Soil, Plant and Atmosphere

Klaus Reichardt • Luís Carlos Timm

Soil, Plant and Atmosphere

Concepts, Processes and Applications

Springer

Klaus Reichardt
Centro de Energia Nuclear na Agricultura
and Escola Superior de agricultura
"Luiz de Queiróz"
University of Sao Paulo
Piracicaba, São Paulo, Brazil

Luís Carlos Timm
Rural Engineering Department
Faculty of Agronomy
Federal University of Pelotas
Capão do Leão, Rio Grande do Sul
Brazil

Translation from the Portuguese language edition: "Solo, planta e atmosfera" by Klaus Reichardt and Luís Carlos Timm, 2nd Edition

ISBN 978-3-030-19321-8 ISBN 978-3-030-19322-5 (eBook)
https://doi.org/10.1007/978-3-030-19322-5

This Springer imprint is published by the registered company Springer Nature Switzerland AG
The registered company address is: Gewerbestrasse 11, 6330 Cham, Switzerland

To my wife, Ceres; my sons, Roberto (in memoriam)
and Gustavo; and my daughter, Fernanda.

Klaus Reichardt

To my parents, Ely and Edemar Timm (in memorium).
To my wife, Cristiane; my daughter, Ana Clara;
and my two sons, Luís Augusto and José Henrique.

Luís Carlos Timm

Foreword

It is a Thursday morning in Davis, California, on September 7, 1995, and I am pretty nervous because in about 2 h, I will have to give an oral presentation on behalf of the Vadose Zone Hydrology Conference, which is held to honor the career achievements of Donald R. Nielsen and James W. Biggar. While waiting in the foyer of the conference hall, a person with a very friendly appearance greets me and introduces himself with "I am Klaus!" So, this is Klaus Reichardt, I think to myself. Finally, I have the pleasure to meet him in person. As a graduate student more than 20 years ago, he was already a pioneer in introducing scaling concepts into soil physics. And returning to Brazil, his unprecedented creative sampling of fields to learn more about symbiotic nitrogen fixation greatly affected my entire career. As he is talking to me with a radiating smile and positive attitude, my nervousness immediately subsides. Later during the conference, I learn about the great artistic ability of Klaus when I admire his hand-painted picture of the main building of the campus of the University of São Paulo, Piracicaba, attractively surrounded by the names of Don's students and scholars.

Looking back over the years as I became more thoroughly acquainted with Klaus and Ceres, his wife, I have learned more and more about his tremendous achievements in his scientific career. One day, one of my colleagues said: "It seems that soil physics is one of the strongest agricultural disciplines in Brazil. Do you know why?"—"I do. Because of Klaus Reichardt." Klaus is one of the most influential scientists worldwide. The many awards he has received do not do justice to celebrate his outstanding capacity, creativity, and personality. One of his greatest achievements was the foundation of the National Center of Research and Development of Agricultural Instrumentation (CNPDIA) in which agronomists and soil physicists collaborate to improve productivity and further develop Brazil's agriculture and ag-related science. But there is also one award that deserves to be mentioned here as an expression of Klaus' special personality and citizenship: in 2013, he became elected citizen of his hometown, Piracicaba.

No more than 5 years after our first meeting, I had the unique pleasure to interact with Klaus' graduate student, Luís Carlos Timm, who worked with me for half a year at ZALF in Germany. During this time, he worked intensively on the analysis of internal drainage experiments and on many spatial data sets that he had collected in Brazil. Today, Timm, who carries degrees in Agricultural Engineering and Agronomy and teaches soil physics and hydrology, is a professor at the Federal University of Pelotas, is one of the leading experts in state-space analysis, and has applied this stochastic modeling approach to a variety of data sets that he and his students have been accumulating over the years. For many years, Professors Reichardt and Timm have taught soil physics and spatial and temporal statistics to students and young scientists from around the world at the International Centre for Theoretical Physics in Trieste, Italy, causing everlasting inspirations for these young scientists to enhance education, research, and practical agriculture in each of their countries. The memories of my first visit in Piracicaba on the occasion of the "First Brazilian Soil Physics Conference" in September of 2011, which is organized by Professor Quirijn de Jong van Lier, when Klaus, Ceres, and I met again, are still fresh in my mind. The experiences of this conference and the following one in Rio de Janeiro, which is organized by Marcos Ceddia and Marta Vasconcelos Ottoni in 2013, were inspirational for me in many ways: the strong link between applied and basic sciences as a result of the high level of education and, among individuals, the respect for each other. These visits were a wonderful initiation of continuing our existing and the foundation of new friendships and interaction.

With this translation of their newest book *O SOLO, A ÁGUA, A PLANTA E A ATMOSFERA* (*Soil, Water, Plant and Atmosphere*), Professors Reichardt and Timm reach out to a broader audience. Especially for those readers without a strong background in agronomy and earth sciences, this book is a great "primer" and provides insights and understanding of the role of human beings in managing soils, water, crops, and atmosphere in sustainable ways. While introducing many complex topics and presenting state-of-the-art knowledge of many agro-ecosystem processes, this book is written in a way that can easily be understood by laymen, i.e., without many equations and with very explanatory illustrations.

Reichardt and Timm walk the readers through 19 chapters in which they emphasize the importance of knowledge about natural systems and about how food and fiber can be produced efficiently while sustaining the resources and important ecosystem-regulating functions. With their first chapter, the authors introduce and illustrate the major challenge of a growing population on earth and the increasing scarcity of resources. In Chap. 2, the fundamental properties of water and its role in the environment, its national abundance, and its relevance for agricultural management are explained. In Chap. 3, the authors present the fundamentals of soils, beginning with a profile description, exploring the details of clay mineral structure, and then transitioning to an overview of soil types in Brazil. Crop vegetative growth, different phenological stages, and aspects of the plant life cycle are elucidated in Chap. 4. An overview of gas composition of the atmosphere and fundamentals of pressure laws, temperature, and radiation is given in Chap. 5. In the next chapter, the reader is able to grasp some principles of thermodynamics and is introduced to the capillarity of the soil pore system. A nice overview on the instruments for the measurement of fundamental soil state variables related to soil moisture is provided in Chap. 6. The logical next step is the introduction to principles of water transport in soils. Of course, when we consider water transport in soils, we have to account for solutes that are transported with water, while their electrical charges determine the interaction with the solid phase and between solutes. A thorough insight in the thermal regime of soils is given in Chap. 10, and the reader has the opportunity to learn about theoretical aspects and practical implications of soil water infiltration in Chaps. 11 and 12, before the attention is directed toward evaporation, evapotranspiration, and crop coefficient. Chapter 14 is focused on the soil-plant-atmosphere continuum. The aspects of the soil water balance are illustrated in Chap. 15, which ends with the impact of agricultural production on the water balance. Plant nutrient uptake is the topic of Chap. 16, and Chaps. 17 and 18 are devoted to the study of the variability of measurements within the Soil-Plant-Atmosphere System, an important subject involving statistics and space and time data series. They close the book with Chap. 19, on dimensional analysis. This book is an important resource for those who want to make themselves familiar with the basic mass and energy transfer processes in agricultural ecosystems. It is a great stimulation for those who decide to study the relevant processes of the water cycle, the crop production, and the processes that govern the cyclic alterations of climate in further detail.

Klaus Reichardt and Luís Carlos Timm deserve a great appreciation for this extremely well-written book and for their successful attempt to present complex matters in a digestible way to the readers who do not yet have a huge scientific background or those at the entrance to agricultural and geosciences in Brazil as one of the strongest nations in terms of agricultural production and other countries. This book is truly inspirational, and with its great educative value, it will help to replace emotion-driven beliefs and assumptions with qualified knowledge, leading to rationale solutions for food supply for our society and clean, sustainable management of resources.

University of Kentucky
Lexington, KY, USA

SSSA
Madison, WI, USA

Ole Wendroth

Preface

One of the goals of this book is to give a broad and detailed view of soil physics applied to the Soil-Plant-Atmosphere System (SPAS). Because of the relative depth in physics, the text involves a great number of mathematical concepts, which chase away agronomists and environmentalists, who reject reading texts full of mathematical expressions. Therefore, we chose to write the chapters in a very comprehensive way, progressing slowly with the use of math. Only in Chap. 3 we start introducing derivatives and integrals. Thereafter, math becomes more and more present, culminating with the solution of differential equations needed for the discussion of most processes that occur in the SPAS. So, if the reader goes directly to Chap. 11 because he is interested in infiltration, he probably will soon give up. In this Preface, the authors intend to illustrate the thoughts made above, placing a parallel between the *exact sciences*, used to describe here the Soil-Plant-Atmosphere System, which are predictable and Cartesian, and the *human sciences* that involve all of us, like art, love, smell, nostalgia, and envy, all considered inexact, difficult to quantify, and almost always unpredictable.

For this parallel, we begin with the academic career of a starting individual, which, for sure, involves exact aspects intertwined with human aspects, and each one, whether student, researcher, or teacher, evolves according to his unique path, with successes and stumblings, to conquer his place in the scientific world. The pathway is long, in which each one develops his own resources for a self-assertion and arrival to a destiny that never comes. Daniel Hillel (1987) was able to formulate an interesting model to describe what he calls the *flow of scientific development* through the interaction of processes addressed by the exact sciences, interconnected by those who addressed the human sciences. The following figure, adapted from this author, illustrates this flow, imagining a researcher (you, the reader) who, at the beginning of his academic career, takes his sailboat and, departing from a point A, navigates in the River of Science, against the water current and the wind, in the directions B, C, D, etc., taking part of the scientific development of humankind. On the right hand, the bank of the River of Science is the Margin of the Theory, where theoretical aspects are mainly dealt with, and on the left hand side, the Data Margin or that of Practice where scientific experimentation prevails. For the reader not versed in the art of sailing, it has to be said that the only alternative to sail against the wind is the zigzag path, which, in our scenery, is from one margin to the other. The journey starts from point A, passing through B, C, etc., toward Z (placed in infinity that is never reached). We might consider that Einstein, Freud, Newton, and so many more have attained Z, but in fact, each of them has made his own journey. In this travel, the researcher is going through obstacles, with voluntary deviations of route or not, entering in the rivers, tributaries, and lakes or beating on the rocks. The navigation from the Data Margin to the Theory Margin symbolizes the *inductive processes* that, based on some experimental and particular data, lead to the establishment of generalizations and general conclusions or theories. From the Theory Margin to the Data Margin, the sailor implements the *deductive processes*, which take a theory to its verification, validation, or proof by means of experimental observations.

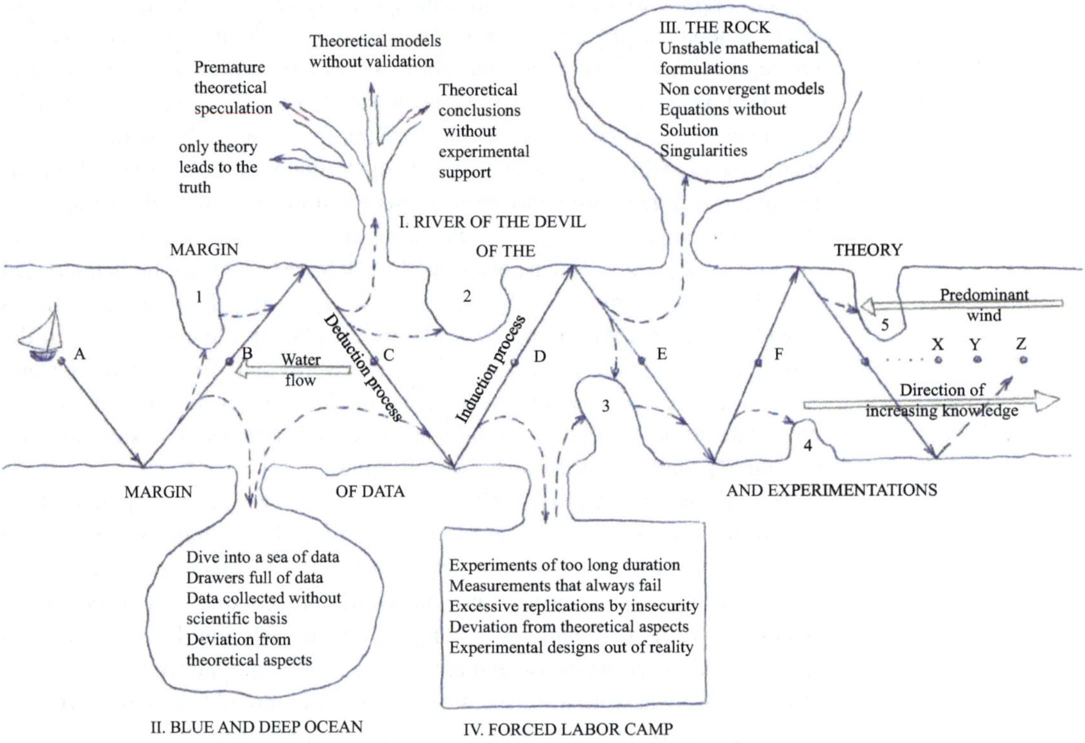

The flow of scientific development along the River of Science, going from A to Z. I, II, II, and IV, indicates the deviations from the main route; 1, 2, 3, 4, and 5 indicate trajectory obstacles hit during travel. Source: Adapted from Hillel, 1987

The ideal trajectory is indicated by the solid zigzag line with arrows, without hitting obstacles (1, 2, 3, etc.) and without route deviations into the rivers or lakes. Few are able to follow this clean route; each one is at the mercy of his destiny and makes his own trajectory. Some get lost in the River of the Devil on the margin of the theory, where they believe that only the theory brings us to the truth, using models without validation and concluding without experimental evidence. Others hit on the rock, using unstable formulations, non-converging models, and equations without solution and becoming lost in singularities. Still others, on the margin of data and practice, get lost in the

Blue and Deep Ocean, submerging in a sea of data, storing enormous volumes of data sometimes collected without scientific basis, and deviating abruptly from the theory, or do not abandon the Forced Labor Field, conducting experiments of unlimited duration, employing methods that do not work, and making a great number of unnecessary replicates only because of insecurity, with experimental designs out of reality.

During his travel along the River of Science, the sailor might also hit obstacles: (1) delay by conventional wisdoms; (2) institutional administration; (3) financing agencies; (4) publication policies and peer reviews; and (5) judgment committees, concurrence, family barriers, retirement, etc. What is important is that, when making its own way, by dribbling the most varied obstacles, the sailor ends up contributing with its share to the scientific development. During the sailing, it is important to be open to changes of course, to innovations, to partnerships, and to collaborations. The worst obstacle one can find is the resistance to change. You never know if what seems bad today is definitely evil forever.

Time is a fundamental coordinate during the navigation on the River of Science, for both the exact and the human sciences. Albert Einstein certainly represents the scientists in the field of exact sciences, who contributed most to the definition of time t, recognizing it as the fourth dimension, next to the spatial coordinates x, y, and z, and showing that their interconnection is so intense as to make time "shrink" or "swell" with the relative velocity between object and observer. All demonstrated by well-founded theories and equations but difficult to understand for a simple mundane.

In the human sciences, which also involve emotion, we can cite the Brazilian composer Chico Buarque de Holanda as the representative of the lovely and poetic interpretation of time. In most of his verses, time stands out, if not directly, at least as a backdrop; he shows how time passes through our lives while everything happens and showing the ephemeral character of time, which often leads us to miss the "trains" of life.

To close this Preface, we will state that we intend with this book to take in hands each student or scientist along the arduous path of the scientific development to advance little by little, chapter by chapter, through the Soil-Plant-Atmosphere System. At the end of this journey, hoping to become mature, ready to lead the frontier of knowledge, with the whish that older career colleagues will have their accomplishments recognized and that sad, insecure, timid colleagues turn around, react, and have strength to enter the academic dance and, finally, recognize that great colleagues, those who have achieved affirmation and recognition, learn to be young again, smile, be humble and tolerant, and aware that there is always something more to be learned and that the end never comes to an end.

São Paulo, Brazil — Klaus Reichardt
Rio Grande do Sul, Brazil — Luís Carlos Timm

Contents

About the Authors

Klaus Reichardt holds a degree and a PhD in Agronomy (1963 and 1965, respectively), a "Frei Docent" degree in Physics and Meteorology (1968), and a full professor in Physics and Meteorology (1981) all from the University of São Paulo and a PhD in Soil Science from the University of California (UC, Davis) (1971). He is currently retired senior professor of the Center for Nuclear Energy in Agriculture (USP) and crop science advisor at the graduate program in Crop Production of the Agricultural College, ESALQ, USP, both in Piracicaba, SP, Brazil. Furthermore, he has experience in physics applied to agronomy, acting mainly on the following subjects: soil water, neutron probe, soil hydraulic conductivity, soil tomography, agricultural balance of water and nitrogen in agricultural crops, spatial variability of soil parameters with emphasis in geostatistics and regionalized variables, and the use of isotopes as markers in agricultural experiments, agrometeorology, soil physics, and crops like corn, soybean, sugarcane, and coffee. In 2019, he has eight graduate students for PhD and MS.

Luís Carlos Timm holds a degree in Agricultural Engineering from the Federal University of Pelotas (1991), a master in Irrigation and Drainage from Federal University of Viçosa (1994), and a PhD in Agronomy from the University of São Paulo (2002). He is currently associate professor of the Department of Rural Engineering, Faculty of Agronomy, Federal University of Pelotas. In addition, he has experience in soil physics and statistics, acting mainly on the following subjects: soil water, neutron and capacitance probes, soil hydraulic conductivity, balance of agricultural crops, spatial and temporal variability of soil and crop parameters with emphasis in time series analysis, and geostatistics.

Man and the Soil–Plant–Atmosphere System

1

"Death of soil is becomming water, death of water is becomming air, air in fire, and vice-versa"

Heraclitus of Efeso (544–484 a.C.)

1.1 Introduction

The twentieth century has undergone changes never before seen in the evolution of mankind, especially in what refers to scientific and technological advances. Plants, animals, and microorganisms that live in a certain area and constitute a biological community are interconnected by a complex network of functional relationships that includes the environment in which they exist. The set of physical, chemical, and biological components, interdependent with each other, is what biologists call the **ecosystem**. This concept is mainly based on the functional relationships between living organisms and the environment in which they live.

The biosphere as a whole can be considered an ecosystem because it represents an envelope that is remarkably small relative to the dimensions of our planet and sustains the only known life form in the universe. About 400 million years ago, favorable conditions for plant development allowed an enrichment of the atmosphere to a mixture of approximately 20% oxygen plus nitrogen, argon, carbon dioxide, and water vapor. With incalculable precision, this mixture was practically kept constant throughout the millennia by plants, animals, and microorganisms that used it and reconstituted it in equal rates. The result was a closed system, a balanced cycle in which nothing is lost, where everything is taken advantage of, in a **dynamic equilibrium**. Dynamic equilibrium is a form of equilibrium very much found in this text and, therefore, deserves an explanation. It is an equilibrium in which there is movement, but "things" are invariant in time. For example, a water tank receives 5 L of water per minute continuously from a tap, and 5 L per minute are also lost by an outlet. The water level in the tank remains constant, indicating still water, while the water moves and is constantly renewed. To maintain this **steady state**, all ecosystems require four basic elements: (1) inorganic substances (gases, minerals, ions); (2) producers (plants) which convert these inorganic substances into food; (3) consumers (animals) using the food; and (4) decomposers (microorganisms) that transform the protoplasm into substances that can be reused by producers, consumers, and even by the decomposers, thus closing the cycle. Only producers have the ability to use solar energy, producing living tissue. That is, the Vegetable Kingdom sustains the Animal Kingdom, and both leave their remains to the decomposers. The efficient use of decomposition products by nature is also a fundamental factor in

K. Reichardt, L. C. Timm, *Soil, Plant and Atmosphere*, https://doi.org/10.1007/978-3-030-19322-5_1

the formation of a soil. And the process is so delicate and complex that it is estimated that the formation of a few inches of fertile soil takes centuries.

Ecosystems are governed by a number of fundamental laws to maintain balance and life. One is adaptation: each species finds a precise place in the ecosystem that provides it with food and environment. At the same time, all species have the defensive power to multiply faster than their own mortality rate. As a result, predators become necessary to keep the population within the limits of its food availability. A jaguar hunting an antelope is necessary for the maintenance of the community of antelopes, even if this does not seem fair to the eliminated individual. Diversity is also another necessary law. The more different species there are in an area, the smaller the chance of one proliferating and dominating the area. Diversity is the survival tactic in nature. Therefore the importance of today's slogan **biodiversity**. The Soil–Plant–Atmosphere System (SPAS), as part of the biosphere, is subject to all these laws and principles. From man's point of view, the biosphere is the supplier of inorganic substances, the producer of his food, and the decomposer of his wastes, that allows the cycle to close. However, man has violated all the laws of balance and has threatened both nature and his very existence on the planet. The main factor of imbalance is the **demographic explosion** (the adaptation and biodiversity laws mentioned above failed). It is estimated that the population of *Homo sapiens* rose from 5 million 8000 years ago to 1 billion in 1850, which clearly demonstrates that in this long period there was a reasonable balance between man and nature. However, from 1850 onward, the time needed for the doubling of the world population has declined significantly. In 1930, the world population reached 2 billion, but in 1991, it already surpassed the 5 billion. In Fig. 1.1, we can see this exponential growth. Today, in 2018, according to the Unites States Census Bureau (USCB), we have already reached the figure of 7.2 billion. Fortunately, in the first years of the twenty-first century, there may already be some slowdown of this growth. Even so, estimates for 2100 are: (1) Pessimistic, ever increasing, reaching 16 billion; (2) average, stabilizing at about 10 billion; (3) Optimistic, growing slower until 2040 and then decreasing to 6 billion in 2100.

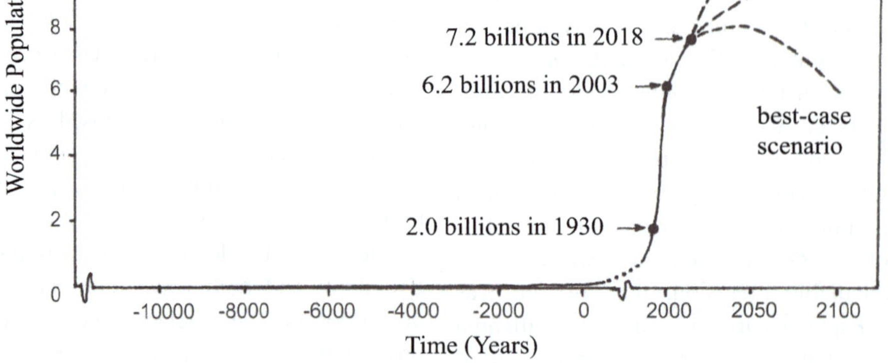

Fig. 1.1 Distribution of the worldwide population

In the Soil–Plant–Atmosphere System, each constituent suffers a typical influence of man. In the first place, we will deal with the water. Essential to life, it is found on the surface of the earth in more quantities than any other pure substance. According to Stikker (1998), about 97.5% are salty and 2.5% sweet, of which 69% represent eternal glaciers and snow, 30% groundwater, 0.9% other reservoirs not readily available, and only 0.3% is in lakes and rivers readily available to man. Of these, 65% are used in agricultural activities, 22% by industry, and 7% by municipalities, and the remaining 6% are lost for human use, which is why scarcity of drinking water can already be felt for long time. The 65% freshwater used in agriculture goes almost entirely to irrigation, but attention, to produce food and fiber. Therefore, a more rational management of irrigation can lead to the economy of large volumes of water. It is important to note that to produce 1 kg of potato we spend 133 L of water, 2500 L for rice, 3700 L for chicken, and 17,000 L for beef and that is why agriculture consumes a lot of water. Today, in the export/import market of agricultural commodities, we are already talking about the "cost" of the water needed to produce them (called virtual water), when they are exported. This is the case of the export from one country to another of corn, soybeans, and beef, which reaches thousands of tons. The water problem is a major global problem, a challenge for the twenty-first century. In a book on soil water management, published 40 years ago, Reichardt (1978), already worried about this situation, stated:

> I devote these pages to the common sense of men, hoping that in the near future the crystalline and potable water will once again be the most abundant natural resource on the surface of the earth.

Water pollution can occur in the SPAS with the most varied agents: (1) biodegradable products and organic substances in general; (2) chemicals (minerals, heavy metals, acids, bases); (3) non-degradable organic products (plastics, detergents, pesticides, and other petrochemical products). These pollutants enter the food chains of ecosystems at certain stages and can reach man. Serious problems are intoxications with heavy metals, such as lead, mercury, arsenic, and cadmium.

In the case of still or semi-still water, such as lakes and dams, it is common to use the term **eutrophication** to indicate an increase of the concentration of ions in water, especially nutrients, as substances containing nitrogen, N, and phosphorus, P. This increase in N and P levels of organic (industrial, urban, or agricultural) or inorganic (industrial) origin causes an imbalance in these ecosystems. Certain species of plants, such as algae, develop in an alarming way in relation to other species, modifying oxygenation conditions, light penetration, temperature, fauna, and flora. Eutrophication is a virtually irreversible process and, in the few cases where something could be done, large sums were spent on its recovery.

The **biochemical (or biological) oxygen demand** (BOD) is used as an index to assess the action of biodegradable agents with respect to their pollution potential. These organic agents are toxic indirectly, because for their biological decomposition, they consume the dissolved oxygen in the water. Thus, the greater the BOD of a product released in a watercourse, the more oxygen is withdrawn from it. This decrease in the level of dissolved oxygen in water has a great influence on fauna and flora. This is the case of the dumping of urban sewage and waste from the paper and sugar refinery industries and alcohol mills. **BOD** is the amount of oxygen required for the decomposition of biodegradable material, under aerobic conditions, by biological action. Thus, a polluting waste water at a rate of BOD $= 6000$ mg L^{-1} (or ppm) refers to such a waste water that every liter discharged into a river will cause 6000 mg of dissolved oxygen to be consumed per liter of river water. Urban sewages have BOD between 200 and 400 ppm and the sugar mills between 15,000 and 20,000 ppm. Logically, the amount of material released in relation to the body volume of water is also of great importance in decreasing the dissolved oxygen content. In general, water sources have dissolved oxygen content in the order of 12 ppm. Most fish require a minimum of 4–6 ppm. The release of debris with high BOD in water bodies causes values of dissolved oxygen close to zero to be reached, with a dramatic effect on the aerobic fauna.

Soil pollutants can also be classified as in the case of water. Here, we will discuss three aspects of **soil pollution**: irrigation, fertilization, and the use of agrotoxics. It is not any water that is suitable for irrigation. Irrigation depends on both the quantity and the quality of the dissolved material in the water. The quality of the water has been neglected in the past for many years, because of the abundance of good quality and easy use of the water sources. The concentration of water in mineral salts is of great importance, both qualitatively and quantitatively. Quantitatively, the saline concentration of irrigation water is evaluated by the **electrical conductivity** (EC). The waters are classified according to their feasibility and risk for irrigation. For example, in the USA, waters with electrical conductivity below 0.75 mmhos cm^{-1} (at 25 °C) are considered to be of first class, very suitable. The US classification is followed in Brasil and water quality for irrigation should be analyzed taking into accountsix basic parameters: (1) total concentration of soluble salts, given by the EC, due to the risk of making the soil saline; (2) relative proportion of sodium, Na, calcium, Ca, and magnesium, Mg, given by the **Sodium Adsorption Ratio** (SAR), to be discussed in detail in Chap. 8, to avoid the risk of alkalinization or soil sodification; (3) concentration of toxic elements in plants; (4) concentration of bicarbonates since these tend to precipitate Ca and Mg increasing the proportion of Na; (5) health aspects; and (6) aspects of clogging emitters in localized irrigation systems. A disadvantage of the criterion of electrical conductivity is the fact that it does not take into account the quality of the ion, as it is a measure of total ion concentration. Ca, Na, Mg, and potassium, K, have distinct effects on the soil, with Na being the most problematic ion. Many waters available for irrigation have EC well above 0.75, and uncontrolled irrigations can cause dispersion of the soil colloidal system—significantly altering its physical properties, determining its salinization, and turning the soil infertile. Such irrigations may also contaminate groundwater reservoirs. In California, USA, for example, the Imperial Valley—one of the most productive regions on the globe—in the 1960s and 1970s was threatened by gradual salinization as a result of the adopted irrigation practices. The recovery of salinized areas is difficult and expensive, especially because it requires a lot of good water and the construction of drainage systems.

In many parts of the world, excessive rates of fertilizers have been used, especially N, whose productivity response is compensatory but whose use has been exaggerated. In several regions, groundwater is condemned because of the high concentration of nitrate (NO^{-3}). Another problem is the use of insecticides and herbicides. The demand for food for growing populations has increased so much that the use of pesticides in increasing doses has become inevitable. Among these organic substances, many are not biodegradable and are very resistant to decomposition by any other process. Typical examples are DDT, BHC, 2,4-D, and glyphosates, which, absorbed by plants and insects, are carried by water and, at a given moment, enter the food chain of ecosystems through a variety of ports.

Air pollution is brutal, resulting mainly from industrial activities, such as paper, steel, oil, and chemical industries in general, and from internal combustion engine's waste gases. Agriculture contributes mainly with the burning of both forests and crop residues, and with cattle raising. Among the main pollutants in the atmosphere are the oxides of carbon, sulfur, nitrogen, organic substances, and heavy metals. Carbon monoxide combines with hemoglobin in the blood, making it unable to carry oxygen. The consequence is suffocation and heart and lung problems. Likewise, nitrogen oxides also reduce the oxygen-carrying capacity of the blood, while sulfur oxides contribute to the appearance of lung diseases. Another important effect of air pollution is the modification of the physicochemical properties of the atmosphere. As we shall see in Chap. 5, of the atmosphere, millennia passed for this gaseous layer to enter into dynamic equilibrium, presenting characteristic concentrations of its different constituents, concentrations of about 340 ppm for CO_2, which allowed the establishment of life on the planet. Air pollution, mainly due to the combustion of fossil fuels such as oil and coal, has increased to values of the order of 390 ppm by 2018. Particularly in localized areas, such as urban and industrial centers, of major importance is the modification of the quality and amount of solar energy that reaches the ground. In some areas, pollution already reduces solar energy by 40% of its normal value. Qualitatively, certain wavelengths are absorbed (mostly by CO, CO_2) almost completely, so that a spectrum of different characteristics reaches the surface of the ground. These changes in the energy balance affect the distribution of temperature, atmospheric pressure, wind, rain, etc. As a consequence, there are problems of visibility, thermal inversion, and the **greenhouse effect**. This effect, which will be discussed in more detail in Chap. 5, is a natural effect of the greatest importance, without which there would be no conditions for life on Earth. Through it the atmosphere acts as a "filter" of solar radiation, allowing only part of the radiation coming directly from the Sun to reach the surface of the soil, warming it. This soil surface in turn heated emits terrestrial radiation that is partially blocked by the atmosphere. Thus, the balance between incoming solar radiation and the outcoming terrestrial radiation is controlled by the greenhouse effect. The so-called **greenhouse gases (GG)**, especially CO_2, CH_4, and N_2O, change the characteristics of the atmosphere and affect the radiation balance which is mostly positive. In the late 1990s, concern about **global change**—in the Earth's atmosphere—increased tremendously. The Kyoto Protocol, signed in 1997, asks for the reduction of the CO_2 emissions, especially to the great polluters of the developed world. There is a fundamental difference between the CO_2 originated from the combustion of fossil fuels (non-replaceable sources) such as petroleum and coal, and that from existing biological materials (renewable sources) such as alcohol, biodiesel, and wood. The former represent an addition of CO_2 to the atmosphere coming from a carbon source that would otherwise remain in the Earth's depths, and the latter are from C sources that participate in biosystems where they cycle between fixed forms and atmospheric CO_2. In any case, all burnings contribute to air pollution and need to be minimized or mitigated. The term **carbon sequestration** is used for processes or management systems that fix carbon, and, among them, emphasis is given to those that occur in agriculture. Correct soil management can contribute to the sequestration of atmospheric carbon, as recent research shows. Currently, 25 million ha are cultivated in Brazil by the **no-tillage system**, and this area is responsible for approximately 13 million tons of CO_2 sequestered from the atmosphere per year. The soil fixes about three times more CO_2 than it exists in the atmosphere. Today carbon credits, also called certified CO_2 emission reductions, are certificates issued to a person or company that has reduced its GG emissions. By convention, one ton of CO_2 corresponds to one carbon credit. Kutilek and Nielsen (2010), in their book *Facts About Global Warming*, discuss mainly the recent increases in CO_2 content and air temperatures in different parts of the globe.

Plastic, indirectly (because it pollutes more when burned), is among the greatest polluters of the atmosphere, because there is no microorganism capable of taking advantage of the existing energy in the plastic and, physically–chemically, it is very resistant, at least under environmental conditions. Few people are aware that the plastic in which meat and vegetables are wrapped in the supermarket cannot be destroyed without harm to nature, and if left in the environment, it will remain intact for generations. The easiest way to destroy it is by combustion, hence its importance as a polluter of the atmosphere. Today, the use of so-called biodegradable plastics is more common, such as the POLY ECO which is made of biodegradable resin and is recyclable, which can be found in some supermarkets.

The influence of man on the environment here analyzed in very general terms and, in particular, the influence of man on the Soil–Plant–Atmosphere System, makes the importance of knowing in detail the processes that control this system, become evident. The increased demand for food due to population increase, problems of environmental pollution, storage and treatment of waste, recharge of groundwater reservoirs, and effective control of the natural properties of the Soil–Plant–Atmosphere System make it essential for man to study basic physical–chemical processes responsible for any change in the dynamic equilibrium state of this system.

At the end of the 1980s, a phase of great ecological awareness began, recognizing that the current model of agricultural production needed profound changes aiming at the greater conservation of the environment. The term in fashion today is "**sustainable agriculture**"—one that would make agricultural production possible in balance with the environment for generations. This is a major challenge that must be faced in the twenty-first century, and overcoming it depends on the deep knowledge of the processes that govern the dynamics of the Soil–Plant–Atmosphere System, with which this book aims to contribute.

1.2 Exercises

1.1. What is meant by an ecosystem?
1.2. List agents that pollute water, soil, plant, and atmosphere.
1.3. What is BOD?
1.4. What are transfer processes?
1.5. What is the greenhouse effect?
1.6. What is known about the hole in the ozone layer?
1.7. What is sustainable agriculture?
1.8. What is meant by carbon sequestration?
1.9. What global changes does the Kyoto Protocol (1997) address?

References

Kutilek M, Nielsen DR (2010) Facts about global warming. Catena, Cremlingen-Destedt
Reichardt K (1978) A água na produção agrícola. McGraw-Hill, São Paulo
Stikker A (1998) Water today and tomorrow: prospects for overcoming scarcity. Futures 30:43–62

2 Water, the Universal Solvent for Life

2.1 Introduction

Water is one of the most important substances of the earth's crust for both vital and physicochemical processes. In the liquid and solid phases, it covers more than two-thirds of our planet and, in the gaseous form, is constituent of the atmosphere, being present in every part. Without water, life would not be possible as we know it. Living organisms were originated in aqueous media and became absolutely dependent on it in the course of evolution. Water is a constituent of protoplasm in proportions that can reach 95% or more of the total weight of living organs. In the protoplasm, it participates in important metabolic reactions, such as photosynthesis and oxidative phosphorylation. It is the universal solvent and makes possible most of the chemical reactions to occur. In plants, water has also the function of maintaining the cellular turgor, responsible for the vegetative growth. Thus, knowledge of its physicochemical properties is essential for the study of its functions in nature, in particular its behavior in the Soil–Plant–Atmosphere System as a whole.

2.2 Molecular Structure and Phase Change of Water

The chemical formula of water is H_2O, that is, it consists of two hydrogen atoms and one of oxygen. In nature, there are three isotopes of hydrogen (^1H = hydrogen, ^2H = deuterium, and ^3H = tritium) and three oxygen isotopes (^{16}O, ^{17}O, and ^{18}O). These different atoms in terms of weight enable 18 different combinations in the formation of a water molecule; however, ^2H, ^3H, ^{17}O, and ^{18}O are poorly abundant in nature. These 18 different isotopes behave in the same way from the chemical and biological points of view, having the same properties but differing only in their weight. Hydrogen ^1H and deuterium ^2H or D are stable isotopes, the first with an abundance of 99.98% and the second 0.0026–0.018%. Tritium (^3H) is a naturally occurring radioisotope because it is constantly produced in the atmosphere by nuclear reactions with cosmic radiation, with an abundance of traces only. It emits beta radiation, but at such low concentrations, it does not impair life. The Oxygen Isotopes 16, 17, and 18 are present in

K. Reichardt, L. C. Timm, *Soil, Plant and Atmosphere*, https://doi.org/10.1007/978-3-030-19322-5_2

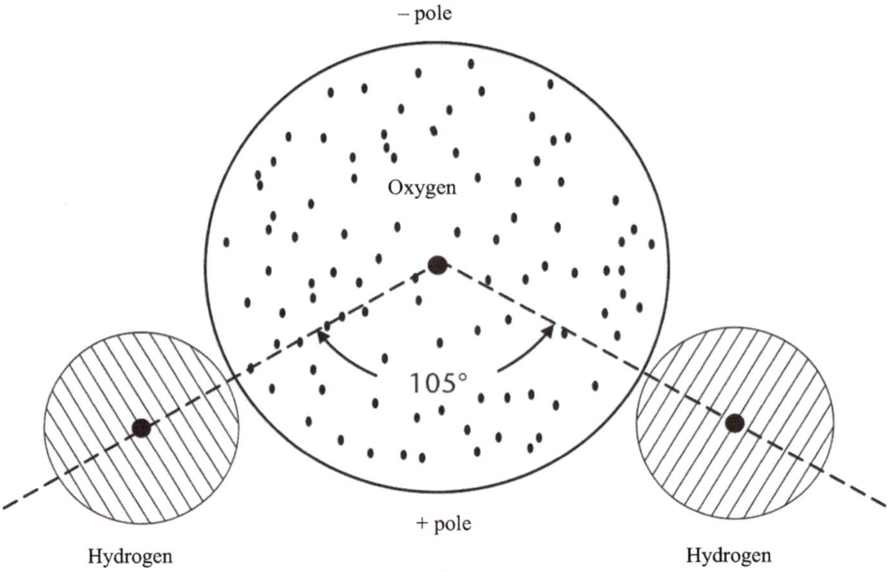

Fig. 2.1 Schematic presentation of the water molecule showing its asymmetry

nature with abundances of 99.762, traces, and 0.238%, respectively.

The mean diameter of the water molecule is approximately 3 Å (1 Å = 3×10^{-10} m), and the two hydrogen atoms are bound to the oxygen atom forming an angle of about 105° (see Fig. 2.1), which is responsible for the space imbalance of electrical charges in the molecule.

This asymmetric charge distribution creates an electric dipole responsible for a number of physicochemical properties of the **water molecule**. Due to this polarity, the water molecules orient themselves forming structures. Polarity is also the reason why water is a good solvent, and it is adsorbed on solid surfaces or hydrates ions and colloids.

Each hydrogen of a molecule is attracted by the oxygen of the neighboring molecule, with which it forms a secondary bond, called the hydrogen bond. The hydrogen bond has a much weaker bonding energy than the intra-molecular bond of oxygen to the two hydrogen atoms. As a result, water is made up of a chain of molecules bound by hydrogen bonds (polymer). This structure has fewer faults when the water is in the solid state (**ice**). Under these conditions, each molecule is bound to four neighboring molecules forming a relatively open hexagonal structure. With the melting of the ice, this structure is partially destroyed, so that other molecules can enter the intra-molecular spaces. For this reason, each molecule can have more than four neighboring molecules, and the density of water in the liquid state becomes slightly greater than that of ice and floats on liquid water. This is an interesting exception since in pure materials during the melting, the solid phase is denser than the liquid and sinks into the liquid. If ice would not float in the water (like icebergs with 1/9th of their volumes out of water and 8/9 in the liquid), it would sink and accumulate in the bottom of the oceans and over very long time the ice would be accumulating up to the surface, without the possibility of having liquid water for the formation of life.

For water in the liquid state, a structure of the same type as that of ice continues to exist, but this structure is not rigid and permanent, but flexible and transient. In the gaseous state, this structure disappears completely and the molecules have maximum freedom.

In the transition from the solid state to liquid and gaseous states, the hydrogen bonds are ruptured, whereas in the reverse passages, they are re-established. Thus, in the melting process of 0.001 kg of ice, 335.0 J of energy need to be supplied (**latent heat of fusion**), and in the solidification of 0.001 kg of water, the same amount of energy is released from the system. The **melting point** of the water under normal atmospheric pressure is 0 °C, while the **boiling point** is 100 °C. In this temperature range, the water is in the liquid state and its specific heat is 4186 J kg^{-1} $°C^{-1}$. This value is very high compared to ice (−10 °C): 2093.0; aluminum: 900.0; iron: 447.9; mercury: 138.1; oxygen: 920.9. Because of this, water behaves as a great buffering system for the energy available in the atmosphere, that is, a lot of energy is needed to raise its temperature slightly. This property of the water makes biological systems (whose percentage in water is very high) more resistant to variations in temperature.

At the boiling point, the water passes from the liquid to the gaseous state (or vice versa), and the heat involved in the phase change is 2.26×10^6 J kg^{-1} (**latent heat of vaporization** or condensation). The water may also be brought to the gaseous state at temperatures less than 100 °C, but such vaporization, called **evaporation**, requires a greater amount of heat. Thus, for example, at 25 °C the latent heat of vaporization or latent heat of evaporation is 2441×10^6 J kg^{-1} (see Table 2.1). In this table, it is seen that even the ice can pass to the vapor state, which is the case of **sublimation**. According to the kinetic theory of gases, the molecules of a liquid are in continuous motion, a movement that is an expression of their thermal energy. Molecules often and repeatedly hit each other and absorb enough energy to escape from the liquid and enter into the air. Its kinetic energy is dissipated during the passage through the potential energy barrier caused by the intermolecular attraction on the liquid surface (measured by surface tension). Escaping the liquid, the molecule becomes part of the gas phase. In the same way, molecules of the gas phase can return to the liquid phase.

The rate at which transfers of molecules occur from liquid to vapor, and vice versa, depends on the concentration of water vapor in the atmosphere in contact with the liquid surface. An atmosphere in equilibrium with the surface of the water is considered saturated with water vapor (the same number of molecules that leaves the liquid and passes to the gas phase returns to the liquid phase). The vapor pressure of the air in equilibrium with the water surface depends on the pressure and temperature of the system. In general, we can say that under normal pressure conditions the air can retain more vapor the higher its temperature. In Chap. 5, about the atmosphere, this subject will be discussed in more detail.

As mentioned above, water can also pass directly from the solid state to the gaseous (or vice versa), a phenomenon called sublimation. The latent heat of sublimation is equal to the sum of the latent heats of melting and vaporization. When rainfall falls through cold regions below

Table 2.1 Some physical properties of the water

Temperature (°C)	Density ($kg\ m^{-3}$)	Specific heat ($J\ kg^{-1}$)	Latent heat of vaporization ($\times 10^6\ J\ kg^{-1}$)	Surface tension ($kg\ s^{-2}$)	Viscosity ($kg\ s^{-1}\ m^{-1}$)
−5	999.18	4227.86	2.511	0.0764	–
0	999.87	4215.30	2.919	0.0756	0.001787
4	1000.00	4206.93	2.491	0.0750	0.001567
5	999.99	4202.74	2.489	0.0748	0.001519
10	999.73	4190.19	2.477	0.0742	0.001307
15	999.13	4186.00	2.465	0.0732	0.001139
20	998.23	4181.81	2.453	0.0727	0.001002
25	997.08	4177.63	2.441	0.0719	0.000890
30	995.68	4177.63	2.430	0.0711	0.000798
40	992.25	4177.63	2.406	0.0695	0.000653
50	988.07	4181.81	2.382	0.0679	0.000547

0 °C, hail is formed by solidification, which is ice that is amorphous. But when the water vapor in regions below 0 °C forms snow by sublimation, we have crystals of geometric structure of extraordinary beauty.

2.3 Surface Tension

Surface tension is a typical phenomenon of a liquid–gas interface. Most liquids behave as if they were covered by an elastic membrane, under tension, with a permanent contraction tendency (trying to assume minimal area). This happens because the cohesive forces acting on each molecule of water are different if the molecule is inside the liquid or at the surface (Fig. 2.2).

Molecules within the liquid are attracted in all directions by equal forces, while surface molecules are drawn into the denser liquid phase with forces greater than the forces with which they are pushed to the less dense gaseous phase. These unbalanced forces cause the molecules of the surface to tend toward the interior of the liquid, that is, they present a tendency to contract the surface.

If we take an arbitrary line of length L on the surface of the liquid, a force F will be acting on both sides of the line, trying to contract the surface. The ratio F/L (which is a constant, since the greater L, greater F), is called surface tension σ, whose dimension is force per unit length (d cm^{-1} in the CGS system and N m^{-1} in the international system). The same phenomenon can be described in terms of energy. Increasing the surface of a liquid requires energy expenditure which remains stored on the enlarged surface and which can perform work if the surface contracts again. Energy per unit area has the same units as force per unit length. Thus, the surface tension can also be expressed in erg cm^{-2} in the CGS system and J m^{-2} in the international system.

The surface tension is then the measure of the resistance to the formation of the elastic membrane which is formed in a liquid–gas interface. It depends on the temperature: it usually decreases with its increase (see Table 2.1). The decrease in surface tension is further accompanied by an increase in vapor pressure. Substances dissolved in water lead to changes in surface tension in both directions. Electrolytes generally increase surface tension because the affinity between the

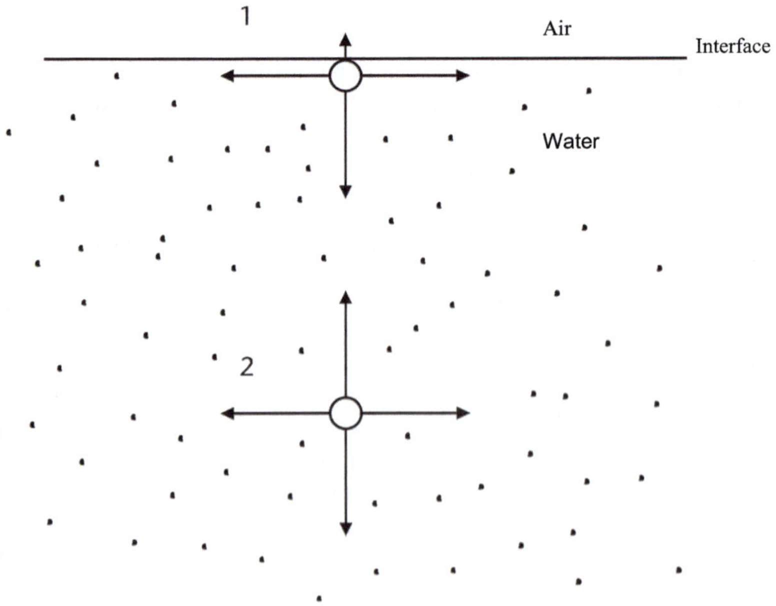

Fig. 2.2 Forces acting on water molecules: (1) Water molecule at air–water interface; (2) Water molecule inside the liquid phase

electrolyte ion and a water molecule is greater than the affinity between water molecules, and as a result, the solute tends to penetrate the solvent. Otherwise, that is, when the affinity between the solute and the solvent is less than the affinity between solvent molecules, the solute tends to be pushed outward and be concentrated on the surface of the liquid, reducing its surface tension. Such is the case of organic solvents, in particular detergents.

For flat liquid surfaces such as lakes, dams, and class A evaporation tank, there is no pressure difference between immediately higher and lower points at the liquid–gas interface. However, for curvilinear surfaces, such as drops and *menisci* in capillaries, there is a pressure difference, responsible for a series of capillary phenomena. These phenomena will be studied in Chap. 6, which will analyze the interactions between solid (soil), liquid (soil solution), and gas (soil air).

2.4 Viscosity

When water is in motion, other properties are still important. A fluid, when moving, can be imagined made up of superposed sheets that slide over each other (laminar flow, Fig. 2.3). It has been found that for **Newtonian fluids** (water and gases in general), the force (F) required to move the sheets is proportional to the *modulus* of the gradient of the velocity v ($\mathrm{d}v/\mathrm{d}x$), plotted perpendicular to the direction x of the movement of the fluid and the contact area A between the slices. Therefore:

$$F = \eta A \frac{\mathrm{d}v}{\mathrm{d}x} \tag{2.1}$$

The coefficient of proportionality η is called **absolute viscosity**, and Eq. (2.1) is also known as Newton's viscosity equation. **Viscosity** can be seen as the property of the fluid that measures its slipping resistance or internal friction. The viscosity η is defined in the CGS system as the force per unit area (F/A) required to maintain a speed difference of 1 cm s^{-1} between two unit ($A = 1\ \mathrm{cm}^2$) parallel sheets separated by a distance of 1 cm. It is easy to verify in Eq. (2.1) that $F/A = \eta$, when $\mathrm{d}v/\mathrm{d}x = 1$. The absolute viscosity dimensions are $\mathrm{M\ L^{-1}\ T^{-1}}$. The viscosity varies with temperature (see Table 2.1) and is also affected by the type and concentration of solutes. In addition to the absolute viscosity coefficient, in practice the **kinematic viscosity** coefficient $\mathrm{L^2\ T^{-1}}$ is obtained by dividing η by the specific mass of the fluid.

Kutilek and Nielsen (1994) and Lal and Shukla (2004) go into detail in relation to water structure, surface tension, and viscosity. These are texts that need to be consulted for a deeper understanding of these concepts. There are also

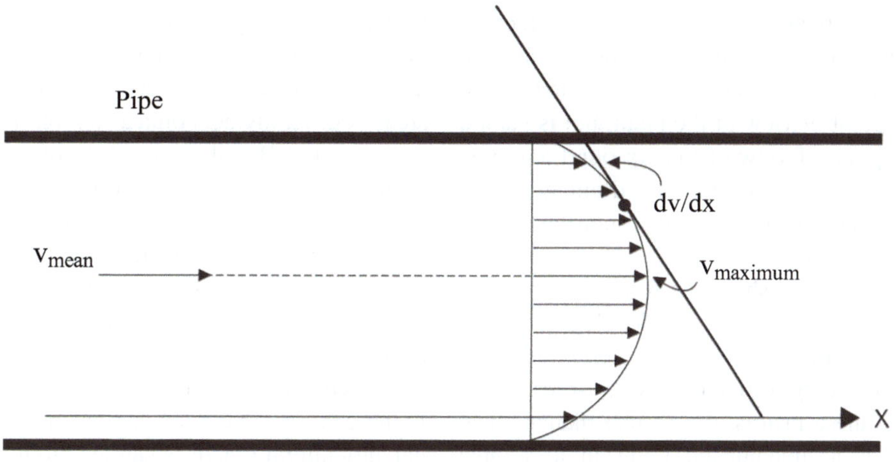

Fig. 2.3 Profile of the velocities of water in the sheets (laminar flow) indicating that the maximum value is in the center of the pipe, and that theoretically the velocity at the pipe wall is zero

extensive texts on these topics in the area of Fluid Mechanics. To understand phenomena that prevail in the Soil–Plant–Atmosphere System, what we have seen on η and σ is sufficient.

2.5 The Importance of Water for Agricultural Production

Water is a key factor in plant production. Its lack or excess decisively affects the development of plants, and therefore, the rational management of water is imperative in the maximization of agricultural production.

Any crop during its development cycle consumes a huge volume of water, and up to about 98% of this volume only passes through the plant, reaching the atmosphere by the transpiration process. This flow of water, however, is essential and necessary for plant development, and for that reason, its rate must be kept within the appropriate limits for each crop. The water fixed by photosynthesis (Chap. 4) is minimal in relation to transpiration, and is incorporated in the formation of sugars, and finally into dry matter.

The water reservoir is the soil that, temporarily, stores water, and it can return the water to the plants according to their needs. As the natural recharge of this reservoir (rain) is discontinuous, the volume of water available to the plants is variable. When rainfall is excessive, its storage capacity is exceeded and large losses can occur. These losses are possible by surface runoff, causing soil erosion, or by deep percolation, being lost to the groundwater. This percolated water is lost from the point of view of the plant, but is gained from the point of view of underground aquifers. One negative point is that the drainage water carries along all soluble salts (nutrients) and organic compounds.

When the rain is short, the soil acts as a reservoir of water essential to plant development. The exhaustion of this reservoir by a crop requires its artificial reloading, which is the case of irrigation.

Due to these factors, the correct management of water is a fundamental point in a rational agriculture. In arid and semi-arid regions, proper management implies in water-saving policies and care practices with salinity problems. In superhumid regions, the fundamental problem is the leaching of soil materials and drainage. In regions where rainfall is sufficient, there are usually distribution problems that lead to periods of water shortage. In these areas, it is of utmost importance to obtain the highest possible efficiency in the use of water for crops as well as the use of supplementary irrigation.

We take Brazil as an example due to its territorial extension and climate diversity that presents all sorts of situations. The North, represented by the Amazon Basin, is a super-humid region, with average precipitation above 2000 mm per year. The soils are mostly poor, and leaching problems are of major importance. With the recent development of this area, much has been learned so that viable and productive cultivation methods can be implemented. Next to the Amazon, to the east side, we have the Northeast with mostly semi-arid areas, where productive agriculture can only be developed at the expense of irrigation. Many national irrigation projects have already been implemented in these areas, which aim to use water from dams and, especially, the São Francisco River. The project to transpose the waters of São Francisco to the northeast is of fundamental importance, but due to political and environmental problems, it has not yet been consolidated. As we have said, special precautions should be taken with regard to the quality of water in these irrigation projects. Waters apparently good for irrigation can, over the years, turn saline-extensive areas, rendering them unproductive. The recovery process of such areas is generally economically prohibitive.

In Central Brazil, about 25% of the national territory consisted of Cerrado (a specific type of savannah), a peculiar Soil–Plant–Atmosphere System, whose main characteristic is the low fertility of the soil. The Cerrado is, however, strengthened to extinction due to the expansion of agriculture in an erratic way. It is one of the ecosystems of greatest biodiversity of the globe. In these areas, it has been demonstrated that rational agricultural practices can lead to highly compensating productivities. These practices involve, in particular, correction of soil pH (liming), proper fertilization, and correct water management. Dry spells can affect productivity in many cases, requiring additional irrigation. An important example is the west of the Bahia state, where the availability of surface water for irrigation is good, in which irrigations and fertirrigations (fertilizer application via irrigation water) are successfully carried out by means of central pivots, which can cover areas as large as 100 ha each. In the South and Center-South regions, rainfall generally meets the needs of agriculture. Disturbance problems, however, can be fatal on many occasions. Thus, water management must be adequately conducted to maximize production and minimize problems such as erosion, deep percolation, and surface and groundwater pollution.

The discussion made above is related to **rational water management** and synonymous expressions. What do these concepts represent, anyway? In general terms, they represent the most appropriate use of the available natural resources with respect to the different Soil–Plant–Atmosphere Systems. For this, basic knowledge is needed and to be applied by those responsible for agricultural projects. In this text, we present—quite objectively, to be accessible even to those not directly involved with water management problems—essential basic knowledge for understanding the processes that occur with water in the Soil–Plant–Atmosphere System.

2.6 Exercises

2.1. Write the 18 types of water molecules using the three isotopes of H and the three isotopes of O.

2.2. 100 g of water, 100 g of aluminum, 100 g of mercury, and 100 g of air, all at 20 °C, receive 100 cal. The specific heats of Al, Hg, and air are, respectively, 0.2, 0.03, and 0.172 cal g^{-1} $°C^{-1}$. What are the respective temperature increases (ΔT)?

2.3. Considering the densities of the materials in exercise 2.2 as 1, 2.7, 13.6, and 0.0013 g cm^{-3}, respectively, calculate (ΔT) when 100 cm^3 of each material receives 100 cal.

2.4. A flat water surface receives 1.2 cal cm^{-2} min^{-1}, and all energy is used in the evaporation process. How many grams of water are evaporated per hour per m^2 surface when the water is at 10, 20, or 30 °C?

2.7 Answers

2.1. $^1H^1H^{16}O$; $^1H^1H^{17}O$; $^1H^1H^{18}O$; $^1H^2H^{16}O$; ......; $^1H^3H^{17}O$; ......; $^3H^3H^{18}O$.

2.2. $\Delta T = 1, 5, 33.3$, and 5.8 °C.

2.3. $\Delta T = 1, 1.9, 2.4$, and 4472 °C.

2.4. 1217, 1229, and 1241 g.

References

Kutilek M, Nielsen DR (1994) Soil hydrology. Catena, Cremlingen-Destedt

Lal R, Shukla MK (2004) Principles of soil physics. Marcel Dekker Inc, New York

The Soil as a Water Reservoir for Plants

3

3.1 Introduction

The term soil refers here to the outer and agriculturally usable layer of the earth's crust. Its origin is the rock that, through the action of physical, chemical, and biological processes of disintegration, decomposition, and recombination, has become, during the geological ages, a porous material of peculiar characteristics. Five factors are recognized in soil formation: original material (rock) M; time (age) I; climate (C); topography (T); and living organisms (O). Using the mathematical language, one can say that:

$$\text{Soil} = f(M, I, C, T, O) \qquad (3.1)$$

From the combination of the last four **soil-forming factors** acting at different intensities on the same original material M, a great diversity of soil types can result.

By making a vertical cut in the profile of a typical soil, a series of overlapping layers called **soil horizons** is obtained. The set of layers receives the name of **soil profile** (Fig. 3.1). A complete soil is formed by four horizons—A, B, C, and D—which can be further subdivided. The **soil horizon A** is the top surface layer of the soil, exposed directly to the atmosphere. It is known as the **eluviation horizon**, a horizon that is more susceptible to losses of chemical elements by successive washes with rainwater. It is subdivided in A_{00} (superficial layers in forest soils with large amount of non-decomposed organic material: branches, leaves, fruits, and animal wastes); A_0 (just below A_{00}, is made up of decomposed organic material, that is humified, a concept we will see later); A_1 (already a mineral horizon, but with high percentage of humified organic matter that gives it a dark color); A_2 (the typical A horizon, which is lighter in color, and corresponds to the zone of maximum loss of mineral elements, i.e., eluviation); and A_3 (transition horizon between A and B, with characteristics of both). The **soil horizon B** is known as the horizon of illuviation, that is to say, a horizon that is more susceptible to the gain of chemical elements coming from horizon A. It is subdivided into B_1 (transition between A and B, but has more characteristics of B); B_2 (formed by the zone of maximum illuviation, i.e., accumulations of leached material of A, composed mainly of Fe, Al, and Ca); and B_3 (transition between B and C). The **soil horizon C** is formed by the material that gave origin to the soil, still in a state of decomposition, sometimes called subsoil, and the **soil horizon D**, formed mainly by the matrix rock. The thickness of the horizons is variable, and the lack of certain horizons in particular soils is quite common. All this depends on

K. Reichardt, L. C. Timm, *Soil, Plant and Atmosphere*, https://doi.org/10.1007/978-3-030-19322-5_3

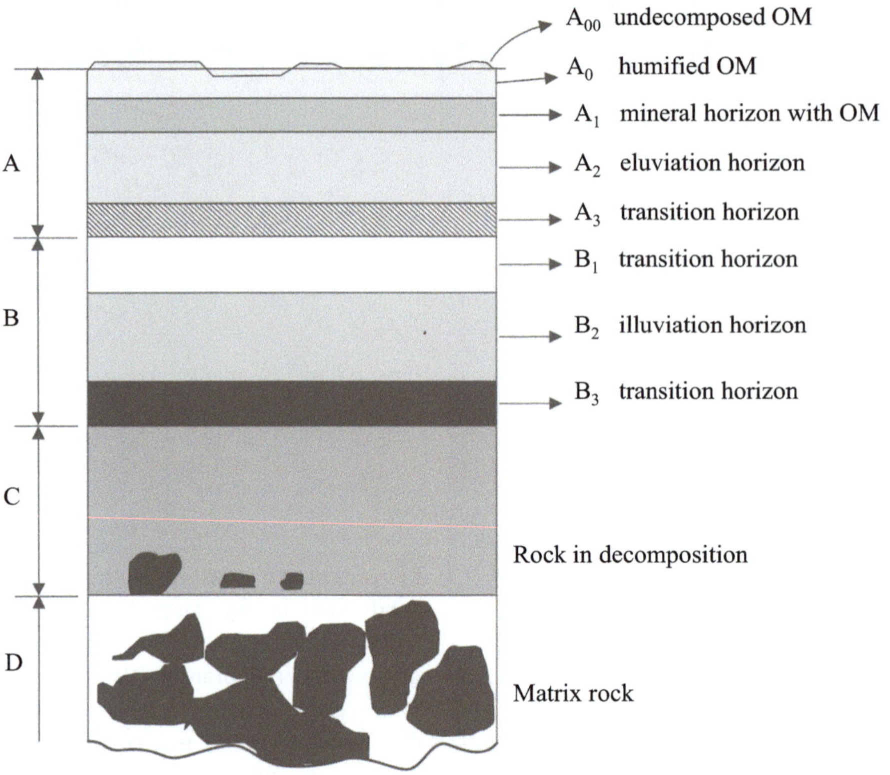

Fig. 3.1 A schematic view of a complete soil profile. *OM* organic matter

the intensity of the action of the formation factors I, C, T, and O on M.

As the soil is a porous material, it consists of three phases: solid, liquid, and gas. The solid part of the soil consists of mineral and organic material. The mineral part comes from the rock of which the soil was formed and is called primary when it has the same structure and composition of the minerals that make up the rock. It is called secondary when the matter is new, transformed, in composition, and with different structures, produced during the process of soil formation. Primary materials are fragments of rock or minerals, such as quartz and feldspar. Secondary materials are, e.g., montmorillonite and kaolinite clay minerals, and calcium carbonate. The organic part consists mainly of CHO compounds, in many stages of decomposition, ending up as humus. Living organisms and microorganisms also take part of the soil.

The liquid part of the soil consists of a solution of mineral salts and organic components whose concentration varies from soil to soil and certainly with its water content.

The gaseous part consists of air, with composition slightly altered in relation to the air that circulates above soil surface, also varying according to a large number of factors. In general, the amount of O_2 is reduced compared to air over the soil surface, and the amount of CO_2 is higher as a consequence of the biological activities occurring in the soil. Its relative humidity in natural conditions is almost always saturated or very close to saturation.

3.2 The Solid Fraction of the Soil

The **solid soil particles** of the soil vary greatly in quality and size. In relation to the size, some are large enough to be seen with the naked eye, while others are so tiny not to be seen individually, representing colloidal properties. The term **soil texture** refers to the distribution of soil particles

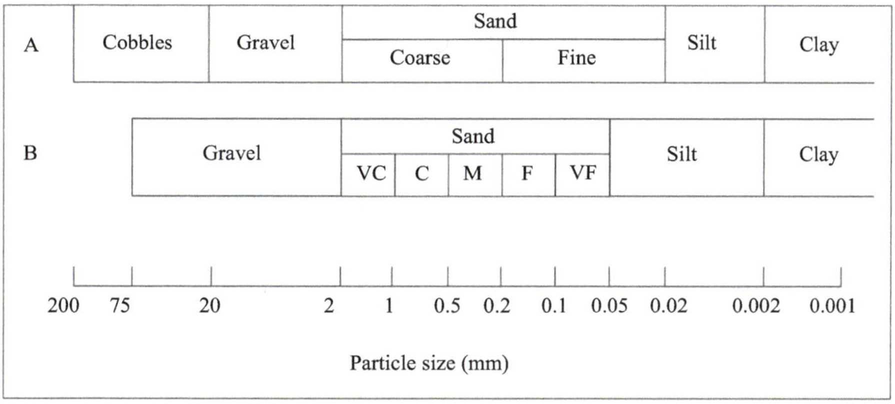

Fig. 3.2 Two commonly used soil textural classifications: (**a**) Atterberg and (**b**) US Department of Agriculture (Soil Survey Staff 1951). *VC* very coarse, *C* coarse, *M* medium, *F* fine, *VF* very fine

only as to their size, not taking into account their quality. Each soil receives a designation referring to its **texture**, which gives us an idea of the size of the most frequent constituent particles. Traditionally, soil particles are divided into three size fractions, called **textural fractions**: **sand**, **silt**, and **clay**. There is still no agreement on the definitions of these classes. In Fig. 3.2, the two most commonly used classification schemes are shown.

Once the mechanical or textural analysis (separation in sizes) of a soil is made, the soil receives a designation relative to the quantities of the three fractions—**sand**, **silt**, and **clay**—its textural class. Thus, soils with different proportions of sand, silt, and clay receive different designations.

The determination of soil particle size distribution is known as **soil mechanical analysis**. Traditionally, this analysis is performed on sieved dry soil samples using a 2 mm sieve, that is, particles with diameters lees than 2 mm. Particle separation is generally done by sieving the soil sample in water (to destroy aggregates that could erroneously be considered as larger particles) with a sieve sequence down to a particle diameter of about 0.05 mm. In order to separate the particles of smaller diameter, the sedimentation method is used, which consists of dispersing an aqueous suspension of a soil sample and measuring the settling (or sedimentation) velocities of the particles of different sizes.

A particle in free fall under vacuum does not find resistance to motion and, therefore, increases its velocity along its trajectory in accelerated motion, according to gravity g. A spherical particle falling inside a fluid encounters resistivity (friction) proportional to its radius r, velocity v, and fluid viscosity η. By **Stokes law** (1851), the friction force f_r is given by:

$$f_{\mathrm{r}} = 6\pi v r \eta \tag{3.2}$$

Due to this resistance, after some time the particle reaches a constant rate of fall, i.e., without acceleration. Under these conditions, the sum of all forces acting on the particle must be zero (v = constant or acceleration = 0). In addition to the downward directed friction force, the particle is subjected to a buoyancy force f_b (Archimedes) also directed from the bottom up, which is given by the weight of the volume of the displaced fluid which, for a sphere of radius r, is:

$$f_{\mathrm{b}} = \frac{4\pi}{3} r^3 \rho_{\mathrm{f}} g \tag{3.3}$$

where ρ_{f} is the density of the fluid, g is the acceleration of gravity, and $4\pi r^3/3$ is the volume of the particle. The last force acting on the particle is its weight f_w, directed from top to bottom and given by the product of its mass by the acceleration of gravity. Since mass is equal to volume multiplied by density, we have:

$$f_w = \frac{4\pi}{3} r^3 \rho_g g \quad (3.4)$$

where ρ_g is the particle density.

Making the balance of these forces, we obtain:

$$-6\pi v r \eta - \frac{4\pi}{3} r^3 \rho_f g + \frac{4\pi}{3} r^3 \rho_g g = 0$$

or

$$v = \frac{2}{9} \frac{r^2 g}{\eta} \left(\rho_g - \rho_f\right) = \frac{d^2 g}{18\eta} \left(\rho_g - \rho_f\right) \quad (3.5)$$

remembering that the particle diameter = $2r$. This is because the mechanical analysis is made by separating particles by diameter.

Assuming that the equilibrium velocity is reached almost instantaneously, which in this case is true the larger the particle, one can calculate the time required for a particle of diameter $d = 2r$ to fall a height h, knowing that velocity is space traveled by unit of time ($v = h/t$):

$$t = \frac{18 h \eta}{d^2 g \left(\rho_g - \rho_f\right)} \quad (3.6)$$

Considering soil particles are spherical and of uniform density, that they move individually, that the flow of the fluid around them is laminar, and that the particles are large enough not to be affected by the thermal movements of the molecules of the fluid, Stokes law, represented by Eq. (3.5) or (3.6), can be used to determine the distribution of the falling particles. For this purpose, 0.05–0.1 kg of soil is passed through a 2 mm sieve, dispersed in 1 L of water, in a laboratory cylinder. Various soil dispersants are used, which are necessary to keep the soil particles separated (not agglutinated, i.e., dispersed), such as soda or Calgon. Wishing to find the amount of suspended material in solution, having a diameter d less than a chosen value, we let the soil solution to decant for a time t (calculated by Eq. (3.6)), so that a height h (from the top liquid surface) is free of particles of diameters greater than d. This is because the rate of fall is proportional to the mass and the heavier particles (larger diameter than d) already passed the depth h. In general, the height h is chosen as 0.10 m, a height sufficient for a densimeter to be introduced (between 0 and h), thus determining the density of the suspension. Examples will clarify the procedure:

Example 1: For how much time we have to wait so that the first 10 cm of a liquid soil suspension becomes free of (a) sand particles; (b) sand and silt particles. For the diameters of these particles, consult Fig. 3.2. Consider the properties of the solution equal to those of pure water at 20 °C, and the soil particle density as 2.65 g cm^{-3}.

Solution:

$$\text{(a)}\ t_a = \frac{18 \times 10 \times 0.01}{(0.002)^2 \times 980 \times (2.65 - 1.0)} = 278.3\ \text{s} = 4.64\ \text{min}$$

which means that after 4.64 min the first 0.10 m of the suspension does not have sand particles, but still has silt and clay particles.

$$\text{(b)}\ t_b = \frac{18 \times 10 \times 0.01}{(0.0002)^2 \times 980 \times (2.65 - 1.0)} = 27{,}829\ \text{s} = 7.73\ \text{h}$$

that is, after 7.73 h, we find only clay in the first 0.10 m.

Example 2: In the solution of the previous example, the concentration of suspended solids (in the upper layer of 0.10 m) was measured by a densimeter, resulting $C_a = 30$ g L^{-1} and $C_b = 18$ g L^{-1} obtained at instants t_a and t_b, respectively. What is the textural class of the soil, knowing that 50 g of soil were dispersed in 1 L of water?

Solution:

The initial concentration C_0 is 50 g L^{-1}, and this includes sand, silt, and clay. C_a includes silt and clay and C_b only clay. In this way:

$$\text{Sand} = \left(\frac{C_0 - C_a}{C_0}\right) \times 100 = \left(\frac{50 - 30}{50}\right) \times 100 = 40\%$$

$$\text{Silt} = \left(\frac{C_a - C_b}{C_0}\right) \times 100 = \left(\frac{30 - 18}{50}\right) \times 100 = 24\%$$

$$\text{Clay} = \left(\frac{C_b}{C_0}\right) \times 100 = \left(\frac{18}{50}\right) \times 100 = 36\%$$

According to the classic **textural triangle**, the soil of this example belongs to the textural class: clay loam (see Fig. 3.3).

The results obtained through mechanical analysis of a soil are generally presented as a table or graph (percentage of particles with diameter smaller than d as a function of the log of the diameter), as can be seen in Fig. 3.4. More details on mechanical soil analysis can be seen in Dane and Topp (2002).

In 1992, Vaz et al. published a new method for mechanical soil analysis that measures concentrations C by attenuation of a gamma radiation beam passing through the soil–water suspension. This method is continuous and does not interfere in the sedimentation process, avoiding the introduction of a densimeter or pipette. This method has been further improved by Oliveira et al. (1997). The sedimentation vessel is prismatic, with the thickness (x) being traversed by the gamma radiation beam (see the principle of gamma attenuation in Chap. 6), and the C concentrations of suspended particulates can be estimated at any time t and depth h (the position of the radiation beam), by the equation:

$$C = \frac{\ln(I_0/I)}{x(\mu_p - \mu_w d_w/d_p)} \tag{3.7}$$

where I_0 and I are the incident and emergent intensities of the radiation beam passing through the vessel, μ_p and μ_w are the attenuation coefficients of the particles and water, and d_w and d_p the water and particle densities. The method has become much faster because the authors have recognized that in Eq. (3.6) the ratio t/h is constant, which means that there are infinite combinations of t in which the volume of water above h is free of particles larger than d. They abandoned the traditional depth $h = 0.10$ m and made the beam of radiation fall on variable heights, starting from the deepest part of the sedimentation vessel. Thus, it was possible to measure the coarser particles (sands) at greater h with

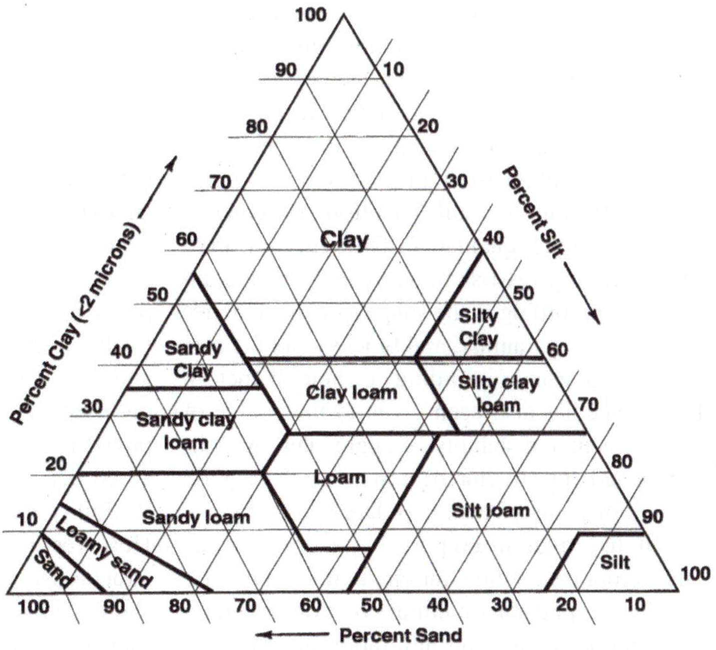

Fig. 3.3 Soil textural triangle and textural classes. (Source: Plant & Soil Sciences eLibarr^{PRO}) (http://passel.unl.edu/pages/informationmodule.php?idinformationmoule=1130447039&topicorder=2&maxto=10)

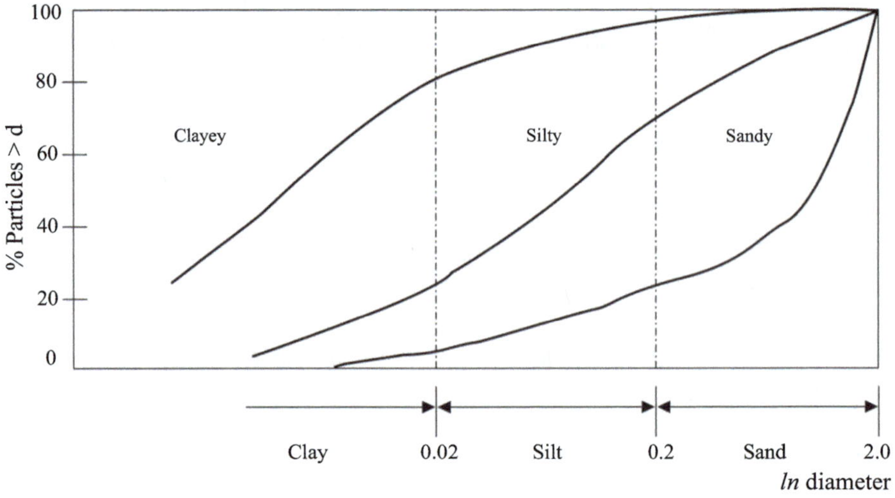

Fig. 3.4 An example of particle distribution made for three typical Brazilian soils

more discrimination and, when the beam was slowly elevated along the vessel and almost reached the surface of the liquid ($h = 1$ cm), they could measure the clay in a much shorter time (of the order of 20 min). As a result, the method was automated in an apparatus patented by CNPDIA/Embrapa, São Carlos, SP, Brazil.

The term (and the concept) **soil structure** is used to describe the soil in terms of the arrangement of the solid particles, orientation, and organization. The structure also defines the geometry of porous spaces. Since the arrangement of soil particles in general is too complex to allow for any simple geometric characterization, there is no practical way of measuring (or attributing a number) the structure of a soil. Therefore, the concept of soil structure is qualitative. The **aggregation** of soil particles gives rise to aggregates, which are classified according to the shape (prismatic, laminar, columnar, granular, and blocky) and the size of the aggregates (according to their diameter). Importance has also been given to the degree of development and stability of aggregates. A well-aggregated (or structured) soil presents a good amount of pores of relatively large size. We say that it has high macroporosity, a quality that affects root penetration, air circulation (aeration), handling from the agricultural point of view (cultivation operations), and infiltration of water. An indirect evaluation of soil structure is made by means of penetrometers, instruments that measure the resistance of a soil to penetration, a subject discussed in greater detail at the end of Chap. 6.

In recent years, the standard methods used as indicators of **soil structural quality** have been complemented by simple methods of visual field evaluation. Mueller et al. (2009) cite that several methods of visual description of the soil are available and may differ in many respects, including the considered depth of the soil, emphasis on particular characteristics of soil structure, and the scoring scales. Among these, the methods known as "Visual Soil Assessment" (VSA), described in Shepherd (2009), and "Visual Evaluation of Soil Structure" (VESS), described in Ball et al. (2007), have been widely used in the evaluation of soil structure in the field. Garbout et al. (2013) mention that the advantages of soil physical assessment directly in the field are: relatively short time consumption, immediate availability of results, use of simple equipment, observation of rapid changes in conditions of physical properties of the soil that can be difficult to be determined by any other means, and the flexibility to handle a wide range of situations. However, it is necessary that indicators of soil structural quality measured directly in the field should be correlated with soil properties determined by standard methodologies. Mueller et al. (2009) report that, in general, there are significant

Table 3.1 Approximated chemical composition of the main primary materials in the soil

Mineral	SiO_4	Al_2O_3	Fe_2O_3	CaO	MgO	K_2O	P_2O_5
Quartz	100	–	–	–	–	–	–
Orthoclase	64	19	–	1.5	–	12	–
Albita	65	23	–	4.5	–	2	–
Anorthite	42	33	–	15	–	1	–
Muscovite	45	35.5	1	–	1.5	9.5	–
Biotite	34.5	21	10	1	11	7.5	–
Hornblenda	48	9.5	3	7.5	14	1	–
Angita	50	6.5	3	21	13	–	–
Olivina	39	–	1.5	–	40	–	–
Apatite	–	–	–	54.5	–	–	41
Magnetite	–	–	69	–	–	–	–

correlations between soil properties (soil resistance to penetration, soil bulk density, and total porosity) measured by standard procedures and visual field evaluation parameters.

Guimarães et al. (2013), who evaluated in a field soil located in the municipality of Maringá, Paraná state, Brazil, the correlation between the VESS and the standard method of determining the **least limiting water range** (LLWR, see also Chap. 14) in a very clayey soil under conditions of no-tillage for a long period of time. With the objective of evaluating the use and the ability of the VSA and VESS field methods to describe the structural quality of two soils with contrasting textures and under two different management practices (conventional and no-tillage), and to identify the similarities between them and some soil physical-water properties (aggregate stability, soil bulk density, total soil porosity, and saturated and unsaturated soil hydraulic conductivities), Moncada et al. (2014) found that both methods are reliable and promising, and can provide rapid semiquantitative measurements of soil physical quality indicators as well as provide farmers with a simpler and faster method of assessing soil structural quality.

From the qualitative point of view, we will separate the solid fraction of the soil into four subfractions: **primary matter**, oxides and salts, organic matter, and **secondary matter**. Table 3.1 gives a qualitative description of the species of primary materials in the soil. As can be seen, the nutrients (see Chap. 4) for plants Ca, Mg, K, and Fe are abundant in the primary materials. Phosphorus P is less frequent, with the absence of the N and S and the micronutrients (Cu), (Zn), (Co), (Mo), and (B).

Table 3.2 Composition of the main oxides found in tropical and temperate climate soils

Oxides	Tropical soil	Temperate soil
SiO_4	3–30	60–95
Al_2O_3	10–40	2–20
Fe_2O_3	10–70	0.5–10
CaO	0.05–0.5	0.3–2
MgO	0.1–3	0.05–1
K_2O	0.01–1	0.1–2
P_2O_5	0.01–1.5	0.03–0.3

Oxides and salts, especially carbonates and sulfates, may contain nutrients such as Ca, Mg, S, and Fe. However, carbonates, when present, are probably more important as pH buffers. Iron oxides are important as cementing agents (formation of aggregates) between particles. In Table 3.2, we present the composition of the main oxides found in tropical and temperate climate soils. Oxides and hydroxides of Fe and Al (aluminum) exist in many soils, both in crystalline form and as amorphic material. Depending on the external pH and the saline concentration of the soil solution, they dissociate H^+ or OH^- groups, becoming electrically charged, being able to adsorb cations and anions at negative and positive charge points, respectively, thus contributing to the ion exchange capacity, which will be studied in detail in Chap. 8.

The **organic matter** is one of the main sources of N, S, and P to the plant. Its content, in most mineral soils, varies between 1 and 10%. The organic matter also has a large specific surface area, which is reactive due to the dissociation of the COOH, OH, and NH_2 groups, producing complexes with Fe, Mn, Ca, Mg, and others. **Soil organic matter** (OM) is the part of the solid fraction composed of solid organic compounds of vegetal or animal origin, in its most varied degrees of transformation. The most advanced stage of transformation is called **humus**, formed by the action of microorganisms, whose typical characteristics are: colloidal state, dark color, and high soil stability.

The composition of raw OM is very variable, being grouped into ten categories related to their facility of transformation:

1. Carbohydrates: sugars, starch, cellulose, and hemi-cellulose;
2. Lignin;
3. Tannins;
4. Glycosides;
5. Organic acids, salts, and esters;
6. Fats, oils, and waxes;
7. Resins;
8. Nitrogen compounds: proteins, amino acids, etc.;
9. Pigments: chlorophyll, xanthophyll, etc.; and
10. Mineral constituents: salts, acids, bases etc.

Their decomposition in the soil is made by action of a large number of native micro-organisms, following the reaction:

$$C_{org} + O_2 \xrightarrow[\text{Microorganisms}]{\text{Enzymes}} CO_2 + H_2O + \text{Energy}$$

which is, in essence, the inverse of the photosynthesis process (see Chap. 4).

An important characteristic of OM for agriculture is its **C/N ratio**, which varies greatly according to its origin. Leguminous residues are more proteic (they fix atmospheric N_2 by symbiosis) and have a low C/N ratio, that is, between 20/1 and 50/1. Cereal straws have values between 50/1 and 200/1 and woods between 500/1 and 1000/1. The interesting thing is that, whatever is the C/N ratio of the plant residue, its transformation in the soil leads to an exponential decrease of the ratio, finally reaching values of the order of 10/1 to 12/1, a typical value for humus. The transformation time is variable for each type of waste. According to this reaction, the microorganisms draw the carbon energy and use the nitrogen in their metabolism. The latter is also released by mineralization processes, reaching the soil solution in the form of NO_3^-. The ideal C/N ratio for decomposition is 30/1; the microorganisms consume at least two thirds of the organic matter to obtain energy and fix one third in their tissues. For example, a C/N = 31/1 residue decomposes as:

31 kg residue ↗ 10 kg C + 1 kg N → humus 10/1
31 kg residue ↘ 20 kg C → CO_2

This relation 30/1 is that of the maximum exploitation of C and N of the residue. Any other relationship, both higher and lower, will lead to excessive losses of C or of N. For example, a high ratio C/N = 101/1:

101 kg residue ↗ 10 kg C + 1 kg N → humus 10/1
101 kg residue ↘ 90 kg C → CO_2

In this case, it can be seen that 9/10 of C are lost as CO_2, and that the amount of humus produced is relatively lower, and the decomposition time is higher because N has become a limiting factor. Now, a low ratio of C/N = 15/1, which characterizes an excess of N, the remainder of which is eliminated by microorganisms in the form of NH_3^-. To compare this case with that of the 30/1 ratio, let us convert a residue from 15/1 to 30/2:

32 kg residue ↗ 10 kg C + 1 kg N → humus 10/1
32 kg residue → 20 kg C → CO_2
32 kg residue ↘ 1 kg C → NH_3

Animal manure has low C/N ratios, and it is therefore advisable to mix it with cane straw or other crop residues of high C/N ratio. From what we have seen, the decomposition of OM leads to the production of: (1) energy; (2) simple products (CO_2, H_2O, mineral salts containing, above all, N and P); and (3) humus, composed of numerous compounds of high molecular weight (2000–4000). The presence of humus in the soil increases the cation exchange capacity (CEC) (because of the large number of free radicals present in its structure), increases nutrient availability, increases buffering power, and tends to increase the pH of acidic soils and Al toxicity to plants. An example of application of decomposition of dry matter in relation to nitrogen is discussed in Dourado-Neto et al. (2010), in which the isotope ^{15}N is used as the tracer (you find details of the use of tracers in Chap. 16).

The soil has pores of various shapes and sizes, which condition a peculiar behavior to each soil. The fraction of the soil that most decisively determines its physicochemical behavior is the clay fraction, which is **secondary matter**. It has the largest specific area (area per unit mass) and, therefore, is the most active fraction in physicochemical processes occurring in the soil. Clay particles retain water and are responsible for the processes of expansion and contraction when a soil retains or loses water. Most of them are negatively charged and thus form an "electrostatic double layer" with ions in solution and with water molecules that are dipoles. Sand and silt have relatively small specific areas and consequently do not show much physicochemical activity. They are important for the macroporosity of the soil where capillary phenomena predominate, when the soil is close to saturation. Along with clay, silt and sand form the solid matrix of the soil, also called the mineral fraction of the soil.

The clay, consisting of particles smaller than 2 μm (10^{-6} m), comprises a large group of minerals, some of which are amorphous, but most of them are microcrystals of colloidal size and of defined structure. Among these crystals, or **clay minerals**, are the **aluminosilicates**.

Basically, the aluminosilicates are composed of two structural units: a tetrahedron of oxygen atoms involving a silicon atom (Si^{4+}), and an octahedron of oxygen atoms (or OH^- hydroxyl group), involving an atom of aluminum (Al^{3+}). The tetrahedra and octahedra are joined by their vertices by means of oxygen atoms that are shared. For this reason, layers of tetrahedra and octahedra are formed, which can be seen in section, in Fig. 3.5. There are two main types of aluminosilicates, depending on the relationship between layers of tetrahedra and octahedra. In 1:1 minerals, such as kaolinite, a layer of octahedra shares oxygen from a layer of tetrahedra. In 2:1 minerals, like montmorillonite, a layer of octahedra shares oxygen from two layers of tetrahedra.

These described structures, also called **clay mineral micelles**, are ideal and are electrically neutral. In nature, however, substitutions of atoms (**isomorphic substitutions**) occur during their formation, which produce a non-balance of electric charges. Thus, it is common to substitute Si^{4+} by Al^{3+} in the tetrahedra and the substitution of Al^{3+} by Mg^{2+} and/or by Fe^{2+} in the octahedra. These substitutions occur because, as far as the atomic size is concerned, these atoms can perfectly substitute each other in the crystalline lattice, and as a result, negative charges of oxygen remain unbalanced, making the micelles electrically charged surfaces.

Another source of unbalanced charges in clay minerals is the incomplete neutralization of the atoms at the ends of the crystals and of organic materials. The unbalanced charges of the clays are externally neutralized by the aqueous solution of the soil, that is, by exchangeable ions (Ca^{2+}, H^{1+}, Mg^{2+}, $H_2PO_4^{1-}$, NO_3^{1-}, PO_4^{3-}, etc.) or even by water dipoles. These ions also penetrate between overlapping micelles in order to neutralize the charges caused by isomorphic substitution. A typical case is potassium, an essential element of plants, which, when penetrating between the layers of some aluminosilicates, becomes practically unavailable. These electrically adsorbed ions are not part of the crystalline structure and may be "exchanged" or replaced by others. This substitution phenomenon is called **ion exchange** and has vital importance in soil physicochemistry because it affects the retention or release of nutrients for plants, for minerals, and for the flocculation and dispersion processes of soil colloids.

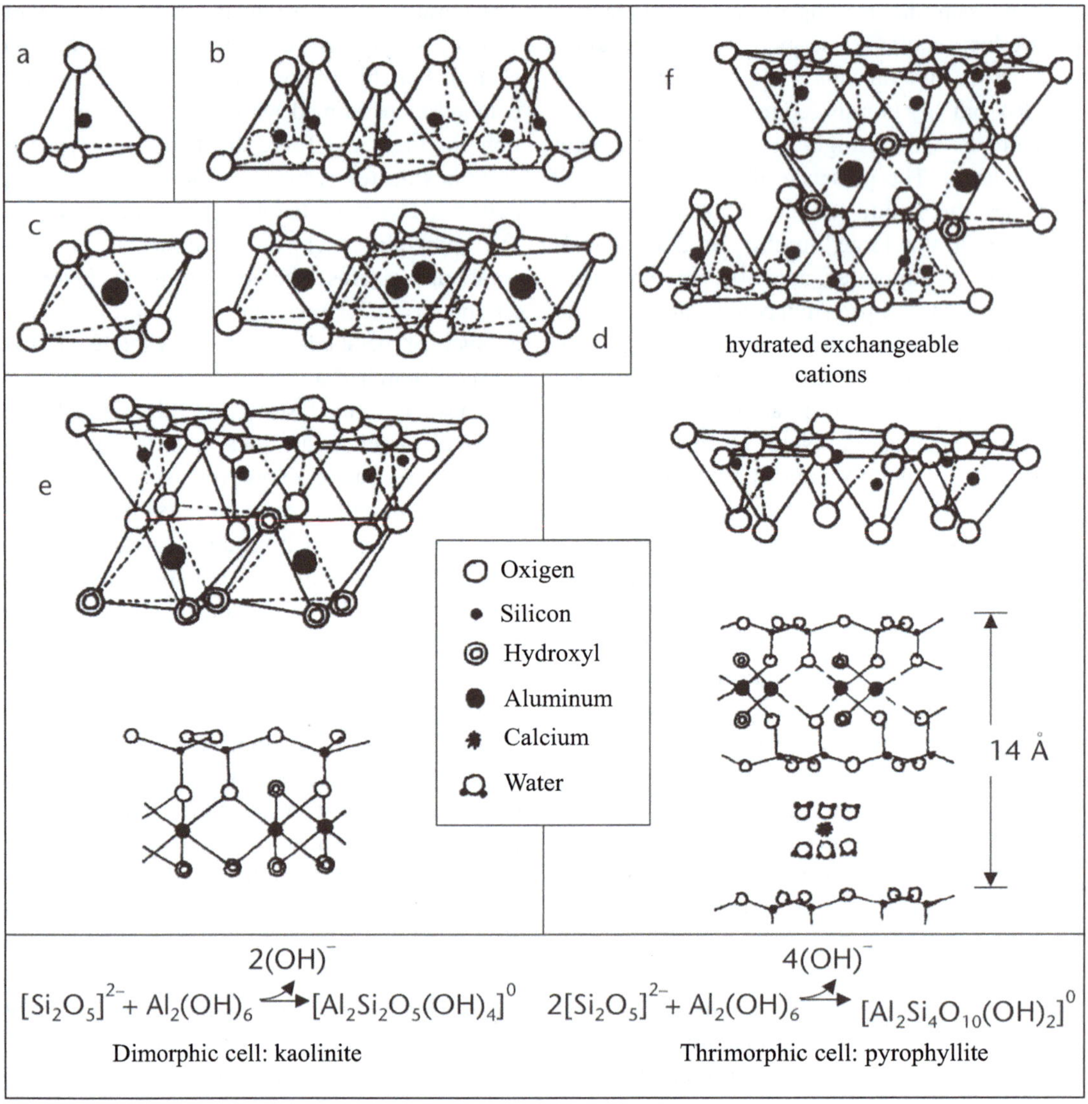

Fig. 3.5 Structure of aluminosilicates: (a) silicon tetrahedron; (b) layer of tetrahedra; (c) aluminum octahedron; (d) layer of octahedra; (e) 1:1 micelle; (f) 2:1 micelle

The unbalanced charge of the clay micelles is generally described as a **surface charge density** (net number of exchange positions per unit of micelle area), but the characterization of a soil by the **cation exchange capacity** (CEC) is more common. The exchange capacity will be studied in detail in Chap. 8. It is usually given in centimoles of charge per cubic decimeter of soil ($cmol_c\ dm^{-3}$). Thus, a soil with CEC of 15 $cmol_c\ dm^{-3}$ has the ability to retain 15 cmol of any cation in every dm^3 of soil. The CEC depends on a number of factors, distinguishing between them the pH of the soil solution. In the case of cations, pure montmorillonite samples at pH = 6 have a CEC of around 100 $cmol_c\ dm^{-3}$, whereas kaolinite has only 4–9 $cmol_c\ dm^{-3}$. These differences are mainly due to differences in surface charge density and specific **soil surface area**, which is defined as the total surface area of the solid particles per unit mass of the soil. It is usually given in $m^2\ g^{-1}$. Montmorillonite and pure kaolinite present, approximately and respectively, 800 and

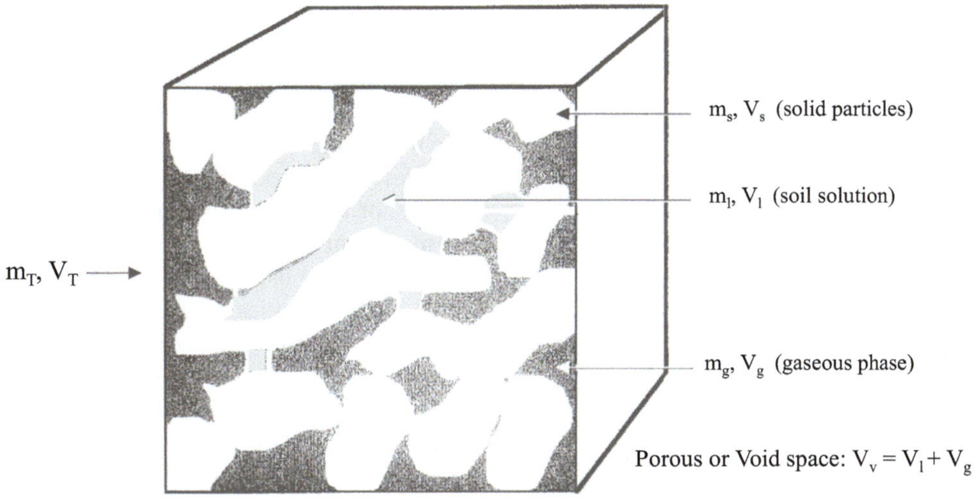

Fig. 3.6 Schematic soil sample indicating solid, liquid, and gaseous fractions

100 $m^2\ g^{-1}$. By way of comparison, the specific surface of the sand is not larger than 1 $m^2\ g^{-1}$.

Organic matter also contributes significantly to the cation exchange capacity. The CEC of humic acids ranges from 250 to 400 $cmol_c\ dm^{-3}$, i.e., three times greater than montmorillonite, but depending largely on pH. Organic matter also exerts great influence on the structure of a soil.

Some mass–volume relationships are commonly used to describe the three soil fractions: solid, liquid, and gaseous, and their interrelationships. If we take a soil sample (sufficiently large to contain the three fractions and represent a certain portion of the profile—see Fig. 3.6—say a soil clod of 0.100–0.500 kg), we can discriminate the masses and the volumes of each fraction:

$$m_T = m_s + m_l + m_g \quad (3.8)$$

$$V_T = V_s + V_l + V_g \quad (3.9)$$

where m_T is the total mass of the sample; m_s is the mass of the solid particles; m_l is the mass of the soil solution, which, because being diluted is taken as pure water; m_g is the mass of the gaseous phase, i.e., soil air, which is a negligible mass with respect to m_s and m_l. V_T is the total sample volume; V_s is the volume occupied by the solid particles; V_l by water; and V_g is the volume of the gases (not negligible as in the case of their mass).

The following definitions related to the solid fraction are important and frequently used in soil physics:

1. **Specific mass of particles**, in the soil science literature called **particle density** (which was in the past also called real soil density):

$$d_p = \frac{m_s}{V_s}\ (\text{kg m}^{-3}) \quad (3.10)$$

In physics, the correct term for this concept is **soil-specific mass**, with the dimensions given in Eq. (3.10), because density is a relative and dimensionless quantity, in this case it would be the relationship between the specific mass of the soil and the specific mass of the water at 25 °C. As the use of the term density with the indicated units is embedded in soil science, in this book, we will use it also.

2. **Specific soil mass** or **soil bulk density** (which was also called apparent soil density):

$$d_s = \frac{m_s}{V_T}\ (\text{kg m}^{-3}) \quad (3.11)$$

If we take a soil clod with $m_s = 0.335$ kg, $V_s = 0.000126\ m^3$, and $V_T = 0.000255\ m^3$, we will have:

$$d_p = \frac{0.335}{0.000126} = 2658.7 \text{ kg m}^{-3}$$

$$d_s = \frac{0.335}{0.000255} = 1313.7 \text{ kg m}^{-3}$$

The particle density depends on the constitution of the soil and, as it varies relatively little from soil to soil, it does not vary excessively between different mineral soils. The density of the particles is similar to those of the rocks that formed the soil. The quartz has $d_p = 2650$ kg m^{-3}, and since it is a frequent component in soils, the density of soil particles oscillates around this value. The average for a wide variety of soils is 2700 kg m^{-3}. If the formation of the soil is very different, as in the case of peat soils (with a lot of organic matter), its value may be lower. Some examples are:

Argisols, sandy fraction:

$$d_p = 2650 \text{ kg m}^{-3}$$

Nitosols:

$$d_p = 2710 \text{ kg m}^{-3}$$

The determination of the particle density is done by weighing a sample of dry soil (m_s), that is, after a constant weight in an oven at 105 °C (about 24–48 h), and by the measurement of their volume V_s, obtained by the change in volume registered when m_s is immersed in a liquid. For example, 458 g dry soil is placed in a beaker containing 200 cm^3 of water. The final volume is 369 cm^3. Of course, $V_s = 369 - 200 = 169$ cm^3 and therefore the $d_p = 458/169 = 2.710$ g cm^{-3} (= 2710 kg m^{-3}). In this methodology, special vials called **pycnometers** are used, which allow precise determination of volumes. A special care that must be taken is to avoid air bubbles in aggregates or micropores, which lead to error in measurement. Therefore, alcohol or water is often used under vacuum to remove the dissolved air.

Soil bulk density, having the total volume of sample V_T in the denominator of Eq. (3.11), varies according to the V_T. When compressing a soil sample, m_s remains constant and V_T decreases; therefore, d_s increases. Soil bulk density is therefore an indicator of the degree of **soil compaction**. For coarse and sandy soils, the particle arrangement possibilities are not very large, and therefore the compaction levels are also not large. Due to the fact that they have larger particles, the porous space is also constituted, mainly, of larger pores denominated, arbitrarily, as **macropores**; in an apparently paradoxical way, in these soils, the total volume of pores is small. The densities of sandy soils oscillate between 1400 and 1800 kg m^{-3}. For a single sandy soil, this range of variation, at different levels of compaction, is much lower.

For fine, clayey soils, the possibilities of arrangement of the particles are much greater. Their porous space consists essentially of **micropores**, and the volume of pores V_g is relatively large, which is why they present a slightly higher range of soil densities (900–1600 kg m^{-3}).

Directly linked to the definition of density is that of **porosity**, a measure of the porous space of the soil. The total porosity α, also called the **total soil pore volume** (TPV), of a soil is defined by:

$$\alpha = \frac{V_p}{V_T} = \frac{V_T - V_s}{V_T} \qquad (3.12)$$

It is dimensionless (m^3 m^{-3}/m^3 m^{-3}) and is generally expressed as a percentage. For the example of the clod used to exemplify Eqs. (3.10) and (3.11), we have:

$$\alpha = \frac{0.000255 - 0.000126}{0.000255} = 0.506 \text{ or } 50.6\%$$

which means that 50.6% of the sample can be occupied by air and/or water.

The total porosity is logically also affected by the level of compaction. For the same soil, the greater the d_s, the smaller the α is. As discussed for soil bulk density, a distinction is made between macro and micropores. This definition is arbitrary, but logical. The larger pores, represented by the macroporosity, are the most important in soil aeration and fast water flows, while smaller ones, represented by the microporosity, contain the most retained water in the soil. Examples of three Brazilian soils are shown in Table 3.3. The boundary between macro and microporosity will

Table 3.3 Values of soil bulk density (d_s), particle density (d_p), and total porosity (α) for some soils of Minas Gerais, Brazil (Freire et al. 1980)

Soil class	d_s (kg m^{-3})	d_p (kg m^{-3})	α (%)
Argisol	1200	2600	53.8
Nitosol	1000	2700	62.9
Latosol	1100	2700	59.2

be discussed in connection with Chap. 6, as it involves the concept of soil water retention.

It can also be shown that:

$$\alpha = \left(1 - \frac{d_s}{d_p}\right) \quad (3.12a)$$

A formula much used to estimate α from d_s and d_p data.

The visualization or feeling of the porosity of soils can be improved by a simple model in which a soil is represented by n^3 equal spheres of radius r, arranged in the cubic system, placed in a cubic box of side $L = 2rn$, as indicated in Fig. 3.7 for $n = 4$, that is, four spheres fit exactly along an edge of the cube:

1. Volume of the spheres: $n^3\left(\frac{4\pi}{3}\right)r^3$
2. Volume of the box: $L^3 = (2nr)^3$
3. Porosity:
   $\alpha = \frac{n^3(4\pi/3)r^3}{8n^3r^3} = \frac{4\pi}{24} = 0.5236$ or 52.36%

This simplified soil model shows that the porosity of this simplified and homogeneous model is independent of n, that is, of the number of spheres placed in the box, provided that the cubic arrangement is respected. Even with only one sphere is placed in the box, the porosity is 52.4%. In this arrangement, if we place in 1 L n spheres of the same size within the sand interval, the porosity will be around 52.4%. Exchanging for equal spheres within the silt range, or in the clay band, the porosity does not change, it will be the same 52.4%. In real soils, of course, we do not have particles of the same size and all are found in different numbers. The interesting thing is that the porosity values of most soils vary around 50%, which corresponds to a soil with $d_s = 1325$ kg m^{-3} and $d_p = 2650$ kg m^{-3}, according to Eq. (3.12a). If this soil is compacted

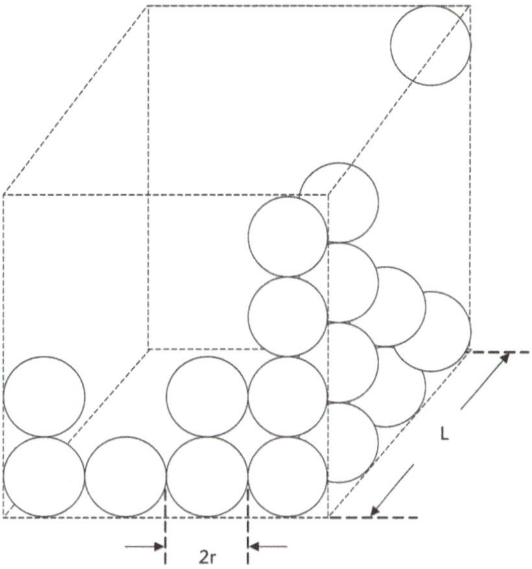

Fig. 3.7 Cubic arrangement of spheres of radius r placed in a cubic box of side L

to $d_s = 1725$ kg m^{-3}, α becomes 34.91%. On the other hand, if it is softened to $d_s = 925$ kg m^{-3}, α becomes 65.09%. These would be about the extreme porosity values for this soil. Kutilek and Nielsen (1994) discuss in more depth the modeling of porosity.

In determining the soil bulk density, the major problem is in the measure of the total volume of the sample, V_T. A clod has an irregular shape and is porous, and the pores are part of V_T. One common method of measuring V_T is the **paraffin method**, in which the dry mass m_s of a soil clod, suspended by a very thin sting, is immersed in water after being covered with paraffin (the paraffin must be very viscous not to enter the pores, just covering the surface of the clod making it impermeable to water). Its volume can then be determined by buoyancy (Archimedes' Principle). Another very common technique is that of the volumetric ring: a ring of internal volume V_T, with sharp edges, is introduced into the soil until it is completely filled with soil. After that, the excess of soil is eliminated with a spatula. This determination, however, must be done with care, in order not to compact the soil during ring introduction into the soil (many times done with a rubber hammer), which is very difficult.

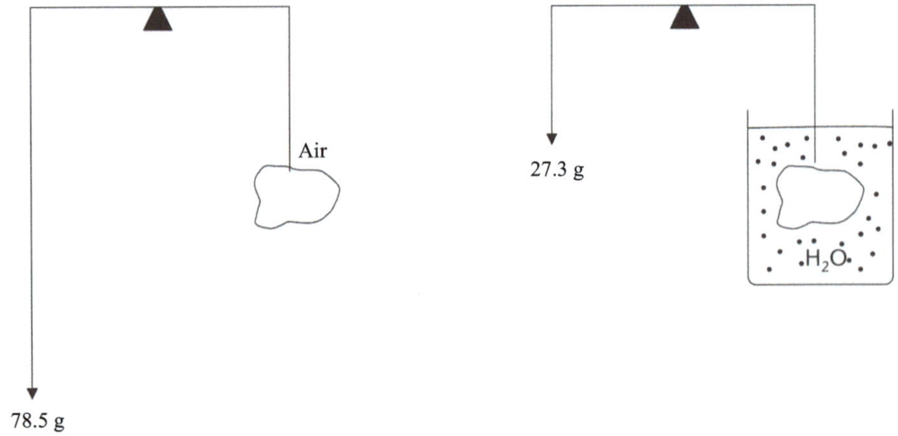

Fig. 3.8 Illustration of the paraffin method for the measurement of soil bulk density

Figure 3.8 explains the paraffin method:

1. Mass of the dry clod: 75.0 g;
2. Mass of the dry clod covered with paraffin: 78.5 g;
3. Mass of paraffin: 3.5 g (at immersion the paraffin should not be too warm, neither to cold, but liquid, so that paraffin does not enter the pores);
4. Mass of the clod with paraffin suspended in water: 27.3 g;
5. Volume of the clod covered with paraffin: (78.5 − 27.3) = 51.2 g = 51.2 cm^3 (since after Archimedes the buoyancy force is the weight of the displaced liquid, in the case, the water of specific mass 1 g cm^{-3});
6. Volume of paraffin: assuming a specific mass of 0.9 g cm^{-3}, results in 3.9 cm^3;
7. Volume of the clod: 51.2 − 3.9 = 47.3 cm^3; and finally
8. Soil bulk density: d_s = 75.0/ 47.3 = 1.585 g cm^{-3} or 1585 kg m^{-3}.

Another problem in determining soil bulk density is the water content. The definition involves m_s obtained with dry soil and, when drying a soil, most samples contract, decreasing V_T. This is of high significance in expansive, clayey soils with type 2:1 clays, such as Vertisols. In these cases, it is best to indicate the soil water content at which V_T was determined.

In Chap. 6, we discuss in detail the methodology used to measure soil bulk densities and water contents. More information on sampling, number of samples, and sampling sites can still be found in Reichardt (1987) and Webster and Lark (2013).

3.3 Liquid Fraction of the Soil

The liquid fraction of the soil is an aqueous solution of mineral salts and organic substances, the mineral salts being of major importance. In general, the soil solution is not the main reservoir of plant nutrient ions, except for chlorine and perhaps sulfur, which are not adsorbed by the solid fraction or incorporated into the organic matter. When the plant withdraws ions from the soil solution, its concentration may vary with time for each nutrient and each special environmental condition. There is a constant interaction between the solid fraction (ion reservoir) and the liquid fraction, which is a complex interaction, governed by solubility products, equilibrium constants, etc. For this reason, the description of the soil solution concentration becomes difficult and only average and approximate values can be obtained. As examples, Reisenauer (1966) presents mean values for a number of soils in the state of California, USA (Table 3.4), and Table 3.5 shows soil composition data, according to Fried and Shapiro (1961).

For Brazilian conditions, Malavolta (1976) presents the concentration ranges of the main nutrients (Table 3.6).

Observing the data from these tables, there is a great difference between the authors, but in average terms, the main nutrients, except phosphorus, are present in concentrations between 10^{-4} and 10 M. Phosphorus usually has the lowest concentration, usually between 10^{-5} and 1 M.

The presence of solutes grants to the solution what in the past was called **osmotic pressure** (P_{os}). The osmotic pressure of a solution expresses the potential difference of water in the solution over that of pure water. When a solution is separated from a volume of pure water by means of a **semipermeable membrane** (allows the solvent to pass through and does not let the solute pass), the water will tend to diffuse through the membrane in order to occupy the lower energy state in solution (Fig. 3.9).

The osmotic pressure is the pressure to be applied to the solution in order to prevent the passage of water into the solution. In Fig. 3.9, the solution was initially under pressure P_0 (equal to atmospheric pressure) and, with the passage of water through the membrane, its pressure increases to ($P_0 + \rho g h$), where ρ is the density of the solution. In the equilibrium condition, $\rho g h$ is a measure of the osmotic pressure of the solution.

For diluted solutions, P_{os} can be estimated by **Van't Hoff's equation**, which is empirical and, paradoxically, uses concepts of gas thermodynamics to describe liquid solutions:

$$P_{os} = -RTa \tag{3.13}$$

where R is the universal gas constant (0.082 atm L K^{-1} mol^{-1}); T, the absolute temperature (K), and $\boldsymbol{a}$, the **activity of the solution** (see Chap. 8), which in the case of dilute solutions can be replaced by the concentration C.

The osmotic pressure is actually the osmotic potential of water due to the presence of the solutes and is an energy, more specifically energy per unit volume, which is a pressure (see Chap. 6). The negative sign in Eq. (3.13) shows that the energy of water in the presence of solutes is smaller than the energy of pure water, considered standard and equal to zero. Hence, the spontaneous tendency of the ions to move from a higher concentration to a smaller one, and vice versa, and the water moves to regions with a higher salt concentration.

Example 1: Calculate the osmotic pressure of a 1 M solution of sucrose and that of a 0.01 M $CaCl_2$, both at 27 °C?

Solution:

(a) Sucrose:

$$P_{os} = -0.082 \times 300 \times 1 = -24.6 \text{ atm}$$

Table 3.4 Mean nutrient concentration value for a number of soils in the state of California, USA (Reisenauer 1966)

Nutrient	Concentration ($mmol_c$ L^{-1})
Calcium (Ca^{2+})	1.870
Magnesium (Mg^{2+})	3.086
Nitrogen (NO_3^-)	8.929
Phosphorus (PO_4^{-3})	0.0003
Potassium (K^+)	1.023
Sulfur (SO_4^{-2})	1.558

Table 3.5 Soil solution composition data ($mmol_c$ L^{-1}) (Fried and Shapiro 1961)

Element	Range values for all soil types	Acid soil	Calcareous soil
Ca	0.5–38	3.4	14
Mg	0.7–100	1.9	7
K	0.2–10	0.7	1
Na	0.4–150	1.0	29
N	0.16–55	12.1	13
P	0.001–1	0.007	0.03
S	0.1–150	0.5	24
Cl	0.2–230	1.1	20

Table 3.6 Soil solution composition data ($mmol_c\ L^{-1}$) for some Brazilian soils (Malavolta 1976)

	Concentration ($mmol_c\ L^{-1}$)	
Nutrient	All soil types	Acid soils
N	0.16–55	12.1
P	0.0001–1	0.007
K	0.2–10	0.7
Mg	0.7–100	1.9
Ca	0.5–38	3.4
S	0.1–150	0.5
Cl	0.2–230	1.1
Na	0.4–150	1.0

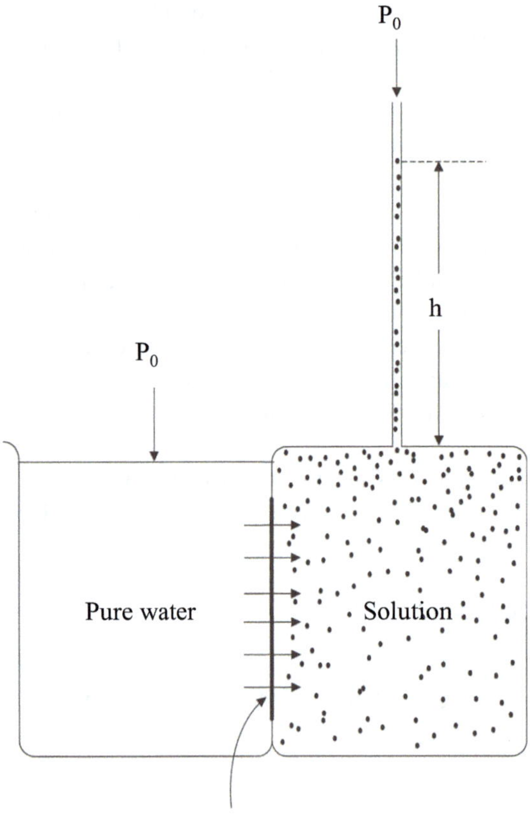

Fig. 3.9 Illustration of the osmotic pressure using a semipermeable membrane

(b) Calcium chloride:

$$P_{os} = -0.082 \times 300 \times (0.01 + 0.02)$$
$$= -0.74 \text{ atm}$$

Example 2: Calculate the osmotic pressure of a nutrient solution containing: KNO_3, $a = 0.006$ M; $Ca(NO_3)_2{\cdot}4H_2O$, $a = 0.004$ M; $NH_4H_2PO_4$, $a = 0.002$ M; $MgSO_4{\cdot}7H_2O$, $a = 0.001$ M; and other micronutrients of negligible concentration. $T = 27$ °C.

Solution:

$$C = (0.006 + 0.006 + 0.004 + 0.008 + 0.002 + 0.002 + 0.001 + 0.001) = 0.030 \text{ M}$$

$$P_{os} = -0.082 \times 300 \times 0.030 = -0.738 \text{ atm}$$

In the scheme of Fig. 3.9, h would, in this case, be approximately 7 m high, since each atmosphere is equivalent to about 10 m of water column.

The quantitative determination of the liquid fraction of a soil, which does not take into account the solutes, or simply refers to pure water, is made in several ways. As we are going to see along the following chapters, that the interest of knowing the water content of a soil sample is, most the times based on volumes. For plants, specifically for seeds, the interest is directed to weights (or masses) of water (see item *g* of Water in the plant, Chap. 4).

1. **Soil water content based on mass *u*:**

$$u = \frac{m_l}{m_s} = \frac{m_T - m_s}{m_s} \quad (3.14)$$

where m_T, m_l, and m_s were already defined in Eq. (3.8) and Fig. 3.6.

The water content u is dimensionless ($kg\ kg^{-1}$), but its units should be kept not to be confused with the volume-based soil water content, which is also dimensionless but numerically different. The term moisture is colloquial and, whenever possible, should be replaced by the expression soil water content to represent the liquid fraction of the soil. The soil water content u is also often given as a percentage, in relation to m_s. The measurement is quite simple: the sample is weighed wet $= m_T$ and then left in an oven at 105 °C, until constant weight m_s (24–48 h or until constant weight). The difference between m_T and m_s is m_l, the mass of the water. Under these conditions, although the soil still contains crystallization water, from an agronomic point of view, the soil is considered to be dry, with $u = 0$. The sample for u measurement can be of any size, provided it is not too small, nor too large (ideal 0.050–0.500 kg), and may have a deformed (not-preserved) structure, like soil sampled with an auger. For the determination, therefore, samples taken in the field with any instrument (auger, shovel, hoe, spoon, etc.) are used, but care must be taken not to let the water evaporate before wet weighing.

2. **Soil water content on volume basis θ**

$$\theta = \frac{V_l}{V_T} = \frac{m_l}{V_T} = \frac{m_T - m_s}{V_T} \tag{3.15}$$

The soil water content on volume basis θ is also dimensionless ($m^3\ m^{-3}$) and is often presented as a percentage. For oven dry soil samples, we also consider $\theta = 0$ and when it is saturated $\theta = \theta_0 = \alpha$. The measurement of θ is more difficult because it involves measuring V_T and, therefore, the sample cannot be deformed (preserved structure). We assume $V_l = m_l$ considering the density of the soil solution as 1000 kg m^{-3} (= 1 g cm^{-3}). V_T is the most difficult to be measured. The most common technique is the use of volumetric rings, identical to those used to measure soil bulk density.

It is not obvious that u is different than θ for the same soil sample. It can be demonstrated that:

$$\theta = \frac{(u \cdot d_s)}{1000} \tag{3.16}$$

u given in kg kg^{-1}, d_s in kg m^{-3}, resulting θ in m^3 of H_2O m^{-3} of soil. One convenient way to obtain θ is measuring u first and thereafter multiplying the result by d_s. Logically d_s has to be known, but since under field conditions, d_s does not vary too much in time (but yes in space!), this procedure is well taken. The bulk density changes a lot as a result of management practices, like plowing, subsoiling, and harrowing. But, in general, the largest changes in d_s occur in the 0–0.30 m surface layer. For deeper layers, d_s is taken as constant.

Example: A soil sample having a volume of 150 cm^3, with a wet mass of 258 g and a dry mass of 206 g. In this way:

$$u = \frac{258 - 206}{206}$$
$$= 0.252\ \text{g g}^{-1} \text{ or kg kg}^{-1} \text{ or } 25.2\%$$

$$\theta = \frac{258 - 206}{150}$$
$$= 0.347\ \text{cm}^3\ \text{cm}^{-3} \text{ or m}^3\ \text{m}^{-3} \text{ or } 34.7\%$$

Note that for the same sample, u is different from θ, hence the need to maintain the units, even though both values are dimensionless.

$$d_s = \frac{206}{150} = 1.373\ \text{g cm}^{-3} \text{ or } 1373\ \text{kg m}^{-3}$$

$$\theta = \frac{(u \cdot d_s)}{1000} = \frac{0.252 \times 1373}{1000} = 0.346\ \text{m}^3\ \text{m}^{-3}$$

It is, therefore, seen that only for the particular case of $d_s = 1$ g cm^{-3}, $\theta = u$, which is the case of a very fluffy soil.

Still using the average value of 2650 kg m^{-3} for the density of the particles, we have:

$$\alpha = 1 - \frac{d_s}{d_p} = 1 - \frac{1373}{2650}$$
$$= 0.482\ \text{m}^3\ \text{m}^{-3} \text{ or } 48.2\%$$

In Reichardt (1987), we can find criteria on how, where, and when to do soil sampling for determinations of u, θ, and d_s. In general, measurements are made with several replicates and at various depths, depending on what you are interested in.

Besides u and θ, the **degree of saturation** S is also used in several situations:

$$S = \frac{\theta}{\alpha} = \frac{0.347}{0.482} = 0.72 \text{ or } 72\% \quad (3.17)$$

S will be 100% when $\theta = \alpha$, which indicates that all porous space is filled with water. A soil under these conditions is called saturated soil. The degree of saturation will be 0% when $\theta = 0$, that is, when the soil is dry (constant weight in oven at 105 °C). Thus, S indicates the fraction of porous space occupied by water. The advantage of using S lies in the fact that it is dimensionless and varies from 0 to 1 for any type of soil (see Chap. 19). The **relative degree of saturation** S_{re} is defined as a function of the residual water content (θ_r) of a very dry soil sample, e.g., the water content of an air dry soil sample. It is defined by:

$$S_{re} = \frac{(\theta - \theta_r)}{(\theta_s - \theta_r)} \cdot 100 \quad (3.17a)$$

However, S will be 0 when $\theta = \theta_r$, not when $\theta = 0$. The use of S_{re} has been quite extensive in the development of mathematical models to represent the water retention curve in the soil (will be discussed in Chap. 6) and to calculate the hydraulic conductivity of an unsaturated soil (Chap. 7).

From the agronomic point of view, it is of fundamental importance to know the amount of water stored in a soil layer or in a soil profile at a given moment. Given the values of soil water content, which are punctual, how do you determine the amount of water stored in a soil layer?

Traditionally, quantities of water are measured by height. Thus, it is said that in Piracicaba, São Paulo (SP) state, Brazil, it rains on average 1200 mm per year. What does that represent? Rainwater is measured by rain gauges, water-collecting containers exposed to weather. They have a catchment area S (m^2) (cross section of their "mouth") and collect a volume V (m^3) of water during a rainfall. The rainfall height is h (m) = V/S, which can be converted to mm (Fig. 3.10). The interesting thing is that h is independent of the size of the mouth of the rain gauge, since a 2S mouth gauge will collect twice the volume, that is, 2V, resulting in the same h. The meaning of h can then be better visualized for the case of $S = 1\ m^2$, i.e., h equal to the volume of water falling on the unit surface.

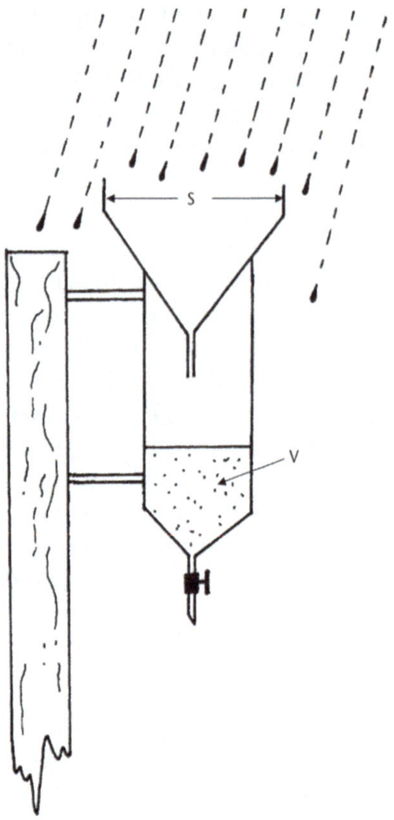

Fig. 3.10 Illustration of a rain gauge

If we throw 1 L of water on a flat, leveled, and impermeable surface of 1 m^2, we obtain a water film of height 1 mm. Thus, 1 mm of rain corresponds to 1 L m^{-2} and therefore the 1200 mm that fall over Piracicaba in 1 year represents 1200 L m^{-2}. Which means that if all the water that precipitated in Piracicaba did not infiltrate, nor evaporate, and does not runoff, at the end of a year we would have 1.2 m of water distributed throughout the area of 1 m^2, or the whole area of Piracicaba. Water supplied by irrigation, water lost by evaporation, etc., are all measured in mm. It would be interesting, therefore, to also measure soil water in mm. This is the concept of **soil water storage**, which will be seen below.

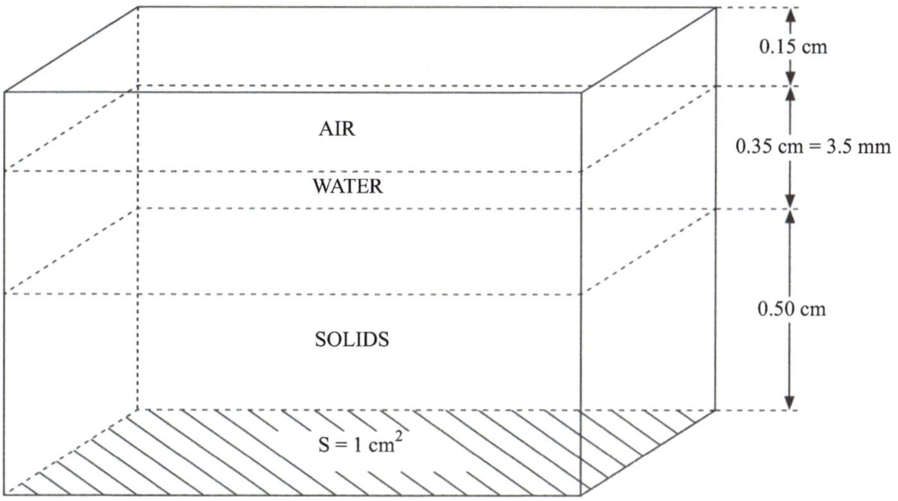

Fig. 3.11 Schematic view of 1 cm^3 of a soil with volumetric water content 0.350 cm^3 cm^{-3}, separating "artificially" the three fractions, solids, water, and air

As in the case of rainfall, the height of water stored by the soil is independent of the area and so, for a unitary surface, $h = V$. For this concept to be better visualized, we use the centimeter as the unit for our calculations. Let us take a unit soil surface of 1 cm^2 and consider the first cm of soil depth. In this case, $V = 1$ cm^3 of soil with soil water content θ_1 (cm^3 of H_2O cm^{-3} of soil—see Eq. (3.15)) and $S = 1$ cm^2. We then have a volume of water V equal to θ_1 cm^3 of water on an area of 1 cm^2 and calculating V/S we have $h_1 = \theta_1$. Let's see an example: if 1 cm^3 of soil has $\theta = 0.35$ cm^3 cm^{-3}, this means that in that soil cube, whose base is 1 cm^2, we have 0.35 cm^3 of water. Therefore, the water height is 0.35 cm or 3.5 mm (Fig. 3.11).

Following the same reasoning in depth, the second cm of soil with moisture θ_2 will have a water height $h_2 = \theta_2$, and so on, so that the nth cm of water with θ_n will have a height $h_n = \theta_n$. It is logical, therefore, that up to a depth of L cm, the stored water height is the sum of all the 1 cm layers down to L. Let the amount of water stored to the depth L be equal to S_L, then:

$$S_L = \sum_{i=1}^{n} \theta_i \tag{3.18}$$

In this reasoning, it is assumed that the soil water content does not vary horizontally, only in depth.

For example, we take a soil in which the water content varies with depth according to Table 3.7.

From what we have been seen, each layer has a water height in cm equal to θ. Thus, the water stored from 0 to 5 cm, according to Eq. (3.18), is:

$$0.101 + 0.132 + 0.154 + 0.186 + 0.201 = 0.774 \text{ cm}$$

or

7.74 mm of water

The water stored up to 10 cm will be: 23.16 mm.

Using higher calculus, the sum of Eq. (3.18) can be replaced by an integral and thus we obtain the definition of **soil water storage**, S_L.

$$S_L = \int_0^L \theta \mathrm{d}z \tag{3.19}$$

where z is the variable that represents the depth in the soil and varies from 0 (soil surface) to L (arbitrary depth of interest); dz represents an

Table 3.7 An example of the calculation of soil water storage

Soil depth z (cm)	Soil water content θ (cm^3 cm^{-3})	S_5 (mm)	S_{10} (mm)
0–1	0.101		
1–2	0.132		
2–3	0.154		
3–4	0.186		
4–5	0.201	7.74	
5–6	0.222		
6–7	0.263		
7–8	0.300		
8–9	0.358		
9–10	0.399		23.16

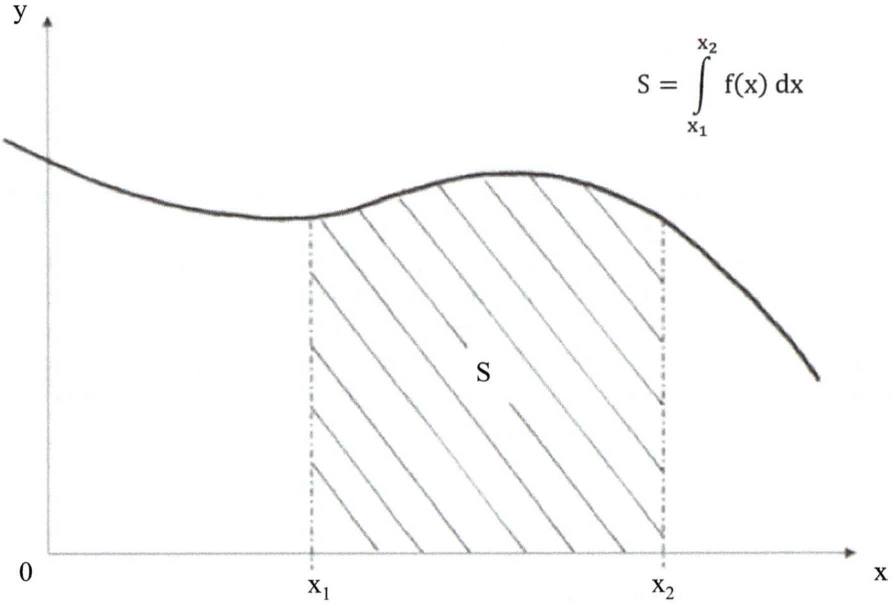

Fig. 3.12 Graphical representation of a definite integral, between limits x_1 and x_2

infinitesimal value of z, i.e., soil layer of thickness as small as desired and which, in the previous example, was taken as 1 cm.

Equation (3.19) is the correct definition of S_L. Comparing Eqs. (3.18) and (3.19), it is worth noting that in Eq. (3.18) dz does not appear because the increment of depth Δz was taken as unitary (1 cm) and in Eq. (3.19) it is an arbitrary infinitesimal dz. We also remind the reader that an integral is the sum of increments. In the case of Eq. (3.18), the increments are finite, covering 1 cm of depth, and in the case of Eq. (3.19), (more exact) they are infinitesimal.

The reader should remember that given a function $y = f(x)$, the integral of y with respect to x, in a given interval, represents the area under the curve, as shown in Fig. 3.12.

Equation (3.19) is an integral of the same type, just replace y by θ and x by z. In Eq. (3.19), θ represents $f(x)$ and it would be better to substitute θ for $\theta(z)$. So, as S can be determined graphically in Fig. 3.12, S_L can also be obtained from a graph

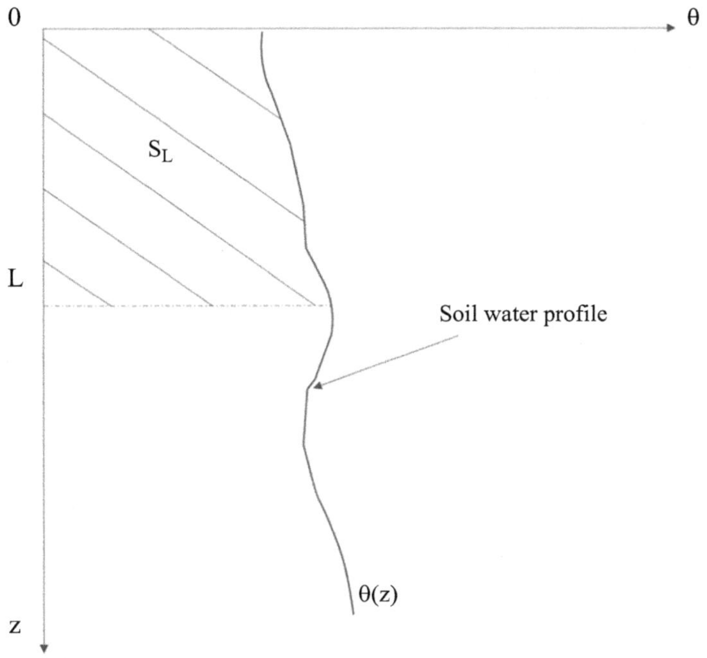

Fig. 3.13 Graphical representation of the soil water storage S_L, from soil surface to depth L

of θ versus z, which, in general, is presented as in Fig. 3.13, the abscissa being the soil water content θ and the soil depth z being taken, for convenience, as positive from top (soil surface taken as reference) downward.

In Fig. 3.13, the graph $\theta(z)$ is called the **soil water profile**. It is seen that, in order to determine the water storage of a profile, the ideal is to know the function $\theta(z)$ that defines the water profile and, thus, to carry out an analytical integration. Since $\theta(z)$ is a function of time and can assume the most varied forms, it is practically impossible to make analytic integrations. We then call upon numerical integrations and, for this, data of θ as a function of z are required to draw the soil water content profile. It is also logical that the more data available, the better the profile and therefore the better the storage calculation.

When a few soil water content data are available, or when the data originate from samples covering soil layers, one way is to transform the curve $\theta(z)$ into a histogram, as shown in Fig. 3.14 (trapezoidal method).

In this case, the storage S_L can be approximated by a sum of rectangles of base θ_i and height Δz, that is:

$$S_L = \theta_1 \Delta z + \theta_2 \Delta z + \cdots + \theta_n \Delta z$$

or:

$$S_L = (\theta_1 + \theta_2 + \cdots + \theta_n) \Delta z$$

where θ_1, θ_2,... θ_n are the θ values for the equidistant depths Δz.

If we multiply and divide the second member by the number n of layers of thickness Δz, we have:

$$\begin{aligned} S_L &= \left[\frac{(\theta_1 + \theta_2 + \ldots + \theta_n)}{n} \right] n \cdot \Delta z \\ &= \overline{\theta} \cdot L \end{aligned} \tag{3.20}$$

where $\overline{\theta}$ is the average soil water content of the layer 0–L, where $n{\cdot}\Delta z = L$. It can, therefore, be seen that the water height contained in a soil layer of thickness L is equal to the product of the depth by the average water content of the layer.

Storage does not necessarily have to be defined from the surface. For a layer extending from a depth L_1 to L_2, the storage will be:

$$S_{(L_2 - L_1)} = \int_{L_1}^{L_2} \theta \mathrm{d}z = \overline{\theta} \cdot (L_2 - L_1) \tag{3.21}$$

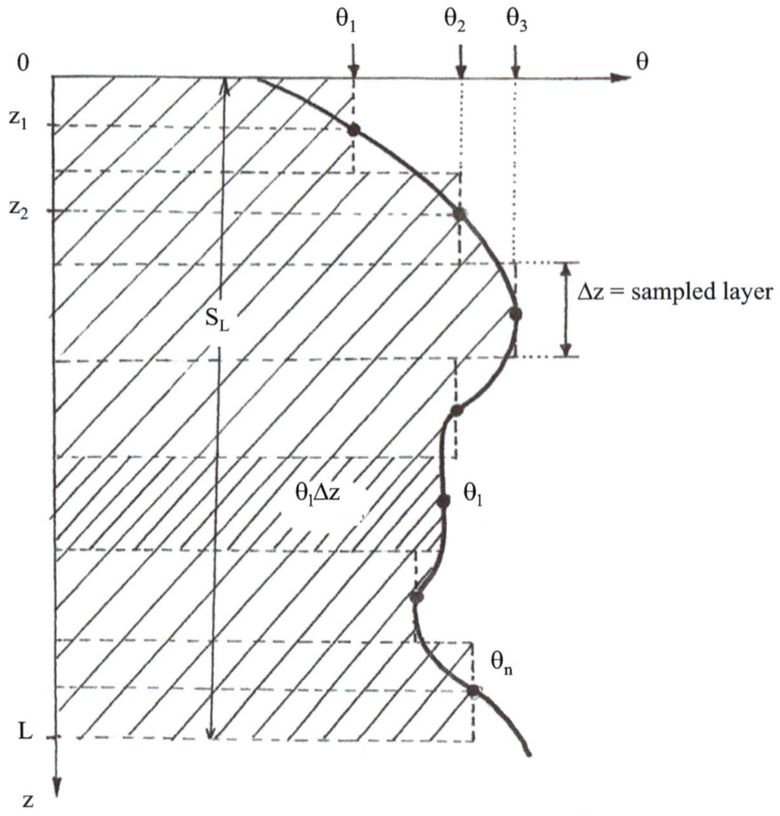

Fig. 3.14 Illustration of the trapezoidal method of integration applied to soil water storage. Observe that each point of θ represents a soil layer Δz

Now $\overline{\theta}$ is the average between L_1 and L_2.

Other ways of determining S_L include the use of a planimeter or the measurement of the area S_L by drawing the graph θ versus z on a good paper in terms of granature, cutting the area S_L, which is weighed on a precision scale and compared to the weight of known area (say a $10 \times 10\ cm^2$ square) the same paper. The profile θ versus z can also be adjusted by models for later analytical or numerical integration.

Exercise: At a given time, soil samples were collected in a sugarcane crop, and the results presented in Table 3.8 were obtained.

Calculate the storages in the 0–0.45 m; 0–0.90 m; 0–1.20 m; 0.45–1.20 m; and 0.15–0.30 m layers. To do this, first transform the data of u into θ by Eq. (3.16) and compute the storages by Eq. (3.20) or (3.21).

Answers by trapezoidal method: $S_{0\text{–}0.45} = 75.6$ mm; $S_{0\text{–}0.90} = 159.3$ mm; $S_{0\text{–}1.20} = 201.6$ mm; $S_{0.45\text{–}1.20} = 126.0$ mm; $S_{0.15\text{–}0.30} = 25.8$ mm

Table 3.8 Soil bulk density and gravimetric water content of a sugarcane field in Piracicaba, SP, Brazil

z (m)	d_s (kg m^{-3})	u (%)
0–0.15	1250	12.3
0.15–0.30	1300	13.2
0.30–0.45	1300	13.8
0.45–0.60	1150	15.2
0.60–0.75	1100	18.6
0.75–0.90	1100	16.3
0.90–1.05	1050	13.7
1.05–1.20	1000	13.7

Note: As discussed, these results depend on the method of calculation and, therefore, one should not expect to obtain identical results as those presented here.

Turatti and Reichardt (1991) conducted a detailed study of soil water storage in a dark red Latosol (Nitosol), using different integration methods and studying the spatial variability of the concept.

Of great importance for the analysis of the behavior of a crop during growth are the changes of soil water content and, consequently, of water storage. The variations are a reflection of the evapotranspiration rates, rainfall, irrigation, and water movements in the soil profile. These processes will be studied in detail from Chaps. 6 to 10; here, we shall be concerned only with the calculation of these variations.

Soil water profiles, such as those shown in Figs. 3.13 and 3.14, are representative for a given and fixed time t. If there is movement of water in the soil, additions by rain or irrigation, and withdrawals by evapotranspiration, these profiles change in shape, and, logically, the water storage also changes. The variable θ is therefore a function of the depth z and, in terms of depth, function of time t, that is, $\theta = \theta(z,t)$. Thus, the storage definition given by Eqs. (3.19), (3.20), and (3.21) includes the notion that it is an integral of θ as a function of z, for a fixed time. We can, however, study storage variations for a fixed depth as a function of time.

Table 3.9 Soil water content profiles collected in a corn field during a period without rainfall

	Soil water content (m^3 m^{-3})			
Soil depth (m)	05/01	09/01	13/01	17/01
0–0.20	0.351	0.292	0.249	0.202
0.20–0.40	0.325	0.276	0.232	0.200
0.40–0.60	0.328	0.260	0.226	0.203
0.60–0.80	0.315	0.296	0.275	0.266
0.80–1.00	0.316	0.316	0.315	0.314

Let's consider a corn crop, in full development, for which the soil water profiles indicated in Table 3.9 and in Fig. 3.15 were determined during a period without rainfall. The profiles were determined at 8 o'clock in the morning on days 5, 9, 13, and 17 of January 1988. Just by observing the profiles of Fig. 3.15, one can verify that soil water content varies as a function of depth. These changes are a result of differences in water-related properties of the soil and, above all, to the distribution of the root system of the crop that extracts water. In this case, from Eq. (3.20), the storage variation between two dates t_i and t_j is given by:

$$\Delta S_L = S_L(t_j) - S_L(t_i)$$
$$= \left[\overline{\theta}(t_j) - \overline{\theta}(t_i)\right]L \qquad (3.22)$$

where $S_L(t_i)$ and $S_L(t_j)$ are the soil water storages of the layer 0–L (fixed), at times t_i and t_j, respectively. $\overline{\theta}(t_j)$ and $\overline{\theta}(t_i)$ are the average soil water

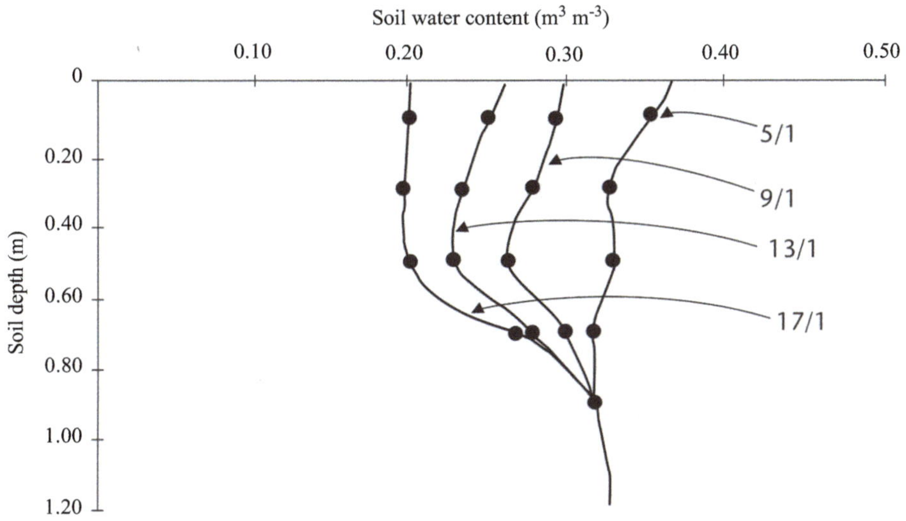

Fig. 3.15 Graphical representation of soil water content profiles shown in Table 3.9

contents of the layer 0–L, at the same times t_i and t_j, respectively.

Mathematically, we say that θ is a function of t and z, and we write: $\theta = \theta(t,z)$. Since θ is a function of two variables, its derivative in relation to one of them is called **partial derivative**. The variation of θ with t is called partial derivative of θ with respect to t, keeping z constant, and we write:

$$\left(\frac{\partial\theta}{\partial t}\right)_z$$

Note the use of the symbol ∂ instead of **d** in the derivative, this is because it is a partial derivative.

This exact concept of partial derivative can be approximated, for practical purposes, by a ratio of finite differences of θ and t:

$$\left(\frac{\partial\theta}{\partial t}\right)_z \equiv \left(\frac{\Delta\theta}{\Delta t}\right)_z = \left(\frac{\theta_j - \theta_i}{t_j - t_i}\right)_z \quad (3.23)$$

with the soil having a water content θ_i at time t_i (previous) and θ_j at the instant t_j (later), but at the same depth z.

Thus, for example, using the data in Table 3.9, $\partial\theta/\partial t$ for $z = 0.10$ m, during the period from January 5 to 9, is:

$$\left[\frac{\partial\theta}{\partial t}\right]_{0.20} = \frac{0.292 - 0.351}{9 - 5} = -0.0148 \text{ m}^3\text{ m}^{-3}\text{ day}^{-1}$$

performing the same calculation over the same period for $z = 0.30$; 0.50; 0.70; and 0.90 m, we have, respectively: −0.0122; −0.0170; −0.0005; and 0.00 $\text{m}^3\text{ m}^{-3}\text{ day}^{-1}$. The same calculation can also be done for the other periods. $\partial\theta/\partial t$ represents the rate at which the water of the soil varies with time and is a direct reflection of the distribution of the root system at the respective depth. With additions of water (rain, irrigation), $\partial\theta/\partial t$ is positive. Recalling the concept of derivative, which is the tangent to the curve at the point, this can be done for any depth.

In the same way as we proceeded for θ, we can repeat the reasoning for the soil water storage S_L, if we have the function S_L versus t. But we can approximate the partial derivative of S_L as a function of time by:

$$\frac{\partial S_L}{\partial t} \equiv \frac{\Delta S_L}{\Delta t} = \frac{[\overline{\theta}(t_j) - \overline{\theta}(t_i)]L}{t_j - t_i} \quad (3.24)$$

In this way, for $z = 0.20$ m, for the period of January 5–9, we have $S_{Li} = 70.2$ mm and $S_{Lj} = 58.4$ mm, so that:

$$\frac{\partial S_{0.20}}{\partial t} = \frac{58.4 - 70.2}{4} = -2.95 \text{ mm day}^{-1}$$

If we calculate the storage changes in the different periods, for the different depths, we obtain the following results, all in mm day^{-1} (Table 3.10):

Table 3.10 Soil water loss rates in mm day^{-1} for the corn crop illustrated in Fig. 3.15

z (m)	5–9	9–13	13–17
0–0.20	2.95	2.15	2.35
0–0.40	5.40	4.35	3.95
0–0.60	8.80	6.05	5.10
0–0.80	9.75	7.10	5.55
0–1.00	9.80	7.15	5.60

This table gives us a detailed idea of the extraction of water by the crop, for the different layers in the different periods. The contribution of the last layer, 0.80–1.00 m, is insignificant, probably because there are no roots at this depth. Logically, the total water lost by the crop to the depth of 1.00 m throughout the period, from the data in Table 3.10, is:

$$(9.80 \times 4) + (7.15 \times 4) + (5.60 \times 4) = 90.2 \text{ mm}$$

The daily average loss for the 0–1.00 m layer is:

$$(9.80 + 7.15 + 5.60)/3 = 7.52 \text{ mm day}^{-1}$$

This is a typical value of evapotranspiration in the month of January in Piracicaba, SP, of a growing crop with good water supply.

Exercise: We suggest that the reader recalculates the $\partial S_L/\partial t$ data presented for all depths and periods. Again we draw attention to the fact that exactly equal values will not be obtained, since they depend on the integration method employed.

An example of an estimate of water root extraction is given by Silva et al. (2009) for a coffee crop. Another example is Rocha et al. (2010) who tested a macroscopic model of soil water extraction by the root system based on the microscopic scale process, describing the results of an experiment with plants whose root system was divided in layers of soil with contrasting hydrological properties.

Another factor that significantly affects the estimation of soil water storage is its spatial variability. Soil water contents vary spatially, horizontally, and in depth, due to variations in the pore space arrangement and variations in the quality and size of the matrix particles. Thus, questions arise about the form and number of samples required to obtain a representative value of water content and, consequently, of storage. Kachanoski and De Jong (1988) and Turatti and Reichardt (1991) address these aspects. Furthermore, Reichardt et al. (1990) and Silva et al. (2006) discuss how the spatial variability of storage can affect the estimation of water balances.

The concepts of soil water content and storage we have seen up to now refer strictly to soil water. When our interest is in the solutes, that is, in the solution of the soil, the "salts" need to be included in the definitions. The most common procedure is to use solute C concentration in the **soil solution**, which is done when the solution is extracted from the soil (by filtration, centrifugation, suction, etc.) and the concentration is expressed as follows:

$$C = \frac{\text{Mass of solute}}{\text{Volume of solution}} \tag{3.25}$$

In units like: g Cl^- cm^{-3}; mg NO_3^- cm^{-3}; g $H_2PO_4^-$ L^{-1}.

The important thing is to know that, in this case, the volume is of solution and not of soil. The case is different in the field when considering the soil as a whole. For example, a cm^3 of soil containing 25 mg Cl^- and has a water content of 0.31 cm^3. Therefore, the Cl concentration in the water inside the soil is:

$$\begin{aligned} C &= \frac{25\ \text{mg Cl}^-}{0.31\ \text{cm}^3\ \text{of soil solution}} \\ &= 80.6\ \text{mg Cl}^-\ \text{cm}^{-3}\ \text{of soil solution} \end{aligned}$$

and the concentration in the soil as a whole is:

$$C' = \frac{25\ \text{mg Cl}^-}{1\ \text{cm}^3\ \text{of soil}} = 25\ \text{mg Cl}^-\ \text{cm}^{-3}\ \text{of soil}$$

It is clear that C is different from C'. As we said, the most common is to use C, which is the concentration of the solution that is moving inside the pores of the soil. However, for practical purposes, it is often desired to know the amount of a salt per unit volume (or mass) of soil, which is C'. It is easy to verify that:

$$C' = \theta C \tag{3.26}$$

From the data of the previous example, it can be seen that $25 = 0.31 \times 80.6$. If we want to express C' in terms of soil mass, just divide by soil density:

$$\frac{C'}{d_s} = \frac{\text{g salt cm}^{-3}\ \text{of soil}}{\text{g soil cm}^{-3}\ \text{of soil}} = \text{g salt g}^{-1}\ \text{of soil}$$

In this case, if d_s is 1.5 g cm^{-3}, we have $C'/d_s = 25/1.5 = 16.7$ mg Cl^- g^{-1} of soil.

It is also important to know the total salt inside a soil profile. It is a concept similar to that of water storage. This would be the case of an "SS, salt storage":

$$SS_L = \int_0^L (\theta C)\mathrm{d}z \tag{3.27}$$

with units:

$$\frac{\text{Mass of solute}}{\text{cm}^3\ \text{of soil}} \cdot \text{cm of soil} = \frac{\text{Mass of solute}}{\text{cm}^2\ \text{of soil}}$$

and which can easily be transformed in kg ha^{-1}. It is implicit that this mass lies in the volume of the layer of depth L. If mass of solute per unit mass of soil is desired:

Table 3.11 Example of nitrate concentration data collected under field conditions

Soil depth (m)	mg L^{-1} NO_3 in the soil solution	u (kg kg^{-1})	d_s (kg m^{-3})
0–0.10	25	0.18	1310
0.10–0.20	21	0.22	1350
0.20–0.30	18	0.26	1340

$$SS_L = \int_0^L \left(\frac{\theta \cdot C}{d_s}\right) dz = \int_0^L (u \cdot C) dz \qquad (3.28)$$

Of course, to integrate Eqs. (3.27) and (3.28), it is necessary to know how C, θ, and/or d_s vary as a function of z, and this can often be complicated. A reasonable simplification is the trapezoidal method already used for the calculation of water storage, using mean values of θ, u, or d_s (up to depth L) and taking them as constant and independent of z. Like this:

$$SS_L = \bar{\theta} \cdot \bar{C} \cdot L \text{ or } \bar{u} \cdot \bar{C} \cdot L \qquad (3.29)$$

Let's see an example: data of Table 3.11 belong to a given soil:

$$\bar{C} = \frac{(25 + 21 + 18)}{3} = 21.33 \text{ mg L}^{-1}$$
$$= 21.33 \times 10^3 \text{ mg m}^{-3}$$

$$\bar{u} = \frac{(0.18 + 0.22 + 0.26)}{3} = 0.22 \text{ kg kg}^{-1}$$

$$\bar{d}_s = \frac{(1310 + 1350 + 1340)}{3} = 1333.33 \text{ kg m}^{-3}$$

$$\bar{\theta} = \frac{0.22 \times 1333.33}{1000} = 0.29 \text{ m}^3 \text{ m}^{-3}$$

and so:

$$SS_{0.30} = 0.29 \times 21.33 \times 0.30$$
$$= 1.86 \times 10^3 \text{ mg m}^{-2} = 18.6 \text{ kg ha}^{-1}$$

i.e., 1 ha to the depth of 0.30 m contains 18.6 kg of NO_3.

As for the case of water storage, Eqs. (3.27), (3.28), and (3.29) can be applied to any layer of thickness L_2–L_1; in this case, the mean values of C, θ, u, or d_s must be taken in the same layer.

3.4 Gaseous Fraction of the Soil

The gaseous fraction of the soil is constituted by the air of the soil or the atmosphere of the soil. Its chemical composition is similar to that of the free atmosphere, close to the surface of the soil (see Table 5.1 of Chap. 5), but it shows differences mainly in the O_2 and CO_2 contents. Oxygen is consumed by microorganisms and by the root system of the higher plants, so that its concentration is lower than in the free atmosphere. On the contrary, CO_2 is released in metabolic processes that occur in the soil and, therefore, its content is in general higher. In cases of fertilization with urea, ammonium sulfate, etc., NH_3 levels in the soil atmosphere can increase significantly. Other organic and inorganic gases may also have their compositions altered, depending on the biological activities of the soil. From the point of view of soil water vapor, soil air is almost always very close to saturation. This will be discussed in Chap. 6, which deals with soil water equilibrium.

Soil air occupies porous space not occupied by water. We have already seen that for a dry soil, all the empty space given by porosity α (Eq. 3.12) is occupied by air. When a soil has a water content θ (Eq. 3.15), only the difference between α and θ can be occupied by air. This means that wetting and drying of a soil diminishes or increases the space for the air, respectively. The difference between α and θ is called **water-free porosity** (or aeration porosity) β:

$$\beta = (\alpha - \theta) \text{m}^3 \text{ of air m}^{-3} \text{ of soil} \qquad (3.30)$$

For a saturated soil: $\theta = \alpha$ and $\beta = 0$, and for a dry soil, $\beta = \alpha$.

Soil aeration is the dynamic process of β variations. Flooded soils or soils after long periods of heavy rain or irrigation are poorly aerated and the lack of oxygen for biological activities impairs crop growth and development. In very dry soils, the aeration is very good, but there is lack of water for the plants. According to Kiehl (1979), an "ideal" soil has a solid fraction that occupies 50% of the volume, being $\alpha = 50\%$, occupied half by water ($\theta = 0.25\ \text{m}^3\ \text{m}^{-3}$) and half by air ($\beta = 0.25\ \text{m}^3\ \text{m}^{-3}$).

As discussed in the case of solutes, the concentration of gas in the soil can be made based on the volume occupied by the fluid alone (soil air) or on the basis of soil volume. Therefore:

$$C = \frac{\text{Volume of gas}}{\text{Volume of air}} \tag{3.31}$$

with units $\text{m}^3\ \text{O}_2\ \text{m}^{-3}$ air or mg L^{-1} of air. The important thing is that C is based on volume of air, and C' on the soil:

$$C' = \frac{\text{Volume of gas}}{\text{Volume of soil}} \tag{3.32}$$

and, in the same way,

$$C' = \beta C \tag{3.33}$$

Let's take a soil that has 20% of O_2 on volume basis, with porosity of 45% and water content $0.23\ \text{m}^3\ \text{m}^{-3}$. In this case:

$$\beta = 0.45 - 0.23 = 0.22\ \text{m}^3\ \text{of air m}^{-3}\ \text{of soil}$$

$$C = 20\% = \frac{0.20\ \text{m}^3\ \text{O}_2}{1\ \text{m}^3\ \text{air}} = 0.20\ \text{m}^3\ \text{O}_2\ \text{m}^{-3}\ \text{of air}$$

$$C' = 0.22 \times 0.20 = 0.044\ \text{m}^3\ \text{O}_2\ \text{m}^{-3}\ \text{of soil}$$

Great care is required with these ways of expressing gas concentration. In the literature, the confusion is enormous!

Just as we did for water and for solutes (Eq. 3.27), we could speak of a gas storage in a soil profile, which would be the integration of βC to a depth L. This part is left as an exercise for the reader.

The soil aeration process is of great importance in soil productivity. Equation (3.9) shows the relationship between V_T, V_s, V_l, and V_g for a soil sample. In the soil profile, that is, in the field, these volumes can vary greatly in time and space, but within limits. In most cases, d_s is considered constant in time for a given situation, which implies in V_T, V_s, and α constants, varying only V_l and V_g, whose sum is always α and what implies in variations of θ and β. θ includes available water for the plants and β soil aeration, both essential for plant development and high agricultural productivity.

When $V_g = 0$, V_l is equal to α and the soil is saturated, $\theta = \theta_0$. This is the case of flooded soils of rice. When $V_l = 0$, V_g is equal to α and the soil is completely dry. The most general case is $\theta_0 > \theta > 0$ and when θ is in the available water range and plants develop well.

In the case of d_s varying, all volumes V vary. When soil is compacted, V_T decreases to a lesser or greater extent, through agricultural management practices. The traffic of machinery compacts the soil and plowing/harrowing decreases d_s, making it more "fluffy." With the increase of d_s, α decreases and the water/air relationships change. The total porosity α is subdivided into **macroporosity** (Ma) and **microporosity** (Mi), in an attempt to distinguish large pores from small pores. The pore distribution is continuous, and, therefore, the boundary between Ma and Mi is arbitrary (see Chap. 6), but in general, Ma's larger pores conduct fluids better, in this case water and air. The external forces responsible for soil compaction act more intensely on Ma. This fact affects the infiltration of water into the soil by reducing the hydraulic conductivity (it will be studied in Chap. 7) and the aeration of the soil by the reduction of β. The relation between θ and β is complex, one increases with the decrease on the other, compromising or aeration (problem related to drainage) or lack of water (problem related to irrigation).

Studies on the **soil compaction** phenomenon led to the establishment of a 10% ($0.10\ \text{m}^3\ \text{m}^{-3}$) critical macroporosity (Ma_{crit}) (Grable and Siemer 1968) below which compaction effects begin to affect agricultural productivity (Hakansson and Lipiec 2000). Although this limit widely used as an indicator of soil compaction, Stolf et al. (2011, 2012) present a way on how Ma can be estimated through a regression with only d_s and sand content of the soil:

$$Ma = 0.693 - 0.465\, d_s + 0.212\ \text{sand}$$

For example, if $d_s = 1.38$ g cm^{-3} and sand content = 0.12 kg kg^{-1}, results in Ma = 0.077 cm^3 cm^{-3} or 7.7%, below Ma_{crit}, it is recommended that this soil should be de-compacted. This can be done with deep plowing or even subsoiling operations. Repeated plowing at the same depth may also induce compacted layers at the plowing depth. There are soils that present genetic compaction, due to the formation of the soil, such as textured B horizons, hard pans, also known as hardsetting horizons.

Soil compaction also affects root penetration. Compacted areas do not allow satisfactory root development, and a reduced soil layer will be available for plant exploitation, affecting water relations and nutrient availability. One equipment that has been used to evaluate **soil mechanical resistance to penetration** of the root system is the so-called **penetrometer**, which expresses soil resistance to root penetration in terms of pressure (kPa or MPa). Several types of penetrometers are available for this measurement, among them Stolf et al. (2012) (Chap. 6). In the literature, penetration resistance values of up to 2.0 MPa have been considered as non-restrictive for an adequate development of the root system of most crops (Taylor et al. 1966; Silva et al. 1994).

In most plants (with the exception of specialized plants, such as irrigated rice), the transfer of oxygen from the atmosphere to the roots needs to be in sufficient proportions to meet their needs. Adequate root growth requires oxygen (aeration) in such a way that gas exchanges between atmosphere and soil are sufficient and at a sufficient rate to avoid O_2 deficiency (or excess CO_2) in the root zone. Microorganisms also require ideal conditions for their development. Measures of oxygen uptake by plant roots (Hawkins 1962) show that approximately 10 L of O_2 per m^2 of crop, per day, are required. Considering a 20% water-free porosity, 20% O_2 air, and a root zone 1 m deep, the amount of O_2 per m^2 of surface is 40 L, that is, four times the demand for roots. This means that to maintain good aeration conditions, 25% of the soil air needs to be renewed daily. The calculations performed have the sole purpose of giving an idea of the proportions of the problem. Under actual conditions, aeration rates are likely to range between more extreme limits. According to Erickson and Van Doren (1960), plant growth depends more on the occurrence of O_2 shortage periods than on average aeration conditions.

Gas exchange and gas movement in the soil can occur: (1) in the gas phase (diffusion or mass transport), in pores not occupied by water, interconnected and in communication with the atmosphere; and (2) dissolved in water. As the diffusion of gases in the air is generally higher than in the liquid phase, the water-free porosity becomes very important in the aeration process.

The air composition of the soil (Van Bavel 1965; Jury and Horton 2004) depends on the aeration conditions. In a soil with good aeration, it does not differ significantly from atmospheric air, except for a relative humidity which is almost always close to saturation (even in a dry air soil, the **relative humidity** is about 95%) and a higher concentration of CO_2, that is, 0.2–1.0% compared to atmospheric air, which is 0.03% (today, 2018, almost reaching 0.04%). Under limited aeration conditions, the CO_2 concentration may increase and the O_2 concentration decrease, both drastically.

The concentration of **gases diluted in water** generally increases with pressure and decreases with temperature. According to **Henry's law**, the dissolved gas concentration is proportional to the partial pressure P_i of the considered gas, thus:

$$C = s\frac{P_i}{P_0} \tag{3.34}$$

where s is the **gas coefficient of solubility** in water and P_0 is the atmospheric pressure.

Coefficients s in water for N_2, O_2, CO_2, and CO_2-free air are given in Table 3.12.

Table 3.12 Solubility coefficients of some gases diluted in water (g L^{-1})

T (°C)	N_2	O_2	CO_2	Air without CO_2
0	0.0235	0.0489	1.713	0.0292
10	0.0186	0.0380	1.194	0.0228
20	0.0154	0.0310	0.878	0.0187
30	0.0134	0.0261	0.665	0.0156
40	0.0118	0.0231	0.530	–

Example: A water surface is in equilibrium with the atmosphere (20 °C and 1 atm), with an oxygen partial pressure of 0.2 atm. What is the concentration of O_2 in water?

$$C = 0.0310 \times \frac{0.20}{1.0} = 0.0062 \text{ g L}^{-1} \text{ or } 6.2 \text{ mg L}^{-1}$$

Dissolved O_2 in water is essential for aerobic life in water. Ideal values are 8–10 mg L^{-1} for fishes. Organic pollutants consume this oxygen and threaten aquatic life. This was already discussed in Chap. 1 related to the **biological oxygen demand** (BOD) of polluting effluents. These dissolved gases in the water move according to the movement of the water. Thus, water flow can be important in the transport of gases. With rain, water infiltrates into the soil and occupies the spaces occupied by gases which are partially dissolved in water. With the evaporation and the absorption of water by the roots, the gases return to occupy the volume occupied by the water. These phenomena are important for gas exchange in the soil.

3.5 Thermal Properties of the Soil

Of importance in the thermodynamic and agronomic studies of the soil are its thermal properties, especially the **soil-specific heat** and the **soil thermal conductivity**. The first refers to the soil as a heat reservoir and the second for heat transfer.

Specific heat is, by definition, the amount of sensible heat given or received by the unit of mass or the volume of soil (during a cooling or heating event) when its temperature varies by 1 °C. Since soil water content is variable, the specific heat of the soil, in its natural state, is not a characteristic of the soil only, but of the soil–water system. For dry soils, it can be considered constant. From soil to soil, it varies, depending on the quality and proportions of mineral and organic matter.

The specific heat or thermal capacity per unit volume of soil can be determined by adding the thermal capacities of the different constituents in 1 cm^3. In general, the heat capacity of the gas fraction inside the soil can be neglected. So, we have:

$$c_s = (1 - \alpha)c_p + \theta \cdot c_w \qquad (3.35)$$

where c_s is the specific heat of the soil (cal cm^{-3} °C^{-1} or J m^{-3} K^{-1}), α the porosity, c_p the specific heat of the soil particles (cal cm^{-3} °C^{-1} or J m^{-3} K^{-1}), θ the volumetric soil water content, and c_w the specific heat of water (1 cal cm^{-3} °C^{-1} or J m^{-3} K^{-1}). According to Kersten (1949) and Reichardt et al. (1965), an average value of 0.16 cal g^{-1} °C^{-1} can be assumed for the specific heat of the particles of most mineral soils. Also taking an average value of 2.65 g cm^{-3} for the particle density, we have an average value of 0.4 cal cm^{-3} °C^{-1} for c_p. For the organic matter present in a fraction f, we can use an average value of c_{om} 0.6 cal cm^{-3} °C^{-1}. Thus, Eq. (3.35), for units in cal cm^{-3} °C^{-1}, can be rewritten:

$$c_s = 0.4(1 - \alpha) + \theta + 0.6f \qquad (3.36)$$

Example: Given soil with OM 5%, α = 49%, u = 13%, and d_s = 1.3 g cm^{-3}, which is its specific heat?

Solution:

$$c_s = 0.4(1 - 0.49) + 0.13 \times 1.3 + 0.6 \times 0.05$$
$$= 0.403 \text{ cal cm}^{-3\circ}\text{C}^{-1}$$

Soil thermal conductivity K can be defined according to the equation of Fourier equation for heat transfer, which states that the **soil heat flux density** q (cal cm^{-2} s^{-1} or J m^{-2} s^{-1} or still W m^{-2}) is proportional to the temperature gradient dT/dx (see definition of gradient in Chap. 7), that is:

$$q = -K\frac{dT}{dx} \quad (3.37)$$

In addition to depending on the composition of the solid fraction of the soil and, in particular, the soil water content, soil thermal conductivity K is also a function of the soil bulk density. As an example, Decico (1967) presents the following equations for calculating K as a function of u (%) and d_s (g cm^{-3}) for two Brazilian soils:

Soil series "Luiz de Queiroz" (Latosol):

$$K = 10^{-4}[1.275 \cdot \log u - 0.71]10^{1.077 \cdot d_s} \text{ cal cm cm}^{-2}\text{ s}^{-1}\text{°C}^{-1}$$

Soil series "Quebra-Dente" (Argisol) sandy phase:

$$K = 10^{-4}[0.945 \cdot \log u - 0.445]10^{1.365 \cdot d_s} \text{ cal cm cm}^{-2}\text{ s}^{-1}\text{°C}^{-1}$$

The soil samples from the "Luiz de Queiroz" series presented 34% sand, 28% silt, and 38% clay, and the "Quebra-Dente" series presented 79% sand, 19% silt, and 2% clay.

Another parameter of great importance in the thermal characterization of a soil is the **soil thermal diffusivity** D. It is defined by the quotient:

$$D = \frac{K}{c_s} \quad (3.38)$$

It is easy to see that D is also a function of θ, d_s because of K, and of the composition of the soil. Diffusivity is of great use in heat flux studies, which will be seen in more detail in Chap. 10, which deals with the flow of heat in the soil and in which these properties are used in some practical problems. A recent text that covers solutions of differential equations related to heat flux in the soil is Prevedello and Armindo (2015).

3.6 Soil Mechanics

Besides the aspects seen so far for the soil system, soils are also studied from the point of view of Engineering. Mandatory courses in Engineering have disciplines on **Soil Mechanics**, which deal with aspects of soil as a support for engineering projects. As an example, the reader should look for the text of Dias Junior (2003). The topics covered involve tensions and deformations in soils, overloads, compressibility and elasticibility, compaction, soil hydraulics, etc. Examples of applications in our environment are those of Lima et al. (2012), Dias Junior et al. (2005, 2007), and Ajayi et al. (2009a, b, c). The approach of soil mechanics differs from the agronomic viewpoint, except in cases of agricultural mechanics, when studying soil/implement interaction.

3.7 Exercises

3.1. To determine the sand, silt, and clay fractions of a soil, 80 g of soil were suspended in 1 L of a dispersing solution at 20 °C. By means of a densimeter, it was found that after 4.64 min and 7.73 h, the concentrations of the suspensions, in the first 0–10 cm depth in the sedimentation cylinder, were 32 and 18 g L^{-1}, respectively. What are the percentages of sand, silt, and clay?

3.2. What are 2:1 clay minerals?

3.3. One soil has a CEC of 12 meq 100 g^{-1} (12 cmol$_c$ dm^{-3}) and is completely saturated with K^+. How many grams of K^+ are adsorbed on 1 kg of soil?

3.4. A soil sample was collected at a depth of 0.60 m, with a volumetric ring of diameter and height of 0.075 m. The wet weight of soil was 0.560 kg, and after 48 h in an oven at 105 °C, its weight remained constant and equal to 0.458 kg. What is the bulk density of the soil? What is its water content on the basis of weight and volume?

3.5. What is the total and water-free porosity of the soil sample of Problem 3.4?

3.6. In the same place where the sample of problem 3.4 was collected, five more samples were collected at the same depth and with the same volumetric ring. The data obtained are shown in the table below. Determine, for each one, d_s and θ.

Sample	m_u (kg)	m_s (kg)	d_s (kg m^{-3})	θ (m^3 m^{-3})
1	0.560	0.458	1382	0.308
2	0.581	0.447		
3	0.573	0.461		
4	0.555	0.457		
5	0.561	0.452		
6	0.556	0.463		

3.7. Using the data from the previous problem, determine the means, standard deviations, and coefficients of variation for d_s and θ.

3.8. A soil sample collected with a volumetric ring of 0.0002 m^3 at a depth of 0.10 m, with $m_u = 0.332$ kg and $m_s = 0.281$ kg. After the collection, a soil compacting test was performed in the field, passing a compressor roller on soil surface. A new sample collected with the same ring at the same depth with $m_u = 0.360$ kg and $m_s = 0.305$ kg. Determine d_s, u, and θ before and after compaction.

3.9. In Problem 3.8, why did u not change in the two cases and θ not? And with the porosity, what happened?

3.10. A researcher needs exactly 0.100 kg of a dry soil and has a wet sample with $\theta = 0.250$ m^3 m^{-3} and $d_s = 1200$ kg m^{-3}. How much wet soil should you weigh to get the desired dry soil weight?

3.11. Given an area of 10 ha, considered homogeneous in terms of soil density and water content up to the 0.30 m depth, what is the weight of dry soil in tons in the 0–0.30 m layer? The soil water content is 0.200 kg kg^{-1} and its bulk density 1700 kg m^{-3}. How many liters of water are retained in the same soil layer?

3.12. A 0.40 × 0.40 m rectangular water box contains 9 L of water. What is the height of water in millimeters?

3.13. The average soil water content of a soil profile up to the depth of 0.60 m is 38.3% on the volume basis. What is the height of water stored in this layer?

3.14. One soil absorbed 15 L of water in each m^2. What is the height of the absorbed water?

3.15. Soil water content (% weight) and soil bulk density (kg m^{-3}) were measured, and the following data were obtained:

Soil layer (m)	d_s (kg m^{-3})	u (%)
0–0.10	1350	22.3
0.10–0.20	1430	24.6
0.20–0.30	1440	26.1
0.30–0.40	1470	27.0
0.40–0.50	1500	27.7

Determine the water storage in the 0–0.50; 0.10–0.30; and 0.40–0.50 m layers.

3.16. Draw the soil water profile of Exercise 3.15 in the same way as in Fig. 3.13.

3.17. In a sugarcane crop, soil water content measurements were made in % by volume:

Soil layer (m)	10 March	13 March	17 March	20 March
0–0.15	32.5	30.1	26.7	44.3
0.15–0.30	33.4	31.2	28.8	41.2
0.30–0.45	34.1	32.6	29.3	36.8
0.45–0.60	36.8	35.9	33.6	32.1
0.60–0.75	35.4	35.5	34.3	34.0
0.75–0.90	37.8	37.9	37.2	36.0

Determine how many millimeters of water the crop lost or gained in each period, in the 0–0.90 m layer.

3.18. What are the water loss or gain rates in Exercise 3.17?

3.19. Plot θ versus z with the data from Exercise 3.17, indicating the storage variations, as was done in Fig. 3.15.

3.20. Soil solution was extracted from a soil, and the NO_3 concentration was measured in the extract to give 5.3 mg L^{-1}. Knowing that the soil had a water content of 0.35 m^3 m^{-3}, what is the NO_3 concentration per m^3 of soil?

3.21. The following table indicates $H_2PO_4^-$ concentrations in the soil solution:

Depth (m)	$H_2PO_4^-$ (mg L^{-1})	Soil water content (m^3 m^{-3})
0–0.15	20.1	0.43
0.15–0.30	16.3	0.39
0.30–0.45	5.7	0.37

How much P in the form $H_2PO_4^-$ is found in 1 ha of this soil, considering the 0–0.45 m layer?

3.22. A soil with a bulk density of 1450 kg m^{-3} and particle density 2710 kg m^{-3} has a water content of 0.22 kg kg^{-1}. What is its water-free porosity?

3.23. In the previous problem, the soil air has 18% O_2 at the volume basis. What is the concentration of O_2 in the soil in kg m^{-3} soil?

3.24. What is the specific heat of a 56% porosity mineral soil and 32% volumetric water content?

3.25. The soil of the previous problem, when subjected to a temperature gradient of 2.3 °C cm^{-1}, allows a flow of 9.3×10^{-3} cal cm^{-2} s^{-1}. What is its thermal conductivity in this situation?

3.26. Calculate for the same soil, in the same situation, its thermal diffusivity.

3.27. Check deeply the difference between thermal conductivity and thermal diffusivity.

3.8 Answers

3.1. 60% sand; 17.5% silt and 22.5% clay.

3.2. See in text.

3.3. 120 meq K^+ or 4.68 g K^+ in 1 kg soil.

3.4. d_s = 1382 kg m^{-3}; u = 22.3%; $\theta = 0.308$ m^3 m^{-3}.

3.5. $\alpha = 47.8\%$; $\beta = 17.1\%$.

3.6. Use Eqs. (3.11) and (3.16) to calculate d_s and θ.

3.7. $d_s = 1378$ kg m^{-3}; $s = 17.8085$ kg m^{-3}; CV = 1.3%. $\theta = 0.326$ m^3 m^{-3}; $s = 0.0439$ m^3 m^{-3}; CV = 13.4%.
This is the typical variability that can be found for a given field situation. These variances are due to differences in compaction, texture, structure, etc., and also to the sampling methodology.

3.8. Before: d_s = 1405 kg m^{-3}; $u = 0.181$ kg kg^{-1}; $\theta = 0.254$ m^3 m^{-3}
After: $d_s = 1525$ kg m^{-3}; $u = 0.180$ kg kg^{-1}; $\theta = 0.275$ m^3 m^{-3}

3.9. In the second case, we have more mass and more water in the same volume of 0.0002 m^3. The porosity α decreases.

3.10. $m_u = 0.121$ kg.

3.11. 51,000 tons of soil and 1.02×10^7 L of water.

3.12. 56.25 mm.

3.13. 0.2298 m or 229.8 mm.

3.14. 15 mm.

3.15. 184.2 mm; 72.8 mm; 41.6 mm.

3.16. From 10 to 03/13: −9.9 mm; 13 to 03/17: −19.8 mm; from 17 to 03/20: +52.3 mm.

3.17. From 10 to 03/13: −9.9 mm; from 13 to 03/17: −19.8 mm; from 17 to 03/20: +52.3 mm.

3.18. −3.30 mm day^{-1}; −4.95 mm day^{-1}; +17.43 mm day^{-1}. In the period from March 17 to 20, it must have rained and the rain may have fallen in a single day. Therefore, the average gain rate equal to 17.43 mm does not have much meaning.

3.19. The procedure to elaborate Fig. 3.19 is the same made to elaborate Fig. 3.15.

3.20. 1855 mg NO_3 m^{-3} of soil.

3.21. 25.6 kg ha^{-1}.

3.22. 0.146 m^3 air m^{-3} soil.

3.23. As the percentage of O_2 is by volume, 1 m^3 has 0.18 m^3 of O_2. Under normal pressure and temperature conditions, 1 mol of any gas occupies the volume of 22.4 L (22.4×10^{-3} m^3), in 0.18 m^3 we have 8 mol. Since the mol of O_2 is 0.032 kg, we have 0.256 kg of O_2. Therefore, 1 m^3 air with 18% O_2 has 0.256 kg O_2. As the free-soil porosity is 0.146 m^3 air m^{-3} soil, the response is: 0.0374 kg O_2 m^{-3} soil.

3.24. 0.496 cal cm^{-3} °C^{-1}.

3.25. Apply Eq. (3.37) disregarding the sign. The result is $K = 4 \times 10^{-3}$ cal cm^{-1} s^{-1} °C^{-1}.

3.26. Apply Eq. (3.38). The answer is 8.1×10^{-3} cm^2 s^{-1}.

3.27. The difference is clear by the units. Notice that in D the cal disappears.

References

Ajayi AE, Dias Junior MS, Curi N, Araújo Junior CF, Aladenola OO, Souza TTT, Inda Júnior AV (2009a) Comparison of estimation methods of soil strength in five soils. Braz J Soil Sci 33:487–495

Ajayi AE, Dias Junior MS, Curi N, Araújo Junior CF, Aladenola OO, Souza TTT, Inda Júnior AV (2009b) Strength attributes and compaction susceptibility of Brazilian Latosols. Soil Tillage Res 105:122–127

Ajayi AE, Dias Junior MS, Curi N, Gontijo I, Araújo Junior CF, Vasconcelos Júnior AI (2009c) Relation of strength and mineralogical attributes in Brazilian latosols. Soil Tillage Res 102:14–18

Ball BC, Batey T, Munkholm LJ (2007) Field assessment of soil structural quality – a development of the Peerlkamp test. Soil Use Manag 23:329–337

Dane JH, Topp GC (eds) (2002) Methods of soil analysis, Part 4, Physical methods. Soil Science Society of America, Madison

Decico A (1967) Condutividade térmica dos solos: equações para o cálculo da condutividade térmica de alguns solos em função da densidade e umidade. PhD Thesis, Escola Superior de Agricultura Luiz de Queiroz/Universidade de São Paulo, Piracicaba, São Paulo state, Brazil.

Dias Junior MS (2003) A soil mechanics approach to study soil compaction. In: Achyuthan H (ed) Soil and soil physics in continental environment. Allied Publishers Private Limited, Chennai, pp 179–199

Dias Junior MS, Leite FP, Lasmar Júnior E, Araújo Junior CF (2005) Traffic effects on the soil preconsolidation pressure due to eucalyptus harvest operations. Sci Agr 62:248–255

Dias Junior MS, Fonseca S, Araújo Junior CF, Silva AR (2007) Soil compaction due to forest harvest operations. Pesq Agr Bras 42:257–264

Dourado-Neto D, Powlson D, Bakar RA et al (2010) Multiseason recoveries of organic and inorganic nitrogen-15 in tropical cropping systems. Soil Sci Soc Am J 74:139–152

Erickson AE, Van Doren DM (1960) The relation of plant growth and yield to oxygen availability. Trans Int Congr Soil Sci 3:428–434

Freire JC, Ribeiro MAV, Bahia VG, Lopes AS, Aquino LH (1980) Maize production under greenhouse conditions as a function of water levels in soils from the Lavras region (Minas Gerais State). Braz J Soil Sci 4:5–8

Fried M, Shapiro RE (1961) Soil-plant relationships and ion uptake. Annu Rev Plant Phys 12:91–112

Garbout A, Munkholm LJ, Hansen SB (2013) Tillage effects on topsoil structural quality assessed using X-ray CT, soil cores and visual soil evaluation. Soil Tillage Res 128:104–109

Grable AR, Siemer EG (1968) Effects of bulk density, aggregate size, and soil water suction on oxygen diffusion, redox potential and elongation of corns roots. Soil Sci Soc Am J 32:180–186

Guimarães RML, Ball BC, Tormena CA, Giarola NFB, Silva AP (2013) Relating visual evaluation of soil structure to other physical properties in soils of contrasting texture and management. Soil Tillage Res 127:92–99

Hakansson I, Lipiec J (2000) A review of the usefulness of relative bulk density values in studies of soil structure and compaction. Soil Tillage Res 53:71–85

Hawkins JC (1962) The effects of cultivation on aeration, drainage and other factors important in plant growth. J Sci Food Agric 13:386–391

Jury WA, Horton R (2004) Soil physics, 6th edn. Wiley, Hoboken, NJ

Kachanoski RG, De Jong E (1988) Scale dependence and temporal persistence of spatial patterns of soil water storage. Water Resour Res 24:85–91

Kersten SM (1949) Thermal properties of soil. University of Minnesota (Institute of Technology), Minnesota

Kiehl EJ (1979) Manual de edafologia: relações solo-planta. Agronômica Ceres, São Paulo

Kutilek M, Nielsen DR (1994) Soil hydrology. Catena, Cremlingen-Destedt

Lima CLR, Miola ECC, Timm LC, Pauletto EA, Silva AP (2012) Soil compressibility and least limiting water range of a constructed soil under cover crops after coal mining in Southern Brazil. Soil Tillage Res 124:190–195

Malavolta E (1976) Manual de química agrícola: nutrição de plantas e fertilidade do solo. Agronômica Ceres, São Paulo

Moncada MP, Penning LH, Timm LC, Gabriels D, Cornelis WM (2014) Visual examinations and soil physical and hydraulic properties for assessing soil structural quality of soils with contrasting textures and land uses. Soil Tillage Res 140:20–28

Mueller L, Kay BD, Hu C, Li Y, Schindler U, Behrendt A, Shepherd TG, Ball BC (2009) Visual assessment of soil structure: evaluation of methodologies on sites in Canada, China and Germany: Part I: Comparing visual methods and linking them with soil physical data and grain yield of cereals. Soil Tillage Res 103:178–187

Oliveira JCM, Vaz CMP, Reichardt K, Swartzendruber D (1997) Improved soil particle-size analysis by gamma-ray attenuation. Soil Sci Soc Am J 61:23–26

Prevedello CL, Armindo RA (2015) Física do Solo com problemas resolvidos. Sociedade Autônoma de Estudos Avançados em Física do Solo, Curitiba

Reichardt K (1987) A água em sistemas agrícolas. Manole, Barueri

Reichardt K, Salati E, Freire O, Cruciani DE (1965) Propriedades térmicas de alguns solos do Estado de São Paulo. In: Congresso Latino Americano, Congresso Brasileiro de Ciência do Solo. Sociedade Brasileira de Ciência do Solo, Piracicaba

Reichardt K, Libardi PL, Moraes SO, Bacchi OOS, Turatti AL, Villagra MM (1990) Soil spatial variability and its implications on the establishment of water balances. Trans Int Congr Soil Sci Kyoto Japan 1:41–46

Reisenauer HM (1966) Mineral nutrients in soil solution. In: Alman PL, Dittmer DS (eds) Environmental biology. Fed Am Soc Exp Biology, Bethesda, pp 507–508

Rocha MG, Faria LN, Casaroli D, van Lier QJ (2010) Evaluation of a root-soil water extraction model by root systems divided over soil layers with distinct hydraulic properties. Braz J Soil Sci 34:1017–1028

Shepherd TG (2009) Visual soil assessment, Pastoral grazing and cropping on flat to rolling country, vol 1, 2nd edn. Horizons Regional Council, Palmerston North, New Zealand

Silva AP, Kay BD, Perfect E (1994) Characterization of the least limiting water range. Soil Sci Soc Am J 58:1775–1781

Silva AL, Roveratti R, Reichardt K, Bacchi OOS, Timm LC, Bruno IP, Oliveira JCM, Dourado-Neto D (2006) Variability of water balance components in a coffee crop grown in Brazil. Sci Agric 63:105–114

Silva AL, Bruno IP, Reichardt K, Bacchi OOS, Dourado-Neto D, Favarin JL, Costa FMP, Timm LC (2009) Soil water extraction by roots and Kc for the coffee crop. Agriambi 13:257–261

Soil Survey Staff (1951) Soil texture. In: Soil Survey Staff (ed) Soil survey manual, United States Government Printing Office, Washington, DC, pp. 205–213.

Stokes GG (1851) On the effect of the lateral friction of fluids on the motion of pendulums. Trans Cambr Philos Soc 9:8–106

Stolf R, Thurler AM, Bacchi OOS, Reichardt K (2011) Method to estimate soil macroporosity and microporosity based on sand content and bulk density. Braz J Soil Sci 35:447–459

Stolf R, Murakami JH, Maniero MA, Silva LCF, Soares MR (2012) Integration of ruler to measure depth in the design of a Stolf impact penetrometer. Braz J Soil Sci 5:1476–1482

Taylor HM, Roberson GM, Parker Junior JJ (1966) Soil strength–root penetration relations to medium to coarse-textured soil materials. Soil Sci 102:18–22

Turatti AL, Reichardt K (1991) Soil water storage variability in "Terra Roxa Estruturada". Braz J Soil Sci 13:253–257

Van Bavel CHM (1965) Composition of soil atmosphere. In: Black CA (ed) Methods of soil analysis. American Society of Agronomy, Madison, pp 315–318

Vaz CMP, Oliveira JCM, Reichardt K, Crestana S, Cruvinel PE, Bacchi OOS (1992) Soil mechanical analysis through gamma-ray attenuation. Soil Technol 5:319–325

Webster R, Lark M (2013) Field sampling for environmental science and management. Routledge, New York

4 Plant: The Solar Energy Collector

4.1 Introduction

Plants develop in the atmosphere close to the ground, most having as physical support the soil. For their development, they use solar energy from the atmosphere, and the root system absorbs water and nutrients, which can be subdivided into:

Macronutrients:

1. Nitrogen, N: mostly absorbed as NO_3^- and NH_4^+;
2. Phosphorus, P: $H_2PO_4^-$ and HPO_4^{2-};
3. Potassium, K: K^+;
4. Calcium, Ca: Ca^{2+};
5. Magnesium, Mg: Mg^{2+}; and
6. Sulfur, S: SO_4^{2-}

which are **essential elements** for the growth and development of plants and are so called not because they are the most important but because they are the ones absorbed (and needed) in larger quantities.

Micronutrients:

7. Zinc, Zn: Zn^{2+};
8. Copper, Cu: Cu^{2+};
9. Manganese, Mn: Mn^{2+};
10. Iron, Fe: Fe^{2+};
11. Boron, B: boric acid H_3BO_3;
12. Molibdenum, Mo: MoO_4^{1-};
13. Chloride, Cl: Cl^-.

Micronutrients are also essential but absorbed and needed in very low quantities. Their lack, however, does not allow the plants to fully complete their life cycle.

Carbon, C; the oxygen, O; and hydrogen, H, are also essential elements, thus closing the list of the 16 essential elements. In addition, there are some other elements which are useful and found in plant tissues, like cobalt, Co (Co^{2+}), important for legumes; silicon, Si (SiO_3^-); and nickel, Ni (Ni^{2+}). These elements, however, are not essential to all plants and, therefore, are not included in the list of 16 essential elements. Havlin et al. (2014) evaluated the nutritional status of plants, describing important details about nutrient dynamics and absorption by plants.

Through the aerial part of the plant, specifically through the stomata, enters the carbon dioxide, CO_2, which participates in **photosynthesis**, a synthesis of sugars carried out at the expense of solar energy. The water needed for the process comes from the soil passing from the roots (through the xylem) to the stem and then to the leaves. The process by which green plants transform radiant (electromagnetic) energy into chemical energy is

K. Reichardt, L. C. Timm, *Soil, Plant and Atmosphere*, https://doi.org/10.1007/978-3-030-19322-5_4

essential for all live on Earth. The general formulation of photosynthesis is given by:

$$CO_2 + H_2O + \text{Solar energy} \xrightarrow{\text{Green plant}} O_2 + \text{Organic matter} + \text{Chemical energy}$$

and the resulting chemical energy built into the organic materials formed is employed by the leaf cell in various metabolic processes. The organic matter produced is the carbohydrate, symbolized by $(CH_2O)_n$. The water molecule, shown in the above equation, is retained in the carbohydrates, and therefore in plant tissues, and is a very small proportion of the absorbed water by the plant, in relation to the transpiration water that only passes through the plant, but with the very important function of bringing nutrients from the soil to the plant. The sunlight-absorbing structures in the leaf are pigments that are found in chloroplasts of higher plants (and some algae) (Table 4.1). The solar radiation absorption happens in peaks of the solar spectrum, in the bands of blue (420–480 nm) and red (643–660 nm), with the central green being reflected, hence the green color of the plants. Therefore, the range of the solar spectrum ranging from 400 to 700 nm is called photosynthetic active, which stands for **PAR** (**photosynthetic active radiation**), discussed in more detail in Chap. 5.

Photosynthesis operates in biochemical cycles. The main photosynthetic cycle of carbohydrate production is that of Calvin, with the formation of PGA, a sugar with three carbons. Therefore, plants that follow this cycle are called C_3 plants. Another group of plants, including some grasses (or Poaceae family), and other plant species adapted to the arid climate, follow a variation of the **Calvin cycle**, discovered by Hatch and Slack, and produce malate, a four-carbon carbohydrate, and are termed C_4 plants. Ehlers and Goss (2016) make a comparison of the physiology of plants C_3 and C_4, ranging from differences in the capacity of transference of vascular system, resistance of stomata to CO_2 flow, differences in the efficiency of nitrogen utilization in assimilation processes, to physiological adaptations to water shortage. Additionally, several plant species that inhabit arid and warm environments present another system of fixation of the CO_2, denominated crassulacean acid metabolism, that is, the **CAM plants**. Among them are cactus, pineapple, and orchids.

To grow and develop, plants also consume energy through the **respiration process**, in which sugars produced by photosynthesis are "burned" by O_2 resulting in CO_2 and H_2O. It is the reverse process of photosynthesis. In order for plants to accumulate dry matter (DM) (growth), it is necessary that the production of sugars by photosynthesis surpasses their burning by respiration. Figure 4.1 shows schematically the rates of carbon assimilation (CAR, photosynthesis) and respiration (RR, respiration) as a function of air temperature T, since plant growth and development are directly related to T_{air}, i.e., the climate. CAR increases with increasing T, passes through a maximum, and then decreases when the air becomes very hot, so that photosynthesis slows down. RR increases linearly with T. For low temperatures, lower than T_{bl} (see Fig. 4.1 and Fig. 5.5 in Chap. 5), RR > CAR, the net carbon assimilation rate (NCAR = CAR − RR) is negative and the plant consumes its own energy and,

Table 4.1 Pigments found in higher plants (and some algae) which take part in photosynthesis

Pigment	Peaks of the solar spectrum range of maximum absorbing sunlight (nm)
Chlorophyll a (green-bluish)	430 and 660
Chlorophyll b (green)	453 and 643
α Carotene (yellow-orange)	420; 440; 470
β Carotene (yellow-orange)	425; 450; 480
Luteol	425; 445; 475
Violoxanthol	425; 450; 475

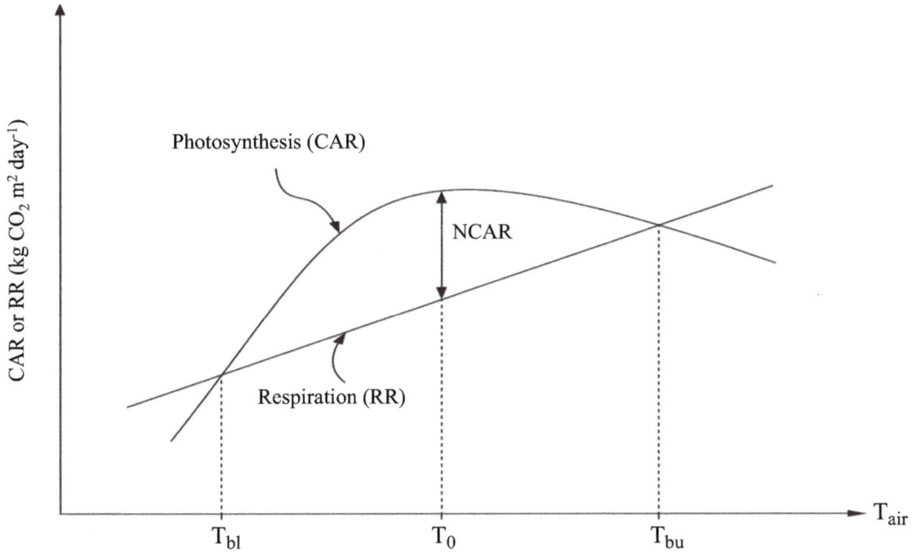

Fig. 4.1 Illustration on how photosynthesis (CAR) and respiration (RR) vary as a function of air temperature, showing the lower base temperature T_{bl}, upper base temperature T_{bu}, and the optimum temperature T_0 at which the net carbon assimilation rate (NCAR) is maximum

consequently, does not grow. For temperatures above T_{bu}, the same occurs. Within the temperature range T_{bu}–T_{bl}, the plant accumulates sugars and grows. At temperature T_0, the NCAR is maximum, and this is the optimum temperature for plant growth. All these temperatures, T_{bl}, T_0, and T_{bu}, are characteristic of each plant species and well known for agricultural crops.

As shown, for the growth and development of plants, most of the water in the soil passes through them and ends up in the atmosphere. Since in this process practically no vital energy enters, the plant is often seen as the link between soil water and the water of the atmosphere, occupying a preponderant role in the **water cycle**.

Water consumption by agricultural crops refers to all water lost by plants (transpiration and guttation) and by the soil surface, plus water retained in plant tissues. The percentage of water in plant tissues is very high, but even so, it is generally about less than 1% of the total evaporated during the plant growth cycle. Therefore, the water consumption of plants usually refers only to water lost by the transpiration of plants and by evaporation of the soil surface. In such studies, the following definitions are important:

(a) **Transpiration**: loss of water in the form of vapor, through the surface of the plant (mainly leaf), strongly linked to the leaf area.
(b) **Evaporation**: loss of water in the form of vapor, through the surface of the soil.
(c) **Evapotranspiration**: is the sum of transpiration with evaporation. In practice, the evapotranspiration of a crop is also called the **consumptive use of water** by a crop.

The water consumption of a crop is of fundamental importance from the agricultural point of view, since, in general, the available water resources are limited. Of importance in the evapotranspiration is the architecture of the plant, since, in this aspect, the leaf area is of great importance because it is the evaporating surface exposed to the atmosphere. The **leaf area** of a crop is the sum of the areas of all leaves. The **leaf area index** (LAI), the relationship between leaf area of one plant and the area of soil occupied by this plant, is widely used. Thus, for a corn population of

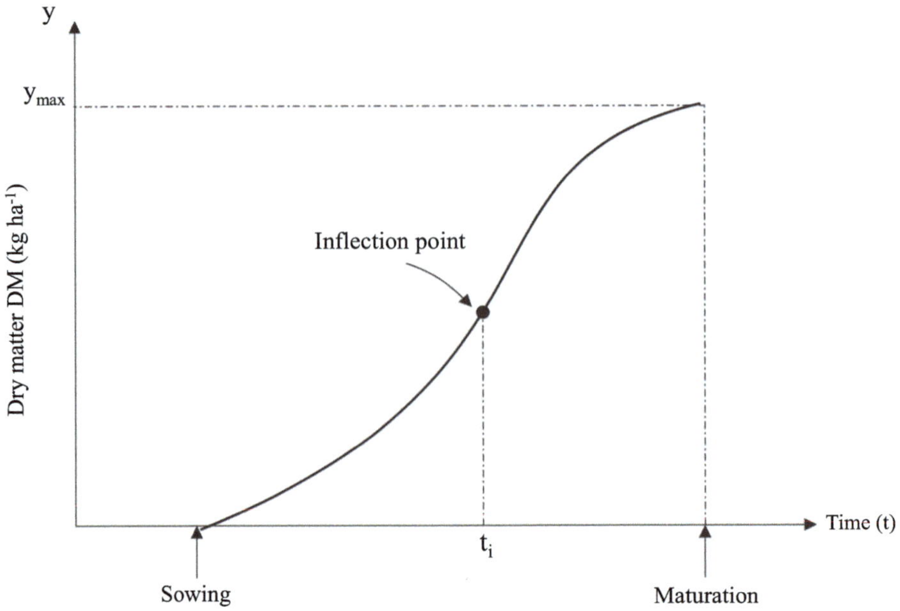

Fig. 4.2 Schematic view of the dry matter (DM) accumulation according to a sigmoid curve showing the moment t_i at which the accumulation rate is maximum (inflection point of the curve)

65,000 plants per ha, we have 6.5 plants in 1 m^2 and a plant occupies 0.154 m^2 of soil and if it has a leaf area of 0.85 m^2, the IAF is 0.85 m^2/0.154 m^2 = 5.5. This means that the transpiration of water occurs over an area 5.5 times larger than the evaporation of soil water. For calculation purposes, evapotranspiration is estimated per m^2 of cropped soil, that is, in L m^{-2} = mm.

LAI varies with the development of the crop. At sowing time, there is neither plant nor leaf and, with the passage of time, the plants grow and can cover totally the ground, becoming a green cover, or a green canopy. After flowering, fruiting, and maturation, the plants lose leaves by **senescence**, and the leaf area decreases. In Fig. 4.7, the variation of LAI is shown schematically along the development of a corn crop that follows in a similar way as the growth of the leaf area.

In the case of total **dry matter** (DM) accumulation, for most cases, a sigmoidal model, such as that presented in Fig. 4.2, is the one that best describes its behavior. A sigmoid is S-shaped, and in the first part of the sigmoid, DM is increasing at always greater rates until in the central part of the curve the rate (first derivative) becomes maximum, and then continues to grow but with decreasing rates until arriving at a minimal rate and an accumulation of the total DM tending to a constant maximum value. After this, DM decreases due to senescence, but this part is not covered by the sigmoid.

An equation that fits well with this sigmoid model is part of a sinusoid, with its beginning in the fourth quadrant, that is, at the angle of 3π/2 rad (Fig. 4.3). Observe the similarity between the sigmoid of Fig. 4.2 and the solid black part of the sinusoid of Fig. 4.3.

A model of this type was suggested by Garcia y Garcia (2002), in which (y) represents the dry phytomass of a crop, say in kg of DM ha^{-1}, as a function of time (t) during its growth cycle:

$$y = y_{max}\left\{\frac{1}{2}\left[\sin\left(\frac{3\pi}{2} + \pi\frac{t}{t_{max}}\right) + 1\right]\right\}^{\alpha} \quad (4.1)$$

y_{max} being the maximum dry biomass reached at maturation and t_{max} the time at maturation, considering $t = 0$ the emergence of the plants. Note that the relation t/t_{max} is dimensionless and since at the beginning of the cycle $t = 0$, we have

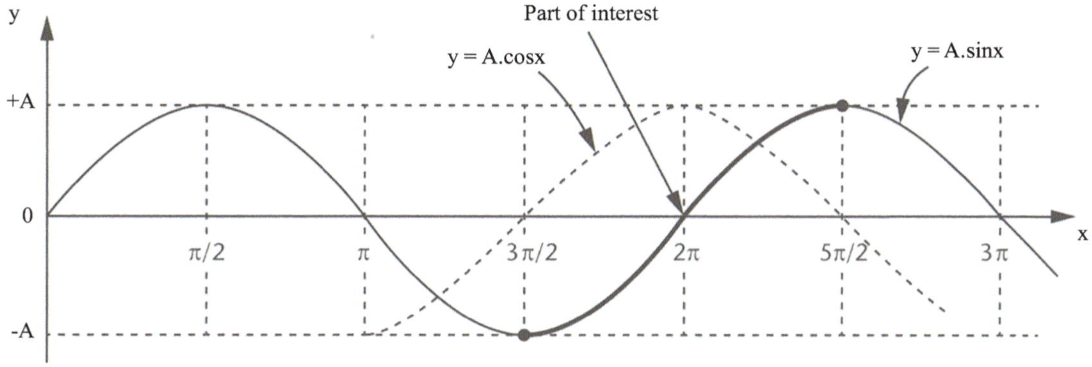

Fig. 4.3 Graph of a sigmoid or a sine curve $y = A{\cdot}\sin x$, showing the part of interest in solid black, going from an angle $x = 3\pi/2$ to $5\pi/2$, with the inflection point at 2π

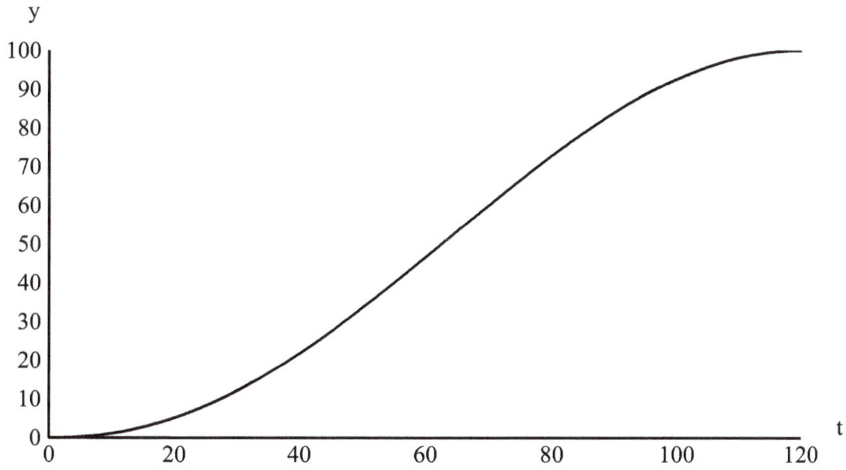

Fig. 4.4 Graph of Eq. (4.1) for $y_{max} = 100$, $t_{max} = 120$, and $\alpha = 1.1$

$t/t_{max} = 0$, and at the end of the cycle $t_{max}/t_{max} = 1$. Such a time scale is very useful because for any crop cycle, $t = 0$ at the beginning and $t = 1$ at the end. To understand Eq. (4.1), note that in the bracket of the sine of $3\pi/2$ appears a time lag of $\pi\ t/t_{max}$, which indicates that for $t/t_{max} = 0$, the sine starts at $3\pi/2$ (third quadrant), starting at -1, as shown in Fig. 4.2. This result, added to the +1 of the equation, results in 0, that multiplying the other factors leads to $y = 0$, which is the phytomass at the start of the cycle. For $t/t_{max} = 1$ (end of cycle), we have sine of $3\pi/2 + \pi = 5\pi/2 = 1$. Now, 1 + 1 = 2, multiplied by 1/2 of the equation results in 1, or $y = y_{max}$. The parameter α is a form factor, that is, it modifies the S of the sinusoid, which is pure only for $\alpha = 1$. According to the value of α , the S shape of the curve lengthens or shrinks, changing in its form and adapting to particular growth curve of the considered crop. The value of α is obtained by fitting Eq. (4.1) to experimental data of y and t by the minimum squares method.

A simulation made with the EXCELL[R] sheet, using Eq. (4.1), $y_{max} = 100$, $t_{max} = 120$, and $\alpha = 1.1$, resulted in the line presented in Fig. 4.4, just showing how well a part of a sine function can represent a sigmoidal function.

Of importance, also, are the growth rates of the plants, which would be the derivatives of the curves of Figs. 4.2, 4.3, and 4.4 or Eq. (4.1),

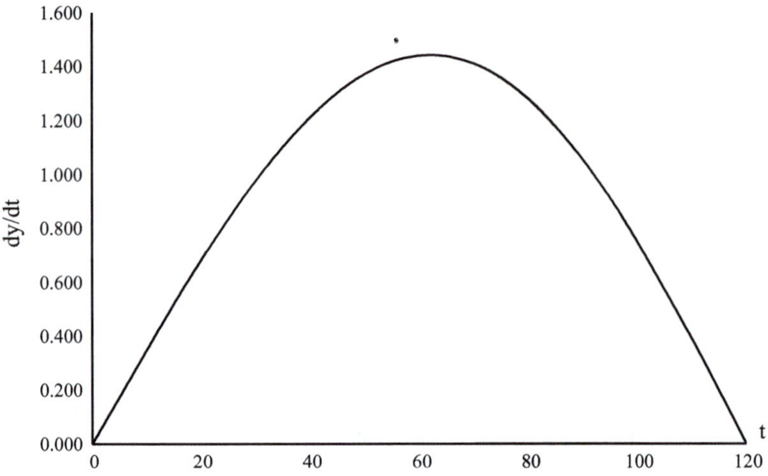

Fig. 4.5 Growth rate of a crop (dy/dt) as a function of time or the development of the crop

that is, dy/dt, that in our example is given in kg ha^{-1} day^{-1}. In the first branch used in the sinusoid (from $3\pi/2$ to 2π) of Fig. 4.3, y is increasing, and the derivatives are positive and increasing faster and faster. At point 2π, which is an inflection point, the second derivative d^2y/dt^2 is zero, which indicates that from there on (from 2π to $5\pi/2$), the derivatives remain positive but decrease in time. The result is that the growth rate dy/dt, as a function of time, becomes part of a cosine curve (remember that the first derivative of sine is cosine), with a maximum, as shown in Figs. 4.3, 4.4, and 4.5. Figure 4.5 was constructed using the derivative of Eq. (4.1), which the reader can find in the solution of Exercise 4.6, at the end of this chapter.

Another model of a sigmoidal curve (Fenilli et al. 2007) is that of Eq. (4.2), adapted for total dry matter (DM) of the aerial part of coffee plants.

$$\mathrm{DM} = a + \frac{b}{\left[t + e^{(t-c)/d}\right]} \tag{4.2}$$

Just to think about, observe how two very different Eqs. (4.1 and 4.2) are able to describe the same phenomenon, in the case the cumulative growth of a crop. Figure 4.6 shows, through the full line, the accuracy by which Eq. (4.2) adjusts to the experimental data of a coffee crop. The adjustment was done by the "Table Curve" program, which provides the parameters of the equation: $a = 313.2$; $b = 2604.5$; $c = 178.0$; and $d = 64.1$. The dotted line is the growth rate (derived from Eq. (4.2), dDM/dt), which passes through a maximum at $t = 172$ days after flowering of the coffee. The figure also shows the four nitrogen fertilizations made at days 1, 63, 105, and 151 after flowering, indicated by means of arrows, during the period of time the accumulation rates of DM were positive and the plants needed more N.

So far, we have been talking about **growth and development**, which are important concepts in crop science. They are not synonyms, they are quite distinct but inseparable. As the plant grows, it develops. Growth refers more to plant size, more properly in dry matter accumulation. Development involves differentiation and the plant goes through several stages until closing the reproductive cycle, producing seeds that will perpetuate the species. In a generic form, we speak about vegetative phases like flowering, fruiting, maturation, senescence, and so on. These phases are periods of the cycle that cannot be confused with stages, which are moments. The description of the phases and stages is called **phenology**. Maize (*Zea mays* L.), from the Poaceae (old Gramineae) family, one of the most studied plants, presents ten stages: (0) emergence (2 leaves); (1) 4 leaves; (2) 8 leaves; (3) 12 leaves; (4) bolting; (5) flowering; (6) milky grains; (7) pasty grains; (8) farinaceous grains; (9) hard grains; and (10) point of **physiological maturity**. Each stage is considered reached when 50% of the plants of a crop present the symptoms of the

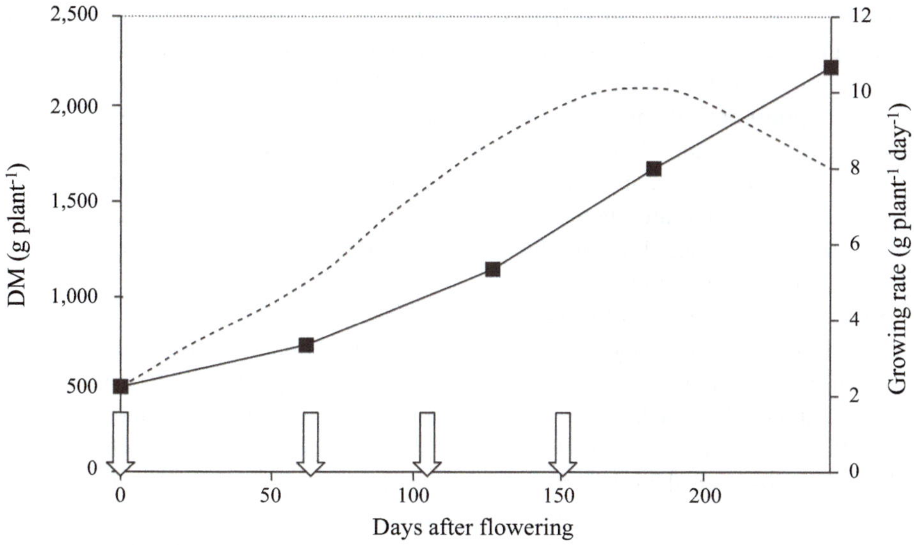

Fig. 4.6 Cumulative growth curve (solid line) and growing rate (dotted line) of a coffee crop indicating through arrows the fertilizer application times

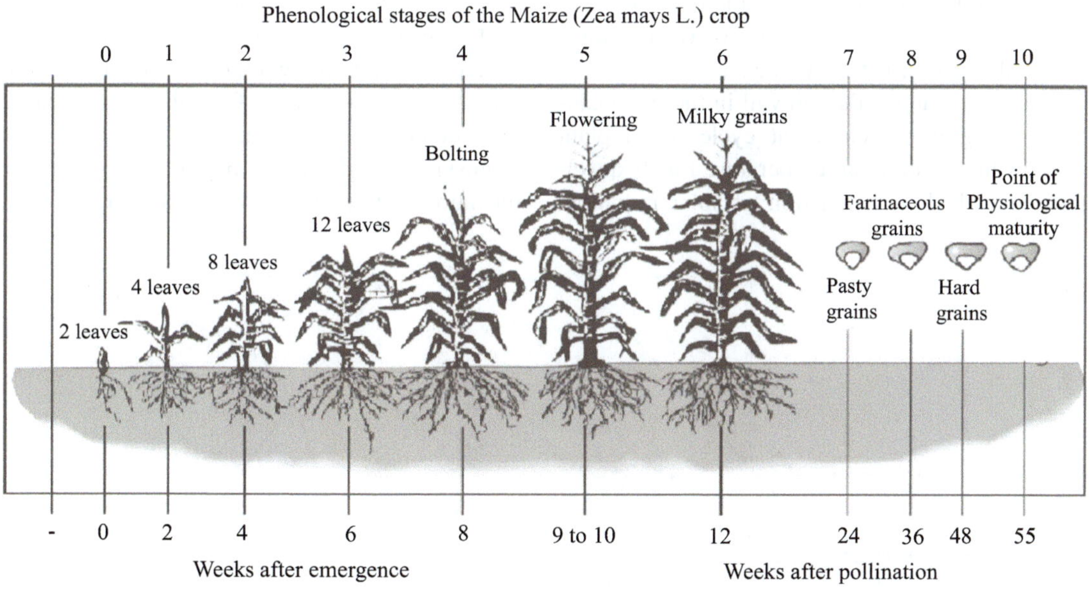

Fig. 4.7 Schematic view of the physiological stages of the corn plant (Fancelli and Dourado-Neto 2000)

specific stage. Figure 4.7 illustrates the ten stages of corn development.

Important is the term DAE (days after emergence), widely used to accompany the development cycle of plants. Instead of Julian days *t*, DAE is commonly used, for example, 50 DAE or DAE_{max}, which is the duration of the cycle.

The duration of the cycle is reasonably constant for a given variety or cultivar, provided that the environmental conditions are adequate, that is, availability of nutrients and water, luminosity, adequate air and soil temperature, and so on. There are early, mid-cycle and late varieties. In relation to luminosity, the duration of the day stands out (see *N*—possible number of hours of sunshine on a given day or length of the day—in Chap. 5, item solar radiation) that leads to the phenomenon of **photoperiodism**. Induction of flowering is affected by photoperiod or day length in many species. There are plants that are not photosensitive and those that are photosensitive, among which are those of long days and those of short days. In sugarcane cultivation, for example, in which production is represented by stalks, the induction of flowering is undesirable. In any case, sugarcane flowers when the photoperiod is between 12.0 and 12.5 h, which in the state of São Paulo, Brazil, occurs between February 25 and March 20. However, there is a combined effect of temperature, that is, flowering is only induced if the maximum air temperature is less than 31 °C (which is rare at this time of year) or greater than 18 °C.

Plants, to grow and develop, use energy that comes from the sun. A practical way of quantifying it is by the concept of **degree days** (DD), detailed in the next chapter. It is based on the air temperatures that prevail in the plant canopy during its development cycle. Each plant species has an optimal temperature for its development, which is a function of solar radiation. There is also a minimum temperature in terms of growth (Fig. 4.1), called the lower base temperature T_{bl}, below which the crop hardly develops, and an upper base temperature T_{bu}, above which the development of the crop is impaired. Thus, the optimal range for growth and development of a crop is (T_{bu}–T_{bl}), having in its center the optimum temperature T_0. The concept of degree days (DD) analyzed in Chap. 5 is based on these temperatures.

Another important aspect of growth and development is the **partition of carbon** (C) from photosynthesis and that is allocated differently to each part or organ of the plant. The total dry matter growth rate, shown in Fig. 4.4, can be unfolded into organs. Figure 4.8 shows schematically this unfolding in percentage terms, while the total sugars of CAR (see Fig. 4.1) are allocated to the root, stem, leaf, and reproductive organs during an ideal cycle of maize cultivation, from the emergence of seeds until physiological maturity. In the period between sowing and emergence, called germination, the main source of C is the seed because there are no leaves. At seed emergence, when the photosynthetic processes begin, the percentage allocation of C is about 1/3 to each organ: root,

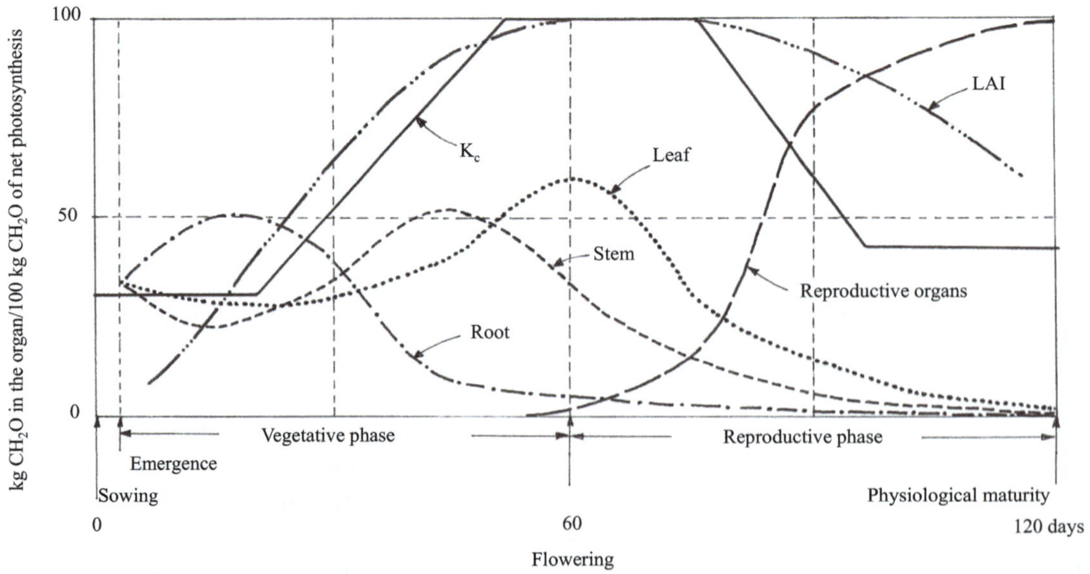

Fig. 4.8 Carbon partition in corn for roots, stem, leaves, and reproductive organs as a function of plant development. K_c is the crop coefficient for ET_0, and LAI is the leaf area index

stem, and leaf. During the vegetative phase, the allocations vary as shown in the example of Fig. 4.8 and decelerate after flowering, when the main sinks of C become the reproductive organs. The knowledge of these C fluxes during the development cycle of the plant is of great importance in crop management, as they determine the best moments or phases in which plants need nutrients and water. Figure 4.8 also shows schematically the crop coefficient (K_c) (see Chap. 13) and the leaf area index (LAI). Both are related to leaf growth. Due to the difficulties of K_c measurements, in the initial phase when there is still little leaf, it is taken as constant and around 0.3. It then is assumed to grow linearly until the peak of leaf growth when it assumes values of the order of 1.0 or more, 1.1–1.2. Values greater than 1.0 characterize conditions when plants evapotranspirate more than ET_0. In the reproductive phase, when the older leaves begin to lose photosynthetic activity, K_c decreases linearly until approximately 0.4 and taken as constant thereafter. LAI accompanies leaf growth and also decreases in the final stages due to loss of photosynthetic activity.

In the plant, the movement of water, from its entrance at the root surface, to the leaves, is made in special pathways and, therefore, some notions of plant anatomy become indispensable for the understanding of the process. These notions are presented below.

4.2 Plant Anatomy

The water-absorbing zone of the root extends by a few centimeters from its meristematic end (growth zone), to the point where epidermic suberization of the root becomes evident. **Root hairs** (epidermal cells with an elongated end as shown in Fig. 4.9) are generally present in the **root absorption zone** and increase the contact area between root and soil (for water and nutrient absorption) by of a factor 3 or 4. A cross section of a root in the water absorption zone is shown in Fig. 4.9.

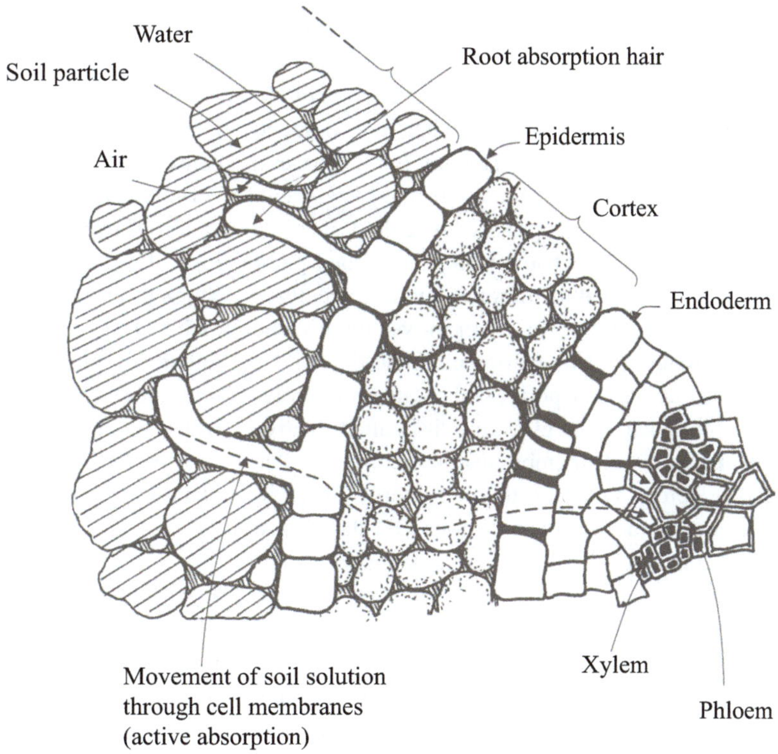

Fig. 4.9 Schematic view of a root cross section showing the movement of water from the soil to the xylem: (1) through cell walls (apoplast) reaching the endosperm ending up in the xylem (solid black line starting at the upper root hair and following cell walls in the cortex); (2) Through the protoplasm (symplast), starting at the lower root hair and ending up in the xylem (dotted line). The black areas between the endosperm cells are the *de Caspari* bands

Root cells are differentiated in layers, and water (and nutrients) pass through these cells or intercellular spaces until they reach the cells of the **xylem** (conductive element) located in the central cylinder of the root. The epidermis consists of a layer of cells, after which the cortex begins, which usually has 5–15 layers of parenchyma cells. After the cortex begins the endoderm, also consisting of a single cell layer. A feature of the endoderm is that part of the cell walls is suberized, in such a way that all movement of water (and nutrients) in the endoderm occurs only through the cells and not through intercellular spaces or cell walls. The suberized barrier of the endodermis is called the *de Caspari* band. After the endoderm is the pericycle, which consists of one or two layers of thin-walled cells, and after the pericycle are the vascular tissues: xylem and **phloem**. The first takes the crude sap (water and mineral salts) to the aerial part of the plant, and the second leads the elaborated sap (solution of organic materials produced by photosynthesis) from the leaves to the roots.

The movement of the **soil solution** can occur in two ways at the root. The first pathway is through the cell walls and intercellular spaces, together called as outer space. In this case, the water moves due to potential differences or gradients, and the solutes either are dragged (mass flow), or move by diffusion. In this process, energy from the plant metabolism is not spent, and that's why this process is called inactive transfer or **passive absorption**. The second pathway is through cell membranes and living cells or inner space. In this case, water moves mainly due to differences in osmotic potential and the solutes move by active transport, that is, transport involving biological energy, and is called **active absorption**. Cell membranes are semipermeable and selective, in which there is energy expenditure during the transport of ions. To date, most surveys indicate that water movement is primarily inactive.

The xylem extends from the roots to the leaves, through the stem. When the vascular bundle of the xylem penetrates the leaf, it divides into a series of branches until it is progressively composed of simple cells. These are in contact with the cells of the laconic parenchyma, a foamy tissue with large amount of intercellular spaces, where the water evaporates, that is, it passes from the liquid state to the vapor state. A cross section of a leaf is shown diagrammatically in Fig. 4.10.

The water vapor of the intercellular spaces reaches the atmosphere by two paths: the cuticle, in smaller quantity, and mainly through the stomata.

The cuticle is a suberized layer that covers the cells of the epidermis and, therefore, the water losses through the cuticle are very small. Stomata are holes 4–12 μm wide by 10–14 μm in length, found on the surface of the leaf (upper, lower, or both, depending on the species) by which the major gas exchanges occur between the plant and the atmosphere. Through them, the water vapor leaves the leaf, reaching the atmosphere, and the carbon dioxide enters the intercellular spaces, being used in the photosynthetic process. They consist of two cells between which is a hole of variable dimensions (Fig. 4.11). The structure of the stomatal apparatus can vary considerably from plant to plant. Their average number is about 10,000/cm^2. Changes in their opening are due to differences in the potential of the water within the **guard cells**, a function of several factors. An increase in the volume of guard cells causes the opening of the stomata. These volume variations may be due to variations in the translocation of water in the plant and to the intensity of loss of water from the leaf to the atmosphere. Stomata are also sensitive to light, temperature, and CO_2 and potassium concentrations. It is a rather complex operation that will not be dealt with here. More details on the movement of water in the plant can be found in Ehlers and Goss (2016). Other important aspects of plant physiology can be seen in Taiz and Zeiger (2010).

4.3 Water in the Plant

In later chapters (Chaps. 7–16), we will study the movement of water in the Soil–Plant–Atmosphere System as a whole. For this, it becomes

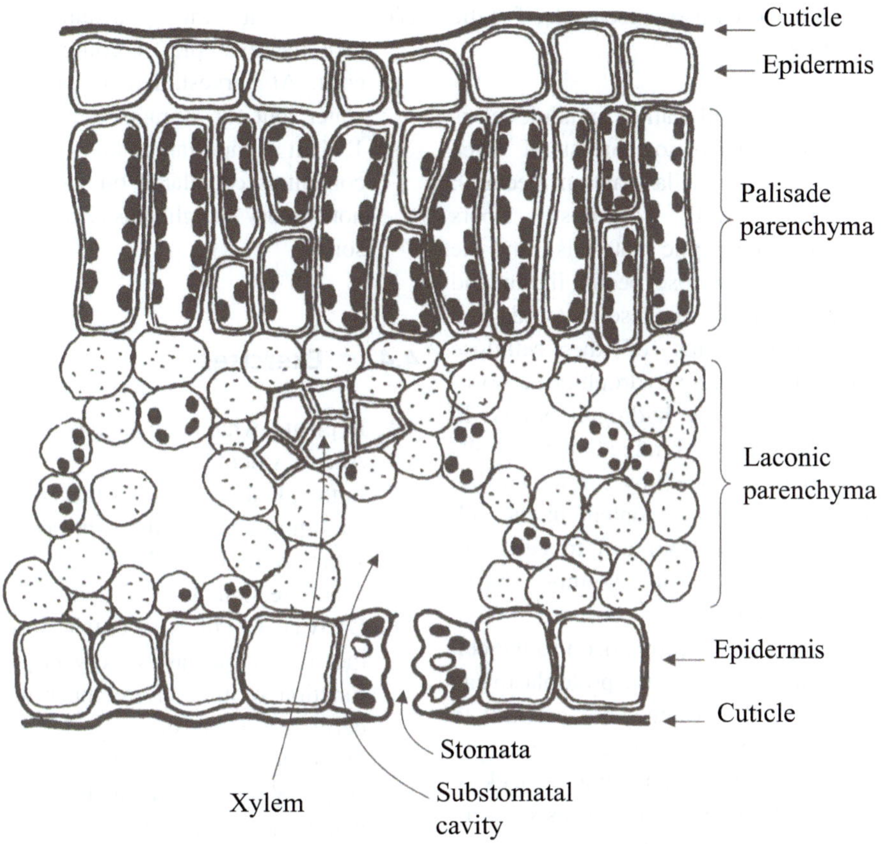

Fig. 4.10 Schematic view of a leaf cross section showing the different layers. For the movement of water from the plant to the atmosphere, of importance is the substomatal cavity in the laconic parenchyma, the stomata, and the cuticle

Fig. 4.11 Leaf surface showing stomata

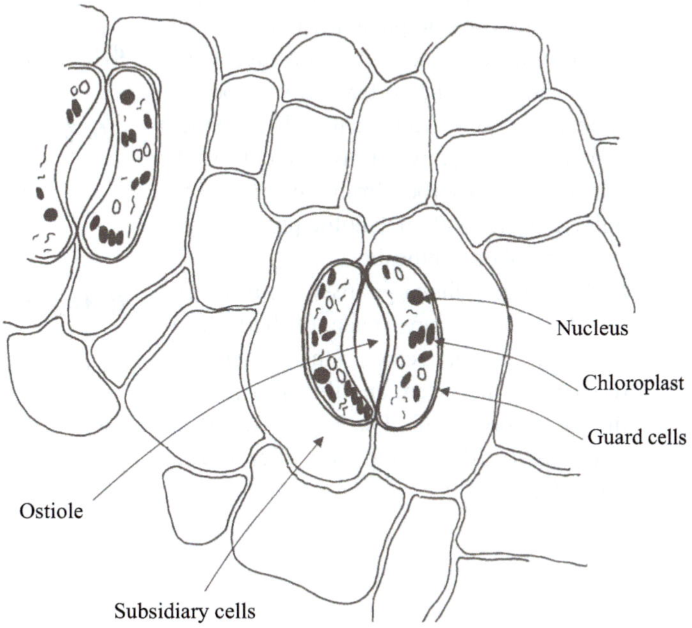

important to know how water presents itself in the various plant tissues:

(a) In cell walls: the cell wall of an adult cell is generally considered to consist of three parts: the central lamella (calcium pectate); the primary wall (cellulose fibers impregnated with peptic materials); and the secondary wall (cellulose, pectin, lignin, and cutin). The surfaces of these materials and the hydroxyl groups of the cellulose molecules are strongly hydrophilic, absorbing water, mainly by hydrogen bonds. In turgid cells, water is retained in the walls (pores of fibrous tissues) by surface tension phenomena. The water content in the cell wall varies greatly from cell to cell, and may reach 50% (on volume basis, $m^3\ H_2O\ m^{-3}$ cell wall).

(b) In the protoplasm: in comparison to the cell wall, the water content in the protoplasm can reach 95% of the volume, but can fall to much lower levels when cell inactivity occurs due to extreme temperature, lack of water, etc. The protoplasm consists essentially of proteins and water.

(c) In the vacuole: here its contents can reach more than 98% of the volume; the remaining 2% are sugars, organic acids, and minerals. All these components are of great importance in the osmotic phenomena in the cell.

(d) In the vascular system: it is the sap, whose composition in the xylem and phloem is generally quite different. The xylem contains, mainly, water as solvent of the mineral salts absorbed from the soil and the phloem, water with metabolite products produced by photosynthesis.

(e) In the wood: in the large trees, the trunk and woody branches are made of bark (dead material, suberized), the sapwood (containing the xylem and the phloem), and the core (wood itself). The "green" wood can have up to 50% water and, after drying, 10–15%.

(f) In seeds: it varies a lot and, therefore, we take the example of commercially grown corn. At harvest, it can have 15–20% of water, but for storage, it is dried to about 13%. It is important to know that plant water content is calculated on wet weight basis, not on dry weight basis, as it is done for soils.

4.4 Exercises

4.1. In plants, what is understood as phloem?
4.2. And as xylem?
4.3. In the plant, where does the water pass from the liquid phase to the vapor phase?
4.4. What are root hairs?
4.5. What are stomata?
4.6. Derivate Eq. (4.1) to obtain an equation of the rate of increase of dry phytomass as a function of time. Remember that the sine derivative is a cosine.
4.7. Does the leaf area index vary from crop to crop? With the spacing between lines and plants, how LAI develops with the growth stage of a crop?
4.8. What are the differences between growth and development?

4.5 Answers

4.6. $\frac{dy}{dt} = \frac{\alpha y_{max} \pi}{2 t_{max}} \left\{ \frac{1}{2} \sin \left[\left(\frac{3\pi}{2} + \frac{\pi t}{2} \right) + 1 \right] \right\}^{\alpha - 1}.$
$\cos \left(\frac{3\pi}{2} + \frac{\pi t}{t_{max}} \right)$

References

Ehlers W, Goss M (2016) Water dynamics in plant production, 2nd edn. CABI, London
Fancelli AL, Dourado-Neto D (2000) Produção de milho. Agropecuária, Guaíba
Fenilli TAB, Reichardt K, Dourado-Neto D, Trivelin PCO, Favarin JL, Costa FMP, Bacchi OOS (2007) Growth,

development and fertilizer N-15 recovery by the coffee plant. Sci Agric 64:541–547

Garcia y Garcia A (2002) Modelos para área foliar, fitomassa e extração de nutrientes na cultura de arroz. PhD Thesis, Escola Superior de Agricultura Luiz de Queiroz, Universidade de São Paulo, Piracicaba, São Paulo state, Brazil

Havlin JL, Tisdale SL, Nelson WL, Beaton JD (2014) Soil fertility and fertilizers: an introduction to nutrient management, 8th edn. Pearson, Upper Saddle River

Taiz L, Zeiger E (2010) Plant Physiology, 5th edn. Sinauer Associates Inc, Massachusetts

5 Atmosphere: The Fluid Envelope That Covers the Planet Earth

5.1 Introduction

Because of its dimensions and the physicochemical and biological processes that have occurred over a long time, the planet Earth has a gaseous layer that surrounds it and constitutes the **atmosphere**, being essential to the forms of life that evolved here and that helped to shape it. During the geological ages, the chemical **composition of the atmosphere** must have varied greatly, having reached a dynamic equilibrium in the last 200 million years. For purposes of meteorological studies and definitions, its average composition is given in Table 5.1.

The total mass of the gas layer that constitutes the atmosphere corresponds to 0.001% of the total mass of the planet and is practically concentrated in the first 10 km of altitude (which is called the **troposphere**), corresponding to a very thin layer when compared with the radius of the planet, which is approximately 6000 km. Due to the dimensions of the planet (gravitational force), to gas density, and to heat exchange processes, the atmosphere has a characteristic vertical structure, which is indicated in Fig. 5.1. The lowest layer, near the ground, is called the troposphere and is characterized by the decrease of the temperature with altitude. This layer contains about 80% of the total mass of the atmosphere and is the layer that is most strongly influenced by the energy transfers occurring on the Earth's surface. These processes create temperature and pressure gradients, which produce the atmospheric movements responsible for transporting water vapor and heat. The thickness of the troposphere varies, ranging from 16 to 18 km in the tropics and from 2 to 10 km in the polar regions.

The troposphere has as limit a layer called **tropopause**, where atmospheric movements are reduced, hence its name. Above the tropopause lies the **stratosphere**. In this layer, the temperature increases and, at about 50 km of altitude, the temperature assumes values corresponding to the one of the surface of the Earth. This increase in temperature is associated with the absorption of ultraviolet radiation by ozone, present in high concentration (relatively) at altitudes ranging from 20 to 50 km. Above the stratosphere, there is a layer where the temperature passes through a maximum, called **stratopause**. Above it extends the **mesosphere**, characterized by the decrease of the temperature with the altitude. The lower temperature region, which is the upper limit of the mesosphere, is called the **mesopause**. Above 80–90 km, the temperature seems to increase continuously with the altitude, until reaching temperatures of the order of 1500 K to 500 km. This region is called the **thermosphere** or **ionosphere**.

K. Reichardt, L. C. Timm, *Soil, Plant and Atmosphere*, https://doi.org/10.1007/978-3-030-19322-5_5

Table 5.1 Composition of the atmosphere

Gas	Volume
Nitrogen (N_2)	780,840 ppmv[a] (78.084%)
Oxygen (O_2)	209,460 ppmv (20.946%)
Argon (Ar)	9340 ppmv (0.934%)
Carbon dioxide (CO_2)	390 ppmv (0.039%)
Neon (Ne)	18.18 ppmv (0.001818%)
Helium (He)	5.24 ppmv (0.000524%)
Methane (CH_4)	1.79 ppmv (0.000179%)
Krypton (Kr)	1.14 ppmv (0.000114%)
Hydrogen (H_2)	0.55 ppmv (0.000055%)
Nitrous oxide (N_2O)	0.3 ppmv (0.00003%)
Carbon monoxide (CO)	0.1 ppmv (0.00001%)
Xenon (Xe)	0.09 ppmv (9×10^{-6}%)
Ozone (O_3)	0.0–0.07 ppmv (0% to 7×10^{-6}%)
Nitrogen dioxide (NO_2)	0.02 ppmv (2×10^{-6}%)
Iodine (I)	0.01 ppmv (10^{-6}%)
Ammonium (NH_3)	Traces
Gases not included in the high atmosphere (sample without water):	
Water vapor (H_2O)	~0.40% at all atmosphere, normally between 1 and 4% at atmosphere surface

Source: Google—Composition of the terrestrial atmosphere. https://pt.wikipedia.org/wiki/Atmosfera_terrestre (accessed 04/23/2018)

[a]ppmv: parts per million per volume (note: the volume fraction is equal to the molar fraction only for ideal gases)

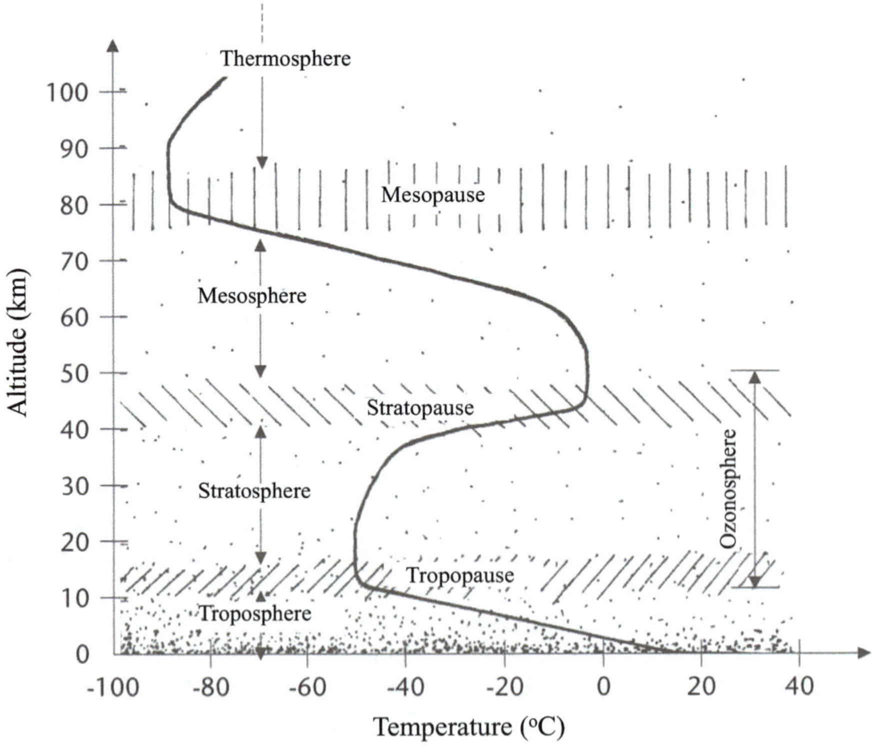

Fig. 5.1 Structure of the atmosphere for the first 100 km

5.2 Thermodynamic Characteristics of the Air Close to Soil Surface

This gaseous envelope that surrounds the Earth, with due approximations, can be considered an ideal gas, in which each element of mass or volume is well characterized by the equation of state of the **ideal gases**, applied to a given volume, say 1 L, 1 m^3 or a given mass, 100 g, 1 kg:

$$\frac{PV}{T} = nR \quad (5.1)$$

where P is the pressure (local **atmospheric pressure**) inside this volume V; T, its absolute temperature; n, the number of moles of air in the element under consideration; and R, the **universal ideal gas constant** ($R = 0.082$ atm L mol^{-1} K^{-1}, or 8.31 J mol^{-1} K^{-1}). This equation relates the three **thermodynamic coordinates** of a gaseous system, that is, P, V, and T. Knowing any two variables, the third is defined by Eq. (5.1).

Since the atmosphere has no defined volume, the volume of the unit mass, called **specific volume of air** $v = V/m$ (m^3 kg^{-1}), is generally taken. The number of moles n is the relationship between the mass of the gas and its molecular mass M. Thus, from Eq. (5.1), we can write, remembering that $n = m/M$:

$$\frac{PV}{T} = \frac{m}{M}R \quad \text{or} \quad \frac{Pv}{T} = \frac{R}{M}$$

Note also that the inverse of v is the specific mass of gas d (kg m^{-3}), also called as **air density** by meteorologists, and as gases expand and contract with temperature variations, both v and d depend on T.

The temperature T (K) is the absolute air temperature measured in meteorological stations as t in °C, that is, in the shade and in a ventilated environment. The relationship is T (K) = t (°C) + 273.

The atmospheric pressure $P = P_{atm}$, which enters in the place of P in Eq. (5.1), is the sum of the partial pressures of the constituent elements of atmospheric air (Table 5.1), where each of them acts independently, according to **Dalton's law**. So we have:

$$P_{atm} = P_{N_2} + P_{O_2} + P_{argon} + P_{water\ vapor} + \ldots \quad (5.2)$$

This atmospheric pressure is measured by barometers or barographs also at meteorological stations. Its value oscillates around 760 mmHg at sea level. The atmospheric pressure corresponds to the weight per unit area of a column of air extending from the considered point to the outer boundary of the atmosphere, varying greatly with the altitude of the considered point. For this reason, in the comparison of P_{atm} data taken at different locations, a correction, called reduction to the sea level, is necessary. In Piracicaba, which is 580 m above sea level, the value of the atmospheric pressure oscillates around 712 mmHg. The unit mmHg is still widely used in Meteorology, but the correct way is the use of Pascal (Pa). Conversions of the different units used to express pressure can be made with the aid of Table 5.2.

Since air is a mixture of gases, in order to apply it to Eq. (5.1), it is necessary to know its average molecular mass M_{air}. The best choice is a mass-weighted average, taking the number of moles n in the mass m of air contained in V. The

Table 5.2 Conversion table for different pressure units

	atm	cmHg	cmH_2O	bar	Pa	lb $in.^{-2}$
1 atm	1	76	1033	1.034	101,325	14.696
1 cmHg	1.316×10^{-2}	1	13.6	0.014	1333.2	0.1934
1 cmH_2O	9.681×10^{-4}	7.35×10^{-2}	1	9.81×10^{-4}	98.1	1.423×10^{-2}
1 bar	0.967	73.6	1019	1	98,000	14.214
1 Pa	9.869×10^{-6}	7.5×10^{-4}	1.019×10^{-2}	1.02×10^{-5}	1	1.45×10^{-4}
1 lb $in.^{-2}$	6.805×10^{-2}	5.172	70.292	0.0704	6894.8	1

(weighted) average molecular mass of the atmospheric air can be estimated as follows:

$$M_{air} = \frac{\sum (n_i M_i)}{\sum n_i} \quad (5.3)$$

where M_i is the molecular mass of each component i, and n_i is their respective number of moles in a volume element, respectively.

Adopting the composition of Table 5.1, we will have by Eq. (5.3), approximately:

$$M_{air} = 29 \text{ g mol}^{-1}$$

which is a value close to the molecular mass of the nitrogen, N_2 that is 28 g mol^{-1}, because N_2 is the most expressive component in air.

By dividing the value of the universal gas constant R by the molecular mass of the gas in question, a specific value of R for that gas is obtained. Therefore, the specific R (R_a) value for atmospheric air is given by:

$$R_a = \frac{R}{M_{air}} = \frac{8.31 \text{ J mol}^{-1} \text{ K}^{-1}}{29 \text{ g mol}^{-1}} = 0.287 \text{ J mol}^{-1} \text{ K}^{-1}$$

Example: Calculate the density of an air sample when the atmospheric pressure is 712 mmHg and the temperature 27 °C?

Solution:

$$\frac{Pv}{T} = R_a \quad \text{or} \quad d = \frac{P}{R_a T}$$

$$d = \frac{94{,}925.5 \text{ Pa}}{0.287 \text{ J g}^{-1} \text{ K}^{-1} \times 300 \text{ K}} = 1102.5 \text{ g m}^{-3} = 1.1025 \text{ kg m}^{-3}$$

Equation (5.2), Dalton's law, can also be applied individually to any of the air components. Of importance is its application to water vapor. For water vapor in the atmosphere, Eq. (5.2) can be rewritten as follows:

$$P_{atm} = P_{dry\ air} + P_{water\ vapor}$$

The dry air pressure ($P_{dry\ air}$) being the sum of the partial pressures of all the components minus the water, which will be designated hereafter by P_a. $P_{water\ vapor}$ is the pressure only of water and will be designated as e_a, which is the contribution of the water vapor to the atmospheric pressure. It is simply referred to as the current or actual **partial pressure of vapor**, often erroneously designed as vapor tension. So:

$$P_{atm} = P_a + e_a$$

Since the concentration of water in the air in the normal range of temperatures is variable (from 0% to about 4%), we can have air samples from dry to wet (or saturated). Three basic principles govern the behavior of water vapor in the atmosphere:

1. The actual partial pressure of vapor e_a, that acts at the moment under consideration, is proportional to the mass of vapor (m_v) in the chosen volume element of air. This can be verified by the water vapor state equation (Eq. (5.1) applied only to water vapor, the more vapor the greater n_v, V and T taken as constants):

$$\frac{e_a V}{T} = n_v R \quad \text{or} \quad e_a = \frac{m_v}{M_v} \cdot \frac{RT}{V} \quad (5.4)$$

   or looking at e_a that is proportional to m_v.
2. At a given temperature T, there is a maximum of water vapor that the air can hold or absorb. The partial pressure of vapor e_a, when the air retains the maximum of vapor, is denominated **saturated partial pressure of vapor**, e_s.
3. The higher the temperature of the air, the greater the mass of vapor it can retain, that is, the larger T, the larger e_s.

In Fig. 5.2, the curve of e_s versus t is presented. This curve is a physical feature of water that is part of psychrometric charts. The equation for t in °C (**Tétens equation**), with the result of e_s given in kPa, is:

$$e_s = 0.6108.10^{\left(\frac{7.5 \cdot t}{t+237.3}\right)} \quad (5.5)$$

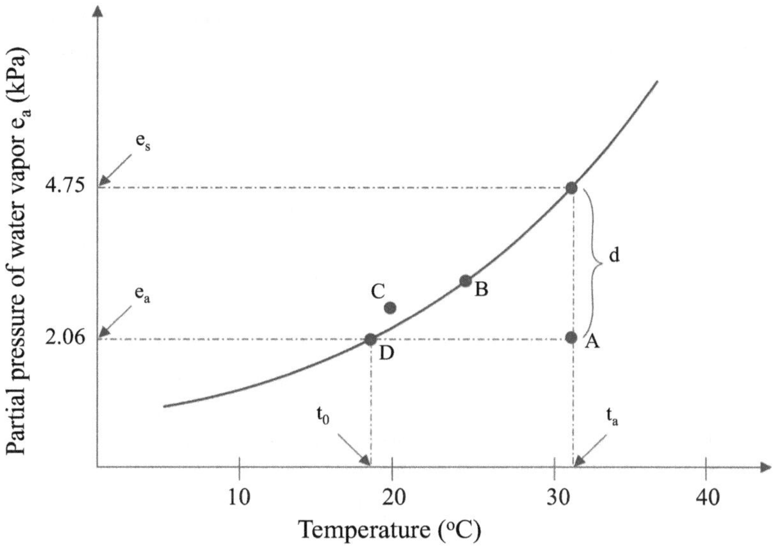

Fig. 5.2 Partial vapor pressure as a function of temperature. This curve following Eq. (5.5) is universal, that is, valid for any air sample

The reader should note that this graph is the P versus T curve of Eq. (5.1), applied to water vapor, where $P = e_a$ and $T = t$. As stated, any two variables of P, V, and T characterize the gas; in this case, e_a and t characterize the state of the water vapor in the atmosphere.

Psychrometers measure the value of e_a; and e_s is given in tables, psychrometric graphs, or even calculated by Eq. (5.5).

A given sample of air with values of e_a and t such that in Fig. 5.2 fall below the saturation curve, as is the case of an air sample characterized by point A, is not saturated with water vapor. The air is called saturated when its values of e_a and t fall on the saturation curve, and in this situation $e_a = e_s$ (Point B). Above the saturation curve, water is present in the liquid phase (Point C). In very special cases, the air may be slightly above the saturation curve because it is supersaturated.

In Meteorology and in Eq. (5.4), $\rho_v = m_v/V$ is called the **actual air humidity**. For a saturated air, ρ_v is maximum (at a given temperature) and is called **saturation air humidity** ρ_{vs}. These humidities can be easily calculated from Eq. (5.4), expliciting m_v/V:

$$\rho_v = \frac{M_v}{R} \cdot \frac{e_a}{T} = \frac{288 \cdot e_a}{273 + t} = \frac{2168.1 \times e_a}{273 + t} \quad (5.6)$$

$$\rho_{vs} = \frac{M_v}{R} \cdot \frac{e_s}{T} = \frac{288 \cdot e_s}{273 + t} = \frac{2168.1 \times e_s}{273 + t} \quad (5.7)$$

In the equations shown above, the constant 288 appeared to obtain ρ_v and ρ_{vs} in grams of vapor m^{-3}, when e_a and e_s are used in mmHg and t in °C. In the case of e_a and e_s in kPa, the constant changes to 2168.1.

The **relative humidity** (RH) of an air sample is defined by the percentage relation between e_a and e_s, at a given air temperature. As seen above, if t changes, e_s also changes, and so RH:

$$\text{RH} = \frac{e_a}{e_s} \cdot 100 = \frac{\rho_v}{\rho_{vs}} \cdot 100 \quad (5.8)$$

In Fig. 5.2, the air represented by A is at a temperature t_a and has a vapor pressure e_a. The saturation pressure at the same temperature is indicated in the graph. The RH is 43.4%. The air represented by B has an RH = 100%. In general, at 100% RH, all excess of vapor is condensed to the liquid phase. A supersaturated air, with relative humidity above 100%, is stable only under special conditions. The **air saturation deficit** (d) is defined as the difference between the saturation pressure and the current pressure of an air sample:

Table 5.3 An example of an air mass that in 5 h has its temperature decreasing from 30 to 5 °C

Time (h)	Temperature (°C)	e_a (kPa)	e_s (kPa)	d (kPa)	RH (%)	ρ_v (g m^{-3})	Condensed vapor (g m^{-3})
0	30	1.71	4.24	2.53	40	12.15	0
1	25	1.71	3.16	1.45	54	12.36	0
2	20	1.71	2.40	0.69	73	12.57	0
3	15	1.71	1.71	0	100	12.79	0
4	10	1.23	1.23	0	100	9.38	3.41
5	5	0.87	0.87	0	100	6.78	2.70

$$d = e_s - e_a \tag{5.9}$$

On the outside of a cold beer glass, droplets of water accumulate. This happens because the air close to the glass reaches low temperatures and cannot anymore hold all the water dissolved in it.

The temperature at which the air would reach saturation by cooling, without varying its humidity, is called the **dew point temperature** (t_O), or simply, dew point. At the dew point, $e_s = e_a$ and $d = 0$. If, for example, the air sample represented by A in Fig. 5.2 is cooled, keeping e_a constant, it reaches saturation at point D. The corresponding temperature is the dew point t_O.

Let's look at another example summarized in Table 5.3: imagine that a certain mass of air, with constant e_a, undergoes a gradual drop in temperature. Let the initial conditions be t = 30 °C and e_a = 1.71 kPa, and the temperature drop is 5 °C h^{-1}. After the first hour, t falls to 25 °C and, as a consequence, e_s drops to 3.16 kPa and the RH rises to 54%. In the second hour, t = 20 °C, e_s = 2.4 kPa, and RH = 74%. In the third hour, the air is saturated, $e_a = e_s$, and RH = 100%, reaching the dew point. In the fourth hour, as the air cooled further, e_s drops to 1.23 kPa and it can no longer retain all the vapor it has. Part condenses, always equilibrating e_a with e_s.

Relative humidity, vapor pressure, air temperature, and various meteorological parameters can be obtained with an instrument called a **psychrometer**. It consists of two thermometers, one of them wrapped in a constantly moistened fabric, called the wet thermometer, and another next to it, only in thermal equilibrium with the air in the shade, the dry thermometer. The wet thermometer receives a constant flow of air through a ventilation system. Because of this, the water of the fabric will be evaporated, subtracting energy from the moist bulb. Its temperature will drop and, when the steady state is reached, it will stabilize. The temperature recorded by the wet thermometer under these conditions is called **wet bulb temperature** t_u. It is considered that state at which the heat flow from the air to the bulb of the thermometer is equal to the energy spent in the evaporation. Under these conditions, it can be shown that:

$$e_a = e_{su} - \lambda\,(t - t_u) \tag{5.10}$$

where

e_a = current partial pressure of the water vapor in the air, kPa;
e_{su} = partial saturation pressure of the air at the temperature of the wet thermometer (t_u), in kPa;
t = air temperature (measured by the dry thermometer), °C;
t_u = wet bulb temperature, °C;
λ = psychrometric constant: 0.062 kPa °C^{-1} for ventilated psychrometers and 0.074 kPa °C^{-1} for common, non-ventilated psychrometers.

Once the **psychotropic depression** ($t - t_u$) is known, we can calculate by Eq. (5.10) the current partial pressure e_a, and thus the other parameters previously seen that characterize air.

Example: In a greenhouse, using a non-ventilated psychrometer, the following data were obtained: t = 25.3 °C and t_u = 19.8 °C. So that:

1. e_s = 3.22 kPa (Eq. (5.5), for t = 25.3 °C);
2. e_{su} = 2.31 kPa (Eq. (5.5), for t = 19.8 °C);
3. e_a = 1.90 kPa (Eq. (5.10), for e_{su} = 2.31 kPa);
4. d = 3.22–1.90 = 1.32 kPa;
5. ρ_v = 13.74 g m^{-3} (Eq. 5.6);
6. RH = (1.90/3.22) × 100 = 59%.

5.3 Solar Radiation

One of the most important chapters of Meteorology is the study of the energy received from the sun by **solar radiation**. Making a completely negligible mistake, the Sun can be considered the only source of energy responsible for the physical, chemical, and biological processes that develop in the atmosphere (Fig. 5.3). Qualitatively, solar radiation is made up of electromagnetic waves of wavelength ranging from about 0.2 to 20.0 μm. Electromagnetic waves carry energy from the Sun to Earth, where the energy E_λ of a ray or **photon** is given by $E_\lambda = h{\cdot}f = h{\cdot}c/\lambda$, where h is the **Plank constant** (6.63 × 10^{-34} J s), f (μm^{-1}) the frequency of the radiation, λ (μm) its wavelength, and c the speed of light in the vacuum (300,000 km s^{-1}). From these definitions, it becomes clear that the higher the frequency or the lower the wavelength, the more energetic the radiation. When talking about a beam of radiation, it can be monochromatic (one wavelength) or polychromatic (various wavelengths), as in the case of solar radiation. When referring to a spectrum of electromagnetic radiation, all the wavelengths of which it is constituted are understood. For each wavelength, therefore, an amount of energy (W m^{-2} μm^{-1}) is associated, whose summation in terms of wavelength gives us the total energy of the spectrum (W m^{-2}). In Figs. 5.3 and 5.4, we can see aspects of the solar energy, and in the last one, the upper curve is the solar spectrum in the absence of the atmosphere, that is, the beam that reaches the top of the atmosphere. In the same figure, we can also see the spectrum of global radiation (the one that reaches the surface of the ground).

All bodies of nature emit radiations in the form of electromagnetic waves, including the Sun and Earth, provided they are at an absolute

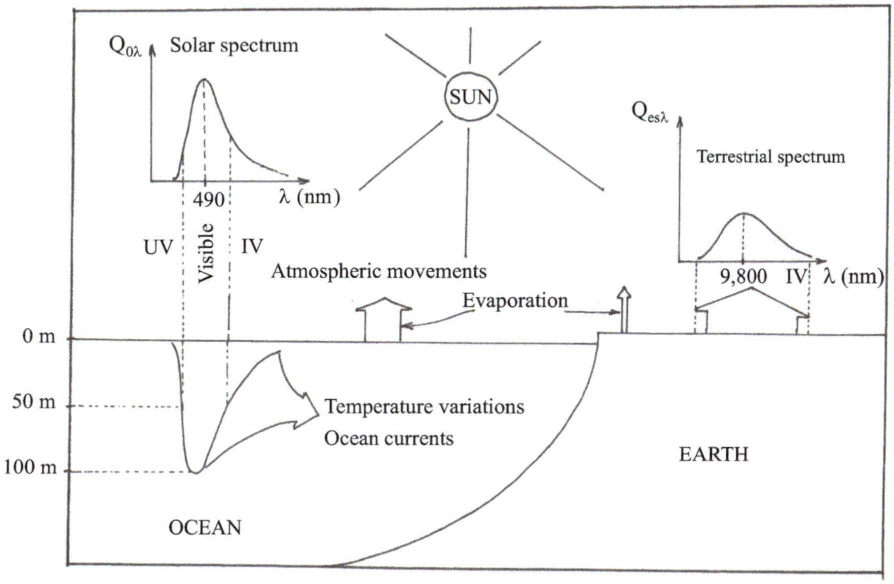

Fig. 5.3 An illustration of the action of the Sun on Earth, including oceans and land

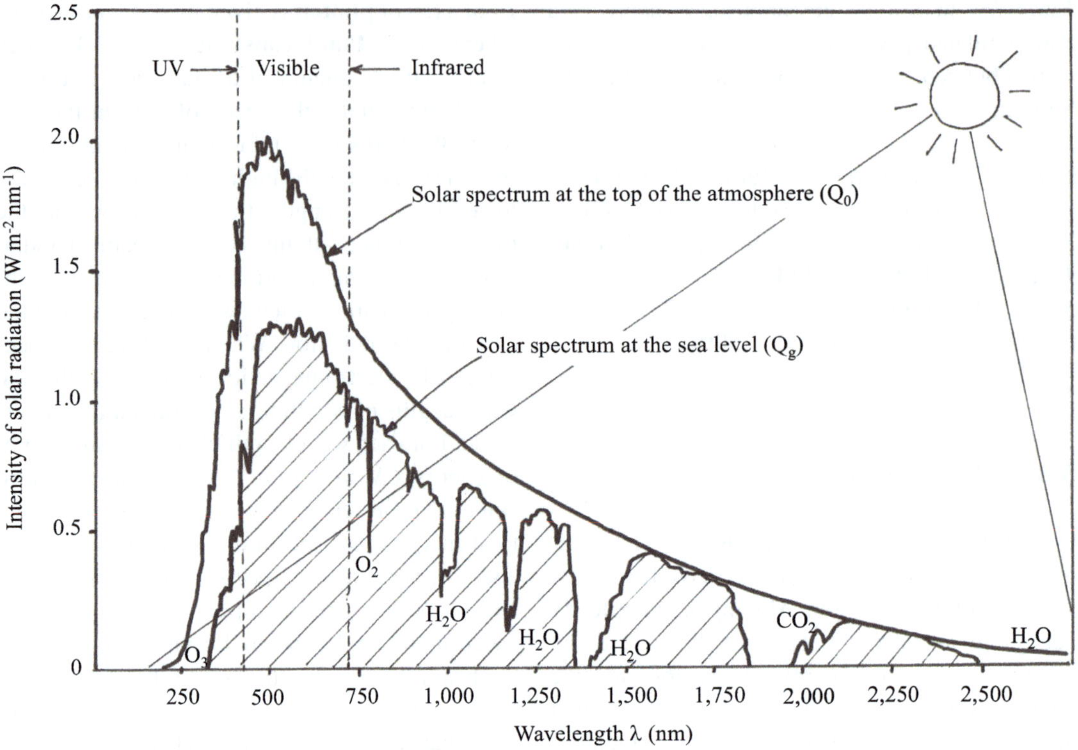

Fig. 5.4 Solar spectrum before and after radiation enters the atmosphere

temperature T greater than 0 K. This statement is based on **Stephan–Boltzmann's law**, whose expression is Eq. (5.18), which shows that the emission of a body is proportional to the fourth power of its absolute temperature T. Thus, the emission of the Sun is proportional to about $(6000)^4$, and the emission of the Earth, to about $(300)^4$. Considering the size of the Sun relative to that of the Earth, its total emission is still much greater. It turns out that only a tiny fraction of this solar energy reaches our planet.

The amount of radiant energy that reaches a unitary surface per unit of time, perpendicular to the solar rays, in the absence of the atmosphere, and at a distance of the Sun equal to the average of the Earth–Sun distances (which are very small differences since the trajectory of the Earth around the Sun is an ellipse of low eccentricity) is called the **solar constant**. Its approximate value is $J_0 \cong 2$ cal cm^{-2} min^{-1} or 1400 W m^{-2}. This solar radiation is called **short-wave radiation**, in contrast to that emitted by Earth, called **long-wave radiation** or **terrestrial radiation** (Fig. 5.5). By **Wien's law**, the most frequent wavelength λ_f in the emission of a body at temperature T is given by:

$$\lambda_f \cdot T = 2.940 \times 10^6 \text{ nm K} \tag{5.11}$$

So that:

$$\lambda_f(\text{Sun}) = \frac{2.940 \times 10^6}{T} = \frac{2.940 \times 10^6}{6000} = 490 \text{ nm} = 0.49\ \mu\text{m}$$

which corresponds to the green color of the solar spectrum, and

$$\lambda_f(\text{Earth}) = \frac{2.940 \times 10^6}{T} = \frac{2.940 \times 10^6}{300} = 9800 \text{ nm} = 9.8\ \mu\text{m}$$

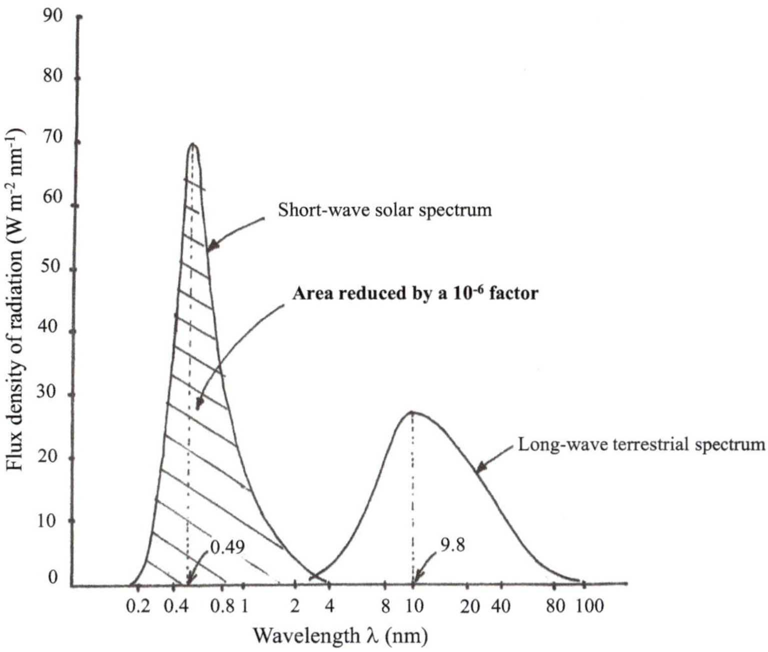

Fig. 5.5 Solar and terrestrial spectra based on Stephan–Boltzman's law. Note that the area of the solar spectrum is very much reduced to fit into the same graph

that falls in the far infrared range (heat), non-visible, therefore called long-wave radiation.

When a radiation beam $I_{00\lambda}$ (W m^{-2}) falls on a transparent medium, it is partially reflected $I_{r\lambda}$; partially absorbed/diffused in the medium $I_{a\lambda}$; and partially transmitted $I_{t\lambda}$. The emergent beam from this medium is $I_{t\lambda}$, which depends on the properties of the medium and follows the **Beer–Lambert** law:

$$I_{t\lambda} = I_{00\lambda} \cdot \exp - (\mu \cdot d \cdot x) \qquad (5.12)$$

where μ is the **radiation attenuation coefficient** of the medium, d the density of the medium, and x its thickness crossed by the beam. Figure 5.6 shows schematically the process, which according to the energy conservation principle can be written as $I_{00\lambda} = I_{r\lambda} + I_{a\lambda} + I_{t\lambda}$. The following characteristics of the translucent material are defined according to the intensity of each component of the beam:

Reflecting power or **reflectivity *r***:

$$r = \frac{I_{r\lambda}}{I_{00\lambda}}$$

Absorbing power or **absorptivity *a***:

$$a = \frac{I_{a\lambda}}{I_{00\lambda}}$$

Transmitting power or **transmissivity *t***:

$$t = \frac{I_{t\lambda}}{I_{00\lambda}}$$

For wavelengths in the visible range, ***r*** tends to 1 in a mirror ($I_{r\lambda} \cong I_{00\lambda}$, $I_{a\lambda} = 0$, $I_{t\lambda} = 0$); ***a*** tends to 1 for bodies painted black ($I_r = 0$, $I_{a\lambda} \cong I_{00\lambda}$, $I_{t\lambda} = 0$), and ***t*** tends to 1 in a pure quartz glass ($I_r = 0$, $I_{a\lambda} = 0$, $I_{t\lambda} \cong I_{00\lambda}$). In the majority of the cases, ***r***, ***a***, and ***t*** take values between 0 and 1, depending on the medium. A **black body** is by definition a body with $a = 1$, being a theoretical definition.

Kirchhoff's law shows that the absorptivity "*a*" is equal to the emissivity "*e*" of a black body, for the same wavelength. In words, the law says that "every good absorber is also a good emitter." The black body is an ideal concept and is so called because for sunlight, dark bodies and bodies painted black have values of "*a*" close to

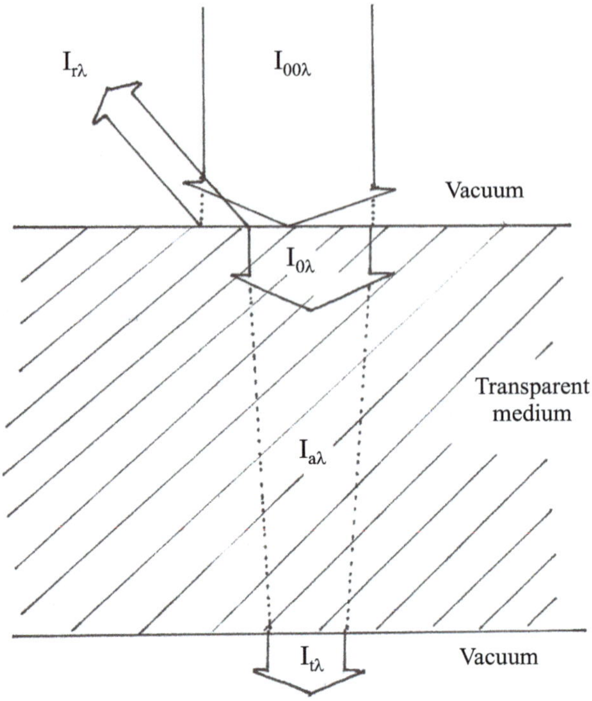

Fig. 5.6 Schematic view of the splitting of an incident radiation beam into a translucid medium

1, that is, they absorb almost every radiation they receive. As a consequence, they reach a higher temperature. By **Stephan–Boltzmann's law** (Eq. (5.18)), black bodies also emit more, for two reasons: emission is proportional to T^4, which is larger, and proportional to the emissivity "*e*," which is equal to 1 by Kirchhoff's law.

The transmissivity *t*, as we saw above, varies greatly with the wavelength. For short waves (those of the Sun), for example, *t* is very high for glass and transparent plastics, letting almost every radiation pass through. But for long waves (those of the Earth), *t* is very low, letting to pass very little. This fact leads to what we call the **greenhouse effect**, which occurs in glass or plastic houses used as nurseries and commonly called greenhouses. In them, the short-wave solar radiation penetrates with great ease during the day, heating its interior. The heated interior emits, by Stephan–Boltzmann's law, far infrared radiation (long wave), which cannot escape the greenhouse, being reflected back into it. In this way, the internal temperature of the greenhouse increases in relation to the external temperature and, at night, the temperature is also maintained at higher levels. In winter, the greenhouses are advantageous; and in summer, the temperature can become so high that forced ventilation or the use of perforated plastic is required. In global terms, the Earth's atmosphere also has a characteristic transmissivity, which allows the arrival of solar radiation to the Earth's surface. Air pollution, however, has changed its transmissivity mainly in relation to long waves, and the overall energy balance is affected, resulting in an increase in the average temperature of the globe (Fig. 5.7).

With certain limitations, Eq. (5.12) can also be applied to a polychromatic beam such as the whole solar radiation spectrum. In this case, μ is an average value for all λ of the radiation beam in relation to all the constituents of the atmosphere, *d* is variable in relation to the altitude, and *x* is the thickness of the atmosphere which is mainly function of the inclination of the solar rays and of the latitude. For the correspondent values of these constants, the term exp $-(\mu dx)$ is also termed the **atmospheric transmissivity *t*** and is equal to the relation Q_g/Q_0. The solar radiation in its path through the atmosphere suffers various losses due to absorption and diffusion

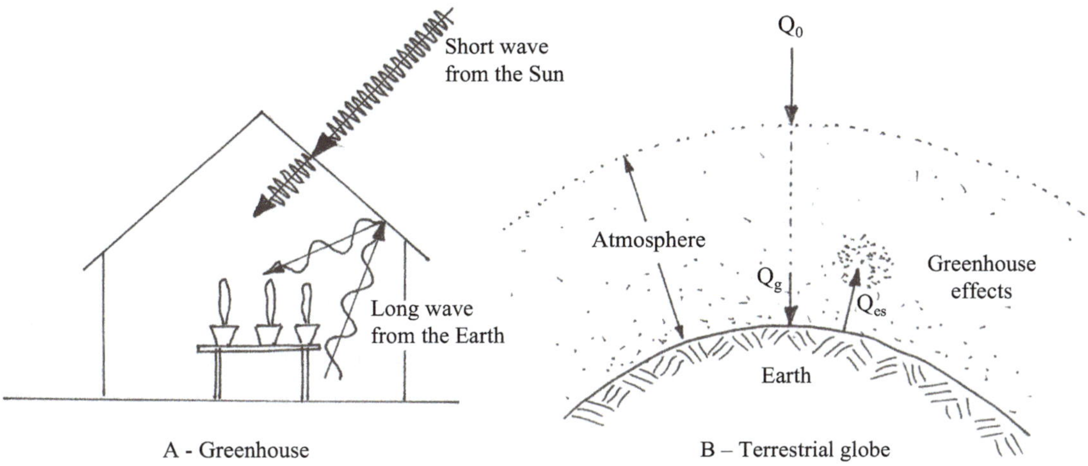

Fig. 5.7 Illustration of the greenhouse effect: to the left a "glass greenhouse" showing that the short waves pass through the glass and the long waves do not; to the right the Earth with its atmosphere showing that the air layer also acts partially according to the greenhouse effect, because the short waves pass easier through the atmosphere than the long waves

phenomena. The main components of the atmosphere responsible for the absorption are: (a) **ozone**, which almost completely intercepts **ultraviolet radiation** UV (the value of μ is very high for UV) and also has soft absorption bands (bands that correspond to wavelengths for which μ is very high and the absorption in this range is high) in the infrared region; (b) carbon dioxide, which has selective absorption at the wavelengths 1.5–2.8–4.3–15 μm, with the highest absorption in the region of the infrared (15 μm); (c) water vapor, which is the most important of the selective sorbents, despite its low atmospheric content (±2% on average), and absorbs predominantly in the infrared region, that is, between 0.8 and 2.7, 5.5 and 7 μm, and 15 μm as shown by the graph of the global radiation in the infrared band (Fig. 5.4).

In addition to the absorption process, the diffusion process of solar radiation, determined by its different constituents, occurs in the atmosphere. This process is responsible for appreciable losses of radiation returning to space. Due to the size of the diffusion-determining particles, two distinct effects must be noted. If the particles are of the order of 0.01–0.1 of the wavelength of the radiation, the diffusion will be proportional to the inverse of the fourth power of the wavelength of the radiation ($1/\lambda^4$). This means that smaller wavelengths are preferably diffused, and the diffusion grows rapidly toward the violet in the visible spectrum. This process of diffusion is responsible for the vivid blue color of the sky. It is termed selective diffusion in order to be distinguished from the diffusion process determined by larger, non-selective particles. In this case, the diffusion is no longer dependent on the wavelength of the radiation, with diffusion at all wavelengths, resulting in the white color of the diffuse light. A typical example is the fog, which is grayish white.

Suppose that a flux of radiation Q_0 (cal cm^{-2} min^{-1} or W m^{-2}) penetrates into the atmosphere at its most extreme limit (Fig. 5.8).

The reader should not confuse the **solar constant** J_0 with Q_0. In the case of J_0, the receiving surface is always perpendicular to the solar rays and, therefore, its value is constant. In the case of Q_0, the surface is parallel to the horizontal plane of the place under consideration and thus the value Q that reaches the area A depends on the inclination of the solar rays, which in turn depends on the time of year, latitude, and time of day. Figure 5.9 illustrates the Law of Cosines in relation to the slope of a ramp receiving sun

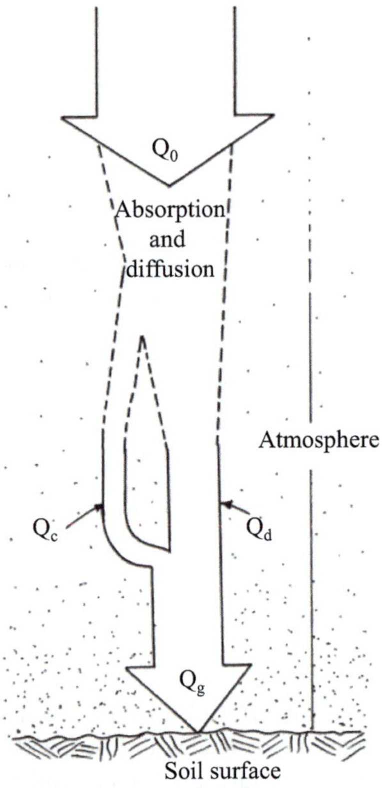

Fig. 5.8 Details of the penetration of the solar radiation into the atmosphere, down to the earth surface

rays. By focusing on a sloped surface at an angle α in relation to the direction of the rays, the beam falls on an area A greater than the corresponding area perpendicular (or normal) to the rays, and as the beam intensity is given in W m^{-2}, it decreases by the factor cosα. Therefore, because the beam is the same, but acting on a larger area, it apparently weakens. By the **Law of Cosines**:

$$Q = Q_0 \cdot \cos\alpha \tag{5.13}$$

Because the movement of the Earth around the Sun occurs with an angle its axis (always parallel to itself) of approximately 23° (degrees) in relation to the plane of translation (ecliptic), the seasons appear with variable **day lengths**. The day has 12 h on the **equinoxes**, which occur on March 21 and September 23, in any latitude, that is, any part of the globe. The day is longer in the summer **solstice** and shorter in the winter solstice, occurring in the Southern Hemisphere on December 21 and June 21, respectively (the opposite in the Northern Hemisphere). On solstices, the higher the latitude the greater the day/night variation. Thus, the values of N and Q_0 vary, as shown in Table 5.4. N is the length of the day, also called the **photoperiod**. When we

Fig. 5.9 Illustration of the cosines law, showing that the normal (perpendicular) area A_N is smaller than the area A receiving radiation at an angle α

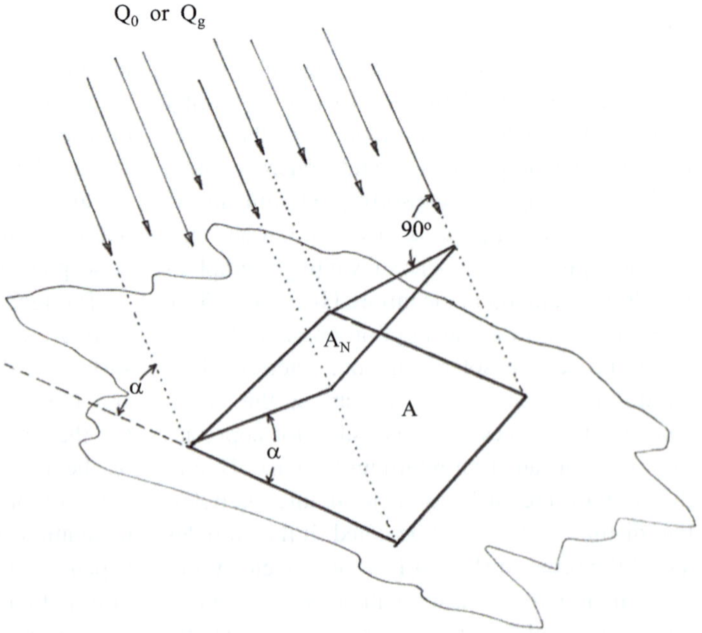

Table 5.4 Monthly changes of the incident radiation Q_0 (cal cm^{-2} day^{-1}) for three latitudes of the southern hemisphere

Latitude	−20°		−22° 30′		−25°	
Month	Q_0	N	Q_0	N	Q_0	N
January	994	13.2	1004	13.4	1013	13.6
February	952	12.8	953	12.8	952	12.9
March	870	12.2	859	12.2	846	12.3
April	748	11.6	726	11.6	701	11.5
May	635	11.2	606	11.1	576	10.8
June	571	10.9	539	10.8	507	10.5
July	586	11.0	555	10.9	523	10.7
August	672	11.4	646	11.3	618	11.2
September	792	12.0	774	12.0	755	11.9
October	897	12.5	891	12.6	883	12.7
November	968	13.2	973	13.2	977	13.4
December	996	13.3	1008	13.5	1018	13.8

talked about the development of the plants in Chap. 4, we showed the importance of N (not nitrogen), which leads to the **photoperiodism** of each species.

During the course of the radiation through the atmosphere, as we have seen, part of it is absorbed, diffused, or transmitted, and finally reaches the Earth's surface (Fig. 5.8). The fraction diffused by the atmosphere or diffuse radiation is constituted of a part that returns to the sidereal space and of another part that reaches the surface of the Earth, that is called **sky radiation** (Q_c). At the surface of the Earth, therefore, we will have the sky radiation plus the fraction of radiation that has not undergone alteration when passing through the atmosphere, called **direct radiation** (Q_d). The sum of these two fractions is called **global radiation** (Q_g):

$$Q_g = Q_c + Q_d \tag{5.14}$$

The global radiation can be measured with special instruments (**actinographs** or **solarimeters**), which record the energy reaching the soil in cal cm^{-2} min^{-1} or W m^{-2}.

Part of the radiation that reaches the ground Q_g is reflected back to the atmosphere. The **reflecting power** of a surface is also called **albedo** (A), and is defined as

Table 5.5 Reflecting power or albedo for different materials in the Soil–Plant–Atmosphere System

Material	Albedo	
Stones	0.15–0.25	
Cultivated soil	0.07–0.14	
Forests	0.06–0.20	
Light sand	0.25–0.45	
Crop	0.12–0.25	
Sun height		
Water	90–40°	0.02
Water	30°	0.06
Water	20°	0.13
Water	10°	0.35
Water	5°	0.59

$$A = \frac{Q_r}{Q_g} \tag{5.15}$$

The albedo depends on the type of surface (topography, coloring, roughness, etc.) on which the light incides. For the surface of the Earth, the average albedo is given in Table 5.5.

The term **net energy** (Q_L) is used to express the difference between the short-wave radiation that arrives at a given plane on the ground and the long-wave radiation leaving that same plane. This plane can be the surface of the ground or any plane, imaginary, at a certain height of the ground. Considering the global solar radiation Q_g that actually reaches a surface in question, it is easy to verify that the total absorbed energy (Q_a) by the soil is:

$$Q_a = Q_g(1 - A) \tag{5.16}$$

This energy absorbed by the surface will warm it up and make it emit (Q_{es}) (Stephan–Boltzmann's law), which is the emission of Earth in the long-wave form. This radiation, in turn, will cause an increase in the temperature of the atmosphere, which will also emit in the form of long wave (Q_{ea}).

The net radiation is then defined by the expression:

$$Q_L = Q_a - Q_{es} - Q_{ea} \tag{5.17}$$

Equation (5.17) is also called the **Radiation Balance** (RB), which is a balance between short-wave (SWB) and long-wave (LWB) balances. Therefore:

$$\text{RB} = \text{SWB} + \text{LWB}$$

The total radiant emission of a **black body** is, according to **Stephan–Boltzmann's law**, proportional to the fourth power of the temperature of its surface and is expressed by:

$$Q_e = e\sigma T^4 \quad (5.18)$$

where

Q_e = radiation emission, in W m^{-2};
e = emissivity of the body, equal to 1 for a black body, non-dimensional;
σ = Stephan–Boltzmann's constant (= 5.67 × 10^{-8} W m^{-2} K^{-4} = 4.903 × 10^{-9} MJ m^{-2} K^{-4});
T = temperature in K.

Therefore, in terms of terrestrial radiation:

$$Q_{es} = e_s\sigma T_s^4 \quad (5.19)$$

and atmospheric radiation:

$$Q_{ea} = e_a\sigma T_a^4 \quad (5.20)$$

T_s being the temperature of the surface (soil, crop, etc.) and T_a the temperature of the air.

Substituting Eqs. (5.16), (5.19), and (5.20) into (5.17), we have:

$$Q_L = Q_g(1 - A) - e_s\sigma T_s^4 - e_a\sigma T_a^4 \quad (5.21)$$

which is a practical expression for the determination of Q_L.

The net radiation can be measured by special instruments called Net Radiometers. Such instruments are not always available and, therefore, Ometto (1968) made a study of the relationships between net radiation, global radiation, and hours of sunshine, in Piracicaba, São Paulo state, Brazil. The expressions obtained by the author allow us to estimate the net radiation in the absence of the specific instrument. They are:

$$Q_L = -12.0 + 0.56 \times Q_g \text{ (spring – summer)}$$

$$Q_L = -23.0 + 0.45 \times Q_g \text{ (autumn – winter)}$$

where Q_L and Q_g are given in cal cm^{-2} day^{-1}. The linear regressions were significant at the 0.1% probability level.

In the case of agricultural crops, the important wavelengths for photosynthesis are those in the range of the blue to red (see Table 4.1). Therefore, it was established that the **Photosynthetically Active Radiation** (PAR) is the band between 400 and 700 nm. PAR is measured through special radiometers sensitive only to this range of wavelengths. As shown in Fig. 5.10, one way to measure the absorbed PAR_a is using three radiometers: one above the crop canopy, turned upward that measures the incoming PAR_0, another one, also above the canopy, but turned downward that measures the reflected radiation by the crop and soil PAR_r, and) the last one turned upward, on soil surface that measures the transmitted PAR_t. The absorbed part is calculated by the difference:

$$PAR_a = PAR_0 - (PAR_r + PAR_t) \quad (5.22)$$

Dividing Eq. (5.22) by PAR_0 and calling PAR_a/PAR_0 as **coefficient of radiation interception** of day i (CRI_i), and making the ratio (PAR_r + PAR_t)/PAR_0 = exp($-K{\cdot}LAI_i$) [following Beer–Lambert's law, Eq. (5.12)], we have:

$$CRI_i = 1 - \exp(-K \cdot LAI_i) \quad (5.23)$$

where K is the light extinction coefficient and LAI_i is the leaf area index of day i. K values rotate around 0.7 for various crops. The relation of CRI with LAI is empirical, but it makes sense because the greater the LAI, the greater the part of PAR_0 that is absorbed. In this way, knowing LAI, PAR_a by the crop can be estimated. This procedure is widely used in modeling the productivity of agricultural crops.

Both Q_g and PAR_0 are available for the physicochemical and biological processes that occur in the canopy of a crop. They are therefore also related to the growth and development of crops discussed in the previous chapter. These concepts are relatively new and, long before them, the

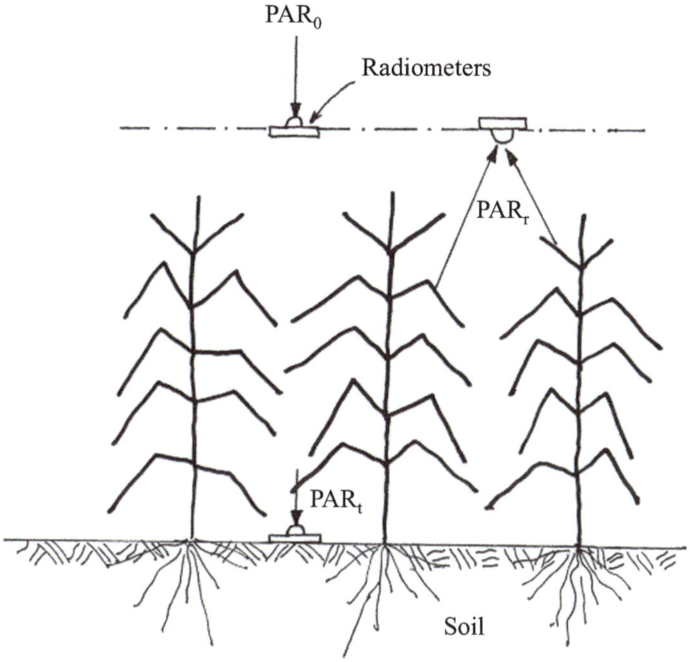

Fig. 5.10 Schematic representation of the measurement of solar radiation in a corn canopy

concept of **Degree Days** (DD) was defined, with similar purposes. The origin of this concept comes from the eighteenth century, suggested by Reaumur in France around 1735. At that time, the simplest (if not the only) measurement related to energy or heat quantities was the measurement of temperature. For this reason, Reaumur assumed that the sum of air temperatures during the vegetative cycle of a species expresses the amount of solar energy it needs to reach maturity. Thus, Reaumur was the precursor of the system of **thermal units** or **Degree Day DD**, currently used to characterize the phenological cycle of agricultural crops. The DDs are still widely used when only temperature data are available.

Figure 5.11 shows schematically the daily variation of the air temperature T_{air} through the function $T(t)$. T_{air} rises from a minimum value along the morning (T_{min}) until it reaches a maximum (T_{max}) around midday, and decreases again to T_{min} during the next night, day after day. Assuming for a given crop, say corn, a lower basal temperature (T_{bl}) and one upper basal temperature (T_{bu}) (also shown in Fig. 4.1) above or below which the plant does not develop or grows at minimal rates, we have a temperature interval (T_{bu}–T_{bl}), during a time interval (t_f–t_i) during which the plant grows and develops. According to the concept of thermal units or DD, the widely dashed area A shown in Fig. 5.11 is proportional to the part of the solar energy received and used for development by the plant on that date, which is part of the net energy given by Eq. (5.24).

The area A is given by the integral:

$$DD_j = A = \int_{t_i}^{t_f} T(t)\mathrm{d}t - B - C \tag{5.24}$$

with unit °C day when t is given in days. The subscript j refers to day j. The name DD comes from this product.

Since the $T(t)$ curve is not always known or is difficult to be modeled, several simple formulas for calculating DD_j were proposed. When we studied soil water storage (Chap. 3), we showed that an integral can also be given by the mean value of the function multiplied by the integration interval. By this criterion, it can be said that:

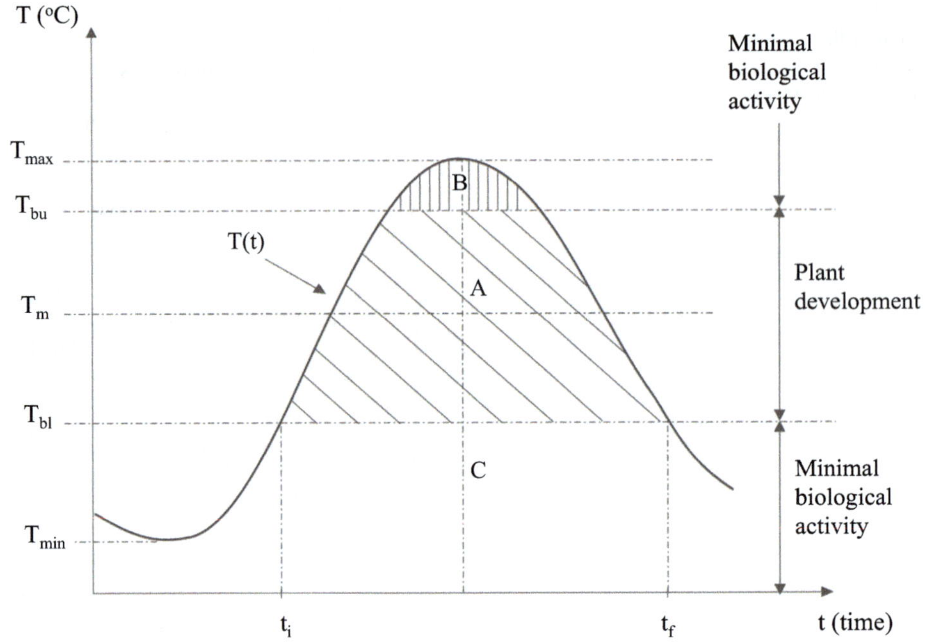

Fig. 5.11 Air temperature distribution along a sunny day. Air temperatures: Maximum T_{max}; upper basal temperature T_{bu}; average temperature T_m; minimum basal temperature T_{bl}; minimum temperature T_{min}

$$DD_j \cong \bar{T}_m(t_f - t_i) - B - C$$

where $\bar{T}_m$ is the average of T between t_i and t_f. In Meteorology, the average air temperature is calculated over a time interval greater than t_f–t_i, that is, over approximately 1 day, and the equation shown above is simplified in:

$$DD_j \cong T_m - B - C$$

Note that, in this case, the question of time is treated with very little rigor, $t_f - t_i$ being considered equal to 1, regardless of the time of year and latitude, that is, of the day length N.

As each crop adapts to a given climate, which is done by agricultural zoning, there are few days in which T_{max} is greater than T_{bu}, and area B is considered null. As for temperatures lower than T_{bl} there is no plant development, Pereira et al. (2002) suggest calculating DD_j by:

$$DD_j = T_m - T_{bl} \tag{5.25}$$

for cases where T_{min} is greater than T_{bl}. In this way, the area C is automatically eliminated. This expression considers that at each degree of temperature above T_{bl}, we account for 1 °C day. For cases where T_{bl} is equal to or greater than T_{min}, and less than T_{max}, Villa Nova et al. (1972) suggest the equation:

$$DD_j = \frac{(T_{max} - T_{min})^2}{2\,(T_{max} - T_{min})} \tag{5.26}$$

The sum of DD_j for the number of days in the crop cycle (emergence to maturation) is called the thermal constant or **Cumulative Degree Days** (CDD):

$$CDD = \sum_{j=1}^{k} DD_j \tag{5.27}$$

Each crop has its CDD value, that is, to develop and complete the cycle, it needs CDD degree days. In milder conditions of temperature, the number of days to reach the CDD is greater, that is, the cycle becomes longer. In warmer conditions, the cycle is shortened. Some examples of CDD are given below, taken from Pereira et al. (2002) (Table 5.6).

Table 5.6 Cumulative Degree Days for various crops

Crop (variety)	T_{bl} (°C)	CDD (°C day)
Rice (IAC-4440)	11.8	1985
Sunflower (Contisol 621)	4.0	1715
Soybean (UFV-1)	14.0	1340
Pea (super-early)	6.0	1225–1525
Pea (early)	6.0	1526–1725
Pea (semiearly)	6.0	1726–2000
Pea (late)	6.0	2000–2275

5.4 Wind

Wind is the air displacement caused by differences in atmospheric pressure and temperature between different positions on the globe. On a larger scale, we have the **Alisios winds**, between the tropics and the equator; the west and east winds, between the tropics and the poles; cyclones and anticyclones; and the "**El Niño**" and "**La Niña**" caused by temperature differences in the Pacific Ocean. In the case of agriculture, local winds are of greater importance, those affecting a region, or even a single crop. One important aspect of the wind is the transport of masses of air of different temperature, pressure, and humidity, from region to region. Therefore, the winds affect air temperatures, rainfall regimes (cloudiness), and evaporation and evapotranspiration processes. Dry air passing through a crop speeds up the evapotranspiration process. In Chaps. 13 and 15, these aspects will be discussed.

The wind is a vectorial quantity, with a modulus (strength), sense, and direction. The last two are characterized by the **wind rose**, with eight fundamental directions: N, NE, NW, S, SE, SW, E, and W. The module or strength is measured by **anemometers**, which record their instantaneous or mean velocity of a period (usually 1 day). In fact, as there is turbulence, the horizontal component is mostly measured in m s^{-1} or km h^{-1}, being 1 m s^{-1} = 3.6 km h^{-1} (Pereira et al. 2002).

As wind affects evapotranspiration, its module is part of several ET models, such as the most used one, suggested by Penman–Monteith (Allen et al. 1998). In these cases, the average daily value of the wind speed, measured at 2 m from the surface of the ground, is used. In Chap. 13, this method will be presented in more detail and the wind factor better appreciated in relation to its effect on evapotranspiration.

In some regions, the wind can also affect the management of agriculture by its mechanical effect on the canopy. In sugarcane, corn, and soybean, for example, the stalks laid down by the wind are a problem at the time of mechanical harvest. In cases of more intense winds, "windbreaks" are used by farmers, constituted of lines of shrubs or even trees, displaced inside the crops or along borders.

5.5 Exercises

5.1. Transform a pressure of 100 cmHg into cmH_2O, atm, Pa, and bar.

5.2. Transform 4.6 kPa in cmH_2O and atm.

5.3. The tire of a car was calibrated at 24 lb $in.^{-2}$. What pressure is this, absolute or relative? What is its value in atmosphere?

5.4. A vacuum gauge indicates "zero" when in contact with atmospheric pressure. In a certain vacuum condition, it reads "−0.5 atm." What pressure is that?

5.5. In a place where the atmospheric pressure is 720 mmHg, we rise to a height of 2000 m. What is the pressure drop? Use the expression: $P_r = P_0 \exp(-h/8.4)$, where h is given in km.

5.6. Under certain conditions, the air temperature is 25 °C and the partial vapor pressure is 12.3 mmHg. What is its relative humidity?
The value of e_s for 25 °C is 23.76 mmHg.

5.7. In what condition is the air of question 5.6? What are its saturation deficit and its dew point?

5.8. An air mass has relative humidity of 85% and is at 35 °C. What is the current partial vapor pressure?

5.9. The air in the previous question is heated to 40 °C without losing or gaining vapor. The value of e_s for 40 °C is 55.2 mmHg. What is its new RH?

5.10. The air in question 5.8 is cooled to 20 °C. What is the new RH?
5.11. A non-ventilated psychrometer indicates $t_u = 23$ °C and $t = 32$ °C. What is the relative humidity of the air?
5.12. Think a little bit about the effect of the inclination of the solar rays, in different seasons of the year, considering the declination of the Sun. What do you think of the sunshine difference of a north-facing slope compared to a south-facing slope? Do not overlook the latitude.
5.13. In a given place of latitude 20°S, in the month of July, the sun shines on average 9.3 h per day. For this location, Eq. (5.13) has as parameters $a = 0.21$ and $b = 0.63$. What is the global radiation at this location in July?
5.14. For the same location of the previous question, the net radiation for the month of July is 313 cal cm^{-2} day^{-1}. How many millimeters of water can be evaporated with this energy, if fully utilized? Assume the water temperature equal to 20 °C.

5.6 Answers

5.1. 1.316 atm; 1360 cmH_2O; 0.133 MPa; and 1.4 bar.
5.2. 0.045 atm and 47.52 cmH_2O.
5.3. This is a positive gauge pressure, that is, above atmospheric pressure; 1.63 atm. If the local atmospheric pressure is 0.9 atm, its absolute value is 2.53 atm.
5.4. Pressure of −0.5 atm, below local atmospheric pressure, also called tension or negative pressure.
5.5. Where $P_0 = 720$ mmHg and $h = 2$ km, we have that $P_r = 567$ mmHg.
5.6. 51.76%.
5.7. It is an unsaturated air. Saturation deficit $d = 11.46$ mmHg and $t_0 = 14.4$ °C.
5.8. 35.84 mmHg.
5.9. 64.8%.
5.10. Since the saturation tension at 20 °C is lower than 27.05 mmHg, the air has reached saturation (RH = 100%) at a temperature greater than 20 °C. To reach 20 °C, it remains saturated and all excess water condenses. Therefore, the air needs to lose water to reach 20 °C.
5.11. 45%.
5.12. In the Southern Hemisphere, the sun spends more time with its parabola facing north, and so the north-facing slopes receive the sun with smaller "inclinations" and they absorb more solar energy.
5.13. 435 cal cm^{-2} day^{-1}.
5.14. 5.3 mm day^{-1}.

References

Allen RG, Pereira LS, Raes D, Smith M (1998) Crop evapotranspiration – guidelines for computing crop water requirements. FAO, Roma

Ometto JC (1968) Estudo das relações entre radiação solar global, radiação líquida e insolação. PhD Thesis, Escola Superior de Agricultura Luiz de Queiroz, Universidade de São Paulo, Piracicaba, São Paulo state, Brazil

Pereira AR, Angelocci LR, Sentelhas PC (2002) Agrometeorologia: fundamentos e aplicações práticas. Agropecuária, Guaíba

Villa Nova NA, Pedro Júnior MJ, Pereira AR, Ometto JC (1972) Estimativa de graus-dia acumulados acima de qualquer temperatura base, em função das temperaturas máxima e mínima. Caderno Ci Terra 30:1–8

The Equilibrium State of Water in the Systems

6

6.1 Introduction

The water in the soil, the plant, the atmosphere, etc., whether it be in the liquid, solid or gaseous states, like anybody in nature, can be characterized by a state of energy. Different forms of energy can determine this state. **Thermodynamics** is the part of physics that studies the energetic relationships that involve the various systems, and that defines as a system the object of study and, as a medium, everything that surrounds it and maintains energetic relations (Fig. 6.1). In any thermodynamic study, it is fundamental to define what will be considered as a **system** and, consequently, as a **medium**, whose limits may or may not be well defined.

In the case of the water in the soil, the most common approach is to consider the liquid water, including everything present in it: ions, molecules, dissolved gases. In the different parts of the Soil-Plant-Atmosphere System, where water is present and whose state of energy we want to define, it is not easy to define the limits of the system of interest. The first difficulty is to choose its limits (size), which is arbitrary, but requires criteria. As we shall see in this chapter, our concepts are, for most part, punctual, of dimensions considered infinitely small. We will speak, for example, of the soil water content at a point 30 cm deep. Of course, we need a minimum volume of soil to define its water content. If this volume is too small, we may be considering only a pore or a grain of sand, which in no way represent the soil. In an arbitrary way, we could say that the minimum volume of soil that represents a point in the profile is 1 cm^3. Our samplings are, however, generally larger, collected by cylinders of heights of 3–10 cm, with diameters of 3–10 cm, resulting in volumes of 20–1000 cm^3. Volumes very large to be considered a point measurement, but that's the way we proceed. We have only to recognize this fact when analyzing our problem. In the mentioned case, our system has even smaller volume, because it is the liquid water contained in the pores of this volume of soil. Its limits are the tangle of water-air and water-solid interfaces, distributed within the pores. At times, we include in the system the soil air water vapor, which, in general, is in equilibrium with the liquid phase. The medium is considered as the solid part or matrix of the soil and the other gases of the soil air, with which the water (system) maintains energetic relations.

For example, the case of point A of Fig. 6.2 it is quite easy to define our system. This point is a portion of the water that lies within the body of water in a dam. The system could be defined by 1 cm^3 of water around point A. Its limit is

K. Reichardt, L. C. Timm, *Soil, Plant and Atmosphere*, https://doi.org/10.1007/978-3-030-19322-5_6

hypothetical and the medium would be the remainder of the dam water, which acts on the system, for example, by the hydrostatic pressure of a water column of height h_A.

In the atmosphere, at a given point, the system could be the water vapor contained in 1 L of air around the point. In the plant, we have different structures and we speak of the leaf, the xylem, the phloem, the seed and, for each case, the system must be defined. Within a leaf, for example, the system could be the water of the vacuole of a cell, or a portion of plant tissue, such as the lacunar parenchyma, shown in Fig. 4.9 of Chap. 4.

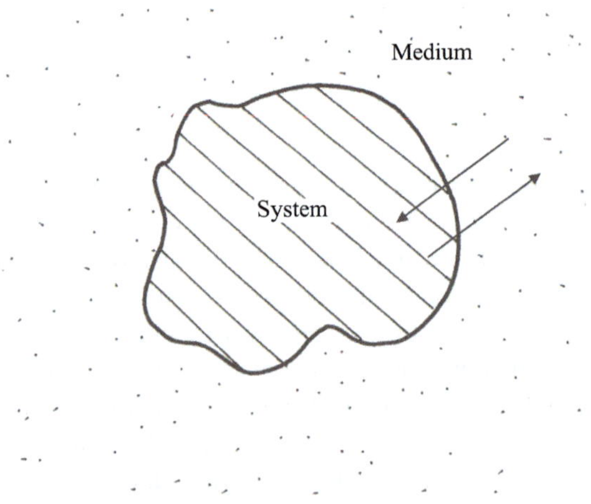

Fig. 6.1 Schematic view of the thermodynamic definition of system and medium

Energy relations between system and medium can be separated into mechanical, which appear by the action of forces, and thermal, by virtue of differences in temperature. The forces give rise to the mechanical works and give movement to the water. As the movement of water in the different parts of the Soil-Plant-Atmosphere System is very slow, its **kinetic energy**—proportional to the square of the velocity *v*—is, in most cases, negligible. This is not the case with running water in channels or pipes, where kinetic energy can be of great importance. Bernoulli's theorem of hydrodynamics (see Chap. 7) involves these aspects. On the other hand, the **potential energy**, function of the position and internal condition of the water at the point under consideration, is of paramount importance in the characterization of its state of energy. For the different water systems mentioned above, we can separately better define the types of energy that act in the system-medium relations. In the case of water vapor, heat (thermal energy) and pressure work are of importance. In the case of liquid water, gravitational work can assume great importance. All interactions between water and the solid matrix of the soil, which involve capillary, adsorption, electric, and

Fig. 6.2 Schematic view of a dam indicating selected systems occupying points A, B, C and D

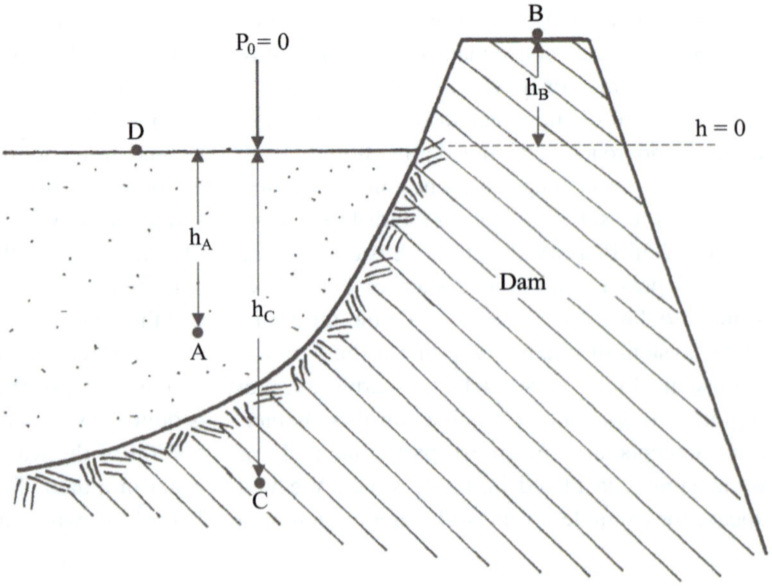

other forces, are, for convenience and simplicity, called matrix forces, which lead to a matrix work. The presence of solutes in the water implies chemical work. As a consequence of these system-medium relations, water assumes a state of energy that can be described by the thermodynamic functions, like Gibbs free energy and other conservative works, which in the Soil-Plant-Atmosphere System receive the particular name of total potential of the water. Differences in water potential between distinct points in the system give rise to its movement. The spontaneous and universal tendency of all matter in nature is to assume a state of minimum energy, seeking equilibrium with the environment. Water obeys this universal tendency and is constantly moving in the direction of diminishing its total potential. The rate of potential decrease along a direction is a measure of the force responsible for movement. Thus, the knowledge of its state of energy at each point in the system can allow us to calculate the forces acting on the water and determine how far it is from the equilibrium state.

The concept of total water potential is therefore of fundamental importance. For the soil case, it eliminates the arbitrary categories in which soil water was classically subdivided: **gravitational water**, **capillary water**, and **hygroscopic water**. In fact, all soil water is affected by the terrestrial gravitational field, so that all of it can be called as "gravitational." In addition, the capillary laws do not fully explain the phenomenon of soil water retention. Finally, it should be emphasized that the water is the same at any position and time within the soil. It does not differ in "form" but rather in the state of energy.

6.2 Thermodynamic Basis of the Soil Water Potential Concept

Thermodynamics is a vast chapter of physics, and it is not the intention of this SPAS text to present a complete treatise on the subject. However, thermodynamics is in each page of this text and is fundamental for the full understanding of its context. The reader, if necessary, should use basic texts of thermodynamics, for a better understanding of the concepts presented here (Rock 1969; Zemansky and Dittman 1997). The authors hope, however, that the following material will be sufficient for the soil science researcher to understand the complex energy relationships to be described in the Soil-Plant-Atmosphere System.

We will begin our study by applying thermodynamic concepts to the simplest system in the SPA, which is that of water vapor in the atmosphere. It is simpler because it is an inert gaseous system. It is suspended in the air because the gravitational force is balanced by the buoyance force. The water vapor is dissolved in the air and becomes a component of it. We have already seen, in Chap. 5, that under special conditions it is at maximum concentration (saturated air) and all excess condenses into the liquid phase. In the presence of tiny condensation nuclei, droplets are formed which, as they grow, undergo the action of gravity which prevails over the other forces, and begin to precipitate in the form of rain.

The characterization of the different states of a system is made using **thermodynamic coordinates**, which are the physical characteristics of the system. Gaseous systems are perfectly characterized by three coordinates: pressure P (Pascal, Pa), volume V (m^3), and temperature T (Kelvin, K). Thus, a state A of a gaseous system is characterized by particular values P_A, V_A, and T_A, these being related by a state equation. The most accepted state equation is that of the **ideal gases** (Eq. (5.1), Chap. 5). The amount of matter in the system is defined by volume V, which contains m kg or n mol of the gas, in the case, vapor.

In the vapor state, the energies involved in the system are only three: **heat** (Q), **mechanical work** of expansion or compression (W), and the **internal energy** (U) of the system. The **first principle of thermodynamics** is the balance of these energies and states that one type can be transformed into the others, while the total energy remains constant:

$$Q - W - \Delta U = 0 \tag{6.1}$$

In Eq. (6.1), the internal energy is presented as a variation ΔU, because its absolute value U is very difficult to be determined. For gases, U is only a function of T, the warmer a gas, the larger U. The signals of each term are conventional, usually we take Q positive as it enters the system (comes from the medium) and negative when it leaves the system (going to the medium). In the case of W, an expansion is considered positive work and it is negative for a compression. The ΔU signal will depend on the sum of Q and W. Heat is only considered as thermal energy during its transfer between system and medium. Once heat has entered the system, one does no longer talk about heat; it becomes internal energy. The heat supply may then result in increased internal energy and/or expansion work. Thus, according to Eq. (6.1), if a system absorbs heat from the medium ($Q = 600$ J) and at the same time is compressed by the medium ($W = -200$ J), its internal energy variation will be 800 J. For infinitesimal variations of energy, the equation can be written in the form:

$$\mathrm{d}U = \mathbf{d}Q - \mathbf{d}W \tag{6.1a}$$

The differential $\mathrm{d}U$ is an **exact differential**, because U is a **point function**, that is, a function that depends only on the initial (i) and final (f) states of a transformation. This means that $\Delta U = U_f - U_i$ has the same value for any "path," "line," or "process" used to take the system from i to f. The symbol d, in this case, represents an infinitesimal amount of heat inputs or outputs. When a system passes from any state i to another one f, the heat involved in the passage depends on the process or path that connected i to f. There are, therefore, infinite values of Q for the passage from i to f. Therefore, Q is called a **line function** (it depends on the line, the process, the path that connects i and f) and its differential $\mathrm{d}Q$ is denoted as **non-exact differential**, whose **d** is shown in bold to differentiate it from the exact differentials. The differential $\mathbf{d}Q$ is equal to $C_\mathrm{v}\mathrm{d}T$ for **isovolumetric processes** (C_v = heat capacity at constant volume, J K^{-1}); $C_\mathrm{p}\mathrm{d}T$ for **isobaric processes** (C_p = heat capacity at constant pressure, J K^{-1}); for the **isothermal process**, $\mathrm{d}U = 0$ because U is only a function of T and if T = constant, $\mathrm{d}T = 0$, and $\mathrm{d}U = 0$; as a consequence of Eq. (6.1a), $\mathbf{d}Q = \mathbf{d}W$, that is, in the isotherm, all heat is transformed into work, or vice versa; in **adiabatic processes**, $\mathbf{d}Q = 0$, and, in general, when the process is not defined, $\mathbf{d}Q$ is equal to $T\mathrm{d}S$ (S = entropy). The entropy S solves the question of the dependence of heat on the path, since it is a function connected to Q, but that is independent of the path that connects i and f, and, therefore, it is also a point function and its differential $\mathrm{d}S$ is exact. The definition of S is given by:

$$S_f - S_i = \int_i^f \frac{\mathbf{d}Q}{T} \tag{6.2}$$

which shows that if each $\mathbf{d}Q$ that is involved in the process that leads the system from i to f is divided by the respective temperature T, its summation (integral) $\Delta S = S_f - S_i$ is independent of the process. Other definitions of S are linked to order and disorder of the system, and natural and spontaneous processes always lead to greater disorder (a pack of playing cards stacked in a logical sequence, as it falls to the floor, changes into a state of greater disorder: its entropy is greater, and ΔS is positive, because to order the cards again, it is necessary to spend energy; the process of ordering the cards again is not spontaneous as the fall, and its entropy is smaller; ΔS is negative).

Entropy is closely linked to the **second principle of thermodynamics**, which can be seen as a restriction to the first principle, which shows that any energy is capable of being transformed into another, however conserving the total. The second principle, with regard to changes of heat into work, which is made by thermal machines, such as alcohol, gasoline, and diesel engines, states that "it is impossible, in cyclical transformations, to transform totally heat into work." Always a part of the heat remains unavailable for the production of work. The theory of the second principle is extensive and to be fully understood needs to be studied in specialized texts.

To close Eq. (6.1a), $\mathbf{d}W$ represents the mechanical work of expansion or contraction, given by $P\mathrm{d}V$ and which, also, is a non-exact differential, since W depends on the process that connects i to f. Once the difference between exact and non-exact differentials is known, we will not use the bolt notation for **d** in this text.

In addition to the internal energy function U, thermodynamics defines other energy functions, termed **thermodynamic potentials**, each of special interest under certain conditions (Rock 1969). They are:

(a) Enthalpy $H = U + PV$ (6.3)

(b) Helmholtz free energy $F = U - TS$ (6.4)

(c) Gibbs free energy $G = H - TS = U + PV - TS$ (6.5)

The **enthalpy**, also called the heat function, is widely used in chemistry, for example, in calculations of heats of combustion, reaction, enthalpy of formation, etc. The **Helmholtz free energy**, also called work function, is used more in applications involving mechanical work. **Gibbs free energy** is also of great importance in chemistry, as in reaction constants. It is also very convenient in describing the energy status of the water. Let us see in more detail its meaning. G is a thermodynamic property of the system, in the same way as entropy and internal energy, and involves energy. It is a point function, that is, its value depends only on the state of the system (of its coordinates, in case P, V, and T), in the same way as the internal energy. This means that if a system in a state A, which has free energy G_A, passes to another state B with free energy G_B, the difference $G_A - G_B$ is identical for all processes that take the system from state A to state B. G_A and G_B characterize states A and B and therefore are properties of the system. This property allows us to calculate $\Delta G = G_B - G_A$ by any process that links states A and B. We can therefore select the most convenient process to calculate it, or even the simplest. As its name implies, Gibbs' free energy is the part of the energy that is available to produce work. Differentiating Eq. (6.5), we have:

$$dG = dU + PdV + VdP - TdS - SdT$$

and since $dU = TdS - PdV$, we have:

$$dG = VdP - SdT \tag{6.6}$$

which is the differential equation of the Gibbs free energy. From Eq. (6.6) we see that G was taken as a function of T and P, so that we can write $G = G(T, P)$. Equation (6.6) can also be written in a more generic form, using **partial derivatives** (see Chap. 3):

$$dG = \left(\frac{\partial G}{\partial T}\right)_P dT + \left(\frac{\partial G}{\partial P}\right)_T dP \tag{6.6a}$$

where:

$$\left(\frac{\partial G}{\partial T}\right)_P = -S \quad \text{and} \quad \left(\frac{\partial G}{\partial P}\right)_T = V$$

and the indices P and T disposed outside the brackets indicate that the respective variable was held constant during the derivation. These equations show that the partial derivatives have a well-defined physical meaning. The entropy S can therefore also be seen as the partial derivative of Gibbs free energy with respect to temperature. The reader should note in Eq. (6.6a) the difference of the symbols used for **total differential** "d" and partial differential "∂" and also for **total derivatives** such as "dy/dx" and **partial derivatives** "$\partial y/\partial x$." When a function depends only on a variable, such as $y = y(x)$, the derivative is total dy/dx. For more than one variable, the partial derivatives appear, as in the case of $y = y(x, t, v)$, we have:

$$dy = \left(\frac{\partial y}{\partial x}\right)_{t,v} dx + \left(\frac{\partial y}{\partial t}\right)_{x,v} dt + \left(\frac{\partial y}{\partial v}\right)_{x,t} dv$$

where dy is the total differential and:

$$\left(\frac{\partial y}{\partial x}\right)_{t,v}$$

is the partial derivative of y with respect to x, with t and v held constant. Let us adapt Eq. (6.6) to our system, which consists of water vapor in an "arbitrary" volume of air. Of all the variables we have seen so far, a distinction can be made into two groups: **intensive variables** and **extensive variables**. The first, with examples P and T, are independent of the "size" or "extension" of the system. If the temperature of a room is 30 °C and the room is divided in half, we will not have 15 °C for each side. If the temperature of the room is stable, whatever the size of the air sample we take, its temperature will be 30 °C. This is no longer the case, for example, with the room volume V, which, when divided, becomes $V/2$. The volume is an extensive variable, depending on the size of the system. Also, U, G, m (mass), and S are extensive; the larger the system, the larger their values. As it is advantageous to work with intensive quantities, it is common to transform extensive variables into intensive. The ratio of two extensive variables results in an intensive one. Thus, m/V is the specific mass or density d. The air density of the room in the previous example does not change with its splitting. The inverse of the density V/m is the specific volume v, also intensive. If Eq. (6.6) is divided by m, we obtain the same equation for intensive variables:

$$\mathrm{d}g = v\mathrm{d}P - s\mathrm{d}T \tag{6.6b}$$

where $g = G/m$ is the specific **Gibbs free energy** and s is the specific entropy.

Following the most commonly used nomenclature in agrometeorology (Chap. 5), for water vapor, the pressure P is the partial vapor pressure $\boldsymbol{e}$ (Eqs. (5.4) and (5.5)), and, by calling the specific free energy g of water potential Ψ, Eq. (6.6b) becomes:

$$d\Psi = v de_{\mathrm{a}} - s\mathrm{d}T \tag{6.6c}$$

As we have already said, G and therefore g and Ψ are point functions, and thus to determine their value between two states, we can choose the most convenient path. Choosing the isothermal ($\mathrm{d}T = 0$), Eq. (6.6c) is reduced to $\mathrm{d}\Psi = v\mathrm{d}e_{\mathrm{a}}$. Assuming, further, that the water vapor of the atmosphere behaves as an ideal gas (see Chap. 5), its state equation is:

$$e_{\mathrm{a}}v = \frac{RT}{M_{\mathrm{v}}}$$

separating the value v from this equation and substituting it in the previous equation, we have:

$$d\Psi = \frac{RT}{M_{\mathrm{v}}}\frac{\mathrm{d}e_{\mathrm{a}}}{e_{\mathrm{a}}}$$

that, after integrating from the standard state e_{s} to the state of the air e_{a}, we have:

$$\Delta\Psi = \int_{e_{\mathrm{s}}}^{e_{\mathrm{a}}} \frac{RT}{M_{\mathrm{v}}}\frac{\mathrm{d}e_{\mathrm{a}}}{e_{\mathrm{a}}} = \frac{RT}{M_{\mathrm{v}}}\ln\frac{e_{\mathrm{a}}}{e_{\mathrm{s}}} \tag{6.7}$$

taking into account that the integral of $\mathrm{d}y/y$ is $\ln y$, because $e_{\mathrm{a}}/e_{\mathrm{s}}$ varies from 0 to 1, Ψ will always be negative. The standard state e_{s} will be discussed later in this chapter. In summary, the absolute value of Ψ is difficult to be obtained and, therefore, we measure $\Delta\Psi$ between a standard state Ψ_0 and the considered state Ψ, so that $\Delta\Psi = \Psi - \Psi_0$. Since the value of Ψ_0 is arbitrary, we choose zero, $\Psi_0 = 0$, and so that $\Delta\Psi = \Psi$.

Then, taking a measure with a **psychrometer** (see example in Chap. 6) of the vapor pressure e_{a} and the temperature T of the air, one can determine Ψ. Let's take an example: calculate the potential of the water vapor in a mass of air whose relative humidity is 50% or 0.5, at a temperature of 27 °C (300 K):

$$\begin{aligned}\Psi &= RT\,\ln\frac{e_{\mathrm{a}}}{e_{\mathrm{s}}} = 0.082\frac{\mathrm{atm\,L}}{\mathrm{mol\,K}} \\ &\times 55.5\frac{\mathrm{mol}}{\mathrm{K}} \times 300\ \mathrm{K} \times \ln(0.5) \\ &= -947\ \mathrm{atm} = -94.7\ \mathrm{MPa}\end{aligned}$$

The constant 55.5 appears because 1 mol of water equals 18 g and in 1 L = 1000 g, we have 55.5 mol of water. As it can be seen, the potential of the water vapor in the atmosphere assumes very negative values (logarithms of numbers smaller than 1 are negative). Using Eq. (6.7), the potential of water in the atmosphere can be calculated for different **relative humidities**, all at 27 °C, as shown in Table 6.1.

It can, therefore, be seen that the potential of water in the atmosphere is, in general, much more negative than it is in the plant and in the soil, as we shall see below. This fact causes the water to move spontaneously from the soil to the plant and from the plant to the atmosphere.

Table 6.1 Values of relative humidity (RH) of air samples with respective values of vapor potential in the atmosphere

Relative humidity	Ψ atmosphere	
%	atm	MPa
20	−2200	−220
50	−947	−95
80	−305	−30.5
90	−144	−14.4
95	−70	−7.0
99	−13.7	−1.4
99.9	−1.4	−0.14
100	0	0

6.3 Total Potential of Water in the Soil

Let us now turn to the more complicated case where our system is the soil water contained in a volume element, say 1 cm^3 of soil, and the amount of water contained in it is θ cm^3 (Eq. (3.15)). This is our system now. Soil particles, which interact with water (matrix potential), are part of the medium. Water vapor that is in equilibrium with liquid water is also part of the system. As we have already said, in addition to $T\mathrm{d}S$ (heat) and $P\mathrm{d}V$ (mechanical work), other works appear in this case, some of a mechanical nature and others of a chemical nature, which means that G is not only a function of T and P but also a function of other variables responsible for other forms of energy. As regards to chemical work, the number of moles n_i of each solute i present in the soil water will participate in the free energy of Gibbs G. Chemical work is performed when there are variations in concentration of each species of solute. The natural tendency of their displacement is from more to less concentrated regions. As we shall see in Chap. 8, the solute/solvent amounts can be expressed in terms of number of moles n, concentration C, activity $\boldsymbol{a}$, and mole fraction N. The index $i = 1$, 2, 3, ... n refers to the n components of the system, including the solvent H_2O. Among them, we can mention H^+ and OH^- (responsible for the pH), Na^+, K^+, Ca^{2+}, Mg^{2+}, NH_4^+, ..., NO_3^-, Cl^-, SO_4^{2-}, $H_2PO_4^-$, ..., organic compounds, humic acids, etc. Each of them contributes to the chemical potential of water, measured against a standard, in this case pure water.

With respect to other mechanical works, we distinguish between gravitational and matrix work, remembering that the former acts on water with constant intensity and whose variable is the relative height z of the system in the gravitational field. This field is conservative, that is, the work to take the system from level A to level B does not depend on the path, it depends only on $z_B - z_A$. The gravitational work ($mg\mathrm{d}z$, where mg is weight, i.e., force, which multiplied by distance results in work or energy), so it is a point function like G and thus can be added to it. Matrix work involves the (also conservative) forces that interact between the system (water) and the medium (solid particles of soil). It is a complex sum of forces, such as capillary, of adsorption, of cohesion, electric, etc., whose separate description is very complicated and, for the purposes of this thermodynamic approach, not necessary. All being conservative, they are enclosed in one, under the denomination of matrix force. It has been verified that its intensity is related to the soil water content θ and, therefore, the matrix work can also be added to G using the variable θ ($m\omega\mathrm{d}\theta$, where $m\omega$ is the matrix potential that is an energy that does not change its characteristics when multiplying by the nondimensional differential $\mathrm{d}\theta$). Thus, for the soil water system, Eq. (6.6a) needs to be complemented by including these mentioned energies or works:

$$\mathrm{d}G = \left(\frac{\partial G}{\partial T}\right)_{P,z,n_i,\theta} \mathrm{d}T + \left(\frac{\partial G}{\partial P}\right)_{T,z,n_i,\theta} \mathrm{d}P + \sum_{i=1}^{n} \left(\frac{\partial G}{\partial n_i}\right)_{P,T,n_j,z,\theta} \mathrm{d}n_i + mg\mathrm{d}z + m\omega\mathrm{d}\theta \tag{6.8}$$

that in terms of energy per unit of mass is:

$$\mathrm{d}g = -s\mathrm{d}T + v\mathrm{d}P + \mu_w \mathrm{d}n_w + g\mathrm{d}z + \omega\mathrm{d}\theta \tag{6.8a}$$

where g is the Gibbs specific free energy per mass of the water in the soil. For $i = 1$ we have the water and $i = 2, 3, 4 ...., n$ we have the other ionic species. Thus, the term n_j in Eq. (6.8) refers to the other ionic species present in the soil solution, and the sum of the chemical work indicated in this equation was simplified for one chemical work only, $\mu_w dn_w$, which represents the performance of $n - 1$ solutes on the water component (solvent). Thus, μ_w represents the **chemical potential of water** and n_w the number of moles in the specific volume. The variation of moles of water by the addition of solutes is very small, but its implication in terms of energy may be considerable. In this chapter, the situation is reversed because we will be emphasizing the solutes and the potential of each of them becomes more important than that of water.

The constitution of Eqs. (6.8) and (6.8a) can be argued as being a "mixture" of **Gibbs free energy** and other works, in particular, gdz which is the work of the gravitational field, external to the system. Equations are presented in this way because of the complexity of the soil water system, in terms of energy. For example, dg of Eq. (6.8a) cannot be estimated by different paths because it is a point function. The terms sdT, vdP, and $\mu_w dn_w$ are dependent on each other by the concept of free energy but not dependent on gdz and not directly on $\omega d\theta$. This means that if we choose the path with T, P, n_w, and θ all constants, dg cannot be calculated by gravitational potential alone. On the other hand, if we take z, T, n_w, and θ constants and consider as a system the water vapor of the soil that is in thermodynamic equilibrium with the liquid water of the soil, dg can be calculated only by vdP, as shown by Eq. (6.7). The soil and plant psychrometers, which will be seen below, operate on this basis.

The variable ω was not well defined. By analogy to μ_w, it could be called **soil matrix potential**, derived from the combination of all the works that appear between the water and the solid matrix of the soil, among them capillary work σdA (σ = surface tension and A = area) and the electric work εde (ε = electric potential and e = electric charge). The variable ω relates to Ψ, and although involving another work, it can be seen as $(\partial G/\partial \theta)_{T,P,z,n_w}$, and being a function $\omega = \omega(\theta)$ it has to do with the **soil water retention curve**, discussed in detail later in this chapter.

The differential Eq. (6.8a) comes from the function g, which we will replace by Ψ, the **total potential of water**:

$$g = \Psi = \Psi(T, P, z, n_w, \theta) \tag{6.9}$$

The total potential of water Ψ_i in a state i, can hardly be determined in absolute form. Its difference between a state taken as **standard state of water** and the considered state i of the system is then determined, as it was done for the particular case of Eq. (6.7). The standard state is the one in which the water system is in normal conditions of T and P, free of mineral salts and other solutes, with a flat liquid-gas interface, situated in a reference position. To this state, we assign the arbitrary value $\Psi_0 = 0$ (see Fig. 6.3). Thus, the potential Ψ_i of water in the state i is given by:

$$\Delta\Psi = \int_{\Psi_0}^{\Psi_i} d\Psi = (\Psi_i - \Psi_0) = \Psi_i \tag{6.10}$$

or, in detail for each variable of Eq. (6.8a), we have:

$$\Psi_i = \underbrace{\int_{T_0}^{T_i} -sdT}_{\Psi_T} + \underbrace{\int_{P_0}^{P_i} vdP}_{\Psi_P} + \underbrace{\int_{0}^{z} gdz}_{\Psi_g} + \underbrace{\int_{n_0}^{n_i} \mu_w dn_w}_{\Psi_{os}} + \underbrace{\int_{\theta_0}^{\theta_i} \omega d\theta}_{\Psi_m} + \tag{6.11}$$

The lower index of the last integral is the water content at saturation θ_0, since the water in a saturated soil system is free of tensions or pressures, such as the free water in the standard state.

Equation (6.11) shows that the total potential of water is the sum of five components: thermal, pressure, gravitational, osmotic, and matrix. The **thermal potential component** Ψ_T is difficult to be measured but generally considered negligible. The relatively small variations of T occurring in the soil imply, most of the time, in negligible variations of this potential, so that the processes

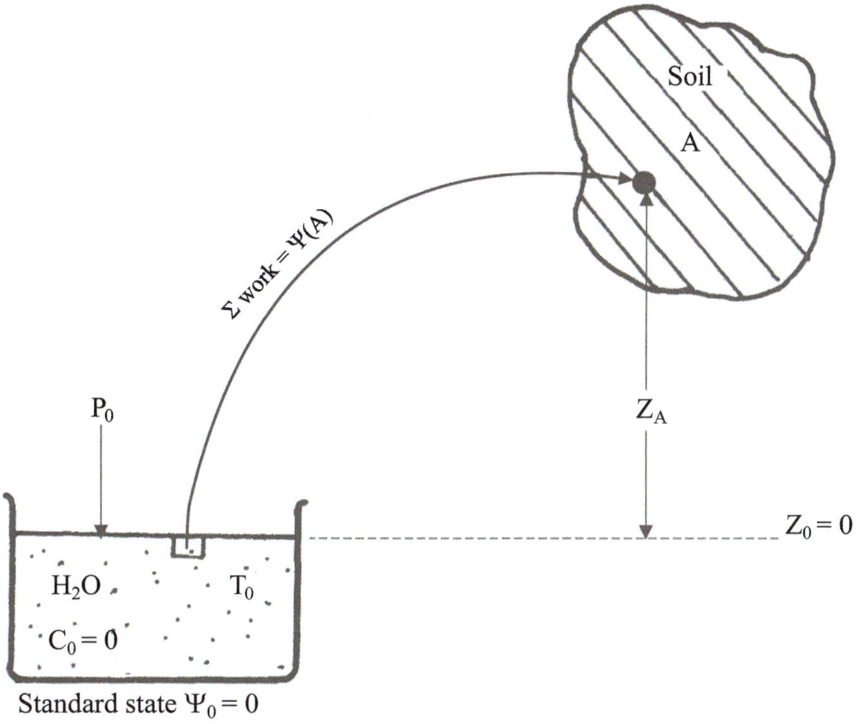

Fig. 6.3 Schematic view of the work made to transport water in the standard state $\Psi_0 = 0$ to a generic point A in the soil Ψ_A

can be considered isothermal. Taylor and Ashcroft (1972) discuss aspects of the thermal potential. In comparison to the other four components that may assume considerable importance from situation to situation, Ψ_T can safely be neglected. Therefore, the other components will be seen in more detail below.

The Gibbs free energy units and of the other works are energy units. Since the energy of a system is an extensive quantity, it is opportune to express it per unit of another quantity proportional to the extension of the system, as already seen. Three forms are mostly used: (a) **energy per unit mass**, (b) **energy per unit volume**, and (c) **energy per unit weight**. These three quantities are "energies," but they have the property of being intensive, that is, they do not depend on the extension or size of the system. They are potentials.

(a) Energy per unit mass: its dimension (see Chap. 19) is $L^2\ T^{-2}$ and the most common units are J kg^{-1}, erg g^{-1}, cal g^{-1}, and atm L mol^{-1}.
(b) Energy per unit volume: it has pressure dimensions, because just as energy can be expressed as a product of pressure per volume, the ratio of energy per volume expresses a pressure. Its size is $M\ L^{-1}\ T^{-2}$ and the most common units are Pa = N m^{-2}, d (dyne) cm^{-2}, and atm.
(c) Energy per unit weight (hydraulic load): it has dimensions of length L, because just as energy can be expressed as a pressure, it can be expressed in terms of column (height) of liquid. For example, a pressure of 1 atm corresponds a column of 1033 cm of water or one of 76 cm of mercury. The most common units are cm of water, m of water, cm of Hg, and mm of Hg.

Example: 10 g of water 10 cm above the reference level have a potential energy E of:

$$E = m \cdot g \cdot h = 0.01 \text{ kg} \times 9.8 \text{ m s}^{-2} \times 0.1 \text{ m} = 0.00098 \text{ J}$$

(a) Energy per unit of mass:

$$\frac{E}{m} = g \cdot h = 9.8 \text{ m s}^{-2} \times 0.1 \text{ m} = 0.98 \text{ J kg}^{-1}$$

(b) Energy per unit of volume:

$$\frac{E}{V} = \rho \cdot g \cdot h = 1000 \text{ kg m}^{-3} \times 9.8 \text{ m s}^{-2} \times 0.1 \text{ m} = 980 \text{ Pa}$$

(c) Energy per unit weight (hydraulic load):

$$\frac{E}{mg} = h = 0.1 \text{ m or } 10 \text{ cm}$$

To transform units, the following relationships, already presented in Chap. 5, Table 5.2, are very useful:

$$1 \text{ atm} = 76 \text{ cmHg} = 1033 \text{ cmH}_2\text{O} = 1{,}013{,}250\ b = 101{,}325 \text{ Pa} = 101 \text{ kPa} = 14{,}696 \text{ lb in.}^{-2}$$

Thus, the total soil water potential (Ψ) can be rewritten as follows (neglecting the thermal component):

$$\Psi = \Psi_P + \Psi_g + \Psi_{os} + \Psi_m \qquad (6.12)$$

where:

Ψ = total soil water potential

Ψ_P = **pressure potential component**, which appears whenever the pressure acting on the water in the soil is different and **greater** than the pressure P_0 acting on the water in the standard state. For example, the water at the bottom of a dam is subject to a pressure equivalent to the water column above it. In saturated soils, there is also a load of water acting on the considered point. These positive pressures are the pressure component. When an expansive soil (nonrigid body) is in saturation or non-saturation conditions, it also weighs on the water itself, and it is referred to as **overburden potential**;

Ψ_g = **gravitational potential component**, which is always present due to the presence of the terrestrial gravitational field;

Ψ_{os} = **osmotic potential component**, which appears because the water in the soil is a solution of mineral salts and other solutes and the water in the standard state is pure;

Ψ_m = **matrix potential component**, which is the sum of all other works involving the interaction between the solid soil matrix and water, such as capillary work, work against adsorption and electric forces, etc. These phenomena lead to water at pressures lower than P_0, which act on water in relation to the standard state. They are, therefore, negative pressures, also denominated as **tensions** or **suctions.**

6.4 Pressure Component

$$\Psi = \boxed{\Psi_P} + \Psi_g + \Psi_{os} + \Psi_m$$

The **pressure component** Ψ_P considers only positive pressures, those that act on the system when $P_i > P_0$, the standard pressure. Negative pressures $P_i < P_0$, like tensions or suctions, will be taken care of by Ψ_m. Its calculation comes from Eq. (6.11):

$$\Psi_P = \int_{P_0}^{P_i} v\text{d}P$$

For example, the dam indicated in Fig. 6.3, where it is desired to determine the water pressure potential component at point A. Under such conditions, as water is incompressible, the specific volume of water around A is constant and:

$$\Psi_P(\text{A}) = \int_{P_0}^{P_\text{A}} v\text{d}P = v \int_{P_0}^{P_\text{A}} \text{d}P = v[P]_{P_0}^{P_\text{A}} = v(P_\text{A} - P_0)$$

and since P_0 is taken as zero ($P_0 = 0$), the reference (standard) state:

$$\Psi_P(\text{A}) = vP_\text{A} \tag{6.13}$$

For the case of measuring Ψ_P in units of energy per volume, we have to divide Ψ_P by v, resulting only P_A. From hydrostatics, we know that $P_\text{A} = \rho g h_\text{A}$, so that $\Psi_P = \rho g h_\text{A}$.

As already seen, Ψ_P(A) can be measured in three units: energy/volume ($\rho g h_\text{A}$), energy/mass ($g h_\text{A}$), and energy/weight (or hydraulic load) (h_A). So, if $h_\text{A} = 5$ m, we have:

$$\Psi_P(\text{A}) = \rho \cdot g \cdot h_\text{A} = 1000 \times 9.8 \times 5$$
$$= 49,000 \text{ Pa} = 49 \text{ kPa}$$

or

$$\Psi_P(\text{A}) = g \cdot h_\text{A} = 9.8 \times 5 = 49 \text{ J kg}^{-1}$$

or

$$\Psi_P(\text{A}) = h_\text{A} = 5 \text{ m or } 500 \text{ cm}$$

Point B of Fig. 6.3 is on the soil surface and submitted to atmospheric pressure P_0, so that its pressure potential is zero:

$$\Psi_P(\text{B}) = 0$$

and also $\Psi_p(\text{D}) = 0$.

Point C is in the saturated soil at the bottom of the dam and, as the hydrostatic pressure of the water is transmitted through the pores of the soil (considered rigid porous material), the pressure on C, although through a tortuous path, is also given by $\rho g h_\text{C}$. So that:

$$\Psi_P(\text{C}) = \rho \cdot g \cdot h_\text{C} \quad \text{or} \quad g \cdot h_\text{C} \quad \text{or} \quad h_\text{C}$$

only considered for positive pressures, that is, above atmospheric pressure. For negative values (tensions), i.e., subatmospheric pressures, as already said, the matrix component Ψ_m which measures capillary tensions, etc., will be considered, as will be seen below. Therefore, the pressure component is only of importance for saturated soils, where hydraulic pressures are present. Thus, for example, in a flooded rice field, as schematized in Fig. 6.4, the pressure potential cannot be neglected.

The values of Ψ_P at the indicated points given in terms of hydraulic load are:

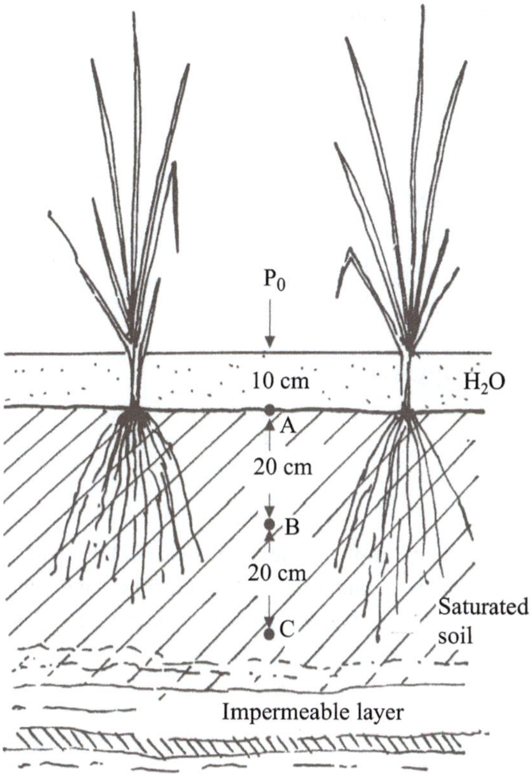

Fig. 6.4 A paddy rice field with a 10 cm load (depth) of water

$$\Psi_P(\text{A}) = 10 \text{ cmH}_2\text{O}$$
$$\Psi_P(\text{B}) = 30 \text{ cmH}_2\text{O}$$
$$\Psi_P(\text{C}) = 50 \text{ cmH}_2\text{O}$$

6.5 Gravitational Component

$$\Psi = \Psi_\text{P} + \boxed{\Psi_\text{g}} + \Psi_\text{os} + \Psi_\text{m}$$

The **gravitational component** Ψ_g is always present and is calculated by the third integral of Eq. (6.11). It can also be measured in energy per unit volume, energy/mass, or energy/weight. It is the potential energy of the gravitational field itself, equal to mgz, where z is measured from an arbitrary referential. In general, it is assumed that $z = 0$ at the surface of the soil and thus z is the vertical position coordinate z itself:

1. Energy/volume:

$$\Psi_g = \int_0^z \rho g \mathrm{d}z = \rho g \int_0^z \mathrm{d}z = \rho g z \tag{6.14}$$

2. Energy/mass:

$$\Psi_g = \int_0^z g \mathrm{d}z = g \int_0^z \mathrm{d}z = g z \tag{6.14a}$$

3. Energy/weight:

$$\Psi_g = \int_0^z \mathrm{d}z = z \tag{6.14b}$$

In the example of the dam (Fig. 6.3), considering $z = 0$ at the free water interface, we have for A, $z = -h_A$; for B, $z = -h_B$; and for C, $z = -h_C$:

$$\Psi_g(\mathrm{A}) = -\rho \cdot g \cdot h_A \quad \text{or} \quad -g \cdot h_A \quad \text{or} \quad -h_A$$

$$\Psi_g(\mathrm{B}) = +\rho \cdot g \cdot h_B \quad \text{or} \quad +g \cdot h_B \quad \text{or} \quad +h_B$$

$$\Psi_g(\mathrm{C}) = -\rho \cdot g \cdot h_C \quad \text{or} \quad -g \cdot h_C \quad \text{or} \quad -h_C$$

$$\Psi_g(\mathrm{D}) = 0$$

Note that using the hydraulic load unit Ψ_g is equal to the depth or height, i.e., the vertical z coordinate itself. Because of this, Ψ_g is indicated, in most scientific works, simply by z.

A simple experiment demonstrates very well the importance of the gravitational potential and is outlined in Fig. 6.5. Take a sponge, put it in a tray, and saturate it with water (state 1).

In this state $z = 0$ (arbitrary reference), the values of the gravitational potential in A and B are 0, and we have gravitational equilibrium, since $\Delta\Psi_g = 0$ between A and B. Lift the sponge carefully, without compressing it, until

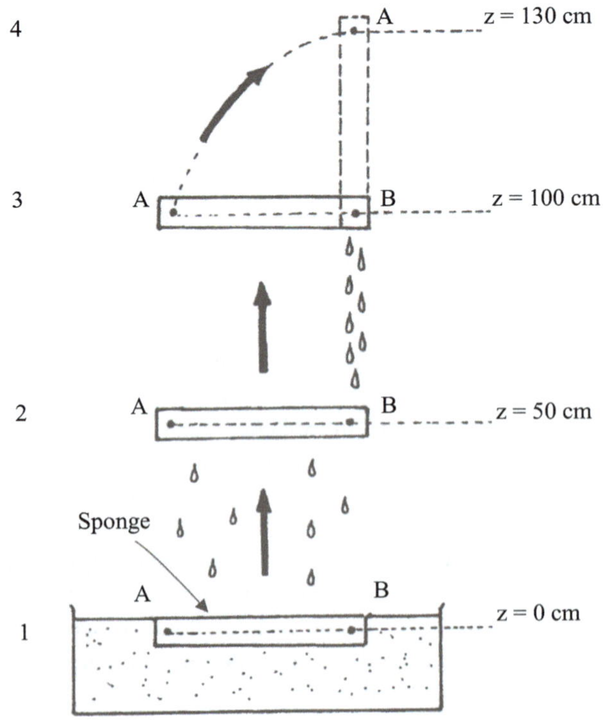

Fig. 6.5 Demonstration of the importance of the gravitational potential. Sponge at state 1 is in equilibrium and saturated. At state 2 still almost saturated and in another state of equilibrium. State 3 almost equal to state 2, but no more drops falling. State 4, after rotation at point B, in nonequilibrium with water dropping by action of gravity

$z = 50$ cm, and let it loose all free water until dripping practically stops (15–30 s). In this state 2, it still retains a lot of water by the action of the matrix potential, which we will see next. In this condition, as $\Psi_g(A) = \Psi_g(B) = 50$ cmH$_2$O, we still have gravitational equilibrium because $\Delta\Psi_g = 0$ between A and B. Lifting the sponge horizontally and carefully up to $z = 100$ cm (state 3), no drip of water leaves the sponge. The gravitational equilibrium continues with $\Psi_g(A) = \Psi_g(B) = 100$ cmH$_2$O and $\Delta\Psi_g = 0$ between A and B. It is seen that the transition from state 2 to state 3 does not imply loss of water. The gravitational potential increases from 50 to 100 cmH$_2$O, which means that water in state 3 has more gravitational energy than in state 2. Since the reference level is arbitrary, we could make $z = 0$ in the state 3. Well, under these conditions $\Psi_g = -50$ cmH$_2$O in state 2 and -100 cmH$_2$O in state 1. The important thing is that in each state $\Delta\Psi_g = 0$ between A and B, there is gravitational equilibrium. If, however, from state 3 we move to state 4 by rotation, keeping point B fixed, without compressing the sponge, we will notice that a large amount of water will emerge from the sponge, only by the action of the gravitational potential. In state 4, $\Psi_g(A) = 130$ cmH$_2$O and $\Psi_g(B) = 100$ cmH$_2$O and as a result $\Delta\Psi_g = 30$ cmH$_2$O between A and B. This difference in gravitational potential causes water to flow.

This illustration demonstrates well the performance of the gravitational potential. Because the gravitational field is always present, the gravitational potential always exists. As we have said, it is a component of the total potential. Its importance in relation to the total potential depends on the magnitude of all other components. In general terms, we can say that in saturated soils, the gravitational component is the one with the greatest quantitative importance and has a significant weight in the total potential. When a soil gradually loses water, the matrix component becomes more important than the gravitational. However, it is important to point out once again that the gravitational component is always present.

6.6 Osmotic Component

$$\Psi = \Psi_p + \Psi_g + \boxed{\Psi_{os}} + \Psi_m$$

Also because soil water is a solution of mineral salts and organic substances, it has an **osmotic potential component** Ψ_{os}, which contributes to its total potential Ψ. In Chap. 3, the subject has already been discussed, and the osmotic pressure there discussed is the very osmotic component of the total water potential. In Eq. (6.8a), the osmotic potential is defined by:

$$d\Psi_{os} = \mu_w dn_w$$

dn_w being the variation of moles of water and μ_w the chemical potential of water, both given the presence of solutes.

In the soil, this potential is difficult to be determined, but there are special instruments to measure it. The previous equation must be integrated (fourth integral of Eq. (6.11)) between the limits n_0 = number of moles of pure water in the element of standard volume and $n = n_i$ (number of moles of water in the soil solution). For this, it is necessary to know the function $\mu_w(n_w)$, which defines how μ_w varies as a function of n_w as a result of the addition of salts, which is very difficult. A simpler way would be to calculate Ψ_{os} by the concept of osmotic pressure P_{os}, the result of the joint action of all the solutes on water, and, being a pressure, it is the osmotic component itself measured in terms of energy per unit volume. In the osmometer of Fig. 3.9 (Chap. 3), the pressure acting on the water in the standard state is $P_0 = 0$, and the pressure acting on the solution is $P_{os} = -RTC$, according to the **equation of van't Hoff**. That is:

$$\Psi_{os} = \int_{n_0}^{n_i} \mu_w dn_w = \int_0^{P_{os}} v dP = v(P_{os} - 0)$$
$$= P_{os} = -RTC \qquad (6.15)$$

considering $v = 1$ cm^3 g^{-1} for soil solution.

The **osmometer** of Fig. 3.9 separates the pure water from the solution by means of a semipermeable membrane which allows water to flow but not the solutes, which leads to the solution to be under osmotic pressure P_{os}. A solution per se, that is, without a semipermeable membrane, is not subjected to a P_{os}, irrespective of the value of the concentration C, since both water and solute move freely, seeking mutual equilibrium. The movement of solute and solvent occurs because of differences in concentration, by diffusion (see Chap. 8). If the reservoirs of the osmometer of Fig. 3.9 are brought into contact without the presence of semipermeable membrane, the tendency is to redistribute solute and solvent until, after a long time, the concentration equals to a value C' in both reservoirs. In fact, this movement is a consequence of osmotic potential differences (at the beginning, $\Psi_{os} = 0$ in pure water and $\Psi_{os} = -RTC$ in the solution, and in the final equilibrium, $\Psi_{os} = -RTC'$ in both reservoirs).

The negative sign of Eq. (6.15) indicates that the larger the C, the smaller (more negative) is Ψ_{os}, which is the osmotic potential of the water due to the presence of salts. For the solute, the opposite happens, and therefore, in the redistribution of "salts," there is movement of water in one direction and movement of solute in the opposite direction.

Under normal soil conditions, the concentration C of the solution is practically constant and, therefore, there is no water movement due to the presence of solutes. Under special conditions, such as localized fertilization, there is movement of water toward the fertilizer. The absence of semipermeable membranes and the small variation of the concentration of the soil solution lead to the osmotic potential not to be considered. In the plant, as we will see below, Ψ_{os} plays an important role due to cell membranes.

6.7 Matric Component

$$\Psi = \Psi_P + \Psi_g + \Psi_{os} + \boxed{\Psi_m}$$

The **matrix potential component** of the total potential, represented by the fifth integral of Eq. (6.11), given to its complexity, cannot be easily calculated as we did for Ψ_P and Ψ_g. It was not yet possible to establish a reasonable, theoretically grounded equation for $\omega(\theta)$, since it would include all interactions between water and the solid matrix of the soil. Because of this, to date the Ψ_m measurement is experimental, made by means of tensiometers or pressure or suction instruments, which will be described next. In the soil, Ψ_m is related to θ, that is, the greater the θ (wetter soil), the higher the Ψ_m (or less negative).

The soil matrix potential was often referred to as **capillary potential**, soil water stress, suction or negative pressure. This potential is the result of capillary and adsorption forces arising from the interaction between water and solid particles, i.e., the soil matrix. These forces attract and "fix" the water in the soil, lowering its potential energy relative to free water. Capillary phenomena that result from the surface tension of the water and its angle of contact with solid particles are also responsible for this potential.

For flat water/air interfaces, there is no pressure difference between immediately higher and lower points at the liquid-gas interface. For curvilinear surfaces, however, there is a difference in pressure responsible for a series of capillary phenomena.

If we place a drop of a liquid on a flat surface of a solid, the liquid will settle onto the solid, acquiring a certain shape (see Fig. 6.6). The tangent of the liquid-gas interface at point (A) and the surface of the solid form an angle (α), characteristic for each liquid-solid-gas combination, called the **contact angle**. A contact angle equal to 0° would represent a complete spreading of the liquid on the solid or a perfect "wetting" of the solid by the liquid. A contact angle equal to 180° would correspond to a "no wetting" or total rejection of the liquid by the solid. The value of α depends on the adsorption forces between the molecules of the liquid and the solid. If these forces between the solid and the liquid are larger than the cohesive forces within the liquid and larger than the forces between the gas and the solid, α tends to be sharp and the liquid is said to "wet" the solid. Otherwise, $\alpha > 90°$, the liquid is said to be repelled by the solid.

The contact angle of a given liquid on solid is generally constant under given physical conditions. It may be different under dynamic

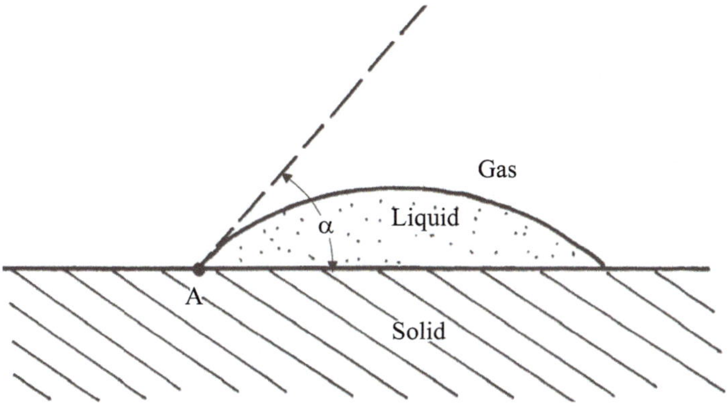

Fig. 6.6 Illustration of the contact angle α between the dashed line and the solid surface. Point A represents the gas, the liquid, and the solid interfaces

conditions, that is, when the liquid moves relative to the solid. The contact angle for pure water on flat, inorganic surfaces is generally close to zero, but roughness or impurities adsorbed by the surface usually make it different from zero. In the case of quartz (glass), the main component of the sands, and water, α is close to 0°.

When a capillary tube is immersed in through the surface of a liquid, the liquid will spontaneously rise through the tube and form a **meniscus** resulting from the material of the tube, its radius, and the angle of contact between the walls of the tube and the liquid. The curvature of the meniscus will be the larger the smaller the internal diameter of the tube and, because of this curvature, a pressure difference is established at the liquid-gas interface of the meniscus. A liquid with an acute α will form a concave meniscus to the side of the gas (water and glass), and a liquid with an obtuse α will form a convex meniscus to the gas side (mercury and glass). In the first case, the pressure P_1 under the meniscus is less than the atmospheric pressure P_0, and in the second, P_1 is greater than P_0 (see Fig. 6.7). As a result, in the first case, the liquid rises in the capillary tube and in the second, the liquid is repelled from the capillary. If the contact angle is zero, the meniscus will be a hemisphere and the radius of curvature of the meniscus R will be equal to the radius of the tube r. For α between 0° and 90°:

$$R = \frac{r}{\cos \alpha} \tag{6.16}$$

as it can be seen in the triangle inside Fig. 6.7.

The pressure difference between the water immediately below the meniscus and the above atmosphere is given by:

$$P = P_1 - P_0 = \frac{2\sigma \cos \alpha}{r} \tag{6.17}$$

As $P_1 < P_0$, for the case of pure water in a glass capillary, the pressure is negative; therefore, it is a subatmospheric pressure, usually referred to as **tension**.

From hydrostatics, we know that $P = \rho g h$ and, therefore, it is easy to verify that in the glass capillary tube the water height is given by:

$$h = \frac{2\sigma \cos \alpha}{\rho g r} \tag{6.18}$$

where ρ is the density of the liquid, g the gravitational acceleration, and σ the surface tension of the liquid.

Example: A glass capillary of 0.1 mm radius is inserted into a flat surface of water. What is the height reached by the water inside the tube? The water is at 30 °C, its density is 1.003 g cm^{-3}, and its angle of contact with the material of the capillary tube is 5°.

Solution:

$$h = \frac{2 \times 71.1 \times \cos 5^\circ}{1.003 \times 981 \times 10^{-2}} = 14.4 \text{ cm}$$

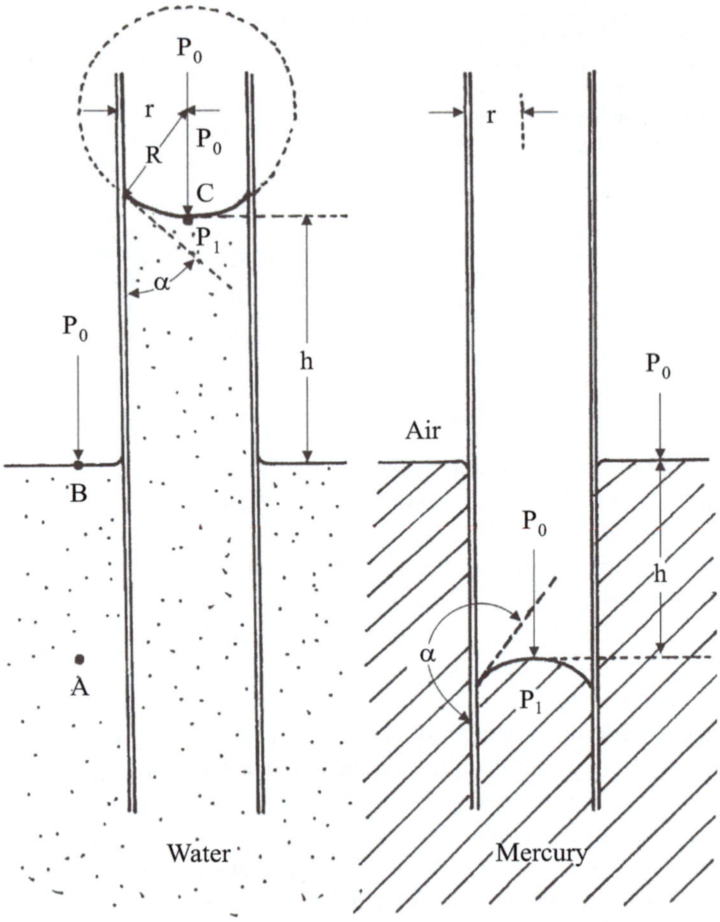

Fig. 6.7 Left hand side: glass capillary immersed in water showing a concave meniscus; right hand side with the same capillary immersed in mercury, with a convex meniscus

For thinner capillaries of radius 0.01 and 0.001 mm, we will have h values of 144 and 1440 cm, respectively.

The soil can be seen as a bunch of capillaries of different shapes, diameters, and arrangements. When water lodges in these capillary spaces, menisci of all sorts are formed. Each solid material has its own contact angle. It is therefore seen that it is difficult to apply formulas of the type 6.18 to the soil. Considering average values of pores and various approximations, something can be done.

Example: To a soil clod saturated with water, a pressure P of 0.3 atm is applied to extract part of its water and we wait for equilibrium. Considering the soil being a bunch of capillaries of diameter r, from which pores has the water been withdrawn by the applied pressure and which pores remain with water? ($T = 30$ °C, $\alpha = 5°$, $\rho = 1.0$ g cm^{-3}).

Solution:
Equation (6.18) indicates that capillaries of radius greater than r can be emptied with a pressure P. Remembering, that 1 atm corresponds to 1.013×10^6 b:

$$r = \frac{2 \times 71.1 \times \cos 5^{\circ}}{0.3 \times 1.013 \times 10^6} = 4.66 \times 10^{-4} \text{ cm}$$

and we can say that pores of radius greater than 4.66×10^{-4} cm were emptied and pores with smaller radius continue with water.

Just as water under a flat surface has a positive pressure potential ($+\rho gh$, see point A in Fig. 6.7), on the surface it has zero potential (point B, in the same figure) and just below the capillary meniscus has potential of negative pressure ($-\rho gh$, at point C). In the soil, it may also be under positive, zero, or negative pressures, its potential being, respectively, positive, null, or negative. For unsaturated soils, due to the presence of menisci (liquid-gas interfaces) and the presence of adsorption surfaces (solid-liquid interfaces), the pressure is negative, which gives it a negative matrix potential. Hence, the designation of water tension in the soil is frequent. In sandy soils, with relatively large pores and particles, adsorption is not very important, while capillary phenomena predominate in determining the matrix potential. For thin textures, the contrary occurs. Variations in potential also occur for the same soil, with different water contents. When relatively moist, capillary forces are of importance and, as the water content decreases, the adsorptive forces take their place.

In general, we can say that the matrix potential is mainly the result of the combined effect of two mechanisms—capillarity and adsorption—that cannot be separated easily. The water in capillary meniscus is in equilibrium with the water of adsorption "films," and the modification of the state of one of them implies the modification of the other. Thus, the older term **capillary potential** is inadequate, and a better term is **matrix potential** or **matric potential** because it refers to the total effect resulting from interactions between the water and the solid matrix of the soil. A work involving all these concepts related to hygroscopicity of a soil is that of Tschapek (1984).

As we have already said, the mathematical description of this matrix potential Ψ_m is quite difficult, and its determination is usually experimental, as will be seen in detail below.

For each sample of a homogeneous soil, Ψ_m has a characteristic value for each water content θ. The graph of Ψ_m as a function of θ is then a characteristic of the sample under consideration and is commonly referred to as a **soil water characteristic curve,** or simply a **retention curve**. For high water contents, in which capillary phenomena are of importance in the determination of Ψ_m, the characteristic curve depends more on the geometry of the sample, that is, on the arrangement and the dimensions of the pores. In this wet range, the curve becomes a function of soil density and porosity. For low water contents, the matrix potential is practically independent of geometric factors, with soil density and porosity being of little importance in its determination. Figure 6.8 shows retention curves for different samples.

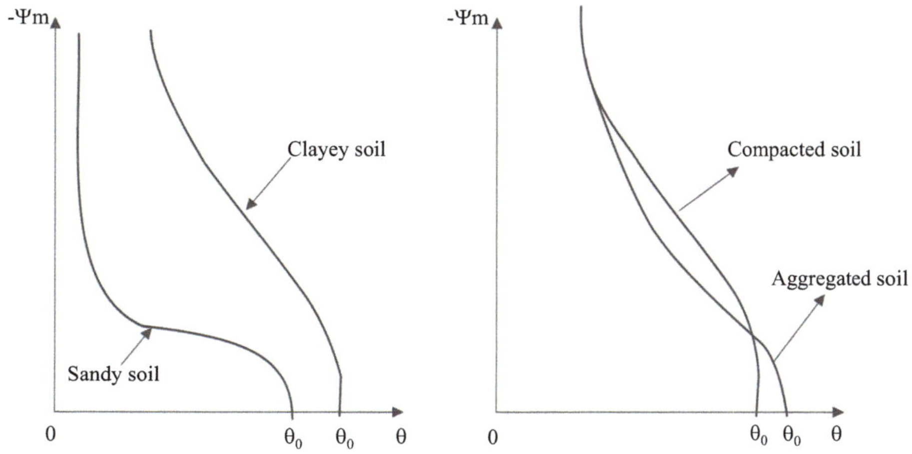

Fig. 6.8 Examples of soil water retention curves. To the left two extreme soils, to the right the same soil in two different conditions

With the retention curve of a soil, one can estimate Ψ_m knowing θ, or vice versa. In practice, the determination of θ is much simpler, such that θ is measured and Ψ_m estimated by the retention curve. As long as the system geometry does not vary with time, the characteristic curve is unique and does not need to be determined for each experiment. As can be seen in Fig. 6.8, the saturation water content θ_0 of a clayey soil is larger than that of a sandy soil. For a compacted soil, θ_0 is also smaller because the compaction decreases the porosity α and $\alpha = \theta_0$. In a saturated soil, in equilibrium with pure water at the same elevation, the matrix potential Ψ_m is zero. By applying a small suction to a saturated soil, water will not produce any water flow until the suction reaches a certain value at which the largest pore empties. This critical suction or tension is called "**air entry value** (or air entry suction)." For coarse-textured soils, this value is small, and for fine-textured soils, this cannot be neglected. By further increasing the tension, more water is withdrawn from the pores that cannot hold the water against the applied tension. Recalling the capillary equation, we can immediately predict that a gradual increase in tension will result in a progressively thinner pore emptying, until at very high tensions only very small pores can retain water. Each tension value corresponds to a value Ψ_m; as for each tension the soil has a water content θ, the curve of θ versus Ψ_m can be determined with ease. As we have already said, it is determined experimentally. To date, there is no satisfactory theory for predicting the characteristic curve Ψ_m versus θ. Today, the presentation of water retention curves is a routine.

The relationship between the matrix potential Ψ_m and the soil water content θ is not usually univocal, which means that its determination depends on how θ is evaluated. This relationship can be obtained in two different ways: (a) by a soil "drying" procedure, that is, taking a sample of soil initially saturated with water and then gradually extracting its water by applying larger and larger suctions and recording the pressures applied (Ψ_m, in energy per volume) at each successive measurement and evaluating the respective value of θ, or (b) by a soil "wetting" procedure, taking a soil sample initially in air dried condition and allowing its gradual wetting by reduction of the pressure Ψ_m. Each method provides a continuous curve, but the two, in most cases, are distinct. The phenomenon is called **hysteresis**. Soil water content in the equilibrium condition, at a given potential, is greater in the "drying" curve than in the "wetting" curve. Figure 6.9 shows the hysteresis phenomenon. As can be seen, θ_0 is the same, since, being the same

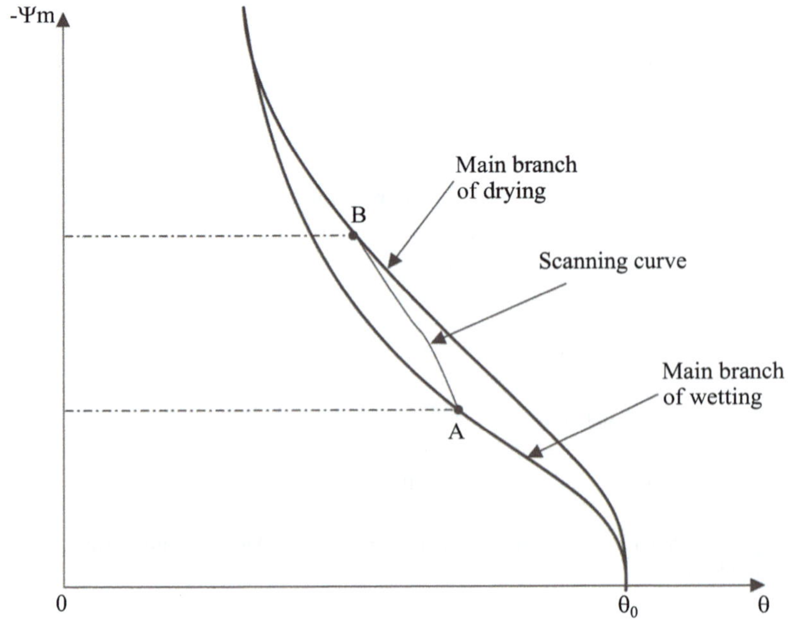

Fig. 6.9 Illustration of hysteresis, showing the main branches for wetting and drying. A scanning curve AB is obtained by reversing the process of drying to wetting or vice versa

soil, when saturated must always have the same water content. If a retention curve is obtained by wetting from a dry soil to the value of Ψ_m at point A of Fig. 6.9 and the soil is again dried by increasing the tension, another curve is obtained, represented by the segment AB. These intermediate curves are called **scanning curves**, and the two full curves are designated as main branches of hysteresis.

Basic studies of soil water hysteresis were carried out extensively in the past, including Haines (1930), Miller and Miller (1955a, b, 1956), Poulovassilis (1962), Philip (1964), Topp and Miller (1966), and Topp (1969). According to these papers, the hysteresis is attributed to the nonuniformity of the individual pores in relation to capillary phenomena, air bubbles that remain fixed in the macropores, and contraction/expansion of clays during drying and wetting.

The hysteresis presents serious problems for the mathematical description of the flow of water in the soil. We will see that the potential gradient is the force responsible for water movement, and if the relation $\Psi_m(\theta)$ is not univocal, the partial derivatives $\partial\Psi_m/\partial\theta$, $\partial\Psi_m/\partial x$, and $\partial\theta/\partial x$ will also not be. The problem can partially be avoided using the wetting curve parameters when dealing with wetting phenomena, as infiltration, and drying curve parameters in drying phenomena, as in the case of evaporation. When the two phenomena occur simultaneously, the problem becomes difficult. Most of the time the hysteresis needs to be neglected. It has to be recalled that for sandier soils the hysteresis can safely be neglected.

Another concern of the researchers is to find adjustment models for water retention curves in soils. Once the curves $\Psi_m(\theta)$ or $\theta(\Psi_m)$ are obtained experimentally, there is a need to define the best curve that fits the experimental data in order to be possible to estimate intermediate values. Among the most varied models, we highlight van Genuchten's (1980) model, based on parameter adjustments. The **equation of van Genuchten**, explicited in terms of θ, is:

$$\theta = \theta_r + \frac{(\theta_0 - \theta_r)}{\left[1 + (\alpha\Psi_m)^n\right]^b} \tag{6.19}$$

which is an S-shaped curve starting at saturation (θ_0) for $\Psi_m = 0$ and going asymptotically to infinity for extremely large (negative!) values of Ψ_m. The adjustment parameters α, n, b, and θ_r are obtained minimizing the deviations.

At this point of discussion, it is important to observe again Eq. (6.11), here shown again:

$$\Psi_m = \int_{\theta_s}^{\theta_i} \omega(\theta)\mathrm{d}\theta$$

when we said, empirically, that the result of this integral is a function of θ, the **soil water retention curve**. As the model of van Genuchten fits to most soils, we could say that $\omega(\theta) = \mathrm{d}\Psi_m/\mathrm{d}\theta$, that is, the derivative of Eq. (6.19), written in the form $\Psi_m(\theta)$. Equation (6.19) in this form is:

$$\Psi_m = \frac{1}{\alpha n}\left[\left(\frac{\theta_s - \theta_r}{\theta - \theta_r}\right)^{1/b} - 1\right]^{1/n} \tag{6.20}$$

$$\omega(\theta) = \frac{d\Psi_m}{\mathrm{d}\theta} = \frac{-1}{\alpha n b}\left[\left(\frac{\theta_s - \theta_r}{\theta - \theta_r}\right)^{1/b} - 1\right]^{\left(\frac{1}{n}-1\right)} \cdot \frac{(\theta_s - \theta_r)^{1/b}}{(\theta - \theta_r)^{\left(\frac{1}{b}+1\right)}} \tag{6.20a}$$

This analytical expression of $\omega(\theta)$ is one of the first attempts to describe the **soil matrix potential**, described above in connection with Eq. (6.8).

As an example, the reader can verify the application of van Genuchten's model to experimental data of Ψ_m and of θ obtained under field conditions with tensiometers and neutron probe, by Villagra et al. (1988). Moraes (1991) presents a study of the spatial variability of the water retention curve, adjusted by the Van Genuchten model. Dourado-Neto et al. (2000) presented software (SWRC—Soil Water Retention Curve software) for the determination of the retention curve, using several models, including van Genuchten, from experimental data. Another important aspect of the water retention curves in the soil is related to the concepts of total porosity α, **macroporosity** (Ma), and **microporosity** (Mi). It was already said that when a soil is saturated, $\theta = \alpha$ and $\Psi_m = 0$. The pore distribution in terms of size is continuous, and therefore it is difficult to find the limit between Ma and Mi, which was done arbitrarily based on the retention curve. Water at potentials close to saturation is poorly retained and therefore more subject to the action of the gravitational potential, and, in very wet profiles, water drains to the deeper layers. On the other hand, water at the negative potentials (e.g., $-1/10$ or $-1/3$ atm, considered as field capacity) and -15 atm, the permanent wilting point, is more retained by the soil matrix. In this way, the limit between Ma and Mi mostly employed is $\Psi_m = -60$ cmH_2O or -6 kPa. This point is taken in the retention curve, and the water between $\Psi_m = 0$ and -60 cm is considered as macropore water. Water retained at potentials less than -60 cm is considered as micropore water.

In addition to the Ma and Mi concepts, the retention curve is also used to define the field capacity (FC) through the soil water content θ_{FC}, corresponding to the matric potential of $-1/3$ atm (-33 kPa), or sometimes $-1/10$ atm (-10 kPa), depending on the soil type (see Chap. 12), and define the permanent wilting point (PWP) point through the θ_{PWP}, corresponding to the potential of -15 atm (-1500 kPa).

6.8 Total Water Potential of the Plant

In the same way as in the soil, the processes that take part of the **plant total water potential** can be considered reasonably isothermal, and the thermal component of the potential has minor importance in the plant. The contribution of the other components also depends on the part of the plant considered. The gravitational component Ψ_g is generally disregarded because the reference level is arbitrary and can always be brought to the same level as the object under study. In addition, there must be continuity in the water system for this potential to act. The water in a glass can be raised at any time, with increasing gravitational potentials, but at each height, it remains in equilibrium. Each of them, individually, can be considered at the gravity reference, $\Psi_g = 0$. For water from a cell vacuole, for example, only the osmotic and pressure components are of importance. So:

$$\Psi(\text{cell}) = \Psi_P + \Psi_{\text{os}}$$

and the pressure component appears because of the positive pressure that the cell walls exert on the cell juice (vacuole) when the cell is turgid. This component is also called a potential wall or **turgor potential**. The osmotic component Ψ_{os} is of great importance because the cell membranes are semipermeable, allowing water to pass through and being selective for various ions. In fact, one component (pressure) is a consequence of the other (osmotic). The cell functions as the osmometer of Fig. 3.9, because there is a higher saline concentration in the cell vacuole, in relation to the saline concentration outside the cell. Therefore, the tendency of water is to enter the cell, which implies in an increase in its volume. Because the walls are elastic only to a certain extent, the pressure inside the cell increases and is positive in relation to pure water, and the plant becomes turgid.

Typical values of Ψ_P (turgor) are +2 to +5 atm (0.2 to 0.5 MPa) and Ψ_{os} from -2 to -10 atm (-0.2 to -1 MPa). These potential values are responsible for cell turgor (erect plants as opposed to wilted) and for plant growth by cell elongation.

For the case of water retained by cell walls, the matrix component Ψ_m might be of importance. In the same way as water is retained by the soil pores, it can be retained by the pores of the fibrous tissues of the cell walls like cellulose and lignin, which also happens with starch agglomerates and other reserve substances in seeds. In these cases:

$$\Psi = \Psi_m + \Psi_{os}$$

Ψ_m and Ψ_{os} can be very negative, dozens of atm. Due to these very negative values, seeds absorb water with great ease, often doubling their volume. The literature on water potential in the plant is extensive. We recommend the works of Zimmermann and Stendle (1978) and Oertli (1984) and, for more details, Kirkham (2014), Ehlers and Goss (2016), Taiz et al. (2018), among others.

A controversial subject is the rise of water in trees much taller than 10 m. As we know, the atmosphere can hold by suction water columns of only about 10 m. Since the potential of water in plant tissues is in general very negative, water rises in the plant against gravity by suction. In non-capillary tubes or pipes, it is known that by aspiration the water can be elevated close to -1 atm (or -10 mH_2O or -0.1 MPa) and, thereafter, the water column ruptures, by the formation of dissolved air bubbles (vapor) which are released in the liquid water breaking the column. This phenomenon occurs in the water of tensiometers, instruments used to measure Ψ_m in soils, which will be seen later in this chapter. If we are to draw water from a well, by suction, the depth of the well cannot be greater than 10 m, this at sea level where $P_{atm} = 1$ atm. The pump raises water by suction, and as this cannot be greater than the local P_{atm}, it does not operate for depths greater than 10 m. What happens is that P_{atm} that acts at the free water level of the well pushes the water up, at the most 10 m. For deeper wells, the pump is lowered inside the well to an appropriate depth. In many cases the pump is immersed in the well water.

In plants, things are different because the "pipes" are capillaries (xylem). Angelocci (2002) deals very well with this subject, when referring to the dynamics of water in the plant, in the liquid phase. According to this author, in plants with a low transpiration rate—which occurs with low atmospheric demand or at night with closed stomata, or leaf loss due to falling or pruning—the raw sap of the xylem turns to be under positive pressures in relation to the atmosphere. With the occurrence of wounds, as by pruning, gutting and sap exudation prove the positive pressure values. The most accepted theory for this forced rise of water assumes that the roots function as an osmometer (like shown in Fig. 3.9), when the transpiration is low. Under conditions of higher atmospheric demand, the transpiration is more intense and the xylem has a structure of low resistance to the transport of the sap, which is due to the very low values of the total potential of the water in the leaf. The rise of the sap is best explained by the theory of adhesion-cohesion, based on the late-nineteenth-century work by Bohem, Askenasi, Dixon, and Joli. The second conceived an assembly that shows that a column of water under tension (negative pressure) caused by a porous capsule adapted at its upper end (simulating a leaf) can be maintained for tensions far more negative than -1 atm, without breaking. The adhesion-cohesion theory assumes that water in the xylem forms a continuous liquid phase, from the root to the leaf. In the micro-capillaries of the cell walls of the leaf mesophyll, the free energy of the water is reduced as a function of the curvature of the menisci, leaving the entire column under tension. This tension is transmitted throughout the xylem at the expense of the high cohesive force between the water molecules, maintaining liquid continuity and the transpiration. These physiological aspects of water-plant relationships are very well addressed by Nobel (1983) and Kramer and Boyer (1995).

Returning to the study of the total potential of water in the Soil-Plant-Atmosphere System as a whole, defined by Eqs. (6.8) and (6.11), let us see the importance of each of the components in several specific cases:

1. **In the soil:**
   (a) Saturated soil, immersed in water:

$$\Psi = \Psi_P + \Psi_g$$

In this case Ψ_g is of great importance and Ψ_P depends on the value of the hydraulic load acting on the considered point; $\Psi_m = 0$, because there are no water/air interfaces and Ψ_{os} is not considered because there is no semipermeable membrane (as long as there is no plant).

(b) Non-saturated soil:

$$\Psi = \Psi_m + \Psi_g$$

Ψ_g is of great importance in the wet range, close to saturation. Ψ_g gets losing its importance as the water content decreases (although maintaining its constant value at every point). This is because Ψ_m is zero at saturation and keeps gaining importance as soil loses water. For very dry soils, $\Psi = \Psi_m$ (very negative) and Ψ_g can be completely disregarded. As there is no free water in the system, $\Psi_P = 0$ and Ψ_{os} is not considered because there is no semipermeable membrane.

2. **In the plant:**

(a) In cells of soft tissue (like leaf cells):

$$\Psi = \Psi_P + \Psi_{os}$$

Ψ_P is the plant turgor (positive pressure), and Ψ_{os} the osmotic potential due to the presence of solutes mainly in the vacuole and of semipermeable membranes. $\Psi_m = 0$ and Ψ_g are not considered, which implies that the gravitational reference is taken to the level of the cells.

(b) Fibrous tissue, woody, or agglomerate (e.g., wood, fibers in the stem, seeds):

$$\Psi = \Psi_m + \Psi_{os}$$

In this case, Ψ_m is the negative potential resulting from the interactions between the water and the plant porous materials. Seeds and other woody tissues in stems, roots, and fruits can present very negative values of Ψ_m. As Ψ_{os} is also negative, the final value of Ψ becomes very negative. Therefore, seeds are eager for water and absorb it quickly, often doubling their volume. Ψ_g is neglected because it is relatively small in relation to the other or because the referential is brought to the system's level.

3. **In the atmosphere:**

$$\Psi = \Psi_P$$

Ψ_m and Ψ_{os} do not enter into consideration since the system consists of water vapor dissolved in the air. Ψ_g is also neglected or considered as zero by taking the gravitational reference to the point under study.

4. **Transfer of water from soil to plant:**

(a) Flooded soil (e.g., paddy rice):

$$\Psi = \Psi_g + \Psi_P + \Psi_{os}$$

The component Ψ_{os} appears because of the semipermeable membranes of the plant.

(b) Aerated soil (e.g., highland rice):

$$\Psi = \Psi_g + \Psi_m + \Psi_{os}$$

By the same reason, Ψ_{os} cannot be neglected.

We see that, from situation to situation, the total water potential Ψ consists of different components, each according to its importance.

6.9 Equilibrium of the Water

As has been said in the introduction to this chapter, water obeys the universal tendency to constantly seek a state of minimum energy. The total potential of water represents its state of energy and is therefore a criterion for knowing its state of equilibrium. We can then say that water is in equilibrium when its total potential is the same at every point in a system. Let's look at some examples, starting from the simplest. This is the case of water in a glass, as shown in Fig. 6.10. Of course, it is an equilibrium system and thus the total water potential Ψ must be the same at any point inside the glass. We arbitrarily chose points A, B, and C to show that Ψ has the same value. Because it is pure water, the components Ψ_{os} and Ψ_m are zero and Eq. (6.12) reduces to:

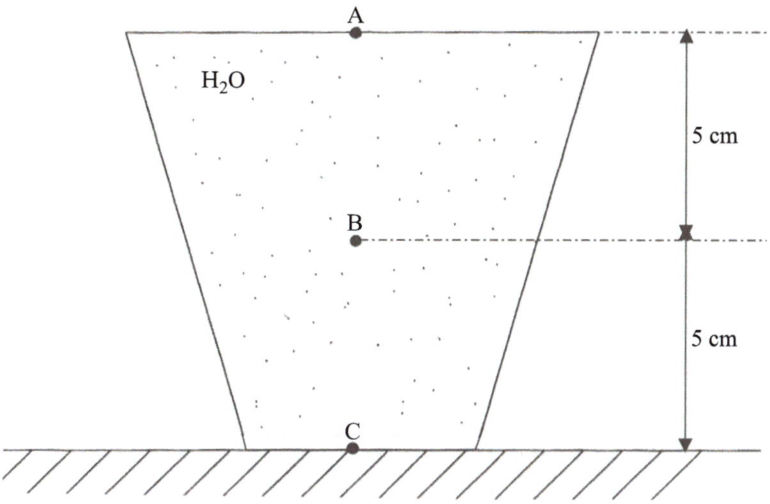

Fig. 6.10 A glass of water in equilibrium

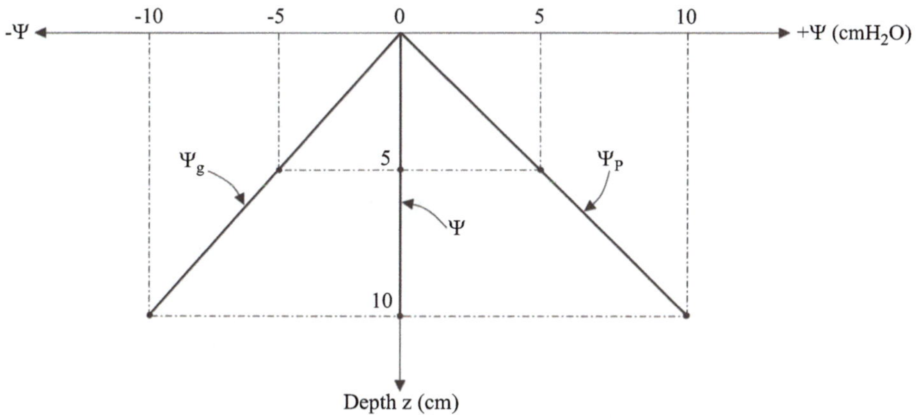

Fig. 6.11 A potential versus height graph for the water glass shown in Fig. 6.10

$$\Psi = \Psi_P + \Psi_g$$

and for points A, B, and C, we have (considering as gravitational reference the free water surface) using the units of energy/volume or hydraulic load:

$$\Psi(\mathrm{A}) = 0 + 0 = 0\ \mathrm{cmH_2O}$$

$$\Psi(\mathrm{B}) = +5 - 5 = 0\ \mathrm{cmH_2O}$$

$$\Psi(\mathrm{C}) = +10 - 10 = 0\ \mathrm{cmH_2O}$$

and we see that Ψ has the same value (= 0), indicating equilibrium.

Graphically we can represent these values as shown in Fig. 6.11.

If we change the gravitational reference to the bottom of the glass, we have:

$$\Psi(\mathrm{A}) = 0 + 10 = 10\ \mathrm{cmH_2O}$$

$$\Psi(\mathrm{B}) = 5 + 5 = 10\ \mathrm{cmH_2O}$$

$$\Psi(\mathrm{C}) = 10 + 0 = 10\ \mathrm{cmH_2O}$$

and one can see that again Ψ is the same (= 10 $\mathrm{cmH_2O}$) in any point, indicating equilibrium. The change in the reference only changes the value of Ψ, which anyway is relative because it depends on the choice of the reference. With this change, Fig. 6.11 is transformed into Fig. 6.12.

If we make a small hole in the bottom of the glass, exactly at point C, we know that water will

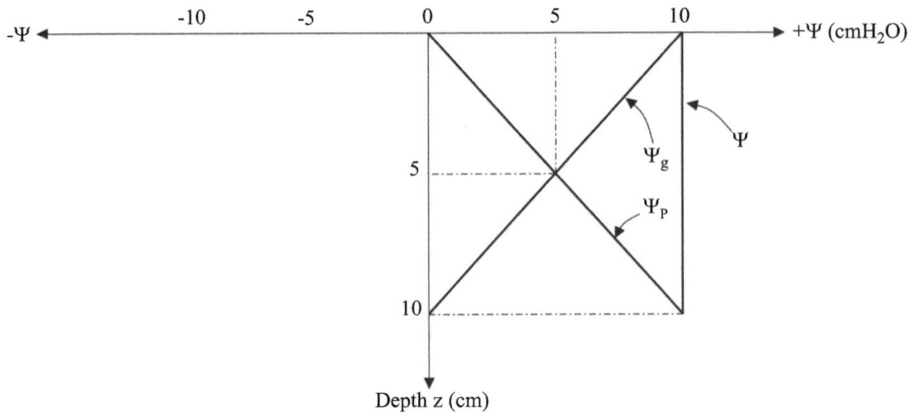

Fig. 6.12 The same graph of Fig. 6.11 with a different gravitational reference

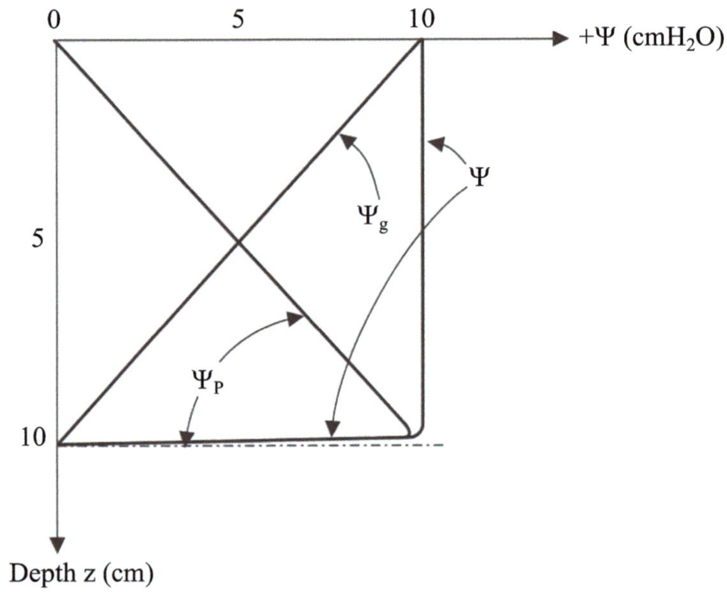

Fig. 6.13 The glass of Fig. 6.10 with a hole at point C

flow through the hole, indicating that it is no longer in equilibrium. When opening the hole in C, the pressure potential drops instantly from $+10\ \mathrm{cmH_2O}$ to zero, as it is subject to atmospheric pressure at the hole. Keeping the gravitational reference in the bottom of the glass, we have:

$$\Psi(\mathrm{A}) = 0 + 10 = 10\ \mathrm{cmH_2O}$$

$$\Psi(\mathrm{C}) = 0 + 0 = 0\ \mathrm{cmH_2O}$$

As we see, now Ψ varies within the glass and, therefore, the water moves. It moves from a larger total potential to a smaller one, in the case from A to C. With the presence of the hole (at the very beginning), the graph of Fig. 6.12 changes to the shape of Fig. 6.13.

Of course, over time, the water level in the beaker decreases and the graph in Fig. 6.13 changes continuously.

Let us now consider a plant tissue with cells, in which $\Psi_P = +5$ atm (cell turgor) and $\Psi_{os} = -7$ atm (cell vacuole saline concentration) were determined. In this case:

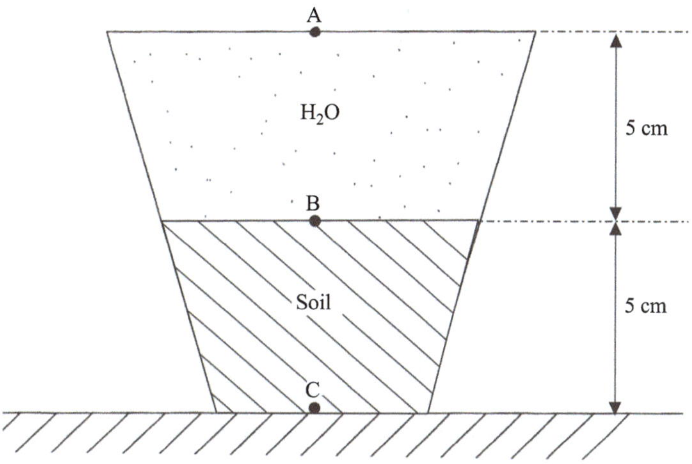

Fig. 6.14 The glass of Fig. 6.10 half filled with soil

$$\Psi(\text{cell}) = \Psi_P + \Psi_{os} = 5 - 7 = -2 \text{ atm}$$

If this tissue is thrown into the pure water glass of the previous example, what would happen? The potential of the water in the glass is equal to 0 or 10 cmH_2O (0.01 atm), depending on the gravitational reference. In any case, we can consider it 0 if the gravitational level is passed through the cell, as in the case of standard water. As the water potential in the beaker is greater than that of the cells ($0 > -2$), water will penetrate the cells, looking for a lower potential. As cells have semipermeable membranes, water enters and salts do not leave the cell, resulting in increased turgor and decreased salt concentration. After some time we will have the balance between the cell and the water in the glass.

$$\Psi(\text{cell}) = \Psi_{\text{water in the glass}} = 0$$

In this new condition, Ψ_P in cells = +6 atm and Ψ_{os} = −6 atm were determined; therefore, $\Psi = +6 - 6 = 0$.

Let us now consider the glass of Fig. 6.10, with 5 cm of soil in its bottom, as shown in Fig. 6.14. What would be the distribution of potentials inside the glass?

As the soil is submerged, all its pores are filled with water, there are no water/air interfaces, and therefore there are no capillary phenomena and $\Psi_m = 0$. The pressure due to the water load propagates through the soil pores and the values of Ψ_P are identical to those found in the beaker with water alone. The graphs of Figs. 6.11 and 6.12 are also valid for this system. This system can be identified with a flooded soil, in which there is a water impermeable layer at a given depth, such as that of paddy rice given in Fig. 6.4.

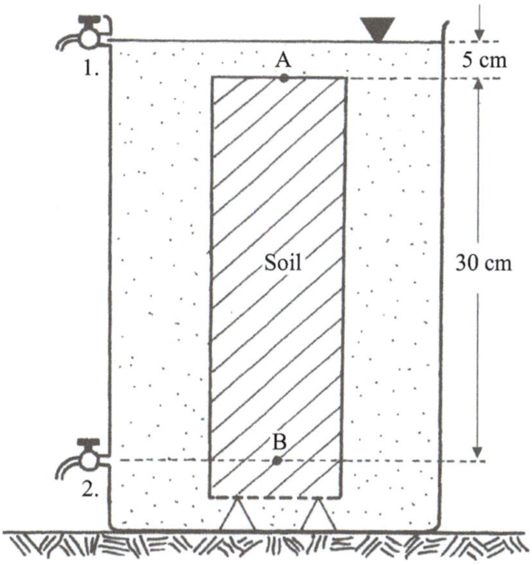

Fig. 6.15 A submerged soil column with two water taps 1 and 2

Now let be the system shown in Fig. 6.15, essentially the same as that of Fig. 6.14.

A column of soil is placed inside an acrylic plastic cylinder with an open bottom for the passage of water, which is totally immersed in a vial of water.

In the condition of immersed soil, that is, with taps 1 and 2 closed, and using point B as gravitational reference, we have:

$$\Psi(A) = \Psi_P + \Psi_g = 5 + 30 = 35 \text{ cmH}_2\text{O}$$

$$\Psi(B) = \Psi_P + \Psi_g = 35 + 0 = 35 \text{ cmH}_2\text{O}$$

demonstrating equilibrium.

If tap 2 at level B is opened, the water in the flask drops quickly depending on that tap water flow, to level B. Inside the soil, the water level drains much slower, because of the tortuousness of the water path in the soil and because of the resistance offered by the soil. Air starts entering the soil and after a long time, without allowing losses by evaporation, the equilibrium is established and, once again, the total potential has to be constant at all points of the soil. The gravitational component does not change, but the pressure component does, because the soil is in the unsaturated condition; water/air interfaces appear and the soil pressure potential becomes negative, i.e., Ψ_m appears. Only a small region near point B is still saturated, although subject to negative pressures. In this region called capillary fringe, the suction of a few cm of water is not enough to empty even the largest pores of the soil. This is the case of the **air entry value**, already discussed previously, in relation to the water retention curve in the soil. In this new condition, if there is no evaporation of water on the soil surface, the following distribution of potentials is obtained:

$$\Psi(A) = \Psi_m + \Psi_g = -30 + 30 = 0 \text{ cmH}_2\text{O}$$

$$\Psi(B) = \Psi_m + \Psi_g = 0 + 0 = 0 \text{ cmH}_2\text{O}$$

The graphs in Fig. 6.16 show the situation before and after reaching the new equilibrium.

Let's see another example: consider a capillary immersed in a flat surface of pure water, as shown in Fig. 6.17. This system is typically in equilibrium. Its total potential Ψ is composed of two components, the gravitational Ψ_g and the pressure Ψ_P. Table 6.2 gives each component at points A to E, expressed as hydraulic load.

As we see, Ψ is constant along z, an essential condition for equilibrium.

Another example, similar to Fig. 6.15, considers a soil column in contact with a water

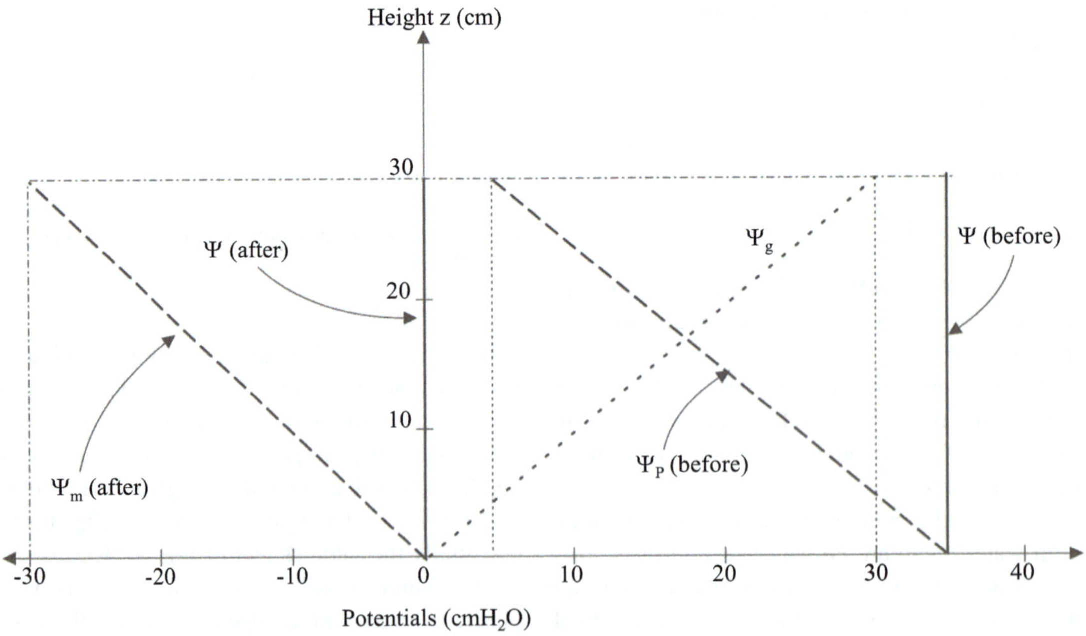

Fig. 6.16 Potential versus height graphs for the case of Fig. 6.15, before opening tap 2 and after opening tap 2

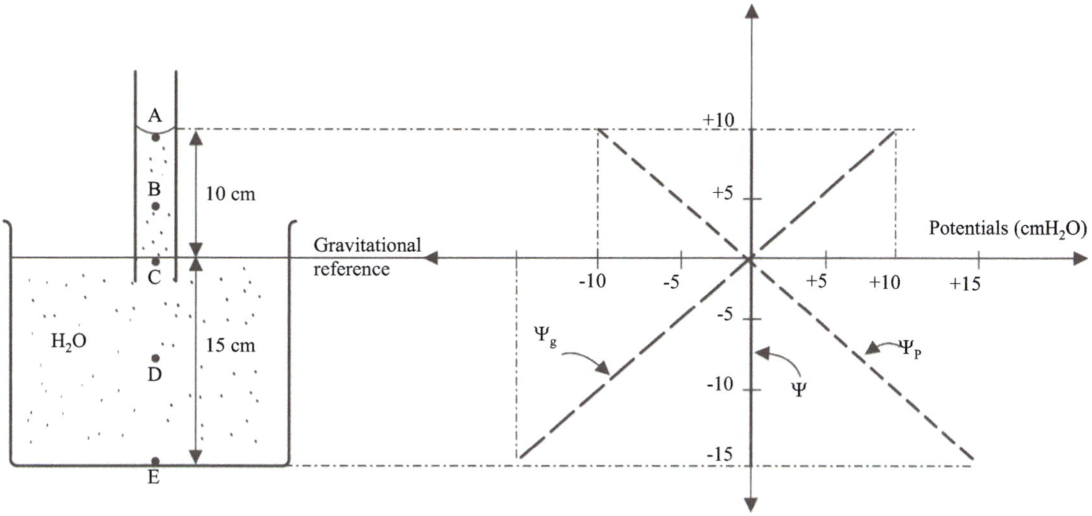

Fig. 6.17 A capillary immersed in water and the respective graph of potential versus height

Table 6.2 Calculation of the total potential of the different points indicated in Fig. 6.17

Points	Ψ_g	Ψ_P	Ψ
A	+10	−10	0
B	+5	−5	0
C	0	0	0
D	−7.5	+7.5	0
E	−15	+15	0

surface, as shown in Fig. 6.18, in which the matrix potential (which was measured) varies according to the curve indicated. The evaporation rate at soil surface is constant.

Let's look at Fig. 6.18. The Ψ_m curve is obtained experimentally by means of porous cups (mini-tensiometers) installed in depth along the soil column. The straight line Ψ_g is obtained by directly measuring the gravitational potential with a ruler, from an arbitrary referential, and the curve Ψ is obtained by adding Ψ_g and Ψ_m at each point. As we can see, Ψ is variable along z. Because Ψ is variable, we can say that the system is not in equilibrium. Therefore, there is water flow within the column, at least in the aerial part of the column, i.e., for z values varying from 10 to 30 cm. In the immersed part, Ψ is constant and there will be no flow. This part of the column is in steady-state equilibrium, resembling the example in Fig. 6.15, but water flows in order to maintain the evaporation flow.

As we have already said, the curves of Ψ_g depend on the reference. By changing the reference, the curves change. As $\Psi = \Psi_g + \Psi_m$, the curve of Ψ also varies. This has no drawback because, for any chosen reference, Ψ will be constant for cases of equilibrium and Ψ will be variable for nonequilibrium cases. The absolute value of Ψ will be different, but this is of little importance, because the potential differences between two points will always be the same. It is suggested that the reader makes new drawings of Figs. 6.17 and 6.18 for other references and verify that the potential difference between two points does not depend on the reference.

Now a field example, as shown in Fig. 6.19. Measurements were made with tensiometers and from the soil water characteristic curve elaborated with samples of the same soil profile (curves of the type of Fig. 6.8 or 6.9), the values of Ψ_m for each θ were determined and also shown in Table 6.3.

Figure 6.19 shows us, first, that the total potential Ψ varies with depth and, therefore, it is a dynamic situation. The approximately 0–50 cm layer is characterized by increasingly negative

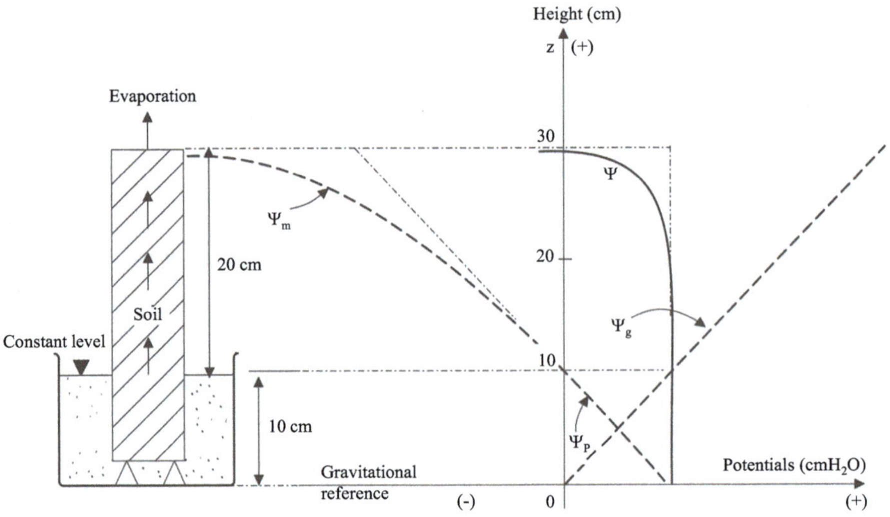

Fig. 6.18 A soil column with a fixed water table and the respective graph of potential versus height

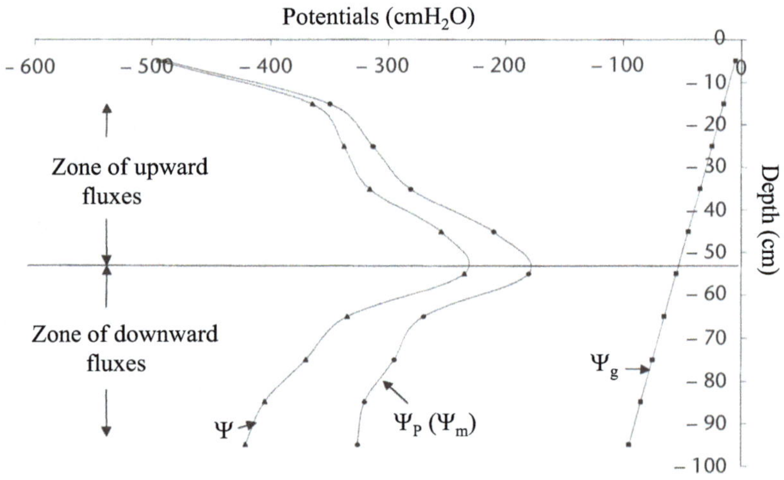

Fig. 6.19 Potential versus depth graph for the data presented in Table 6.3

Table 6.3 Soil water data of a field soil in a condition of upward and downward water movement

Depth (cm)	Soil water content θ ($cm^3\ cm^{-3}$)	Ψ_m (cmH_2O)	Ψ_g (cmH_2O)	Ψ (cmH_2O)
0–10	0.256	−490	−5	−495
10–20	0.295	−350	−15	−365
20–30	0.321	−313	−25	−338
30–40	0.336	−281	−35	−316
40–50	0.345	−210	−45	−255
50–60	0.351	−180	−55	−235
60–70	0.338	−270	−65	−335
70–80	0.330	−295	−75	−370
80–90	0.315	−320	−85	−405
90–100	0.313	−326	−95	−421

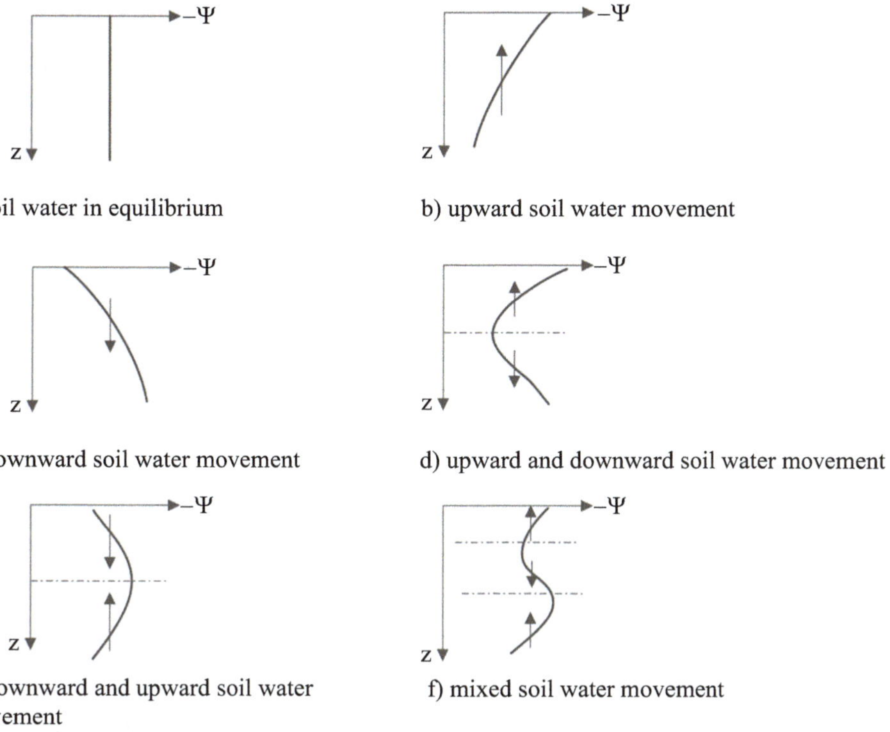

Fig. 6.20 Various situations of water equilibrium and flow, indicated by the arrows

values in the direction of the soil surface. As the water always moves from points of greater Ψ to points smaller Ψ, water moves (in this layer) upward. Below 50 cm, the opposite occurs and the water moves from top to bottom. This deeper part of the profile is under drainage. This is a typical example of what can occur in the field. At about 50 cm depth, the water flux inverts from upward to downward and so many researchers call the plane of this depth as the **zero flux plane**. As time passes without rainfall, the upper layer dries out at the expenses of the stored water, and the upper layer also dries out at the expenses of stored water, by drainage or root uptake. In this process, the zero plane layer might remain at the same depth, but more frequently, it lowers in depth.

Graphs of the type of Fig. 6.19 are important for defining soil water dynamics in the field. Figure 6.20 illustrates the issue for other cases.

In the case of Fig. 6.19, total soil water potentials range from −200 to −500 cmH_2O, or −0.2 to −0.5 atm, approximately. With additions of water (rainfall or irrigation) and with subtractions of water (evapotranspiration or drainage for deeper horizons), the graph of Fig. 6.19 changes continuously. In general, for crops in full development, without water deficit, the values of total soil water potential range from −0.1 to −1.0 atm (−10 to −100 kPa). Under the same conditions, the total potential of the plant varies from −5 to −40 atm (−0.5 to −4 MPa) and, in the atmosphere, from −100 to −1000 atm (−10 to −100 MPa). Hence, the normal movement of water is from the soil to the plant roots and from the leaves to the atmosphere. In Fig. 6.21 this movement is shown from a generic point A in the soil to another B in the root, to another C in the leaf, and finally to D in the atmosphere.

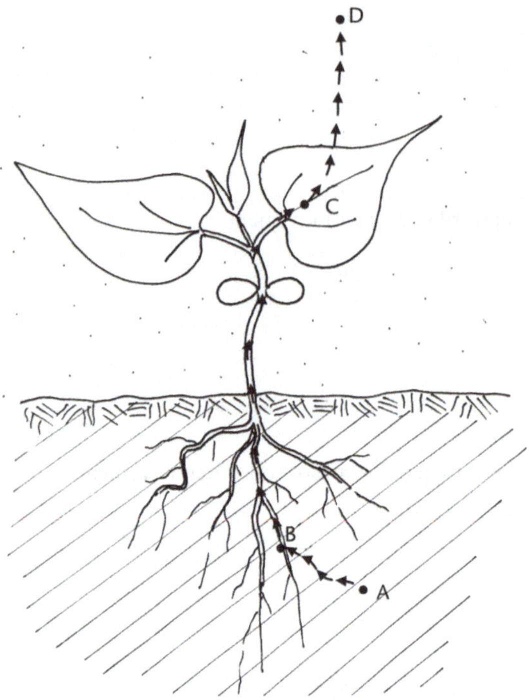

Fig. 6.21 Water flow in the Soil-Plant-Atmosphere System during a normal transpiration condition with available water in the soil

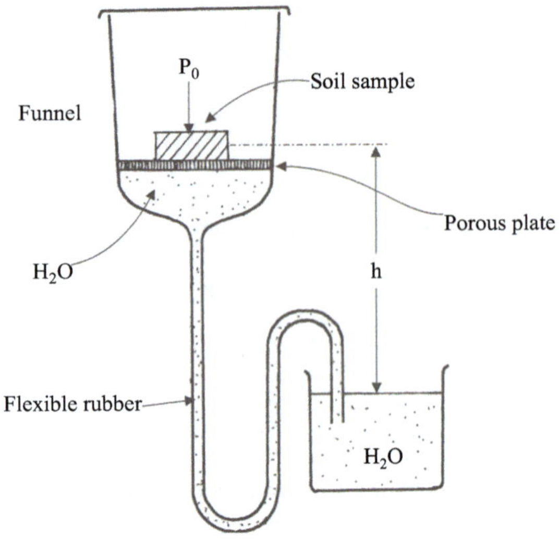

Fig. 6.22 Schematic view of the porous plate funnel showing a soil sample under tension $-h$

6.10 Instruments for Soil Water Measurements

There is a number of instruments used to determine soil water properties. These measurements are made with soil water in equilibrium. We start with the soil water potential, with instruments that provide us with one or more components, depending on the type of instrument.

As we have already seen, the gravitational potential is measured directly with the help of a ruler, since it depends only on the relative position of water in the terrestrial gravitational field.

Pressure and matrix potentials appear only when the soil is subject to a hydraulic load or a suction, and these are also proportional to the height of a fluid column.

The osmotic potential can be estimated by the concentration of the soil solution, as we saw. Serious difficulties are encountered in determining the saline concentration of the soil water. From the macroscopic point of view, it can be considered constant, but, microscopically, it varies greatly because of the adsorption phenomena by solid particles and preferential extraction by plants. In Chap. 8, on soil solution, the subject will be discussed in detail. In practice, as the saline concentration of the soil water generally varies little from point to point, the osmotic potential is neglected. Thus, in most problems of soil physics, the total potential is considered as the sum of the gravitational, pressure, and the matrix potential, under isothermal and isobaric conditions, hence the importance of measuring soil matrix potential.

6.10.1 Porous Plate Funnel

This instrument is shown in Fig. 6.22. It consists of a porous plate adapted to a funnel. The porous plate is saturated, consisting of pores of such dimensions that the suction h applied to its bottom is not sufficient to withdraw the water from the tiny pores. These plates and other porous ceramic materials are manufactured with different granulometry. They behave like an artificial soil of very homogeneous porosity. Thus, ceramic plates and capsules are found of 0.5, 1, 3, 5, and

15 atm, which means that these materials remain saturated to the indicated suctions (or pressures). Potential measurements at low suctions, from 50 to 500 cmH_2O (0.05 to 0.5 atm), are generally made with funnels.

In the condition found in Fig. 6.22, the pore water of the plate (although saturated) is under suction or tension of $-h$ cmH_2O. If the soil sample has good contact with the plate, in the equilibrium condition, the soil matrix potential Ψ_m will be $-h$. In relation to the gravitational potential, the plate thickness and the sample height are considered to be negligible, i.e., $h \gg z$.

Porous plate funnels are used to make water retention curves in the wet range of soils. After placing the soil sample in contact with the plate, raise the free water vessel to the upper level of the plate. Under these conditions $h = 0$ and the soil becomes saturated. θ_0 can thus be determined, logically after the equilibrium has been reached. Then, the water containers lowered to a preset pressure h, say $h = 30$ cm. In equilibrium, the soil will have $\Psi_m = -h = -30$ cmH_2O. In this condition, the new θ is determined which, logically, will be smaller than θ_0. From what we have seen previously, the pores of the soil whose diameter is larger than the diameter of a pore with potential -30 cmH_2O are emptied. By lowering the container to a new h, say 60 cm, we have in equilibrium $\Psi_m = -60$ cmH_2O and a new value of θ. The same is true for several values of h, measuring the respective values of θ, until the working limit of the plate is reached. If the limit is exceeded, some pores are emptied from the plate and the air begins to enter the chamber below the plate, causing the water column to break, and the instrument stops working.

The procedure previously seen, although simple, the greater the negative potential ($-h$), the more it takes to reach equilibrium. The procedure also implies in a series of problems that the operator only learns in practice. The main ones are:

(1) Is the contact between the soil sample and the plate good?

(2) Should deformed or un-deformed samples be used?

(3) How to avoid water loss through evaporation?

(4) Keep the temperature of the laboratory reasonably constant?

(5) Should the same sample be used for each (potential) point or not?

(6) What is the ideal number of replicates?

(7) Is hysteresis negligible?

6.10.2 The Water Tensiometer with Mercury Manometer

This tensiometer consists of a ceramic cup in contact with the soil, connected to a manometer, by means of a PVC tube completely filled with water, as shown in Fig. 6.23. It, in fact, is not much different from the porous plate funnel seen in the previous item. Imagine only the flat plate transformed into a cup and, reversing the positions, that is, the cup goes down and is

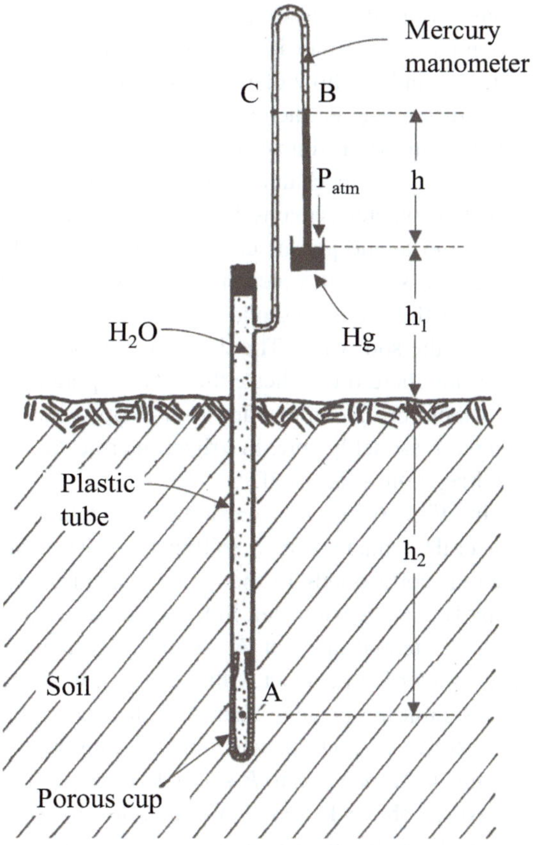

Fig. 6.23 Schematic view of a water tensiometer with a mercury manometer

introduced into the soil, and the water vessel rises above the surface of the soil. The simple inversion would imply in positive pressures in the capsule, but this is avoided by the proper construction, seen below. In the tensiometer, the water does not come into direct contact with the atmospheric pressure, as in the case of the funnel. It is hermetically sealed and P_{atm} acts only through the mercury manometer (Fig. 6.23).

When placed in the soil, the water from the tensiometer comes in contact with the soil water through the pores of the porous cup and the equilibrium tends to be established. At first, that is, before placing the instrument in contact with the soil, its water is at atmospheric pressure. Soil water, which is generally under subatmospheric pressures, withdraws a certain amount of water from the cup, causing a drop in the hydrostatic pressure inside the instrument. Once the equilibrium has been established, the potential of the water within the tensiometer cup is equal to the potential of the water in the soil around the capsule Ψ_m and the flow of water ceases. The pressure difference is indicated by a manometer, which can be a simple U-tube with water or mercury, or a mechanical or electric indicator. The tensiometer remains in the soil for a long time, and as the porous capsule is permeable to water and salts, the water in the tensiometer will be with the "same" composition and concentration of the soil water. The difference in pressure does not, therefore, indicate the osmotic potential. It is therefore recommended to use tap water as long as reasonably pure. In irrigation projects, use the irrigation water itself.

In Fig. 6.23, the tensiometer reading is $-h$ cmHg, made with a millimeter ruler. This reading corresponds to the water tension at the point B of the tensiometer. Since we are interested in the potential at point A ("point at which the tensiometer is in contact with the soil"), we need to discount the positive charge of the water column between A and B, which is the water height from C to A, or equal to $h + h_1 + h_2$ cmH_2O. The pressures above B and C are annulled by each other. Thus, the matrix potential at A, in unit of hydraulic load (cmH_2O), is given by:

$$\Psi_m(\mathrm{A}) = -13.6 \times h + h + h_1 + h_2$$

or

$$\Psi_m(\mathrm{A}) = -12.6 \times h + h_1 + h_2 \qquad (6.21)$$

in which:

h = reading in cm of Hg that is transformed into a water height through the factor 13.6 (specific mass of Hg) (see Fig. 6.24)

h_1 = height of the open mercury surface (P_{atm}) in relation to soil surface

h_2 = depth of installation of the ceramic cup (from soil surface down to the center of the cup), which is about 5 cm long

Figure 6.25 shows two tensiometers in equilibrium with the soil around them:

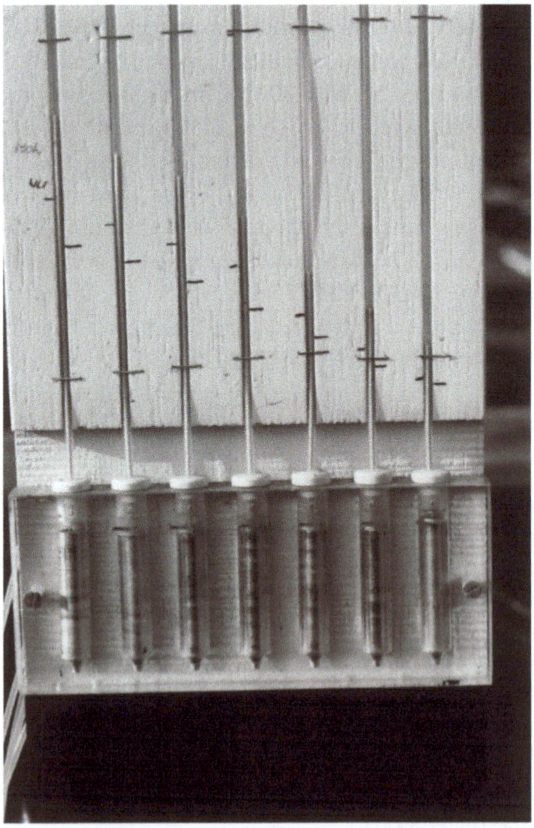

Fig. 6.24 A battery of mercury gauges (manometers) showing different heights h and, consequently, different soil water potentials

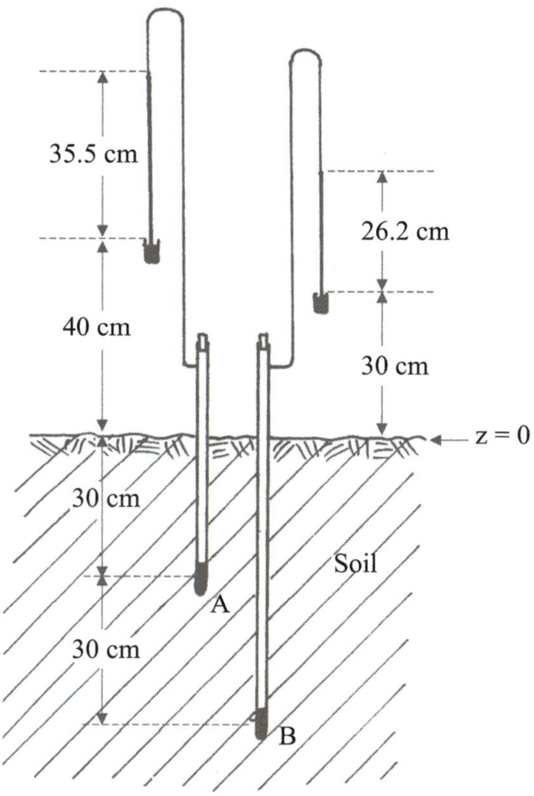

Fig. 6.25 Two tensiometers installed in the field at depths A and B

Tensiometer A:

$$\Psi_m(A) = -12.6 \times 35.5 + 40 + 30$$
$$= -377.3\ cmH_2O = -36.5\ kPa$$

Tensiometer B:

$$\Psi_m(B) = -12.6 \times 26.2 + 30 + 60$$
$$= -240.1\ cmH_2O = -23.2\ kPa$$

The total potential Ψ, at the two points, taking as gravitational reference the soil surface, will be:

$$\Psi(A) = \Psi_m(A) + \Psi_g(A) = -377.3 - 30$$
$$= -407.3\ cmH_2O$$

$$\Psi(B) = \Psi_m(B) + \Psi_g(B) = -240.1 - 60$$
$$= -300.1\ cmH_2O$$

It is, therefore, concluded that $\Psi_A < \Psi_B$ and that there is upward water movement in this soil profile. When $\Psi_A = \Psi_B$, soil water is in equilibrium (**field capacity**), and when $\Psi_A > \Psi_B$, the profile is under drainage and there might be solute leaching. The measurement of the matrix potential by tensiometer is, in general, limited to values lower than 1 atm, because of the local values of P_{atm}. This is because the manometer measures gauge pressures (vacuum) relative to the external atmospheric pressure. When the tension reaches high negative values, close to −1 atm, air bubbles appear in the water that interfere with the equilibrium, even interrupting the water column. This process is minimized by the use of deaerated water when filling the tensiometers and can be minimized by fluxing the instrument from time to time. The fluxing is the operation that forces a flow of water through the manometer tube, eliminating air bubbles.

In practice, the useful range of the tensiometer is from $\Psi_m = 0$ (saturation) to $\Psi_m = -0.8$ atm, approximately. This range of limited potentials, measurable by the tensiometer, is not as limited as it seems. It is a small part of the total range of potentials but, in the field, covers the main range of soil water contents of importance for agricultural practices. Reichardt (1987) details further these aspects. A key text on tensiometers is that of Cassel and Klute (1986). An advance in the simplification of manometers used in tensiometers is presented by Villa Nova et al. (1989) and Villa Nova et al. (1992). It should be noted that the use of mercury tensiometers has been quite limited in practice due to environmental issues related to mercury pollution.

6.10.3 The Polymer Tensiometer

Polymer tensiometers are a true advancement in the field measurement of Ψ_m because they operate in the full agricultural range of the matric potential. They are devices developed by Bakker et al. (2007) at the University of Wageningen, in the Netherlands, with the aim of circumventing problems encountered in the use of common tensiometers. This tensiometer is basically composed of a solid ceramic disk, a stainless steel cover where a polymer is placed (the polymer has a great expansion capacity when receiving

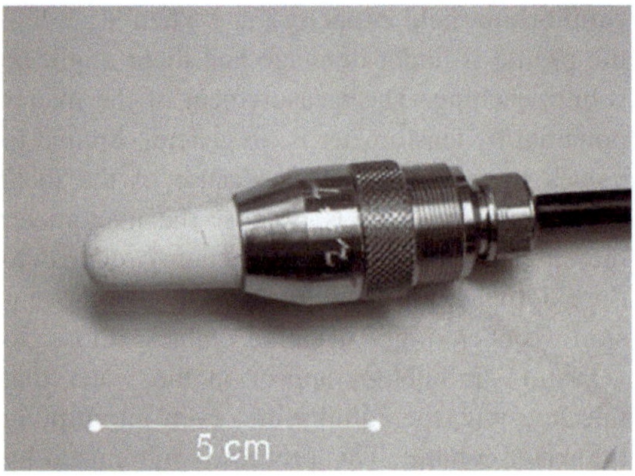

Fig. 6.26 The probe of a polymer tensiometer

Fig. 6.27 Sensor (Watermark®) for the determination of soil water potential based on electrical resistance

water and retraction when losing water) and a pressure transducer (to measure the variations pressure due to the water inlets and outlets through the ceramic disc) with a temperature sensor (Fig. 6.26). The sensors are connected to an individual datalogger and the data is collected continuously in short intervals, say every 15 min. More details can be found in Bakker et al. (2007).

These tensiometers filled with polymers are able to measure a much greater range of water potentials in the soil, being only less accurate near the saturation. According to Durigon et al. (2011) and Durigon and De Jong van Lier (2011), this new tensiometer has the capacity to measure the tension of water retained in the soil (Ψ_m) from near saturation to the point where $\Psi_m = -200$ mH_2O, surpassing the permanent wilting point (around -150 mH_2O). In this way, this equipment comes as an efficient alternative for carrying out measurements of Ψ_m in the field. One negative point is their cost.

6.10.4 Electric Resistance Sensors

Recently, the category of soil resistance reading sensors (Fig. 6.27) has also been used to monitor soil water potential. Resistance values are converted into Ψ_m readings by means of a calibration equation inserted in a digital meter (datalogger) that enables the storage of individual data. Its operating range is 0–200 centibars (200 kPa) and are commercially known as Watermark®.

6.10.5 Richard's Pressure Membrane

The Richards membrane or pressure plate apparatus is outlined in Fig. 6.28.

This apparatus consists, in synthesis, of a pressure chamber connected to the atmosphere by means of a plate (or cellophane membrane), on which the soil sample is placed. The soil water is extracted by applying pressure to the chamber. It is similar to the porous plate funnel seen above, but in which soil water is extracted by suction. The instrumental arrangement is such that the underside of the plate is continuously under atmospheric pressure $P_{atm} = 0$. In this equipment, the soil water is withdrawn by pressure, which is an advantage as high pressure values can be reached, in the case 15 atm (1.5 MPa) or even more.

The sample is placed on the plate which is already saturated and the soil is saturated with water for a period of 24 h. Then, a pressure P (0.1–2 atm for one type of instrument and 1–20 atm for another) is applied to the chamber. By virtue of the applied pressure, the water is pushed out of the soil until equilibrium is established and, under such conditions, the soil will have a water content θ retained at a matrix potential Ψ_m. In the equilibrium condition:

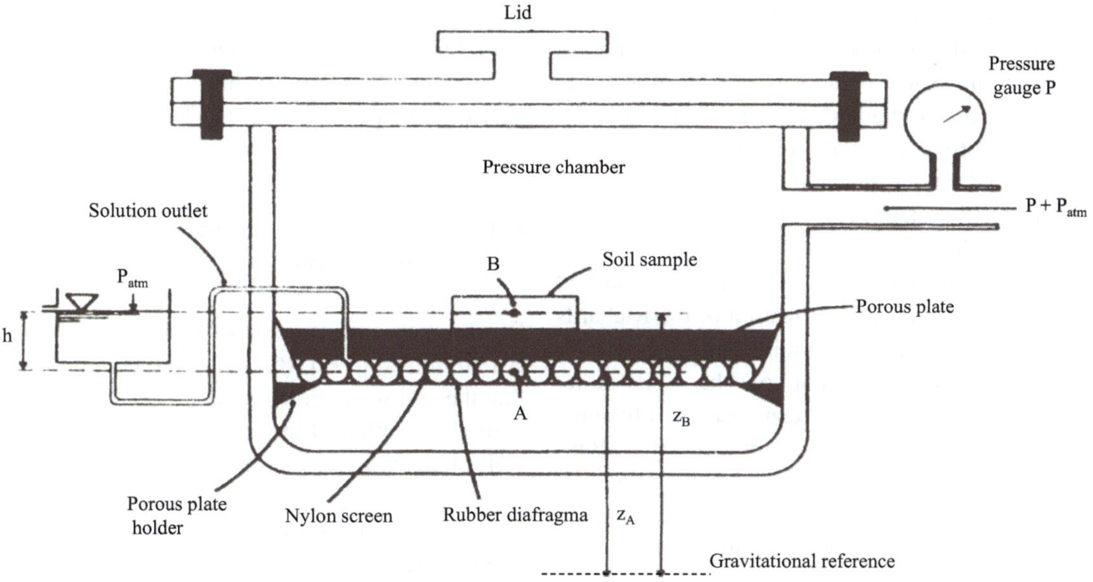

Fig. 6.28 Schematic view of a Richards pressure plate apparatus [with verbal authorization of Libardi (2012)]

$$\Psi_m\ (\text{soil}) = P$$

and the manometer reading directly provides the soil water matrix potential in that equilibrium condition. The operation is repeated for so many P values required to obtain a good **water retention curve**. As the relation Ψ_m versus θ is exponential, it is important to have more values of P in the wet range and values more spaced in the dry range. An example of choices for P is 0, 0.06, 0.1, 0.33 (1/3 atm which is the classic FC field capacity, see Chaps. 12 and 14), 0.5, 1, 3, 5, and 15 atm (the latter being the classic PWP permanent wilting point, see Chap. 14). A typical example is given in Table 6.4.

Table 6.4 Example of soil water retention data showing the importance of using logarithms (absolute value) when making Ψ_m versus θ graphs

$P_i = \Psi_m$ (atm)	$\log \Psi_m$	θ ($m^3\ m^{-3}$)
0 (saturation)	$-\infty$	0.575
0.06	−1.221	0.551
0.1	−1.000	0.530
0.33 (FC)	−0.418	0.415
0.5	−0.301	0.399
1.0	0	0.380
3.0	0.477	0.362
5.0	0.699	0.355
15.0 (PWP)	1.176	0.351

As can be seen, while θ ranges from 0.575 (saturation) to 0.351 $m^3\ m^{-3}$, Ψ_m varies exponentially from 0 to 15 atm. Therefore, when displaying the data in a graph, we use log $|\Psi_m|$ or ln$|\Psi_m|$. Because $\log 0 = \ln 0 = -\infty$, this point does not enter the graph.

The limitation of these instruments is on the porosity of the plate or the porous membrane. In the used pressure range P, the membrane (when wet) must be impermeable to air and permeable to water. It is impermeable to air because the pressure P is not enough to eliminate water from its capillaries. There are two types of apparatus: the so-called pressure cooker, for pressures $0 < P < 2$ atm, and another, with a high microporosity plate, called a Richards plate or membrane, for pressures $1 < P < 20$ atm.

Another equipment that has recently been used to determine soil water potential is the so-called HYPROP®, making measurements during soil evaporation, developed by the German company UMS. Nowadays, UMS and Decagon merged to the Meter Group (https://www.metergroup.com). In a summarized way, it consists in evaluating the soil mass and the matrix potential in soil samples during the drying process caused by the evaporation of the water contained in a sample. The principle of the method was first formulated by Wind (1966), and its simplification was proposed by Schindler (1980), which is implemented in the HYPROP® system. The system also allows the determination of soil water flux density by using two mini-tensiometers at two depths of the soil sample as long as the application of the Richards equation (see Chap. 7) is valid. Thus, the hydraulic conductivity function of the soil under non-saturated conditions can also be obtained. The instrument includes a software (HYPROP-DES®) that provides the user with the choice of seven models of soil water retention curves and four functions of hydraulic conductivity.

6.10.6 Psychrometer for Air Water Potential

Psychrometers shortly described in Chap. 5, which operate according to Eqs. (5.4), (5.5), and (5.10), are schematically shown in Fig. 6.29. They make quick measurements of the relative humidity (RH) and through the application of Eq. (6.7) the potential of the water vapor in the air can be calculated (see also Table 6.1).

6.10.7 Psychrometer for Soil Matric Potential

Psychrometric measurements of soil water potential are linked to Eq. (6.7). This equation tells us that the soil water potential is proportional to the natural logarithm of the air relative soil humidity RH (obtained by psychrometric techniques). The idea of measuring soil water potential through its relative humidity is not recent. The main difficulty of measuring is technical. RH of the soil air, when saturated at 20 °C, is 100% and its potential is zero. For a soil in which the water has a matrix potential of −15 atm (−1.5 MPa), which corresponds to the permanent wilting point (PWP), the relative humidity is still very high, 98.88% at 20 °C. From this, it becomes clear that, from the agronomic point of view, the useful range of RH is between 99% and 100%. Hence,

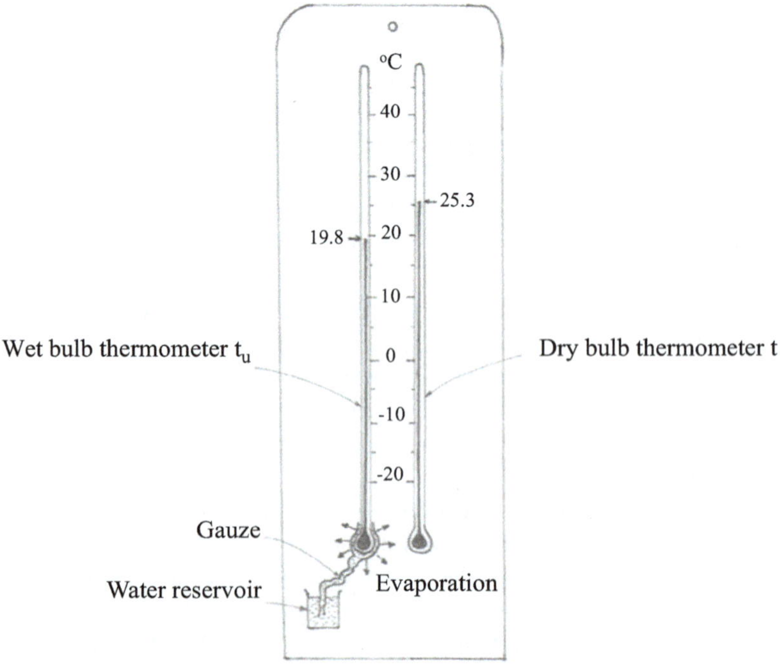

Fig. 6.29 Psychrometer for air relative humidity measurement. The wet bulb thermometer has a gauze that conducts water from the reservoir to the wet bulb. The temperatures t_u = 19.8 °C and t = 25.3 °C are those of the example of Eq. (5.10) in Chap. 5

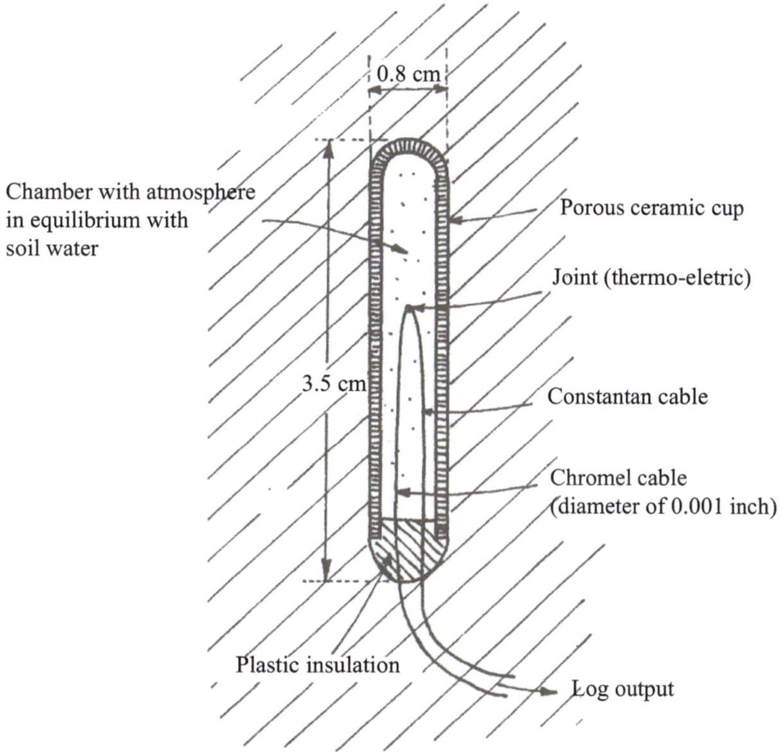

Fig. 6.30 Psychrometer for soil matric potential

the technical difficulties of measuring such tiny differences in RH. The theory of such methods was approached by W. H. Gardner and his team (Rawlins 1966; Campbell and Gardner 1971; Wiebe et al. 1971), and there are now in commerce instruments based on psychrometry.

A **soil psychrometer** is shown schematically in Fig. 6.30. The operating principle is the same as the psychrometers used for relative atmospheric humidity measurement described in Chap. 5. The psychrometric depression ($t - t_u$) is measured with a thermocouple and the vapor pressure is calculated by Eq. (5.10).

Due to the practical limitation of the use of water/mercury tensiometers previously mentioned, HYPROP® has been used in conjunction with the WP4® equipment (https://www.metergroup.com/environment/articles/create-full-moisture-release-curve-using-wp4c-hyprop/) whose principle is also based on the equilibrium of the water in the liquid phase of a soil sample with the water in the vapor phase in the airspace above the soil sample in a sealed chamber. Within this chamber, the dew point temperature of the fresh air is measured in a mirror cooled by means of a thermal sensor, and the temperature of the sample is measured by an infrared thermometer.

6.10.8 Measurement of Soil Bulk Density and Water Content

Soil water content has already been defined in Chap. 3 by Eqs. (3.14) and (3.15), which represent, respectively, mass-based water content u and soil water content based on volume θ. It is worth recalling at this point that $\theta = u \cdot d_s$, where d_s is the soil bulk density, defined by Eq. (3.11).

6.10.9 Soil Bulk Density d_s

The soil bulk density can be determined by any process that allows us to determine the mass m_s of the material contained in a volume V of soil. Two methods are the most common (Fig. 6.31). The

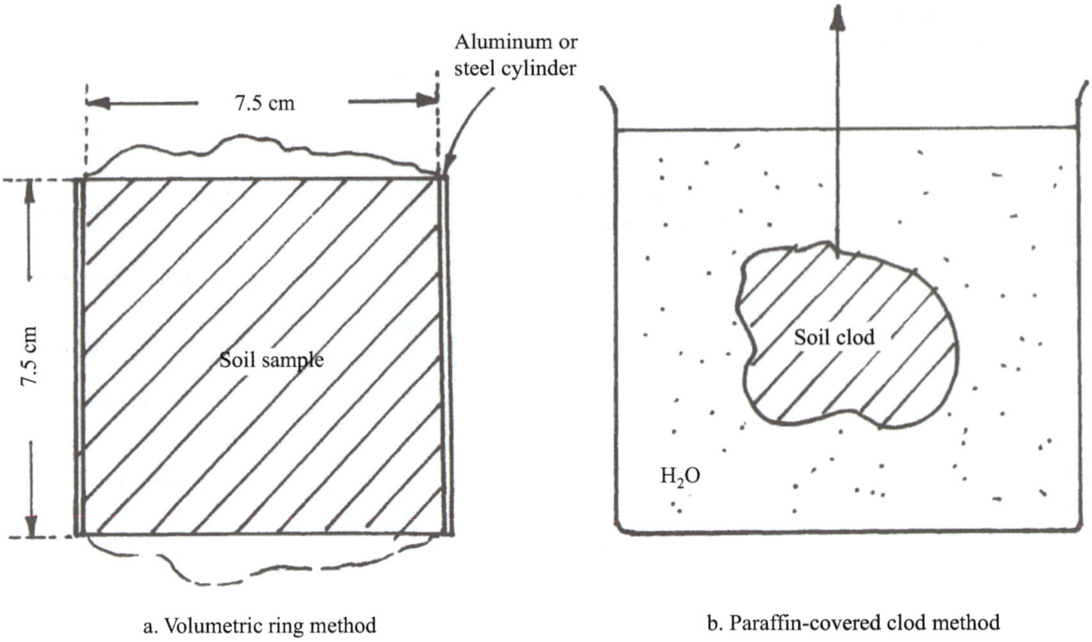

Fig. 6.31 Illustration of a soil bulk density determination: on the left the volumetric ring and on the right the paraffin clod

first, called the **volumetric ring method** (a), consists of the introduction of a cylinder of volume V into the soil (Uhland cylinder); after being removed from the soil, the excess soil is cut at the ends in order to make sure that the soil occupies only the volume V; the sample is brought to an oven at 105 °C for 48 h to eliminate the water or until constant weight, and m_s is determined and d_s calculated. The diameters and heights of the most used rings vary between 3 and 10 cm. The second, **paraffin-covered clod method** (b), consists of collecting soil clods of different sizes (dry mass of 50–200 g), which are air-dried. The clods are then immersed in liquid paraffin so that they are covered by an impermeable layer, and the volume V of the clods is determined by their buoyancy when immersed in water. The volume of the paraffin is generally not negligible and should be determined by its weight and density. In Chap. 3, a numerical example is given. Residual moisture needs to be taken into account. To avoid this, and if the soil allows, it is best to use ovendried clods. Details of these methodologies can also be found in Blake and Hartge (1986).

Other more sophisticated methods of determining soil bulk density are based on the principle of interaction of a gamma radiation beam with matter. As sources of gamma radiation, sources of ^{60}Co, ^{137}Cs, and ^{241}Am have been used, with activities ranging from a few mCi to 300 mCi (1 mCi = 3.7×10^7 Bq, and 1 Bq = 1 disintegration/s). Two principles are used in these measurements: the absorption and scattering of gamma radiation by matter. If we introduce into the soil a set, such as that shown in Fig. 6.32, which is a depth-probe for the determination of soil bulk density, the gamma radiation emitted by the source cannot cross the lead barrier directly and reach the radiation detector. On the other hand, the radiation penetrating the soil is scattered (spread) by the Compton effect in all directions, and the number of scattered radiation reaching the detector is proportional to the soil bulk density (including water) surrounding the probe. An apparatus of this nature must be calibrated (empirically) for each type of soil.

There are also gamma-neutron surface probes that are used to measure soil density and water

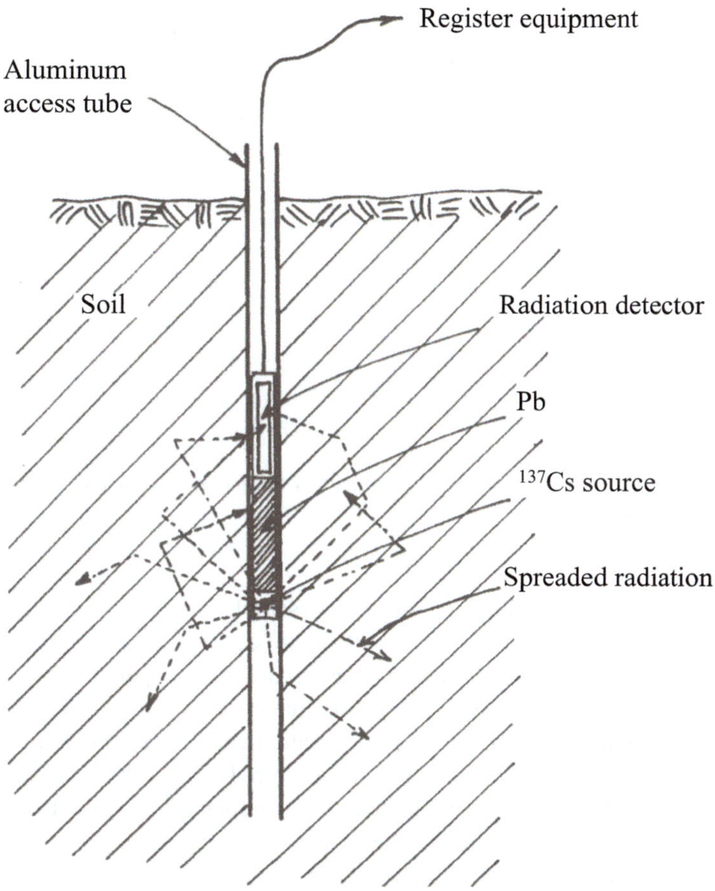

Fig. 6.32 Schematic view of a gamma ray density probe introduced in a soil through an aluminum access tube

content for surface layers down to about the 30 cm depth. Cássaro et al. (2000) used these probes to diagnose compacted soil layers in the range $z = 0$ to $z = 30$ cm. Tominaga et al. (2002) studied changes of soil water content in the 0–15 cm layer, in a sugarcane crop with straw burning after harvest and straw left on the soil surface, also after harvest. The same surface probe was used by Dourado-Neto et al. (1999) to study the relationship between soil water content and temperature in the same sugarcane experiment. Timm et al. (2006) evaluated the structure of spatial and temporal variability of soil density and water content data along a 200 m transect in a coffee plantation, using the same surface probe.

In the laboratory, a collimated beam of gamma radiation can be produced, as shown in Fig. 6.33. By collimated beam we understand a radiation beam of almost parallel rays, produced when a radiation source is protected in an appropriate lead shield with a hole that allows radiation to escape. In this case, the absorption of the beam of intensity I_0 (before crossing the soil) is proportional to the density of the soil d_s, the water content θ, and the thickness of the soil sample x. The principle governing the absorption or attenuation of radiation is **Beer's law** given by Eq. (6.22):

$$I = I_0 \cdot \exp - [(\mu_s d_s + \mu_a \theta)x] \qquad (6.22)$$

where I is the intensity of the emerging beam and μ_s is a coefficient called the **mass absorption coefficient** of the soil and μ_a of the water. For dry soil and for radiations in the 661 keV energy range (^{137}Cs), the value of μ_s is practically independent of the soil type, being about 0.07 cm^2 g^{-1}.

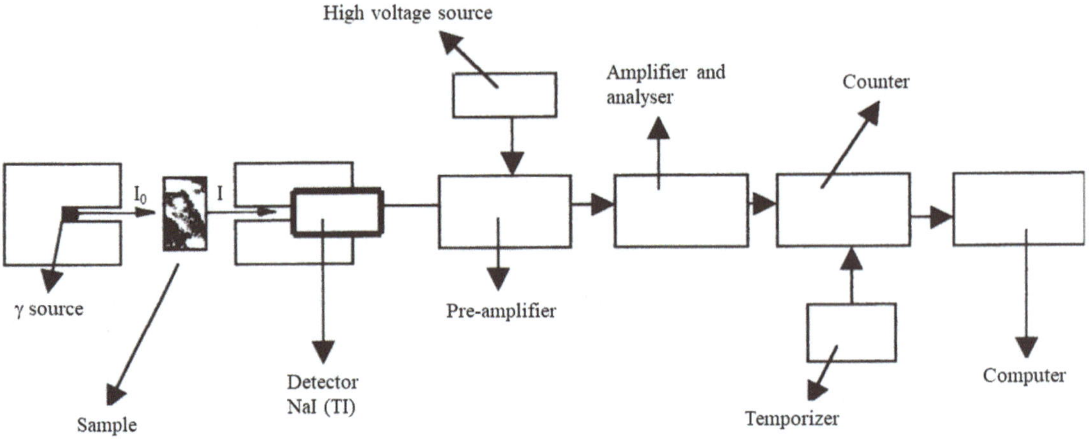

Fig. 6.33 The layout of the instrumentation for gamma ray attenuation measurement of soil bulk density and water content

To make a measure of d_s, a sample of ovendried soil at 105 °C ($\theta = 0$) is crossed by the beam, whose thickness x (traversed by the beam) must be known. A measurement of I_0 without the soil sample and a measurement of I with the soil sample are made. Applying Eq. (6.22), d_s can be calculated. Measurements of I are obtained in a few minutes. The technique as described here seems rather simple, but a number of difficulties arise as to the precision, sample thickness, radiation energy, and so on. Detailed discussion of the method can be found in Davidson et al. (1963), Reichardt (1965), Gardner and Calissendorff (1967), Gardner et al. (1972) and Ferraz (1983).

Soil compaction is related to high values of soil bulk density and low total porosity, consequently to a difficulty of root penetration and also of agricultural implements. Another way to evaluate the compaction is by means of **penetrometers**, devices to determine the resistance of the medium in which roots penetrate. Its agricultural use refers more to the determination of the compaction of the soil and the evaluation of the thickness and depth of compacted layers. Penetrometers are composed of a metal rod with conical end, which is introduced into the soil by continuous movement or by impacts. In conventional penetrometers, the resistance, while the tip advances in the soil, can be read or recorded, by means of a dynamometer. In this case, the apparatus being nominated as a penetrograph. Another type is the impact penetrometer, in which the dynamometer and the recorder are replaced by a constant stroke weight, which causes the rod to penetrate the ground by means of impacts. The resistance of the cone is measured in terms of pressure, in kgf cm^{-2}, that is, the ratio of the resistance force (measured by the dynamometer in kgf) and the area of the cone base (cm^2) that comes into contact with the soil. In the impact penetrometer, the resistance of the terrain is measured in terms of the number of impacts/10 cm of depth. Stolf (1992) states that a major constraint in relation to the technique has been the lack of information on the transformation of the practical unit N (impacts/10 cm) into the theoretical unit R (kgf cm^{-2}), and author suggests a relationship:

$$R = 5.6 + 6.89 \times N$$

Using this type of penetrometer, Stolf et al. (1998) studied the mechanical impedance in a clayey soil with gravel.

In the penetrograph, resistance variations are automatically recorded in relation to depth, in a standard abacus, allowing the farmer to have, in a few seconds, a graphical visualization of the various degrees of soil compaction. By means of the curve obtained in the abacus, one can recommend the most suitable implement for the rupture of compacted beds.

Vaz and Hopmans (2001) combined a penetrometer with time domain reflectometry (TDR), allowing simultaneous measurement of resistance to soil penetration and soil water content.

The compaction evaluation is also done through soil bulk density and macroporosity (Ma) assessments. Ma is a soil attribute used for studies of soil compaction degree (Stolf et al. 2011). Low values of Ma may result in poor drainage, low aeration, and increased soil resistance to penetration.

According to Hakansson and Lipiec (2000), many studies indicate the value of 10% (0.10 $m^3 m^{-3}$) as a critical limit for soil aeration, and these authors hypothesized that the value of soil relative bulk density (d_r) of 0.87 corresponds to Ma = 10%. The d_r is the relation between the density of a given soil d_s and its maximum density (d_{smax}), defined as the soil density when Ma = 0; hence, its maximum value is 1. The d_{smax} is measured by the compression of (200 kPa) in a cylinder designated **Proctor cylinder** (Hakansson 1990). Using samples collected in the field, Stolf et al. (2011) obtained models for the estimation of Ma and Mi through the sand content and the soil bulk density, which greatly simplifies these measures. Their models were compared with data from the literature and showed high precision. More about this subject you can find in Chap. 17 of this book.

In a book by the Brazilian Society of Soil Science (SBCS), Silva et al. (2010) carried out a review on indicators of **soil quality**, in which compaction receives special attention. Silva et al. (2014) present a current text on the different indicators to evaluate the physical quality soil for the growth of the crops, from the simplest to the most complex to be determined.

6.10.10 Soil Water Content (u and θ)

The traditional method (gravimetric) for soil water content is that on weight basis. A sample is collected, its wet mass m_u is determined, and thereafter it is placed in a ventilated oven at 105 °C until constant weight, to determine the dry mass m_d, and the calculations are made according to the equations seen in Chap. 3:

$$u = \frac{(m_u - m_d)}{m_d} \times 100$$

$$\theta = \left[\frac{(m_u - m_d)}{m_d} \times d_s\right] \times 100$$

For the u measurement, the sample can be deformed, like those collected with augers. For the determination of θ, the sample has to have its natural structure, and therefore the most common procedure is using the volumetric cylinders, like those used to determine d_s. Reichardt (1987) provides more details on these procedures, also discussing ways of sampling, number of replicates, etc.

Another way to determine soil water content is by instruments whose electrical resistance varies with soil water content. The electrical resistance of a soil volume element depends not only on its water content but also on its composition, texture, and concentration of salts in the soil solution. On the other hand, the electrical resistance of a porous body placed into a soil and in balance with the soil water can often be calibrated as a function of soil water content. These instruments, called **electrical resistance blocks**, contain a pair of electrodes inside a block of gypsum or nylon (or fiberglass). Already in the 1940s, Colman and Hendrix (1949) presented an electric instrument for measuring soil moisture that was constituted of fiberglass element.

When inserted into the soil, these instruments tend to come into equilibrium and, under these conditions, the soil water potential is equal to the potential of the solution ($CaSO_4$, in the case of gypsum) inside the block. For each equilibrium condition, which corresponds to a value of Ψ_m or of θ of the soil, also corresponds a value of R (electric resistance between the electrodes). Then, for a given soil, we can correlate R with θ or Ψ_m. The calibration curve can thus be established. Example of this type of instrument was previously presented in Fig. 6.27.

The main problems of the electric resistance blocks are:

(a) They are affected by the hysteresis.

(b) Contact between resistance block and soil.

(c) Variation of the hydraulic properties of the resistance block with time.

(d) Blocks made of inert material, such as fiberglass, are highly sensitive to small variations in the saline concentration of the soil solution [for plaster (gypsum) blocks, this does not happen because the solution inside the block has a constant concentration and practically equal to of a saturated solution of calcium sulfate].

(e) And gypsum blocks deteriorate with time given their solubility.

As a result of these factors, the determination of θ with blocks has limitations. As long as all care is taken, they are instruments that can perfectly be utilized. Their working range extends to much drier soils, in which tensiometers cease to function. The main advantage is that they can be connected to dataloggers, enabling continuous readings in the field.

Since the 1960s, the method of **neutron moderation** by the use of **neutron probes** (see Fig. 6.34) has been successfully applied in determining soil water contents in the field. When a source of fast neutrons (energy around 2 MeV) is introduced into the soil, emitted neutrons penetrate radially into the soil, where they encounter several atomic nuclei with which they collide elastically. The loss of neutron energy by collisions is, on average, maximal when colliding with a nucleus of mass close to its own. Such light nuclei are, in particular, those of hydrogen from water. The number of collisions needed to make a fast neutron (2 MeV) to slow down (0.025 eV) can be seen in Table 6.5 for several nuclei of isotopes present in the soil.

Fig. 6.34 View of a neutron probe placed on an aluminum access tube, ready for a measurement

Table 6.5 Number of elastic collisions needed for a fast neutron reduce its energy to slow or moderate neutron

Isotope	Number of collisions
^{1}H	18
^{2}D	25
^{4}He	43
^{7}Li	68
^{12}C	115
^{16}O	152
^{238}U	2172

The neutron moderation process represents the loss of energy by neutrons, going from fast to slow (or moderate). In addition to the process of moderation, there is also the capture process, by which the neutron is captured by an atomic nucleus, processing a nuclear reaction, which results in the formation of a stable or radioactive isotope. The probability of capture is measured by the cross-section coefficient, and this depends on the energy of the neutron and the nucleus bombarded by the neutron. Neutrons are also unstable particles, disintegrating with a half-life of 13 s.

Thus, when a source of fast neutrons is introduced into the soil, neutrons are moderated, captured, or disintegrate. Because of these three processes, the number of slow neutrons around

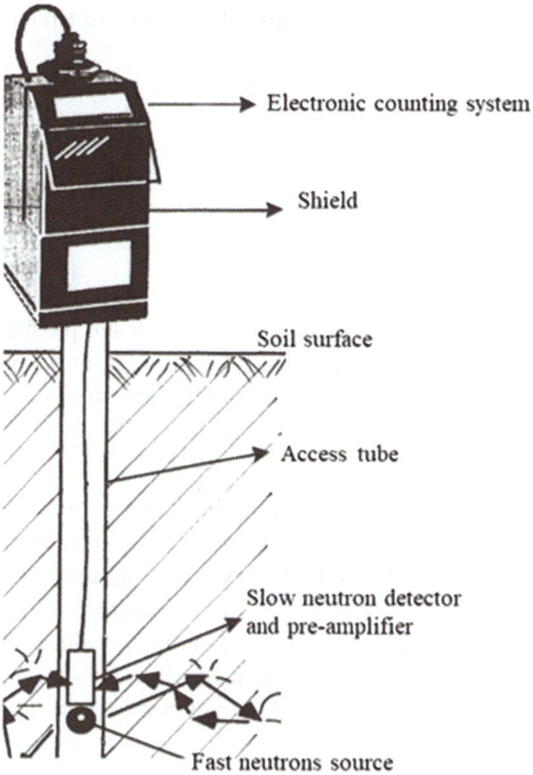

Fig. 6.35 Depth neutron probe diagram at working position

the source quickly reaches equilibrium. In practice, it was found that the number of slow neutrons present around the source is proportional to the concentration of hydrogen in the soil. These slow neutrons diffuse at random in the soil, forming a "cloud" around the source. If a specific detector for slow neutrons is placed in this cloud, a count can be made. These specific counters for slow neutrons are boron trifluoride (BF_3) counters and Li crystal counters. A neutron moderation instrument scheme is shown in Fig. 6.35.

These instruments can then be calibrated to measure soil water content θ. Figure 6.36 shows typical calibration curves for different soil bulk densities.

Details of the art in neutron moderation can be found in Gardner and Kirkham (1952), van Bavel et al. (1956), IAEA (1976), Greacen (1982), Bacchi et al. (2002), among others.

In general, the main problems of the neutron moderation technique are:

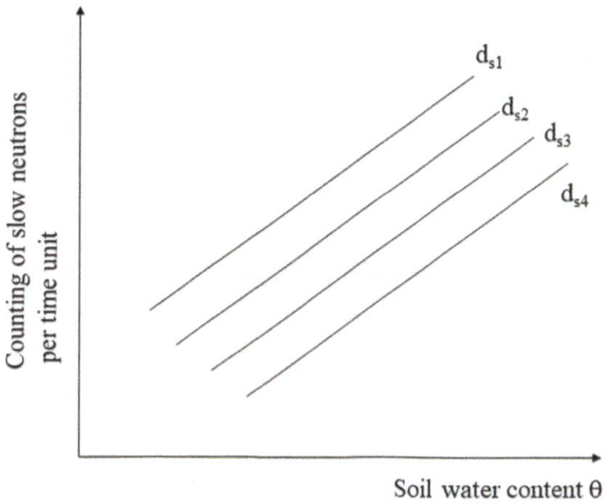

Fig. 6.36 Typical calibration soil water content (θ) curves for different soil bulk densities

(a) The neutron cloud (sphere of radius R) that represents the size of the analyzed sample varies with the soil water content (R equal to approximately 10 cm for moist soils, increasing to 40 cm for extremely dry soils).

(b) The equipment cannot be used on or near the surface by virtue of its radius of action (there are corrections that can be made and even surface instruments).

(c) In general, it can be said that in order to obtain absolute values of θ, the instruments can bring to great errors. This is due to the difficulty of calibration. On the other hand, variations of θ that occur in a profile due to evaporation, drainage, etc. can be measured with optimum accuracy. This happens because the calibration curves, although varying with d_s, are straight lines of the same slope, so that their slope d(cpm)/dθ is constant; therefore, water content variations can be measured even if the absolute value of θ is not known.

(d) And difficult calibration. Reichardt et al. (1997) discuss the problem of calibration of neutron probes in relation to spatial variability of soils, and Carneiro and De Jong (1985) present an interesting method of calibration under field conditions. Figure 6.37 illustrates the soil sampling operation for the calibration of the neutron probe in the field.

The great advantage of the technique is the possibility of measuring changes of θ with time

Fig. 6.37 Field collection of soil samples using volumetric rings to obtain data for calibration curves

and depth in the field. Once the access tubes are installed, periodic measurements can be made without disturbing the system (crop, clean field, forest, etc.). This advantage is extremely important for studies of spatial variability. Kirda and Reichardt (1992) compare neutron moderation with classical soil moisture measurement techniques.

Another technique for measuring soil water contents θ is based on the **attenuation of gamma radiation**, already described by Eq. (6.22).

As can be seen from the above, for wet soil determinations, we have an equation with two unknowns: d_s and θ. The situation is solved in three ways:

1. d_s is initially measured in the dry soil and, once d_s is known at each measurement point, any water movement within the soil can be studied by using Eq. (6.22).
2. When one is not interested in the measurement of d_s, I_0 is measured with the column of dry soil and then the absorption measured by I will be the result of only water and Eq. (6.22) becomes:

$$I' = I'_0 \cdot \exp - (x\mu_a\theta) \qquad (6.22a)$$

where I'_0 is the intensity of the beam passing through the dry soil; it is easy to verify, by comparing Eq. (6.22) with (6.22a), that:

$$I'_0 = I_0 \cdot \exp - (x\mu_s d_s)$$

3. If the radiation beam is composed of two gamma ray energies (e.g., ^{137}Cs—0.661 MeV and ^{241}Am—0.060 MeV), two simultaneous equations of type (Eq. 6.22) can be obtained in which μ_s and μ_a have different values for each energy. In this way, it is possible to measure d_s and θ simultaneously in any situation.

This technique of measuring soil water content by attenuation of gamma radiation was studied in detail and applied by several researchers. References to the subject are found in Davidson et al. (1963), Reichardt (1965), Gardner and Calissendorff (1967), Gardner et al. (1972), Jensen and Somer (1967), and Ferraz (1983).

The major problem with this technique is the exact measurement of the thickness of soil x traversed by the beam of radiation. Small variations in x lead to large errors in d_s and θ. The most recent advance in this area is the introduction of **computed soil tomography** in soil science, which virtually eliminated the need to measure the thickness x.

Tomography can be used for soil samples of any shape, even a clod of soil, and the result is two-dimensional. In a tomographic measurement, the sample moves in relation to a fixed beam of radiation and a series of I/I_0 measurements is taken, without the concern of knowing x. Several parallel measurements (with a certain spacing or linear pitch of the order of millimeters) are taken along a plane or "cut" of the sample. Then a rotation of an angle α (angular pitch) is given and another series of measurements is made in the same plane. With several rotations, a "sweep" is made in that plane and the information set is processed by appropriate computer program. For this reason, we speak of computed tomography CT. In first-generation tomographs, those used in soils, this scanning takes hours and, therefore, obtaining a tomographic cut is still time-consuming. In medicine, third-generation CT scanners provide real-time imaging! The tomographs also need to be calibrated, since the calibration curve relates the CT tomographic units to the attenuation coefficients of a standard material. There is a correspondence between tomographic units (TU) and a scale of grays, ranging from white to black, with increasing the attenuation coefficient or density of the object. In the image, density and/or water content information is obtained for small areas, called **pixels**, each with its tone of gray. The CNPDIA of Embrapa, São Carlos, Brazil, built a minitomograph with a pixel of 131 μm. The cost of tomographs is still high, but price tends to decrease, so that real-time measurements can be made for water dynamics studies. By way of illustration, Fig. 6.38 shows a tomographic section in a soil sample.

One of the worldwide pioneering works on soil tomography is that of Crestana et al. (1985). Vaz et al. (1989), Macedo et al. (2000), and Pires et al. (2002) are examples that used tomography to study soil compaction. The 25 years of soil tomography in Brazil are presented in Pires et al. (2010). Vaz et al. (2014) present new opportunities for the application of the X-ray micro-tomography (advanced benchtop X-ray MicroCT) technique in the field of soil science with emphasis on research in the unsaturated zone of the soil profile (vadose zone).

Time domain reflectometry (TDR) is a technique used to measure soil water content θ based on the effect of θ on the microwave propagation velocity v (electromagnetic waves of frequency in the 50 MHz to 10 GHz range) in conductive (metallic) cables (rods) that are introduced into the soil in the region where it is desired to measure θ. The velocity v depends on the medium that surrounds the rod, that is, its permissivity (or dielectric constant) k, which depends on the proportion of solid matter ($k_s \cong 3$), water ($k_{water} = 80$), and air ($k_{air} = 1$). This large difference between water and other soil components allows a (nonlinear) relationship between k and θ, called the calibration curve. The instrument

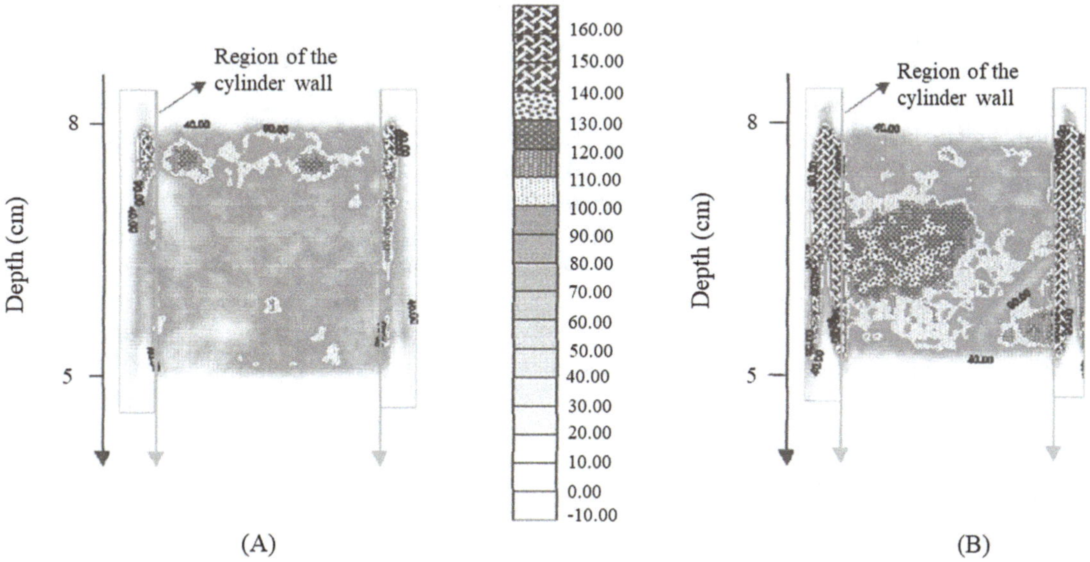

Fig. 6.38 View of soil tomographies obtained by Pires et al. (2002). Today's resolution of such images improved a lot

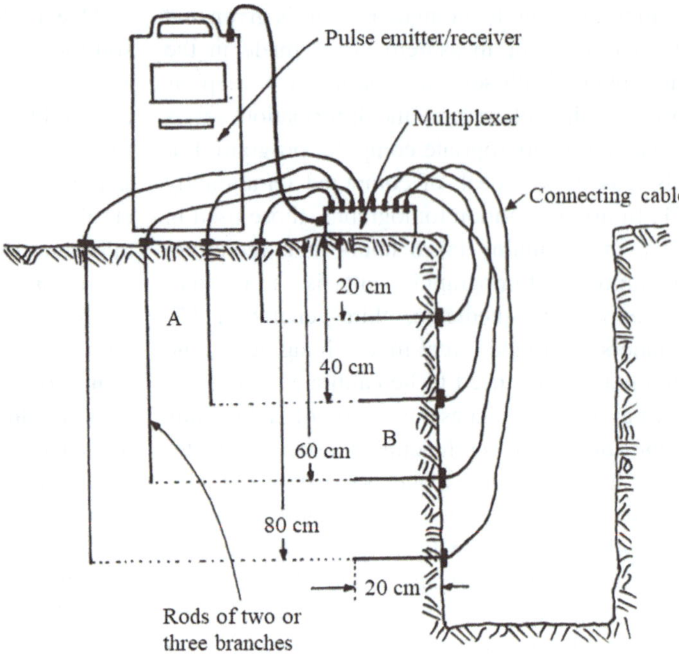

Fig. 6.39 Schematic view of a TDR apparatus with several rods installed in the field

sends a microwave pulse through the rod of length L, which is reflected (traversing, thus, a distance $2L$) and is detected. The time t of propagation in the rod is proportional to the permissivity of soil k, which, in turn, depends on θ, hence the name time domain reflectometry. As shown in Fig. 6.39, the rods are inserted into the soil (at the desired locations and depths), and instantaneous measurements are made by closing the circuit with the wave emitter/receiver apparatus. Rods of 20, 40, 60, 80, and 100 cm are available and the measurements of θ refers to corresponding soil layers, i.e., 0–20, 0–40, 0–60 cm.... The water content of an intermediate layer, such as 40–60 cm, is obtained by difference. Multiple rods were also constructed for simultaneous measurements at various depths (Serrarens et al. 2000). The method is still expensive because of the price of the emitter/pulse detector, but the prospects of cheapening are predicted for the next 10 years. Important works on the technique are those of Topp et al. (1980, 1982), Topp and Davis (1985), and Vaz and Hopmans (2001).

The **frequency domain reflectometry** (FDR) technique is also used to measure soil moisture θ, based on the soil dielectric constant change response. It is often confused with the TDR technique, since both measure the dielectric constant of the soil. However, it is based on the charge time of a capacitor, and this is a function of the dielectric constant of the surrounding soil.

The probe is composed of a pair of electrodes that function as an electronic capacitor. Once activated, the dielectric medium of the capacitor is formed through the soil-water-air matrix around the FDR probe access tube. The capacitor is connected to an LC oscillator circuit (L = inductor; C = capacitor), in which the frequency changes of the circuit depend on the capacitance changes in the ground matrix. The capacitance rises as the number of free water molecules increases, and their dipoles respond to the dielectric field created by the capacitor. For this to happen, the area of the electrodes and the distance between them should be fixed on the probe (Paltineanu and Starr 1997; Sentek 2001).

The FDR technique offers some advantages over other methods, such as obtaining a large number of soil moisture measurements continuously and without damaging soil properties, speed of data collection, equipment is easy to transport, has no radioactivity, and low cost compared to other equipment. However, it also requires calibration.

Like most equipment, it also has a factory calibration equation, but Paltineanu and Starr (1997) emphasize the need for local calibrations, which can improve the accuracy of equipment, even if they are labor-intensive and costly.

In order to obtain the calibration curves of the FDR equipment, which relate the values of relative frequency (FR) and soil water content θ, it is necessary to know the frequency in both liquid medium (Fw), which is water, and air (Fa). From the values of these frequencies, FR is calculated by:

$$\mathrm{FR} = \left(\frac{\mathrm{Fa} - \mathrm{Fs}}{\mathrm{Fa} - \mathrm{Fw}}\right) \qquad (6.23)$$

where Fs is the frequency reading in the soil with the probe inserted inside the access tube installed in the soil.

From the calculation of the relative frequency (FR), the water content values are calculated by:

$$\theta = a \times \mathrm{FR}^b \qquad (6.24)$$

where:

θ = soil water content based on volume ($m^3\ m^{-3}$), a and b = coefficients of the equation (dimensionless), and FR = relative frequency calculated by the Eq. (6.23). An example of this technique is Hu et al. (2008), who used it successfully in the measurement of soil surface water content.

6.11 Exercises

6.1. What is meant by energy? What are its dimensions?

6.2. Transform potentials of −0.1, −0.33, −1, and −15 atm in cmH_2O and Pascal.

6.3. Given the function $u = x^2y + xz + at$, determine $\partial u/\partial x$, $\partial u/\partial y$, $\partial u/\partial z$, and $\partial u/\partial t$.

6.4. In the previous problem, what is the value of $\partial u/\partial x$ when $x = 1$, $y = 1$, $z = 2$, and $t = 3$?

6.5. In a soil the water storage S (mm) varies linearly in-depth z (cm) and exponentially in time t (days), according to the equation:

$S = (3z + 15)e^{-0.1t}$,in the range $0 < z < 50$ and $0 < t < 5$.

(a) Determine $\partial S/\partial z$ and $\partial S/\partial t$.

(b) For $z = 30$ cm and $t = 1$ day, what is the value of $\partial S/\partial t$?

6.6. What is a point and line function?

6.7. What is the potential pressure at a depth of 1.5 m in a pool?

6.8. In a flooded rice crop, the water depth above the soil surface is 15 cm. What is the pressure potential at a point in the soil, 15 cm below the surface of the soil?

6.9. To measure the depth of the water table in a flooded soil, a piezometer (5 cm diameter PVC pipe with lateral perforations) was installed, and the groundwater table was found to be 80 cm below the surface of the soil. What is the water pressure potential at a point 1.5 m from the soil surface?

6.10. Determine the gravitational potential of water at three points A, B, and C located 30, 60, and 120 cm, respectively, below the surface of an unsaturated soil. Give the results in cmH_2O and atm.

6.11. Determine the gravitational potential of the points in Exercises 6.8 and 6.9.

6.12. Considering T = 27 °C, calculate the osmotic potential of 10^{-6}, 10^{-5}, 10^{-4}, 10^{-3}, and 10^{-2} M concentration solutions. Give the responses in atm and cmH_2O.

6.13. The solution of a given soil is 1.5×10^{-3} M. What is its osmotic potential at 27 °C?

6.14. The cell juice of a plant has a concentration of 0.31 M. What is its osmotic potential at 27 °C?

6.15. A glass capillary tube has a radius of 0.02 mm. Until what height does the water rise in this capillary, knowing that the contact angle is 11° and that the water is at 30 °C?

6.16. How do you fill a glass capillary tube with water?

6.17. The capillary tube of Exercise 6.15 is completely immersed in the bottom of a tank at a depth of 1 m. Is there a capillary phenomenon? What is the potential of the water inside it?

6.18. The capillary tube identical to Exercise 6.15 has a length of only 20 cm. If it were longer, the water would rise to 70.9 cm. What happens in this shorter tube?

6.19. For a given soil, the following water retention curve was obtained:

Matric potential (atm)	Soil water content ($cm^3\ cm^{-3}$)
0	0.541
−0.1	0.502
−0.3	0.546
−0.5	0.363
−1.0	0.297
−3.0	0.270
−5.0	0.248
−10.0	0.233
−15.0	0.215

Try to make the graphs of the retention curve, first using the values as shown in the table and then taking the log or ln of the modulus of the matric potential values.

6.20. Explain, by looking at Fig. 6.8, why water retention curves differ for soils of different textures and structures.

6.21. A mercury manometer tensiometer is installed in the soil at a depth of 20 cm. Its reading is 37.3 cmHg, and the free mercury level is 40 cm above the soil surface. What is the soil matric potential at this depth?

6.22. Considering that the curve of Exercise 6.21 is equal to that of the soil of Exercise 6.19, what is the soil water content at the point where the tensiometer is installed?

6.23. After 3 days, the tensiometer of Exercise 6.21 has a reading of 51.1 cmHg. What is the new soil water content?

6.24. For two soils, the following water retention data were obtained:

	θ ($cm^3\ cm^{-3}$)	
$-h$ (cmH_2O)	Soil A	Soil B
0	0.556	0.491
10	0.540	0.398
100	0.430	0.257
300	0.403	0.236
500	0.391	0.227
1000	0.382	0.209
3000	0.375	0.198
10,000	0.359	0.195
15,000	0.343	0.191

(a) Make the retention curves of both, on the same paper, using plain and semi-log paper.

(b) Which is the sandier soil?

(c) What is the water content of the soils to a potential of −0.7 MPa?

(d) A tensiometer installed in soil A provides a reading of 28.8 cmHg. The mercury free level is 33 cm from the soil and porous cup is at a depth of 25 cm. What is the soil water content of the soil?

6.25. A soil sample with a dry mass of 105.6 g was placed on a porous plate funnel. The sample was saturated ($h = 0$), and then the potential h was varied according to the following table, waiting for equilibrium for each h.

$h = -\Psi_m$ (cmH_2O)	Mass of the sample (g) = soil + water
0	146.6
50	144.9
100	141.9
150	135.6
200	129.3
250	125.1
300	121.1

The soil density is 1.41 g cm^{-3}. Make the water retention curve of this soil, for the range 0–300 cmH_2O.

6.26. The relative humidity of the air above a maize crop is 74% and the temperature of the same air is 28 °C. What is the potential of water in the air above corn crop?

6.27. For the cases given in the table below, plot the total potential of the water in the soil and indicate the equilibrium and flow regions (indicating the direction), as in Fig. 6.20.

Depth	Total potential H (cmH_2O)					
z (cm)	A	B	C	D	E	F
0	−150	−350	−60	−60	−1500	−700
20	−130	−300	−70	−82	−600	−550
40	−115	−260	−95	−110	−300	−480
60	−107	−230	−125	−118	−250	−450
80	−103	−210	−125	−105	−220	−510
100	−100	−190	−125	−93	−210	−580
120	−100	−180	−125	−84	−205	−620
140	−100	−175	−125	−77	−200	−650

6.12 Answers

6.1. ML^2T^{-2}.

6.2. −103.3, −340.9, −1033, and −15,495 cmH_2O; −0.01, −0.033, −0.101, and −1.52 MPa.

6.3. $\partial u/\partial x = 2yx + z$

$\partial u/\partial y = x^2$

$\partial u/\partial z = x$

$\partial u/\partial t = a$.

6.4. $\partial u/\partial x = 4$

6.5. (a) $\partial S/\partial z = 3e^{-0.1t}$

$\partial S/\partial t = -0.1e^{-0.1t}(3z + 15)$

(b) $\partial S/\partial t = -9.5$ mm day^{-1}.

6.6. See in text

6.7. +150 cmH_2O.

6.8. +30 cmH_2O.

6.9. +70 cmH_2O.

6.10. −30, −60, and −120 cmH_2O; −0.029, −0.058, and −0.116 atm, considering the soil surface as reference.

6.11. −15 and −150 cmH_2O, considering soil surface as reference.

6.12. −0.025, −0.25, −2.54, −25.4, and −254 cmH_2O; -24.6×10^{-6}, -24.6×10^{-5}, -24.6×10^{-4}, -24.6×10^{-3}, and -24.6×10^{-2} atm.

6.13. −0.037 atm or −38.2 cmH_2O.

6.14. −7.63 atm or −7877 cmH_2O.

6.15. 70.9 cm.

6.16. Simply put one end of the tube in contact with the water, and it rises spontaneously to the height h, which in the case of exercise 6.15 is 70.9 cm.

6.17. If there is no water-air interface, there is no capillarity. Its potential is +100 cmH_2O.

6.18. The water rises to 20 cm and stops, but with a meniscus less concave than that of problem 6.15.

6.19. See figures, as examples, in text

6.20. See in text

6.21. −410 cmH_2O.

6.22. 0.418 cm^3 cm^{-3}, approximately, depending of the interpolation made.

6.23. 0.350 cm^3 cm^{-3}.

6.24. (b) Soil B is probably sandier than A because it has lower total porosity and curvature has a more pronounced inflection.

(c) 0.363 and 0.205 cm^3 cm^{-3}, respectively.

(d) 0.403 cm^3 cm^{-3}.

6.25. See figure, as examples, in text

6.26. −412.5 atm = −41.2 MPa.

6.27. A: upward flow in the 0–100 cm layer and equilibrium below 100 cm.

B: upward flow in any sampled depth.

C: downward flow in the 0–60 cm layer and equilibrium below 60 cm.

D: descending water from 0 to 60 cm and rising water from 60 to 140 cm.

E: ascending water in the whole profile.

F: ascending water from 0 to 60 cm and descending water from 60 to 140 cm.

References

Angelocci LR (2002) Água na planta e trocas gasosas/energéticas com a atmosfera: introdução ao tratamento biofísico. Angelocci LR, Piracicaba

Bacchi OOS, Reichardt K, Calvache M (2002) Neutron and gamma probes: their use in agronomy. International Atomic Energy Agency, Vienna

Bakker G, van Der Ploeg MJ, de Rroij GH, Hoogendam CW, Gooren HPA, Huiskes C, Koopal LK, Kruidhof H (2007) New polymer tensiometers: measuring matric

pressures down to the wilting point. Vadose Zone J 6:196–202
Blake GR, Hartge KH (1986) Bulk density. In: Klute A (ed) Methods of soil analysis. American Society of Agronomy; Soil Science Society of America, Madison, pp 363–375
Campbell GS, Gardner WH (1971) Psychrometric measurement of soil water potential: temperature and bulk density effect. Soil Sci Soc Am Proc 35:8–12
Carneiro C, De Jong E (1985) In situ determination of the slope of the calibration curve of a neutron probe using a volumetric technique. Soil Sci 139:250–254
Cássaro FAM, Tominaga TT, Bacchi OOS, Reichardt K, Oliveira JCM, Timm LC (2000) The use of a surface gamma-neutron gauge to explore compacted soil layers. Soil Sci 165:665–676
Cassel DK, Klute A (1986) Water potential: tensiometry. In: Klute A (ed) Methods of soil analysis. American Society of Agronomy; Soil Science Society of America, Madison, pp 563–596
Colman EA, Hendrix TM (1949) Fiberglass electrical soil moisture instrument. Soil Sci 67:425–438
Crestana S, Mascarenhas S, Pazzi-Mucelli RS (1985) Static and dynamic three dimensional studies of water in soil using computed tomographic scanning. Soil Sci 140:326–332
Davidson JM, Nielsen DR, Biggar JW (1963) The measurement and description of water flow through Columbia Silt Loam and Hesperia Sandy Loam. Hilgardia 34:601–617
Dourado-Neto D, Nielsen DR, Hopmans JW, Reichardt K, Bacchi OOS (2000) Software to model soil water retention curves (SWRC, version 2.00). Sci Agric 57:191–192
Dourado-Neto D, Timm LC, Oliveira JCM, Reichardt K, Bacchi OOS, Tominaga TT, Cassaro FAM (1999) State-space approach for the analysis of soil water content and temperature in a sugarcane crop. Sci Agric 56:1215–1221
Durigon A, de Jong van Lier Q (2011) Determinação das propriedades hidráulicas do solo utilizando tensiômetros de polímeros em experimentos de evaporação. Rev Bras Cienc Solo 35:1271–1276
Durigon A, Gooren HPA, de Jong van Lier Q, Metselaar K (2011) Measuring hydraulic conductivity to wilting point using polymer tensiometers in an evaporation experiment. Vadose Zone J 10:741–746
Ehlers W, Goss M (2016) Water dynamics in plant production, 2nd edn. CABI, Croydon
Ferraz ESB (1983) Gamma-ray attenuation to measure soil water content and/or bulk densities of porous media. In: IAEA Symposium, Aix-en-Provence, France, pp 449–460
Gardner WH, Calissendorff C (1967) Gamma-ray and neutron attenuation measurement of soil bulk density and water content. In: IAEA and FAO Symposium. Isotope and radiation techniques in soil physics and irrigation studies, Istanbul, pp 101–113
Gardner WH, Campbell GS, Calissendorff C (1972) Systematic and random errors in dual gamma energy soil bulk density and water content measurements. Soil Sci Soc Am Proc 36:393–398
Gardner WR, Kirkham D (1952) Determination of soil moisture by neutron scattering. Soil Sci 73:391–401
Greacen EL (1982) Soil water assessment by the neutron method. CSIRO, Adelaide
Haines WB (1930) Studies of the physical properties of soils: V. The hysteresis effects in capillary properties and the modes of moisture distribution associated. J Agric Sci 20:97–116
Hakansson I (1990) A method for characterizing the state of compactness of the plough layer. Soil Tillage Res 16:105–120
Hakansson I, Lipiec J (2000) A review of the usefulness of relative bulk density values in studies of soil structure and compaction. Soil Tillage Res 53:71–85
Hu W, Shao MA, Wang QJ, Reichardt K (2008) Soil water content variability of the surface layer of a loess plateau hillside in China. Sci Agric 65:277–289
IAEA (1976) Tracer manual on crops and soils. International Atomic Energy Agency, Vienna
Jensen PA, Somer E (1967) Scintillation techniques in soil-moisture and density measurements. In: IAEA and FAO Symposium. Isotope and radiation techniques in soil physics and irrigation studies, Istanbul, pp 31–48
Kirda C, Reichardt K (1992) Comparison of neutron moisture gauges with non-nuclear methods to measure field soil water status. Sci Agric 49:111–121 (special number)
Kirkham MB (2014) Principles of soil and plant water relations, 2nd edn. Academic, Oxford
Kramer PJ, Boyer PJ (1995) Water relations of plants and soils. Academic, New York
Libardi PL (2012) Dinâmica da água no solo, 2nd edn. EDUSP, São Paulo
Macedo A, Vaz CMP, Naime JM, Jorge LAC, Crestana S, Cruvinel PE, Pereira JCD, Guimarães MF, Ralisch R (2000) Soil management impact and wood science – recent contributions of Embrapa Agricultural Instrumentation Center using CT imaging. In: Cruvinel PE, Colnago LA (eds) Advances in agricultural tomography. Embrapa Agricultural Instrumentation, São Carlos, pp 44–54
Miller EE, Miller RD (1956) Physical theory of capillary flow phenomena. J Appl Phys 27:324–332
Miller EE, Miller RD (1955a) Theory of capillary flow: I. Practical implications. Soil Sci Soc Am Proc 19:267–271
Miller EE, Miller RD (1955b) Theory of capillary flow: II. Experimental information. Soil Sci Soc Am Proc 19:271–275
Moraes SO (1991) Heterogeneidade hidráulica de uma Terra Roxa Estruturada. PhD Thesis, Escola Superior de Agricultura Luiz de Queiroz, Universidade de São Paulo, Piracicaba, São Paulo, Brazil

Nobel PS (1983) Biophysical, plant physiology and ecology. W.H. Freeman & Company, New York

Oertli JJ (1984) Water relations in cell walls and cells in the intact plants. Z Pflanz Bod 47:187–197

Paltineanu IC, Starr JL (1997) Real-time soil water dynamics using multisensor capacitance probes: laboratory calibrations. Soil Sci Soc Am J 61:1576–1585

Philip JR (1964) Similarity hypothesis for capillary hysteresis in porous materials. J Geophys Res 69:1553–1562

Pires LF, Borges JAR, Bacchi OOS, Reichardt K (2010) Twenty-five years of computed tomography in soil physics: a literature review of the Brazilian contribution. Soil Tillage Res 110:197–210

Pires LF, Macedo JR, Souza MD, Bacchi OOS, Reichardt K (2002) Gamma-ray computed tomography to characterize soil surface sealing. Appl Radiat Isot 57:375–380

Poulovassilis A (1962) Hypothesis of pore water, an application of the concept of independent domains. Soil Sci 93:460–463

Rawlins SL (1966) Theory for thermocouple psychrometers used to measure water potential in soil and plant samples. Agric Met 3:293–310

Reichardt K (1965) Uso das radiações gama na determinação da umidade e da densidade do solo. PhD Thesis, Escola Superior de Agricultura Luiz de Queiroz, Universidade de São Paulo, Piracicaba, São Paulo, Brazil

Reichardt K (1987) A água em sistemas agrícolas. Manole, Barueri, Brazil

Reichardt K, Portezan-Filho O, Bacchi OOS, Oliveira JCM, Dourado-Neto D, Pilotto JE, Calvache M (1997) Neutron probe calibration correction by temporal stability parameters of soil water content probability distribution. Sci Agric 54:17–21 (special number)

Rock PA (1969) Chemical thermodynamics: principles and applications. The Macmillan Company, Toronto

Schindler U (1980) Ein schnellverfahren zur messung der wasserleitfähigkeit im teilgesättigten boden an stechzylinderproben. Arch Acker-u Pflanzenbau u Bod 24:1–7

SENTEK (2001) Calibration of Sentek soil moisture sensors. Sentek Pty Ltd, Stepney, Australia

Serrarens D, Macintyre JL, Hopmans JW, Bassoi LH (2000) Soil moisture calibration of TDR multi-level probes. Sci Agric 57:349–354

Silva AP, Bruand A, Tormena CA, da Silva EM, Santos GG, Giarola NFB, Guimarães RML, Marchão RL, Klein VA (2014) Indicators of soil physical quality: from simplicity to complexity. In: Teixeira WG, Ceddia MB, Ottoni MV, Donnagema GK (eds) Application of soil physics in environmental analysis: measuring, modelling and data integration. Springer, New York, pp 201–221

Silva AP, Tormena CA, Dias Junior MS, Imhoff S, Klein VA (2010) Indicadores da qualidade do solo. In: De Jong van Lier Q (ed) Física do solo. Sociedade Brasileira de Ciência do Solo, Viçosa, pp 241–282

Stolf R (1992) Teoria e teste experimental de fórmulas de transformação dos dados de penetrômetro de impacto em resistência do solo. Rev Bras Cienc Solo 15:229–235

Stolf R, Cassel DK, King LD, Reichardt K (1998) Measuring mechanical impedance in clayey gravelly soils. Braz J Soil Sci 22:189–196

Stolf R, Thurler AM, Bacchi OOS, Reichardt K (2011) Method to estimate soil macroporosity and microporosity based on sand content and bulk density. Braz J Soil Sci 35:447–459

Taiz L, Zeiger E, Moller IM, Murphy A (2018) Fundamentals of plant physiology. Oxford University Press, Oxford

Taylor SA, Ashcroft GL (1972) Physical edaphology: the physics of irrigated and non-irrigated soils. W.H. Freeman & Company, New York

Timm LC, Pires LF, Roveratti R, Arthur RCJ, Reichardt K, Oliveira JCM, Bacchi OOS (2006) Field spatial and temporal patterns of soil water content and bulk density changes. Sci Agric 63:55–64

Tominaga TT, Cássaro FAM, Bacchi OOS, Reichardt K, Oliveira JCM, Timm LC (2002) Variability of soil water content and bulk density in a sugarcane field. Aust J Soil Res 40:605–614

Topp GC (1969) Soil water hysteresis measure in a sandy loam and compared with the hysteresis domain model. Soil Sci Soc Am Proc 33:645–651

Topp GC, Davis JL (1985) Measurement of soil water content using time domain reflectometry (TDR): a field evaluation. Soil Sci Soc Am J 49:19–24

Topp GC, Davis JL, Annan AP (1980) Electromagnetic determination of soil water content: measurements in coaxial transmission lines. Water Resour Res 16:574–582

Topp GC, Davis JL, Annan AP (1982) Electromagnetic determination of soil water content using TDR. I. Applications to wetting fronts and steeps gradients. Soil Sci Soc Am J 46:672–678

Topp GC, Miller EE (1966) Hysteresis moisture characteristics and hydraulic conductivities for glass-bead media. Soil Sci Soc Am Proc 30:156–162

Tschapek M (1984) Criteria for determining the hydrophilicity-hydrophobicity of soil. J Plant Nutr Soil Sci 147:137–149

Van Bavel CHM, Underwood N, Swanson RW (1956) Soil moisture measurement by neutron moderation. Soil Sci 82:29–41

Van Genuchten MT (1980) A closed-form equation for predicting the conductivity of unsaturated soils. Soil Sci Soc Am J 44:892–898

Vaz CMP, Crestana S, Mascarenhas S, Cruvinel PE, Reichardt K, Stolf R (1989) Using a computed tomography miniscaner for studying tillage induced soil compaction. Soil Technol 2:313–321

Vaz CMP, Hopmans JW (2001) Simultaneous measurement of soil penetration resistance and water content with a combined penetrometer-TDR moisture probe. Soil Sci Soc Am J 65:4–12

Vaz CMP, Tuller M, Lasso PRO, Crestana S (2014) New perspectives for the application of high-resolution benchtop X-ray MicroCT for quantifying void, solid and liquid phases in soils. In: Teixeira WG, Ceddia MB, Ottoni MV, Donnagema GK (eds) Application of soil physics in environmental analysis: measuring, modelling and data integration. Springer, New York, pp 261–281

Villa Nova NA, Oliveira AS, Reichardt K (1992) Performance and test of a direct reading 'air-pocket' tensiometer. Soil Technol 5:283–287

Villa Nova NA, Reichardt K, Libardi PL, Moraes SO (1989) Direct reading "air-pocket" tensiometer. Soil Technol 2:403–407

Villagra MM, Matsumoto OM, Bacchi OOS, Moraes SO, Libardi PL, Reichardt K (1988) Tensiometria e variabilidade espacial em Terra Roxa Estruturada. Rev Bras Cienc Solo 12:205–210

Wiebe HH, Campbell CS, Gardner WH, Rawlins SL, Cary JW, Brown W (1971) Measurement of plant and soil water status. Utah Agricultural State, Logan

Wind GP (1966) Capillary conductivity data estimated by a simple method. In: International Association for Scientific Hydrology. Wageningen Symposium, Water in the unsaturated zone, Wageningen, pp 181–191

Zemansky MW, Dittman RH (1997) Heat and thermodynamics: an intermediate textbook, 7th edn. McGraw-Hill, New York

Zimmermann U, Stendle E (1978) Physical aspects of water relations of plant cells. Adv Bot Res 6:45–117

7 The Movement of Water in the Systems

7.1 Introduction

Water moves in the Soil-Plant-Atmosphere System in any of its phases. In the soil and in the plant, the main movements occur in the liquid phase, although the vapor flow may assume great importance when the soil is "drier," and in certain parts of the plant, as is the case with the stomata chambers in the leaf. In the atmosphere, the main movement occurs in the gaseous phase (water vapor), but also in the liquid (rain) and solid phases (hail or snow), and can assume important proportions.

In this chapter, emphasis will be given to the movement in the liquid phase, which occurs in response to differences in total water potential Ψ. We saw in Chap. 6 that whenever Ψ is constant in the system, there is equilibrium, and every time Ψ is variable in space, there is movement. However, under these considerations, it is necessary to discuss the problem of "semipermeable membranes," structures that allow the passage of water, but not of solutes. In the Soil-Plant-Atmosphere System, the main membranes are found in plant cells and water-air interfaces, such as the soil surface, in which water passes in the form of vapor, leaving behind liquid water and solutions. When there are no membranes, solutes move with water, and even in the existence of osmotic potential differences from one region to another, the movement of water due to osmotic potential is considered negligible; the movement of salts is more important, and these move looking for equilibrium. Therefore, in the case of water movement in the liquid phase, without the presence of membranes, the total potential of the water does not include the osmotic component (even if it is not negligible). With membranes, the osmotic component becomes the most important and needs to be included.

The **hydraulic potential** H, which is the **total water potential** Ψ, is then defined for the soil, without the inclusion of the osmotic component. Under these conditions, Eq. (6.12) is simplified in:

$$\Psi = H = \Psi_P + \Psi_m + \Psi_g \qquad (7.1)$$

Since both Ψ_P and Ψ_m refer to pressures, the first to positive and the second to negative, they can be grouped into a single component $h = \Psi_P + \Psi_m$, which covers the whole pressure range. The gravitational component Ψ_g can be expressed in terms of height, and, if the surface of the soil is taken as a reference, it is identified with the depth z. Thus, the most common form of presenting Eq. (7.1) in the literature of soil physics and in terms of hydraulic load is:

K. Reichardt, L. C. Timm, *Soil, Plant and Atmosphere*, https://doi.org/10.1007/978-3-030-19322-5_7

$$H = h + z \tag{7.2}$$

With this notation we will develop, from now on, the equations referring to the water flow in the liquid phase, in systems without the presence of semipermeable membranes.

7.2 Water Movement in the Soil

7.2.1 The Darcy Equation

Water in the liquid state moves whenever there are differences in hydraulic potential H at different points in the system. This movement occurs in the direction of decreasing potential H, that is, water always moves from points of higher potential to points of lower potential. Darcy (1856) was the first to establish an equation that would allow the quantification of water movement in saturated porous materials. The **equation of Darcy** states that the water flow density is proportional to the hydraulic potential gradient in the soil. His equation was later adapted to unsaturated soils (Buckingham 1907), now called the **Darcy-Buckingham equation**, and, despite its limitations, it is the equation that best describes the flow of water in the soil. In a general form, it can be written as:

$$q = -K \cdot \nabla H = -K \cdot \text{grad}\, H \tag{7.3}$$

in which q is the flux density of the water (L m^{-2} day^{-1} = mm day^{-1}), ∇H or grad H the hydraulic potential gradient (m m^{-1}), and K the hydraulic conductivity of the soil (mm day^{-1}). To understand this equation, let us look at the meaning of each term separately.

The **water flux density** q is a vector quantity and should be symbolized by having modulus, sense, and direction. Its module is the volume of water V, which passes per unit of time and by the unit of cross-sectional area (perpendicular to the movement). That is:

$$q = \frac{V}{A \cdot t} = \frac{L^3}{L^2 \cdot T} = L \cdot T^{-1} \tag{7.4}$$

Obs.: do not confuse the L of the dimensional analysis (a length; see Chap. 19) with L = liter (volume).

Thus, if 10 L of water cross 5 m^2 of soil in 0.1 day, the flow density will be 20 mm day^{-1}, since 1 mm = 1 L m^{-2}.

Although this flow has speed dimensions (space per time), it does not represent the speed at which water moves in the soil. The actual velocity of the water in the soil is the volume of water V that passes per unit of time by the area available for the flow within the soil, i.e., the cross section of pores occupied by the water. For a saturated soil, this cross section of pores is assumed to be the product of the effective area A by the porosity (see Eq. 3.12) of the soil.

If the previous example refers to the movement of water through an area of 5 m^2 of cross section without the presence of soil (Fig. 7.1), we will have $q = v = 20$ mm day^{-1}. The flow rate Q, which is defined by V/t, is also 100 L day^{-1}. If the area is reduced to half of the cross section, i.e.,

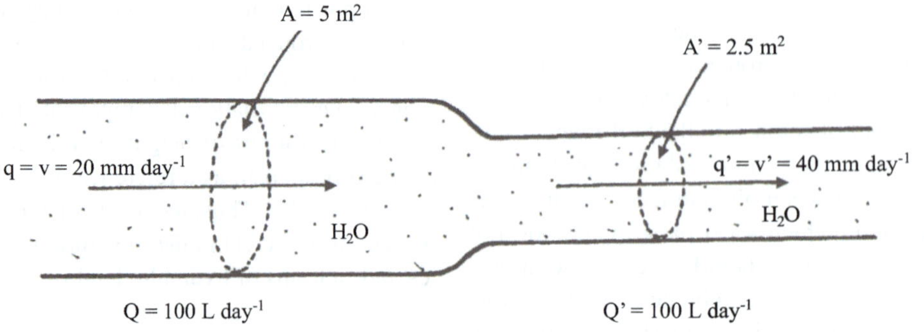

Fig. 7.1 Illustration of a pipe with water flowing that at a given position has its cross section reduced to half. As the area diminishes by half, the velocity doubles

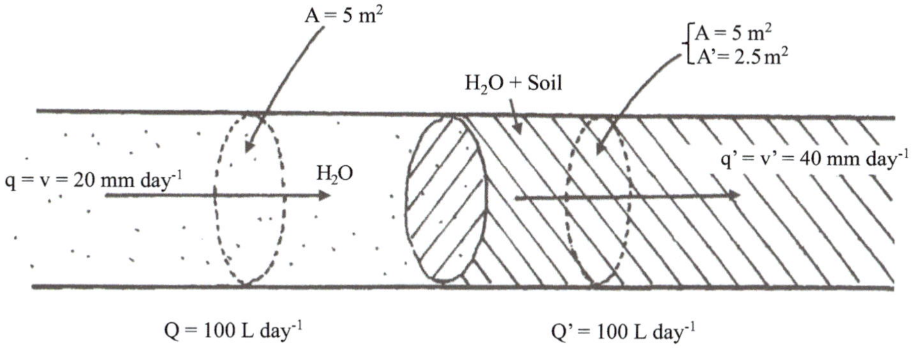

Fig. 7.2 Demonstration that in a pipe of the same diameter, when part is filled by soil, the result is also a reduction of the area available for flow

$A' = 2.5\ \text{m}^2$, it is easy to verify that the flow Q remains the same (equation of continuity: $Q = Av = A' \cdot v' = \ldots$) and that the flow density doubles to $q' = V/A't = 40\ \text{mm day}^{-1}$. Even so $q' = v'$.

It is easy to verify that $q \cdot \text{A} = q' \cdot A'$.

The reduction of area available to the flow can also be made by introducing soil into the pipeline. If the cross-sectional tube of $A = 5\ \text{m}^2$ is filled with soil of porosity ($\alpha = 0.5\ \text{m}^3\ \text{m}^{-3}$), it can be shown that the area A available to the stream is reduced by $A' = \alpha \cdot A = 0.5 \times 5 = 2.5\ \text{m}^2$ (see Fig. 7.2).

Since A (not A') is measured on the soil, the flow density q remains the same, equal to 20 mm day^{-1}, but the **pore velocity** v changes to v'. Therefore, the velocity of the water in the pore v' is different from q. And also therefore, the soil water flow density, which has velocity dimensions, is not equal to the water velocity in the pores. That is:

$$v = \frac{V}{A \cdot \alpha \cdot t} = L \cdot T^{-1} \qquad (7.5)$$

$$v = \frac{q}{\alpha}$$

If the soil is not saturated, the area available for water flow is even smaller, $A' = A \cdot \theta$ (where θ is the volume-based soil water content defined by Eq. 3.15), because water "travels" only through the pores filled with water, and:

$$v = \frac{q}{\theta} \qquad (7.6)$$

Due to variations in the shape, direction, and width of the pores, the current velocity of the water in the soil is highly variable from point to point, and it is not possible to speak of a single velocity of the liquid but, at best, at an average of the real velocity. In the previous example, the average real velocity of water in the soil pores is 40 mm day^{-1}, and the flow density is 20 mm day^{-1}.

Tortuosity of a porous medium is defined as the square of the ratio between the distance actually traveled by a molecule of water and the distance (advance) in a straight line. This parameter is dimensionless and generally varies from 1 to 2. Also, due to this fact, q differs from v.

Thus, the definition of q in the Darcy equation is now clear. Often it is called simply as a rate. If a soil is losing 5 L of water by evaporation in each m^2 each day, its evaporation rate is 5 mm day^{-1}. This is the evaporation flux density. A soil can also lose water by drainage, for example, 2 mm day^{-1}. This is a drainage rate, another example of q.

Let us now see in the Darcy equation (Eq. 7.3) the meaning of the **gradient of H** or ∇H. The potential gradient ∇H is also a vector quantity and should be symbolized by $\vec{\nabla} H$. It is defined in the Cartesian system of three dimensions (see coordinate systems in Chap. 19), by the equation:

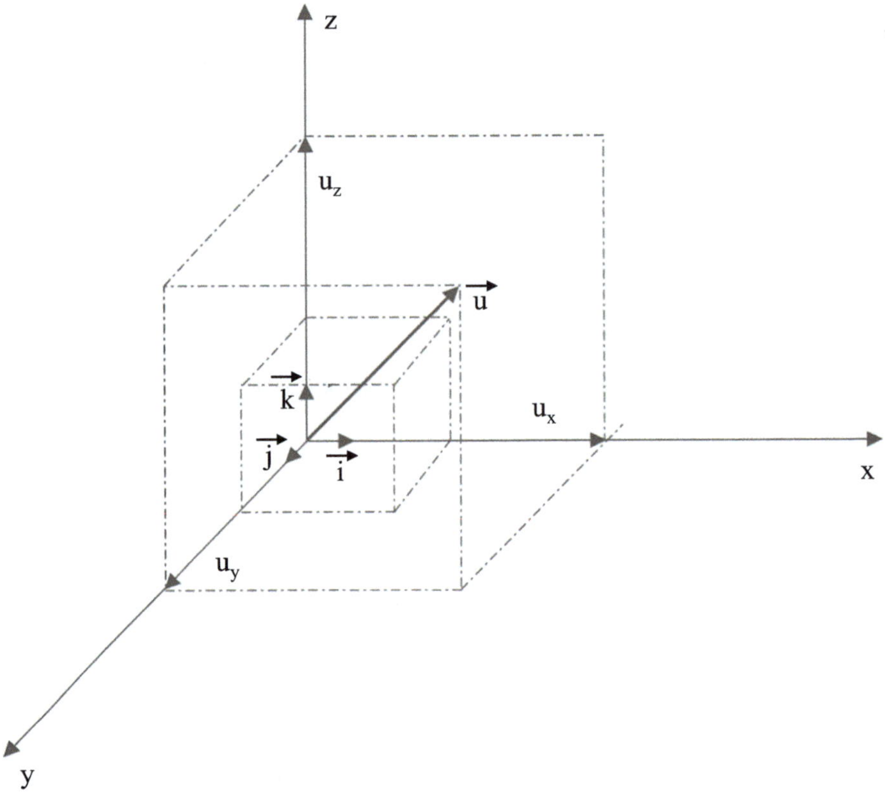

Fig. 7.3 Decomposition of a vector u in its three orthogonal components

$$\text{grad } H = \vec{\nabla} H = \frac{\partial H}{\partial x}\vec{i} + \frac{\partial H}{\partial y}\vec{j} + \frac{\partial H}{\partial z}\vec{k} \tag{7.7}$$

Let's look better at the definition of the gradient. In the Cartesian system, any vector $\vec{u}$ of given sense and direction can be decomposed in three orthogonal components, the vectorial sum of them being the vector:

$$\vec{u} = u_x\vec{I} + u_y\vec{J} + u_z\vec{k}$$

where u_x, u_y, and u_z are the modules of the components and $\vec{i}$, $\vec{j}$, and $\vec{k}$ are vectors of unit module of directions x, y, and z, respectively (Fig. 7.3).

The operator $\vec{\nabla}$ (in Greek nabla) is a vector operator that when decomposed in the directions x, y, and z is expressed as:

$$\vec{\nabla} = \frac{\partial}{\partial x}\vec{i} + \frac{\partial}{\partial y}\vec{j} + \frac{\partial}{\partial z}\vec{k}$$

When $\vec{\nabla}$ operates on a scalar entity, the result is a gradient. Scalar entities, in contrast to vectors, have only modulus for their complete definition, no need of sense and direction. Examples are mass, temperature, volume, concentration, etc. Let's take, for example, the scalar entity temperature T. In this case $\vec{\nabla} T$ is the temperature gradient, which becomes a vector:

$$\vec{\nabla} T = \text{grad } T$$

To make the operation, we apply the operator $\vec{\nabla}$ over T, that is:

$$\vec{\nabla} T = \frac{\partial T}{\partial x}\vec{i} + \frac{\partial T}{\partial y}\vec{j} + \frac{\partial T}{\partial z}\vec{k} = \text{vector}$$

It is then seen that the gradient is the result of the operation of $\vec{\nabla}$ on a scalar quantity and the result is a vector quantity. It is not possible to obtain the gradient of a vectorial entity. In the example we have seen, the temperature is a scalar entity, and the temperature gradient is a vector quantity, with modulus, sense, and direction.

The total potential of water H is a scalar (energy). Its gradient $\vec{\nabla}H$ is a vector (force), with sense and direction.

In several texts of less mathematical rigor, as it is the case of soils, the notation is simplified, and the arrows $\rightarrow$ are not used, even in the unit vectors i, j, and k, even though their modules are unitary. Often the gradient in one single direction is desired.

The Darcy equation can then be presented in the following forms, which are equivalent:

$$q = -K\frac{\partial H}{\partial x} = -K \cdot \text{grad}\, H = -K \cdot \nabla H$$

Dimensionally, the water potential gradient refers to a force, since it represents an energy per unit length: J m^{-1} = (N m) m^{-1} = N. When H is expressed in water height, dimension L, the grad H has dimensions $L\,L^{-1}$, that is, it is dimensionless. It should not be forgotten that, nevertheless, grad H is a force. It is the force responsible for the movement of water in the soil. When the gradient is 0, there is no force, and consequently there is no water movement: equilibrium.

The Darcy-Buckingham equation (Eq. 7.3) tells us only that the flux density q is proportional to the force acting on the water, that is, the potential gradient. The coefficient of proportionality K is the **hydraulic conductivity of the soil**. A negative sign also appears in the equation, which indicates that the direction (sense of the arrow) of the flux density is the inverse of the gradient. The direction of the gradient is, by definition, taken as the one in which the potential field grows, that is, from a smaller value of H to a value greater than H. As we have already said, water moves from a point with greater H to another of smaller H, that is, in the opposite direction of the gradient, hence the inclusion of the negative sign in the Darcy-Buckingham equation. The hydraulic conductivity can therefore be defined by the relationship between the flux density and the gradient:

$$K = \frac{q}{\nabla H} = \frac{L \cdot T^{-1}}{L/L} = L \cdot T^{-1} \tag{7.8}$$

with dimensions equal to those of the flow, $L{\cdot}T^{-1}$(L T^{-1}), when the potential H is measured in energy per unit weight or hydraulic load (cm or mH_2O).

The hydraulic conductivity depends on the properties of the fluid and the porous material. It has experimentally been found that for a rigid porous material:

$$K = \frac{k \cdot \rho_e \cdot g}{\eta} \tag{7.9}$$

where:

k = soil property called **intrinsic permeability** (m^2), dependent of the geometric arrangement of the soil particles and of soil water, which define the available cross section of the flow

ρ_e = fluid-specific mass (water), in kg m^{-3}

g = gravitational acceleration, in m s^{-2}

η = viscosity of the fluid, in kg m^{-1} s^{-1}

The viscosity and density of the soil solution depend on the temperature, pressure, concentration of soluble salts, soil water content, etc. With the exception of expanding and contracting soils, the value of k of a soil is taken as constant for each sample at a given water content. For practical purposes, it is assumed that ρ_e, g, and η are constants for a given experiment, and k varies only with soil water content (useful area for flow). We have already said that for a saturated soil, this area is proportional to the porosity α and that for an unsaturated soil, the useful area for the flow is proportional to the water content θ.

Thus, we can say that, with the mentioned conditions, the hydraulic conductivity of a soil sample is a function of only θ, or $K = K(\theta)$.

Normally, the hydraulic conductivity of a saturated soil is symbolized by K_0. This is the maximum value of K for that sample. It decreases rapidly with decreasing θ (or matrix potential h), because as $h = h(\theta)$ (soil water retention curve), the conductivity can also be expressed in terms of

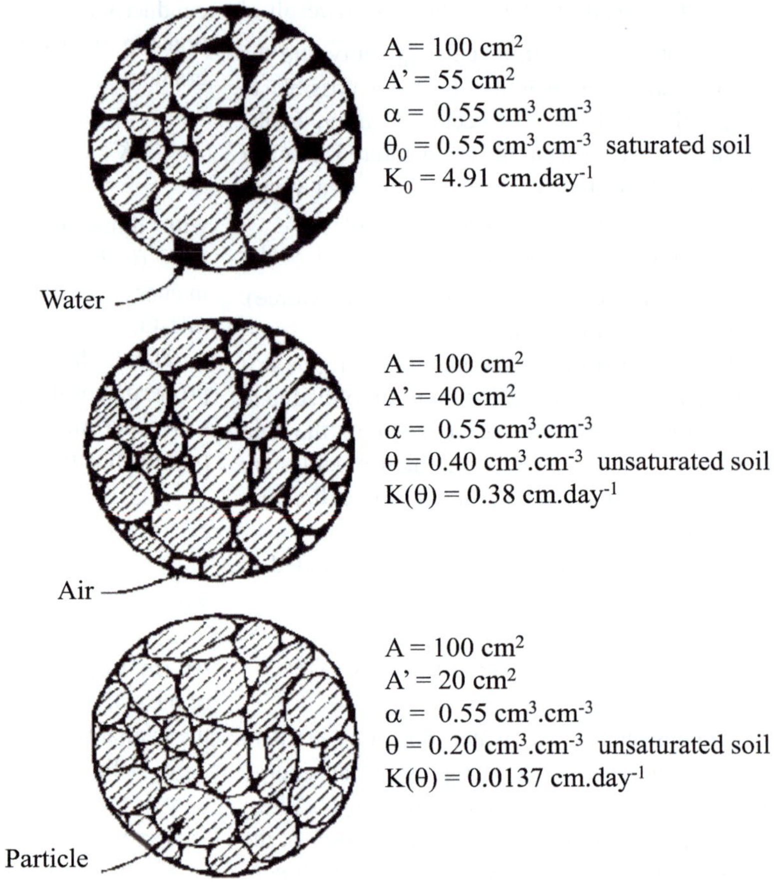

Fig. 7.4 Illustration of the change of the soil hydraulic conductivity with soil water content. Top, I saturated soil; center, II intermediate condition; below, III a very dry condition

the matrix potential h, $K = K(h)$. Figure 7.4 shows K values for a given soil, taken as an example in three soil water conditions.

In addition to the decrease of the useful area due to the reduction of θ, the tortuosity of the soil and the water retention phenomena cause K to decrease drastically with θ. Therefore, in order to represent the curve $K(\theta)$, logarithms are applied to the data of K and not to θ which vary in a much smaller proportion. Graphs of this type are called semi-log plots.

Since K also depends on the pore space geometry, it varies greatly from soil to soil and also for a soil with structural variations, compaction, and so on. Thus, it is convenient for a given soil to plot the logarithm of K versus θ (as shown in Fig. 7.5). Figure 7.5 was obtained with data of K and of θ of Fig. 7.4, in which one can see that the data fits well to a straight line. Under these conditions, using decimal logarithms (with base 10, $\log_{10}$), we have:

$$\log_{10} K(\theta) = a + b\theta$$

the linear coefficient a of the line **a** is the log K value for $\theta = 0$ (dry soil, K_{d}), which in the present case is equal to -3.34. Thus, log $K_{\mathrm{d}} = -3.34$ and its anti-log $K_{\mathrm{d}} = 0.00046$ cm day^{-1}. Note that we are using the subscript **d** in K_{d} to denote the value of K for dry soil. The slope **b** is the slope of the line or the tangent of the angle that the line makes with the axis $x = \theta$, which in the present case is equal to Eq. (7.28) and is dimensionless. Therefore, the above equation can be written as:

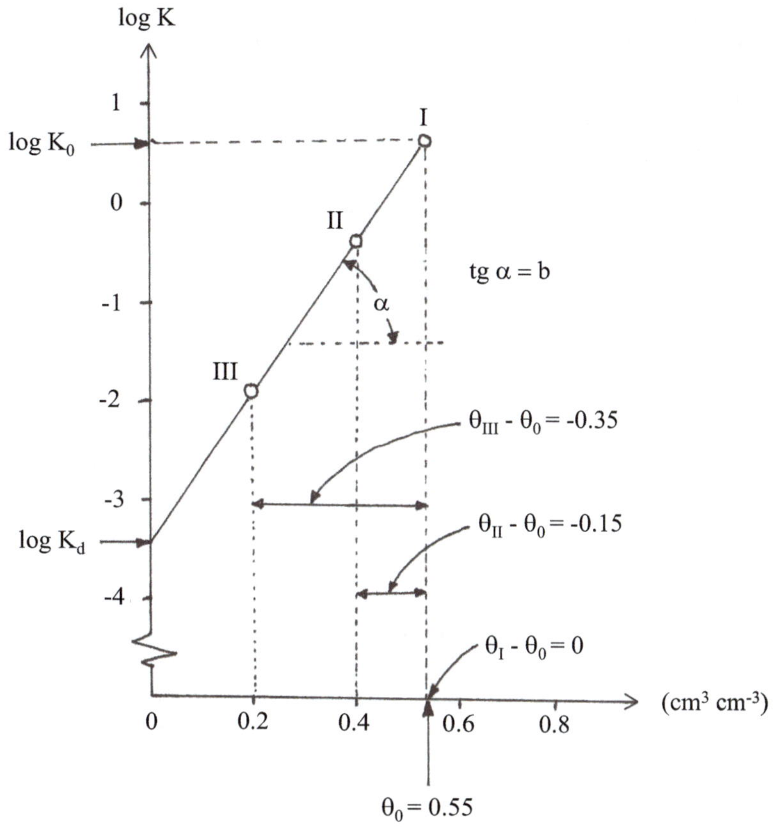

Fig. 7.5 Log_{10} hydraulic conductivity data of Fig. 7.4 plotted as a function of θ. Alfa (α) is the slope of the straight line (not the porosity!), that is, $\tan(\alpha) = b$

$$\log_{10} K(\theta) = -3.34 + 7.28 \times \theta$$

or:

$$K(\theta) = 0.00046 \times 10^{7.28\times\theta} \qquad (7.10)$$

To understand the step between these two equations, take Eq. (7.10), and apply $\log_{10}$ to both members. Remember that the log of a product is a sum, that the log of an exponent is the exponent times the log of the base, and finally that $\log_{10} 10 = 1$.

Figure 7.5 could also be made using Naperian or natural logarithms (ln), with base $e = 2.718\ldots$ (Fig. 7.6). In this case applying ln to the K data of Fig. 7.4, we would have:

$$\ln K(\theta) = a' + b' \cdot \theta$$

or:

$$\ln K(\theta) = -7.68 + 16.8 \cdot \theta$$

or still:

$$K(\theta) = 0.00046 \times e^{16.8\times\theta} \qquad (7.11)$$

in many cases represented as:

$$K(\theta) = 0.00046 \cdot \exp(16.8 \times \theta)$$

Note that in Eqs. (7.10) and (7.11), where in the former, the base of the exponent is 10 and in the second is **e**, K_d is the same and $b \neq b'$; nonetheless, they represent the equation $K(\theta)$ of the same soil, and, for any value of θ, they have to yield the same value of K. For example, if $\theta = 0.5$, then $K = 2.05$ cm day^{-1}, by both equations, approximately, of course, depending on the approximations of decimal places used in the logarithms.

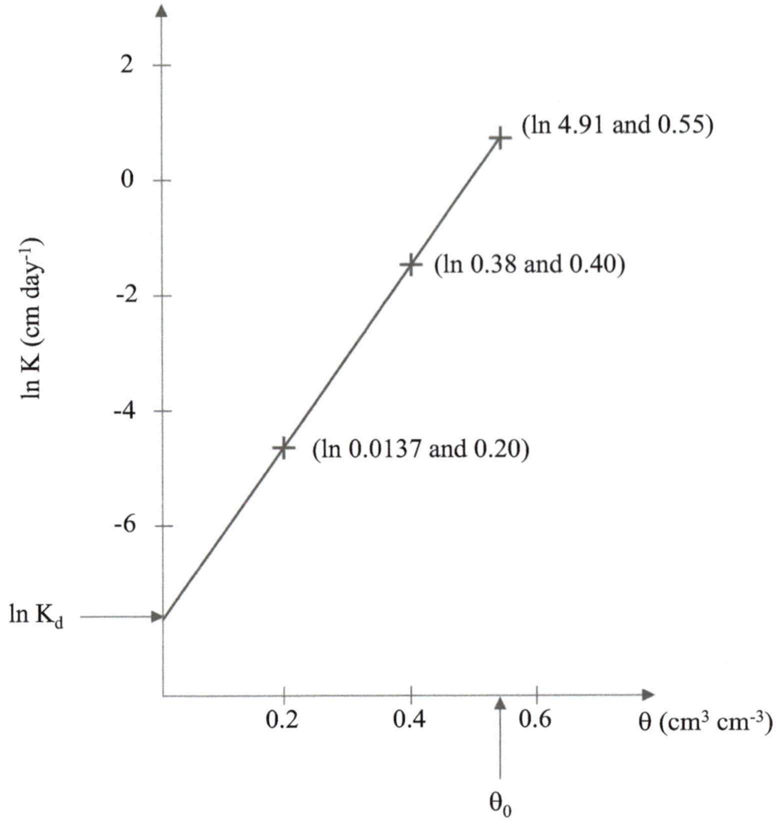

Fig. 7.6 Ln of hydraulic conductivity data of Fig. 7.4 plotted as a function of θ. The slope of the straight line will no longer be b but b' because the numerical values of $\log_{10}$ are different from those of ln

As it can be verified, Eqs. (7.10) and (7.11) include the parameter K_d = hydraulic conductivity of the dry soil. This parameter does not have a useful physical meaning, because if the soil is totally dry, we cannot speak of movement of water. As we have seen, K_d is a value obtained by extrapolation, and, therefore, the equations $K(\theta)$, most of the time, are written including another parameter, the K_0 (**saturated soil hydraulic conductivity**), an important parameter in practice and of well-defined physical meaning. This can be done by introducing a new variable $(\theta - \theta_0)$ in the place of θ, where θ_0 is the saturation soil water content, which is also an important parameter. This new variable, null (at saturation) when $\theta = \theta_0$, is negative for the other values of θ, since $\theta < \theta_0$. This passage from θ to $(\theta - \theta_0)$ implies only in a translation of the ln K coordinate from one place to another. For example, for the points in Fig. 7.5, we will have (I) $\theta - \theta_0 = 0$; (II) $\theta - \theta_0 = -0.15$; and (III) $\theta - \theta_0 = -0.35$, and the graph of ln K versus $(\theta - \theta_0)$ is shown in Fig. 7.7.

In this case, we have:

$$\ln K(\theta) = \mathrm{a}'' + \mathrm{b}'' \cdot (\theta - \theta_0)$$

or:

$$\ln K(\theta) = 1.59 + 16.8 \times (\theta - \theta_0)$$

It is important to note that $\mathbf{a}''$ (ln K_d) is different from $\mathbf{a}'$(ln K_0) and that b'' is equal to $\mathbf{b}'$, because we deal with a simple translation on the θ axis. If both members are made exponents of **e**, we will have:

$$K(\theta) = 4.91 \times e^{16.8 \times (\theta - \theta_0)} \tag{7.12}$$

Therefore, we find in the literature equations of the type:

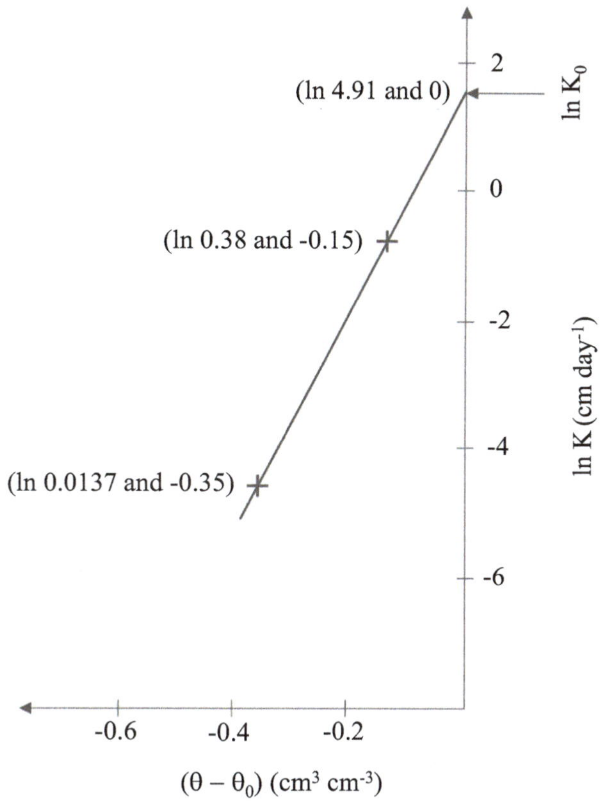

Fig. 7.7 Figure 7.6 translated from abscissa θ to $\theta - \theta_0$

$$K(\theta) = K_0 \cdot e^{\gamma \cdot (\theta - \theta_0)}$$

or:

$$K(\theta) = K_0 \cdot \exp\left[\gamma(\theta - \theta_0)\right]$$

or, when using the decimal log:

$$K(\theta) = K_0 \cdot 10^{\beta(\theta - \theta_0)}$$

These equations have the function of providing us data of K of the soil in question for any value of θ in the interval 0–θ_0 from the observed exponential model. In addition, since θ is a function of h (characteristic or retention curve), and vice versa, it is often more convenient to express K as a function of h, that is, to establish the function $K(h)$. Its equation will depend on the function $h(\theta)$. A model that is very common when K data is not available is the combination of the van Genuchten (1980) model for the retention curve $h(\theta)$ and the Mualem (1976) model for $K(\theta)$ using soil water in a dimensionless way:

$$\Theta = \frac{\theta - \theta_r}{\theta_0 - \theta_r} \tag{7.13}$$

$$\Theta = \left[1 + |\alpha h|^n\right]^{-m} \tag{7.14}$$

$$K(\Theta) = K_0 \Theta^l \left[1 - \left(1 - \Theta^{\frac{1}{m}}\right)^m\right]^2 \tag{7.15}$$

where θ_r is the residual water content, of the air-dried soil; α (not the porosity!), m, and n are the parameters of the van Genuchten curve; and l another empirical parameter called pore connectivity. In order to better understand Eq. (7.13), see in Chap. 19 the item non-dimensional. Equation (7.14) differs somewhat from Eq. (6.19) because it is written in terms of the dimensionless soil water content (Eq. 7.13) and in which b has been replaced by m. Equation (7.15) is the expression of Mualem for $K(\theta)$, which was deduced based on the parameters of the van Genuchten curve and which is not an exponential as in the examples above. In the literature, there are numerous models for $K(\theta)$. From the combination

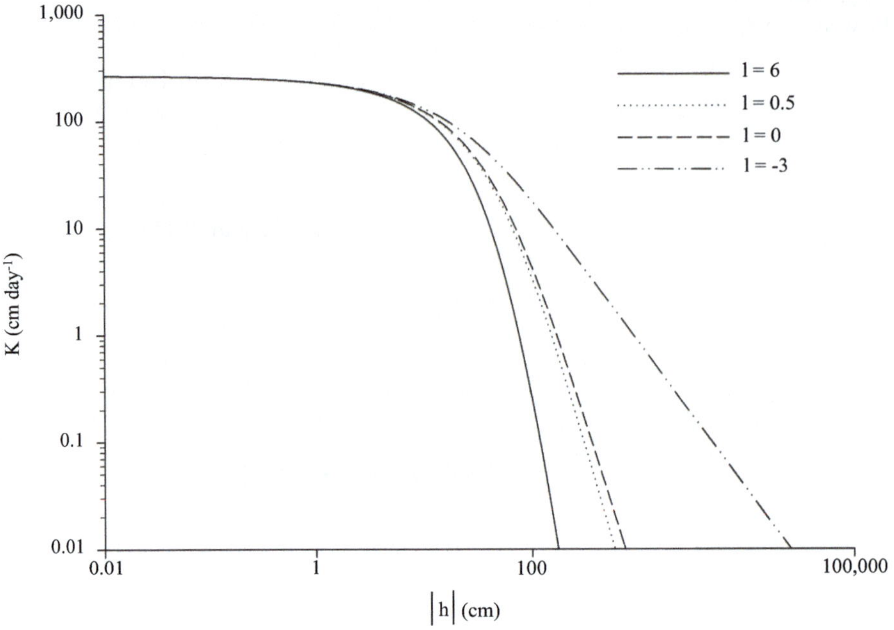

Fig. 7.8 An example of log K versus log h curves showing that the factor l becomes important only in the dry range

of Eqs. (7.14) and (7.15), a curve $K(h)$ is shown for several values of l by way of example (Fig. 7.8). Since both amplitudes of K and of h are large, the figure is presented in as a log-log plot, in which the scales are logarithmic and any data of K or of h can be entered without applying the logarithm.

With these considerations, we rewrite the Darcy-Buckingham equation in the form by which we will use it intensively, that is, for a dimension x (horizontal) or z (vertical):

$$q = -K(\theta)\frac{\partial h}{\partial x} = -K(\theta)\frac{\partial H}{\partial z} \qquad (7.16)$$

in the horizontal case, we use only h since gravity does not act, and in the vertical case, we have H, equal to $h + z$.

In the following examples, we will employ soil columns mounted in plastic or acrylic tubes with diameters of 5–10 cm, in the laboratory, usually filled with air dry soil, sifted through a 2 mm sieve. To hold the soil inside the column, porous plates are used at their ends, which let the water pass. In Fig. 7.9, the water enters with positive pressure, and therefore the porous plate is of very porous texture, not to interfere in the flow of water inside the soil. The same is true at the outlet at which atmospheric pressure is operating. The water inlets and outlets are made by funnels connected to the soil column by means of flexible rubber hoses. By a device not shown, the water level is kept constant, thus maintaining the constant hydraulic loads. The inverted black triangles shown in the figures, placed on the free water level, indicate the constancy of the water level. In this way, we have a flow of water in **dynamic equilibrium**, in which the same amount of water that enters leaves the system and is the water that passes through the soil. In Fig. 7.11, the funnels are below the soil column, to apply negative pressures or suctions. In this case, the porous plates at the ends of the column need to have finer porosity so that they can hold the negative pressures. However, they also cannot have a too fine structure in order not to interfere with the flow of water in the soil.

Example: In Fig. 7.9, we have a soil column mounted vertically with the same soil of Fig. 7.4, through which water flows at a rate that

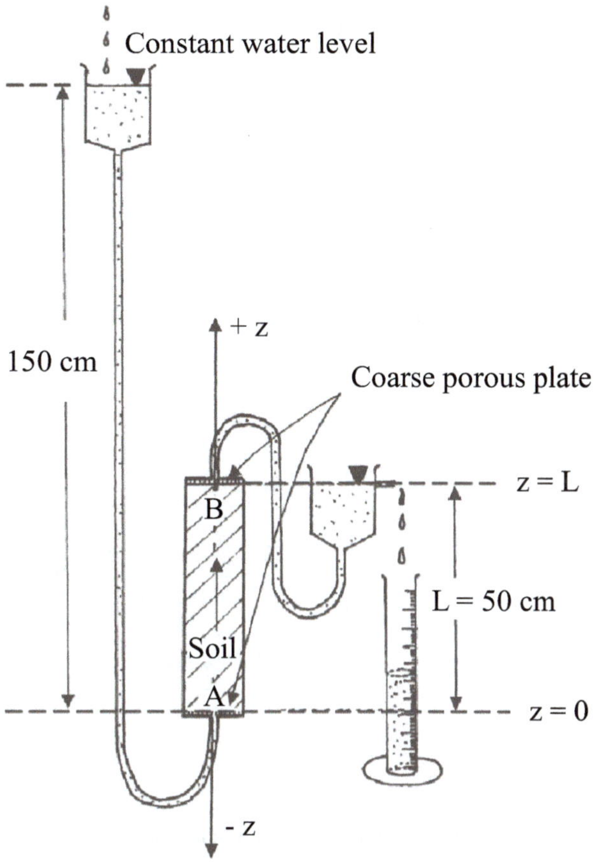

Fig. 7.9 A soil column under steady-state vertical up water flow

is measured in the graduated cylinder. The soil column has cross section $A = 100\ \text{cm}^2$ and length $L = 50$ cm. A volume $V = 982\ \text{cm}^3$ is collected in 24 h. What is the saturated conductivity of the soil?

Answer:
Since the soil is saturated, we will determine K_0. In the Darcy equation, the gradient $\partial H/\partial z$ can be approximated by a finite difference $\Delta H/\Delta z$ or by $(H_B - H_A)/L$, and Eq. (7.16) becomes:

$$\frac{V}{A \cdot t} = -K_0\left(\frac{H_B - H_A}{L}\right)$$

$$q = \frac{V}{A \cdot t} = \frac{982}{100 \times 1} = 9.82\ \text{cm day}^{-1}$$

$$H_A = z_A + h_A = 0 + 150 = 150\ \text{cm}\ H_2O$$

$$H_B = z_B + h_B = 50 + 0 = 50\ \text{cm}\ H_2O$$

$$9.82 = -K_0\left(\frac{50 - 150}{50}\right)$$

$$K_0 = 4.91\ \text{cm day}^{-1}$$

In the previous operations, we see that the grad H has negative sign that, with the minus sign of the equation of Darcy, led to a positive value of K_0. In this case, we choose q as a positive from the bottom-up. K_0 always needs to be positive as it is a property of the soil. The grad H has negative signal because, in calculating it, we chose $H_B - H_A$. If we had chosen $H_A - H_B$, the grad H would be positive and K_0 negative. The best criterion to follow is to first choose the sign of q according to the convenience of the problem in question and then calculate grad H such that K_0 is always positive (the absolute value will always be correct).

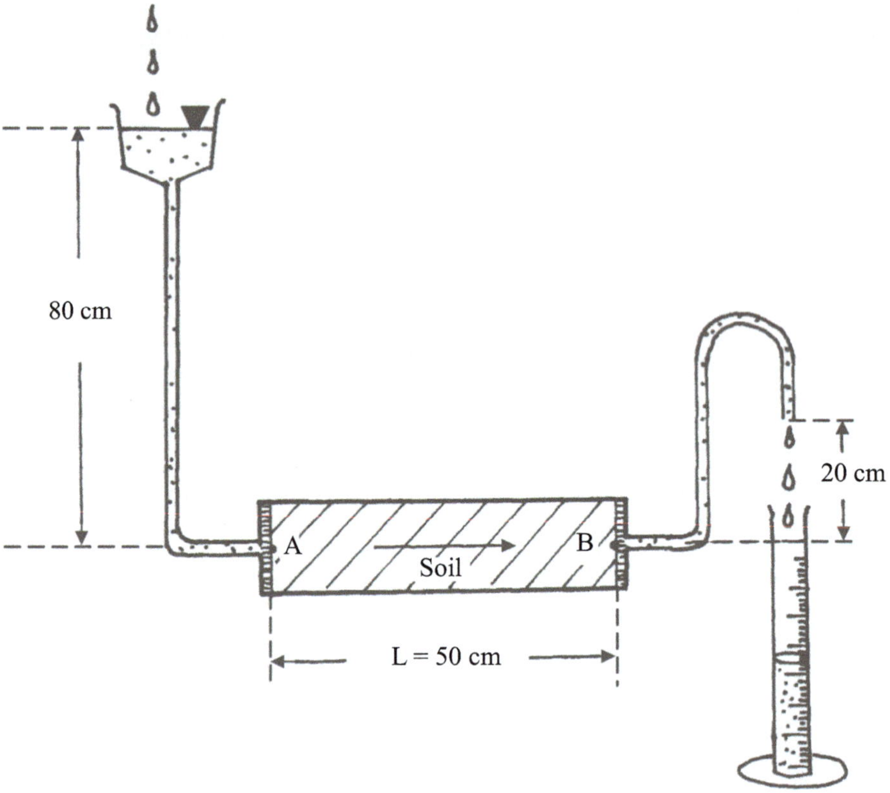

Fig. 7.10 A soil column under steady-state horizontal water flow

In the previous example, we obtained $K_0 = 4.91 \text{ cm day}^{-1}$, which is the hydraulic conductivity of the saturated soil and therefore a characteristic of the sample, not depending on the experimental arrangement. For example, if the height of the upper vessel is reduced to 100 cm, the gradient decreases, but the water flow density decreases proportionally, and as a result, the same K_0 is obtained. In this new situation, 488 cm^3 was collected in 1 day. That is:

$$\text{grad } H = \frac{50 - 100}{50} = -1 \text{ cm cm}^{-1}$$

$$q = \frac{488}{100 \times 1} = 4.88 \text{ cm day}^{-1}$$

$$K_0 = \frac{4.88}{1} = 4.88 \text{ cm day}^{-1}$$

which is very close to the value 4.91 cm day^{-1} obtained in the previous case. The difference, in this case, lies in the experimental error of the volume measurement of 488 cm^3 collected in 1 day.

In the scheme of Fig. 7.9, we could put the soil column in another position, in the horizontal, for example (Fig. 7.10):

$$H_A = z_A + h_A = 0 + 80 = 80 \text{ cmH}_2\text{O}$$

$$H_B = z_B + h_B = 0 + 20 = 20 \text{ cmH}_2\text{O}$$

$$\nabla H = \frac{20 - 80}{50} = -1.2 \text{ cm cm}^{-1}$$

$$q = \frac{588}{100 \times 1} = 5.88 \text{ cm day}^{-1}$$

(measured in the graduate cylinder)

$$K_0 = \frac{5.88}{1.2} = 4.90 \text{ cm day}^{-1}$$

In this case, we see that a similar result was again obtained. Highlighting, K_0 is a property of

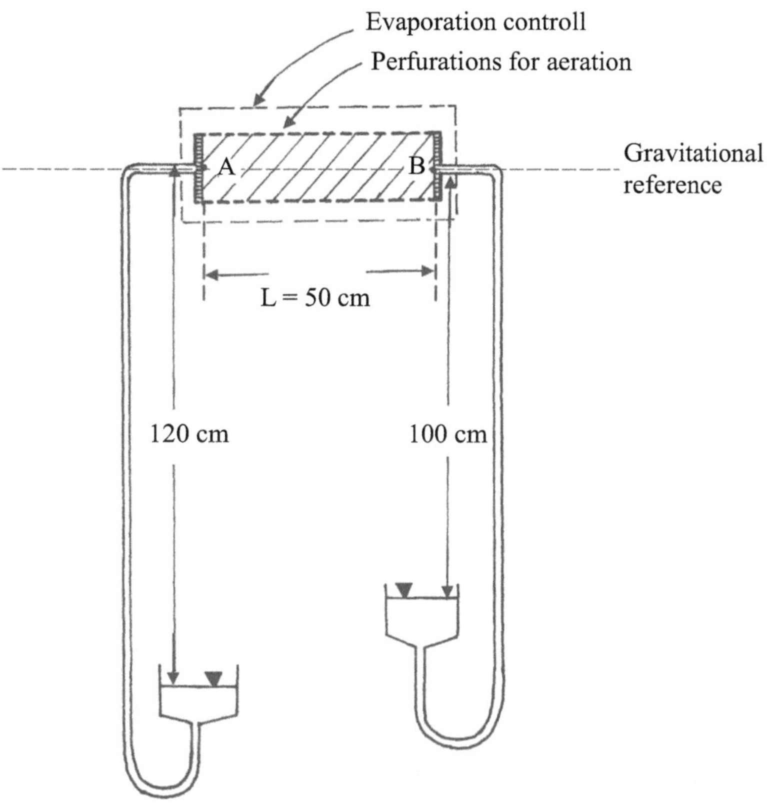

Fig. 7.11 A soil column under steady-state horizontal and unsaturated water flow because water enters at the right side under a tension −100 cm and leaves the soil at the left side under a tension of −120 cm

the sample and independent of the experimental arrangement of measure.

The three previous examples are related to saturated soil and, therefore, we obtain K_0. All are cases of steady state, where the amount of water passing through A is equal to that which passes through B. We could also have an unsaturated soil condition if we applied suctions at A and B, as shown in Fig. 7.11, also with the same soil as in Fig. 7.4. This experiment is difficult to perform since the soil is aerated and the evaporation needs to be controlled. Initially, the right-hand water container is raised so that there is a positive pressure at A, and the left container is held practically at the height of B. So, the soil is wetted, almost saturated, and the water flow is established. The vessels are then lowered to the positions shown in Fig. 7.11.

As the soil is not saturated, the water flow (now from B to A) is much slower. In this type of experiment, the equilibrium is reached only after a long time (weeks, months), and it is necessary to control losses by evaporation and the development of microorganisms. In any case, it also evolves into a dynamic equilibrium case. The soil is not saturated because we have applied suctions on both sides. Since the suction at A is greater than at B, the water moves from B to A. Gravity does not affect the process. If the column was on the vertical position, gravity would act to transport water. As $h_A = -120$ cm H_2O and $h_B = -100$ cm H_2O, the soil should be slightly more humid in B than in A. For calculation purposes, we will use the mean water content θ, which was measured at the end of the experiment, obtaining $\theta = 0.481$ cm^3 cm^{-3}. In this case, we will have:

$$H_A = 0 - 120 = -120 \text{ cm } H_2O$$

$$H_B = 0 - 100 = -100 \text{ cm } H_2O$$

$$\text{grad } H = \frac{[-120 - (-100)]}{50} = -\frac{20}{50} = -0.4 \text{ cm cm}^{-1}$$

Because the water movement is very slow, only 420 cm^3 was collected in a week. Therefore the flow density rate is:

$$q = \frac{420}{100 \times 7} = 0.6 \text{ cm day}^{-1}$$

and:

$$K(\theta) = K(0.481) = \frac{0.6}{0.4} = 1.5 \text{ cm day}^{-1}$$

a very similar value to the one substituting $\theta = 0.481$ in Eqs. (7.10) or (7.11) or still (7.12).

Another interesting example of the application of the Darcy-Buckingham equation in a dynamic equilibrium case is the one outlined in Fig. 6.18. For this case, we will consider a dynamic evaporation equilibrium condition equal to 5 mm day^{-1}, in which the following data were verified:

In point A (located in the soil at the water-air interface level):

$\theta_A = \theta_0 = 0.52$ cm^3 cm^{-3} (saturation)
grad $H = -1.0$ cm cm^{-1} (only gravity acts)
$K(\theta_A) = K_0 = 5$ mm day^{-1} (the evaporation rate)
$q_A = -5$ mm $day^{-1} \times (-1.0$ cm $cm^{-1}) = 5$ mm day^{-1}

At point B (in the center of the column):

$\theta_B = 0.50$ cm^3 cm^{-3} (measured value)
grad $H = -1.4$ cm cm^{-1}
$K(\theta_B) = 3.57$ mm day^{-1}
$q_B = -3.57$ mm $day^{-1} \times (-1.40$ cm $cm^{-1}) = 5$ mm day^{-1}

At point C (slightly below soil surface):

$\theta_C = 0.42$ cm^3 cm^{-3}
grad $H = -5.38$ cm cm^{-1}
$K(\theta_C) = 0.93$ mm day^{-1}
$q_C = -0.93$ mm $day^{-1} \times (-5.38$ cm $cm^{-1}) = 5$ mm day^{-1}

It is thus seen that at any point, the flow is constant (5 mm day^{-1}) and that with the decrease of the water content θ, the hydraulic conductivity drops abruptly. The fall of K is compensated by an increase in the gradient of H, and, as a result, the flow remains constant.

We could also consider the field soil situation as shown in Fig. 7.12, where we see two tensiometers installed at points A and B located horizontally. It is desired to know the water flow between A and B. The soil is the same one whose K curve is presented in Eqs. (7.10)–(7.12) and whose characteristic curve is presented in Fig. 7.13.

Solution:
The matrix potential h of water at points A and B can be calculated by Eq. (6.21) of the tensiometers (note! the symbol h in the tensiometer has another meaning):

$$h(\text{A}) = -13.6 \times h + h + h_1 + h_2 = -212 \text{ cmH}_2\text{O}$$

$$h(\text{B}) = -13.6 \times 30 + 30 + 15 + 30 = -333 \text{ cmH}_2\text{O}$$

The hydraulic potential at points A and B, using as gravity referential the line A–B, will be:

$$H_A = h_A + 0 = -212 \text{ cm H}_2\text{O}$$

$$H_B = h_B + 0 = -333 \text{ cm H}_2\text{O}$$

From the characteristic curve of the soil (Fig. 7.13), it is verified that the water contents at points A and B, corresponding to h_A and h_B, are:

$$\theta_A = 0.50 \text{ cm}^3 \text{ cm}^{-3}$$

and

$$\theta_B = 0.45 \text{ cm}^3 \text{ cm}^{-3}$$

For these values of θ, the corresponding values of hydraulic conductivity (apply Eqs. 7.10 or 7.11 or 7.12) are:

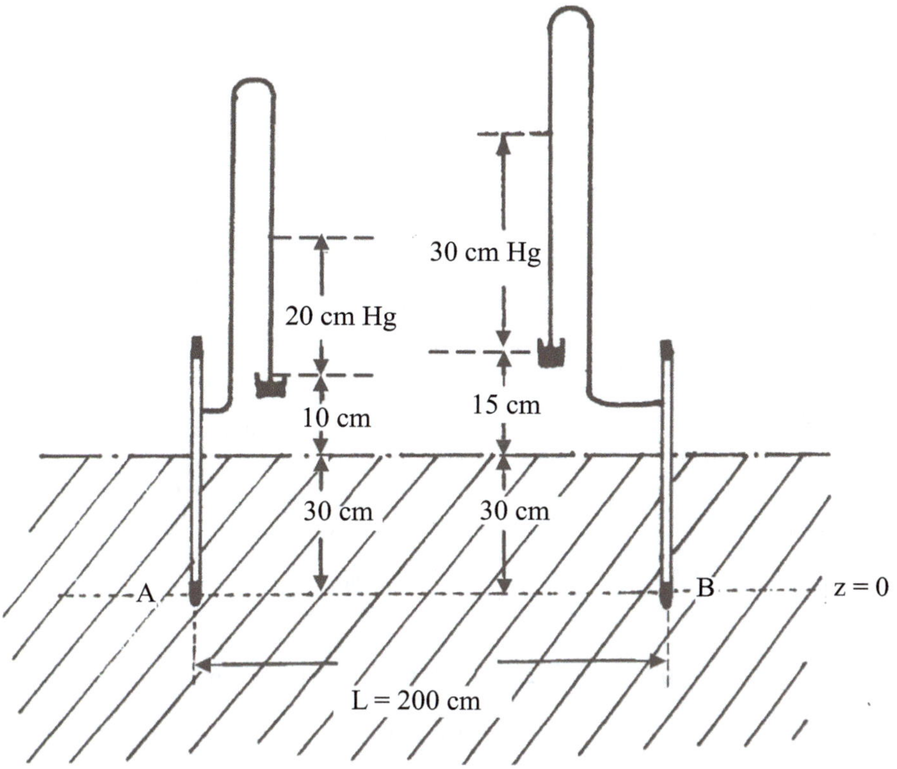

Fig. 7.12 A homogeneous field soil with tensiometers installed at points A and B (30 cm depth), separated by 200 cm, with readings shown on the figure

$$K(\theta_A) = K(0.50) = 4.91 \times \exp[16.8(0.50 - 0.55)] = 2.12 \text{ cm day}^{-1}$$

$$K(\theta_B) = K(0.45) = 4.91 \times \exp[16.8(0.45 - 0.55)] = 0.92 \text{ cm day}^{-1}$$

and the average value is:

$$\bar{K}_1 = \frac{[K(\theta_A) + K(\theta_B)]}{2} = 1.52 \text{ cm day}^{-1}$$

It is important to note that the mean value of K could be calculated differently, first by calculating the mean value of θ and then that of K:

$$\bar{\theta} = \frac{\theta_A + \theta_B}{2} = 0.475 \text{ cm}^3 \text{ cm}^{-3}$$

and:

$$\bar{K}_2 = K(\bar{\theta}) = 4.91 \times \exp[16.8(0.475 - 0.55)] = 1.39 \text{ cm day}^{-1}$$

As you can see, this difference becomes larger with increasing difference in θ. The choice of each of the procedures will depend on the judgment of each researcher.

Thus, the mean water flow between the two tensiometers is:

$$\bar{q} = -\bar{K}_1 \frac{(H_B - H_A)}{L} \quad \text{or} \quad -\bar{K}_2 \frac{(H_B - H_A)}{L}$$

$$\bar{q} = -1.52 \times \frac{[-212 - (-333)]}{200} = 0.92 \text{ cm day}^{-1}$$

In addition to the hydraulic conductivity, there is another soil water parameter, called **soil water diffusivity**. This new parameter was introduced as follows: for horizontal flow $H = h$, since the gravitational component does not come into play.

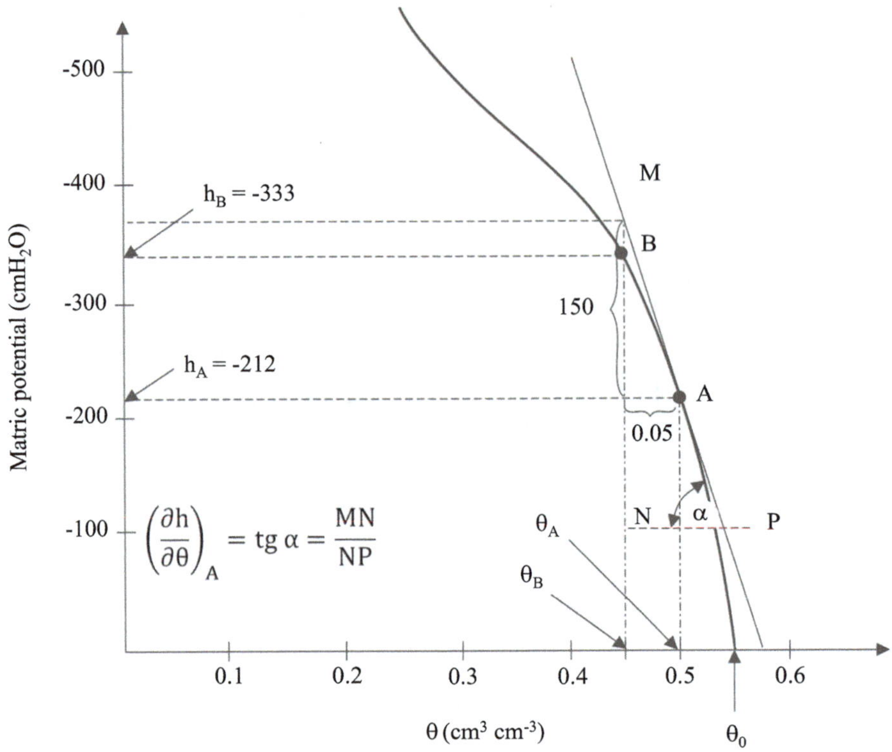

Fig. 7.13 Soil water retention curve (soil of Fig. 7.12) showing the procedure of evaluating the derivative dh/dθ

Thus, the equation of Darcy-Buckingham (Eq. 7.16) becomes:

$$q = -K\left(\frac{\partial h}{\partial x}\right)$$

and since $h = h(\theta)$ (soil water retention curve), we can introduce its derivative, based on the property of a function of a function, because as $h = h(\theta)$ and $\theta = \theta(x)$, the derivative $\partial h/\partial x$ can be replaced by [dh/dθ] × [$\partial\theta/\partial x$] and so:

$$q = -K\left(\frac{dh}{d\theta}\right)\left(\frac{\partial \theta}{\partial x}\right) = -D\left(\frac{\partial \theta}{\partial x}\right) \quad (7.16a)$$

where:

$$D = K\left(\frac{\partial h}{\partial \theta}\right) \quad (7.17)$$

Since $h(\theta)$ is a characteristic, its derivative dh/dθ is also characteristic. D is the diffusivity of the water in the soil, sometimes called, erroneously, **hydraulic diffusivity**, defined by Eq. (7.17). It is then the product of K (at a given value of θ) by the tangent to the characteristic curve (at the point corresponding to the same value of θ).

The Darcy-Buckingham equation in Eq. (7.16a) is often preferable in the case of horizontal movement because the soil water gradient $\partial\theta/\partial x$ is more easily determined than the potential gradient $\partial h/\partial x$. The problem is that Eq. (7.16a) involves **hysteresis** (see Chap. 6) and $\partial h/\partial\theta$ is not unique for a given value of θ. In general, the hysteresis is neglected, or, at best, the "wetting" curve is used in wetting cases and the "drying" curve in the case of drying.

The D parameter was called diffusivity because the Darcy-Buckingham equation, in form Eq. (7.16a), is identical to the Fick equation for diffusion of heat or ions.

In the case of vertical flow, D can also be introduced in the equations:

$$q = -K\frac{\partial H}{\partial z} = -K\frac{\partial}{\partial z}(h+z)$$
$$= -K\frac{\partial h}{\partial z} - K = -K\frac{\partial h}{\partial \theta}\frac{\partial \theta}{\partial z} - K$$
$$= -D\frac{\partial \theta}{\partial z} - K$$

in summary, the two forms used for vertical fluxes are:

$$q = -K\left(\frac{\partial h}{\partial z} + 1\right) \quad (7.18)$$

$$q = -D\frac{\partial \theta}{\partial z} - K \quad (7.19)$$

and for the case of the example of Fig. 7.12, we have:

$$D(\theta_A) = K(\theta_A)\left(\frac{\partial h}{\partial \theta}\right)_A$$
$$\cong 2.12 \times \frac{150}{0.05} = 6360\ \text{cm}^2\ \text{min}^{-1}$$

$$D(\theta_B) = K(\theta_B)\left(\frac{\partial h}{\partial \theta}\right)_B$$
$$\cong 0.92 \times \frac{108}{0.05} = 1987\ \text{cm}^2\ \text{min}^{-1}$$

The average value of the diffusivity is:

$$\bar{D} = \frac{[D(\theta_A) + D(\theta_B)]}{2} = 4174\,\text{cm}^2\ \text{min}^{-1}$$

and the flux density by Eq. (7.16a):

$$\bar{q} = -\bar{D}\left(\frac{\theta_A - \theta_B}{L}\right) = -\bar{D}\left(\frac{0.45 - 0.50}{30}\right)$$
$$= 6.96\,\text{cm}\,\text{min}^{-1}$$

which is 7.6 times greater than that obtained in the example using conductivities. Theoretically the values should be the same (in practice, similar), but, as we can see, the largest source of errors in these calculations, with the Darcy-Buckingham equation, is in the estimates of K, D, and $dh/d\theta$, and this error increases with increasing gradient. The physical characteristic of K and D varying greatly with small variations of θ or h introduces large errors in these calculations, that is, a small error in the measure of θ or of h leads to large errors in the estimation of K or D.

Coming back to the discussion on the Darcy-Buckingham equation (Eq. 7.3), it is important to stress that the water flux density in the soil, being the product of hydraulic conductivity of the soil by the hydraulic potential gradient, depends on the combination of these two quantities. A very small conductivity in the presence of a large gradient may result in a reasonable flow. A high conductivity and a small gradient can also allow a reasonable flux density. When the two are relatively large, the flux density assumes great proportions, and when both are small, the flux density becomes negligible. An impermeable soil layer has $K = 0$ and, under such conditions, always $q = 0$, even in the presence of a non-zero gradient. On the other hand, a null gradient also implies $q = 0$, even if K is large.

Since K and D decrease drastically with θ, for the same gradient $\partial\theta/\partial x$, the smaller the θ, the smaller the flux density. Therefore, the movement of water in a dry soil is generally much lower than in a moist soil.

When the water infiltrates into a dry soil, the upper layer is almost saturated, and K is maximum ($K = K_0$). In addition, the potential gradient between dry and wet soil is enormous. We have then a very large flow. Hence, it is thought that water moves faster into a dry soil. When water infiltrates into moist soil, K is large, but the gradient is small, and infiltration (flow) is small when compared to infiltration into dry soil.

7.3 Equation of Continuity

Only the knowledge of the flux density q by the application of the Darcy equation is not enough in dynamic studies of soil water. In fact, what interests us most is to know, at a given point M, in the soil profile, how the water content varies with time as water flows. For any situation, we would like to have an equation of the type $\theta = \theta(x, y, z, t)$, that is, an equation that allows us to determine θ for any value of chosen x, y, and z (in any position) and for any value of t (at any time). The **continuity equation** will give us the

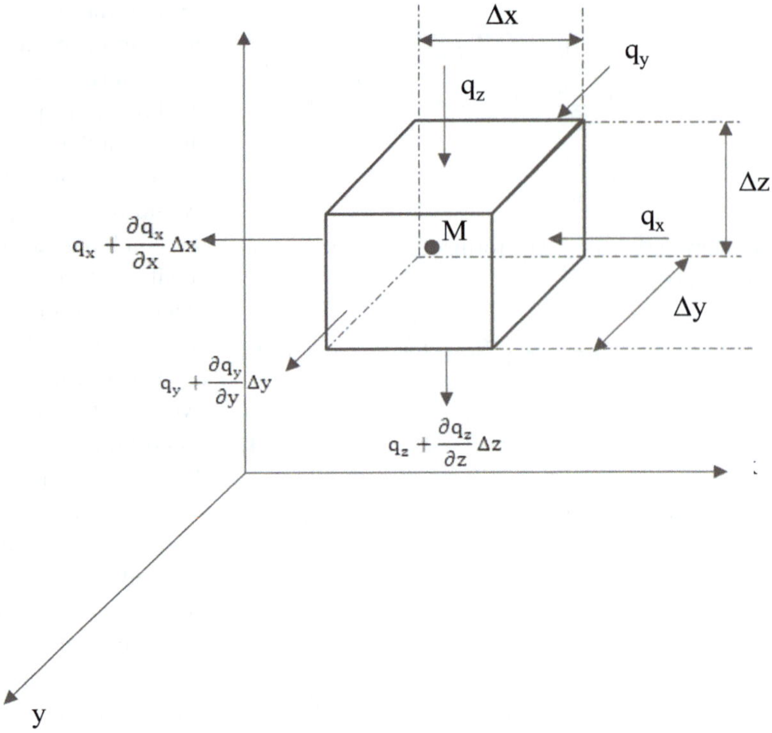

Fig. 7.14 Schematic view of a point M located somewhere in the soil profile

means to establish a differential equation of θ (dependent variable), whose solving for each particular problem is the function $\theta = \theta\ (x, y, z, t)$.

Given a volume element (ΔV) of soil around the generic point M, located in the soil profile, at which we want to study the changes of soil water content, as indicated in Fig. 7.14, we will study changes in θ. The water flux density q that enters the volume element is a vector of any direction, not shown in the figure. As already said, q being a vector, it can be decomposed into the three orthogonal directions x, y, and z, resulting in the modules q_x, q_y, and q_z. Thus, q_x is the flux density of water entering the volume element in the x direction (volume of water per unit time and area, L m^2 day^{-1}, or mm day^{-1}).

The amount of water Q_x (flow) entering the face $\Delta y \cdot \Delta z$ of the element volume (perpendicular to x) per unit of time Δt is then $q_x \cdot \Delta y \cdot \Delta z$ (volume of water per unit time; see definition of flow density, Eq. 7.4). Therefore:

$$\frac{Q_x}{\Delta t} = q_x \cdot \Delta y \cdot \Delta z$$

Considering that along the direction x there can be a change in the flux density q_x, equal to $\partial q_x/\partial x$, the flux density q'_x that leaves the opposite side of the volume element in the direction x will be:

$$q'_x = q_x + \left(\frac{\partial q_x}{\partial x}\right) \cdot \Delta x$$

It is easy to see that $\partial q_x/\partial x$ is the variation of q_x per unit of x and that the total variation along Δx is the product $(\partial q_x/\partial x) \cdot \Delta x$. If, for example, $\partial q_x/\partial x = 0.01$ cm day^{-1} cm^{-1} and $\Delta x = 5$ cm, the total variation is 0.05 cm day^{-1}.

The amount of water Q'_x that leaves the opposite face, also of area $\Delta y \cdot \Delta z$, in the unit of time Δt, is then:

$$\frac{Q'_x}{\Delta t} = \left(q_x + \frac{\partial q_x}{\partial x}\right) \cdot \Delta y \cdot \Delta z$$

The change of the amount of water in the volume element per unit time $\partial Q_x/\partial t = \Delta Q_x/\Delta t$, in the x direction, is the difference between the incoming quantity and the quantity that leaves the element (balance):

$$\left(\frac{Q_x}{\Delta t} - \frac{Q'_x}{\Delta t}\right) = \frac{\Delta Q_x}{\Delta t} = \frac{\partial Q_x}{\partial t}$$

So:

$$\frac{\partial Q_x}{\partial t} = q_x \cdot \Delta y \cdot \Delta z - \left(q_x + \frac{\partial q_x}{\partial x}\Delta x\right) \cdot \Delta y \cdot \Delta z$$

or simplifying:

$$\frac{\partial Q_x}{\partial t} = -\frac{\partial q_x}{\partial x} \cdot \Delta x \cdot \Delta y \cdot \Delta z = -\frac{\partial q_x}{\partial x} \cdot \Delta V$$

because $\Delta x \cdot \Delta y \cdot \Delta z = \Delta V$, the volume of the element chosen around M.

Using the same reasoning in the directions y and z, we obtain similar equations:

$$\frac{\partial Q_y}{\partial t} = -\frac{\partial q_y}{\partial y} \cdot \Delta y \cdot \Delta x \cdot \Delta z = -\frac{\partial q_y}{\partial y} \cdot \Delta V$$

$$\frac{\partial Q_z}{\partial t} = -\frac{\partial q_z}{\partial z} \cdot \Delta z \cdot \Delta x \cdot \Delta y = -\frac{\partial q_z}{\partial z} \cdot \Delta V$$

and the total change $\partial Q/\partial t$, in the element ΔV, will be the sum of the changes in the three directions:

$$\frac{\partial Q}{\partial t} = -\left(\frac{\partial q_x}{\partial x} + \frac{\partial q_y}{\partial y} + \frac{\partial q_z}{\partial z}\right) \cdot \Delta V$$

Since the size of ΔV was not defined, it is now convenient to calculate the change of the quantity of water per unit volume, dividing both sides of the equation by ΔV, and so, the left-hand side becomes $\partial \theta/\partial t$, since θ is the quantity of water per unit volume:

$$\frac{\partial \theta}{\partial t} = -\left(\frac{\partial q_x}{\partial x} + \frac{\partial q_y}{\partial y} + \frac{\partial q_z}{\partial z}\right) \quad (7.20)$$

This is the **equation of continuity** that can be applied to cases of water moving in porous materials. Let us see how one can understand it. For this, we will rewrite it in one dimension only:

$$\frac{\partial \theta}{\partial t} = -\frac{\partial q_x}{\partial x} \quad (7.20a)$$

This equation tells us that, at a chosen point M in the soil, the change in water content θ with time t, due to horizontal flow only, is equal to the variation of the flow q_x in the x direction. This means that only when the flux varies along x, θ varies with time. Logically, if q_x varies along x, more water enters ΔV than leaves (and θ increases) or less water enters ΔV than leaves (and θ decreases). If the same input quantity also leaves the volume, it is because q_x did not change along x, i.e., q_x = constant and $\partial q_x/\partial x = 0$ and $\partial \theta/\partial t = 0$; there is no change in water content over time. The latter case is the steady-state case.

By the Darcy-Buckingham equation (Eq. 7.3), we know that:

$$q_x = -K(\theta)_x \frac{\partial H}{\partial x}$$

$$q_y = -K(\theta)_y \frac{\partial H}{\partial y}$$

$$q_z = -K(\theta)_z \frac{\partial H}{\partial z}$$

where the indexes x, y, and z in the function $K(\theta)$ indicate that K can be different in the three directions.

Substituting these values in Eq. (7.20), we have:

$$\frac{\partial \theta}{\partial t} = -\left\{\frac{\partial}{\partial x}\left[K(\theta)_x \frac{\partial H}{\partial x}\right] + \frac{\partial}{\partial y}\left[K(\theta)_y \frac{\partial H}{\partial y}\right] + \frac{\partial}{\partial z}\left[K(\theta)_z \frac{\partial H}{\partial z}\right]\right\} \quad (7.20b)$$

which is the most general differential equation of soil water movement. Prevedello and Reichardt (1991) is a very good example of the application of Eq. (7.20b).

This equation is often written as:

$$\frac{\partial \theta}{\partial t} = \nabla \cdot K\nabla H = \mathrm{div}(K \cdot \mathrm{grad}\, H) = \mathrm{div}\, q$$

where **div** represents the **divergent**, which is also a vector operator. When the operator ∇ operates on a vector ($K \times$ grad H is a vector), the result is the divergent. For example, v = velocity is a vector. So, ∇v is the divergent of the velocity.

$$\vec{\nabla} \,.\, \vec{v} = \text{div } \vec{v}$$

where the dot (.) indicates the **scalar product** of two vectors. Just to mention, there is also a **vectorial product** of two vectors, which is also a vector.

To make the operation, we simply make a scalar product of ∇ and v. To do that we separate the vectors in their components:

$$\vec{\nabla} \cdot \vec{v} = \left(\frac{\partial}{\partial x}\vec{i} + \frac{\partial}{\partial y}\vec{j} + \frac{\partial}{\partial z}\vec{k}\right) \cdot \left(v_x \vec{i} + v_y \vec{j} + v_z \vec{k}\right) = \frac{\partial}{\partial x}\vec{i}\; v_x \vec{i} + \frac{\partial}{\partial x}\vec{i}\; v_y \vec{j} + \frac{\partial}{\partial x}\vec{i}\; v_z \vec{k} + \frac{\partial}{\partial y}\vec{j}\; v_x \vec{i} \ldots$$

Let's recall the scalar product of two vectors m and n. This product is a scalar of the product of the modulus of m by the modulus of n and by the cosine of the angle between the vectors m and n. For example, the scalar product of a force (vector) by a distance (also a vector since it has a direction) is mechanical work or energy (scalars). So, in the above equation, $i \cdot i = j \cdot j = k \cdot k = 1$, because they are products of unit vectors of the same direction, and $\cos 0° = 1$, and since $i \cdot j = i \cdot k = k \cdot j = \ldots = 0$ (product of perpendicular vectors, $\cos 90° = 0$), the result is:

$$\vec{\nabla} \,.\, \vec{v} = \frac{\partial v_x}{\partial x} + \frac{\partial v_y}{\partial y} + \frac{\partial v_z}{\partial z}$$

We see, then, that the divergent is the result of the operation of ∇ on a vector quantity and the result is a scalar quantity. For the velocity example, its divergent is a measure of the sum of the changes of its components along the x, y, and z directions.

Since the gradient of a scalar is a vector, we can obtain the divergent of a gradient. For example:

T = scalar
grad T = vector
div (grad T) = scalar

In the same way, the divergent of the flux density of water q is the change of the water content in time, as indicated by Eq. (7.20b).

An **isotropic material** for water flow is that one for which K does not change in any direction, so that $K(\theta)_x = K(\theta)_y = K(\theta)_z$. Otherwise, the material is anisotropic. Layered soils are examples of anisotropic materials, mainly in the vertical direction.

In the x dimension, Eq. (7.20b) will be:

$$\frac{\partial \theta}{\partial t} = \frac{\partial}{\partial x}\left[K(\theta)_x \frac{\partial H}{\partial x}\right] \tag{7.20c}$$

which is also called the **Richards equation**.

Three particular cases of the use of Eq. (7.20c) can now be distinguished:

Case (a) **steady-state flow**, also referred to as **permanent regime**, in which the flux density q is a constant and consequently its components q_x, q_y, and q_z are also constant. Since the derivative of a constant is 0, so $\partial\theta/\partial t = 0$. The permanent regime is characterized by the invariability of the system with respect to time, but with a variability with respect to position. In this case, θ does not change with t ($\partial\theta/\partial t = 0$) but varies with x ($\partial\theta/\partial x \neq 0$), and this water content gradient determines the constant flux density q. A steady-state system was discussed in Fig. 7.9. The amount of water entering the soil in A and leaving through B and the soil water content does not change over time.

In the steady state, Eq. (7.20c) becomes:

$$\frac{\partial}{\partial x}\left[K(\theta)\frac{\partial H}{\partial x}\right] = 0$$

In the case of $K(\theta_0)$ constant $= K_0$ (saturated soil), we can further simplify the equation:

$$\frac{\mathrm{d}}{\mathrm{d}x}\left(\frac{\mathrm{d}H}{\mathrm{d}x}\right) = \frac{\mathrm{d}^2H}{\mathrm{d}x^2} = 0 \cdot K_0 = 0 \quad (7.21)$$

It is convenient to remember that Eq. (7.21) is independent of time, hence the total differentials **d** and not the partial ∂, since H is only a function of x. Soil water content θ or potential H varies in space, but not in time. Because there is flux, it is necessary that θ and H vary in space, and this variation is the gradient responsible for the flow. Hence the name dynamic equilibrium.

In three dimensions, $H = H\ (x,\ y,\ z)$ and Eq. (7.21) becomes:

$$\frac{\partial^2 H}{\partial x^2} + \frac{\partial^2 H}{\partial y^2} + \frac{\partial^2 H}{\partial z^2} = 0$$

or:

$$\nabla^2 H = 0; \quad \mathrm{div}\ q = 0$$

these last equations are called **equations of Laplace** is.

Case (b) variable flow or **transient regime**: it is the most general case from which the potentials can vary with time and, of course, with position. In this case, the differential equations used are Eqs. (7.20b) for three dimensions and (7.20c) for one dimension.

Case (c) no flow, thermodynamic equilibrium: in this case the system is static, $\partial\theta/\partial t = 0$, or the gradient, or even $K(\theta)$ is 0.

7.4 Saturated Soil Water Flux

When studying the flow of water in the soil, it is convenient to distinguish between water flow in a saturated soil, which we will simply call saturated flow, and water flow in unsaturated soil. In the first case, θ is not variable, it is constant and equal to the porosity α ($\theta_0 = \alpha$), and K is also constant, assuming the value K_0. For the case of water flow in unsaturated soil, this does not happen, θ and K vary, and everything gets complicated. In the saturated flow, only the gravitational and pressure components of the total water potential are considered. With the soil saturated, water will always be under positive or 0 pressure, never negative.

Because the soil is saturated:

$$\alpha = \theta = \theta_0(\text{saturation}) = \text{constant}$$

$$\frac{\partial\theta}{\partial t} = 0$$

$K = K_0 = \text{constant}$:

$$\Psi_T = \Psi_g + \Psi_P \text{ or } H = z + h$$

$$q = -K_0\frac{\mathrm{d}h}{\mathrm{d}x}\ (\text{horizontal flux})$$

$$q = -K_0\frac{\mathrm{d}H}{\mathrm{d}z}\ (\text{vertical flux})$$

Since θ is constant, $\partial\theta/\partial t = 0$ and we have:

$$\frac{\mathrm{d}^2H}{\mathrm{d}x^2} = 0\ (\text{horizontal flux})$$

$$\frac{\mathrm{d}^2H}{\mathrm{d}z^2} = 0\ (\text{vertical flux})$$

Example 1: Consider the saturated soil of Fig. 7.10. The differential equation that describes the flow in the x direction is:

$$\frac{\mathrm{d}^2H}{\mathrm{d}x^2} = 0$$

In that example we succeeded to calculate q and K_0; however, we have no information of $H = h$ inside the column, that is, we look for a solution of the type $H(x)$. At the ends of the soil column, A and B, called boundaries, we know H_A and H_B. Our problem lies in one direction, x passing through A and B. If x at $A = 0$ and at $B = 50$ cm, we have $H = +80$ cm H_2O at $x = 0$ and $H = +20$ cm H_2O at $x = 50$ cm. Which would be the value of H at any point between A and B? We do not have this information. It will be given by the solution of the problem, which is the solution of the differential equation of H, which in this case is the equation $\mathrm{d}^2H/\mathrm{d}x^2 = 0$. This is the generic form of the continuity equation applied to our problem, which is a particular case of dynamic equilibrium. Our problem is, therefore, to find its solution, that is, a function $H = H(x)$, satisfying the condition that its second derivative is 0. This function will allow us to determine H for any x, that is, for any point between A and B.

The reader must remember that the solution of differential equations (Churchill 1963) is always made by attempts. Several methods are used until a solution is found. Often the solution is not found and the problem can only be solved numerically. Our case, however, is one of the simplest. We know that the function whose second derivative is null is a straight line. So we attempt for a solution of the type:

$$H = a \cdot x + b \tag{7.22}$$

because:

$$\frac{dH}{dx} = a \quad \text{and} \quad \frac{d^2H}{dx^2} = 0$$

Equation (7.22) is called the **general solution**, because the values of a and b are not defined. It actually represents infinite lines and just tells us that H varies linearly along x. It's already something.

The constants a and b appeared in Eq. (7.22) because, although tentatively found, their origin lies in the integration of Eq. (7.21). Since Eq. (7.21) is a second derivative, two integrations were required, and in each integration, an indefinite constant appears. In the first integration, **a** appears, and in the second, **b** appears.

The determination of the constants a and b for our particular problem transforms the general solution into the **particular solution**, which is a well-defined line, valid only for the problem of Fig. 7.9. For this, we need the **boundary conditions**, already mentioned:

At A: $x = 0$ cm; $H = 80$ cm H_2O (first condition)
At B: $x = 50$ cm; $H = 20$ cm H_2O (second condition)

Since the general solution is valid for any x, it is also valid for A and B:

At A: $80 = a \cdot 0 + b$ $(a\,.\,0 + b)$
At B: $20 = a \cdot 50 + b$ $(a\,.\,50 + b)$

So that:

$a = -1.2$ and $b = 80$

and the particular solution is:

$$H = -1.2x + 80 \tag{7.22a}$$

If the reader wants to test if Eq. (7.22a) is the correct solution, simply apply the boundary conditions to it, and check if it works. Thus, for $x = 0$, it indicates $H = 80$; for $x = 50$ indicates $H = 20$, so it is the correct solution.

Equation (7.22a) is the particular solution of our problem. With it we can calculate H at any point in the soil, without making direct measurements. For example, what is the value of H at $x = 10$ cm? Applying Eq. (7.22a), we have $H = 68$ cm H_2O.

What is the flow density of water in the soil?

$$\frac{dH}{dx} = \frac{d}{dx}(-1.2 \times x + 80) = -1.2\frac{\text{cm H}_2\text{O}}{\text{cm of soil}}$$

K_0, already previously calculated, is 4.91 cm day^{-1}, so that:

$$q = -4.91 \times (-1.2) = 5.89 \text{ cm day}^{-1}$$

This example, although simple, is a typical example of the solution of **boundary value problems** (BVPs). These problems consist of a differential equation (Eq. 7.21), boundary conditions (conditions involving the position coordinate at the "ends" of the system under analysis), and initial, intermediate, or final conditions (not present, in this case, because it is a dynamic equilibrium case with no beginning and no end). The BVP solution is a mathematical equation that indicates how the variable of interest, taken as dependent, is a function of the independent variables, space, and time. In the case of the problem in question is Eq. (7.22).

In general, the number of conditions necessary for the solution of a BVP depends on the order of the largest partial derivative. In our example (Eq. 7.21), we only have a partial second-order derivative in relation to space, so we needed two boundary conditions. Yet, an equation of the type:

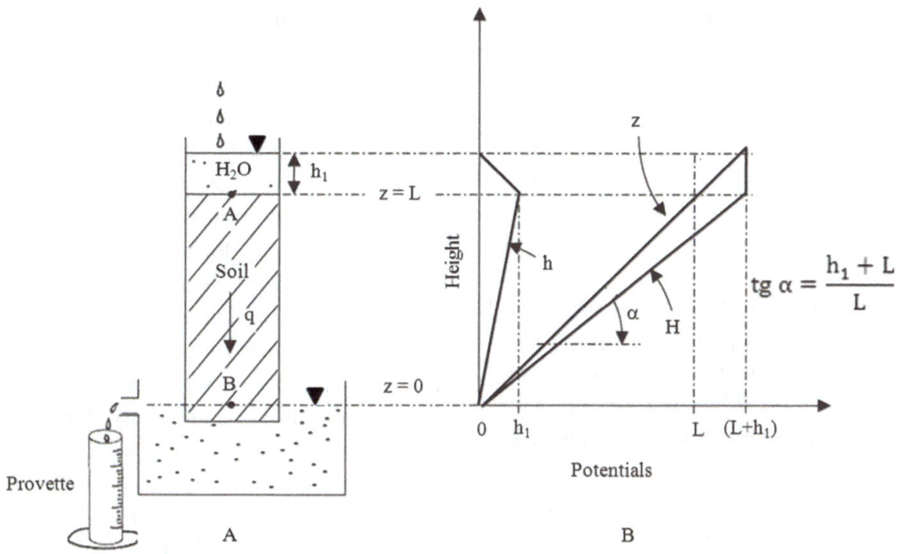

Fig. 7.15 A soil column with water infiltrating vertically in steady-state condition

$$\frac{\partial \theta}{\partial t} = \frac{\partial}{\partial t}\left[D(\theta)\frac{\partial \theta}{\partial x}\right]$$

demands one condition in time (initial condition) and two conditions in x (boundary conditions), because, although not showing explicitly, the second member is a second derivative of θ in relation to x.

Example 2: Here we have the case of saturated flow in a vertical column of soil, as shown in Fig. 7.15. The differential equation will be the same as in the previous problem:

$$\frac{d^2H}{dz^2} = 0$$

with a general solution:

$$H = a \cdot z + b$$

with the boundary conditions, we have:

$$H_A = h_1 + L$$

$$H_B = 0 + 0$$

Substituting these values in the general solution, we have:

$$h_1 + L = aL + b$$

$$0 = a \cdot 0 + b$$

or

$$b = 0$$

$$a = \frac{(h_1 + L)}{L}$$

and the particular solution becomes:

$$H = \frac{(h_1 + L)}{L} \cdot z \tag{7.23}$$

The graph of the distribution of potentials is schematized in Fig. 7.15B. These graphs are discussed in Chap. 6. The flow density will be:

$$q = -K_0\frac{\partial H}{\partial z} = -K_0\left(\frac{h_1 + L}{L}\right)$$
$$= -K_0\frac{h_1}{L} - K_0 \tag{7.24}$$

The negative sign indicates (by our convention) that the flow is from top to bottom.

Imagine, now, that with this soil an experiment was done varying h_1 and measuring q in a provette. The values obtained, for $L = 50$ cm, are shown in the following table:

h_1 (cm)	q (cm min^{-1})
10	−0.60
20	−0.71
30	−0.79
40	−0.90

Of course, the higher the water depth h_1, the greater the flow. If we plot q as a function of h_1 (which in this experiment is a variable), we obtain a straight line (see Eq. 7.24), whose slope must be equal to K_0/L and whose linear coefficient is K_0. This graph is shown in Fig. 7.16.

In this figure, it is verified that the linear coefficient is 0.5 and the slope is:

$$tg\ \alpha = \frac{0.3}{30} = 0.01$$

and $K_0/L = 0.5/50 = 0.01$.

This is a more accurate method of determining K_0. The same can be done with a horizontal column, as seen in the previous example. One can also plot the flow density q as a function of the gradient of H, and in this case, the slope is directly K_0. In Fig. 7.17, this graph is presented for extreme textures: sandy soil and clayey soil.

In a stable (rigid) soil, the hydraulic conductivity K_0 is a constant characteristic of the material. The hydraulic conductivity is obviously affected by soil structure and texture, being higher in highly porous, fractured, or aggregated soils and lower in dense and compacted soils. The conductivity depends not only on the total porosity (α) but, in particular, on the pore size and the activity of the present clays.

For example, a sandy soil in general has a higher hydraulic conductivity than a clayey soil, although the former has less total porosity than the latter.

If, for the example of Fig. 7.15, we have a **stratified soil or a layered soil** with layers of different hydraulic conductivities, logically the smaller K_0 will limit the flow. This is the case in Fig. 7.18, with soil 1 that is sandier on top of soil 2, more clayey.

Since we have a steady-state condition:

$$q = q_1 = q_2 = -K_{01}\frac{\partial H}{\partial z} = -K_{02}\frac{\partial H}{\partial z}$$

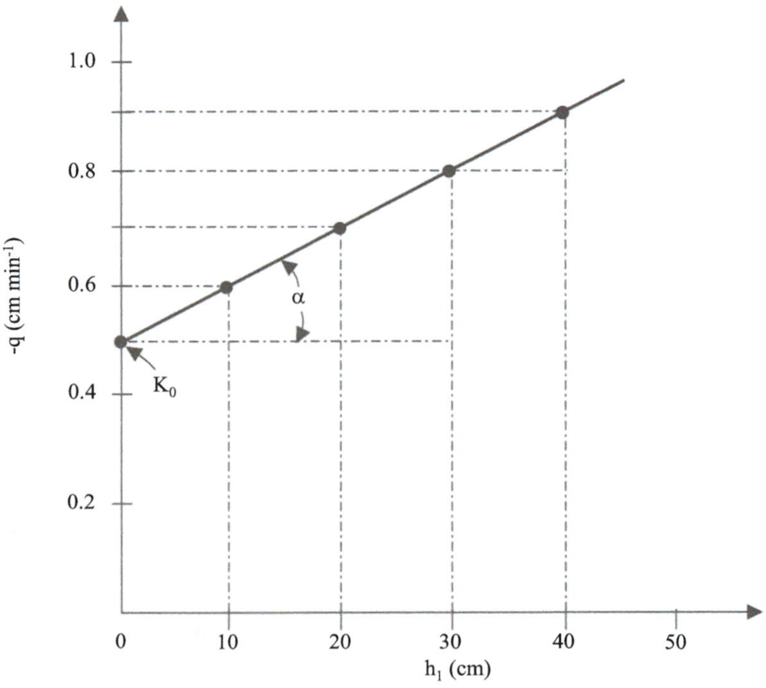

Fig. 7.16 Regression line between water flow q and water depth for experiments made with the soil column of Fig. 7.15

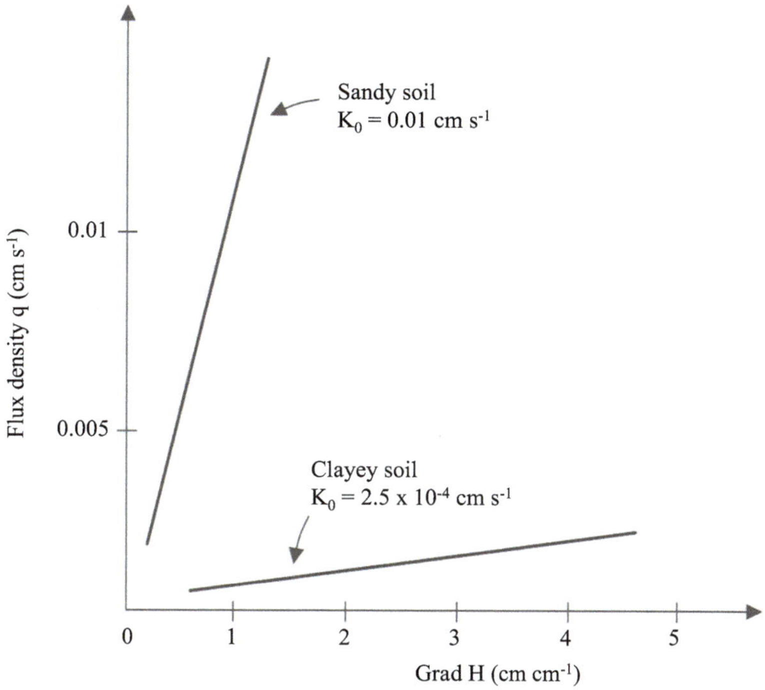

Fig. 7.17 Regression between soil water flux density q and the gradient (Grad H) responsible for the flow

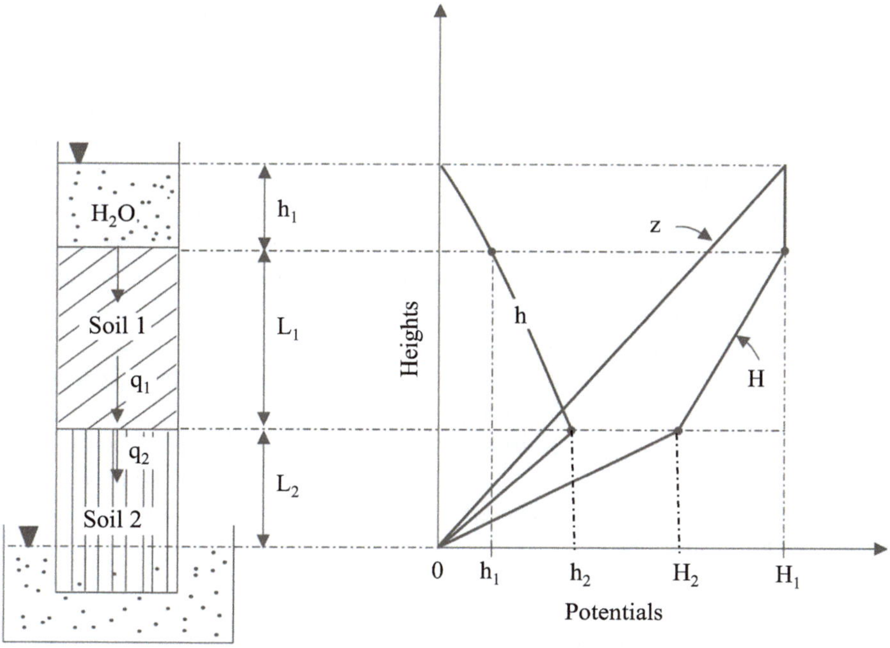

Fig. 7.18 Steady vertical infiltration into a layered column of soil, soil 1 is sandier and soil 2 is more clayey, indicating the potential distributions

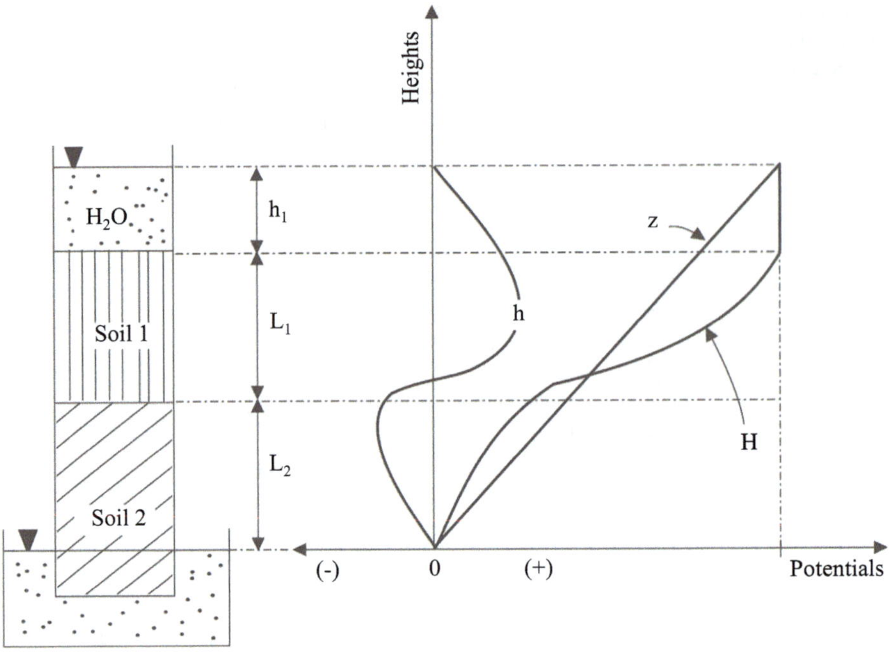

Fig. 7.19 Steady vertical infiltration into a layered column of soil, soil 1 is more clayey and soil 2 is sandier, indicating the potential distributions

If, for example, $K_{01} = 3 \cdot K_{02}$, then $\partial H/\partial z$ in soil 1 must be three times smaller than in soil 2, $(\partial H/\partial z)_1 = (\partial H/\partial z)_2/3$. It is the case of Fig. 7.18, of a more permeable soil (soil 1) on top of a less permeable soil (soil 2). Inverting the situation, as in Fig. 7.19, in which the less permeable soil is above, a suction (negative values of h) may even occur in the most permeable soil below.

The higher conductivity of soil 1 allows a higher flow density, which leads to a suction at the bottom interface of soil 2, which can extend a few centimeters into soil 2. As a result, soil 1 cannot be saturated. The correct graph of h, consequently, of H, will depend on each experimental arrangement.

The Darcy equation is not universally valid for all conditions of fluid movement in porous materials. It has long been recognized that the linearity of the flux-gradient relationships (Figs. 7.16 and 7.17) fails for very low and very high values of the gradient of H. Details of the limitations of the Darcy equation can be found, among others, in Hubert (1956), Swartzendruber (1962), Miller and Low (1963), and Klute (1986).

Methods for measuring soil hydraulic conductivity were reviewed by Klute (1986) and Reichardt (1996). In the laboratory, the hydraulic conductivity of a saturated soil is measured in permeameters, outlined in Fig. 7.20, and in the field, the most convenient methods are those of the cavity of Luthin (1957) and of the piezometer by Johnson et al. (1952). Later, in the application chapters, this subject will be addressed again.

7.5 Non-saturated Soil Water Flux

The water flow is called unsaturated when it occurs in the soil at any water content condition θ below the saturation value (θ_0). Most processes involving soil water movement, either in- or outside a crop, occur with the soil under unsaturated conditions. These unsaturated flow processes are generally complicated and of difficult quantitative description. Variations of soil water content during its movement involve complex functions between the variables θ, H, and K or D, which can be affected by hysteresis. The formulation

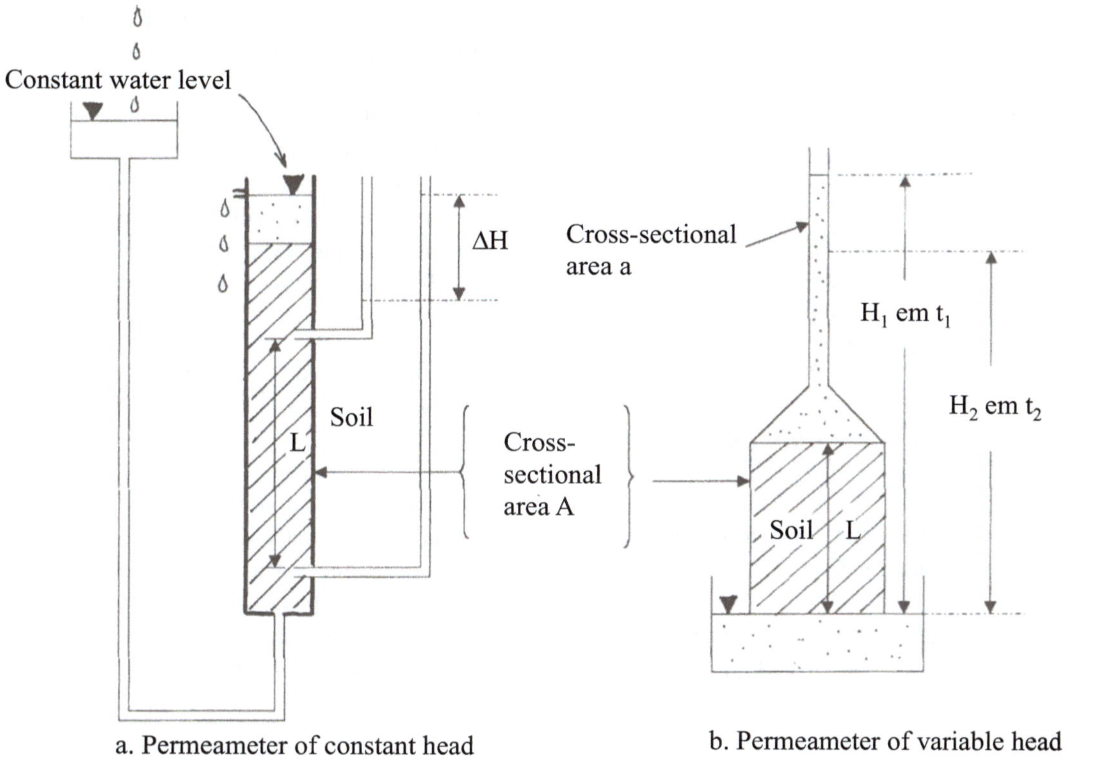

$$K_0 = \frac{V.L}{A.t.\Delta H}$$

$$K_0 = \left[\frac{2.3 \times a \times L}{A(t_2 - t_1)}\right]\left(\log\frac{H_1}{H_2}\right)$$

Fig. 7.20 Schematic views of permeameters: left-hand side, constant head permeameter; right-hand side, variable head permeameter

and resolution of unsaturated flow problems often require the use of complex methods of mathematical analysis, and numerical approximation techniques are often required.

For the case of unsaturated flow, without the presence of semipermeable membranes, only the matrix component h and the gravitational z are of importance. The equations used for the unsaturated flow are:

$\theta_0 > \theta > 0$
$H = h + z$
$h = h(\theta)$, experimental or modeled soil water retention curve
$K = K(\theta)$ or $K(h)$, experimental or modeled functions
$D = D(\theta)$, experimental or modeled functions

For the horizontal flux cases in the x direction, we have:

$$H = h$$

$$q = -K(\theta)\frac{\partial h}{\partial x}$$

$$q = -D(\theta)\frac{\partial \theta}{\partial x}$$

$$\frac{\partial \theta}{\partial t} = \frac{\partial}{\partial x}\left[K(\theta)\frac{\partial h}{\partial x}\right]$$

$$\frac{\partial \theta}{\partial t} = \frac{\partial}{\partial x}\left[D(\theta)\frac{\partial \theta}{\partial x}\right]$$

and for the vertical cases:

$$H = h + z$$

$$q = -K(\theta)\frac{\partial H}{\partial z} = -K(\theta)\frac{\partial h}{\partial z} - K(\theta)$$
$$= -K(\theta)\left(\frac{\partial h}{\partial z} + 1\right)$$

$$q = -D(\theta)\frac{\partial H}{\partial z} - K(\theta)$$

$$\frac{\partial \theta}{\partial t} = \frac{\partial}{\partial z}\left[K(\theta)\frac{\partial H}{\partial z}\right]$$

$$\frac{\partial \theta}{\partial t} = \frac{\partial}{\partial z}\left[D(\theta)\frac{\partial \theta}{\partial z} + \frac{\partial K}{\partial \theta}\cdot\frac{\partial \theta}{\partial z}\right]$$

In the above equations, when convenient, $K(\theta)$ can be substituted by $K(h)$.

Example 1: Let us consider the unsaturated flow in steady state, as shown in Fig. 7.21. It is not saturated because the soil is subject to suction or negative pressure at both ends. To initiate such an experiment, the water container C is lifted above point A, and hence the soil saturates, and the saturated flow is obtained. Thereafter, the container C is lowered, and equilibrium is expected. The column has to be drilled with tiny holes to allow aeration, and evaporation must be controlled. In equilibrium, that takes time to be reached, we have:

$$\frac{\partial \theta}{\partial t} = 0 = \frac{\partial}{\partial x}\left[K(h)\frac{\partial h}{\partial x}\right] \quad (7.25)$$

To solve this equation, it is necessary to know $K(h)$ which is normally determined experimentally. At the beginning of the chapter, we have seen exponential types of $K(\theta)$, and we may also have $K(h)$ functions. Imagine, for example, that for the soil under consideration, in the range

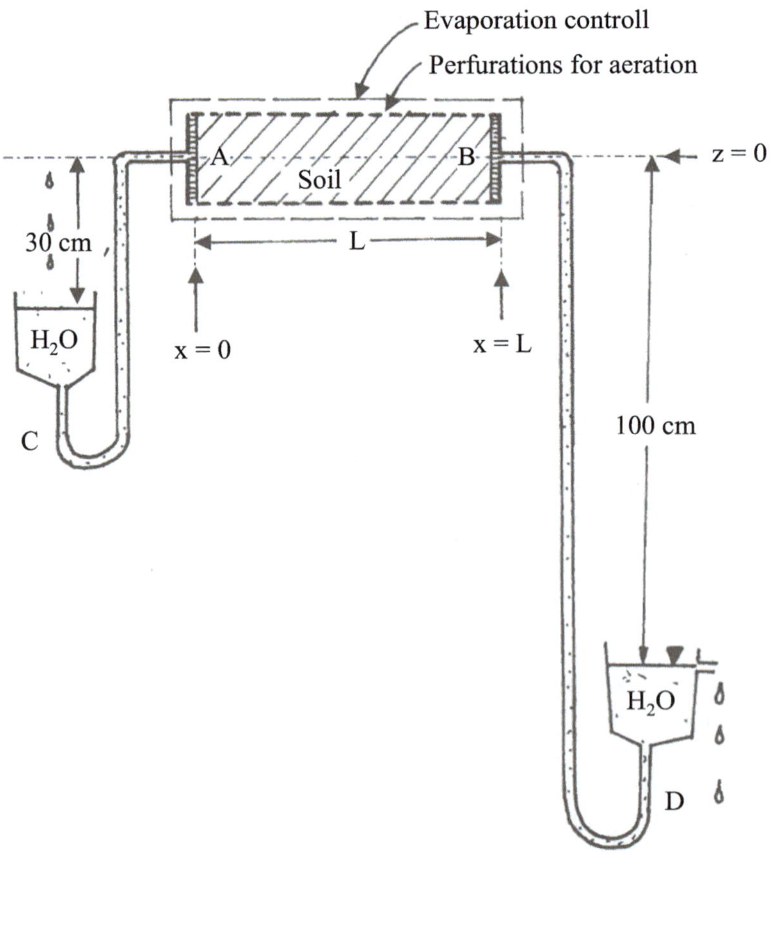

Fig. 7.21 An unsaturated soil column in the horizontal position, used to measure soil hydraulic conductivity

$-30 > h > -100$ cm of water, the relation K versus h is given by:

$$K(h) = \left(1 + \frac{h}{200}\right) \text{ cm h}^{-1} \quad (7.26)$$

For example, at point A, $K(-30) = 1 + (-30/200) = 0.85$ cm h^{-1}; at point B, $K(-100) = 1 + (-100/200) = 0.5$ cm h^{-1}; and at a point where $h = -70$, $K(-70) = 1 + (-70/200) = 0.65$ cm h^{-1}.

In practice, a separate experiment is carried out to determine $K(h)$. Having several values of K for each h, one can make the graph of K versus h. By numerical techniques, it is possible to adapt an equation to the experimental points of the graph. This is the case of Eq. (7.26). Substituting (7.26) in (7.25), we have:

$$\frac{d}{\mathrm{d}x}\left[\left(1 + \frac{h}{200}\right)\frac{\mathrm{d}h}{\mathrm{d}x}\right] = 0 \quad (7.27)$$

The partial derivatives (∂) have been replaced by the totals (d) because h is only a function of x and not of t.

We must now find the solution of Eq. (7.27), looking at the conditions of the problem. It is a bit more complicated than in the case of saturated flow, where K was a constant and the differential equation simplified to $\mathrm{d}^2H/\mathrm{d}x^2 = 0$, whose solution is a straight line. However, since Eq. (7.27) is a second-order differential equation, two integrations with respect to x will be required:

$$\int \frac{\mathrm{d}}{\mathrm{d}x}\left[\left(1 + \frac{h}{200}\right)\frac{\mathrm{d}h}{\mathrm{d}x}\right]\mathrm{d}x = C_1$$

where C_1 is the first integration constant. Thus, since the integral of a derivative is the function itself:

$$\left[\left(1 + \frac{h}{200}\right)\frac{\mathrm{d}h}{\mathrm{d}x}\right] = C_1$$

Separating the variables:

$$\left(1 + \frac{h}{200}\right)\mathrm{d}h = C_1 \cdot \mathrm{d}x$$

and integrating again:

$$\int \left(1 + \frac{h}{200}\right)\mathrm{d}h = C_1 = \int \mathrm{d}x + C_2$$

C_2 being the second integration constant.

$$\int \mathrm{d}h + \int \frac{1}{200} h \cdot \mathrm{d}h = C_1 \int \mathrm{d}x + C_2$$

$$h + \frac{h^2}{400} = C_1 x + C_2$$

multiplying both sides by 400 and adding $(200)^2$, we have:

$$400h + h^2 + 200^2 = 200^2 + 400\,C_1 x + 400\,C_2$$

Since $400C_1$ and $400C_2$ are also constant and C_1 and C_2 were arbitrarily introduced, they can be considered new constants:

$$(h + 200)^2 = 200^2 + C_1 x + C_2$$

or finding the roots of a second-order equation:

$$h = -200 + \sqrt{200^2 + C_1 x + C_2} \quad (7.28)$$

Equation (7.28) is the general solution to the problem. The particular solution is obtained by determining the values of C_1 and C_2, using the boundary conditions. For $x = 0$, $h = -30$, and for $x = L$, $h = -100$, then:

$$-30 = -200 + \sqrt{200^2 + C_1 \cdot 0 + C_2}$$

$$-100 = -200 + \sqrt{200^2 + C_1 \cdot L + C_2}$$

and:

$$C_1 = \frac{-18{,}900}{L} \quad \text{and} \quad C_2 = -11{,}100$$

and the particular solution will be:

$$h = -200 + \sqrt{40{,}000 - \frac{18{,}900}{L}x - 11{,}100} \quad (7.29)$$

This equation allows us to determine h at any point in the column ($L > x > 0$). Its graph is shown in Fig. 7.22 for a column of soil $L = 100$ cm. In this figure it is verified that the matrix potential

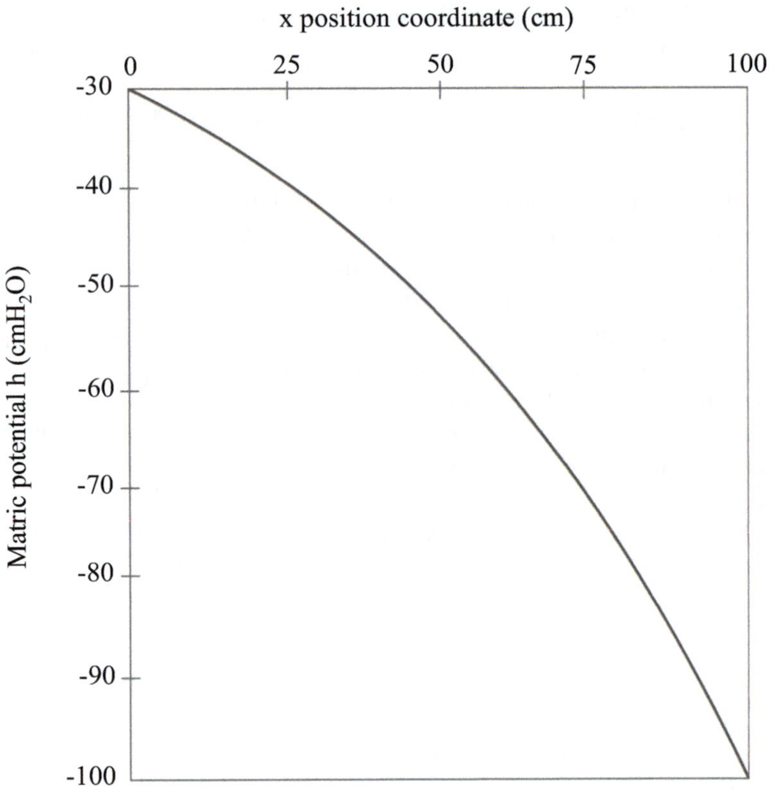

Fig. 7.22 Graph of Eq. (7.29) for the matric potential h versus position x

gradient (dh/dx) varies along x. From Eq. (7.26), it can also be seen that K varies from 0.85 to 0.5 cm h^{-1} along the column. Despite this, the flux density q is a constant. This is because the decrease in conductivity is counterbalanced by an increase in the gradient. Because of this, the flow density q has to be calculated taking into account the whole column:

$$q = -K(h)\frac{dh}{dx}$$

Separating the variables and integrating:

$$q\int_0^L dx = -\int_{-30}^{-100}\left(1+\frac{h}{200}\right)dh$$

$$q[x]_0^L = \left[h+\frac{h^2}{400}\right]_{-30}^{-100}$$

$$qL = -\left[-100+\frac{(-100)^2}{400}\right]+\left[-30+\frac{(-30)^2}{400}\right]$$

$$= 47.25$$

so that $q = 0.4725$ cm h^{-1}, for $L = 100$ cm, once L was not defined.

Example 2: A given soil has a conductivity following Eq. (7.30):

$$K(h) = 2\times e^{0.01\times h}\left(\text{cm day}^{-1}\right) \tag{7.30}$$

where **e** is the basis of the Neperian logarithms. Two tensiometers installed at the same depth and 20 cm apart record $h_A = -350$ cm H_2O and $h_B = -300$ cm H_2O. What is the horizontal water flow density between the tensiometers?

$$q = -K(h)\frac{dh}{dx}$$

Substituting K (Eq. 7.30) into this equation and separating the variables:

$$q \cdot \mathrm{d}x = -2 \times \mathrm{e}^{0.01 \times h} \cdot \mathrm{d}h$$

and integrating between the respective limits:

$$q \int_{0}^{20} \mathrm{d}x = -2 \int_{-350}^{-300} \mathrm{e}^{0.01 \times h}\, \mathrm{d}h$$

$$q[x]_0^{20} = -2\left[0.01 \times \mathrm{e}^{0.01 \times h}\right]_{-350}^{-300}$$

where $q = 3.96 \times 10^{-4}$ cm day^{-1}.

We have already said that K varies from soil to soil and that its values are determined experimentally. The solution of problems of the type seen in the last two examples depends on the analytical function $K = K(h)$. If this is very complicated function or of a certain type, the integration cannot be carried out analytically. In this case, there is no theoretical solution to the problem, but a numerical solution can be found, as will be shown in Chap. 11.

The functions $K(h)$, $K(\theta)$, or $D(\theta)$ are generally exponential for most soils; hence integration is possible in most cases. When one does not have the analytical expression of these functions, methods of numerical analysis can still solve the problem. An introduction to numerical methods is given by Reichardt and Godoy (1972), applied to the equation of water flow in the soil. A more in-depth analysis of numerical methods can be seen in Carnaham et al. (1969).

In the two examples previously seen, the flow density is constant, since these are cases of dynamic equilibrium. Although θ varies with x, θ does not vary with t ($\partial\theta/\partial t = 0$). The most general case is the transient regime, when the flux varies, and hence $\partial\theta/\partial t \neq 0$. Hence, the differential equation to be used is of the type of Eq. (7.20b), whose solution will be of type $\theta = \theta(x,t)$. These problems are, in general, much more difficult from a mathematical point of view, and, for the most part of them, it is not possible to determine the function of $\theta = \theta(x,t)$. Some examples of this kind of solution will be discussed in detail in Chap. 11 of the applied part of this book.

Laboratory methods for determination of hydraulic conductivity and diffusivity were described and discussed by Klute (1986). Among these methods, we highlight Gardner's (1956), which we will study in more detail below, because it is a great example of the application of the equations so far seen and gives us an idea of how complicated the analytical treatment can be. It is important for the reader to appreciate Gardner's mathematical development, although it is outdated in view of all computational tools available today.

At this point, it is opportune that the reader takes in hands a copy of the work of Gardner (1956). Gardner uses a slightly different symbology in the equations we have seen here, but the reader must immediately recognize his first equations. His method of determining K is based on the Richards pressure chamber described in Chap. 6. Consider a soil sample of volume V, cross section S, and height L, located on the porous plate (see Fig. 6.28). At a pressure P_i, the system is in equilibrium, and then at time $t = 0$, the pressure in the chamber is raised by a value ΔP, and the final pressure will be $P_f = P_i + \Delta P$. This increase causes some soil water to escape until a new equilibrium is established. This is the procedure used to determine the water retention curve in the soil, and Gardner had the idea to use the same procedure to determine the hydraulic conductivity, based on the fact that more permeable or wetter soils reach equilibrium more rapidly. For this process of transient flow of water leaving the soil by the action of ΔP, in which θ varies in space and time, we have:

$$\frac{\partial\theta}{\partial t} = \frac{\partial}{\partial z}\left[K(\theta)\frac{\partial H}{\partial z}\right] \tag{7.31}$$

in which the total potential H is given by:

$$H = h + z$$

Gardner disregards the gravitational component z of the water potential because the height

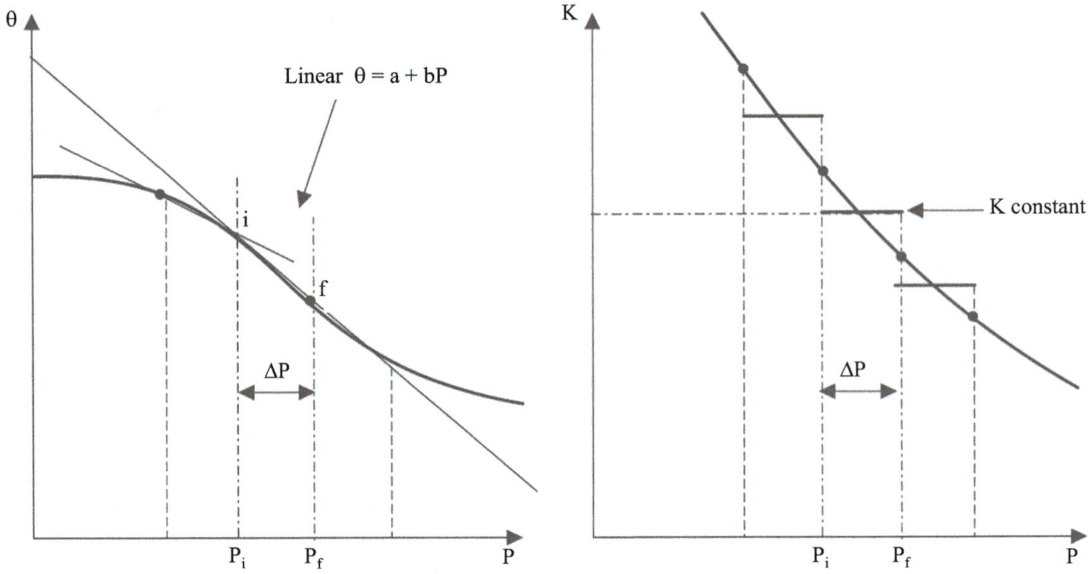

Fig. 7.23 Left-hand side shows that as an approximation the soil water retention curve can be taken as straight line segments of small ΔP intervals; right-hand side, that K can be taken as constant also for small ΔPs

L of the sample is a few centimeters and therefore negligible. This is not to say that z disappears as a coordinate of position; it only disappears as a potential. Since h is the pressure P itself, Eq. (7.31) becomes:

$$\frac{\partial \theta}{\partial t} = \frac{\partial}{\partial z}\left[K(\theta)\frac{\partial P}{\partial z}\right] \tag{7.32}$$

which is Eq. (4) of Gardner's paper. This equation is nonlinear and difficult to solve because it depends on the form of $K(\theta)$. Under certain experimental conditions, one can assume simplifications that linearize the equation and make a solution possible. Gardner assumes that (1) for small ΔP, K can be considered constant during the process of soil water extraction and that (2) the soil retention curve (characteristic) can be considered linear in the same interval ΔP (Fig. 7.23), mainly because $\Delta\theta$ is small for each ΔP step.

1. K = constant in the interval $P_f - P_i = \Delta P$
2. $\theta(P) = a + b \cdot P$, soil water retention curve in the interval $P_f - P_i = \Delta P$

Under such conditions, differentiating $\theta(P)$ to obtain $\partial\theta/\partial t$, we have:

$$\frac{\partial \theta}{\partial t} = \frac{\partial}{\partial t}(a + bP) = b\frac{\partial P}{\partial t}$$

and making the substitutions in Eq. (7.32):

$$\frac{\partial P}{\partial t} = K\frac{\partial^2 P}{\partial z^2} \tag{7.33}$$

in which K is the mean constant value of the curve $K(\theta)$ in the interval ΔP and which includes the constant b, that is, it is a new K, also constant, since b is constant.

The next step is to solve Eq. (7.33). The solution obtained by Gardner was by the classical method of separable variables. Since Eq. (7.33) has a derivative with respect to t and two with respect to z, three integration constants will appear, and, with their determination, the particular solution of the problem is obtained.

We then need three particular conditions of the problem, one with respect to the time (initial condition) and two with respect to z (the boundary conditions). The solution of Eq. (7.33) will be an equation of the type $P = P(z, t)$, and, for this reason, we write the conditions in the same form:

First condition (initial condition): at the beginning of the experiment ($t = 0$) when we apply instantaneously a ΔP, therefore, for $t = 0$, $P = \Delta P$:

$$P(z, 0) = \Delta P$$

or we can write this condition in the form:

$$P = \Delta P; \quad z > 0; \quad t = 0 \qquad (7.34)$$

that is read in this way: P equal to ΔP for any z greater than 0 (inside the soil sample), at the beginning of the experiment.

Second condition (boundary condition): the soil water pressure on its lower surface ($z = 0$) is atmospheric because it is always in contact with the free water of the lower chamber (see Fig. 6.28). So:

$$P(0, t) = 0$$

or $P = 0$ for $z = 0$ for all times t:

$$P = 0; \quad z = 0; \quad t > 0 \qquad (7.35)$$

Third condition (boundary condition): in the upper part of the soil sample ($z = L$), there is no water flux density q; since K is not 0, the gradient must be 0:

$$\left(\frac{\partial P}{\partial z}\right) = 0 \qquad (7.36)$$

As already said several times, the differential Eq. (7.33) and conditions (7.34)–(7.36) constitute a BVP that needs to be solved. Thus, to define the problem's solution, which must be of the type $P = P(z, t)$, Gardner (1956) does not give the details of the solution. This solution can be seen in Reichardt (1985) and will not be repeated here. It is important, however, for the reader to study the mathematical details of this solution. The end result is:

$$P(z,t) = \frac{4\Delta P}{\pi}\sum_{n=1}^{\infty}\frac{1}{(2n-1)}\left[e^{\frac{(2n-1)^2\pi^2 Kt}{4L^2}}\right] \times \left[\text{sen}\frac{(2n-1)\pi z}{2L}\right] \qquad (7.37)$$

in which n is an integer varying from 1 to ∞, i.e., $n = 1, 2, 3, \ldots, \infty$.

Equation (7.37) is Eq. (9) in Gardner (1956). Let's have a discussion about it. The equation seems complicated, but in reality, it is simply a function of the type of separate variables that allows us to calculate P at any z (or point in the soil sample placed in the Richards chamber) at any time t. For example, if $L = 0.4$ cm and we want to determine P for $z = 0.2$ cm and $t = 10$ s or calculate P (0.2; 10), it is necessary to introduce these values of z and t into Eq. (7.37), for n ranging from 1 to ∞, and then add up all the results. It is a time-consuming job, only possible with computers, not very much available at Gardner's time. It turns out, however, that the sum of Eq. (7.37) is rapidly convergent, and we do not have to vary n greatly. In some cases, 2–3 of n are sufficient, that is, $n = 1$, 2, and 3. This means that the contribution to the sum is negligible for terms of $n > 3$, and the problem is simplified. This is not always true for such summations.

Once Eq. (7.37) is understood, we will substitute it in the linear retention curve assumed above to obtain the soil water content changes in our sample as a function of t and z:

$$\theta = a + \frac{4\Delta P}{\pi} b \sum_{n=1}^{\infty}\frac{1}{(2n-1)}\left[e^{\frac{(2n-1)^2\pi^2 Kt}{4L^2}}\right] \times \left[\text{sen}\frac{(2n-1)\pi z}{2L}\right] \qquad (7.38)$$

which is a function of the type $\theta = \theta(z, t)$.

The total water storage (volume) in the sample $W(t)$ is the product of the water storage $S_L(t)$ multiplied by the sample cross section:

$$W(t) = S\int_0^L \theta \mathrm{d}z$$

and:

$$W(t) = S\int_0^L \left\{a + \frac{4\Delta P}{\pi} b \sum_{n=1}^{\infty}\frac{1}{(2n-1)} \times \left[e^{\frac{(2n-1)^2\pi^2 Kt}{4L^2}}\right]\left[\text{sen}\frac{(2n-1)\pi z}{2L}\right]\right\}\mathrm{d}z$$

The result of this integral is:

$$W(t) = aV + \frac{8 \cdot b \cdot \Delta P \cdot V}{\pi^2} \times \sum_{n=1}^{\infty} \frac{1}{(2n-1)^2} \left[e^{\frac{-(2n-1)^2 \pi^2 Kt}{4L^2}} \right]$$

and the initial volume of water at $t = 0$ will be:

$$W(0) = (a + b\Delta P)V$$

because $e^0 = 1$ and

$$\sum_{n=1}^{\infty} (2n-1)^{-2} \cong \frac{\pi^2}{8}$$

The final volume at $t = \infty$, i.e., at equilibrium, will be:

$$W(\infty) = aV$$

because $e^{-\infty} = 0$.

The amount of water $\Delta W(\infty)$ that leaves the sample after having applied ΔP and waiting for the new equilibrium is:

$$\Delta W(\infty) = W(0) - W(\infty) = bV\Delta P$$

and then we can determine the value of b, since the soil water retention curve assumed at the beginning is not yet known:

$$b = \frac{\Delta W(\infty)}{V\Delta P}$$

On the other hand, the amount of water that left the sample $\Delta W(t)$ will be:

$$\Delta W(t) = W(0) - W(t)$$

or:

$$W(t) = W(\infty) \left[1 - \frac{8}{\pi^2} \sum_{n=1}^{\infty} \frac{1}{(2n-1)^2} e^{\frac{-(2n-1)^2 \pi^2 Kt}{4L^2}} \right]$$

Since this sum is also rapidly convergent, Gardner neglected the terms with $n > 1$, and the equation was simplified to:

$$W(t) = W(\infty) \left[1 - \frac{8}{\pi^2} e^{\frac{-\pi^2 Kt}{4L^2}} \right]$$

which after applying ln to both sides becomes:

$$\ln [W(t) - W(\infty)] = \ln \frac{8W(\infty)}{\pi^2} - \frac{\pi^2 Kt}{4L^2} \tag{7.39}$$

Equation (7.39) shows that the graph of the logarithm of the difference $W(t) - W(\infty)$, as a function of time, is linear and that K can be calculated from the slope β. It is just enough to measure W from time to time and $W(\infty)$, which in practice is obtained between 2 and 7 days depending on the soil, and plot the graph of Fig. 7.24.

Recalling that K includes b, which is equal to $\Delta W(\infty)/V \cdot \Delta P$, we have:

$$K = \frac{4 \cdot L^2 \cdot \Delta W(\infty) \cdot \tan \beta}{\pi^2 \cdot V \cdot \Delta P}$$

This value of K corresponds to each step ΔP (see Fig. 7.23), and the same operation is repeated for all other intervals, thus obtaining the complete curve of K.

Gardner (1956) obtained good K values by this method. Bruce and Klute (1956) also present an important method of determining hydraulic diffusivity in the laboratory. This method will be presented in the chapters of the applied part, because it depends on the theory of the process of infiltration of water into the soil, which will be seen in detail in those chapters. This also happens with the hydraulic conductivity determination methods proposed by Rose et al. (1965), Gardner (1970), Hillel et al. (1972), Libardi et al. (1980), Sisson et al. (1980), and Reichardt et al. (2004).

This detailed example of Gardner's method was presented here with the purpose of showing the reader that solving differential equations or BVPs is not a simple matter for non-mathematicians. The solution given by Eq. (7.37), presented in more detail in Reichardt (1985), requires knowledge about differential equations, and, in most cases, one must seek help of colleagues in the area of exact sciences. The important thing is that the person interested in soil science understands the "philosophy" of the process and does not worry about mathematic

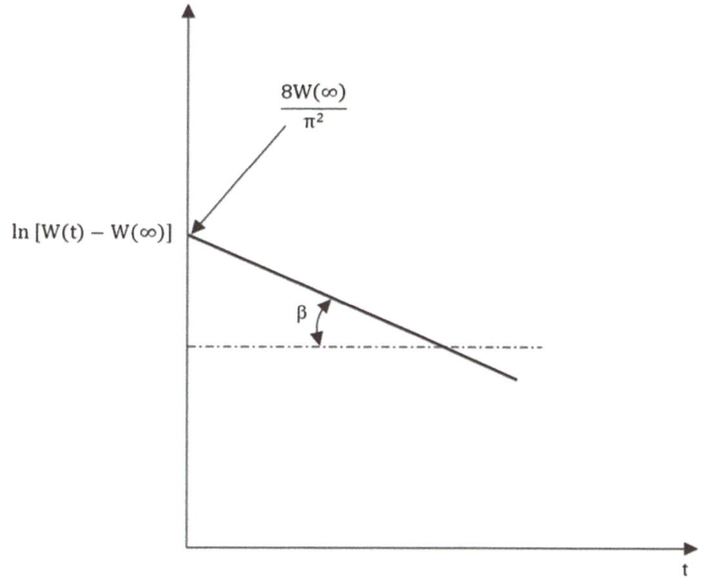

Fig. 7.24 Schematic view of the water volumes *W* drained from the sample as a function of time *t*

details. It is important to recognize a "BVP, boundary value problem," which in the present case constitutes differential Eq. (7.33), subject to conditions (7.34)–(7.36), with its solution 7.37. Most published scientific papers do not provide details of their solution, but the reader needs to understand the BVP. A more focused textbook for the mathematics of solutions is that of Prevedello and Armindo (2015), whose reading is recommended. With the advancement of information technology, there are currently programs that solve differential equations. They have a bank of the classical solutions already known, containing a large number of differential equations, and, if none serves to the user, the program seeks a solution via numerical processes. The interested party enters with the differential equation, and the computer presents the solution or indicates the impossibility of obtaining it. Such a program is "Maple V" which is often presented in new versions. In order to facilitate the determination of the $K(\theta)$ curve or even in the absence of this function, one can use models that are based on physical properties of easier determination and on retention curve models (Chap. 6). Some of the main ones are those proposed by Burdine (1953), Mualem (1976), Brooks and Corey (1964), and van Genuchten (1980). The latter presents restrictions with respect to the parameters of the retention curve, solved by Dourado-Neto et al. (2011).

Another approach was presented by Raats (1970) who suggested the definition of the **matric flux potential** (M) to describe the flow of unsaturated water under conditions where the gravitational potential is negligible or nil. This is a common condition when θ is equal to or less than the θ_{cc}, the **field capacity**, and plants draw water from the soil by the transpiration process. M is defined by the equation:

$$M(h) = \int_{h_{PWP}}^{h} K(h)\mathrm{d}h = \int_{\theta_{PWP}}^{\theta} D(\theta)\mathrm{d}\theta \qquad (7.40)$$

To explain Eq. (7.40), we take the **Darcy-Buckingham equation** (Eq. 7.16), and considering $H = h$, we make $K(h)\mathrm{d}h = \mathrm{d}M$, in a way to combine flow characteristics $K(h)$ with the matric retention characteristics h (or dh). In this way, we have a flux q equal to the gradient of M or the matric potential flux M:

$$q = \nabla M = \frac{\partial M}{\partial x} \qquad (7.41)$$

It is logical that if $\mathrm{d}M = K(h)\mathrm{d}h$, its integral is Eq. (7.40) and that it is the area under the curve $K(h)$ in the limits h and h_{PWP}, as shown in Fig. 7.25.

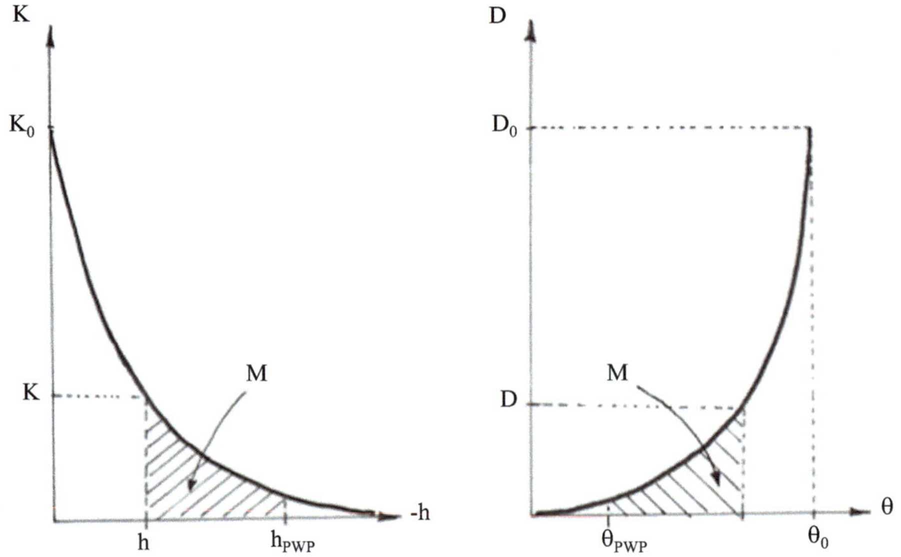

Fig. 7.25 The concept of matric potential M shown on the left by the labeled integration area for the $K(h)$ function and on the right for the function $D(\theta)$

The upper limit h_{PWP} was chosen because it is the limit of the available water for the plants. The dimensions of M are mm day^{-1} m^{-1}. Given the function $M(h)$, the water flow q can be more easily calculated by Eq. (7.41). De Jong van Lier et al. (2009) deal with transpiration modeling using the concept of matrix flow potential.

7.6 Water Movement from the Plant to the Atmosphere

In the same way as we discussed in the case of soil water, water in the plant is in equilibrium when the total potential Ψ is the same at all points (A, B, C ....) of the system:

$$\Psi_A = \Psi_B = \Psi_C = \ldots$$

Here, since there are semipermeable membranes, Ψ is used, which includes the osmotic component and not H, which does not include it.

When there is potential difference between two points, there will be water movement. The water flow can also be described by the **Darcy equation** (Eq. 7.3), substituting H for Ψ. Plant physiologists, however, present this equation differently given the difficulty of measuring potential gradients within the plant and also in the Soil-Plant-Atmosphere System as a whole.

If we substitute H for Ψ and K for $1/r$, where r is a **hydraulic resistivity** (inverse of the conductivity), we shall have:

$$q = -K\frac{\partial \Psi}{\partial x} = -\frac{1}{r}\frac{\partial \Psi}{\partial x} \quad (7.42)$$

The dimensions of K and r depend on the units used for q and Ψ. The flow density is generally expressed in volume per unit area and unit of time, resulting in m s^{-1}, cm s^{-1}, or mm day^{-1}. If Ψ is expressed as a hydraulic load, cmH_2O, the gradient is non-dimensional, and K will have the same units of q, that is, m s^{-1}, cm s^{-1}, or mm day^{-1}. Of course, the resistivity will have inverse units, s m^{-1}, s cm^{-1}, or day mm^{-1}.

If we write Eq. (7.16) in the form of finite differences, we will have:

$$q = -K\frac{\Delta \Psi}{\Delta x} = -\frac{1}{r}\frac{\Delta \Psi}{\Delta x}$$

and since Δx (path traveled by the water inside the plant) is tortuous and difficult to be measured, it can be incorporated into K or r, resulting in conductance or resistance. Thus, physiologists use Eq. (7.16) or (7.42) in the form:

$$q = -K\frac{\Delta\Psi}{\Delta x} = -\frac{1}{r}\frac{\Delta\Psi}{\Delta x}$$

$$q = -\frac{\Delta\Psi}{R}, \text{ where } R = r \cdot \Delta x$$

Logically the units of R will be $(\text{s m}^{-1}) \times \text{m} = \text{s}$.

In the literature, this subject of units of resistance is confusing. A reference text for the question is that of Nobel (1983).

In essence, resistivity r (s m^{-1}) is a point property of the medium that is of transferring water (in the same way as copper resistivity is a point property of the copper medium for the transmission of electricity). The resistance R (s) is a property of a "layer" of thickness Δx (just as the electrical resistance of a conductor depends on its dimensions, even if it is made of copper).

The combination of the Darcy equation with the continuity equation in order to study variations in the water content in the plant, as was done for the soil case, is usually not done. This is because the variations of the water content in the plant are relatively small, and it is possible to consider the flow as being saturated, that is, $\partial\theta/\partial t = 0$. The water lost through transpiration is replaced by root absorption, in a steady-state process.

The vapor flow in the atmosphere is also described by a Darcy-like equation. The phenomenon, however, is more complicated, because the vapor moves due to the potential gradients of Ψ and also due to the atmospheric turbulence (winds). In fact, the equation becomes empirical, and a K_m coefficient is used which includes all vapor transfer processes. In this case, the equation is:

$$q = -K_m\frac{\partial\Psi}{\partial x}$$

The study of vapor transfer in the atmosphere is quite complex and will not be seen in detail here. Rose (1966) gives a good summary of this study. In the atmosphere, the movements are turbulent due to the presence of the wind, and the transport of water in the form of vapor becomes quite complex, involving the definition of wind profiles, which are graphs of the wind speed "u" as a function of the lògarithm of the height z, inside and above the crop canopy. Rosenberg et al. (1983) and Allen et al. (1998) discuss the subject in detail. In Chap. 14, we will return to the study of these flows in the Soil-Plant-Atmosphere System.

7.7 Water Movement in Open Channels and Pipes

This subject is more related toward hydraulics, but since it is part of water management in agriculture, especially in irrigation, we will make an introduction to the subject. In previous cases of water movement in soil and plant, the velocity v of water, i.e., its kinetic energy $E_c = mv^2/2$, was not taken into account because v was very small; on the other hand, in open channels and pipes, E_c cannot be overlooked.

In the same way as we did for the water potential (Chap. 6), we will express the energies of water in terms of heights or hydraulic loads, that is, energy per unit weight. In the case of channels and pipes, three energies are the most important:

Potential energy:

$$\frac{mgh}{mg} = h_g(m) : \text{hydraulic load}$$

Pressure energy:

$$\frac{PV}{mg} = \frac{P}{\gamma} = h_P(m) : \text{pressure load}$$

Kinetic energy:

$$\frac{mv^2}{2mg} = \frac{v^2}{2g} = h_k(m) : \text{kinetic load}$$

where γ is the specific weight of water, equal to the product of water density ($d = 1000$ kg m^{-3}) by the acceleration of gravity ($g = 10$ m s^{-2}).

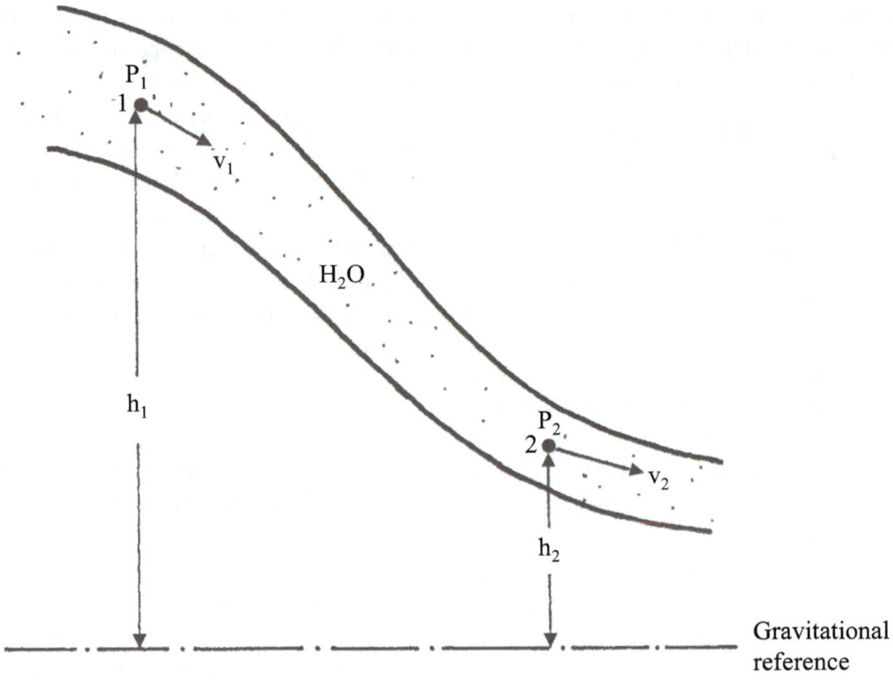

Fig. 7.26 Schematic view of water flow in a pipeline

In the schematic pipeline in Fig. 7.26, the total energy needs to be conserved, but one form of energy can be transformed into another. **Bernoulli's theorem** expresses this conservation for any points in the system:

Consider point 1 of Fig. 7.26 with coordinates h_1 (geometric load), P_1 (pressure load), and v_1 (kinetic load) and point 2 (h_2, P_2, v_2).

We have already seen at the beginning of this chapter that in this situation, the flow at points 1 and 2 is equal $Q_1 = Q_2$ and that, when the cross section changes, the water velocities are different. Thus, in terms of energy per weight and considering the water as a **perfect liquid**, we have:

$$h_1 + \frac{P_1}{\gamma} + \frac{v_1^2}{2g} = h_2 + \frac{P_2}{\gamma} + \frac{v_2^2}{2g} \qquad (7.43)$$

It is noteworthy that perfect liquids do not exist in practice, that is, in nature there is no liquid without viscosity and incompressibility.

Example 1: If at point 1 $P_1 = 200{,}000$ Pa, $h_1 = 100$ m, and $v_1 = 5$ m s^{-1} and at point 2 $P_2 = 300{,}000$ Pa and $h_2 = 10$ m, what is the velocity v_2 at point 2, considering $\gamma = 1000$ kg m^{-3} × 10 m s^{-2} = 10,000 N m^{-3} and $g = 10$ m s^{-2} (see units in Chap. 18)?

Solution

$$100 + \frac{200,000}{10,000} + \frac{(5)^2}{2 \times 10} =$$
$$10 + \frac{300,000}{10,000} + \frac{v_2^2}{2 \times 10}$$

where $v_2{}^2 = 1625$ and $v_2 = 40.3$ m s^{-1}. As we can see, the velocity of the liquid increased at the expenses of the other types of energy.

Example 2: In Fig. 7.27, we have a steady-state water flow, and we would like to know what is the velocity of water coming out at the bottom hole of the container. This example simulates a water container with a tap at depth h_2.

Solution:

Taking points 1 and 2, we have:

Point 1: $h_1 = h$; $P_1 = P_{atm}$; $v_1 = 0$

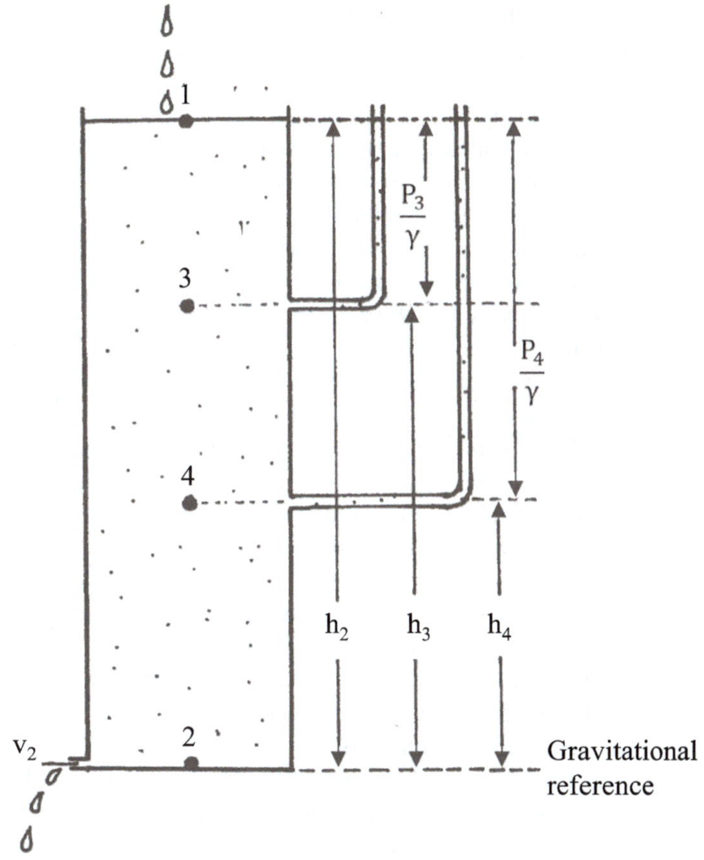

Fig. 7.27 A steady-state condition of a water container with water running at depth h_2

Point 2: $h_2 = 0$; $P_2 = P_{atm}$; $v_2 = ?$

$$h_1 + \frac{P_{atm}}{\gamma} + 0 = 0 + \frac{P_{atm}}{\gamma} + \frac{v_2^2}{2g}$$

$$v_2^2 = 2gh \text{ and } v_2 = \sqrt{2gh}$$

Points 3 and 4 were included in Fig. 7.27 to better visualize the concept of pressure load, which in 3 is equal to the water column $h_2 - h_3$ and in 4 $h_2 - h_4$.

The applications of Bernoulli's theorem are numerous, and their study is part of fluid mechanics, approached by hydraulics or hydrodynamics. Thus, in the use of the water of a dam for the production of electric energy, the water velocity and the flow are of fundamental importance for the rotation of the turbine. This is the classic case of transforming potential energy into electrical energy, passing through kinetics. In practice, we have to introduce the dissipative forces of friction between the water and the walls of the tube and between the water molecules themselves, which is measured by the viscosity. There are, therefore, load losses along pipelines, as is the case in sprinkler irrigation, which need to be considered.

7.8 Exercises

7.1. Through a cross section of soil of 5 m^2, 22 L of water passes per day. What is the water flow density in this soil?

7.2. Considering that the soil water content of Problem 7.1 is equal to 36% by volume, what is the water velocity in the soil pores?

7.3. Inside a 20 cm^2 cross-sectional water pipe pass 150 cm^3 of water in 8 min. What are the flow rate, the flow density, and the water velocity?

7.4. Through a soil column of cross section of 100 cm^2 passes a flow density of 1.5 cm day^{-1}. What is the flow of water?

7.5. In the previous case, the column of soil suffers a reduction, and its cross section passes to 80 cm^2. What is the new flux density?

7.6. A soil with a water flow density of 1.5 mm day^{-1} has a soil water content of 0.421 cm^3 cm^{-3}. What is the velocity of water in the soil pores?

7.7. In Problem (6.27), determine the potential gradient for all cases at $z = 30$ and $z = 110$ cm.

7.8. In Problem (7.1), the gradient of H is 2.2 cm cm^{-1}. What is the hydraulic conductivity of the soil?

7.9. In the previous problem, what is the intrinsic permeability of the soil?

7.10. The hydraulic conductivity of a soil was determined as a function of the water content, and the following results were obtained:

Soil water content θ (cm^3 cm^{-3})	K (cm day^{-1})
0.510 (saturation)	5.42
0.463	1.71
0.405	0.41
0.366	0.157
0.273	0.016
0.214	0.0037

Plot the graphs:

(a) K versus θ.

(b) log K versus θ.

(c) ln K versus θ.

(d) log K versus $(\theta - \theta_0)$, $\theta_0 =$ saturation water content.

(e) ln K versus $(\theta - \theta_0)$.

(f) Make linear regressions of the plots *b*, *c*, *d*, and *e*.

7.11. What is the water flux in the depths 30 and 60 cm of a soil of the following profile:

Depth z (cm)	Soil water content θ (cm^3 cm^{-3})
15	0.320
30	0.341
45	0.375
60	0.396
75	0.420
90	0.452

The soil is that of Problem (7.10) and its retention curve is:

Matric potential h (cm H_2O)	Soil water content θ (cm^3 cm^{-3})
0	0.510
−50	0.501
−100	0.485
−150	0.448
−200	0.407
−250	0.375
−300	0.352
−350	0.326
−400	0.310

7.12. With the data of Problems (7.10) and (7.11), construct the plot K versus h.

7.13. A soil has the following $K(\theta)$ relation:

$$K(\theta) = 3.1 \times 10^{-1} \cdot e^{14.5 \cdot \theta} \, \text{cm day}^{-1}$$

and the saturation soil water content is θ_0 is 0.511 cm^3 cm^{-3}. Find the parameters K_0, α, β, and γ for the following equations of the same soil:

$$K(\theta) = K_0 \cdot \exp[\alpha(\theta - \theta_0)]$$
$$K(\theta) = \beta \cdot 10^{\gamma \times \theta}$$

7.14. For the soil column mounted in the laboratory, as shown in Fig. 7.28, the saturated hydraulic conductivity K_0 is evaluated varying the water depth I and measuring the flow Q. The soil cross section is 200 cm^2.

I (cm)	Q (cm^3 h^{-1})
10	19.9
20	25.5
30	30.2

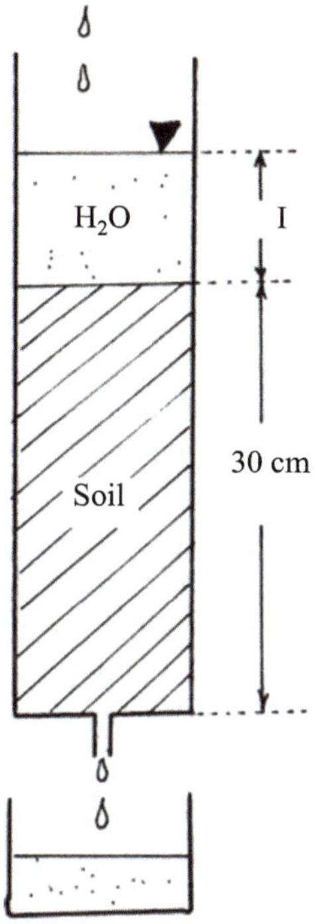

Fig. 7.28 Determination of K_0 for the vertical column of Exercise 7.14

Calculate K_0 for the three values of the table and also through the graph q versus I. Why are the values a little different?

7.15. The same soil sample from the previous problem is placed horizontally, as shown in Fig. 7.29. What are the values of h_1 and h_2, knowing that $Q = 28.8\ \text{cm}^3\ \text{h}^{-1}$?

7.16. What is the water diffusivity value for $\theta = 0.475\ \text{cm}^3\ \text{cm}^{-3}$ of the soil of Problems (7.10) and (7.11)?

7.17. Between two points in the soil, horizontally, there is a total potential gradient of $1.5\ \text{cm cm}^{-1}$, and a water content gradient is $1.7 \times 10^{-3}\ \text{cm}^3\ \text{cm}^{-1}$. The water flow is $0.26\ \text{cm day}^{-1}$. What is the average hydraulic conductivity in this soil region, and also what is the diffusivity of water in the soil of the same region?

7.18. The same soil of Problem (6.26) has equation:

$$K(\theta) = 4.58 \exp\left[-10\,(\theta - 0.15)\right]\ \text{cm day}^{-1}$$

What is the diffusivity value D for $\theta = 0.450\ \text{cm}^3\ \text{cm}^{-3}$?

Fig. 7.29 Determination of K_0 in a vertical soil column

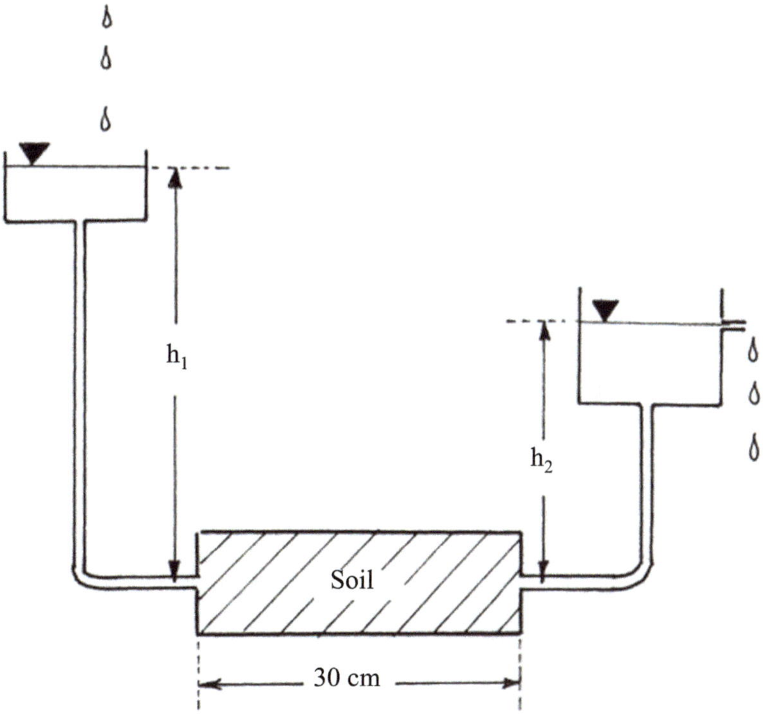

7.19. In a cubic volume element of soil (1 cm^3) with flow only in the vertical direction, a flow of 1.56 mm enters per day on the upper face, and a flow of 1.61 mm exits per day on the bottom face. What is the water content change in 1 day in the volume element?

7.20. For the same soil of Exercise 7.19, in another situation, $q_z = 2.43$ mm day^{-1} enter, and $q'_z = 2.32$ mm day^{-1} leave. The initial θ of the volume element being equal to 0.341 cm^3 cm^{-3}, what is the new water content after 1 day?

7.21. For the same soil from Exercise 7.19, in another situation, the soil water content in the volume element does not change in time. What happens to the water flow?

7.22. In the scheme of Fig. 7.30, calculate the hydraulic conductivity of the saturated soil, whose cross section is 100 cm^2. Under this condition, the soil water content is 0.511 cm^3 cm^{-3}.

7.23. For the same soil of Problem (7.22), h_1 was modified to 150 cm and h_2 to 0 cm. In this new situation, $V = 1477$ cm^3 in 1 day. Determine the K_0, and verify that your determination is independent of h_1 and h_2.

7.24. The same soil of the previous problem is submitted to an average suction of 110 cm H_2O, as indicated in Fig. 7.31. In this condition, the soil has an average soil water content of 0.481 cm^3 cm^{-3}, less than saturation which is 0.511 cm^3 cm^{-3}. What is the average hydraulic conductivity of the unsaturated soil in this condition, since it is known that $V = 138$ cm^3 in 1 week?

7.25. In the scheme of Fig. 7.32, determine K_0 of the soil, and plot the potentials h, z, and H as a function of height. The inside diameter of the soil is 6.5 cm.

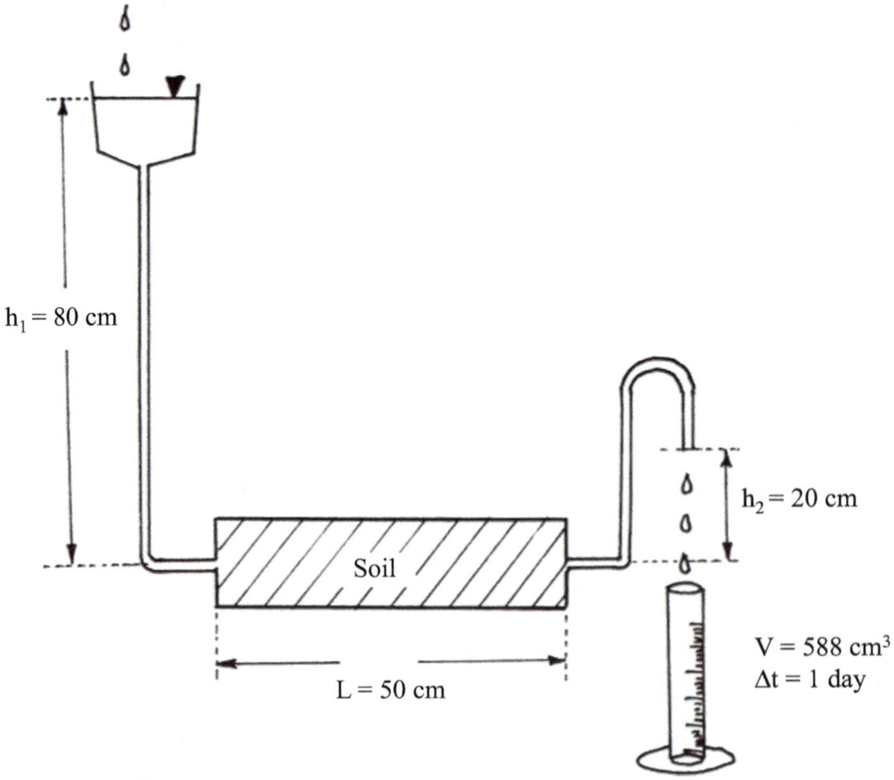

Fig. 7.30 Determination of K_0 in a horizontal soil column

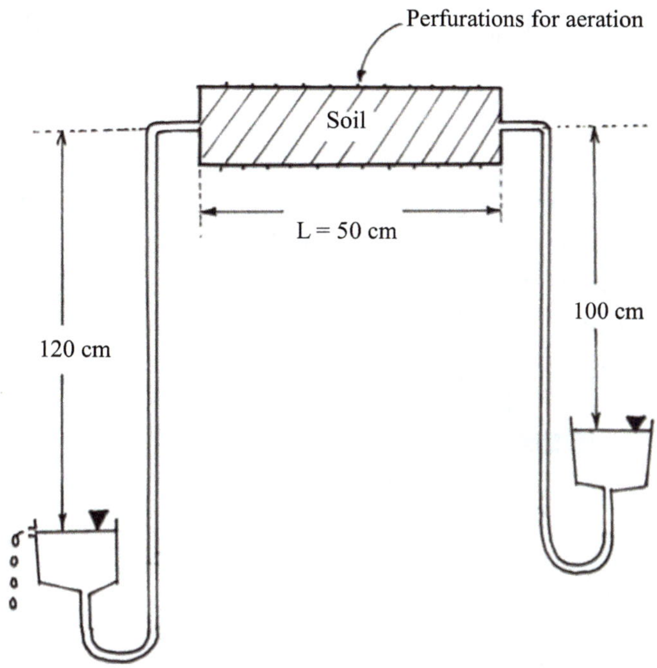

Fig. 7.31 Determination of $K(\theta)$ in a horizontal soil column

7.26. For the experiment outlined in Fig. 7.33, the following data were obtained for the measurement of K_0:

h (cm)	V/At (mm day^{-1})
10	22.0
30	31.0
45	40.5

Which is the average value of K_0?

7.27. In a deep layer of a soil (around 150 cm), the hydraulic conductivity of the soil is given by the equation $K(h) = 128.27 \exp(0.039\ h)$. Two tensiometers installed at depths of 135 and 165 cm measure the potential gradient. The first has a reading of matrix potential of -75 cm H_2O and the second of -88 cm H_2O. What is the drainage flow in this layer?

7.9 Answers

7.1. 4.4 mm day^{-1}.
7.2. 12.2 mm day^{-1}.
7.3. Flow 18.75 cm^3 min^{-1}; flux density 0.94 cm min^{-1}; water velocity 0.94 cm min^{-1}.
7.4. 150 cm^3 day^{-1}.
7.5. 1.875 cm day^{-1}.
7.6. 3.563 cm day^{-1}.
7.7.

z	A	B	C	D	E	F
30	−0.75	−2.00	+1.25	+1.40	−15.00	−3.50
110	0	−0.50	0	−0.45	−0.25	+2.00

Results in cm H_2O cm^{-1} soil or simply cm cm^{-1}. The negative sign indicates upward flux.

7.8. 2.0 mm day^{-1}.
7.9. Solution: apply Eq. (7.9) (e.g., at 25 °C)

$$k = \frac{K\eta}{\rho g} = \frac{0.2\ \text{cm day}^{-1} \times 0.0089\ \text{g cm}^{-1}\ \text{s}^{-1}}{86{,}400\ \text{s day}^{-1} \times 1\ \text{g cm}^{-3} \times 981\ \text{cm s}^{-2}} = 2.1 \times 10^{-11}\ \text{cm}^2$$

7.10. $K(\theta) = 0.0000195 \cdot e^{24.58 \cdot \theta}$; $K_d = 0.0000195$ and $\beta = 24.58$.
$K(\theta) = 5.42 \cdot e^{-24.58 \cdot (\theta - 0.51)}$; $K_0 = 5.42$ and $\gamma = 24.58$.

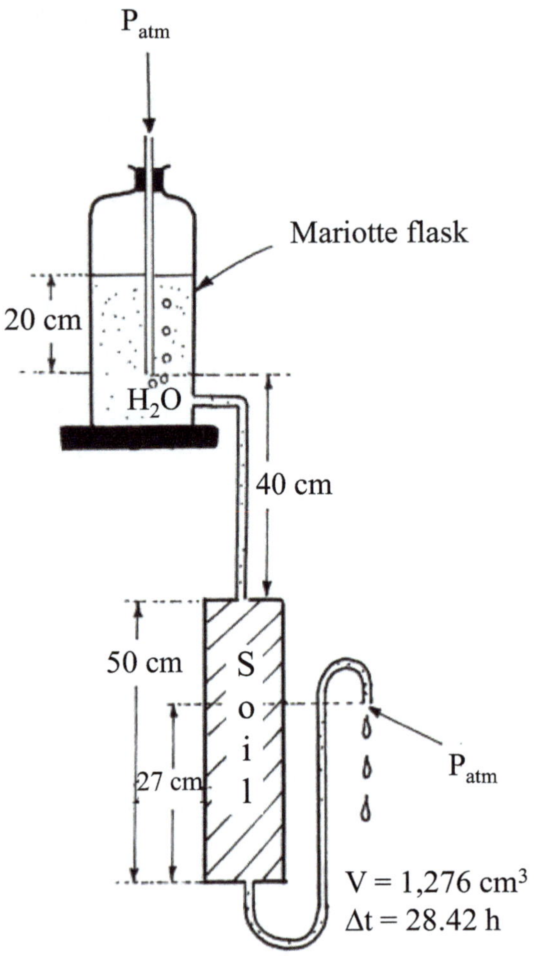

Fig. 7.32 Arrangement for the determination of K_0 in a vertical soil column

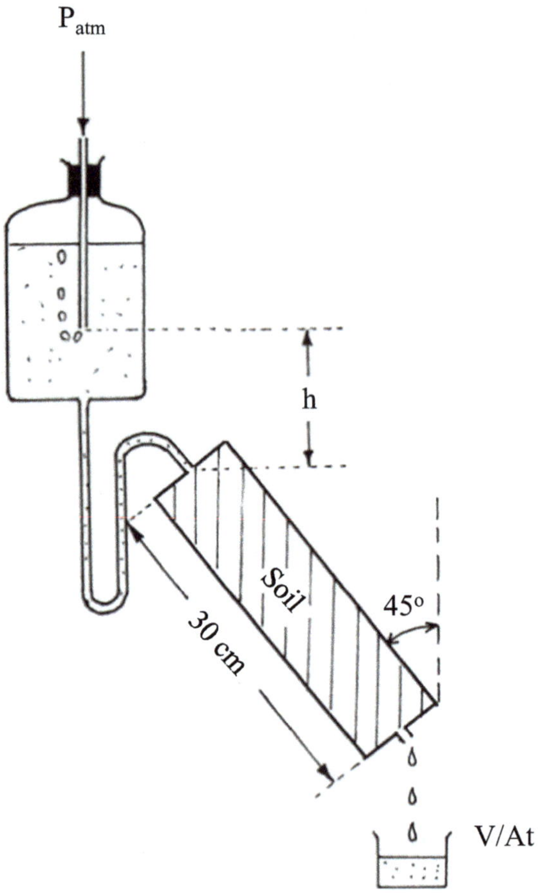

Fig. 7.33 Determination of K_0 in an inclined soil column

$K(\theta) = 0.0000195 \cdot 10^{10.68 \cdot \theta}$; $K_d = 0.0000195$ and $\delta = 10.68$. $K(\theta) = 5.42 \times 10^{-10.68 \cdot (\theta - 0.51)}$; $K_0 = 5.42$ and $\rho = 10.68$.

7.11. Draw the characteristic curve, and transform it with the values of water content (for each depth) into matrix potential. Calculate the total potential, and plot the graph as a function of z. In this graph, determine the potential gradients at 30 and 60 cm, drawing the tangent to the curve at the points. The results are gradient = −3 cm H_2O cm^{-1} soil and gradient = −0.8 cm H_2O cm^{-1} soil. Using any of the equations of $K(\theta)$ of the previous problem, calculate the K at depths 30 and 60 cm, using the respective values of θ. Results K (0.341) = 0.085 and K (0.396) = 0.329 cm day^{-1}. Calculate the flows by multiplying K by grad Ψ. It results in $q_{30} = -2.56$ mm day^{-1} and $q_{60} = -3.86$ mm day^{-1}.

7.12. In many situations, it is more convenient to use the relation $K(h)$ instead of $K(\theta)$. The graph can be plotted using log or ln since K and h have a wide range of variation.

7.13. $K_0 = 0.512$ cm day^{-1}.
$\alpha = 14.5$ (dimensionless).
$\beta = 3.1 \times 10^{-4}$ cm day^{-1}.
$\gamma = 6.28$ (dimensionless).

7.14. Approximately 0.075 cm h^{-1}. The values are slightly different due to experimental errors, especially of the Q measurement. If

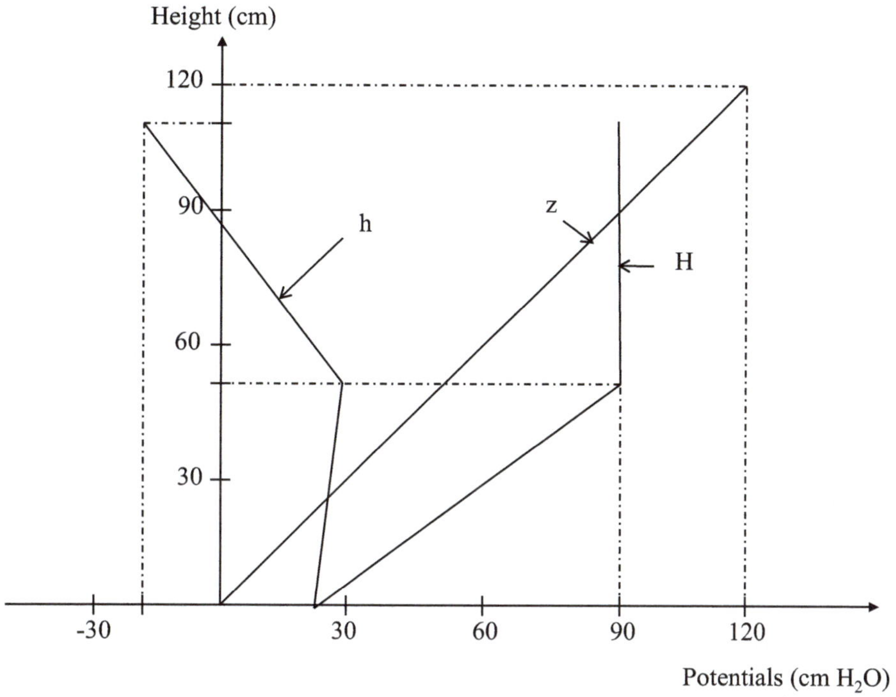

Fig. 7.34 Schematic of the potentials h, z, and H as a function of height

the experiment is too long, due to reduced flow, soil accommodation, anaerobic reactions, temperature, water quality, etc., Q does not fully stabilize for a given I. If I is too small, its measure becomes difficult and the Q flow small. If I is very large, the Darcy equation may no longer be valid.

7.15. Any value, provided that the difference $h_1 - h_2$ is 58 cm.

7.16. Calculate K for $\theta = 0.475$ with one of the equations in Problem (7.4): $K\ (0.475) = 2.29$ cm day^{-1}. In the characteristic curve of Exercise 7.5, calculate dh/dθ at the point $\theta = 0.475$. The result is 1.265 cm H_2O/cm^3 cm^{-3} and $D\ (0.475) = 2.903$ cm^2 day^{-1}.

7.17. $K(\theta) = 0.173$ cm day^{-1} and $D\ (\theta) = 152.9$ cm^2 day^{-1}.

7.18. $D(\theta) = 1470$ cm^2 day^{-1}.

7.19. -5.0×10^{-3} cm^3 cm^{-3} day^{-1} (try to understand the continuity equation, Eq. 7.20a).

7.20. $\theta = 0.352$ cm^3 cm^{-3}.

7.21. Either there is equilibrium and $q = 0$ or it is a case of dynamic equilibrium where $q_z = q_{zs}$, and the value of q_z can be any, within the limits of Darcy's law.

7.22. 4.90 cm day^{-1}.

7.23. 4.92 cm day^{-1}.

7.24. 0.49 cm day^{-1}.

7.25. 25.77 cm day^{-1}; see Fig. 7.34.

7.26. 18.24 mm day^{-1}.

7.27. Apply Eq. (7.16) for resolution of the exercise.

References

Allen RG, Pereira LS, Raes D, Smith M (1998) Crop evapotranspiration – guidelines for computing crop water requirements. FAO, Roma

Brooks RH, Corey AT (1964) Hydraulic properties of porous media. Colorado State University, Fort Collins, CO

Bruce RR, Klute A (1956) The measurement of soil moisture diffusivity. Soil Sci Soc Am Proc 20:458–462

Buckingham E (1907) Studies of movement of soil moisture. United States Department of Agriculture Bureau, Washington, DC

Burdine NT (1953) Relative permeability calculation from size distribution data. Trans AIME 198:71–78

Carnaham B, Luther HA, Wilkes JO (1969) Applied numerical methods. John Wiley & Sons, New York, NY

Churchill RV (1963) Fourier series and boundary value problems. McGraw Hill, New York, NY

Darcy H (1856) Les fontaines publique de la Ville de Dijon. Victor Dalmont, Paris

De Jong van Lier Q, Dourado-Neto D, Metselaar K (2009) Modeling of transpiration reduction in van Genuchten–Mualem type soils. Water Resour Res 45:1–9

Dourado-Neto D, De Jong van Lier Q, van Genuchten MT, Reichardt K, Metselaar K, Nielsen DR (2011) Alternative analytical expressions for the general van Genuchten-Mualem and van Genuchten-Burdine hydraulic conductivity models. Vadose Zone J 10:618–623

Gardner WR (1956) Calculation of capillary conductivity from pressure plate outflow data. Soil Sci Soc Am Proc 20:317–320

Gardner WR (1970) Field measurement of soil water diffusivity. Soil Sci Soc Am Proc 34:215–238

Hillel D, Krentos VD, Stylianou Y (1972) Procedure and test of an internal drainage method for measuring soil hydraulic characteristics in situ. Soil Sci 114:395–400

Hubert MK (1956) Darcy's law and field equations of the flow of underground fluids. Am Inst Min, Metal Petr Eng Trans 207:222–239

Johnson HP, Frevert KR, Evans F (1952) Simplified procedure for the measurement and computation of soil permeability below the water table. Agri Eng 33:283–289

Klute A (1986) Methods of soil analysis. Part I: Physical and mineralogical methods, 2nd edn. American Society of Agronomy, Soil Science Society of America, Madison, WI

Libardi PL, Reichardt K, Nielsen DR, Biggar JW (1980) Simplified field methods for estimating the unsaturated hydraulic conductivity. Soil Sci Soc Am J 44:3–6

Luthin JN (1957) Drainage of agricultural lands. American Society of Agronomy, Madison, WI

Miller EE, Low PF (1963) Threshold gradient for water flow in clay systems. Soil Sci Soc Am Proc 27:605–609

Mualem Y (1976) A new model for predicting the hydraulic conductivity of unsaturated porous media. Water Resour Res 12:513–522

Nobel PS (1983) Biophysical, plant physiology and ecology. W.H. Freeman & Company, New York, NY

Prevedello CL, Armindo RA (2015) Fisica do solo com problemas resolvidos, 2nd edn. Prevedello CL, Curitiba

Prevedello CL, Reichardt K (1991) Modelo tridimensional para medida da condutividade hidráulica de solos não saturados. Rev Bras Ciênc Solo 15:121–124

Raats PAC (1970) Steady infiltration from line sources and furrows. Soil Sci Soc Am Proc 34:709–714

Reichardt K, Godoy CM (1972) Solução numérica de equações diferenciais parciais. Centro de Energia Nuclear na Agricultura. Universidade de São Paulo, Piracicaba

Reichardt K (1996) Dinâmica da matéria e da energia em ecossistemas. 2a ed. Escola Superior de Agricultura Luiz de Queiroz. Universidade de São Paulo, Piracicaba

Reichardt K (1985) Processos de transferência no sistema solo-planta-atmosfera. Fundação Cargill, Campinas

Reichardt K, Timm LC, Bacchi OOS, Oliveira JCM, Dourado-Neto D (2004) A parameterized equation to estimate hydraulic conductivity in the field. Austr J Soil Res 42:283–287

Rose CW (1966) Agricultural physics. Pergamon Press, Oxford

Rose CW, Stern WR, Drummond JE (1965) Determination of hydraulic conductivity as a function of depth and water content in situ. Austr J Soil Res 3:1–9

Rosenberg NJ, Blad BL, Verma SB (1983) Micro-climate: the biological environment. John Wiley & Sons, New York, NY

Sisson JB, Ferguson AH, van Genuchten MT (1980) Simple method for prediction drainage from field plots. Soil Sci Soc Am J 44:1147–1152

Swartzendruber D (1962) Non Darcy behavior in liquid saturated porous media. J Geophys Res 67:5205–5213

Van Genuchten MT (1980) A closed-form equation for predicting the conductivity of unsaturated soils. Soil Sci Soc Am J 44:892–898

Soil Water as a Nutrient Solution

8

8.1 Introduction

Soil water is never free of solutes as considered in Chaps. 6 and 7. Being in close contact with the solid soil fraction, it is a solution of ions, molecules, and organic components. As described in Chap. 3, it has a variable composition due to a series of dynamic processes between the solid and liquid phases of the soil, addition to inputs, absorption of nutrients by the roots, and subsequent transport to the aerial part of the plants. Fried and Broeshard (1967) summarized these processes for plant nutrients by a general equation of form:

$$\text{M (solid)} \leftrightarrow \text{M (solution)} \leftrightarrow \text{M (root)} \leftrightarrow \text{M (aerial part)} \tag{8.1}$$

wherein M represents any nutrient, such as Ca, P, N or K; M (solid) represents the nutrient in the solid phase of the soil (crystalline, amorphous precipitate, organic matter, etc.) or absorbed to the solid phase; and M (solution) the nutrient that is in the liquid phase of the soil, immediately available to plants, mainly in the forms Ca^{2+}, $H_2PO_4^-$, NH_4^+, NO_3^-, or K^+; M (root) and M (aerial part) represent the nutrient absorbed by the root and translocated by the plant to the stems, leaves, and reproductive organs, respectively. The arrows indicate processes in both directions, whose intensities depend on a variety of equilibrium constants from nutrient to nutrient, compound form, temperature, pH, and an endless series of factors. Steady state is rarely achieved due to the continuous water absorption by plants. In general, the predominant direction of nutrient movement is from the soil to the aerial part of the plant. Any blockage in nutrient transfer, in any of the phases indicated by Eq. (8.1), can lead to a deficiency of that nutrient in the plant. If the rate at which a particular nutrient changes from M (solid) to M (solution) is small in relation to the absorption and need of the plant, a deficiency will also appear. The detailed study of each physicochemical transfer process indicated by Eq. (8.1) is quite extensive, complicated and, in many cases, still in the research phase. In this chapter, we will be more concerned with the characterization of M (solution) which, as we have seen, represents a nutrient, be it a molecule or an ion in solution.

8.2 Soil Solution Thermodynamics

In order to describe our M (solution) system, we will again use the **Gibbs free energy** thermodynamic function, defined by Eqs. (6.5)–(6.8), which involves the concept of chemical potential μ_i. Substance i in this case is a given nutrient M and j, all other nutrients also present in the

K. Reichardt, L. C. Timm, *Soil, Plant and Atmosphere*, https://doi.org/10.1007/978-3-030-19322-5_8

soil solution, including the solvent water. Thus, the **chemical potential** of an ionic species is given by:

$$\mu_i = \left(\frac{\partial G}{\partial n_i}\right)_{P_e, T, y_i, n_j} \quad (8.2)$$

If T and P_e are constant and there is no contribution from "other works" then Eq. (6.8) for the soil solution only becomes:

$$dG = \sum_{i=1}^{k} \mu_i dn_i \quad (8.3)$$

and the index i represents each of the ionic species present in the soil solution and other inorganic and organic molecules. In the case of essential plant nutrients, the most important chemical species are shown at the beginning of Chap. 4.

There will be equilibrium in the system when $G =$ constant or $dG = 0$, i.e., developing the sum of Eq. (8.3), we have:

$$dG = \mu_1 dn_1 + \mu_2 dn_2 + \ldots + \mu_i dn_i + \ldots + \mu_n dn_n = 0$$

On the other hand, if a nutrient is found in the system in two phases, for example, a saturated solution of $CaSO_4$ (a salt of very low solubility) with an excess of solid precipitate, in the equilibrium, G has also to be constant in both phases:

$$dG = \mu^s dn^s + \mu^l dn^l = 0$$

where μ^s and μ^l are the chemical potentials of $CaSO_4$ in the solid and liquid phases, respectively, and n^s and n^l are the numbers of moles of the same component in the solid and liquid phases, respectively. Therefore, in equilibrium, the quantity that leaves one phase necessarily passes to the other:

$$dn^s = -dn^l$$

and so:

$$\mu_i^s = \mu_i^l$$

and this equation can be generalized to more complex systems with greater number of phases. For example, liquid l, solid s, adsorbed a, gaseous g, etc., are all in equilibrium among themselves:

$$\mu^s = \mu^l = \mu^a = \mu^g = \ldots \quad (8.4)$$

Let us now imagine a very simple system consisting of a volume of any solvent (water, for example), to which a certain amount of only one salt i (e.g., NaCl) has been added. In this system, we have five components: H_2O molecules and H^+, OH^-, Na^+, and Cl^- ions (considering NaCl completely soluble), and each component i has a specific volume v_i and an **osmotic pressure** P_i (see Chap. 3). Adding an infinitesimal amount of component i maintaining the temperature T constant, and if there is no "other work" contribution, we have:

$$dG = d\mu_i = v_i dP_i \quad (8.5)$$

as it was done in Chap. 6, for the osmotic component of soil water.

On the other hand, the state equation describing such a system, provided it is a dilute solution, is similar to the general equation of **perfect gases** (Eq. 5.1, where $V/n = v =$ molar specific volume):

$$P_i v_i = RT \quad (8.6)$$

and substituting Eq. (8.6) into (8.5), we have:

$$d\mu_i = \frac{R\,T\,dP_i}{P_i}$$

which after integration from a standard state (μ_0, P_0, pure water without ion i) to a given state (μ_i, P_i, in which i is the ion responsible for the osmotic pressure P_i), becomes:

$$\mu_i - \mu_0 = R\,T\,\ln\left(\frac{P_i}{P_0}\right) \quad (8.7)$$

remembering that the integral of dy/y is $\ln y$.

This equation gives us the chemical potential of component i in the system, relative to a standard state μ_0. It is worth pointing out to the reader that if component i is water, $\mu_i = \Psi$, and Eq. (8.7) is identical to Eq. (6.7), where P_i and P_0 are partial pressures of water vapor in the air (e_a and e_s, respectively).

For a solution, the P_i/P_0 ratio is called **ion activity** of the component in the solution (a_i). So that:

$$\mu_i - \mu_0 = R\,T\,\ln\,a_i \quad (8.8)$$

The very name of the activity indicates that this property of the solution shows how active the ion is in the solution and this activity is logically related to its energy, i.e., to the **chemical potential** that is the **Gibbs free energy**.

For ideal solutions, to which the dilute solutions resemble each other, the activity a_i can be considered equal to the C_i, the concentration of ion i in the solution. This is because there is little interaction between molecules or ions in dilute solutions, and in this case:

$$\mu_i - \mu_0 = R\,T\,\ln\,C_i \quad (8.9)$$

which means that the higher the concentration C_i, the more the solution differs from the standard (pure water).

In Eqs. (8.8) and (8.9), a and C_i are "operated" by the operator ln. This operator, as well as the exponential (exp or **e**), can only operate on dimensionless elements. Therefore, a_i and C_i must be dimensionless. Thus, it was assumed that for real solutions:

$$a_i = \gamma_i C_i \quad (8.10)$$

where γ_i is a proportionality coefficient (Sposito 1989) of inverse dimensions of C, such that a becomes dimensionless, called **activity coefficient**, which depends on a series of factors, especially on interactions between components.

Under these conditions:

$$\mu_i - \mu_0 = R\,T\,\ln\,(\gamma_i C_i) \quad (8.8a)$$

It is easy to verify that $\gamma = 1$ for extremely dilute solutions, called **ideal solutions** and, when γ is different from 1, γ is a measure of how much a **real solution** differs from an ideal solution.

When studying the osmotic component of the water potential in Chap. 6, we presented Eq. (6.15), $\Psi_{os} = -RTC$, of the **Van't Hoff equation**. This Eq. (6.15), although not having the logarithm, can be used in the place of Eq. (8.7) or (8.8a). It is important to note that in Eq. (6.15), Ψ_{os} is the osmotic potential of water due to the inclusion of ions in concentration C (case a below). Equation (8.8a) is the inverse, the potential of the ion in aqueous solution (case b below). It is important to clarify this question:

Case (a) Osmotic potential of water due to the presence of ions in concentration C:

$$\Psi_{os} = -R\,T\,C$$

If $C = 0$, there are no ions (standard state), we have $\Psi_{os} = 0$. If, for example, we have a solution of NaCl $= 10^{-3}$ M, we will have also 10^{-3} M in Na^+, so:

$$\begin{aligned}\Psi_{os} &= -0.082\text{atm L mol}^{-1}\,\text{K}^{-1} \\ &\quad \times 300\,\text{K} \times 10^{-3}\,\text{mol L}^{-1} \\ &= -0.0246\,\text{atm}\end{aligned}$$

which is the osmotic potential of water due to the presence of Na^+. In this case, as R was given in atm L mol^{-1} K^{-1}, C must be given in mol L^{-1}.

Case (b) We could also use Eq. (8.8) to calculate the same osmotic potential of the water arising from the presence of Na^+, considering the component i being water itself, that is, $\Psi_{os} = -RT\ln a_i$. Since we do not have a_i, it can be estimated using the concept of molar fraction N_i, defined by:

$$N_i = \frac{n_i}{n_i + \sum n_j} \quad (8.11)$$

where n_i is the number of moles per liter of component i and $\sum n_j$ is the sum of moles per liter of the other components j. In the previous example of the NaCl solution, if i is Na^+ (10^{-3} mol L^{-1}), the others are: Cl^- (10^{-3} mol L^{-1}); water (55.5 mol L^{-1} or 1000 g/18 g); H^+ and OH^-, both negligible with 10^{-7} mol L^{-1} if the pH was 7.

$$N_{Na^+} = \frac{10^{-3}}{10^{-3} + 55.5 + 10^{-3}} = 0.18 \times 10^{-4}$$

and if i is the solvent, we have:

$$N_{H_2O} = \frac{55.5}{55.5 + 2 \times 10^{-3}} = 0.9999819$$

From these results of N we can see that only for very dilute solutions N_i tends to C_i

(N_{Na+} = 0.18 × 10^{-4} and $C_{Na+} = 10^{-3}$ not too close!), making Eq. (8.9) valid. Therefore, it is important to stress the limitations of Eq. (8.9). By the definition of activity itself (Eq. 8.7), a_i varies from 1 (when $P_i = P_0$) to 0 (when $P_i = 0$). On the other hand, C_i can assume "any" value, even greater than 1, as for example 5 mol L^{-1}. Equation (8.8a) is therefore assumed to be much more correct when written in terms of molar fraction N_i:

$$\mu_i - \mu_0 = R\,T\,\ln(\gamma_i N_i) \qquad (8.9a)$$

For the previous result, considering $\gamma_i = 1$, the osmotic potential of water due to the presence of Na^+ will be:

$$\mu_i - \mu_0 = 0.082\text{atm L mol}^{-1}\,\text{K}^{-1} \times 300\,\text{K} \times \ln 0.9999810$$
$$= -4.4299 \times 10^{-4}\,\text{atm L mol}^{-1}$$

or, to compare with the previous result given in atm:

$$-4.4299 \times 10^{-4}\text{atm L mol}^{-1} \times 55.5\,\text{mol L}^{-1} = -0.0246\,\text{atm}$$

which is the same result as obtained in case (a) with the equation $\Psi_{os} = -RTC$.

It can even be shown that the two equations give equal results. For the previous example, if we consider only i = Na^+ and j = H_2O, we have:

$$N_i + N_j = 1$$

which can be seen by summing the results obtained previously, N_{Na+} = 0.18 × 10^{-4} with N_{H2O} = 0.999998, resulting 1. Thus:

$$N_j = 1 - N_i$$

applying ln to both members and using the known series below

$$-\ln N_j = -\ln(1 - N_i) = N_i + \frac{1}{2}N_i^2 + \frac{1}{3}N_i^3 + \ldots$$

that is infinite but very convergent. Neglecting the terms with power greater than 1 when N is very small:

$$-\ln N_j = N_i$$

and so Eq. (8.9a) is simplified to:

$$\mu_i - \mu_0 = R\,T\,N_i$$

and since:

$$N_i = \frac{n_i}{n_i + n_j} \cong \frac{n_i}{n_t} \cong C$$

where $n_i + n_j = n_t$, so that:

$$\mu_i - \mu_0 = \Psi_{os} = R\,T\,C \qquad (8.8b)$$

The importance of the concept of ion activity of an ion in solution can be seen in Eqs. (8.8)–(8.11). The relation between **chemical potential** and **ion activity** is clearly seen in them, the former being a measure of the "free energy" of the component in question. Activity a represents the amount of the ion in question that is actually active for any physicochemical process. It is convenient to say here that the plant, when extracting nutrients from the soil, must "respond" to the activity of a certain nutrient in the solution of the soil and not to its concentration.

8.3 Activity of an Electrolytic Solution

We just defined ion activity of one single component through Eq. (8.10). Let's now see how we measure the activity of a salt in solution.

Consider a salt composed of a positive ion M and a negative ion N, so that its formula is $M_{\nu+}A_{\nu-}$. For example, NaCl, M = Na^+; $\nu^+ = 1$; A = Cl^-; $\nu^- = 1$. Or $Al_2(SO_4)_3$, M = Al^{3+}; $\nu^+ = 2$; A = SO_4^{2-}; $\nu^- = 3$. Logically, for each mol of the salt MA dissolved and dissociated we will have ν^+ mol of M^+ and ν^- mol of A^-. The chemical potential of a salt is the sum of the dissociated cations and anions:

$$M_{\nu^+}A_{\nu^-} \rightarrow \nu^+M^+ + \nu^-A^-$$

$$NaCl \rightarrow Na^+ + Cl^-$$

$$Al_2(SO_4)_3 \rightarrow 2Al^{3+} + 3SO_4^{2-}$$

$$1 \text{ mol L}^{-1}\ Al_2(SO_4)_3 \rightarrow 2 \text{ mol L}^{-1}\ Al^{3+} + 3 \text{ mol L}^{-1}\ SO_4^{2-}$$

$$RT\ \ln\ a_{salt} = v^+\ RT\ \ln\ a_+ + v^-\ RT\ \ln\ a_-$$

eliminating RT and remembering that a sum of logs is the log of their product:

$$\ln\ a_{salt} = \ln\left[a_+^{v+} \cdot a_-^{v-}\right]$$

so that:

$$a_{salt} = a_+^{v+} \cdot a_-^{v-} \quad (8.12)$$

where:

a_{salt} = activity of the salt in solution;
a^+ = activity of the dissociated cation;
a^- = activity of the dissociated anion;
v^+ = number of cations resulting from the dissociation;
v^- = number of anions resulting from the dissociation.

Substituting Eq. (8.10) into Eq. (8.12), we have:

$$a_{salt} = (\gamma_+ C_+)^{v+} \cdot (\gamma_- C_-)^{v-} = \gamma_+^{v+} \gamma_-^{v-}\ C_+^{v+}\ C_-^{v-}$$

Considering an average **activity coefficient** $\gamma_\pm$ that works for both ions and anions, we have:

$$a_{salt} = \gamma_\pm^{v}\ C_+^{v+}\ C_-^{v-} \quad (8.13)$$

where v is the sum of v^+ and v^-.

Take for example a solution of 0.1 M Na_2SO_4:

$$Na_2SO_4 \rightarrow 2Na^+ + SO_4^=$$

it is easy to verify that since this salt is a strong electrolyte:

$$C = 0.1 \text{ M};\ C_+ = 0.2 \text{ M} \quad \text{and } C_- = 0.1 \text{ M}$$

$$v = 3;\ v^+ = 2 \quad \text{and } v^- = 1$$

$$a_{Na_2SO_4} \rightarrow (\gamma_\pm)^3 \times (0.2)^2 \times (0.1)^1 = (\gamma_\pm)^3 \times 0.004$$

If $\gamma = 1$, we would have the Na_2SO_4 salt activity = 0.004 M ($a = C$). Since γ is not 1, the problem is reduced to determining the average activity coefficient for Na_2SO_4. These coefficients are difficult to determine. They can be calculated with satisfactory precision by **Debye–Huckel's theory**, as will be discussed later in this chapter and can be studied in detail in Harned and Owens (1958). This theory is based on the principle of **ionic strength** proposed by Lewis and Randall (1923). According to this principle: "In diluted solutions the average activity coefficient of a given electrolyte is the same in all solutions of the same ionic force S". This principle is extremely useful for mixed solutions, such as soil solution, since the average activity coefficient of an electrolyte can be estimated from a pure solution of the same electrolyte, provided that it has the same ionic strength S defined by:

$$S = \frac{1}{2}\sum_{i=1}^{k} z_i^2 C_i \quad (8.14)$$

in which z is valence. For our Na_2SO_4 solution, the ionic strength is:

$$S = \frac{1}{2}\left[(1)^2 \times (0.2) + (2)^2 \times (0.1)\right] = 0.3$$

From the value of S we can estimate γ for Na_2SO_4 by the Debye–Huckel theory. Rewriting Eq. (8.8a), we have:

$$\mu_i - \mu_0 = R\,T\ \ln\ \gamma_i + R\,T\ \ln\ C_i$$

When $\gamma_i = 1$, $\ln \gamma_i = 0$, and we have an ideal solution. The Debye–Huckel theory is based on the fact that γ_i is different from 1 for real solutions due to electrostatic interactions among ions. In this theory, we calculate the part of the chemical potential ($\mu_i - \mu_0$) that accounts for the electrostatic interactions of the ions, as a result of the ionic force S (Babcock 1963). A simplified form of the obtained equation is:

$$\log_{10} \gamma_\pm = -0.509\ z_+ |z_-| \sqrt{S} \quad (8.15)$$

where z_+ is the valence of the cation and $|z_-|$ the module of the valence of the anion. In our Na_2SO_4 case, we have:

$$\log_{10} \gamma_\pm = -0.509 \times 1 \times 2 \times \sqrt{0.3}$$

so that:

$$\gamma_\pm = 0.277$$

and the activity of the solution of Na_2SO_4 will be (Eq. 8.13):

$$a_{Na_2SO_4} = 1.1 \times 10^{-3}\ \mathrm{M}$$

We have thus seen that a solution prepared to be 0.1 M Na_2SO_4 (which means that we have 14.2 g of salt per liter) actually has an activity of only 1.1×10^{-3} M. This is due to the interaction between the ions. For more dilute solutions, the difference is less, until for the very diluted, $\gamma = 1$. However, according to Eq. (8.13), a_{salt} is less than C, even if $\gamma = 1$. Equation (8.13) is not valid for more concentrated solutions than $C = 1$ M.

8.4 The Theory of Donnan

The **Donnan theory** of membrane systems developed by Donnan (1911) and generalized by Donnan and Guggenheim (1932) has been applied extensively to colloidal systems of organic or inorganic origin, such as those in the soil. Although very old, the Donnan theory explains to date several colloidal behaviors. These colloids are electrically charged surfaces, such as humus and clay minerals, described in Chap. 3, present in most soils. Also, organic matter has unbalanced electrical charges. In this chapter, we will study in detail these electrically charged surfaces.

The system is shown in Fig. 8.1, consisting of two phases: phase I = water + colloid (clay, for example) + electrolyte; and phase II = water + electrolyte. The phases are separated by a semipermeable membrane, which prevents the passage of colloid from I to II, but not of electrolytes.

In this case, the colloid clay has a negative net charge. From Eqs. (8.4) and (8.8) we can say that, in an equilibrium condition:

$$a^{\mathrm{I}}_{\mathrm{electrolyte}} = a^{\mathrm{II}}_{\mathrm{electrolyte}}$$

in which the upper indices indicate the phase. Thus, according to Eq. (8.13) we can write the following relation for the electrolyte, which in the case of Fig. 8.1 is sodium chloride:

$$\left(\gamma^{\nu}_{\pm}\right)^{\mathrm{I}}\left(C_+{}^{\nu+}C_-{}^{\nu-}\right)^{\mathrm{I}} = \left(\gamma^{\nu}_{\pm}\right)^{\mathrm{II}}\left(C_+{}^{\nu+}C_-{}^{\nu-}\right)^{\mathrm{II}}$$

or rearranging:

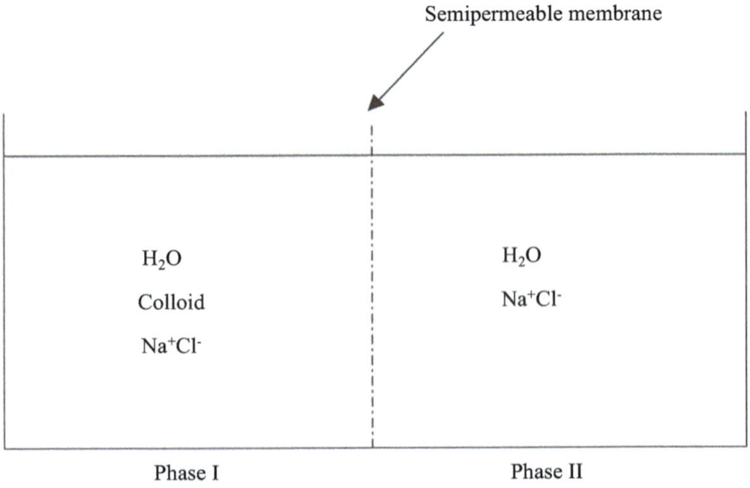

Fig. 8.1 Illustration of a Donnan system consisting of two phases separated by a semipermeable membrane

$$\frac{\left(C_{+}^{\nu+}\right)^{\mathrm{I}}}{\left(C_{+}^{\nu+}\right)^{\mathrm{II}}} = \Omega^{-\nu}\frac{\left(C_{-}^{\nu-}\right)^{\mathrm{II}}}{\left(C_{-}^{\nu-}\right)^{\mathrm{I}}} \tag{8.16}$$

where Ω^{ν} is the ratio of the average coefficients:

$$\Omega^{\nu} = \frac{\left(\gamma_{\pm}^{\nu}\right)^{\mathrm{I}}}{\left(\gamma_{\pm}^{\nu}\right)^{\mathrm{II}}}$$

On the other hand, since each phase should be electrically neutral:

$$N_{+}^{\mathrm{I}} = N_{-}^{\mathrm{I}} \text{ and } N_{+}^{\mathrm{II}} = N_{-}^{\mathrm{II}}$$

where N is the number of equivalents of the referred ion per unit volume (normality).

$$N_{+}^{\mathrm{I}} = z_{+}C_{+}^{\mathrm{I}}$$

$$N_{-}^{\mathrm{I}} = |z_{-}|C_{-}^{\mathrm{I}} + C_{\mathrm{c}}$$

where C_{c} is "concentration" of colloid, also given in number of equivalents per unit volume. The inclusion of C_{c} in the sum of the negative charges is due to the fact that the colloids are composed of surfaces whose net charge is negative, which also have to be taken into account in the charge balance. Thus, in phase I:

$$z_{+}C_{+}^{\mathrm{I}} = |z_{-}|C_{-}^{\mathrm{I}} + C_{\mathrm{c}}$$

or rearranging:

$$C_{+}^{\mathrm{I}} = \frac{|z_{-}|C_{-}^{\mathrm{I}}}{z_{+}} + \frac{C_{\mathrm{c}}}{z_{+}} \tag{8.17}$$

Substituting Eq. (8.17) into (8.16), we have:

$$\Omega^{-\nu}\left[\frac{|z_{-}|C_{-}^{\mathrm{I}}}{z_{+}} + \frac{C_{\mathrm{c}}}{z_{+}}\right] = \frac{\left(C_{+}^{\nu+}\right)^{\mathrm{II}}\left(C_{-}^{\nu-}\right)^{\mathrm{II}}}{\left(C_{-}^{\nu-}\right)^{\mathrm{I}}} \tag{8.18}$$

which is the fundamental relationship of the Donnan equilibrium. With it, the concentrations of the ions in the two phases can be determined. Of great importance for colloidal systems is to know the relationship between anion concentrations in phases I and II. This relationship of anions is usually called anion ratio α:

$$\alpha = \frac{C_{-}^{\mathrm{II}}}{C_{-}^{\mathrm{I}}} \tag{8.19}$$

α provides an idea of the repulsion of the anions of the colloidal phase by virtue of the same signal charges of the colloids. Values of α greater than 1 indicate what is termed **negative adsorption**, i.e., anion repulsion due to the presence of colloid. In addition to α, a factor β called **anionic repulsion** is defined as:

$$\beta = \frac{|z_{-}|\left(C_{-}^{\mathrm{II}} - C_{-}^{\mathrm{I}}\right)}{C_{0}} \tag{8.20}$$

which is a measure of anions repulsion in terms of the quantity of colloids present in the system.

To understand fully the above discussion, let's see a very simple example with NaCl in solution with phase II: $N^{\mathrm{II}+} = N^{\mathrm{II}-} = 500$ meq L^{-1} in equilibrium with phase I containing clay in a concentration $C_{\mathrm{c}} = 1000$ meq L^{-1}. In this case:

$$\nu^{+} = \nu^{-} = 1, \nu = 1 + 1 = 2$$

and Eq. (8.18) becomes:

$$\Omega^{2}\left[\frac{1 \times C_{-}^{\mathrm{I}}}{1} + \frac{C_{\mathrm{c}}}{1}\right] = \frac{\left(C_{+}^{1}\right)^{\mathrm{II}}\left(C_{-}^{1}\right)^{\mathrm{II}}}{\left(C_{-}^{1}\right)^{\mathrm{I}}}$$

For dilute solutions, because Ω is the relation between two γ not too different from each other, it can be considered as unity, and so:

$$\left(C_{-}^{\mathrm{I}}\right)^{2} + C_{\mathrm{c}} \cdot C_{-}^{\mathrm{I}} - C_{+}^{\mathrm{II}} \cdot C_{-}^{\mathrm{II}} = 0$$

which is a second-grade equation in relation to C^{I}. So:

$$C_{-}^{\mathrm{I}} = \frac{-C_{\mathrm{c}} \pm \sqrt{(C_{\mathrm{c}})^{2} - 4C_{+}^{\mathrm{II}} \cdot C_{-}^{\mathrm{II}}}}{2}$$

in which, substituting the values of C_{c}, C_{+}^{II} and C_{-}^{II}, remembering that for monovalent ions $C = N$, results $C_{-}^{\mathrm{I}} = 207$ meq L^{-1}. Calculating α, we have:

$$\alpha = \frac{C_{-}^{\mathrm{II}}}{C_{-}^{\mathrm{I}}} = \frac{500}{207} = 2.4$$

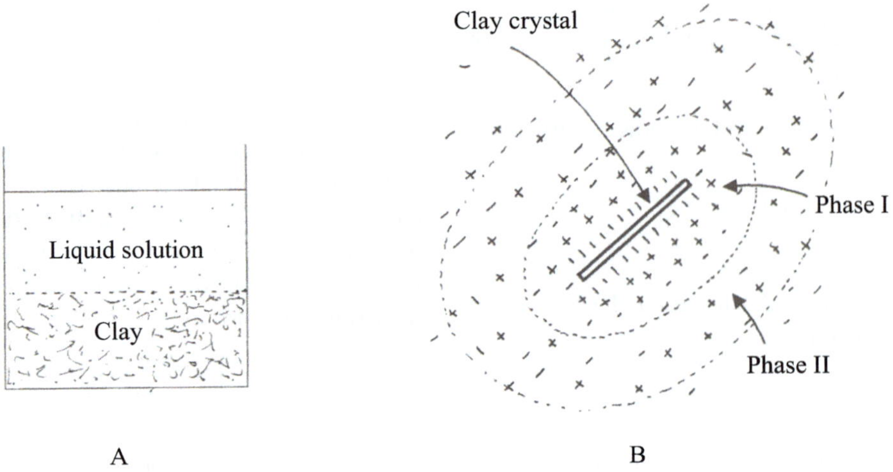

Fig. 8.2 Simplified Donnan systems: (**A**) Beaker with clay in solution after a very long time, until the upper part becomes translucid; (**B**) A single crystal with negative charges immerged in soil solution

$$\beta = \frac{|z_-|(C_-^{II} - C_-^{I})}{C_0} = 0.29$$

Since:

$$(a_{NaCl})^{I} = (a_{NaCl})^{II}$$

or

$$C_+^{I} \cdot C_-^{I} = C_+^{II} \cdot C_-^{II}$$

Because was considered $\Omega = 1$, the values of γ are equal and are cancelled. So:

$$C_+^{I} = \frac{C_+^{II} \cdot C_-^{II}}{C_-^{I}} = \frac{500 \times 500}{207} = 1.207 \text{ meq L}^{-1}$$

This result shows that the clay caused an "expulsion" of Cl^- and that its concentration close to it is much smaller (207 meq L^{-1}) than that of Na^+ (1207 meq L^{-1}).

It is easy to note that for electrolytes with bivalent and trivalent ions the solution of Eq. (8.18) is greatly complicated by the exponents larger than 2 that would appear in the solution and only approximate numerical solutions can be obtained. On the other hand, the Donnan equilibrium, with monovalent examples, as seen earlier, gives us a very good view on colloidal systems, at least in qualitative terms. The membrane of Fig. 8.1 is completely dispensable. A system as shown in Fig. 8.2a, which is a liquid solution floating over a colloid after a long sedimentation time is described by the same equations as seen here. The cationic concentration is higher in the solution next to the colloid. In microscopic terms, one could also apply Donnan's theory to a clay crystal, a constituent of the colloid. The zone near the crystal (cationic adsorption zone) can be considered phase I and the zone further away from the crystal, such as phase II (Fig. 8.2b).

These facts demonstrate that, even under equilibrium conditions, the concentration of an ion is not the same at all points in the system. This is a serious problem when you want to measure the concentration of an ion in the soil. Then the questions arise: how to sample the soil solution? Under which soil water content conditions? At which point? etc. Reichardt (1977) and Reichardt et al. (1977) discuss the sampling of soil solution by means of porous capsules subjected to vacuum, which is only one of the employed methods.

8.5 The Ionic Double Layer

The concept of **ionic double layer** appears in a system of charged colloidal particles in contact with the ionic solution of the soil. This system is

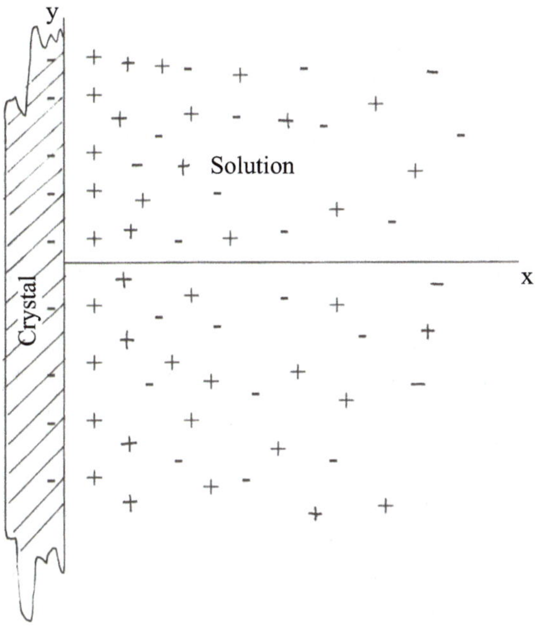

Fig. 8.3 Illustration of a flat crystal surface presenting an excess of negative charges and the adjacent ion distribution of the soil solution, according to Gouy's theory

so called given the presence of two "layers" of charge, one negative—the surface of the colloid (clay)—and the other layer constituted by the adsorbed cations. The knowledge of the distribution of these cations around the colloidal particles is of extreme importance and also of the factors that affect their distribution. The theory of the distribution of ionic concentrations in the vicinity of electrostatically charged particles was developed, in principle, by Gouy (1910). The **Gouy theory** assumes that the clay crystal can be represented by a flat, infinite surface of homogeneous distribution of charges (Fig. 8.3). It also assumes that the charged surface is immersed in an electrolytic solution of uniform dielectric constant, that the charge of the surface is neutralized by an excess of charge of ions of opposite signal and that all the ions are taken as point charges. For such a system, the electric potential along a direction x perpendicular to the charged surface is given by the **Poisson differential equation**:

$$\frac{d^2\Phi}{dx^2} = \frac{4\pi\rho}{D} \quad (8.21)$$

where:

Φ = electric potential;
x = position coordinate;
ρ = net charge in a solution volume element along the direction x (positive minus negative charges);
D = dielectric constant of the medium.

Equation (8.21) is a classical differential equation that can be found in basic Electricity and Magnetism texts. Let us see its solution. Initially, we will calculate ρ that appears on the second member. The cations do not remain "fixed" to the charged surface, there is a dynamic balance between electric attraction and repulsion by virtue of thermal energy (Brownian Motion). The Boltzmann distribution describes this equilibrium through an exponential distribution, whose exponent is the relation between electric and thermal "forces":

$$n_+ = n_0 e^{-\frac{z_+ \varepsilon \Phi}{KT}} \quad (8.22)$$

$$n_- = n_0 e^{-\frac{|z_-| \varepsilon \Phi}{KT}} \quad (8.23)$$

where:

n_+ = number of cations cm^{-3} of solution in the direction x;
n_- = number of anions cm^{-3} of solution in the direction x;
n_0 = number of cations that is equal to the number of anions at an "infinite" distance from the charged surface, where $\Phi = 0$ and the charged particle does not influence the system anymore;
ε = electron charge;
K = Boltzmann's constant;
T = absolute temperature of the solution.

The net electric charge ρ along x is the sum of the positive minus the negative:

$$\rho = \sum |z_i| \varepsilon\, n = (z_+ \,\varepsilon\, n_+) - (|z_-|\, \varepsilon\, n_-) \quad (8.24)$$

Because very close to the negative charged surface the value of n_+ is much greater than n_-, ρ is positive. Far from the surface, in the middle of the solution, there is charge equilibrium and $\rho = 0$, since $n_+ = n_- = n_0$. Substituting Eqs. (8.22) and (8.23) into Eq. (8.24), we have:

$$\rho = n_0 \varepsilon \sum |z_i| e^{-\frac{|z_i| \varepsilon \Phi}{KT}} \quad (8.25)$$

which for the particular case of $z_+ = z_- = z$, we have:

$$\rho = n_0\, z\, \varepsilon \left(e^{\frac{-z\varepsilon\Phi}{KT}} - e^{\frac{z\varepsilon\Phi}{KT}} \right) \quad (8.26)$$

Making,

$$\frac{z\,\varepsilon\,\Phi}{K\,T} = y$$

Equation (8.26) is simplified to:

$$\rho = n_0\, z\, \varepsilon\, (e^{-y} - e^{y}) = 2\, n_0\, z\, \varepsilon \text{ senh } y \quad (8.27)$$

in which sen h y is the hyperbolic sine of y. Substituting Eq. (8.27) into Eq. (8.21), we have:

$$\frac{d^2\Phi}{dx^2} = \frac{8\pi\, z\, \varepsilon\, n_0}{D} \text{ senh } y \quad (8.28)$$

The solution of Eq. (8.28) is not easy and will not be seen here. It can be found in Babcock (1963). So, the function that gives Φ for any x, that is, $\Phi(x)$, is:

$$\Phi = \frac{2\,K\,T}{z\,\varepsilon} \ln \frac{e^{bx} + a}{e^{bx} - a} \quad (8.29)$$

where,

$$b = \frac{8\pi^2 z^2 \varepsilon^2 n_0}{D\,K\,T} \quad \text{and} \quad a = \tan h\left(\frac{z\,\varepsilon\,n\,\Phi_0}{K\,T}\right)$$

Φ_0 is the electric potential on the surface of the colloid, that is, the value of Φ for $x = 0$; $\tan h$ = hyperbolic tangent.

Given the function $\Phi(x)$, one can calculate n_+ and n_-, according to Eqs. (8.22) and (8.23). One can also calculate C_+, C_- and C_0 which are the molar concentrations of each component.

An analysis of Eq. (8.29) reveals that the thickness x [in Angstrom (Å)] of the ionic double layer increases if the concentration of the soil solution (m_0) decreases. In Fig. 8.4A this fact is schematized. The thickness of the double layer can be considered the distance of the crystal ($x = 0$) to the point ($x = x_L$) where $n_+ = n_- = n_0$. It is seen in the particular case of Fig. 8.4A that the thickness is 50 for $m_0 = 0.05$ M and 75 for $m_0 = 0.02$ M. From the analysis of Eq. (8.29) it is also concluded that the potential Φ_0 increases as m_0 increases (Fig. 8.4B) and also, that the thickness of the double layer depends on the valence of the ions (Fig. 8.4C).

Gouy's theory presents a series of problems described by Bolt (1955) but, in general, provides a good picture of what happens in the vicinity of charged particles in the soil. The theory was modified and expanded, including a layer of "fixed" cations, in which the potential varies linearly with x, called the **Stern layer**, followed by the diffuse layer represented by Eq. (8.29). Special texts of soil physicochemistry present more details on the subject. More physical aspects of the issue are presented by Koorevaar et al. (1983) and Sposito (1989).

8.6 Ionic Exchange Capacity

The **ion exchange** phenomenon always occurs in a system in which electrically charged surfaces are present. So far, we have considered only the case of clays presenting negative charges symmetrically distributed on their surfaces. In the Soil-Plant System, clays and organic matter, oxides, root surface, biological tissues, etc., present both positive and negative charges. These charges can be permanent or nonpermanent. The permanent charges are those which appear mainly in the clay crystals as a consequence of the **isomorphic substitution** described in Chap. 3. Nonpermanent charges appear whenever a surface has weak acid or base properties. As the materials indicated above have these properties,

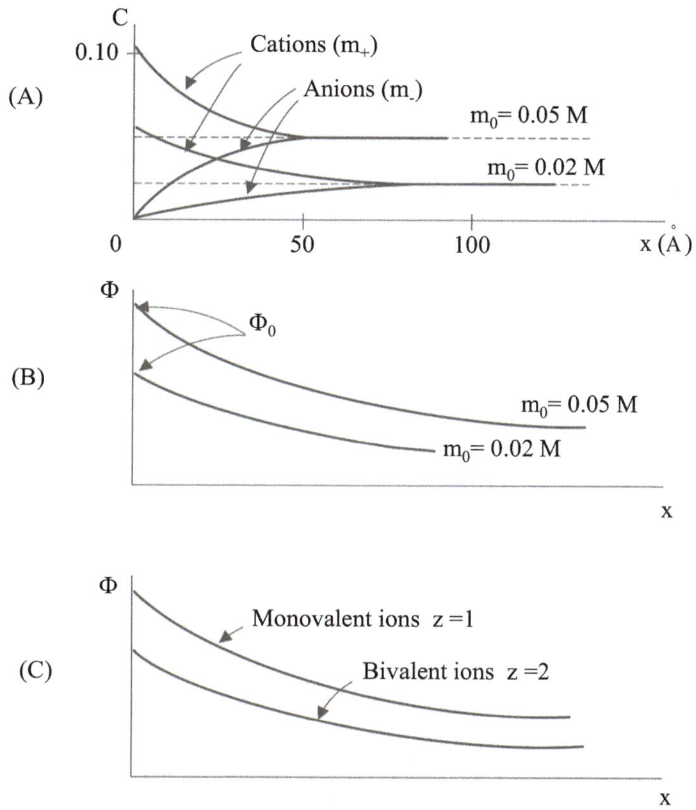

Fig. 8.4 Related to Fig. 8.3 we see distributions of ion concentration *C* and electric potential Φ for different situations

they have nonpermanent charges. These, of course, are dependent on the pH of the solution.

When a surface has excess negative charges, it is called a **cation exchanger**, and when it has excess positive charges, it is called an **anion exchanger**. Thus, for example:

Cation exchangers:

$$\mathrm{R-COOH} \Leftrightarrow \mathrm{R-COO^-} + \mathrm{H^+} \text{ (organic acid)}$$

$$\mathrm{-SiOH} \Leftrightarrow \mathrm{-SiO^-} + \mathrm{H^+} \text{ (silica)}$$

Anion exchangers:

$$\mathrm{-AlOH} \Leftrightarrow \mathrm{-Al^+} + \mathrm{OH^-} \text{(aluminum hydroxide)}$$

$$\mathrm{R-NH_3OH} \Leftrightarrow \mathrm{R-NH_3}^+ + \mathrm{OH^-}$$
(organic radical with ammonia)

To quantify this exchange property of the different materials, we use the **ion exchange capacity**, which is a quantitative measure of the ion exchange phenomenon. It is possible to determine the **cation exchange capacity** (CEC) and the **anion exchange capacity** (AEC), always expressed in $\mathrm{cmol_c\ dm^{-3}}$ or $\mathrm{cmol_c\ kg^{-1}}$. The subscript c after mol indicates charge.

As already mentioned, these values depend on the number of nonpermanent charges, which depends on the pH. There is a pH at which the number of positive charges equals the number of negative charges, with the net charge being 0. This pH characterizes the **isoelectric point** of the system. Van Raij and Peech (1972) discuss these aspects for some tropical soils.

An attempt to describe these relations M(solid) ⇔ M(solution), from Eq. (8.1), could be based on ion exchange equilibrium considerations.

This ion exchange reaction can be treated as a chemical reaction of the type:

$$(\mathrm{soil-A}) + \mathrm{B^-} \Leftrightarrow (\mathrm{soil-B}) + \mathrm{A^+} \quad (8.30)$$

where (soil – A) indicates an ion A adsorbed to a charged surface in the soil. At equilibrium, since the chemical potential of each phase has to be constant (see Eq. 8.4), we can say that:

$$[\mu(A^+) + \mu(\text{soil} - B)] - [\mu(\text{soil} - A) + \mu(B^+)] = 0$$

Applying Eq. (8.8) to each term and re-arranging, we have:

$$RT\ \ln\left[\frac{a_A^+ \cdot a_{\text{soil}}B}{a_B^+ \cdot a_{\text{soil}}A}\right] = \mu_0(A^+) + \mu_0(\text{soil} - B) - \mu_0(\text{soil} - A) + \mu_0(B^+)$$

The second member of this equation is the sum of the chemical potentials at the standard state, so:

$$RT\ \ln\left[\frac{a_A^+ \cdot a_{\text{soil}}B}{a_B^+ \cdot a_{\text{soil}}A}\right] = K'$$

or applying the anti-ln:

$$\left[\frac{a_A^+ \cdot a_{\text{soil}}B}{a_B^+ \cdot a_{\text{soil}}A}\right] = K \qquad (8.31)$$

The reader should note that Eq. (8.31) is an equation that can be applied to any chemical reaction for the calculation of equilibrium constants. This equation shows that, under equilibrium condition, there is a relation between the activities of the different constituents in the different phases. If, for example, the plant withdraws from soil A+ or B−, the reaction (Eq. 8.31) leaves the equilibrium condition. It is restored by the passage of A and/or B from the solid phase to the liquid phase. These ions that "easily" pass from the solid phase to the liquid phase are called available to the plant. The speed of this passage depends on *K*, that is, varies from nutrient to nutrient, from its chemical form, temperature, pH, oxidation conditions, etc.

The exchange of cations between the solid and liquid phase of the soil is a reversible process. The cations "fixed" on the surface of the charged particles of the soil or within the crystals of some mineral species and cations bound to certain organic compounds can be reversibly substituted by those present in the solution. Under normal conditions, a soil usually presents itself as a cation exchanger, and the determination of CEC is of great importance in studies of soil fertility and plant nutrition.

Many methods have been proposed for the determination of CEC and, when applied, the values obtained can vary appreciably. Some reasons for this are the quality and quantity of minerals and organic compounds present in the different soils. These methods can be grouped into different categories:

(a) the soil is "washed" with a dilute solution of acid, for example, HCl. In this process, "all" adsorbed cations to the soil are exchanged for H+ and pass into the effluent. The effluent is then titrated and it is possible to determine, in cmol_c, the amount of each cation extracted from the soil;
(b) the soil is equilibrated with a solution of barium, calcium, or sodium acetate and the amount of cations adsorbed is determined by an appropriate technique;
(c) the soil is washed with calcium acetate to saturate it with Ca. Then it is equilibrated with a calcium nitrate solution containing the radioactive isotope ^{45}Ca. Based on the ^{45}Ca/^{40}Ca isotope ratios, the CEC can be determined. Details of these techniques are found, among others, in Jackson et al. (1986), Van Raij (1987, 1991), and Malavolta (1979).

The great difficulty of applying Eq. (8.31) is in determining the activities of A and B in the solid phase of the soil. By Donnan's theory one can, however, obtain some information about the exchange process. As seen previously, in a system consisting of a crystal immersed in a solution (see Fig. 8.2), the following relationship holds:

$$a^{\text{I}}\ \text{electrolyte} = a^{\text{II}}\ \text{electrolyte}.$$

We will call phase I as the absorbed one (aph) and phase II as external (eph). For an electrolyte of the type M_nA_m, we have:

$$(a_{\text{MnAm}})_{\text{aph}} = (a_{\text{MnAm}})_{\text{eph}}$$

which, according to Eq. (8.12), can be broken in:

$$[(a_M)^n]_{\text{aph}}[(a_A)^m]_{\text{aph}} = [(a_M)^n]_{\text{eph}}[(a_A)^m]_{\text{eph}}$$

re-arranging:

$$\left[\frac{(a_{\mathrm{M}})_{\mathrm{aph}}}{(a_{\mathrm{M}})_{\mathrm{eph}}}\right]^{1/m}=\left[\frac{(a_{\mathrm{A}})_{\mathrm{eph}}}{(a_{\mathrm{A}})_{\mathrm{aph}}}\right]^{1/n} \tag{8.32}$$

Equation (8.32) gives the relation between the activities of an ion in the external and adsorbed phases. For example, for the case of the electrolyte Na_2SO_4, we can say that the following relation is valid in the equilibrium case, where $m = 1$ and $n = 2$.

$$\frac{(a_{\mathrm{Na}})_{\mathrm{aph}}}{(a_{\mathrm{Na}})_{\mathrm{eph}}}=\left[\frac{(a_{\mathrm{SO_4}})_{\mathrm{eph}}}{(a_{\mathrm{SO_4}})_{\mathrm{aph}}}\right]^{1/2}$$

For the soil solution in general, under equilibrium conditions, we could say that:

$$\frac{(a_{\mathrm{H^+}})_{\mathrm{aph}}}{(a_{\mathrm{H^+}})_{\mathrm{eph}}}=\frac{(a_{\mathrm{K^+}})_{\mathrm{aph}}}{(a_{\mathrm{K^+}})_{\mathrm{eph}}}=\frac{(a_{\mathrm{Na^+}})_{\mathrm{aph}}}{(a_{\mathrm{Na^+}})_{\mathrm{eph}}}
=\left[\frac{(a_{\mathrm{Ca^{2+}}})_{\mathrm{aph}}}{(a_{\mathrm{Ca^{2+}}})_{\mathrm{eph}}}\right]^{1/2}=\left[\frac{(a_{\mathrm{Al^{3+}}})_{\mathrm{aph}}}{(a_{\mathrm{Al^{3+}}})_{\mathrm{eph}}}\right]^{1/3}
=\frac{(a_{\mathrm{OH^-}})_{\mathrm{eph}}}{(a_{\mathrm{OH^-}})_{\mathrm{aph}}}=\frac{(a_{\mathrm{Cl^-}})_{\mathrm{eph}}}{(a_{\mathrm{Cl^-}})_{\mathrm{aph}}}=\left[\frac{(a_{\mathrm{SO_2^-}})_{\mathrm{eph}}}{(a_{\mathrm{SO_2^-}})_{\mathrm{aph}}}\right]^{1/2}
=\frac{(a_{\mathrm{H_2PO_4}})_{\mathrm{eph}}}{(a_{\mathrm{H_2PO_4}})_{\mathrm{aph}}}=D$$

all relations being equal to each other, they can only be equal to a constant, in the case D, whose value depends on the ion exchange properties of the soil in question. These relationships allow us to only study the activity of one ion in relation to another, both in solution, and from these relations the classical concepts of "**limestone potential**", "**phosphate potential**", "**K-Ca potential**" and so on, were defined. If we take, for example, the H^+ and Ca^{2+} ions, we can write:

$$\frac{\left[(a_{\mathrm{Ca^{2+}}})_{\mathrm{eph}}\right]^{1/2}}{(a_{\mathrm{H^+}})_{\mathrm{eph}}}=D$$

Applying log we have:

$$\log\left[\frac{1}{(a_{\mathrm{H^+}})_{\mathrm{eph}}}\right]+\frac{1}{2}\log(a_{\mathrm{Ca^{2+}}})_{\mathrm{eph}}=\log D$$
$$=\text{Chemical potential}$$

Using the classical symbols:

$$\log\left[\frac{1}{(a_{\mathrm{H^+}})_{\mathrm{eph}}}\right]=\mathrm{pH}$$

$$\log\left[\frac{1}{(a_{\mathrm{Ca^{2+}}})_{\mathrm{eph}}}\right]=\mathrm{pCa}$$

we have:

$$\textbf{Chemical Potential}=\mathrm{pH}-0.5\times\mathrm{pCa} \tag{8.33}$$

We could also develop other "potentials", as that of phosphate and Ca-K using the ions $H_2PO_4^-$, $Ca^{2+}$and K.

$$\textbf{Phosphate potential}=\mathrm{P}\left[\mathrm{H_2PO_4^-}\right]+0.5\times\mathrm{pCa} \tag{8.34}$$

$$\textbf{Ca}-\textbf{K potential}=\mathrm{pK}-0.5\times\mathrm{pCa} \tag{8.35}$$

These potentials provide the relationship between the activities of one ion in the external phase, and another ion, also in the external phase, in the form of a logarithm. All involve the Ca^{2+} ion because these concepts were developed by researchers in temperate areas, in which Ca^{2+} is the predominant ion in soil solution.

In this text, we do not deal in detail with the applied aspects of ion exchange. This is, however, a matter of great practical importance in terms of liming and fertilizer efficiency. More details can be seen in Havlin et al. (2014).

8.7 Ion Flux in the Soil

So far, we have briefly seen the situation of the equilibrium of ions in the soil. Now let's worry about their transport or movement. As will be seen later in this item, one of the forces responsible for the **ion flux** in the soil is the **activity gradient**. The activity of the ion is evaluated by the free energy of Gibbs, through Eqs. (6.15) and (8.8), already discussed previously. The reader, having gone through the previous items of this chapter, must be convinced of the difficulty of determining the ion activity at any point in the soil. We could take the average activity of the ion in a volume element, but how to measure it? There are techniques, in which the soil is saturated, then extracting the solution and measuring its concentration. It has to be considered that in this extraction of the solution from the soil, first we extract the less concentrated macropores, and later extracting from a drier soil, a more concentrated solution is withdrawn. On the other hand, it is very difficult to remove all the water in the liquid phase from the soil.

This problem could be discussed at length here. It is, however, our intention only to draw the reader's attention to the difficulty of defining and measuring the activity of an ion in the soil. This becomes necessary because the flow equations that we will develop next assume the possibility of measuring the soil solution activity at one point. Finally, it must be said that with electrodes, the activity of an ion in solution can directly be measured (e.g., pH measures the activity of the H^+ ion in solution) and that electrodes sensitive to a series of ions, such as Cu^{+2}, Hg^{+2}, Zn^{+2}, NO_3^-, etc., have already been developed. To date, this technology for sure has improved a lot.

A series of four papers (Sparks 1984; Macher 1984; Baham 1984; and Sposito 1984) was published, which review, discuss, and criticize seriously the concept of ionic activity applied to soil solution. Every researcher interested in this area needs to read these papers.

The ion flux in the soil is mainly due to two processes: **diffusion** and **mass transfer**. Diffusion is the movement of ions due to the activity gradient, and the mass transfer is the movement of ions dragged by the water flow induced by transpiration. We will start with the study of diffusion.

Considering the solution of the soil as a diluted solution, the activity **a** can be replaced by the concentration **C**. It should be stressed, however, that whenever the activity is significantly different from the concentration, one must substitute **C** for **a** in all the equations that we will see from now on.

8.8 Solute Diffusion

The diffusion process of a solute in a given medium is due to differences in its concentration C along one direction. These differences of C are in fact **Gibbs free energy** or chemical potential (activity) differences, as seen at the beginning of Chap. 6. The fundamental equation of solutes diffusion is the **Fick equation** (Crank 1956):

$$j_d = -D_0 \nabla C \tag{8.36}$$

where j_d is the flux density by diffusion of an ion (or compound) by the diffusion process in a given medium, i.e., the amount passing through the cross-sectional unit per unit time, as for example mg of NO_3 m^{-2} day^{-1}; D is the **diffusion coefficient** of the component in the medium under consideration, in the previous example would be the diffusion coefficient of NO_3 in water and ∇C is the ion concentration gradient (see definition of gradient in Chap. 7). Note that the Fick equation is identical to Darcy's equation. In one dimension, we can rewrite it:

$$j = -D_0 \frac{\partial C}{\partial x} \tag{8.37}$$

This equation describes the diffusion of one substance into another, such as the diffusion of NaCl into water. It tells us that the solute will always move in the opposite direction to a concentration gradient, that is, the solute moves from points of highest concentration to those of lower concentration (hence the negative sign in Eq. 8.36). As we have already said, we should speak of activity, because it is proportional to the chemical potential, which is Gibbs free energy for a simple solution.

Diffusion can also be seen only as a casual motion of the solute particles. In Fig. 8.5, we see NaCl diffusing in water. Take two infinitesimal layers **a** and **b** of thickness δ, indicated in the figure.

In an ideal diffusion, there is no preferred direction of particle travel; they move at random. At a given interval of time, on average, half of the

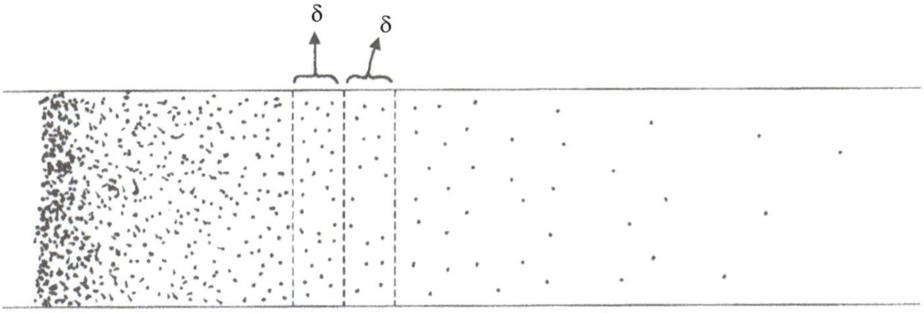

Fig. 8.5 Illustration of NaCl diffusion from a more concentrated part (a) to pure water (b)

ions of (a) leave this region through the central plane and half of those of (b) cross the same plane in the opposite direction. As the mean concentration of (a), C_a, is greater than that of (b), C_b, there is a net flow from (a) to (b). This is one explanation of the diffusion flow from major to minor concentrations. This same reasoning can be applied to the diffusion of one gas into another, diffusion of heat, diffusion of water into water, etc. The **diffusion coefficient** D, in our example, is the diffusion coefficient of NaCl in water. These diffusion coefficients for numerous compounds in different solvents are given in physicochemical tables. Jacobs (1967) is a great text dealing specifically with the diffusion process. For the case of diffusion of a compound in soil water, D_0 of Eq. (8.37) needs to be replaced by the diffusion coefficient D in the soil:

$$D = \theta\, D_0 \left(\frac{L}{L_e}\right)^2 \alpha\, \gamma \tag{8.38}$$

where:

D = diffusion coefficient of a substance i in the soil ($m^2\ s^{-1}$);

D_0 = diffusion coefficient of the same substance in pure water ($m^2\ s^{-1}$);

θ = volumetric soil water content (m^3 of $H_2O\ m^{-3}$ of soil);

$(L/L_e)^2$ = tortuosity, L_e, being the effective path of the substance in the soil, and L the straight line path along a direction x, being a nondimensional factor;

α = nondimensional factor that takes into account the viscosity of the medium, which in turn is a function of θ;

γ = nondimensional factor that takes into account the adsorption of the substance i, i.e., the distribution of i within the pores.

With these considerations, Fick's equation becomes:

$$j_d = -\theta\, D_0 \left(\frac{L}{L_e}\right)^2 \alpha\, \gamma\, \frac{\partial C}{\partial x} = -D \frac{\partial C}{\partial x} \tag{8.39}$$

The volume water content θ is included in the equation because it measures the available area for flow, because the movement occurs only within the solution. In a cross-section of soil A, only θA is available to the solution flow. Similar considerations were made for soil hydraulic conductivity. For the application of Eq. (8.39), D is measured experimentally, avoiding measurements of L/L_e, α, and γ.

Similarly, as it has been said for the flow of water in the soil, it is also very important to study the variations of concentration of an ion at a point in the soil, as a function of time, due to the diffusion process. This requires the **continuity equation**. The reader should now turn to Chap. 7 and see how the continuity equation was established for water flow. It is easy to verify, by the same reasoning used there, that, in the case of ions, in one dimension, the equation becomes:

$$\frac{\partial(\theta C)}{\partial t} = -\frac{\partial j_d}{\partial x} \tag{8.40}$$

which is equivalent to Eq. (7.20a), in the case of water. Here, θ was included because in the development of the continuity equation, a volume element ΔV of soil is taken and the amount of ions contained in that volume is θC. To clarify this point, see in Chap. 3 Eqs. (3.25) and (3.26).

Substituting Eq. (8.39) into Eq. (8.40), we obtain:

$$\frac{\partial(\theta C)}{\partial t} = \frac{\partial}{\partial x}\left(D\frac{\partial C}{\partial x}\right) \tag{8.41}$$

and for cases in which D can be considered constant and independent of C:

$$\frac{\partial(\theta C)}{\partial t} = D\frac{\partial}{\partial x}\left(\frac{\partial C}{\partial x}\right) = D\frac{\partial^2 C}{\partial x^2} \tag{8.41a}$$

Equations (8.41) and (8.41a) are fundamental differential equations for the study of the diffusion of a material in the soil. Their solutions are functions of type C = $C(x, t)$, that is, equations that allow us to determine C at any point x at any time t.

A particular case of importance is the case where the soil water content θ is constant. Under such conditions, D really is constant and θ can be extracted from the derivative of the first member, and Eq. (8.41a) is simplified into:

$$\frac{\partial C}{\partial t} = D'\frac{\partial^2 C}{\partial x^2}$$

where $D' = D/\theta$. We can now establish the same equilibrium cases as those that were established for the case of water flow in the soil:

(a) **steady-state** flow or **permanent regime**:

$$j_d = \text{constant}$$

$$\frac{\partial C}{\partial t} = \text{constant}; \quad \frac{\partial C}{\partial t} = 0$$

(b) variable flow or **transient regime**: which is the most general case in which Eqs. (8.39)–(8.41a) are used:

$$\frac{\partial^2 C}{\partial x^2} = 0 \quad \text{or} \quad \nabla^2 C = 0$$

(c) without flow, thermodynamic equilibrium:

$$j_d = 0, \text{and constant in time and space}$$

$$\frac{\partial C}{\partial x} = 0$$

$$\frac{\partial C}{\partial t} = 0$$

As an example, let's look at a transitional case. In a flooded lowland ($\theta = \theta_0$, saturated and constant) soil and without vegetation, the KNO_3 fertilizer was added to the surface and the surface water concentration was maintained at 0.2 M NO_3^-. Assuming that NO_3^- is not absorbed or consumed by microorganisms, determine the distribution of NO_3^- in the soil profile as a function of time, where $D' = 1.08 \times 10^{-5}$ cm^2 s^{-1}.

The differential equation is then:

$$\frac{\partial C}{\partial t} = D'\frac{\partial^2 C}{\partial z^2}$$

subject to the conditions:

1. $C = 0.2$; $z = 0$; $t > 0$ or $C(0, t) = 0.2$, which means that the concentration will remain constant and equal to 0.2 M at the surface ($z = 0$) for any time t;
2. $C = 0$; $z > 0$; $t > 0$ or $C(z, 0) = 0$, there is no NO_3^- at any depth of the soil profile ($z > 0$) at the beginning of the fertilizer application. Note that z is considered positive downwards;
3. $C = 0$; $z = \infty$; $t > 0$ or $C(\infty, 0) = 0$, assuming a semi-infinite geometry, without limitation for z.

The particular solution of this equation, subject to these three conditions, will not be seen in detail here. It can be shown that it is:

$$C(z, t) = C \cdot \text{erfc}\left(\frac{z}{\sqrt{4D't}}\right)$$

where **erfc** is the **complementary error function**. To understand it, let us first look at the **error function (erf)**, defined by an integral that appears frequently in the solution of differential equations:

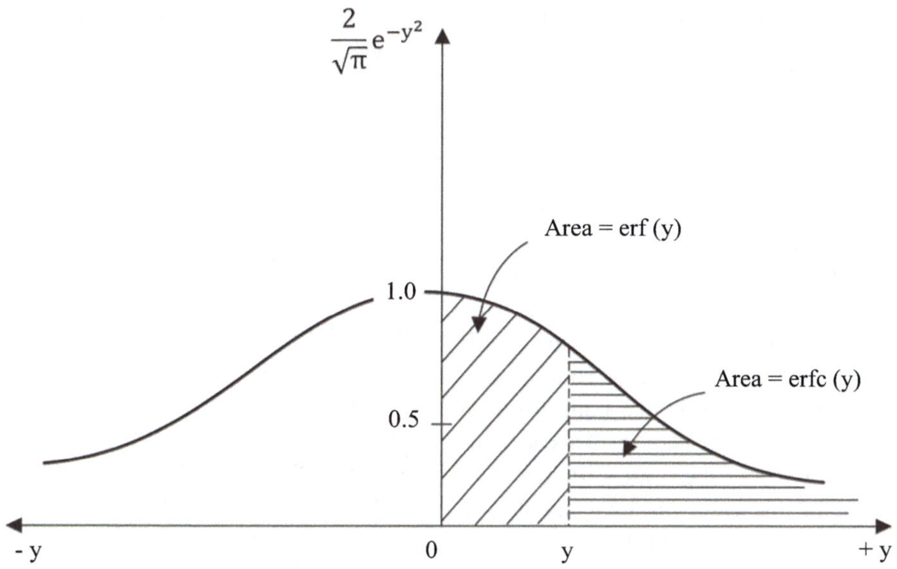

Fig. 8.6 Graph of erf(y) and erfc(y). The sum of both is 1

$$\mathrm{erf}(y) = \frac{2}{\sqrt{y}} \int_0^y \mathrm{e}^{-y^2} \mathrm{d}y \qquad (8.42)$$

and the complementary function is:

$$\mathrm{erfc}(y) = 1 - \mathrm{erf}(y) \qquad (8.43)$$

The reader should not worry about the apparent complexity of these functions. In fact, they are functions like: sine, cosine, logarithm, etc., and as such they are presented in tables for values of y ranging from 0 to ∞. Figure 8.6 shows the graphs of both, and since they are integrals, they represent areas under the curve.

In the above example, if we wish to determine the concentration of $NO_3{}^-$ at 10 cm depth after 1 day (86,400 s), and proceed as follows:

$$y = \frac{z}{\sqrt{4D't}} = \frac{10}{\sqrt{4 \times 1.08 \times 10^{-5} \times 86,400}} = 5.16$$

$$\mathrm{erf}\ (5.16) = \frac{2}{\sqrt{y}} \int_0^{5.16} \mathrm{e}^{-y^2} \mathrm{d}y = 0.02 \qquad \text{(taken from the table)}$$

$$\mathrm{erfc}\ (5.16) = 1 - \mathrm{erf}\ (5.16) = 1 - 0.02 = 0.98$$

and, finally:

$$C(10; 86,400) = 0.2 \times \mathrm{erfc}\ (5.16) = 0.196\ \mathrm{M}$$

It is also important to mention that the diffusion process also appears in the movement of the water from the soil to the root, and at the root interface appear semipermeable membranes, which cause the appearance of concentration gradients.

8.9 Solute Mass Transfer

Let us now see the movement of ions by **mass transfer**. As already said, the solutes of the soil can also move dragged by the flow of water. This process is called mass transfer. If the water flux density in the soil is q (cm^3 of water per cm^2 of cross-section and per day), then qC is the amount of solute carried along with the water:

$$q = 2\ \mathrm{mm\ day^{-1}} = 0.2\ \mathrm{cm^3 cm^{-2}\ day^{-1}}$$

$$C = 0.1\ \mathrm{M\ in\ NO_3^-} = 6.2\ \mathrm{g\ L^{-1}} = 0.0062\ \mathrm{g\ cm^{-3}}$$

and the ion flux j_m by mass transfer will be:

$$j_m = qC = 0.00124 \text{ g of } NO_3^- \text{ cm}^{-2} \text{ day}^{-1} \tag{8.44}$$

In this case, there is no concentration gradient. The concentration is taken as constant. The driving force is the hydraulic gradient ∇H that is embedded in the water flow density q that drags the ion. Much of the plant nutrient uptake is made by mass flow, during transpiration when plants take water from the soil through the root system.

The most common case is that both diffusion and mass flow occur simultaneously. This happens when water moves and there are differences in concentration. In this case, the total ion flux j_t, due to diffusion and mass transfer, is given by the algebraic sum of both (Eqs. 8.39 + 8.44)):

$$j_t = j_d + j_m = -D\frac{\partial C}{\partial x} + q \cdot C \tag{8.45}$$

and the balance of quantities entering and leaving a soil volume element ΔV (continuity equation) is given by:

$$\frac{\partial(\theta C)}{\partial t} = -\frac{\partial j_t}{\partial x} \tag{8.46}$$

$$\frac{\partial(\theta C)}{\partial t} = D\frac{\partial^2 C}{\partial x^2} - \frac{\partial}{\partial x}(q \cdot C)$$

This equation is considerably more complicated for an analytical solution. If q is constant, the problem is less difficult, but, in general, $q = q(x, t)$. Numerical solutions of Eq. (8.46) can today easily be obtained by numerical methods of finite differences (Reichardt and Godoy 1972).

8.10 Solute Sources and Sinks

Other phenomena that affect the balance of quantities entering and exiting a volume element ΔV are cation or anion adsorption phenomena and dissociation of solid phase constituents from the soil and microbial activities that can consume certain compounds and release others. If, for example, we have a soil through which a water flow passes, and at a given instant we add Ca^{++} to this water, calcium moves in the soil by diffusion and mass transfer, but during its movement it is also adsorbed by phenomena described in the item "Ion Exchange Capacity". In this case, Eq. (8.46) becomes:

$$\frac{\partial(\theta C)}{\partial t} = D\frac{\partial^2 C}{\partial x^2} - \frac{\partial}{\partial x}(q \cdot C) \pm \frac{\partial S}{\partial t} \tag{8.47}$$

where $\partial S/\partial t$ represents any "**source**" or "**sink**", that measures the rates of absorption or release of the solute under consideration in the unit volume element. The signal of $\partial S/\partial t$ will depend on each case. In cases of adsorption or desorption, S is called the adsorption or desorption isotherm and, in general, it is a function of C. Again, depending on the complexity of the S function, Eq. (8.47) has no analytic solution. In many cases, the adsorption isotherm is linear or very close to a straight line and can be replaced by a linear regression. In this case:

$$S = a \cdot C + b$$

which means that the adsorbed amount is proportional to the concentration of the solution. For many cases, this is true. Since C is a function of t, deriving S in relation to t, we have:

$$\frac{\partial S}{\partial t} = a\frac{\partial C}{\partial t}$$

and for constant q and θ:

$$(1 + a)\frac{\partial C}{\partial t} = D'\frac{\partial^2 C}{\partial x^2} - v\frac{\partial C}{\partial x}$$

where $v = q/\theta$ (see Eq. 7.6) is the velocity of the water in the pore, and $D' = D/\theta$.Or:

$$\frac{\partial C}{\partial t} = D_1\frac{\partial^2 C}{\partial x^2} - v_1\frac{\partial C}{\partial x} \tag{8.48}$$

where

$$D_1 = \frac{D'}{1 + a} \quad \text{and} \quad v_1 = \frac{v}{1 + a}$$

D_1 is also called **apparent diffusion coefficient** by several authors.

Summarizing these considerations on the flow of ions in the soil, we can separate the following cases, for θ constant:

(a) Diffusion only:

$$\frac{\partial C}{\partial t} = D' \frac{\partial^2 C}{\partial x^2}$$

(b) Mass transfer only:

$$\frac{\partial C}{\partial t} = -\frac{\partial}{\partial x}\left(\frac{q \cdot C}{\theta}\right)$$

(c) Diffusion with mass transfer:

$$\frac{\partial C}{\partial t} = \mathrm{D}' \frac{\partial^2 C}{\partial x^2} - \frac{\partial}{\partial x}\left(\frac{q.C}{\theta}\right)$$

(d) Diffusion and absorption/desorption:

$$\frac{\partial C}{\partial t} = D' \frac{\partial^2 C}{\partial x^2} \pm \frac{\partial S}{\partial t}$$

(e) Mass transfer and absorption/desorption:

$$\frac{\partial C}{\partial t} = -\frac{\partial}{\partial x}\left(\frac{q \cdot C}{\theta}\right) \pm \frac{\partial S}{\partial t}$$

(f) General:

$$\frac{\partial C}{\partial t} = D' \frac{\partial^2 C}{\partial x^2} - \frac{\partial}{\partial x}\left(\frac{q \cdot C}{\theta}\right) \pm \frac{\partial S}{\partial t}$$

Depending on the conditions of each problem, different "components" can be neglected: diffusion, mass transfer, adsorption, etc. For example, if the mass transfer is very significant and the diffusion coefficient too small, the diffusion flux can be neglected. Under equilibrium conditions with respect to water, $q = 0$ and only diffusion is the important mechanism. For nutrients such as phosphorus, highly adsorbed to the solid fraction, the adsorption component $\partial S/\partial t$ becomes very important and for elements such as Cl^-, NO_3^- it is negligible or 0. As an example, Nascimento Filho et al. (1979) studied chlorine fluxes in a Brazilian latosol.

The presentation of the six previous cases, of which "f" is the general case, which encompasses all, can still be more complicated when the nutrient in question undergoes transformations during its movement. It is the case of nitrogen that, besides suffering diffusion and mass transfer, can be transformed by enzymatic or microbiological action:

$$\text{Urea} \rightarrow NH_4 \rightarrow NO_3 \rightarrow N_2.$$

Classical studies that use (and solve) the previous equations for cases of nitrogen are those of Misra et al. (1974), Wagenet et al. (1977), and Wagenet and Starr (1977).

8.11 Miscible Displacement

Several papers presented in the literature (Misra and Misra 1977), under the theme **miscible displacement**, involve diffusion, drag, adsorption, and desorption, and are good examples of the application of the equations seen in this chapter. By miscible displacement we mean the movement of fluids with distinct characteristics that, when crossing a porous medium (mainly the soil) they mix, since they are miscible. To illustrate the question, let us first see the displacement (not turbulent) of two nonmiscible fluids, such as oil and water, through a pipe with velocity v (Fig. 8.7).

If we graph the "concentration" C of the oil, as a function of time, at a point A of coordinate $x = L$, we will have a graph of the type of Fig. 8.8.

For $t = 0$ and t_1 only water passes through A and the oil concentration is 0. During $t_2 - t_1$, only oil passes and $C = 100\%$. For $t > t_2$, C goes back to 0. As the flow is slow, with no turbulence and the liquids are not miscible, one fluid pushes (mass flow) the other as if it were a piston-flow.

If, in the pipe of Fig. 8.7, the oil is exchanged for a solution of NaCl, which is miscible in water, we will have the miscible displacement, which in the case of slow motion will also show the diffusion of NaCl in the water. The graph of Fig. 8.8 changes to that of Fig. 8.9.

Due to diffusion, before t_1, NaCl appears at point A and, for $t > t_2$, NaCl is still present at A. The rectangular graph of Fig. 8.8 becomes a

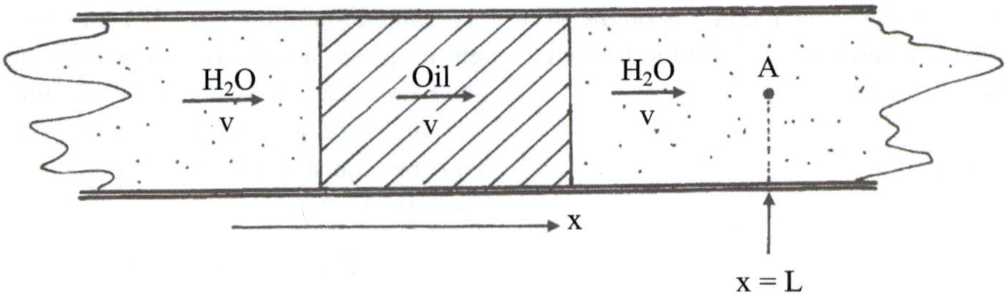

Fig. 8.7 Flow of two nonmiscible fluids in a pipe

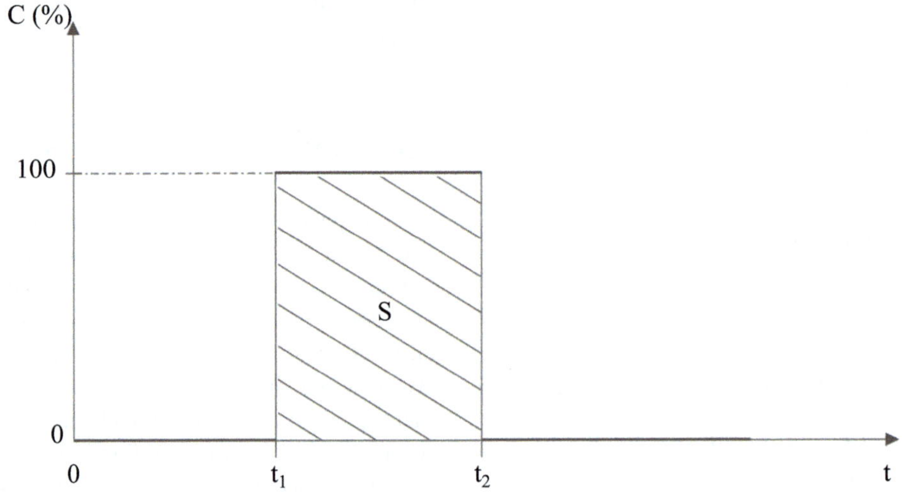

Fig. 8.8 Illustration of a piston flow related to Fig. 8.7 showing the graph of the concentration of the nonmiscible fluid as a function of time

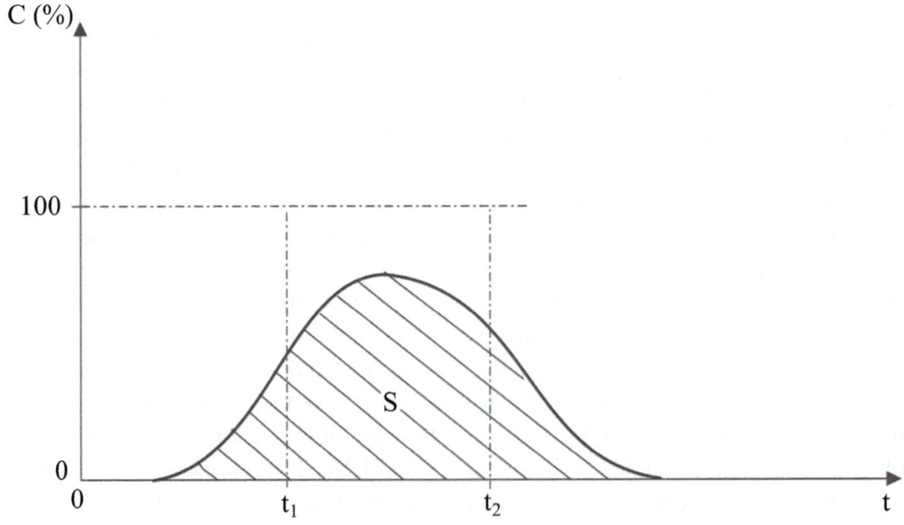

Fig. 8.9 Related to the flow in a pipe (Fig. 8.6), the distribution of NaCl is shown as a function of time. The oil of Fig. 8.7 is replaced by a NaCl solution and the distribution of the concentration is changed to a bell-shaped function due to NaCl diffusion

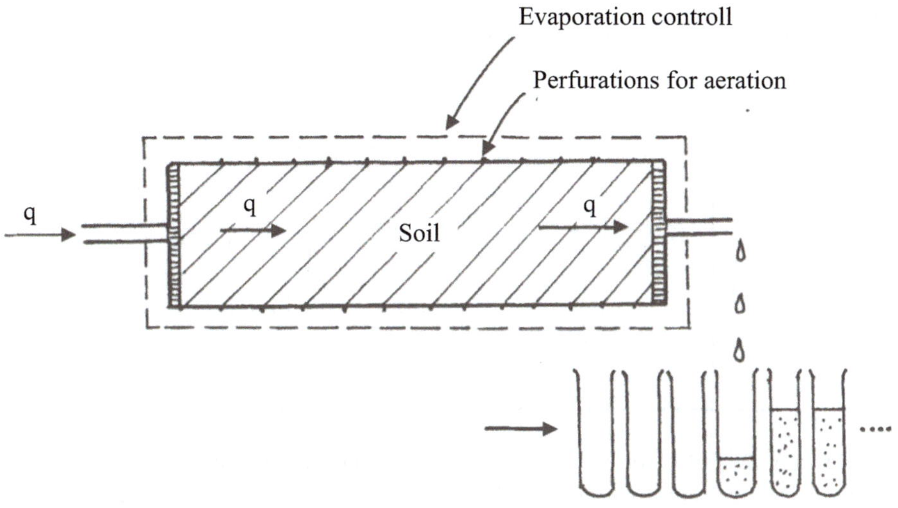

Fig. 8.10 Automatic sample collection during flow of a salt solution through a soil column

"bell" (Fig. 8.9), not always symmetric, with a maximum that does not have to reach 100%, since this depends on v and of the interval $t_2 - t_1$. The important thing is that the areas S of both figures represent the total of the second fluid. Curves of the type in Fig. 8.9 are called **breakthrough curves**.

In the case of soils, many miscible displacement experiments were performed on finite-sized soil columns, as indicated in Fig. 8.10.

A typical example would be the case of a column through which water passes in dynamic equilibrium and, at a given moment, ammonium sulfate is added. In this condition, in relation to the ammonia, we have the drag (mass flow), the diffusion, its adsorption by the negative charges of the soil and its consumption by microorganisms (sinks). The differential equation to be used is Eq. (8.47). In the laboratory, solution samples are continuously collected (in test tubes) to measure the exit concentration C. For this, there are automatic sample changers.

Often, instead of making graphs of C versus t, as in Fig. 8.9, t is changed by the volume of pores V_p. A **pore volume** corresponds to the volume of fluid that fits in the column, that is, its volume V multiplied by the porosity. Thus, a soil column of diameter 5 cm and length 30 cm has a volume of 0.589 L, and if the soil porosity is 0.55 $m^3\ m^{-3}$, the volume of pores V_p is 0.324 L. The use of V_p generalizes the results of the experiment, as it allows the comparison of results of researchers using soil columns of different dimensions. Therefore, a dimensionless concentration, such as C/C_0 or $(C - C_i)/C_0$ is also used (see Chap. 18).

The above-mentioned work by Misra and Misra (1977) deals with the miscible displacement of nitrate [$Ca(NO_3)_2$] and chloride (NH_4Cl) under field conditions. Their Eq. (1), used separately for N-NO_3^- and Cl^- is Eq. (8.47) itself:

$$\frac{\partial C}{\partial t} = D\frac{\partial^2 C}{\partial x^2} - V_s\frac{\partial C}{\partial x} - kC$$

in which x is the vertical coordinate, V_s the solute velocity in the soil pores, and k, a first-order adsorption constant (adsorption of NO_3^- and Cl^- by positive charges). Consider the initial concentration of the soil profile C_i = constant and for a period t_1 apply a pulse of concentration C_0 of N-NO_3^- or Cl^- on the soil surface, followed by pure water. Thus, its boundary conditions are:

$$C - C_i = 0;\ x \geq 0;\ t = 0.$$

$$C - C_i = C_0;\ x = 0;\ 0 < t < t_1.$$

$$C - C_i = 0;\ x = 0;\ t > t_1.$$

and the solution for $t > t_1$ is:

$$\frac{C - C_i}{C_0} = P(x,t) - P[x,(t - t_1)] \qquad (8.49)$$

where

$$P(x,t) = \frac{1}{2}\left\{ \begin{aligned} &\exp\left[\frac{x}{2D}\left(V_s - \sqrt{V_s^2 + 4DK}\right)\right]\cdot \operatorname{erf}\left[\frac{x - t\sqrt{V_s^2 + 4DK}}{\sqrt{4Dt}}\right] + \\ &+\exp\left[\frac{x}{2D}\left(V_s + \sqrt{V_s^2 + 4DK}\right)\right]\cdot \operatorname{erf}\left[\frac{x + t\sqrt{V_s^2 + 4DK}}{\sqrt{4Dt}}\right] \end{aligned} \right\} \qquad (8.50)$$

The experimental data of C for NO_3^- and Cl^- fit well to the theoretical curves represented by this solution. More details need to be analyzed in the paper itself.

Shukla et al. (2002) apply an inspectional analysis (Tillotson and Nielsen 1984) on Eq. (8.47) (which is a type of scaling, see Chap. 18) to miscible displacement experiments with solution identical to that of Misra and Misra (1977). Using scaling factors, they came to a generalized equation, just as Reichardt et al. (1972) for the horizontal movement of water.

8.12 Exercises

8.1. Calculate the activity of a solution 3×10^{-5} M of $CaSO_4$, with an activity coefficient of 0.893?

8.2. For a solution of 2×10^{-2} M NaOH, at 27 °C, calculate:
   (a) The molar fraction of water;
   (b) The molar fraction of NaOH;
   (c) The water potential according to Eq. (8.9a), considering $\gamma = 1$;
   (d) The water potential according to Eq. (8.8b);
   (e) The potential of NaOH through Eq. (8.9a), considering $\gamma = 1$;
   (f) The potential of NaOH through Eq. (8.8b).

8.3. Which are the values of ν_+ and ν_- for $Al_2(SO_4)_3$?

8.4. For a Donnan system (like that of Fig. 8.1) phase I is 1 L of water and 50 g of a soil of CEC 16 meq 100 g^{-1} of soil. Phase II is also 1 L. At equilibrium we measured $N^{II+} = N^{II-} = 300$ meq L^{-1} of KCl. Calculate the anionic repulsion of the system, indicating the values of α and β?

8.5. A soil solution extract shows the following concentrations: $[H^+] = 2.3 \times 10^{-6}$ M; $[H_2PO_4^-] = 5.6 \times 10^{-5}$ M; $[Ca^{2+}] = 7.1 \times 10^{-4}$ M; $[K^+] = 3.2 \times 10^{-4}$ M. Calculate the pH, pCa, $pH_2PO_4^-$, pK, the lime potential, the phosphate potential, and the Ca-K potential.

8.6. The diffusion coefficient of an ion in water is 5×10^{-5} cm^2 s^{-1}. Calculate the diffusion coefficient for a soil with $\theta = 0.386$ cm^3 cm^{-3}; $(L/L_e)^2 = 0.66$; $\alpha = 0.38$ and $\gamma = 0.119$?

8.7. In the soil of Exercise 8.6, we have a concentration gradient of 2×10^{-4} M cm^{-1}. Calculate the ion flux density at this spot of the soil.

8.8. Transform the result of the previous exercise into kg ha^{-1} day^{-1}, considering the mol equal to 96 g.

8.9. At a point of a soil with water content $\theta = 0.425$ cm^3 cm^{-3}, we find a change of the ionic density flux of 2×10^{-12} mol cm^{-1} s^{-2}. Which is the daily change in the ionic concentration at this point in the soil?

8.10. In the paper of Misra and Misra (1977). Miscible displacement of nitrate and

chloride under field conditions. Soil Sci Soc Am J 41:496–499, we ask for:
(a) recognize each term of Eq. (1);
(b) understand the boundary conditions 2a, 2b and 2c;
(c) understand the solution 3;
(d) understand Fig. 8.1 in relation to the solution 3.

8.13 Answers

8.1. Using Eq. (8.10): 2.679×10^{-5} M.
8.2. (a) 0.9996398;
(b) 3.602×10^{-4};
(c) -8.86×10^{-3} atm L mol^{-1} or -0.4919 atm;
(d) -0.492 atm;
(e) -195.05 atm L mol^{-1} or -3.901 atm;
(f) cannot be made because Eq. (8b) is approximate and was developed to calculate the water potential as a function of ion addition.
8.3. $\nu_+ = 2$ and $\nu_- = 3$.
8.4. Solution:

$$C_-^{\mathrm{I}} = \frac{\left[-8 \pm \left(8^2 + 4 \times 300 \times 300\right)^{1/2}\right]}{2} = 296 \text{ meq L}^{-1}$$

$\alpha = 300/296 = 1.0135$ and $\beta = (300 - 296)/8 = 0.5$.

8.5. pH = 5.64; pCa = 3.14; pH_2PO_4 = 4.25; pK = 3.49; lime potential = 4.06; Phosphate potential = 2.68; Ca-K potential = 4.92.
8.6. Through Eq. (8.38): $D = 4.03 \times 10^{-7}$ cm^2 s^{-1}.
8.7. Through Eq. (8.39): $j_d = 8.06 \times 10^{-11}$ mol cm^{-2} s^{-1}.
8.8. 66.8 kg ha^{-1} day^{-1}.
8.9. $\partial(\theta C)/\partial t = \partial j/\partial x = 2 \times 10^{-12}$ mol cm^{-2} s^{-1}; θ considered constant and approximating the derivatives by finite differences: $\theta \Delta C/\Delta t = 2 \times 10^{-12}$ and $\Delta C = 4.06 \times 10^{-7}$ mol cm^{-3} s^{-1} or 4.06×10^{-4} M.
8.10. (a) the left-hand term is the change in the concentration of Cl or NO_3 in time, at a given point; the first term to right: diffusion of Cl or NO_3; second term to the right: mass transport of Cl or NO_3, v_s being the velocity of water in the pores; third term: source or sink of first order.
(b) Equation (2a): at the beginning, at any point there is no Cl or NO_3; (2b): during t_1, the Cl or NO_3, is applied at concentration C_0, at soil surface; (2c): after t_1, the application of Cl or NO_3 is stopped at soil surface.
(c) the solution 3 gives us the concentration of Cl or NO_3 at any depth (x) in the profile, and during times $t > t_1$, in a relative form. To calculate concentrations just apply values of x and t!
(d) the three solid lines of Fig. 8.1 are solution 3 itself for some values of x and t. Therefore, the figure gives us the temporal changes of the concentrations of Cl or NO_3 in some chosen depths.

References

Babcock KL (1963) Theory of chemical properties of soil colloidal systems at equilibrium. Hilgardia 34:417–542
Baham J (1984) Prediction of ion activities in soil solutions: computer equilibrium modeling. Soil Sci Soc Am J 48:525–531
Bolt GH (1955) Analysis of the validity of the Gouy Chapman's theory of the electric double-layer. J Coll Sci 10:206–218
Crank J (1956) The mathematics of diffusion. Oxford University Press, London
Donnan FG (1911) Theorie der membrangleichgewichte und membranpotentiale bei vorhandsein von micht dialysierenden eletrolyten. Z Electrochem 17:572–581
Donnan FG, Guggenheim A (1932) Die genaue thermodynamik der membrangleichgewichte. Z Phys Chem 162(A):346–360
Fried M, Broeshard H (1967) The soil-plant system in relation to inorganic nutrition. Academic, New York, NY
Gouy C (1910) Sur la constituition de la charge electrique a la surface d'un electrolyte. J Phys 9:457–468
Harned HS, Owens BB (1958) The physical-chemistry of electrolytic solutions, 3rd edn. Reinhold Publication Corporation, New York, NY

Havlin JL, Tisdale SL, Nelson WL, Beaton JD (2014) Soil fertility and fertilizers, 8th edn. Prentice Hall, Upper Saddle River

Jackson ML, Lim CH, Zelazny LW (1986) Oxides, hydroxides, and aluminosilicates. In: Klute A (ed) Methods of soil analysis. American Society of Agronomy; Soil Science Society of America, Madison, WI, pp 101–150

Jacobs MH (1967) Diffusion processes. Springer-Verlag, Berlin

Koorevaar P, Menelik G, Dirksen C (1983) Elements of soil physics. Elsevier, Amsterdam

Lewis GN, Randall M (1923) Thermodynamics and free energy of chemical substances. McGraw-Hill, New York, NY

Macher MC (1984) Determination of ionic activities in soil solutions and suspensions: principal limitations. Soil Sci Soc Am J 48:519–524

Malavolta E (1979) ABC da adubação. Agronômica Ceres, São Paulo

Misra C, Misra BK (1977) Miscible displacement of nitrate and chloride under field conditions. Soil Sci Soc Am J 41:496–499

Misra C, Nielsen DR, Biggar JW (1974) Nitrogen transformations in soil during leaching. I. Theoretical considerations. Soil Sci Soc Am Proc 38:289–292

Nascimento Filho VF, Reichardt K, Libardi PL (1979) Deslocamento miscível do íon cloreto em solo Terra Roxa Estruturada (Alfisol) saturado em condições de campo. Rev Bras Cienc Solo 3:67–73

Reichardt K (1977) Extração e análise da solução do solo. Sociedade Brasileira de Ciência do Solo, Campinas

Reichardt K, Godoy CM (1972) Solução numérica de equações diferenciais parciais. Centro de Energia Nuclear na Agricultura. Universidade de São Paulo, Piracicaba

Reichardt K, Libardi PL, Meirelles NMF, Ferreyra FF, Zagatto EAG, Matsui E (1977) Extração e análise de nitratos em solução do solo. Rev Bras Cienc Solo 1:130–132

Reichardt K, Nielsen DR, Biggar JW (1972) Scaling of horizontal infiltration into homogeneous soils. Soil Sci Soc Am Proc 36:241–245

Shukla MK, Kastanek FJ, Nielsen DR (2002) Inspectional analysis of convective-dispersion equation and application on measured breakthrough curves. Soil Sci Soc Am J 66:1087–1094

Sparks DL (1984) Ion activities: an historical and theoretical overview. Soil Sci Soc Am J 48:514–518

Sposito G (1989) The chemistry of soils. Oxford University Press, New York, NY

Sposito G (1984) The future of an illusion: ion activities in soil solutions. Soil Sci Soc Am J 48:531–536

Tillotson PM, Nielsen DR (1984) Scale factors in soil science. Soil Sci Soc Am J 48:953–959

Van Raij B (1991) Fertilidade do solo e adubação. Agronômica Ceres, São Paulo

Van Raij B (1987) Avaliação da fertilidade do solo, 3rd edn. Associação Brasileira de Pesquisa da Potassa e do Fosfato, Piracicaba

Van Raij B, Peech M (1972) Electrochemical properties of some Oxisols and Alfsoils of the tropics. Soil Sci Soc Am Proc 36:587–593

Wagenet RJ, Starr JL (1977) A method for the simultaneous control of the water regime and gaseous atmosphere in soil columns. Soil Sci Soc Am J 41:658–659

Wagenet RJ, Biggar JW, Nielsen DR (1977) Tracing the transformations of urea fertilizer during leaching. Soil Sci Soc Am J 41:896–902

Aspects of the Soil Atmosphere

9

9.1 Introduction

In the study of the "soil atmosphere," that is, the gaseous phase of the soil, the knowledge of the laws and principles governing the movement of gases in the soil is of great importance. Aerobic plants and organisms require certain levels of oxygen in the soil atmosphere, consuming O_2 and releasing CO_2. Because of this, the soil atmosphere generally has a lower concentration of O_2 and higher CO_2, compared to the atmosphere above ground. Gas exchange processes between the upper atmosphere and the soil atmosphere (aeration) can often be limiting for the production of most agricultural crops. This is not true with rare exceptions, as is the case of paddy rice cultivation, which develops adequately in the anaerobic environment. This subject has been briefly discussed in Chap. 3. In Brazil, little has been done related to the dynamics of soil gases. An example on air composition in a bean crop in Piracicaba, SP, is presented by Victoria et al. (1976).

The physical–analytical study of the gas transfer processes in the soil is quite complicated. In addition to the upper atmosphere, which is practically constant (see Table 5.1), there are in the soil "sources" and "sinks" of CO_2, O_2, NH_3, N_2, SO_2, and a number of volatile organic compounds. The renewal of O_2 in the soil comes from the upper atmosphere by diffusion, in solution with water or by mass flow. When it rains, the entrance of water into the porous space of the soil expels a certain amount of air from it, and during evaporation or drainage from the soil, the air is replaced by mass flow. Mass flow is also induced by temperature differences that cause convection currents and establish pressure differences. Despite all these factors, it is believed that the diffusion process is the main process responsible for the transfer of gases in the soil.

9.2 Flow of Gases in the Soil

9.2.1 Gas Diffusion

Let us consider first the case of diffusion of gases in an aerated soil, assuming this process as the main responsible for the flow. Although it can be seen as a random process, as discussed in Chap. 8, the Fick equation states that the force responsible for the diffusion of a compound or gaseous element is its potential gradient, measured by the **Gibbs free energy**, given by Eq. (6.5). In the case of gases, the Gibbs free energy is directly proportional to the partial pressure of the gas in the mixture (see the definition of partial pressure

K. Reichardt, L. C. Timm, *Soil, Plant and Atmosphere*, https://doi.org/10.1007/978-3-030-19322-5_9

in Chap. 5) and also directly proportional to its concentration. Thus, the **gas flux density** by **diffusion**, given by **Fick's Law** or equation, is:

$$j_d = -D_0 \frac{\partial P}{\partial x} = -D'_0 \frac{\partial C}{\partial x} \tag{9.1}$$

where

j_d = gas diffusion flux density (volume or mass of gas per unit area and time).

D_0 = diffusion coefficient of the gas in air or other medium, which is found in tables of physical constants. It is a function of the temperature and practically independent of C, so that in isothermal cases, it can be considered constant.

P = partial pressure of the gas.

C = concentration of the gas.

$D'_0 = a \times D_0$, if $P = a \times C$, that can be considered a proportionality constant for the chosen gas.

Equation (9.1) tells us that the diffused amount of a gas is proportional to the partial pressure gradient of the gas, which in turn is proportional to its concentration. It is necessary for the reader to understand the importance of the gas partial pressure gradient in this question. The total pressure of the gases is almost always the same, equal to the atmospheric pressure, which varies very little. Partial pressures, however, can vary widely, thus establishing huge partial pressure gradients, which lead to considerable flows. We have already said that O_2 is consumed at high rates in the soil by the action of microorganisms and roots, and, consequently, its partial pressure is greatly reduced in relation to the partial air pressure of the upper atmosphere of the soil. This gradient is responsible for the flow of O_2 into the soil, since the diffusion coefficient changes little.

For the case of the diffusion of gases in the soil, as we have already seen in the case of diffusion of ions in solution, the area available for the flow is reduced and the path to be traveled is longer. Since the available space per cm^3 of soil is the water-free porosity β (see Eq. 3.30) and the tortuosity factor is $(L/L_e)^2$ (see Chap. 8), Eq. (9.1) applied to the soil is:

$$j_d = -D'_0(\alpha - \theta)\left(\frac{L}{L_e}\right)^2 \frac{\partial C}{\partial x} = -D\frac{\partial C}{\partial x} \tag{9.2}$$

where D is the **diffusion coefficient** of the gas in the soil, equal to:

$$-D'_0(\alpha - \theta)\left(\frac{L}{L_e}\right)^2$$

If we wish to study the changes of the concentration of a given gas as a function of time, in a given position in the soil, it is again necessary to use the **continuity equation** introduced for the case of water flow in Chap. 7. In the case of gases, the continuity equation can be written as follows:

$$\frac{\partial(\beta C)}{\partial t} = -\frac{\partial j_d}{\partial x} \tag{9.3}$$

For a better understanding of Eq. (9.3), see also Eqs. (3.31)–(3.33). Substituting Eq. (9.2) into (9.3) and remembering that D can be considered constant for isothermal flow, we will have:

$$\frac{\partial(\beta C)}{\partial t} = D\frac{\partial^2 C}{\partial x^2} \tag{9.4}$$

This is the most general differential equation of the diffusion of a gas in the soil. Its application to particular problems with certain boundary conditions can result in a solution of type $C = C(x, t)$, that is, an equation that gives the gas concentration at any time and point in the space under consideration. The same equilibrium conditions, used in Chaps. 7 and 8, can also be presented here:

(a) Flux in steady state or **permanent regime**: q = constant and consequently $\partial(\beta C)/\partial t = 0$. In this case, we have:

$$\frac{d^2C}{dx^2} = 0 \quad \text{or} \quad \nabla^2 C = 0$$

(b) Variable flux or **transient regime**: is the most general case in which we use Eq. (9.4).

(c) Thermodynamic equilibrium: no flux, the case in which the **Gibbs free energy** is constant.

Let us now see a simplified example of case (a). Consider that the concentration of O_2 at the soil surface is equal to that of the atmosphere and considered equal to 20% (2.8×10^{-4} g cm^{-3}) and that at 30 cm depth there is a colony of microorganisms that consume the O_2 as rapidly as it can diffuse into the soil, that is, the colony maintains the O_2 concentration equal to zero at 30 cm. What is the distribution of O_2 in the soil and the flow of O_2 penetrating the soil surface? The soil has a value of $D = 0.018$ cm^2 s^{-1}.

The differential equation, using the vertical coordinate z, will be:

$$\frac{d^2C}{dz^2} = 0$$

subject to the conditions:

$$C(0) = 2.8 \times 10^{-4}$$
$$C(30) = 0$$

The general solution will obviously be a straight line because it satisfies the condition that the second derivative be zero:

$$C(z) = a \times z + b$$

Applying the boundary conditions, we can find the values of a and b:

$$C(0) = a \times 0 + b = 2.8 \times 10^{-4}$$
$$C(30) = a \times 30 + b = 0$$

so that:

$$b = 2.8 \times 10^{-4}$$
$$a = -0.95 \times 10^{-5}$$

and the particular solution will be

$$C(z) = -0.95 \times 10^{-5} \times z + 2.8 \times 10^{-4}$$

This equation gives the distribution of O_2 in the considered soil layer, that is, the 0–30 cm depth. The flow is given by Eq. (9.2), and the gradient can be obtained by derivating the previous function:

$$\frac{dC}{dz} = \frac{d}{dz}\left(-0.95 \times 10^{-5} \times z + 2.8 \times 10^{-4}\right)$$
$$= -0.95 \times 10^{-5}$$

and, therefore, according to Eq. (9.2):

$$j_d = 0.018 \times 0.95 \times 10^{-5}$$
$$= 0.0171 \times 10^{-5} \text{g cm}^{-2}\text{ s}^{-1}$$
$$= 17.1 \times 10^{-8} \text{ g cm}^{-2}\text{ s}^{-1}$$

Let's look at another example:

Consider that the concentration of O_2 is zero within a soil profile (it has only N_2) with $\beta = 0.35$ cm^3 cm^{-3} and that at time $t = 0$, an atmosphere of constant O_2 concentration C_O (20%) is brought into contact with its surface. What will the distribution of O_2 be as a function of time and depth?

$$D = 0.035 \text{ cm}^2\text{ s}^{-1}.$$

In this case, we will have:

$$\frac{\partial C}{\partial t} = \frac{D}{\beta}\frac{\partial^2 C}{\partial z^2}$$

$$C = 0, z > 0, t = 0$$

$$C = C_O, z = 0, t > 0$$

$$C = 0, z = \infty, t > 0 \quad (\text{semi} - \text{infinite geometry})$$

The solution of this problem is identical to that one seen in Chap. 8, therefore:

$$C(z,t) = C_O \times \text{erfc}\left(\frac{z}{\sqrt{4D't}}\right)$$

where $D' = D/\beta$. The reader should check the definition of erfc in Eqs. (8.42) and (8.43).

Given values at z and t in the interval of interest, we can compute C and develop the graph shown in Fig. 9.1.

It is important to remember at this point that at the same time as O_2 diffuses into the soil, N_2 diffuses outward. In these gas movements, currents of opposite senses are always present, because in most cases, the total pressure of the system must remain constant. Therefore, references to **counter-diffusion** are frequently cited in the literature.

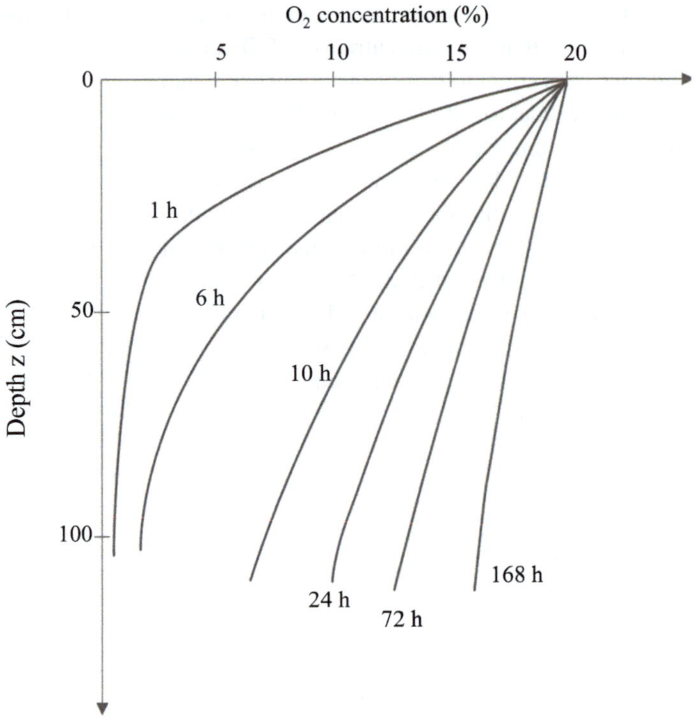

Fig. 9.1 Diffusion of O_2 into a soil profile initially only with N_2

9.3 Sources and Sinks of Gases

Problems of this type are more complicated by the presence of "sources" or "sinks" of the gas in question. In these cases, Eq. (9.4) can be written as:

$$\frac{\partial(\beta C)}{\partial t} \pm A = D\,\frac{\partial^2 C}{\partial x^2} \tag{9.5}$$

where A represents sources and/or sinks. The function A can be extremely complicated and make it impossible to obtain a solution to Eq. (9.5). In general, A must be a function of t and x. In some particular cases, A can be considered constant, as would be the case of microorganisms uniformly distributed in the soil, consuming O_2 at a constant rate. Let us suppose that this is true in a soil depth L, which has an impermeable layer at $z = L$. In this case, under steady-state conditions, Eq. (9.5) becomes:

$$\frac{\partial^2 C}{\partial z^2} = \frac{A}{D} \tag{9.6}$$

subject to the conditions:

$$C = C_0, z = 0 \tag{9.7}$$

$$\frac{\mathrm{d}C}{\mathrm{d}z} = 0, \quad z = L \tag{9.8}$$

The last condition says that the gradient $\mathrm{d}C/\mathrm{d}z$ is zero at $z = L$, which is to say that there is no flow at $z = L$ or that there is an impermeable layer at $z = L$.

Integrating Eq. (9.6) with respect to z gives:

$$\frac{\mathrm{d}C}{\mathrm{d}z} = \frac{A}{D}z + K_1 \tag{9.9}$$

According to condition (9.8), $\mathrm{d}C/\mathrm{d}z = 0$ for $z = L$, we have:

$$\frac{A}{D} \times L + K_1 = 0$$

and

$$K_1 = -\frac{AL}{D} \tag{9.10}$$

Substituting Eq. (9.10) in (9.9) and integrating once more in relation to z, we have:

$$C = \frac{A}{2D}z^2 - \frac{AL}{D}z + K_2 \tag{9.11}$$

According to condition (9.7), $C = C_0$ for $z = 0$, so that:

$$C_0 = 0 + 0 + K_2$$

and the particular solution will be:

$$C = \frac{A}{2D}z^2 - \frac{AL}{D}z + C_0 \tag{9.12}$$

Equation (9.12) then provides the distribution of O_2 in the soil profile $(0 - L)$ for the steady-state case, that is, all O_2 diffused into the soil is consumed by the microorganisms. The final solution of this same problem, when $\partial C/\partial t$ is non-zero (variable flow), is much more complicated. In this case, Eq. (9.5) remains as is, subject to the conditions:

$$C = 0, z > 0, t = 0$$

$$C = C_0, z = 0, t > 0$$

$$\frac{\partial C}{\partial z} = 0, z = L, t > 0$$

and the solution is:

$$C(z,t) = C_0 + \frac{Az}{D}\left(\frac{z}{2} - L\right) + \frac{16AL^2}{\pi^2 D^2}\sum_{n=1}^{\infty}\left[T(t)Z(z) - \frac{4C_0}{\pi}F(t)H(z)\right] \tag{9.13}$$

in which $T(t)$ and $F(t)$ are exponential functions of t and $Z(z)$ and $H(z)$ sinusoidal functions of z.

It is worth noting that for $t = \infty$, the steady state is reached, in which case, $T(t)$ and $F(t)$ are null, and Eq. (9.13) is simplified in Eq. (9.12).

9.4 Gas Mass Transfer

In addition to these complications arising from the presence of sources or sinks, the gases can move by mass flow, that is, transported by moving air in the soil, and the problems become even more complicated. An analytical description becomes feasible only when the mass flow is constant. In this case, as done for the case of water solutes (Eq. 8.45), Eq. (9.4) becomes:

$$\frac{\partial C}{\partial t} = D\frac{\partial^2 C}{\partial x^2} - \frac{\partial}{\partial x}(v \times C) \tag{9.14}$$

where v is the velocity of displacement of the gaseous mass in the soil.

The most complete case is, finally, the one in which there is diffusion, sources, absorbers, and mass flow:

$$\frac{\partial C}{\partial t} = D\frac{\partial^2 C}{\partial x^2} + A(x,t) - \frac{\partial}{\partial x}(v \times C) \tag{9.15}$$

and solving such problems becomes extremely difficult. For the most part, only numerical solutions can be obtained.

By way of example, the reader can analyze the work of Nielson et al. (1984), which uses diffusion equations to study radon fluxes in the soil as a function of the available space. They cite a number of other interesting works. Prevedello and Armindo (2015) and Jury and Horton (2004) also appropriately address the issue of soil gas dynamics.

9.5 Exercises

9.1. In the paper of Nielson et al. (1984):

(a) Understand Eqs. (1), (2), and (3).
(b) Understand Eq. (6).

9.2. To determine the diffusion coefficient D of a gas in the soil, an assembly was constructed as shown in Fig. 9.2.
The soil sample is cylindrical and has an internal diameter of 3.5 cm. The set is placed on an accurate scale (measuring 0.01 g), and at equilibrium, it is found that the mass decreases 5.3 g h^{-1}. The vapor concentration of the volatile liquid when saturated is 0.56 g L^{-1}. Calculate D.

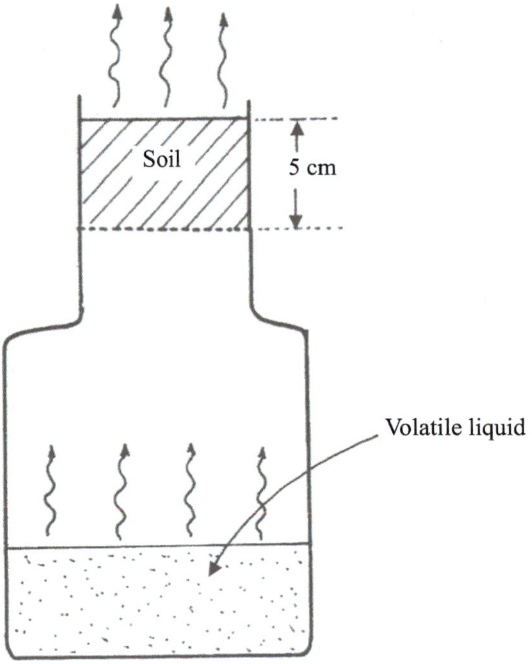

Fig. 9.2 Schematic view of an experiment for the measurement of gas fluxes through a soil sample

9.6 Answers

9.1. (a) Equation (1) is our Eq. (9.1) and therefore refers to the diffusion of radon in the air, without mentioning soil; Eq. (2) corrects the j for flow in the soil. The authors are not clear and define P as total porosity; the correct one would be water-free porosity β, this is the free porous space for gaseous diffusion; Eq. (3) is already the equation for flow in the soil and, therefore, the D of Eq. (3) is different from D of Eq. (1).

(b) The research studies a pore model of the soil using a natural radioactive tracer, radon (Rn), a rare gas in the soil. It is produced from the natural radioisotopes of the series of uranium, found in most soils, in different proportions. The radon is in dynamic equilibrium, hence the right-hand member of Eq. (6) is zero ($\partial C/\partial t = 0$). The first member on the left is the radon diffusion. The second is a sink, which represents the radioactive decay of radon. When radon emits a particle, it is transformed into another isotope, and thus the amount of radon decreases (sink). The third member is the production of radon (source) from the parent elements present in the soil; it is therefore an increase in the amount of radon. Radon, being radioactive, can be detected with special equipment, hence its convenience. Concentration C is proportional to its radioactivity.

9.2. $j = Q/At = (5.3\ \text{g})/(9.62\ \text{cm}^2 \times 3600\ \text{s}) = 1.53 \times 10^{-4}\ \text{g cm}^{-2}\ \text{s}^{-1}$

$\text{grad}\ C = (C_s - 0)/L = [(0.00056 - 0)\ \text{g cm}^{-3}]/(5\ \text{cm}) = 1.12 \times 10^{-4}\ \text{g cm}^{-2}$

$D = j/\text{grad}\ C = 1.366\ \text{cm}^2\ \text{s}^{-1}$.

References

Jury WA, Horton R (2004) Soil physics, 6th edn. John Wiley & Sons, Hoboken, NJ

Nielson KK, Rogers VC, Gee GW (1984) Diffusion of random through soils: a pore distribution model. Soil Sci Soc Am J 48:484–487

Prevedello CL, Armindo RA (2015) Física do Solo com problemas resolvidos, 2nd edn. Prevedello CL, Curitiba

Victoria RL, Libardi PL, Reichardt K (1976) Composição do ar num perfil de solo. Centro de Energia Nuclear para Agricultura. Universidade de São Paulo, Piracicaba

How Heat Is Propagated in the Soil 10

10.1 Introduction

Soil temperature is an important factor in plant growth and development. Many efforts were made to vary the soil temperature in order to create a plant-friendly environment. Various types of mulch, such as straw, aggregates, polyethylene, and so on, were used either to increase or to stabilize soil temperature (Oliveira et al. 2001; Bamberg et al. 2011). The shape of the soil bed can also be adapted in order to increase the heating of the soil next to the plants. Irrigation, too, can be used to modify the thermal behavior of the soil.

Soil temperature affects seed germination, root and plant development, microorganism activity, diffusion of solutes and gases, chemical reactions, and a number of important processes occurring in the field. On the other hand, it is affected by the mineralogical composition of the soil, by the bulk density and water content, by the color of the soil surface, by the structure, by the organic matter, and so on.

It is therefore important to study the **thermal energy transfer processes** in the soil, which can be grouped in three categories:

(a) **radiation**: process of energy transfer by electromagnetic radiation, especially in the visible and infrared region;

(b) **convection**: process of energy transfer by mass flow; and

(c) **thermal conduction** or **diffusion**: process of diffusion transfer of energy from "hot" regions to "cold" regions.

In the soil, the conduction process of heat transfer by diffusion is undoubtedly the main process that occurs. On the surface of the soil and in the atmosphere, the other processes can assume considerable importance. In this chapter, we will study diffusion only. A small introduction about the radiation process in the atmosphere has already been presented in Chap. 5. Convection occurs in fluids, in our case, water and air. In the case of water in the soil, its movement is so slow that heat convection can be neglected. In the atmosphere, the convective movements are of great importance.

10.2 Heat Conduction in Soils

The heat flux density by conduction (Carslaw and Jaeger 1959; Jury and Horton 2004) is given by the **Fourier equation**, which can be written in the form (already seen in Chap. 3):

K. Reichardt, L. C. Timm, *Soil, Plant and Atmosphere*, https://doi.org/10.1007/978-3-030-19322-5_10

Fig. 10.1 A transect of digital soil thermometers installed at the depth of 10 cm in a sugarcane interrow

$$q = -K \frac{\partial T}{\partial x} \quad (10.1)$$

where

q = flux density of heat, equal to the amount of heat (J, cal, erg) per unit area (m^2, cm^2) and time (s, min, day). In the International Unit System, J m^{-2} s^{-1} = W m^{-2};

K = **soil thermal conductivity**, W m^{-1} °C^{-1};

T = soil temperature, °C (see Fig. 10.1);

x = position coordinate, m, cm.

In the same way as it was seen for the flow of water, ions, and gases, it is necessary to use the continuity equation again to study the variation of the amount of heat at a given point in the soil as a function of time.

In this case, we will have:

$$\frac{\partial Q}{\partial t} = -\frac{\partial q}{\partial x} \quad (10.2)$$

where Q is the amount of thermal energy contained in the volume element.

In Thermal Physics, it is shown that the amount of sensible heat or heat energy dQ, stored or lost, per unit volume by a material of isobaric specific heat c (J m^{-3} °C^{-1}), when its temperature varies dT, is given by:

$$\mathrm{d}Q = c\,\mathrm{d}T \quad (10.3)$$

where c is given by Eq. (3.35) or (3.36), for the soil case.

Substituting Eqs. (10.3) and (10.1) into 10.2, we have:

$$c\frac{\partial T}{\partial t} = \frac{\partial}{\partial x}\left(K\frac{\partial T}{\partial x}\right) \quad (10.4)$$

which is the general differential equation of **heat diffusion in soils**. For homogeneous soils of constant composition, density, water content, and porosity, Eq. (10.4) can be simplified, since K and c can be considered constant. Like this:

$$\frac{\partial T}{\partial t} = \frac{K}{c}\frac{\partial^2 T}{\partial x^2} \quad (10.5)$$

where K/c is usually symbolized by D and termed **soil thermal diffusivity**, already described in Chap. 3. Thus:

$$\frac{\partial T}{\partial t} = D\frac{\partial^2 T}{\partial x^2} \quad (10.6)$$

The same equilibrium conditions presented in the previous chapters can be here represented:

(a) steady-state flow: q = constant and, consequently, $\partial T/\partial t = 0$. In this case, Eq. (10.6) is summarized as:

$$\frac{\partial^2 T}{\partial x^2} = 0 \quad \text{or} \quad \nabla^2 T = 0$$

It is important to remember that dT/dx is not null because there is flux.

(b) variable flow: this is the most general case in which Eq. (10.6) is used, as presented above;

(c) thermal equilibrium: there is no flow. The temperature gradient $\partial T/\partial x$ or the thermal conductivity (thermal insulation) is 0.

In order to exemplify the case of heat diffusion in the soil, we will study the variations of soil temperature in the day-night period, at different depths, through a simplified model, which gives us a good idea of the temperature behavior in a soil profile without vegetation. This model was used by Wierenga (1969) and, in Brazil, by Decico (1974).

10.3 Model for the Description of Temperature Changes in the Soil

Consider a homogeneous soil profile with constant density and water content along z, exposed to solar radiation. As it was seen in Chap. 5, the incident radiation is the **global radiation**, part of which is reflected by the soil (see albedo), and the other part is absorbed, which will heat its surface. This heat is then diffused into the soil. The apparent trajectory of the sun relative to the surface varies a lot around the globe, and therefore, we chose in this example a day of **equinox** (duration of night = day, March 21 and September 21 all over the globe), which can be approximated by a sine wave, from its sunrise to the sunset. Thus, the temperature T on the soil surface can also be described by a sine function of the type (Fig. 10.2):

$$T(0,t) = \bar{T} + T_0 \sin \omega t \qquad (10.7)$$

where

$\bar{T}$ = the average temperature around which temperature fluctuates, at 25 °C in the example of Fig. 10.2;

T_0 = amplitude of oscillation, equal to 10 °C;

ω = angular velocity of the Earth ($2\pi/24$ radians/hour). Note that for $t = 0$, we have sin $0° = 0$ and $T = \bar{T}$. For $t = 24$ h, sin $2\pi = 0$.

Note also that in Fig. 10.2, the time t does not coincide with the actual watch time, that is, $t = 0$ coincides with 6 o'clock in the morning, $t = 6$ with midday, and $t = 18$ with midnight.

The limitations of Eq. (10.7) to describe temperature variations on the soil surface are obvious: (1) the apparent trajectory of the sun is not a pure sine function; (2) it varies with the time of year and latitude; (3) the equation applies only to cloudless days and so on. Even so, we will see that this model is very elucidative to understand the diffusion of heat into and out of the soil.

Let us consider that at a theoretically infinite depth (in practice about 1 m is enough), the soil temperature does not vary with time and is equal to $\bar{T}$. So that:

$$T(\infty,t) = \bar{T} \qquad (10.8)$$

This is observed in practice. For example, in the underground cellars where wines are stored

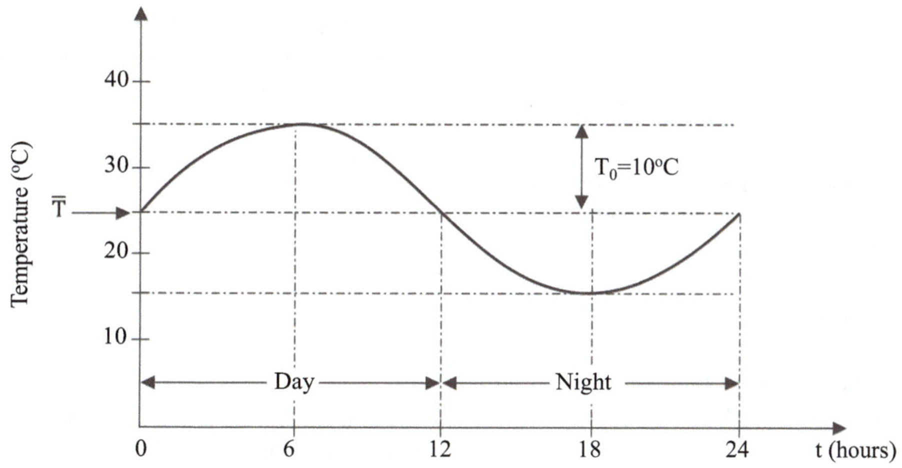

Fig. 10.2 A sine function describing the air temperature changes for an ideal equinox day

for maturation, the temperature is practically constant throughout the year.

An initial condition of type $T(z,0)$ is not necessary here because this **boundary value problem** (BVP—recall in Chap. 7) has no beginning or end. The end of 1 day is the beginning of the other, and the solution is valid for a sequence of days.

Let us also consider that all heat transport in the soil is made only by conduction (diffusion); so that the differential equation that governs the process is:

$$\frac{\partial T}{\partial t} = D \frac{\partial^2 T}{\partial z^2} \tag{10.9}$$

The solution of Eq. (10.9), subject to conditions (10.7) and (10.8), is:

$$T(z,t) = \bar{T} + T_0 \exp^{\left(-z\sqrt{\omega/2D}\right)} \times \sin\left(\omega t - z\sqrt{\omega/2D}\right) \tag{10.10}$$

This solution, whose details can be seen in Decico (1974), tells us that the temperature varies exponentially with depth (the exponential term has only the variable z) and sinusoidally with time. Equation (10.10) is a sinusoid, such as Eq. (10.7). The reader should note that for $z = 0$ (soil surface) it reduces to Eq. (10.7), because exp. (0) = 1 and the time lag $z\sqrt{\omega/2D}$ is 0.

In Eq. (10.10), the amplitude A is only a function of z, given by:

$$A(z) = T_0 \exp\left(-z\sqrt{\omega/2D}\right) \tag{10.11}$$

and, as it can be seen, varies exponentially with the depth z.

Let's consider that the soil has a diffusivity of 5×10^{-3} cm^2 s^{-1}. Under such conditions:

$$\sqrt{\omega/2D} = \sqrt{\frac{2\pi}{86{,}400 \times 2 \times 5 \times 10^{-3}}} = 0.0853$$

and the temperature wave amplitudes for different depths can be calculated:

$A(0) = T_0 \exp(0) = T_0 = 10$ °C (see Fig. 10.1)
$A(10) = T_0 \exp(-10 \times 0.0853) = 4.26$ °C
$A(20) = T_0 \exp(-20 \times 0.0853) = 1.82$ °C
$A(30) = T_0 \exp(-30 \times 0.0853) = 0.77$ °C

....

$A(100) = T_0 \exp(-100 \times 0.0853) = 0.0019$ °C
$A(\infty) = 0$ °C

Figure 10.3 shows the previously calculated data continuously.

It is noted, then, that the most pronounced variations of T occur in the surface layers of the soil. Therefore, when measuring soil temperature, there is no logic in measuring it at many points at great depths. More measurements are required in the vicinity of the soil surface. Good depth choices for the T measurements would be 2, 4, 8, 16, 32, 64 cm; 3, 9, 27, 81 cm; and 5, 25, 100 cm, if possible with measurement on the surface, which is very difficult, since the "bulb" of the thermometer would receive direct solar radiation and since the bulb is a non-soil material, this temperature could not be representative.

The sinusoidal part of Eq. (10.10) also deserves a discussion. In it, we find a sine of a difference that involves space and time:

$$\sin\left(\omega t - z\sqrt{\omega/2D}\right) \tag{10.12}$$

The factor $z\sqrt{\omega/2D}$ is called lag in time, which is greater as the z increases. The temperature wave has, therefore, a delay which increases with z.

Figure 10.2 (top), which corresponds to the soil surface ($z = 0$), shows that for $t = 6$ h, the surface temperature passes through the maximum. At this moment:

$$\text{for } z = 0: \quad \sin\left(\omega t - z\sqrt{\omega/2D}\right) = 1$$

Since $\sin 90° = \sin \pi/2 = 1$.
This is easy to verify,

$$\sin\left(\frac{2\pi \times 6}{24} - 0\right) = \sin(\pi/2) = 1$$

Let us now look at the depth of 10 cm, asking the question: When does the temperature wave go through a maximum at this depth?

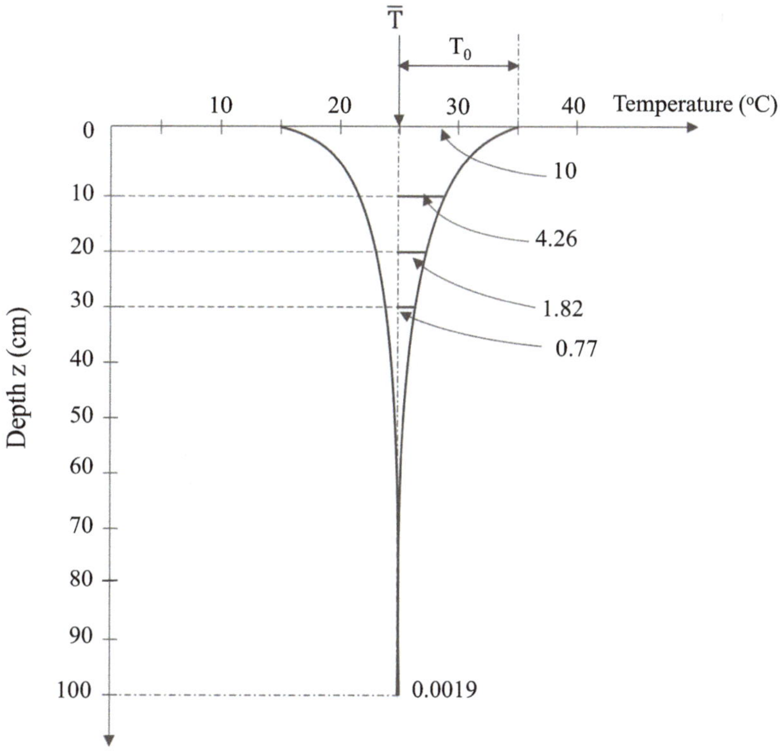

Fig. 10.3 Variation in depth of the amplitude of the temperature wave in a homogeneous soil profile

$$\text{for } z = 10: \quad \sin\left(\frac{2\pi \times t}{24} - 10 \times 0.0853\right)$$

which has to be = 1 so that we have a maximum, because $\sin \pi/2 = 1$. So:

$$\left(\frac{2\pi \times t}{24} - 0.853\right) = \frac{\pi}{2}$$

from which taking the value of t yields 9.25 h (the digit 25 after the comma are not 25 min but 0.25 h), therefore, with a delay of 9.25–6.00 = 3.25 h in relation to the soil surface.

Doing the same for $z = 20$ cm, we get a delay of 6.51 h.

From what we have just seen, we conclude that the temperature wave, when penetrating the soil, has its amplitude diminished with depth and its maximum suffers a delay in relation to the surface, also depending on the depth (Fig. 10.4). As a result, the temperature at higher depths is constant (Condition (10.8)). At these depths, the amplitude becomes minimal, and the maximum of one day is confused with the maximum of the previous day.

This model, although simplified, provides a good idea of the propagation of the temperature wave in the soil. Wierenga (1969) and Decico (1974) also employed this model to determine soil thermal diffusivity under field conditions. Applying Eq. (10.11) to two depths, z_1 and z_2, and dividing one equation by the other, we have:

$$\frac{A(z_1)}{A(z_2)} = \frac{\exp\left(-z_1\sqrt{\omega/2D}\right)}{\exp\left(-z_2\sqrt{\omega/2D}\right)}$$

and, applying ln to both sides, we have:

$$D = \frac{\omega(z_2 - z_1)^2}{2\left[\ln \frac{A(z_1)}{A(z_2)}\right]^2} \tag{10.13}$$

Therefore, by measuring the amplitude A at two depths, z_1 and z_2, we can calculate D. For this, we need to make T measurements throughout the day to measure the amplitudes.

On the other hand, at two depths, z_1 and z_2, the wave will pass through a maximum at different

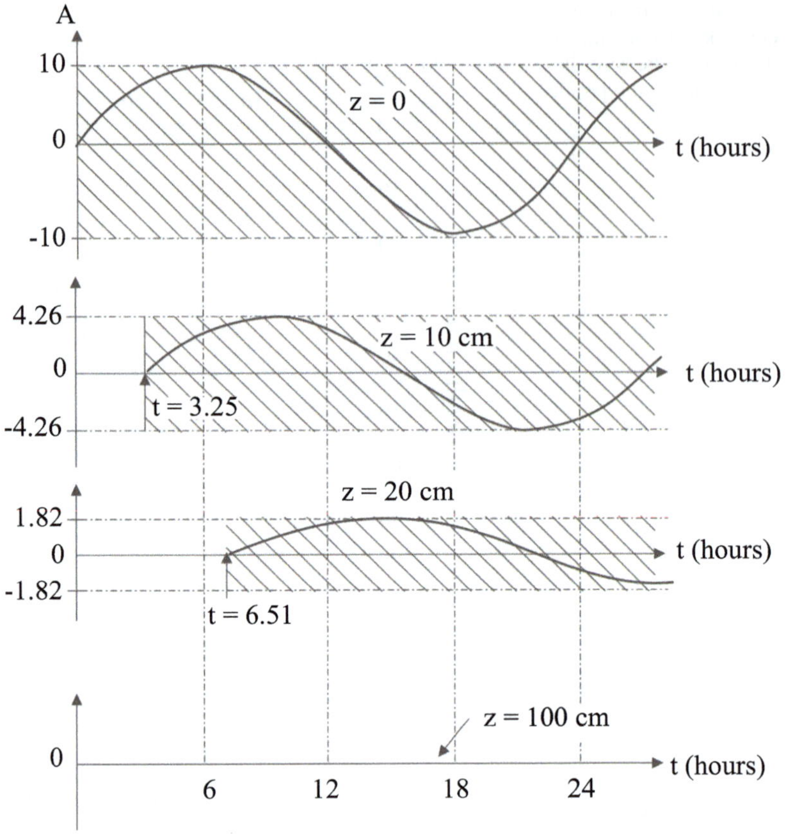

Fig. 10.4 The temperature wave at the soil surface and at two depths, 10 and 20 cm, showing that the amplitude is reduced in depth and that there is a delay in relation to soil surface, also proportional to the depth

times, t_1 and t_2. In this case, according to Eq. (10.12):

$$\left(\omega t_1 - z_1\sqrt{\omega/2D}\right) = \left(\omega t_2 - z_2\sqrt{\omega/2D}\right) = \frac{\pi}{2}$$

or simplifying:

$$D = \frac{1}{2\omega}\left(\frac{z_2 - z_1}{t_2 - t_1}\right)^2 \tag{10.14}$$

It is therefore sufficient to measure the interval of time (t_2 − t_1) between the passage of the maximum through the depths z_1 and z_2, to calculate D.

Equations (10.13) and (10.14) therefore represent methods for determining the thermal diffusivity of a soil.

The data obtained by Decico (1974) show that the model works well for the daytime period and at depths greater than 5 cm.

Of course, in different situations, such as cloudy days, soils with vegetation cover, or mulching, this model has to be adapted. The generalities, however, remain:

1. The amplitude decreases dramatically in depth.
2. There is a delay in the propagation of the heat wave, the greater the depth.

10.4 Exercises

10.1. In a given bare soil, the temperature T was measured at 10 and 20 cm depth, hourly, starting at 6 h. The data are in the following table. Draw the graphs of T versus t for the two depths and calculate the thermal diffusivity of the soil by the amplitude and the lag methods. The average soil profile temperature is 29 °C.

	T (°C)	
t (h)	$z = 10$ cm	$z = 20$ cm
6	29.1	27.9
7	30.1	28.2
8	31.2	28.5
9	31.9	29.0
10	32.5	29.6
11	32.8	30.1
12	33.1	30.6
13	32.7	31.1
14	32.4	31.4
15	31.9	31.5
16	31.2	31.5
17	30.3	31.3
18	29.0	30.8
19	28.2	30.3
20	27.5	29.6
21	27.0	29.1
22	26.7	28.5

10.5 Answers

10.1. For amplitudes $A(10) = 4.2$ °C and $A(20) = 2.4$ °C, taken from the graph. By Eq. (10.13), we have $D = 0.0106$ cm^2 s^{-1}, and for a 3.1 h time lag, also taken from the graph, we have by Eq. (10.14) $D = 0.0051$ cm^2 s^{-1}. We can see that one value is double the other. Explanation: this is not a homogeneous soil, perhaps also not homogeneously humid, and in this case, the model does not describe the process well. What is the best value? You cannot tell. It is best to use the mean value $D = 0.008$ cm^2 s^{-1}. Better yet, repeat the experiment, use other depths and so on.

References

Bamberg AL, Cornelis WM, Timm LC, Gabriels D, Pauletto EA, Pinto LFS (2011) Temporal changes of soil physical and hydraulic properties in strawberry fields. Soil Use Manag 27:385–394

Carslaw HS, Jaeger JC (1959) Conduction of heat in solids. Oxford University Press, London

Decico A (1974) A determinação das propriedades térmicas do solo em condições de campo. Tese (Livre Docência), Escola Superior de Agricultura Luiz de Queiroz. Universidade de São Paulo, Piracicaba

Jury WA, Horton R (2004) Soil physics, 6th edn. John Wiley & Sons, Hoboken, NJ

Oliveira JCM, Timm LC, Tominaga TT, Cássaro FAM, Bacchi OOS, Reichardt K, Dourado-Neto D, Camara GMD (2001) Soil temperature in a sugar-cane crop as a function of the management system. Plant Soil 230:61–66

Wierenga PJ (1969) An analysis of temperature behavior in irrigates soil profiles. PhD thesis, University of California, Davis

11 Water Infiltration into the Soil

11.1 Introduction

Water reaches the surface of the soil (or agricultural crops, vegetation in general) mainly by the processes of rainfall and irrigation. Of less importance (from a quantitative point of view) are hail and dew, and at least for tropical and subtropical regions, snow. This water that comes in contact with the plants and the soil is primarily absorbed by the soil. The process by which this water penetrates the soil surface is called **infiltration**. Depending on the vegetation cover and the slope of the land, some of the water flows over the soil surface. This process is called surface runoff, or simply **runoff**. It is a loss of water for the plants and agriculture, and its intensity determines **soil erosion**. After the process of infiltration, the movement of water continues inside the soil, a process called **redistribution of water**. If this movement reaches greater depths, below the root zone of the plants, the water is lost again from the agricultural point of view. This process is called deep drainage or **internal drainage**. As water carries along ions and soluble compounds, there is also loss of nutrients. The process of chemical losses is called **leaching**. During infiltration, redistribution and internal drainage, water is absorbed by plants and is translocated to all of its parts, and most of it is returned to the atmosphere through leaves by the **transpiration** process. Soil and water bodies (dams, rivers, lakes, etc.) also return water to the atmosphere by the **evaporation** process. For these processes, transpiration and evaporation, energy input is required. It is necessary to bring water from the liquid state to the vapour state and the energy for this obviously comes from the Sun. As in most cases the two processes occur simultaneously, the term **evapotranspiration** is used. Water vapour dissolved in the atmosphere enters the processes of general air circulation and sometimes participates in the formation of clouds. From these it returns to the system in the form of **rain**, **hail** or **snow** and the cycle is closed. For the continuity of the cycle, solar energy enters.

As seen above, the process by which water enters the soil is called **infiltration**, which endures as long as water is available on soil surface. This process is of great practical importance because its rate or velocity often determines the runoff or the excess water that flows over the surface, responsible for the phenomenon of erosion during rainfall or inadequate irrigation management. Infiltration determines the water balance in the root zone and, therefore, the knowledge of this process and its relationship with soil properties is of fundamental importance for the efficient management and conservation of soil and water. Excellent reviews of the infiltration process were made in the past by

K. Reichardt, L. C. Timm, *Soil, Plant and Atmosphere*, https://doi.org/10.1007/978-3-030-19322-5_11

Parr and Bertrand (1960) and Philip (1969). Textbooks encompassing more up-to-date reviews are, among them, Hillel (1980, 1998), Hornberger et al. (1998) and Ehlers and Goss (2016), the latter more focused on hydrology and plant–water relationships, respectively. Libardi (2012) shows that the first models of the infiltration process began in the early twentieth century: (1) Green and Ampt (1911), very simple, but that until the present day produces approximate results quite reasonably; (2) the 1932 Kostiakov equation, which follows the conditions used by Philip and which will be seen below; (3) The Horton equation of 1940, who used a logarithmic model. Also, Kutilek and Nielsen (1994) and Radcliffe and Simunek (2010) deal with the infiltration process. In order to study the infiltration process analytically, we will first take the case of homogeneous soils, in which we will distinguish between the different directions in which infiltration may occur. First we will see the horizontal case, in which the gravitational potential does not come into play, and second the vertical infiltration, in which the gravitational potential may have preponderant participation. Thereafter, we will analyse the process in some field situations, where it can occur in the most varied directions.

11.2 Horizontal Infiltration into Homogeneous Soils

Very rarely water infiltrates into the soil only in the horizontal direction. But because it is a case in which gravitational forces do not act in the process, the analytical description becomes easier. To understand the **horizontal water infiltration** process into a homogeneous soil, we will study it in a controlled laboratory situation. Consider a uniform soil of constant bulk density, packed in a horizontal and transparent acrylic column of constant cross-section and infinite length (quite long to the point that we finish our analysis before the water reaches the end of the column), and with a constant and very dry initial soil water content θ_i. Figure 11.1 illustrates the column. Such columns are mounted as described in Figs. 7.9 and 7.10, except that in this case the constant

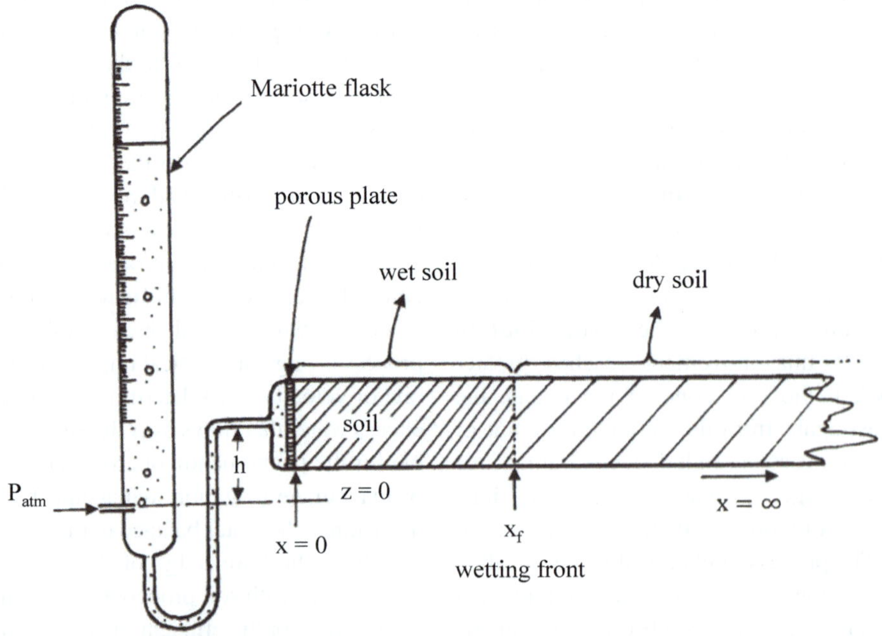

Fig. 11.1 Horizontal infiltration into a homogeneous soil, indicating at time t the position of the wetting front separating wet from dry soil. A graduated Mariotte flask is also shown, allowing water entry into the soil at a negligible pressure (suction) of $-h$

water supply for the infiltration process is made by a **Mariotte flask**. The advantage of this bottle is that the water leaves the bottle at constant pressure, in this case a suction h $(-h)$ and the infiltrating water volumes can be measured. The bottle has a hole where the atmospheric pressure acts, through which air enters as the water comes out and enters the soil. The outside atmospheric pressure P_{atm} is balanced by the internal pressure $P_{\text{atm}} - h_{\text{L}}$ in the bottle such that the h_{L} water column does not act and the water enters the soil with zero pressure. In fact, the setup is arranged in such a way that the small suction h at water entry is slightly larger than the radius of the soil column. If for example the radius of the column is 2.5 cm, we could use $h = -4$ cm, which means that infiltration will occur at a slightly negative pressure, not zero. This negligible suction is important experimentally to avoid water leaks due to positive pressures acting on the water entrance favouring leaking. In terms of soil matrix potential it is very small, and makes it possible water to infiltrate practically at atmospheric pressure, as is the most common case in nature.

At time $t = 0$ (beginning of the infiltration process), the porous plate of negligible resistance to water flow is connected from one side to the Mariotte flask at the $z = -h$ level and at the other side is brought into contact with the column at $x = 0$. Under these conditions the infiltration process begins and the $x = 0$ end of the column is maintained at a saturation water content θ_0 for the entire infiltration time. θ_0 is the saturation water content, if the water in the porous plate is under zero or positive pressure, and will be lower than the saturation humidity if it is under suction, but in our case the lowering of the Mariotte flask to the negligible height—h, we can assume that θ_0 is maintained. We will study here the most common case of infiltration under zero load, or practically nil. The water will penetrate the soil and the soil water content θ will be a function of the point x considered within the column and time t. In this case, the differential flow equation to be used is Eq. (7.20b), which we rewrite here, using the diffusivity D (Eq. 7.17) concept instead of the conductivity K, because it is the horizontal case, in which gravity does not act,

$$\frac{\partial \theta}{\partial t} = \frac{\partial}{\partial x}\left[D(\theta)\frac{\partial \theta}{\partial x}\right] \qquad (11.1)$$

that will be subject to the following conditions:

$\theta = \theta_i$ for any $x > 0$ at time $t = 0$ (initial)

$\theta = \theta_0$ for $x = 0$ at any time $t \geq 0$ (boundary)

$\theta = \theta_i$ for $x = \infty$ at any time $t > 0$ (boundary)

or simply:

$$\theta\,(x, 0) = \theta_i \rightarrow \theta = \theta_i, x > 0, t = 0 \qquad (11.2)$$

$$\theta\,(0, t) = \theta_0 \rightarrow \theta = \theta_0, x = 0, t > 0 \qquad (11.3)$$

$$\theta\,(\infty, t) = \theta_i \rightarrow \theta = \theta_i, x = \infty, t > 0 \qquad (11.4)$$

As already mentioned in Chap. 5, Eqs. (11.1)–(11.4) are a typical **Boundary Value Problem** (BVP), belonging to the classical chapter of mathematics of **solution of differential equations**. Today, with the modern computational tools, solutions of most differential equations subject to boundary conditions can easily be found even on a cell phone. However, for didactic reasons we will go into details of the solution of this BVP. We understand that it is very important for the reader to go in detail through the development of this solution. Condition (11.4) refers to the semi-infinite condition, which indicates that the solution we are looking for will be valid well before reaching infinity, i.e., the experiment in Fig. 11.1 should be terminated before the water reaches the end of the soil column. A classical solution to this BVP is that of Philip (1955), which will not be discussed here since its extension to the vertical case (Philip 1957) will be seen in the case of vertical infiltration. Our horizontal infiltration problem is, then, to find a function $\theta = \theta\,(x,t)$ satisfying Eq. (11.1) subject to conditions (11.2)–(11.4). This function will allow us to calculate θ at any point x in column at any time t. This solution is not easy and was only obtained for some cases in which the function $D(\theta)$ is known, and in such a way as to allow the solution. One of the ways a solution was found was when D is an exponential function of θ. However, a solution of type $\theta(x, t)$ is difficult and, therefore, Swartzendruber (1969) suggests that x is transformed into a dependent variable, that is, we will look for a solution of the form:

$$x = x(\theta, t) \tag{11.5}$$

In practice, this transformation does not disturb anything. For the solution $\theta(x,t)$, we have the soil water content θ at any point x and time t. For the solution $x = x(\theta, t)$ we have the position of any soil water content θ at any instant t.

To make this transformation of variables, we will use rules of the elementary calculus that shows that given Eq. (11.5), $\partial\theta/\partial t$ and $\partial\theta/\partial x$ are given by:

$$-\frac{\partial\theta}{\partial t} = \frac{\partial x}{\partial\theta}\frac{\partial\theta}{\partial x} \tag{11.6}$$

the negative sign in Eq. (11.6), although strange, is correct. It comes from the deduction of the process of transformation of variables, which can be seen in calculus texts. Another part of the transformation is:

$$\frac{\partial\theta}{\partial x} = \frac{1}{\partial x/\partial\theta} \tag{11.7}$$

and, consequently, in the operator form:

$$\frac{\partial}{\partial x} = \frac{1}{\partial x/\partial}$$

Substituting Eqs. (11.6) and (11.7) into Eq. (11.1), we obtain:

$$-\frac{\partial x/\partial t}{\partial x/\partial\theta} = \frac{1}{\partial x/\partial}\left[D(\theta)\frac{1}{\partial x/\partial\theta}\right]$$

or, simplifying:

$$-\frac{\partial x}{\partial t} = \frac{\partial}{\partial\theta}\left[\frac{D(\theta)}{\partial x/\partial\theta}\right] \tag{11.8}$$

It is important to recognize that Eq. (11.8) is identical to Eq. (11.1), only x being the dependent variable. In addition to the mathematical transformations, it is important to understand their physical meaning. To solve Eq. (11.8), we will use the **separable variables technique** (Cain and Meyer 2005) already presented in Chap. 7, during the solution of a Gardner problem. Let then the solution of Eq. (11.8) be given by the product of two functions:

$$x = \eta(\theta)\cdot T(t) \tag{11.9}$$

where $\eta(\theta)$ is a function only of θ, and $T(t)$ a function only of t, that's the reason for the name separable variables. Since Eq. (11.9) is, by hypothesis, a solution of Eq. (11.8), Eq. (11.8) has to be satisfied. So, we will calculate the derivatives contained in Eq. (11.8), starting from Eq. (11.9):

$$\frac{\partial x}{\partial t} = \eta\frac{dT}{dt}$$

Because derivating Eq. (11.9) in relation to t, θ is maintained constant and, consequently, η also assumes a constant value θ.

In a similar way:

$$\frac{\partial x}{\partial\theta} = T\frac{d\eta}{dt}$$

and so:

$$\eta\frac{dT}{dt} = -\frac{d}{d\theta}\left[D(\theta)\frac{1}{T}\frac{d\theta}{d\eta}\right]$$

Separating the variables:

$$T\frac{dT}{dt} = -\frac{1}{\eta}\frac{d}{d\theta}\left[D(\theta)\frac{d\theta}{d\eta}\right] \tag{11.10}$$

Since the left member is a function only of t and the right only of θ, and for any values of t and θ the equality must be observed, the only suitable way is to make each member equal to the same constant. Be it **a**. So:

$$T\frac{dT}{dt} = a \tag{11.11}$$

and

$$-\frac{1}{\eta}\frac{d}{d\theta}\left[D(\theta)\frac{d(\theta)}{d\eta}\right] = a \tag{11.12}$$

Equation (11.11) can easily be integrated to obtain $T(t)$:

$$\frac{1}{a}\int T\mathrm{d}T = \int \mathrm{d}t$$

$$\frac{T^2}{2a} = t + C$$

$$\mathrm{T} = \sqrt{2\mathrm{a}(\mathrm{t}+\mathrm{C})} \tag{11.13}$$

The solution of Eq. (11.12) to obtain $\eta(\theta)$ is not easy and will be discussed ahead. At the moment we will see how $\eta(\theta)$ can be obtained experimentally and so understand its meaning.

Substituting Eq. (11.13) into Eq. (11.9), we have:

$$x = \eta(\theta)\sqrt{2\mathrm{a}(t+C)} \tag{11.14}$$

Since $\eta(\theta)$ is a function of θ only and is not yet known, we can multiply it by any constant without changing its general form. It will only be magnified or reduced, depending on the value of the constant. For this reason, we will group the factor $\sqrt{2\mathrm{a}}$ with $\eta(\theta)$ and call the product as $\lambda(\theta)$, which has the same properties of $\eta(\theta)$, that is:

$$\lambda(\theta) = \sqrt{2a}\,\eta(\theta) \tag{11.15}$$

In this way, Eq. (11.14) is reduced to:

$$x = \lambda(\theta)(t + C)^{1/2} \tag{11.16}$$

and this is the most general solution found to the moment. This solution has also to satisfy the conditions (11.2)–(11.4). Therefore, for $x = 0$ we have $\theta = \theta_0$, and so:

$$0 = \lambda(\theta_0)(t + C)^{1/2} \text{ (condition 11.3)}$$

and since the product of these two functions is equal to zero, it follows that $\lambda(\theta_0) = 0$, because if $t = 0$ the product will never be zero. This means that the function $\lambda(\theta)$, which we do not yet know, assumes the value zero at saturation.

For the initial condition $t = 0$, we have:

$$x = \lambda(\theta_i)(0 + C)^{1/2} \text{ (condition 11.2)}$$

remembering that θ_i is the initial value of θ, present along the whole column, we have:

$$\lambda(\theta_i) = \frac{x}{\sqrt{C}}$$

For finite values of C, $\lambda(\theta_i)$ in the above equation $\lambda(\theta_i)$ will vary with x, which is impossible because θ_i is constant and also $\lambda(\theta_i)$. Therefore, the only choices for C are 0 and ∞, to have $\lambda(\theta_i)$ constant and independent of x, as required by the condition (11.2). For $C = \infty$, $\lambda(\theta_i) = 0$ and we have a trivial solution, i.e., $\lambda(\theta_i) = \lambda(\theta_0)$ and $\theta_i = \theta_0$, indicating there is no infiltration. So, the last alternative is $C = 0$, which implies in $\lambda(\theta_i) = \infty$, but no problem. Therefore, Eq. (11.16) becomes:

$$x = \lambda(\theta)t^{1/2} \tag{11.17}$$

Equation (11.17) is a particular partial solution of the horizontal infiltration problem. Partial, not final, because the function $\lambda(\theta)$, also known as **Boltzmann's transformation**, is still unknown to us. This solution tells us that, in the horizontal infiltration, the advance (x) of θ inside the soil column is proportional to the square root of time. The infiltration, therefore, begins rapidly and decreases with the square root of time—A decelerated process. We will see an illustrative example that makes these facts clearer. Consider an experiment of horizontal infiltration of water into a soil column, as shown in Fig. 11.2. After 1600 min of infiltration, the soil water content was measured along the column in a non-destructive manner (as it is the case with the gamma ray beam absorption technique described in Chap. 6), and the data are shown in Table 11.1. With the data of Table 11.1 the **soil water content profile** of Fig. 11.2 was drawn, showing the saturation value θ_0 at $x = 0$ and the position x_f of the **wetting front**, whose water content is taken as θ_i. In the laboratory, the wetting front is easily observed for columns mounted in a clear acrylic tube, since the wet soil of the region $0 < x < x_f$ is darker and the rest of the column ($x > x_f$) is the region of the dry soil of lighter colour. Our first step is to find experimentally the function $\lambda(\theta)$.

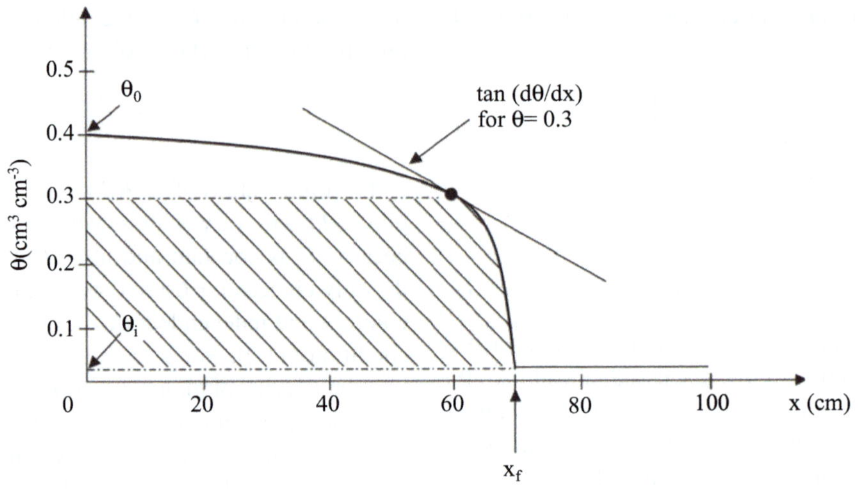

Fig. 11.2 Graph of θ versus x during horizontal infiltration, for $t = 1600$ min, indicating the procedure of Bruce and Klute (1956) for the calculation of $D(\theta)$. The dashed area is the integral used in the method

Table 11.1 Values of θ, λ and x for $t = 1600$ min, for horizontal infiltration into a homogeneous soil

x (cm)	θ (cm^3 cm^{-3})	λ
0	0.39	0
32.5	0.38	0.813
47.1	0.36	1.178
54.8	0.34	1.37
61.1	0.32	1.528
66.5	0.3	1.663
69.9	0.28	1.748
71.1	0.26	1.778
73.8	0.23	1.845
74.5	0.21	1.863
75.1	0.17	1.878
75.5	0.1	1.888
76	0.02	1.9
80	0.02	–
100	0.02	–

Applying the solution (11.17) to our example, we have:

$$\lambda(\theta) = x \cdot t^{-1/2} = \frac{x}{\sqrt{1600}} = \frac{x}{40}$$

which means that the values of x in Table 11.1, after divided by 40, yield respective values of λ. So:

$$\lambda(0.39) = \frac{0}{40} = 0$$

$$\lambda(0.38) = \frac{32.5}{40} = 0.813$$

....

$$\lambda(0.02) = \frac{76}{40} = 1.90$$

Note that $\lambda(\theta_i)$ did not result in ∞ as predicted, in theory, earlier. It is very difficult (if not impossible) to make measures of θ at the exact point where the wetting front encounters the dry soil. In this thin layer, the curve $\theta(x)$ becomes asymptotic in relation to x, which would lead to a $\lambda(\theta_i) = \infty$. Experimentally, however, this detail does not matter and it is assumed that $\theta = \theta_i$ at x_f, which implies that the curve $\theta(x)$ ends abruptly at the point (θ_i, x_f).

With the obtained data, we can draw the graph $\lambda = \lambda(\theta)$. This graph is shown in Fig. 11.3. It is worth noting that the function $\lambda(\theta)$ is the curve θ versus x, for $t = 1$. It is easy to verify this, since for $t = 1$, Eq. (11.17) becomes:

$$\lambda = x$$

which means that the curve θ versus x coincides with θ versus λ for $t = 1$. Thus, to determine $\lambda(\theta)$

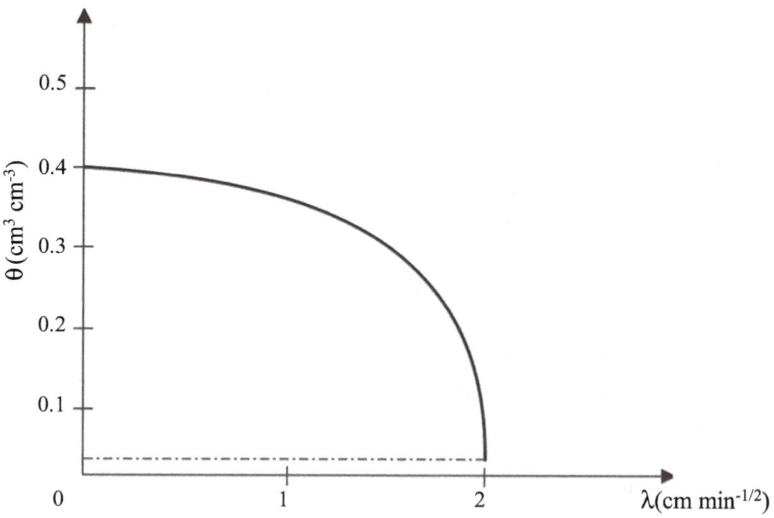

Fig. 11.3 Graph of θ versus λ for the horizontal infiltration (Boltzman's variable)

it would be sufficient to determine $x(\theta)$ for $t = 1$, but this is impossible, experimentally, if the unit of t is seconds or minutes. If the unit of t is hour or day, it would be possible. We will see later that the curve $\lambda(\theta)$ is unique, the same when calculated from any water content profile. Hence, if the curve λ versus θ is known, the problem is solved and we can calculate the soil water content θ at any time t. For example, if we wanted to calculate the distribution of soil water contents in the example of the Fig. 11.2 column, at time $t = 100$ min (which was not measured!), we would have:

$$x = \lambda \cdot t^{1/2} = \lambda \cdot \sqrt{100} = 10 \cdot \lambda \tag{11.18}$$

Attributing arbitrary values to x and calculating λ with Eq. (11.18), respective values of θ are extracted from the curve $\lambda(\theta)$:

$$x = 5 \rightarrow \lambda = 0.5 \rightarrow \theta = 0.38$$

$$x = 10 \rightarrow \lambda = 1 \rightarrow \theta = 0.37$$

$$x = 15 \rightarrow \lambda = 1.5 \rightarrow \theta = 0.36$$

....

$$x = 19 \rightarrow \lambda = 1.9 \rightarrow \theta = 0.02$$

We suggest to the reader to construct and draw the complete θ profile for $t = 100$ min.

The differential Eq. (11.1) and its solution (11.17) will describe the movement of water in the soil, provided that certain conditions are observed:

1. there should be no rearrangement of soil particles during the infiltration process. Therefore, soils that do not have expanding or contracting clays are those to which this theory most adapts;
2. the movement of air should not influence the movement of water;
3. the properties of water, especially density and viscosity, must be the same at any time and position in the soil;
4. the experimental conditions must be isothermal.

When it is desired to verify whether a given soil follows the present theoretical analysis, an experiment like that of the previous example is developed. If the theory is satisfactory, the function $\lambda(\theta)$ must be univocal, that is, independent of the time t at which the water content profile has been determined. This can be verified in two ways. The first is to observe the path of the wetting front x_f in the column and divide the distances by the square root of the times. If these coefficients are constant for all times, they will define (according to Eq. 11.17) the value of λ

corresponding to the water content θ_i that prevails at the wetting front. This also means that the graph of x_f versus the square root of time must be linear. If it starts to deviate from a straight line, the mathematical analysis presented here is no longer valid. The second way to determine if $\lambda(\theta)$ is univocal is to depart from different moisture distribution curves (see Fig. 11.2) and determine the curve $\lambda(\theta)$. If the same curve $\lambda(\theta)$ is obtained for different soil water profiles obtained at different times, the curve $\lambda(\theta)$ is univocal. An important work carried out in this sense is that of Davidson et al. (1963).

Let's look at something else about the horizontal infiltration process. Often it is important to know the amount of water that has penetrated the soil and at what speed. Given the distribution θ versus x (Fig. 11.2), it is easy to note that the area under the curve represents the total water I that penetrated the soil per unit area until time t. This is the **water storage**, defined in Chap. 3, by Eq. 3.18. The reader should verify that doing the $x\mathrm{d}\theta$ integration we obtain the same result as doing the $\theta\mathrm{d}x$ integration. That is:

$$A_L = I = \int_{\theta_i}^{\theta_0} x\mathrm{d}\theta = \int_0^{x_f} \theta\mathrm{d}x \quad (11.19)$$

In which I is the **accumulated infiltration** up to the chosen time t.

Substituting Eq. (11.17) into Eq. (11.19), we obtain:

$$\mathrm{I} = t^{1/2} \int_{\theta_i}^{\theta_0} \lambda(\theta)\mathrm{d}\theta \quad (11.20)$$

The integral of Eq. (11.20) is the area under the curve λ versus θ (Fig. 11.3) and, since this curve is unique, independent of t, its integral is a constant, called **sorptivity** S, here represented by A, thus:

$$I = At^{1/2} \quad (11.21)$$

which means that the accumulated infiltration is also proportional to the square root of time. It starts quickly, decreasing over time.

The **infiltration rate** i, also called instantaneous velocity or infiltration, is defined by:

$$i = \frac{\mathrm{d}I}{\mathrm{d}t} \quad (11.22)$$

and, according to Eq. (11.21), we have:

$$i = \frac{\mathrm{d}}{\mathrm{d}t}\left(At^{1/2}\right) = \frac{1}{2}At^{-1/2} \quad (11.23)$$

In our numerical example, the area A can be calculated graphically (by means of a planimeter), resulting in:

$$A = \int_{0.39}^{0.02} \lambda \cdot \mathrm{d}\theta \cong 0.623$$

and so:

$$I = 0.623\, t^{1/2}\ \mathrm{cm^3\ cm^{-2}\ or\ cm}$$

$$\mathrm{i} = 0.312\mathrm{t}^{-1/2}\ \mathrm{cm^3\ cm^{-2}\ min^{-1}\ or\ cm\ min^{-1}}$$

Assigning arbitrary values to t, we can determine the curve i versus t (Fig. 11.4). This graph shows that i decreases rapidly with t. For very short times, close to zero, i tends to ∞, not representing reality. For very long times ($t \to \infty$), i tends to zero. This is because, after a long time, the water content of the column for small values of x will be practically constant, such that its gradient $\partial\theta/\partial x$ will be practically nil and, according to the Darcy–Buckingham equation of flux density will also be practically zero.

Both I and i can be measured directly on the graduated cylinder (Mariotte bottle) indicated in Fig. 11.1 and the researcher can thus compare experimental data with data calculated by the theory.

Let us return, again, to Eq. (11.12). Although it is not possible to obtain an analytical expression for $\eta(\theta)$, let us look at the following:

$$\frac{\mathrm{d}}{\mathrm{d}\theta}\left[D(\theta)\frac{\mathrm{d}\theta}{\mathrm{d}\eta}\right] = -\eta\mathrm{a} \quad (11.12)$$

Since **a** is an arbitrary constant and does not appear in solution (11.17), we can assign it any

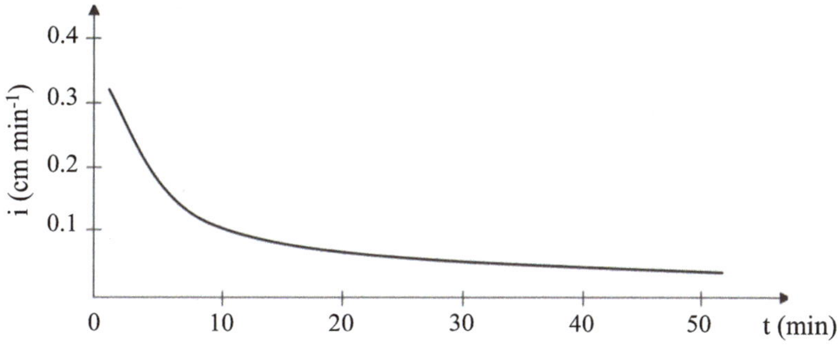

Fig. 11.4 Graph of the infiltration rate i versus time t for horizontal infiltration

value. In order to simplify the equations, we will take $a = 1/2$ and thus Eq. (11.15) becomes:

$$\lambda(\theta) = \eta(\theta)$$

and Eq. (11.12) can be rewritten in the form:

$$\frac{\mathrm{d}}{\mathrm{d}\theta}\left[D(\theta)\frac{\mathrm{d}\theta}{\mathrm{d}\lambda}\right] = -\frac{\lambda}{2}$$

or:

$$\lambda \mathrm{d}\theta = -2\mathrm{d}\left[D(\theta)\frac{\mathrm{d}\theta}{\mathrm{d}\lambda}\right] \tag{11.24}$$

Integrating Eq. (11.24) from θ_i to a generic value of θ, we obtain:

$$\int_{\theta_i}^{\theta} \lambda \mathrm{d}\theta = -2\int_{\theta_i}^{\theta} \mathrm{d}\left[D(\theta)\frac{\mathrm{d}\theta}{\mathrm{d}\lambda}\right]$$

$$\int_{\theta_i}^{\theta} \lambda \mathrm{d}\theta = -2\left[D(\theta)\frac{\mathrm{d}\theta}{\mathrm{d}\lambda} - D(\theta_i)\frac{\mathrm{d}\theta_i}{\mathrm{d}\lambda}\right]$$

Since θ_i is constant, $\mathrm{d}\theta_i/\mathrm{d}\lambda = 0$, and we have:

$$\int_{\theta_i}^{\theta} \lambda \mathrm{d}\theta = -2\left[D(\theta)\frac{\mathrm{d}\theta}{\mathrm{d}\lambda}\right] \tag{11.25}$$

Bruce and Klute (1956) presented a method for the determination of the **soil water diffusivity** $D(\theta)$ based on Eq. (11.25). If we explicit $D(\theta)$ in Eq. (11.25), we have:

$$D(\theta) = -\frac{1}{2}\left(\frac{\mathrm{d}\lambda}{\mathrm{d}\theta}\right)\int_{\theta_i}^{\theta} \lambda(\theta)\mathrm{d}\theta$$

Since $\lambda = xt^{-1/2}$:

$$D(\theta) = -\frac{1}{2}\frac{\mathrm{d}}{\mathrm{d}\theta}\left[xt^{-1/2}\right]\int_{\theta_i}^{\theta} xt^{-1/2}\mathrm{d}\theta$$

and for a chosen fixed time t_j, we have:

$$D(\theta) = -\frac{1}{2t_j}\left(\frac{\mathrm{d}x}{\mathrm{d}\theta}\right)\int_{\theta_i}^{\theta} x\mathrm{d}\theta \tag{11.26}$$

So, once the curve θ versus x (Fig. 11.2) is known for a chosen time t_j, $D(\theta)$ can be calculated for any θ between θ_i and θ_0, applying Eq. (11.26). As an example we calculate $D(\theta = 0.3\ \mathrm{cm^3\ cm^{-3}})$ using the curve θ versus x of Fig. 11.2, where $t = t_j = 1600$ min:

$$D(\theta) = D(0.3) = \frac{1}{2 \times 1600}\left(\frac{\mathrm{d}x}{\mathrm{d}\theta}\right)_{0.3}\int_{0.02}^{0.3} x\mathrm{d}\theta$$

$\mathrm{d}x/\mathrm{d}\theta$ is the tangent to the water content profile at the point $\theta = 0.3$ (note that in Fig. 11.2, θ is the ordinate and x the abscissa), that can be calculated graphically, resulting $(\mathrm{d}x/\mathrm{d}\theta)_{0.3} \cong -190$.

The integral in this calculation is indicated in Fig. 11.2 and can be estimated by several methods, even using a planimeter, resulting 22.5 (pay attention to all units!). So:

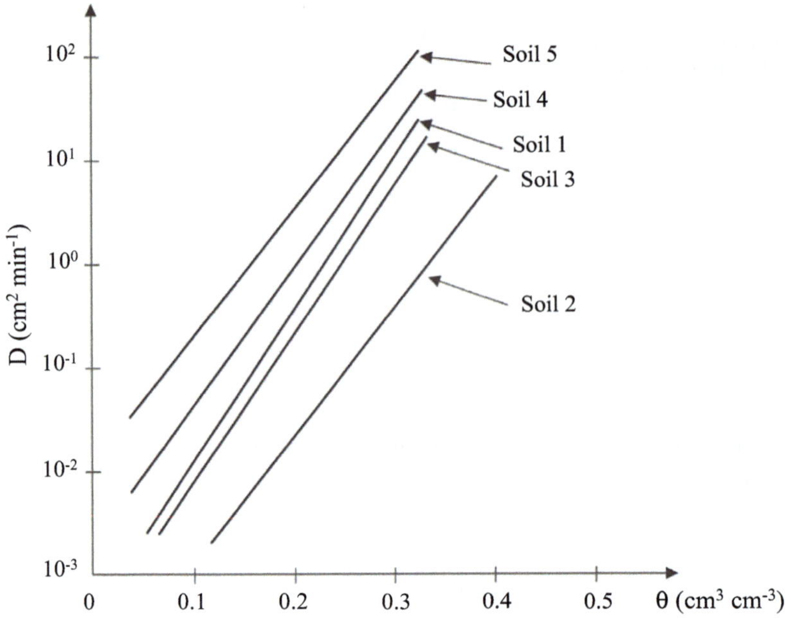

Fig. 11.5 Soil water diffusivity for several soils of the São Paulo state, Brazil. Note on the ordinate the logarithmic scale

$$D(\theta = 0.3) = \frac{1}{2 \times 1600} \times (-190) \times 22.5$$
$$= 1.42 \text{ cm}^2 \text{ min}^{-1}$$

We proceed in the same way to determine $D(\theta)$ for any other value of θ.

As an example, linear regressions of log D versus θ data obtained by this technique by Libardi and Reichardt (1973b) for some soils in the state of São Paulo are presented in Fig. 11.5. Known $D(\theta)$ and the soil retention curve (see Chap. 6), one can determine the **soil hydraulic conductivity** $K(\theta)$. Figure 11.6 also shows log K values for the same soils of Fig. 11.5, obtained by Libardi and Reichardt (1973a). A very intensive database of infiltration was recently published by Rahmati et al. (2018) containing data of soils all over the globe.

Philip (1955) presents a numerical solution of Eq. (11.25), by the finite difference technique, when $D(\theta)$ is known, obtaining $\lambda(\theta)$, thus completing the general solution of the horizontal infiltration problem given by Eq. (11.17). Philip's solution is quite long and difficult at his time, when computing facilities were small. Moreover, we have already seen how one can determine $\lambda(\theta)$ experimentally, which does not differ much from Philip's technique, since the answer of a numerical solution is a table of numbers and not an analytic expression.

At this point, it is also important to make an interpretation of the **Equation of Continuity** (Eq. 7.20) in light of the horizontal infiltration experiment we have just seen. If the volume element with the generic point M (see Fig. 7.14) is placed at an intermediate point on the wet part of the soil ($0 < x_j \leq x_f$), we will have the case illustrated below, and because the soil water content at M increases in time ($\partial\theta/\partial t > 0$), q_e (entry) > q_s (out) and, as a consequence, $\partial q_x/\partial x$ is negative, i.e., q decreases along x. Because q_e is greater than q_s the water content at M increases, giving continuation to the infiltration process. It is also important to note that at $x_j = 0$ (beginning of the soil column) $q_e = q_0 = i$, given by Eq. (11.23). If $x_{j+1} = x_f$, we have $q_{x+\Delta x} = 0$ (the position of the **wetting front**).

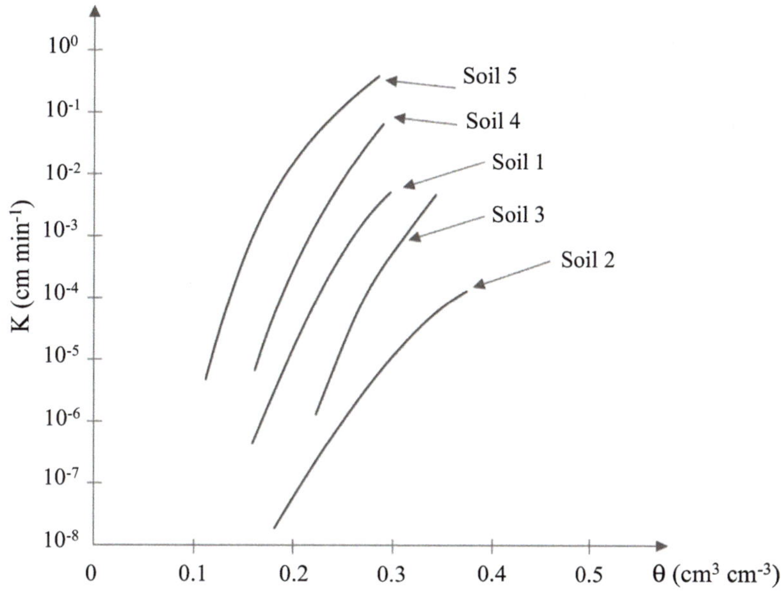

Fig. 11.6 Soil hydraulic conductivity for several soils of the São Paulo state, Brazil. Note on the ordinate the logarithmic scale

Entrance outlet

$$q_e = q_x = D(\theta)\left.\frac{\partial\theta}{\partial x}\right|_j \rightarrow \boxed{M} \rightarrow \left(q_x + \frac{\partial q_x}{\partial x}\Delta x\right) = q_s = D(\theta_{j+1})\left.\frac{\partial\theta}{\partial x}\right|_{j+1}$$

x_j x_{j+1}

We have already seen that in the horizontal infiltration, the graph of the distance to the wetting front x_f versus the square root of time t is a straight line that passes through the origin (Eq. 11.17). In Fig. 11.7, we present this graph for some soils. Of course, the smaller the slope (angular coefficient), the slower the movement of water into the soil. These curves characterize soils with respect to the horizontal infiltration process.

Reichardt et al. (1972), using time graphs such as those shown in Fig. 11.7, were able to generalize the theory of horizontal infiltration, based on the concept of **similar media**, introduced in Soil Physics by Miller and Miller (1956). A generalization in the use of the similar media concept and other scaling techniques is presented by Tillotson and Nielsen (1984) and will be discussed ahead in Chap. 18. In another work, Reichardt and Libardi (1973a) present an equation for estimating the water diffusivity of any mineral soil, knowing only the slope of the line x_f versus the square root of time, and Reichardt et al. (1975) present an equation for the estimation of hydraulic conductivity of a soil by the same technique.

11.3 Vertical Infiltration into Homogeneous Soil

Again, very rarely water infiltrates into the soil only in one direction, neither only the vertical. In the most and general cases, water infiltrates and moves in several directions. In these cases gravitational forces act in the process, and this makes a great difference. However, to better understand the processes, we will study first the pure vertical

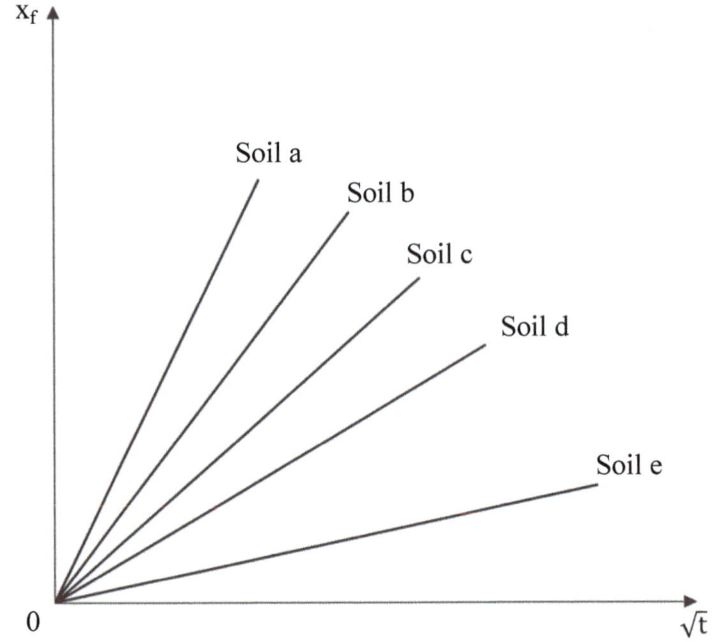

Fig. 11.7 Graphs of the distance to the wetting front x_f versus $\sqrt{t}$ for horizontal infiltration experiments into several homogeneous soils

water infiltration process into a homogeneous soil, and in a controlled laboratory situation, as we did for the horizontal case.

Consider the infiltration of water into a column of homogeneous soil mounted in the vertical position, of constant cross section, constant density, of infinite length and with an initial water content θ_i, similar to the column of Fig. 11.1, as shown in Fig. 11.8. This figure also shows a typical soil water content profile, which in the following figure (Fig. 11.9) is shown for three infiltration times. These profiles are typical and, the sandier the soil or in general for long times, the profile becomes practically vertical and its front is more abrupt. Therefore, the soil layer with θ practically constant was called the **transmission zone**, TZ (which increases over time); the plane where the water meets the dry soil with θ_i, at depth z_f, is the **wetting front**, which also progresses with time; and the layer between z_f and the wetting zone (of almost constant thickness but moving in the downstream direction) is called the **wetting zone** (WZ).

In the same way as explained in the horizontal case, at time $t = 0$, a porous plate is placed in contact with the soil at $z = 0$ and the infiltration process begins. Here we do not need a slightly negative pressure at $z = 0$. In this case, the equation to be used is 7.20, in which $H = h + z$, and which can be rewritten using the hydraulic diffusivity $D(\theta)$ defined by Eq. (7.17):

$$\frac{\partial \theta}{\partial t} = \frac{\partial}{\partial z}\left[D(\theta)\frac{\partial \theta}{\partial z}\right] - \frac{\partial K(\theta)}{\partial z} \tag{11.27}$$

where z is the vertical coordinate, taken as positive downwards. For the geometry of the case, this equation is subject to the following conditions:

$$\theta\,(z,0) = \theta_i \rightarrow \theta = \theta_i, z > 0, t = 0 \text{ (initial)} \tag{11.28}$$

$$\theta\,(0,t) = \theta_0 \rightarrow \theta = \theta_0, z = 0, t \geq 0 \text{ (boundary)} \tag{11.29}$$

$$\theta\,(\infty,t) = \theta_i \rightarrow \theta = \theta_i, z = \infty, t \geq 0 \text{ (boundary)} \tag{11.30}$$

To facilitate the procedure of finding a solution, let us first examine what happens for relatively short values of $t = t_s$ (Fig. 11.9). Rewriting Eq. (11.27) in the form:

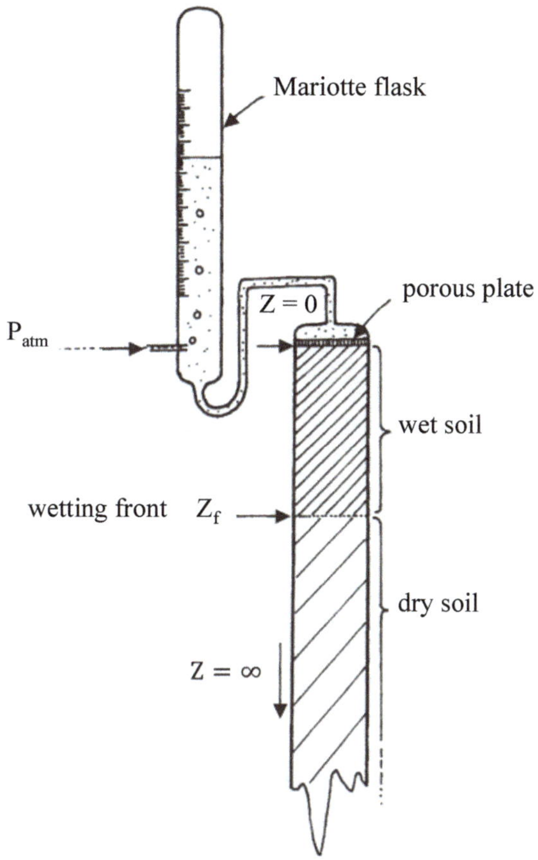

Fig. 11.8 Vertical infiltration experiment showing the wetting front separating the wet from the dry soil. The Mariotte flask can also be seen, with infiltration at pressure 0

$$\frac{\partial \theta}{\partial t} = \frac{\partial}{\partial z}\left[D(\theta)\frac{\partial \theta}{\partial z} - K(\theta)\right]$$

Let us see the order of magnitude of the factors within the square brackets. Since t_s is small (beginning of the infiltration process), the order of magnitude of $\partial\theta/\partial z$ is quite large (θ_0 much larger than θ_i, over a small distance Δz), theoretically the derivative is infinite for $t = 0$. This is because the soil water content at the surface is θ_0 and just below there is dry soil at θ_i (see Fig. 11.9). In addition, the values of $D(\theta)$ are numerically (using the same fundamental units, see Chap. 10) much larger than the values of $K(\theta)$, for the same θ. Thus, product $D(\theta)\partial\theta/\partial z$ is much greater than $K(\theta)$, in such a way that K can safely be neglected within the brackets. K for the soil at these times t_s is very small. In this way, Eq. (11.27) is identical to Eq. (11.1), used for the description of the horizontal flow, already solved above. By analogy, the solution for the vertical case will therefore be of the same type as the horizontal, resembling Eq. (11.17):

$$z = \lambda(\theta)t^{1/2} \text{ (short times)} \qquad (11.31)$$

This means that for relatively short times the vertical infiltration process is identical to the horizontal one. This relatively short time varies from soil to soil. At the end of this chapter, the reader will be more able to perceive it. In general, this short time is somewhat longer for fine textured soils than for coarser textured soils. Equation (11.31) shows that, for short times, the action of gravity is negligible and the vertical movement is identical to the horizontal, that is, the matric potential dominates over the gravitational potential. In terms of gradient, we have $\partial h/\partial z \gg \partial z/\partial z$.

For somewhat longer (medium) times $t = t_m$ (Fig. 11.9) another solution has to be found because both terms $D(\theta)\partial\theta/\partial z$ and $K(\theta)$ are of the same order of magnitude, both are important and none can be neglected. The method of separable variables used for the horizontal case cannot be applied here. A solution can be found assuming an **infinite series**, another technique largely employed in the solution of differential equations (Bronson and Costa 2014). In this way, let the solution of Eq. (11.16)—taking z as the dependent variable as recommended by Swartzendruber (1969) for the horizontal case:

$$\begin{aligned} z(\theta,t) &= f_0 + f_1t^m + f_2t^{2m} + \ldots \\ &= f_0 + \sum_{i=1}^{\infty} f_i t^{im} \end{aligned} \qquad (11.32)$$

in which m is a positive constant and the functions f_i are functions of θ only, for $i = 1, 2, \ldots\infty$. Equation (11.32) being proposed as a solution without restrictions on time t, it should include Eq. (11.31) for short times. The smaller t the smaller the importance of terms of $i > 2$. This can be observed for $t < 1$, because such values of t to the power will be decreasing as i increases.

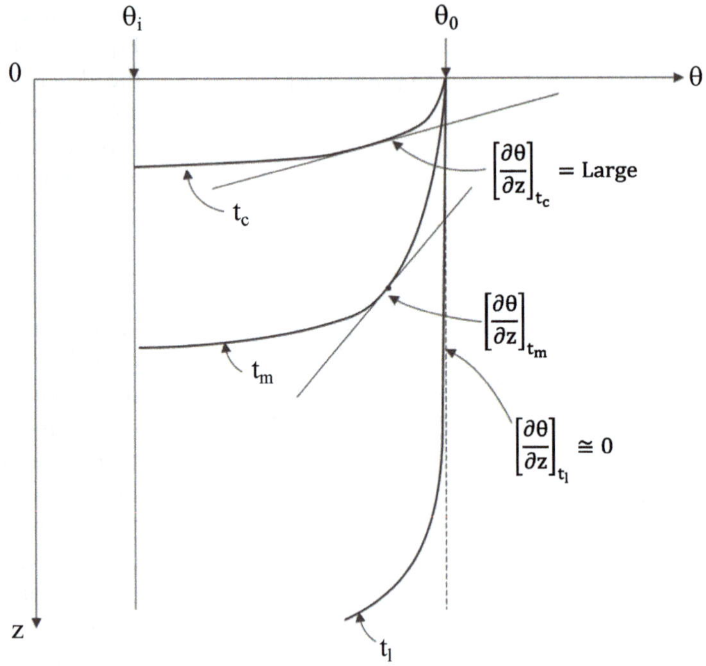

Fig. 11.9 θ versus z soil water content profiles for short times (t_s), medium times (t_m) and long times (t_l). Note the change in time of the derivative of θ with respect to z

Therefore, for very small t we can write Eq. (11.32) neglecting the terms of power $i > 2$:

$$z(\theta, t) = f_0 + f_1 t^m \qquad (11.33)$$

Comparing Eq. (11.33) with (11.31), which are two solutions of the same problem, we come to the conclusion that $f_0 = 0$, $f_1 = \lambda(\theta)$ and $m = 1/2$. This artefact allowed us to find m, and Eq. (11.32) can be rewritten without neglecting the terms of $i > 2$:

$$z(\theta, t) = \lambda(\theta)t^{1/2} + f_2 t + f_3 t^{3/2} + f_4 t^2 + \ldots \qquad (11.34)$$

This is the solution we were looking for. Again Philip (1957) proposes a numeric method for the determination of $f_2(\theta), f_3(\theta), f_4(\theta), \ldots$ which will also not be seen here, for the same reasons mentioned above. The series of Eq. (11.34), however, is not convergent for large values of time, when the solution turns out to be not valid.

For long infiltration times $t = t_l$ (Fig. 11.9), a large part of the soil profile will be at water contents very close to θ_0, so that $\partial\theta/\partial z$ can be assumed to be 0. Since the soil is homogeneous, K is constant, assuming a value K_0 correspondent to θ_0. We can say that for $t = \infty$, the infiltration process tends to a steady-state equilibrium. In this case, Darcy-Buckingham's equation can be written as follows:

$$q = K_0 \frac{\partial H}{\partial z} = K_0 \frac{\partial}{\partial z}(h + z) = K_0 \frac{\partial h}{\partial z} + K_0$$

and since θ is constant in relation to z, h is also constant and $\partial h/\partial z = 0$, so that:

$$q = K_0 \quad \text{and} \quad \frac{\partial \theta}{\partial t} = 0 \qquad (11.35)$$

meaning that the flux q tends to the constant value $K_0 = K(\theta_0)$, for $t = \infty$ and this fact can be used for the determination of K_0, the **saturated soil hydraulic conductivity**, also called **basic infiltration**. Reichardt et al. (1978) present an example for a read latosol (Nitosol) in Brazil.

In Fig. 11.10, the functions $\lambda(\theta)$, $f_2(\theta)$, $f_3(\theta)$ and $f_4(\theta)$ are presented schematically. As shown in the horizontal case, we can now calculate the **accumulated infiltration** I integrating Eq. (11.34) with respect to θ maintaining t constant:

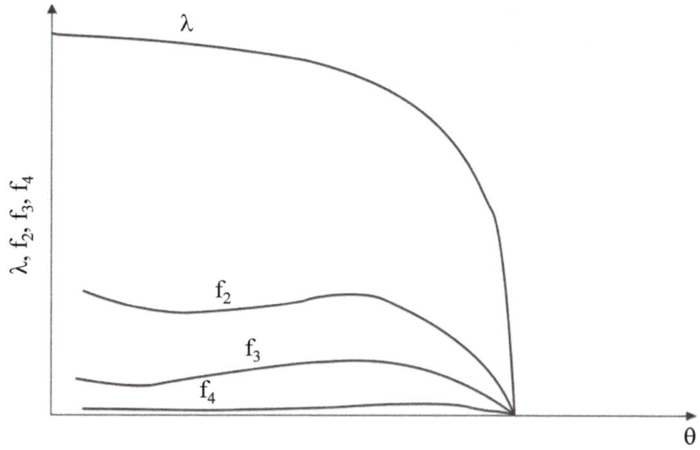

Fig. 11.10 Phillip's functions $\lambda(\theta), f_2(\theta), f_3(\theta)$ and $f_4(\theta)$ of the series solution of the vertical infiltration problem

$$I = \int_{\theta_i}^{\theta_0} z\mathrm{d}\theta = \left\{ \left[\int_{\theta_i}^{\theta_0} \lambda(\theta)\mathrm{d}\theta \right] t^{1/2} + \left[\int_{\theta_i}^{\theta_0} f_2(\theta)\mathrm{d}\theta \right] t + \left[\int_{\theta_i}^{\theta_0} f_3(\theta)\mathrm{d}\theta \right] t^{3/2} \ldots \right\} \tag{11.36}$$

It is easy to see that in Eq. (11.36) the integrals between brackets are numerically equal to the areas under the curves λ, f_2, f_3, f_4 ... presented in Fig. 11.10. Since these functions are characteristic of the soil, their integrals are also constants and Eq. (11.36) can be written in the form:

$$\mathrm{I} = At^{1/2} + \mathrm{B}t + Ct^{3/2} + \ldots \tag{11.37}$$

The **sorbtivity** S is the A in Eq. (11.37) and B, C, ... other constant coefficients.

At this moment, it is important that the reader returns to Eq. (11.21), of the horizontal infiltration, and makes a comparison with Eq. (11.37). Since B, C, ... are positive numbers, it can easily be recognized that the vertical infiltration is more rapid. This is the contribution of the gravitational potential to the process.

The **infiltration rate**, defined by Eq. (11.22), will be:

$$i = \frac{A}{2}t^{-1/2} + B + \frac{3C}{2}t^{1/2} + \ldots \tag{11.38}$$

Above, right after Eq. (11.34), we said that this solution contains a non-convergent series for large values of t. Equation (11.38) shows that clearly because depending on the magnitudes of B, C, ..., i starts increasing in time. In practice, however, it is observed that i also tends to K_0 for $t = \infty$. In Fig. 11.11 we present graphs of i versus t for two extreme soils in terms of texture, one sandy with high K_0 and another one clayey with low K_0. As it can be seen in this figure, i decreases rapidly in time, the sandier soil becoming stable at a higher value as compared to the clayey soil, however at different times.

The infiltration equation of Philip (Eq. (11.37)) is the most complete analytic solution of the existing models for the vertical infiltration. Besides this one, several other models are found in the literature, mainly in the areas of plant–water–soil relationships (Kirkham 2014), hydrology (Hornberger et al. 1998), soil physics and hydrology (Kutilek and Nielsen 1994; Lal and Shukla 2004; Radcliffe and Simunek 2010), micrometeorology (Moene and van Dam 2014) and plant production (Ehlers and Goss 2016). The oldest contribution is that of Green and Ampt (1911), still in use by the irrigation people to estimate the soil depth reached by an irrigation.

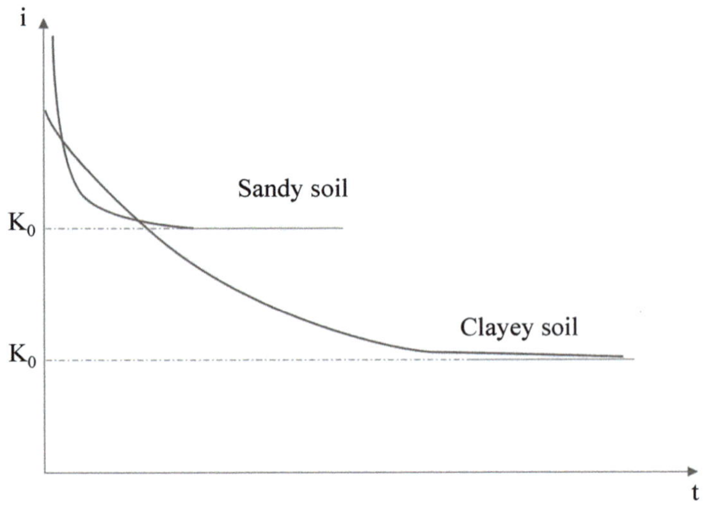

Fig. 11.11 Vertical infiltration rate *i* for two soils of distinct texture

Their method works better for sandy soils, or those in which the **transmission zone** (TZ) is deep and the **wetting zone** (WZ) short, so that water movement can be seen as a piston flow. In the **Green and Ampt model**, the infiltration rate *i* is given by the Darcy–Buckingham equation, the only soil water movement equation known at their time:

$$i = \mathrm{d}I/\mathrm{d}t = K_0(H_0 - H_\mathrm{f})/z_\mathrm{f} \qquad (11.39)$$

in which K_0 is the saturated hydraulic conductivity of TZ, H_0 the hydraulic head at the water entry and H_f at the wetting front. Considering the initial water content θ_i and the value at TZ equal to θ_t, both constant, *i* will be given by the change in water storage *S* due to infiltration, that is equal to the accumulated infiltration $I = (\theta_t - \theta_i)z_\mathrm{f}$, and Eq. (11.39) becomes:

$$\begin{aligned} \mathrm{d}I/\mathrm{d}t &= (\theta_t - \theta_i)\ \mathrm{d}z_\mathrm{f}/\mathrm{d}t \\ &= K_0(H_0 - H_\mathrm{f})/z_\mathrm{f} \\ &= K_0(H_0 - H_\mathrm{f})/(\theta_t - \theta_i)z_\mathrm{f} \end{aligned} \qquad (11.40)$$

where $\mathrm{d}z_\mathrm{f}/\mathrm{d}t$ is the wetting front advance, obtained derivating $I = (\theta_t - \theta_i)z_\mathrm{f}$ in relation to *t*. Reorganizing (11.40) by separating the variables, we have:

$$z_\mathrm{f}\ \mathrm{d}z_\mathrm{f} = K_0[(H_0 - H_\mathrm{f})/(\theta_t - \theta_i)]\mathrm{d}t$$

which integrated results in:

$$z_\mathrm{f}^2/2 = K_0[(H_0 - H_\mathrm{f})/(\theta_t - \theta_i)]t \qquad (11.41)$$

or,

$$z_\mathrm{f} = \{2K_0[(H_0 - H_\mathrm{f})/(\theta_t - \theta_i)]t\}^{1/2} \qquad (11.42)$$

or,

$$I = (\theta_t - \theta_i)\ \{2K_0[(H_0 - H_\mathrm{f})/(\theta_t - \theta_i)]t\}^{1/2} \qquad (11.43)$$

showing that for this model the accumulated infiltration and the depth of the wetting front are proportional to $t^{1/2}$.

Reichardt (1987), among many other authors, describes the most used model in practice, given by:

$$\mathrm{I} = At^n \qquad (11.44)$$

This model is practical, because having data of *I* as a function of *t*, we can, by means of a log *I* versus log *t* graph, estimate the parameters *A* and *n*. Such data are, for example, collected with infiltration rings. The first impression is that Eq. (11.44) disregards the values of *B*, *C*, . . ., of Philip's model, which is not true, since n will be estimated (not necessarily 1/2), resulting, in general, in a value $n > 1/2$, as if it was a mean value between 1/2, 1, 3/2, ... and the new A of Eq. (11.44) is a combination of S, B, C, ... from Eq. (11.38). For most soils, the regression in a log–log type of graph fits very well to a straight line. Applying log to both members of Eq. (11.44), we have:

$$\log I = \log A + n \log t$$

log A and n being the intercept and the slope of the log I versus log t graph. Reichardt (1987) presents an example of such graphs for Brazilian soils.

11.4 Infiltration Direction

The vertical infiltration we have just described is restricted only to the movement of water from top to bottom, in favour of gravity. In addition to this process, we could consider the vertical upward infiltration, from the bottom up, against gravity, commonly referred to as **capillary ascension**. Its differential equation is identical to Eq. (11.27), considering z positive from the bottom up. We will not go into the details of its solution here, but only make a comparison of the graphs of x_f (where x is the coordinate of the position of the wetting front in any direction as a function of $\sqrt{t}$ for infiltration in the three senses discussed here (Fig. 11.12). For the example of this figure it becomes clear that, for times shorter than 25 min, the advance of the wetting front is identical for vertical columns in which the water moves up or down and for horizontal columns. For times greater than 25 min, differences start already to appear. For example: $t = 100$ min, the wetting front is 15 cm for the vertical case down, 10 cm for the horizontal case and 7.5 cm for the case of capillary ascension.

In the field, infiltration can occur in any direction. A typical case can be observed in the furrow irrigation. In Fig. 11.13, the advance of the wetting front is plotted during irrigation by furrows of two different soils in terms of texture, which are initially dry.

As for the clayey soil, gravity becomes important only after a certain time, at the beginning of the infiltration the water advances practically with the same velocity in all directions and the cut shown across the profile, indicated in Fig. 11.13, shows a practically circular advance. In the case of a sandy soil, from the beginning, gravity acts and vertical movement, from top to bottom, is the most pronounced movement. The balance between gravitational and matrix forces can be evaluated by the **Bond Number**, which is a dimensionless parameter (see Chap. 19 and Ryan and Dhir 1993). Similar considerations could be made for drip irrigation and "sub-irrigation" with porous pipes.

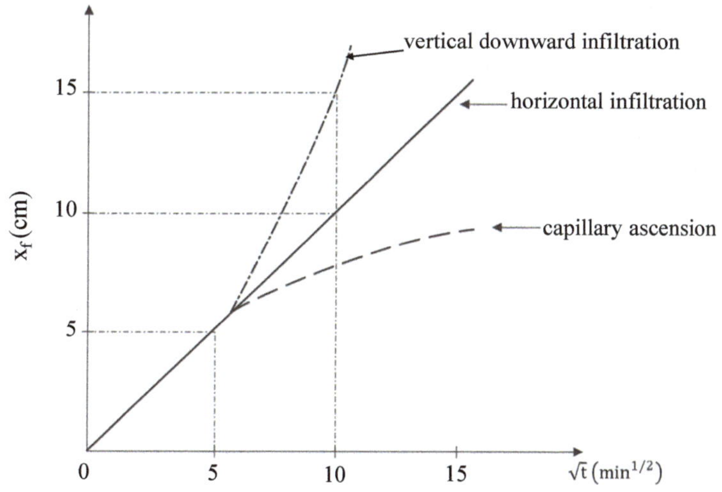

Fig. 11.12 Position of the vetting front x_f versus for different water movement directions into a homogeneous soil

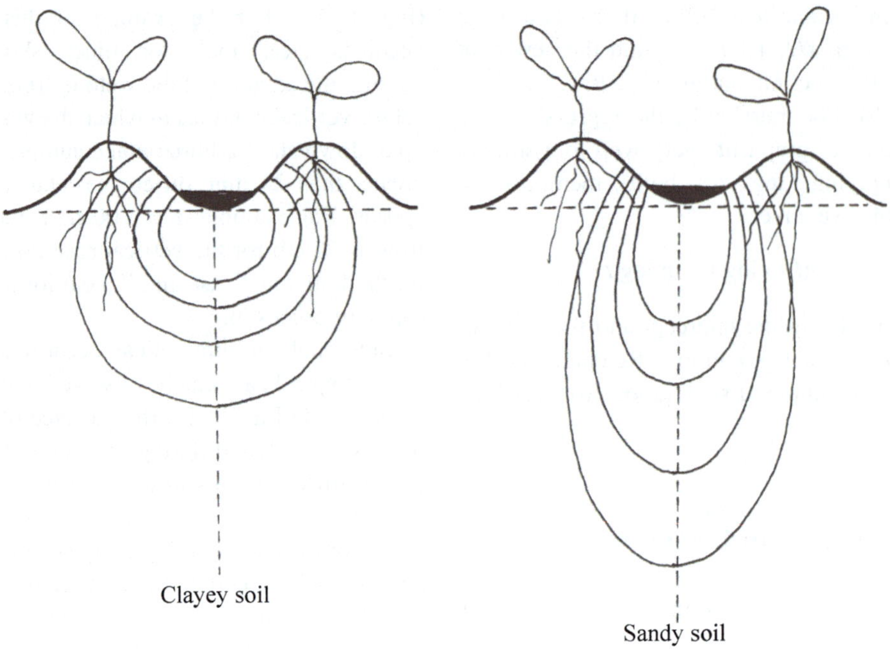

Fig. 11.13 Shapes of wetting front advance during furrow irrigation into soils of different texture

11.5 Infiltration into Heterogeneous Soils

In spite of the fact that the same general principles governing the movement of water in soils can be applied to heterogeneous soils, the analytical description of infiltration in heterogeneous soils is much more difficult. A serious complication that arises is that the hydraulic conductivity of the soil becomes a function of the position coordinate. This fact makes the analytical examination of the problem very difficult, and as a result most of the scientific contributions in this area are based on numerical solutions of Eqs. (11.1) and (11.27). Although several programs for the solution of differential equations are available in the INTERNET, we think that it is important for the reader to understand the basic approaches used in numerical solutions.

Water infiltration in stratified soils in dynamic equilibrium was studied, among others, by Takagi (1960), Swartzendruber (1960) and Srinilta et al. (1969). In general, it has been determined that the water flow is limited by the less permeable layer, regardless of its position in the soil profile. Depending on the texture of the layers, the presence of unsaturated zones in the profile can be observed. Under unsaturated conditions, some important studies are those by Colman and Bodman (1944), Fergusson and Gardner (1962), Whisler and Klute (1966), Philip (1967), Reichardt et al. (1972) and Sisson (1987).

For **heterogeneous soils**, we can write Richards equation including the dependence of K and H on the position coordinate x, in the form:

$$\frac{\partial \theta}{\partial t} = \frac{\partial}{\partial x}\left[K(x,\theta)\frac{\partial H(x,\theta)}{\partial x}\right] \qquad (11.45)$$

and due to the dependence of K and H on the position coordinate, no analytic solution can be found. Equation (11.45) can, however, be solved numerically. We repeat here that, although advanced programs of the solution of differential equations, especially that of Richards, are now available to researchers, it is important to have an accurate notion of what a numerical solution is and how it is obtained. As an example, we will see how Eq. (11.45) can be approximated by the method of finite differences in:

$$\frac{\theta_i^{j+1} - \theta_i^j}{\Delta t} = \frac{\left[K_{i-1/2}^j\left(H_{i-1}^j - H_i^j\right)\right] - \left[K_{i-1/2}^j\left(H_{i+1}^j - H_i^j\right)\right]}{(\Delta x)^2} \quad (11.46)$$

here the lower indexes i refer to position, and the higher indexes j, to time. Δt and Δx represent, respectively, the time and position increments. Knowing the functions $K(x, \theta)$ and $H(x, \theta)$, even experimentally, during each calculation step of Eq. (11.46), the appropriate values of K and H are introduced, and the numerical solution becomes possible. The numerical method of **finite differences** is based on the approximation of the derivatives (partial or total, first or second order) that appear in the differential equation, by finite differences. These differences can be obtained from **Taylor's expansion**. Let $y(x)$ be a continuous function with n continuous derivatives. According to Taylor's expansion, we get:

$$y(y + \Delta x) = y(x) + \Delta x \cdot y'(x) + \frac{(\Delta x)^2}{2!} y''(x) + \ldots + \frac{(\Delta x)^n}{n!} y^{(n)}(x) \quad (11.47)$$

$$y(y - \Delta x) = y(x) - \Delta x \cdot y'(x) + \frac{(\Delta x)^2}{2!} y''(x) - \ldots + \frac{(-1)^n (\Delta x)^n}{n!} y^{(n)}(x) \quad (11.48)$$

From Eq. (11.47), for very small values of Δx, we can obtain an expression for the **first derivative** $y'(x)$, neglecting the terms of power 2 (because Δx is very small, so that Δx^2 is negligible) or higher:

$$y'(x) = \frac{y(x + \Delta x) - y(x)}{\Delta x} \quad (11.49)$$

One should note that the smaller Δx, the more precise is the value of $y'(x)$ and that at the limit ($\Delta x \to 0$) the value is exact and coincides with the classical definition of a derivative.

Adding Eqs. (11.47)–(11.48) and neglecting the terms equal or higher than 3, we can obtain an expression for the **second derivative** $y''(x)$:

$$y''(x) = \frac{y(x + \Delta x) - 2y(x) + y(x - \Delta x)}{(\Delta x)^2} \quad (11.50)$$

In the same way, the smaller Δx, the more precise the value of $y''(x)$.

Such approximations of $y'(x)$ and $y''(x)$ by finite differences are commonly used in the solution of differential equations. We take the example of Eq. (11.1) for horizontal infiltration. Since θ is a function of x and t, the field of variation of the θ values can be divided into equidistant points Δx and the t field in equidistant points Δt, as shown in Fig. 11.14.

To simplify our symbols we will adopt:

$$\theta(x_i, t_j) = \theta_i^j$$

and i always refers to space x and j to time t.

To show how numeric solutions are made, we take first Eq. (11.1) in a simpler form by using a constant soil water diffusivity, and so:

$$\frac{\partial \theta}{\partial t} = D \frac{\partial^2 \theta}{\partial x^2} \quad (11.51)$$

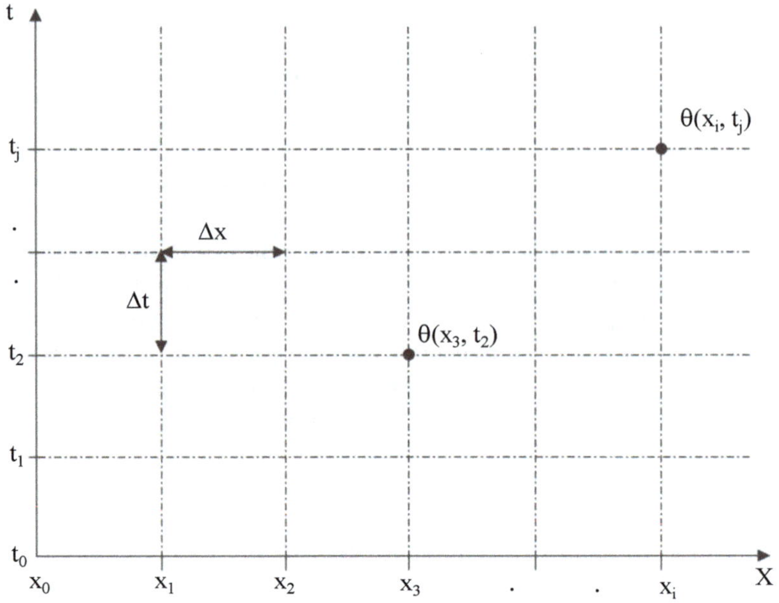

Fig. 11.14 Scheme of numeric integration showing the increments in space and time

and we utilize (11.49) and (11.50) to rewrite 11.51 in finite differences:

$$\frac{\theta_i^{j+1} - \theta_i^j}{\Delta t} = D\frac{\theta_{i+1}^j - 2\theta_i^j + \theta_{i-1}^j}{(\Delta x)^2}$$

The more general way is considering D as a function of θ, and Eq. (11.1) has to be written in the form presented at the beginning of this chapter. Based on the continuity equation (Chap. 7) we know that $q = -D(\theta)\partial\theta/\partial x$, and that:

$$\frac{\partial \theta}{\partial t} = -\frac{\partial q}{\partial x} \tag{11.1a}$$

which in finite differences can be approximated by:

$$\frac{\theta_i^{j+1} - \theta_i^j}{\Delta t} = \frac{q_{i-1}^j - q_i^j}{\Delta x} \tag{11.52}$$

in which

$$q_{i-1}^j = D(\theta_{i-1/2})\frac{\theta_{i-1}^j - \theta_i^j}{\Delta x} \tag{11.53}$$

and

$$q_i^j = D(\theta_{i+1/2})\frac{\theta_i^j - \theta_{i+1}^j}{\Delta x} \tag{11.53a}$$

where $D(\theta_{i \pm 1/2})$ represents the mean value of $D(\theta)$ between $\theta_{i \pm 1}$ and θ_i.

Substituting Eqs. (11.53) and (11.53a) in (11.52) and making θ_i^{j+1} explicit, we have:

$$\theta_i^{j+1} = \theta_i^j + \left[\frac{\Delta t}{(\Delta x)^2}\right]\left[D_{i-1/2}^j\left(\theta_{i-1}^j - \theta_i^j\right) - D_{i+1/2}^j\left(\theta_i^j - \theta_{i+1}^j\right)\right] \tag{11.54}$$

Since this method allows θ_i^{j+1} to become explicit, we have the **numerical explicit method** of solution of differential equations. It can thus be seen from Eq. (11.54) that, knowing θ at positions $i - 1$, and $i + 1$, at time j, it is possible to calculate θ at i at a more advanced time $j + 1$. With the boundary and initial conditions (Eqs. 11.2–11.4) we can begin our calculations using Eq. (11.54), again, for $i = 1, 2, 3, 4, \ldots, \infty$ and $j = 1, 2, 3, 4, \ldots, \infty$. With the results, a table of the type

Table 11.2 Results of the example of numeric integration of Eq. (11.49)

	x_i						
t_j	$x_0 = 0$	$x_1 = 0.2$	$x_2 = 0.4$	$x_3 = 0.6$	$x_4 = 0.8$	$x_5 = 1.0$	$x_6 = 1.2\ldots$
$t_0 = 0$	0.55	0.05	0.05	0.05	0.05	0.05	0.05…
$t_1 = 0.01$	0.55	θ_1^1	θ_2^1	θ_3^1	θ_4^1	θ_5^1	θ_6^1
$t_2 = 0.02$	0.55	θ_1^2	θ_2^2	θ_3^2	θ_4^2		
$t_3 = 0.03$	0.55	θ_1^3	θ_2^3				
$t_4 = 0.04$	0.55						
⋮	⋮						
⋮	⋮						
⋮	⋮						
⋮	⋮						
				θ_{i-1}^j	θ_i^j	θ_{i+1}^j	
				⋮	θ_i^{j+1}	⋮	
t_j				⋮	⋮	⋮	
t_{j+1}				⋮	⋮	⋮	
⋮				⋮	⋮	⋮	
⋮				⋮	⋮	⋮	
⋮				⋮		⋮	

shown in Table 11.2 can be constructed, in our case elaborated for $\Delta x = 0.2$ cm; $\Delta t = 0.01$ s; $\theta_0 = 0.55$ m^3 m^{-3} and $\theta_i = 0.05$ m^3 m^{-3}.

We initiate calculations with θ_1^1 applying Eq. (11.54):

$$\theta_1^1 = \theta_1^0 + \left[\frac{0.01}{(0.2)^2}\right] \left[D_{0.5}^0\left(\theta_0^0 - \theta_1^0\right) - D_{1.5}^0\left(\theta_1^0 - \theta_2^0\right)\right]$$

$$\theta_1^1 = 0.05 + 0.25[1.2(0.55 - 0.05) - 0.1(0.05 - 0.05)] = 0.20$$

where $D_{0.5}^0 = 1.2$ is the average of D values for $\theta_0 = 0.55$ and $\theta = 0.05$, data that have to be known. It is easy to recognize that the calculations of θ_2^1, θ_3^1, … result in zero. Once finished the line t_1 ($j = 1$), we follow to line t_2 ($j = 2$), or, $\theta_1^2, \theta_2^2, \theta_3^2, \ldots$ and so on until filling out the table in the range of interest. Computer programs can easily be built to carry out such calculations or even an EXCEL table can be prepared for that.

For larger values of i calculations can become unstable and yield erroneous data. This is solved adopting stability criteria, in our case being $D\Delta t/(\Delta x)^2 \leq 1/2$. This means that the choices of Δt and Δx, although arbitrary, have to meet the above-mentioned criterion.

Through Table 11.2, we can see that the result of a numeric solution is a table of values of the dependent variable. This is the main difference from the analytic solution that is an equation. Therefore, analytic solutions of the type $\theta = \theta(x,t)$ are preferred. With today's computational facilities, it is easy to construct a program that executes Eq. (11.54), line by line.

Swelling soils by addition or withdrawal of water can be included among the heterogeneous soils. For those, the mathematical analysis becomes even more complicated. Despite the few published infiltration experiments, the theory of infiltration in these soils was well developed (Philip 1968; Smiles and Rosenthal 1968; Philip 1969). The analysis presented by these authors is based on variations of the porous medium resulting from shrinking or expansion of the soil matrix. The mathematical processes employed are extremely complicated.

One way to solve other heterogeneous cases is the use of the concept of similar media, already discussed here and presented by Tillotson and Nielsen (1984). To illustrate some recent developments, we suggest the paper of Prevedello et al. (2008) that introduced a new approach for the solution of the infiltration problem, using similarities in a different way.

As we have seen in the case of horizontal infiltration, the water content profile front for $t = 1$ is the function $\lambda(\theta)$, which contains the information of the shapes of the profiles. The main part of the water retention curve $h(\theta)$ looks similar to $\lambda(\theta)$, one being the mirror image of the other, and so Prevedello included this aspect in his mathematical analysis, adopting a **similarity hypothesis**.

In this context of solutions of differential equations, it is important to mention the existence of numerical calculation programs, such as MAXIMA, http://maxima.sourceforge.net/. and MATLAB®. In addition, Radcliffe and Simunek (2010) employ the HYDRUS model, in the lecturing of principles and applications in Soil Physics, addressing the processes of infiltration, evaporation and percolation in soils of various textures, as well as stratified. The SWAP program is also widely used in water balance simulations (see Chap. 15) of which the infiltration and deep drainage components are of great importance. Pinto et al. (2015) use SWAP in simulations of a coffee crop in Brazil.

11.6 Some Practical Agronomic Implications

We have seen that the **infiltration rate** i decreases along time, tending towards zero in the case of horizontal infiltration and tending to K_0 (**saturated soil hydraulic conductivity**) in the vertical case. During rain or irrigation, if the soil is reasonably dry, at the beginning of the infiltration process, almost all water will infiltrate, irrespective of the amount of rainfall or irrigation. For example, let K_0 be 5 cm h^{-1}. With a rain intensity of 10 cm h^{-1} (which is uncommon), in the first minutes of infiltration, the soil absorbs all water because the soil water potential gradient is very large. With the decrease of i over time, it may happen that the rainfall intensity is greater than the rate of infiltration into the soil **i** and, in that case some of the water flows over the soil surface. It is the surface **runoff**, water lost for the crop and sometimes responsible for soil erosion. Rain cannot be controlled, but irrigation does. Hence K_0 (called the basic infiltration rate) in irrigation projects is an important parameter in the design of irrigation systems. For the sprinkler irrigation process, the irrigation intensity should not be greater than K_0, except for a short time. In this way, one has the assurance that, even if irrigating for long periods, there will be no losses due to superficial runoff.

In cases of furrow and flood irrigations, the criterion is opposite. As the water needs to flow quickly to reach the end of the furrow or cover the pad, it has initially to be administered at rates much higher than K_0. Very large rates cause soil erosion of the bottom and lateral walls of the furrows and very small rates, permit losses by deep infiltration. In such calculations the soil type, terrain slope, furrow spacing, furrow length, etc. are essential, and the most important parameter is K_0.

11.7 Exercises

11.1. For a given soil an experiment of horizontal infiltration was carried out, following the layout of Fig. 11.1. The table below shows the results:

(a) Construct the graph of θ versus x;
(b) Construct the graph x_f versus $t^{1/2}$;
(c) Construct the graph of λ versus θ;
(d) On graph a include the estimated curves of de θ versus x for $t = 200$ min and $t = 500$ min;
(e) Using the $t = 900$ min soil water content profile, determine the function $D(\theta)$ through the method of Bruce and Klute (1956).

	θ (cm^3 cm^{-3})		
x (cm)	$t = 100$ min	$t = 400$ min	$t = 900$ min
0	0.601	0.601	0.601
2	0.585	–	–
4	0.560	0.580	0.590
6	0.523	–	–
8	0.456	0.551	0.570
9	0.390	–	–
10	0.121	0.535	0.559
11	0.101	–	–
12	0.101	0.519	0.550
14	0.101	0.496	–
16	0.101	0.453	–
18	0.101	0.382	0.521
19	0.101	0.320	–
20	0.101	0.120	0.510
21	0.101	0.101	–
22	0.101	0.101	0.490
24	0.101	0.101	0.465
26	0.101	0.101	0.420
28	0.101	0.101	0.315
30	0.101	0.101	0.102

11.2. Imagine that a total of h mm of rain totally infiltrates a deep homogeneous soil profile of average soil water content θ (cm^3 cm^{-3}), below Field Capacity of an average value of θ_{cc} (cm^3 cm^{-3}). Develop an equation that gives the soil depth z_h reached by the rain.

11.3. A soil has the following equation for the accumulated vertical infiltration I:

$$I = 0.52\, t^{0,62} \; (I = \text{cm and } t = \text{min}).$$

We ask for:

(a) How many mm of water infiltrated after 4 h?

(b) Which is the equation for the infiltration rate?

(c) Which is the value of i after 2 h?

(d) Make the graphs of I versus t and i versus t and find the value of K_0 from the graph.

11.8 Answers

11.2. The depth reached by the rain of intensity h is: $Z_h = h/(\theta_{cc} - \theta)$.

So, for a rain of 25 mm infiltrating into a soil of water content of $\theta = 0.245$, and θ_{cc} of 0.326, the depth z_h will be 2.5/(0.326–0.245) = 31 cm.

11.3. (a) 155.5 mm

(b) $i = 0.3224\, t^{-0,38}$

(c) 0.5 mm min^{-1}

(d) approximately 5×10^{-3} mm min^{-1}.

References

Bronson R, Costa GB (2014) Differential equations, 4th edn. McGraw-Hill Education, New York, NY

Bruce RR, Klute A (1956) The measurement of soil moisture diffusivity. Soil Sci Soc Am Proc 20:458–462

Cain G, Meyer GH (2005) Separation of variables for partial differential equations: an eigenfunction approach. Chapman and Hall/CRC, Boca Raton, FL

Colman EC, Bodman WC (1944) Moisture and energy conditions during downward entry of water into moist and layered soils. Soil Sci Soc Am Proc 9:3–11

Davidson JM, Nielsen DR, Biggar JW (1963) The measurement and description of water flow through Columbia Silt Loam and Hesperia Sandy Loam. Hilgardia 34:601–617

Ehlers W, Goss M (2016) Water dynamics in plant production, 2nd edn. CABI, Croydon

Fergusson AH, Gardner WH (1962) Water content measurement in soil columns by gamma-ray absorption. Soil Sci Soc Am Proc 26:243–246

Green WH, Ampt GA (1911) Studies in soil physics: I. The flow of air and water through soils. J Agric Sci 4:1–24

Hillel D (1998) Environmental soil physics. Academic, San Diego, CA

Hillel D (1980) Applications of soil physics. Academic, New York, NY

Hornberger GM, Raffensperger JP, Wiberg PL, Eshleman KN (1998) Elements of physical hydrology. The John Hopkins University Press, Baltimore, MD

Kirkham MB (2014) Principles of soil and plant water relations, 2nd edn. Academic, San Diego, CA

Kutilek M, Nielsen DR (1994) Soil hydrology. Catena Verlag, Cremlingen-Destedt

Lal R, Shukla MK (2004) Principles of soil physics. Marcel Dekker, Inc, New York, NY

Libardi PL, Reichardt K (1973a) Características hídricas de cinco solos do Estado de São Paulo: II. Curvas de retenção de água e condutividade hidráulica. Sociedade Brasileira de Ciência do Solo, Santa Maria
Libardi PL, Reichardt K (1973b) Características hídricas de cinco solos do Estado de São Paulo: I. Difusividade da água no solo. O Solo 65:28–32
Libardi PL (2012) Dinâmica da água no solo, 2nd edn. EDUSP, São Paulo
Miller EE, Miller RD (1956) Physical theory of capillary flow phenomena. J Appl Phys 27:324–332
Moene AF, van Dam JC (2014) Transport in the atmosphere-vegetation-soil continuum. Cambridge University Press, New York, NY
Parr JF, Bertrand AR (1960) Water infiltration into soils. Adv Agron 12:311–363
Philip JR (1968) Kinetics of sorption and volume change in clay colloid pastes. Austr J Soil Res 6:249–267
Philip JR (1969) Theory of infiltration. Adv Hydrosci 5:215–305
Philip JR (1967) Sorption and infiltration in heterogeneous media. Austr J Soil Res 5(1):10
Philip JR (1957) Numerical solution of equations of the diffusion type with diffusivity concentration-dependent. Austr J Phys 10:29–42
Philip JR (1955) Numerical solution of equations of the diffusion type with diffusivity concentration-dependent. Trans Faraday Soc 51:885–892
Pinto VM, Reichardt K, van Dam J, De Jong van Lier Q, Bruno IP, Durigon A, Dourado-Neto D, Bortolotto RP (2015) Deep drainage modeling for a fertigated coffee plantation in the Brazilian savanna. Agric Water Manag 140C:130–140
Prevedello CL, Loyola JMT, Reichardt K, Nielsen DR (2008) New analytic solution of Boltzmann transform for horizontal infiltration into sand. Vadose Zone J 7:1170–1177
Radcliffe DE, Simunek J (2010) Soil physics with hydrus: modeling and applications. CRC Press Taylor & Francis Group, Boca Raton, FL
Rahmati M, Weihermüller L, Vanderborght J et al (2018) Development and analysis of the soil water infiltration global database. Earth Syst Sci Data 10:1237–1263
Reichardt K (1987) A água em sistemas agrícolas. Manole, Barueri
Reichardt K, Libardi PL (1973) A new equation for the estimation of soil water diffusivity. In: IAEA (ed) IAEA and FAO symposium. Isotopes and radiation techniques in soil physics, irrigation and drainage in relation to crop production. IAEA, Vienna, pp 45–51
Reichardt K, Libardi PL, Nascimento Filho VF (1978) Condutividade hidráulica saturada de um perfil de Terra Roxa Estruturada. Rev Bras Cienc Solo 2:21–24
Reichardt K, Libardi PL, Nielsen DR (1975) Unsaturated hydraulic conductivity determination by a scaling technique. Soil Sci 120:165–168
Reichardt K, Nielsen DR, Biggar JW (1972) Scaling of horizontal infiltration into homogeneous soils. Soil Sci Soc Am Proc 36:241–245
Ryan RG, Dhir VK (1993) The effect of soil-particle size on hydrocarbon entrapment near a dynamic water table. J Soil Cont 2:59–92
Sisson JB (1987) Drainage from layered field soils: fixed gradient models. Water Resour Res 23: 2071–2075
Smiles DE, Rosenthal MJ (1968) The movement of water in swelling materials. Austr J Soil Res 6:237–248
Srinilta SA, Nielsen DR, Kirkham D (1969) Steady flow of water through a two-layer soil. Water Resour Res 5:1053–1063
Swartzendruber D (1969) The flow of water in unsaturated soils. In: RJM DW (ed) Flow through porous media. Academic, New York, NY, pp 215–292
Swartzendruber D (1960) Water flow through a soil profile as affected by the least permeable layer. J Geophys Res 67:4037–4042
Takagi S (1960) Analysis of the vertical downward flow of water through two layered soils. Soil Sci 90:98–103
Tillotson PM, Nielsen DR (1984) Scale factors in soil science. Soil Sci Soc Am J 48:953–959
Whisler FD, Klute A (1966) Analysis of infiltration into stratified soil columns. In: UNESCO (ed) Wageningen symposium, water in the unsaturated zone. UNESCO/ International Association for Scientific Hydrology, Wageningen, pp 451–470

12 Water Redistribution After Infiltration into the Soil

12.1 Introduction

When the rain or irrigation ceases and the water reservoir on the soil surface disappears, the infiltration process comes to an end. The movement of water within the profile, however, does not stop and can often persist for a long time. The almost or totally saturated soil layer does not retain all the rain or irrigation water. Part of it moves downward, that is, to deeper layers, especially under the influence of the gravitational potential, and may also move along gradients of other potentials when present. This post-infiltration movement is termed **internal drainage** or **redistribution of water**, and is characterized by increasing water content of deeper layers at the expense of water contained in the initially moistened upper layers.

In some cases, the speed of redistribution decreases rapidly, becoming negligible after a few days, so that one has the impression that the soil retains all this water. Concomitantly, the soil water is evaporated on the surface or removed from the profile by the roots of the plants. In other cases, the redistribution is slow and can continue at an inconsiderable speed, although it decreases with time, for many days and even for weeks.

The importance of the redistribution process should be self-evident because it indicates the amount of water withdrawn at each moment from the different layers of the soil profile, a water which is available to plants. The speed and duration of the process determine the effective capacity of the soil in storing water, a property of vital importance to supply the water needs of the plants. As we shall see later, the field capacity, i.e., the storage of soil water at the end of the drainage period, is not a fixed quantity or a static property of the soil, but rather a temporal phenomenon, determined by the dynamics of the movement of water in the soil.

12.2 Analysis of the Redistribution Process

In the absence of groundwater, and the soil profile being sufficiently deep, the typical soil water profile at the end of the infiltration process consists of an upper wet layer and a non-wetted zone below, as shown in Fig. 12.1, at time $t = 0$ (end of infiltration). The initial velocity of redistribution depends on the depth of the wet layer resulted from the infiltration process, as well as the water content of the deeper dry zone and of the hydraulic conductivity of the soil in the various layers. If the initially wet layer is shallow and the bottom soil is drier, the potential gradients

K. Reichardt, L. C. Timm, *Soil, Plant and Atmosphere*, https://doi.org/10.1007/978-3-030-19322-5_12

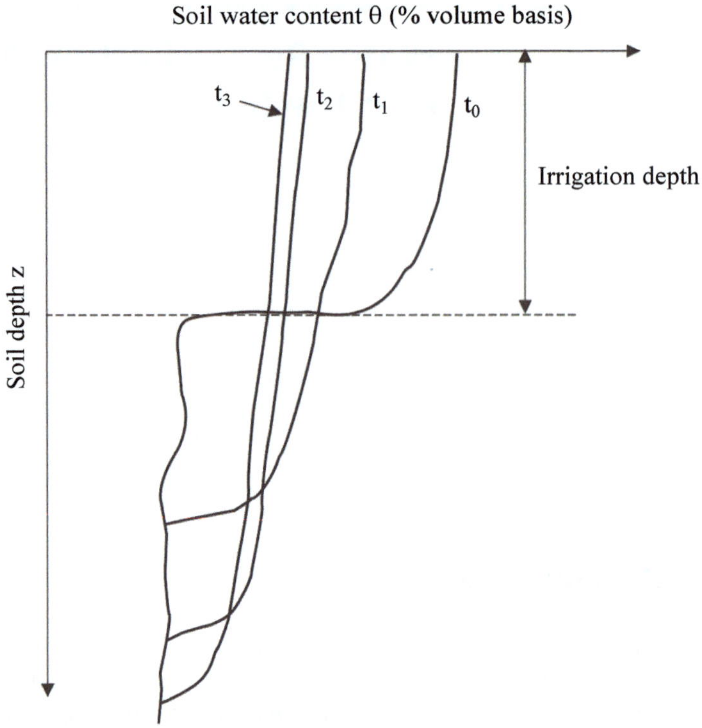

Fig. 12.1 Soil water content distributions at times t_1, t_2, and t_3, after the end of infiltration at time t_0

will be large and the rate of redistribution relatively rapid. On the other hand, if the initially wet layer is deep and if the soil below is relatively moist, the gradient of matrix potential will be small and the redistribution process is slower and occurs mainly under the influence of gravity.

In either case, the rate of redistribution decreases with time, for two reasons:

(a) The gradient of matrix potential between wet and dry zones decreases as the former loses and the latter gains water.
(b) As the wet zone loses water, its hydraulic conductivity drops abruptly.

Both decreasing over time, the flow decreases rapidly. The advance of the wetting front decreases in an analogous way. At the beginning of the infiltration process, the wetting front is well defined, but gradually dissipates during the redistribution process.

The redistribution process involves hysteresis. As the upper part of the soil is in a drying phase and the lower part in a wetting phase, the relation between the matrix potential "$-h$" and the soil water content "θ" will be different depending on the depth and variable over time, even in a homogeneous soil profile. This fact makes it difficult to analyze the redistribution process and makes its mathematical description more complex.

The water flow equation in a vertical profile, already described in Chap. 7, will be rewritten here:

$$\frac{\partial \theta}{\partial t} = \frac{\partial}{\partial z}\left[K(\theta)\frac{\partial H}{\partial z}\right] = \frac{\partial}{\partial z}\left[K(\theta)\frac{\partial h}{\partial z} - K(\theta)\right] \tag{12.1}$$

Because of **hysteresis**, Miller and Klute (1967) suggest that Eq. (12.1) should be written as follows:

$$\left(\frac{\partial \theta}{\partial h}\right)_{hy} \cdot \left(\frac{\partial h}{\partial t}\right)_{hy} = \frac{\partial}{\partial z}\left[K_{hy}(h)\left(\frac{\partial h}{\partial z}\right)_{hy}\right] - \left[\frac{\partial K(h)}{\partial z}\right]_{hy} \tag{12.2}$$

where the subscripts hy indicate the dependence on the hysteresis.

The analytical solution of Eq. (12.2), if not impossible, was very complicated in the 1960s, which is why most of the scientific contributions of the time were based on numerical solutions of this equation. Some examples are the works by Remson et al. (1967), Rubin (1967), Staple (1969), and Gardner (1970). Today, the tools have become so powerful that their solution is quite feasible.

In order to exemplify the redistribution process, we will examine a simplified case used to determine the hydraulic conductivity of a soil in the field, in which, after the infiltration process has ceased, there is no evaporation (surface covered with plastics) or water absorption by roots (without vegetation) and in which the hysteresis is neglected because the analysis will be made in the upper layers that are only in the drying process. In this example, the infiltration process must occur for a very long time, until $i = K_0$ (see Chap. 11) and, in this condition, the profile is wetted in depth and the mathematical analysis is done only on the surface layers that have become saturated and that during redistribution they will only lose water. Figure 12.2 illustrates an experiment in which the water depth over soil surface is measured during the infiltration process. Because the area of 5 × 5 m has a small slope, it was divided into three parcels. In Fig. 12.3, we have the redistribution phase, already covered with plastic. Under the conditions of this experiment, we have

$$\text{(First) } t = 0, z \geq 0, \theta = \theta_0(z) \tag{12.3}$$

$$\text{(Second) } t > 0, z = 0, q = -K(\theta)\frac{\partial H}{\partial z} = 0 \tag{12.4}$$

$$\text{(Third) } t \geq 0, z = \infty, \theta = \theta_i \tag{12.5}$$

The first condition tells us that at the beginning of the redistribution, the water content varies with z according to the function $\theta = \theta_0(z)$, which is a water content profile that can be measured. As the infiltration process was sufficiently long for the upper layer to saturate or almost reach saturation, we have that $\theta(z) = \text{constant} = \theta_0$, logically depending on the homogeneity of the profile. The second condition tells us that the flow is zero at the surface for all times (plastic cover) and, thirdly, that at a fairly large depth the soil is always drier, with a constant water content θ_i.

If we integrate Eq. (12.1) with respect to z, from 0 to L, where L is a layer of arbitrary depth chosen for the determination of $K(\theta)$, we obtain

Fig. 12.2 Infiltration experiment on an area with slope, therefore three flooded areas next to each other, serving as replicates. Inside one can see tensiometers with their mercury manometers in the front. Installed are also neutron probe access tubes for water content measurement

Fig. 12.3 Redistribution plot after infiltration (Fig. 12.2) covered with plastic showing nine neutron probe access tubes covered with beer cans to avoid rain entrance. Black box in the center is the neutron probe for soil water content measurement. Wood boards across the plots are to avoid people to step on the plastic covering a muddy soil

$$\int_0^L \frac{\partial \theta}{\partial t} \mathrm{d}z = \left[K(\theta)\frac{\partial H}{\partial z}\right]_L - \left[K(\theta)\frac{\partial H}{\partial z}\right]_0 \quad (12.6)$$

In the second member of Eq. (12.6), the last term is null, according to the second condition (12.4), so that

$$\int_0^L \frac{\partial \theta}{\partial t} \mathrm{d}z = K(\theta)_L \left(\frac{\partial H}{\partial z}\right)_L$$

and, if interested in the function $K(\theta)$,

$$K(\theta)_L = \frac{\int_0^L \frac{\partial \theta}{\partial t} \mathrm{d}z}{\left(\frac{\partial H}{\partial z}\right)_L} \quad (12.7)$$

The integral of the first member of Eq. (12.6) represents the flow of water passing through the plane $z = L$. This is because it must be equal to the second member, which is the Darcy equation itself, that is, the flow density q_L passing the depth L. If in the layer $0 - L$, $\partial\theta/\partial t$ is independent of z, which means that the water content profiles for different times are parallel, which usually happens in practice for most soils, $\partial\theta/\partial t$ can be withdrawn from the integration signal and the integral becomes equal to L. It turns out, then

$$\frac{\partial \theta}{\partial t} L = K(\theta)_L \left.\frac{\partial H}{\partial z}\right|_L \quad (12.8)$$

The first member of Eqs. (12.6) and (12.8) is numerically equal to the area between two consecutive profiles from 0 to depth L (Fig. 12.4). Let us look at an example—the soil profile shown in Fig. 12.4. For $t = 0$, the water content varies with z, and with good approximation at t_0, until $L = 60$ cm, it can be considered constant and its average value is $\overline{\theta}_0 = 0.40\ \mathrm{cm^3\ cm^{-3}}$. The same happens for $t_1 = 24$ h, with an average value of $\overline{\theta}_1 = 0.35\ \mathrm{cm^3\ cm^{-3}}$. $\partial\theta/\partial t$ can then be approximated by

$$\frac{\partial \theta}{\partial t} = \frac{\overline{\theta}_0 - \overline{\theta}_1}{t_0 - t_1} = \frac{0.050}{24} = 0.002\ \mathrm{cm^3\ cm^{-3}\ h^{-1}}$$

and then

$$\int_0^L \frac{\partial \theta}{\partial t} \mathrm{d}z = L\frac{\partial \theta}{\partial t} = 60 \times 0.002 = 0.120\ \mathrm{cm\ h^{-1}}$$

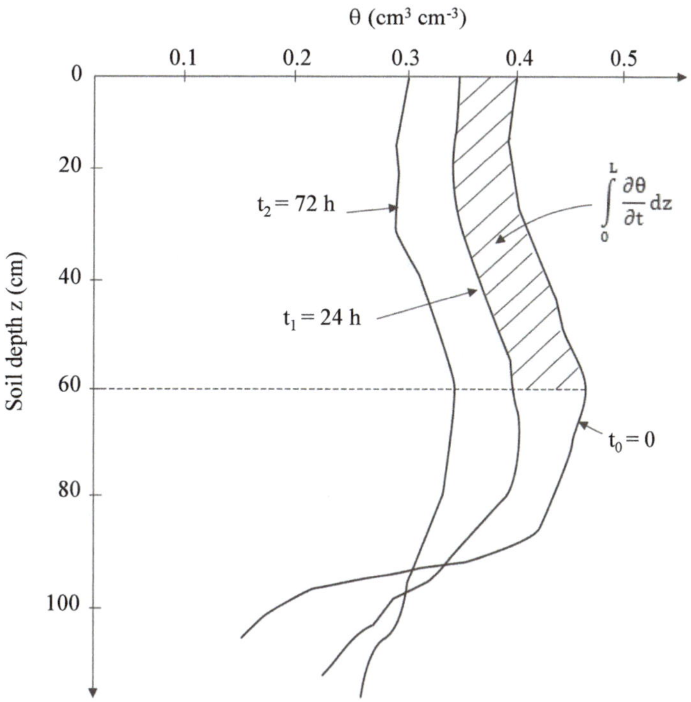

Fig. 12.4 Soil water content profiles during redistribution, indicating (dashed area) the amount of drained water in the first 24 h, down to the depth of 60 cm

Calculating the integral, it is sufficient to divide it by the gradient of the total potential, as indicated in Eq. (12.7), to obtain the conductivity. The value of K obtained is a value of K between the water contents 0.35 and 0.40, i.e., 0.375. This is why we write $K(\theta)$ and we have

$$K(0.375) = 0.120 \text{ cm h}^{-1}$$

If several water content profiles are available, average values of K for other water contents can be calculated.

Hillel et al. (1972) present, in detail, a methodology for determining the hydraulic conductivity in the field, based on Eq. (12.7).

Since the integral of the first member is equal to the variation of the change in storage A_L in time, we can rewrite it in the following way:

$$K(\theta)_L = \frac{\frac{\partial A_L}{\partial t}}{\left(\frac{\partial H}{\partial z}\right)_L} \tag{12.9}$$

For the application of the technique suggested by them, soil water content profiles θ (preferably with neutron probe or TDR equipment or FDR, because they are nondestructive methods—see Chap. 6) should be measured for various drainage times. Profiles of total hydraulic potential H (obtained with tensiometers or indirectly by the use of retention curves) to determine the gradients $\partial H/\partial z$. With the data of $\theta(z,t)$ and $H(z,t)$, the authors presented details for the calculation of the function $K(\theta)$, which made their work become a standard for the determination of soil hydraulic conductivity in the field, the **method of Hillel**. A complete example of $K(\theta)$ is given in Exercise 12.1. Reichardt and Libardi (1974) successfully employed this technique in a Nitossol profile in Brazil. Villagra et al. (1994) have adapted models to describe the functions $A_L(t)$, $H_L(t)$, and $\theta_L(t)$ and, thus, facilitated the application of Eq. (12.7) or (12.9). The internal soil drainage process is decelerated and as the values of A_L, H, and θ stabilize over time, their experimental data fits very well with the following linear–logarithmic models:

$$A_L(t) = a + b \ln t \tag{12.10}$$

$$H_L(t) = c + d \ln t \quad (12.11)$$

$$\theta_L(t) = e + f \ln t \quad (12.12)$$

By means of linear regressions applied to the experimental data, we can estimate the coefficients *a*, *b*, *c*, *d*, *e*, *f*, logically observing the signs of each one. Thus, the numerator of Eq. (12.9) is simplified to *b*/*t*, and using finite differences for the denominator, we have

$$\left(\frac{\partial H}{\partial z}\right)_L = \frac{H_{L+\Delta z}(t) - H_{L-\Delta z}(t)}{2\Delta z} = c' + d' \ln t \quad (12.13)$$

where $c' = [(c_2 - c_1)/2\Delta z]$ and $d' = [(d_2 - d_1)/2\Delta z]$, with c_1, c_2, d_1, and d_2 as the coefficients of the regressions $H_{L+\Delta z}(t)$ and $H_{L-\Delta z}(t)$. The smaller Δz, the better the estimative of the gradient at *L*. So, Eq. (12.9) can be written as

$$K(\theta)_t = \frac{a + b \ln t}{c' + d' \ln t} \quad (12.14)$$

Note the new notation $K(\theta)_t$, since this equation gives us *K* values for different drainage times. It turns out that for each drainage time, a value of θ prevails at depth *L*, given by Eq. (12.12). By calculating $K(\theta)_t$ and $\theta_L(t)$ for the same times, we obtain a Table *K* versus θ, which will give rise to the function $K(\theta)$. Based on Eq. (12.14), Reichardt et al. (2004) developed a parametric method for the determination of the function $K(\theta)$.

In these experiments of $K(\theta)$ determination by internal drainage after infiltration, the hydraulic gradient is theoretically unity, and therefore, it is important to have a consideration of the **unit hydraulic gradient** (Reichardt 1993). For the unit hydraulic gradient to prevail during drainage, it is necessary that in Eq. (12.13) we have $d' = 0$ and $c' = 1$, which is not always true. In this sense, the methodology presented above also allows the experimental verification of the unitary gradient hypothesis.

The **method of Libardi** et al. (1980) for the determination of $K(\theta)$ is based on simplifications of Eq. (12.7). Practical experience shows that *K* (θ) is exponential and that an equation of the type (see Chap. 7)

$$K(\theta) = K_0 \exp[\beta(\theta - \theta_0)] \quad (12.15)$$

describes $K(\theta)$ for a great majority of soils. In Eq. (12.15), β is a positive constant and θ_0, the maximum value of θ for the mentioned soil in the chosen layer under field conditions. θ_0 is very close (or equal) to saturation, because it is very difficult to fully saturate a soil profile under field conditions. K_0 has the same meaning as in Chap. 11, i.e., it is the basic infiltration velocity whose measurement at the soil surface is shown in Fig. 12.5. This model of $K(\theta)$ is the one adopted in this method.

They also consider the unitary hydraulic gradient ($\partial H/\partial z = 1$) and, consequently, ($\partial h/\partial z = 0$, which is reasonable since θ is practically constant for all *z* after excessive infiltration. Substituting Eq. (12.15) in (12.7), the integral can be solved and the final result is

$$\theta - \theta_0 = \frac{1}{\beta} \ln t + \frac{1}{\beta} \ln \left(\frac{\beta K_0}{aL}\right) \quad (12.16)$$

which indicates that soil water content θ (at a chosen depth *L*) is a linear function of ln *t*, since $\beta K_0/aL$ is a constant. Libardi et al. (1980) show that **a** is the regression coefficient $\overline{\theta} = a\theta + b$, between θ at the depth *L* and θ between 0 and *L*. They found that the mean profile water content $\overline{\theta}$ may very well be described by the above linear equation, where θ is the water content at *L*. Therefore, knowing *a* and *b* for a given soil $\overline{\theta}$ can be obtained from measurements at only one depth.

Returning to Eq. (12.16), it is easy to verify that, from graphs $\theta_0 - \theta$ versus ln *t*, obtained during the redistribution process, one can determine β, since it is the slope or angular coefficient of this graph. K_0 can be measured on the surface of the soil instants before time ($t = t_0 = 0$) taken as the beginning of the redistribution process, when the infiltration is under steady state, i.e., $i = K_0$, or can be calculated from the linear coefficient $(1/\beta)\ln(\beta K_0/a\mathrm{L})$. An exercise (solved), presented at the end of this chapter, illustrates this method.

Fig. 12.5 Details of the infiltration measurement with a plastic cylinder, showing in the front a neutron access tube covered with a beer can

The Libardi method was improved by Libardi and Reichardt (2001), which includes the estimation of θ_0 for deeper layers from a θ_0 measurement at the soil surface.

In Brazil, the following studies are applications of the theory seen in this chapter: Reichardt (1974), Reichardt et al. (1979), IAEA (1984), Bacchi and Reichardt (1988), Bacchi et al. (1989), Timm et al. (2000), Reichardt et al. (2004), Silva et al. (2006, 2007, 2009), among others.

A completely different form of calculation of $K(\theta)$ was proposed by Sisson et al. (1980), called the **Sisson method**, applied to the same experimental conditions of the methods described above (Hillel et al. 1972 and Libardi et al. 1980) and also considering a homogeneous soil and the total potential gradient being unitary ($\partial H/\partial z = 1$ or $\partial h/\partial z = 0$). Under these conditions, Eq. (12.1) simplifies into

$$\frac{\partial \theta}{\partial t} = \frac{\partial K}{\partial z} = \frac{\mathrm{d}K}{\mathrm{d}\theta}\frac{\partial \theta}{\partial z} \tag{12.17}$$

since K is not a direct function of z, it is a function of θ, which in turn is a function of z. Sisson et al. (1980) have demonstrated, based on the theory of characteristic value problems, or Cauchy's problems, that another kind of solution to Eq. (12.17) can be given by

$$\frac{z}{t} = \frac{\mathrm{d}K}{\mathrm{d}\theta} \tag{12.18}$$

Equation (12.18) can be seen as follows: since the soil is draining starting at $t = 0$, the water content decreases as a function of time and depth. If we fix a given soil water content θ, this value can be assumed to "move" within the profile, in the downward direction, as if it were a wave, in such a way that, for a given fixed depth, the water contents pass at different times. The longer the time, the lower the water content. In short, a certain water content θ_j can only occur at a given depth z_j after a time t_j. The quotient z_j/t_j of Eq. (12.18) characterizes θ_j and the solution says that it equals $\mathrm{d}K/\mathrm{d}\theta$ for θ_j. Figure 12.6 illustrates the issue.

At $t = t_0 = 0$, all depths are saturated (θ_0). At t_1, layer $0 < z < z_1$ has dried somewhat, and for $z > z_1$, the soil remains saturated. At t_2, a larger layer has dried ($0 < z < z_2$), and for $z > z_2$, the soil remains saturated. One sees, therefore, that the water content θ_0 "shifts" in depth, like a wave. By Eq. (12.18), we have the example of Fig. 12.6 and for the water content θ_0:

$$\frac{z_1}{t_1} = \frac{z_2}{t_2} = \frac{z_3}{t_3} = \ldots = \left(\frac{\mathrm{d}K}{\mathrm{d}\theta}\right)_{\theta_0}$$

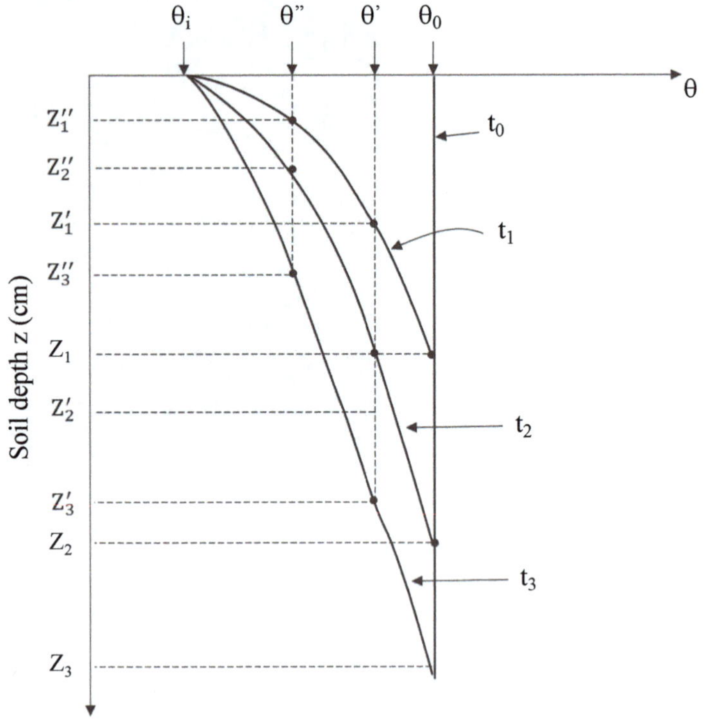

Fig. 12.6 Soil water content profiles according to Sisson's theory showing how chosen water content values move downwards as time passes

The same happens for any water content lower than θ_0, but at slower "speed". Thus, for a water content θ', we will have

$$\frac{z'_1}{t_1} = \frac{z'_2}{t_2} = \frac{z'_3}{t_3} = \ldots = \left(\frac{dK}{d\theta}\right)_{\theta'}$$

and for an even lower water content θ'',

$$\frac{z''_1}{t_1} = \frac{z''_2}{t_2} = \frac{z''_3}{t_3} = \ldots = \left(\frac{dK}{d\theta}\right)_{\theta''}$$

Thus, we obtain a series of values dK/dθ, for several θ. Hence, the step is to reconstruct the equation $K(\theta)$ from its derivatives.

Sisson et al. (1980) present several examples with different models for the $K(\theta)$ curve. Let us consider one of them, which considers the exponential $K(\theta)$ curve (Eq. 12.15), as did Libardi et al. (1980). The derivative of Eq. (12.15) is

$$\frac{dK}{d\theta} = \beta K_0 \exp[\beta(\theta - \theta_0)]$$

and by Eq. (12.18),

$$\frac{z}{t} = \beta\, K_0 \exp[\beta(\theta - \theta_0)] \qquad (12.19)$$

Therefore, if for a fixed depth $z*$ we measure θ as a function of t, we can plot the graph of ln $(z*/t)$ as a function of $(\theta - \theta_0)$ (Fig. 12.7) and, from the regression (that is linear) we can determine β, because

$$\ln\left(\frac{z^*}{t}\right) = \ln(\beta K_0) + \beta(\theta - \theta_0) \qquad (12.20)$$

An elucidative example. For a given soil at $z* = 50$ cm, θ was measured as a function of time and the results were

θ (cm^3 cm^{-3})	t (days)
0.536	0
0.453	1
0.418	2
0.393	4
0.385	7
0.380	10

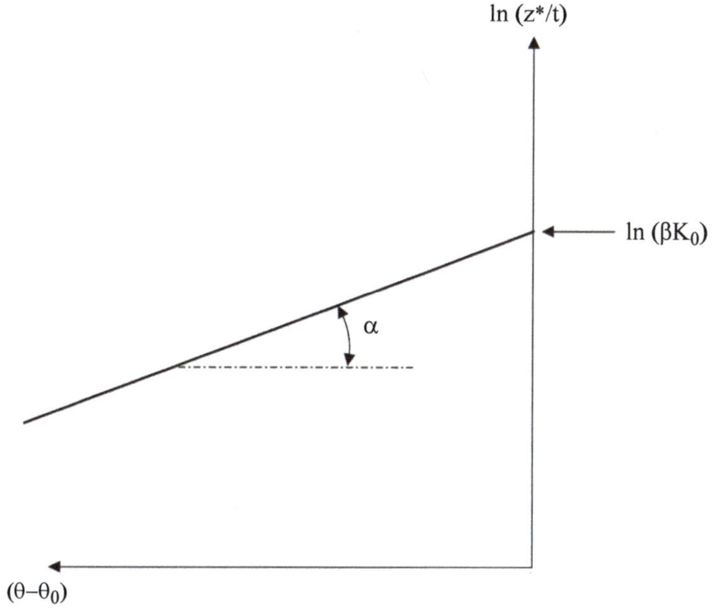

Fig. 12.7 Schematic graph of the data of Sisson's $K(\theta)$ method

Under such conditions, we have

t	$z*/t$	ln $(z*/t)$	$(\theta - \theta_0)$
1	50	3.9120	−0.083
2	25	3.2189	−0.118
4	12.5	2.5257	−0.143
7	7.143	1.9661	−0.151
10	5	1.6094	−0.155

and by linear regression we obtain

$$\ln\left(\frac{z^*}{t}\right) = 5.5 + 31.94(\theta - \theta_0)$$

and so

$$\ln(\beta K_0) = 5.5, \text{ resulting } \beta K_0 = 244.69$$

$$\beta = 31.94 \text{ and } K_0 = \frac{244.69}{31.94} = 7.66 \text{ cm day}^{-1}$$

and finally (Fig. 12.8)

$$K(\theta) = 7.66 \exp[31.94(\theta - 0.536)],$$
$$K(\theta) \text{ expressed in cm.day}^{-1}$$

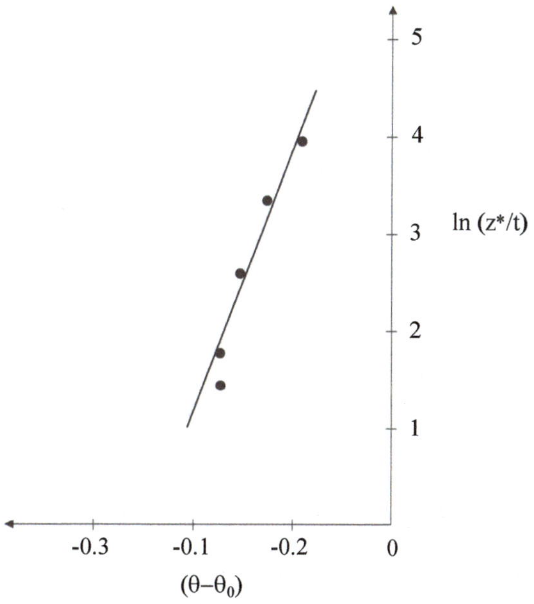

Fig. 12.8 Final graph of Sisson's $K(\theta)$ method

As can be seen, it is a simple and very good method, since it provides values of $K(\theta)$ compatible with other classical methods. The method is convenient for the use of tensiometers when you have a water retention curve representative of the soil in question. In this case, the tensiometer is installed at depth z* and the readings made at various times are converted in values of θ by the retention curve. Sisson (1987) extends this method for cases of stratified soils, where $\partial H/\partial z$

is not unitary. It is, however, a work that requires a more detailed reading and a deeper knowledge of the mathematical procedure.

Bacchi et al. (1991) showed that the methods of Libardi et al. (1980) and Sisson et al. (1980) are essentially the same, that is, Eqs. (12.16) and (12.20) do not differ from each other, considering $a = 1$, that is, homogeneous soil. Further, Reichardt (1993) shows that although it works in practice, the **unit hydraulic gradient** (grad $H = 1$) cannot theoretically exist during the redistribution process. More critical works also show the difficulties of using $K(\theta)$ relationships for soil water flow measurements, such as Reichardt et al. (1998) and Silva et al. (2007). These last studies show the importance of determining the function $K(\theta)$ under field conditions. As shown in Chap. 7, hydraulic conductivity is a physical characteristic of the soil, and therefore, its determination is independent of the experimental arrangement. In that chapter, we have shown examples of measurements of K_0 in the laboratory, with columns arranged in various situations (horizontal, vertical, etc.) always resulting in the same K_0. The above studies show that in the determination of K under field conditions, the hydraulic conductivity increases as a function of depth. The presence of L in the Libardi Eq. (12.16) is evidence of this. Pioneering studies such as Davidson et al. (1963) and LaRue et al. (1968) already affirmed that the hydraulic conductivity of the soil increases in depth and showed this evidence with experimental data. This contradicts what was said above, because if it would be possible to invert the soil profile of an experimental field, the conductivity would decrease in depth. The truth is that experimental methods and data show that K increases in depth and that the explanation for this lies in the boundary conditions that in the field are specific and determine this characteristic for the soil profile.

Dourado-Neto et al. (2007) present software for the determination of $K(\theta)$ for several of these methods presented here, provided field experimental data are available. In this software, the methods of determination of $K(\theta)$ of Hillel et al. (1972), Libardi et al. (1980), and Reichardt et al. (2004) are included.

12.3 Field Capacity

It has been observed from an early date that the water flow and the velocity of soil water content changes decrease with time after the infiltration or rainfall processes. It has been found that the flux becomes negligible or even ceases after a few days. The soil water content at which the internal drainage practically ceases, called **field capacity**, has long been assumed universally as physical property of the soil, characteristic and constant for each soil.

The concept of field capacity was originally derived from inaccurate observations of soil water content in the field, at times when measurements and samplings limited the accuracy and validity of the results. Many researchers have tried to explain it in terms of a static equilibrium. It was commonly assumed that the application of a certain amount of water in the soil would fill the deficit up to the field capacity, down to a well-defined depth, beyond which water would not penetrate. Thus, for example, the amount of water to be applied by irrigation on the basis of the deficit to the field capacity of the soil layer to be irrigated, is commonly calculated.

Recently, with the development of water movement theories in the soil and with more precise experimental techniques of measurement, the concept of field capacity, as defined in its origin, has been considered arbitrary and not as an intrinsic property of the soil, independently of the way of his determination. Veihmeyer and Hendrickson (1949) defined field capacity as "the amount of water retained by the soil after draining its excess, when the velocity of the downward movement virtually ceases, which usually occurs two to three days after rainfall or irrigation, in permeable soils of uniform structure and texture". Richards (1960) discusses this concept and goes so far as to say that "the concept of field capacity has caused more evils than clarification." How can one judge when redistribution has almost ceased or has become negligible? Obviously, the criteria for such determination are subjective, depending greatly on the frequency and accuracy of soil water measurement. The practical definition of field capacity (water content of the layer initially moistened a few days after infiltration) does not take into account factors such as soil water content before infiltration, wetting depth, amount of water applied, profile heterogeneity, etc.

The redistribution process is, in fact, continuous and shows no abrupt interruptions or static levels. Although its velocity decreases with time, the process continues indefinitely and the tendency to equilibrium occurs only after a long period of time.

The soils to which the concept most adapts are soils of coarser texture, in which the hydraulic conductivity decreases rapidly with the decrease of soil water content and the flow becomes very small rapidly. In soils of medium and fine texture, however, the redistribution process can persist appreciably for several days and even months.

The drainage of a given layer of a soil profile depends on its texture, hydraulic conductivity, and the composition and structure of the entire profile, since the presence of a flow-limiting layer at any position within the profile delays the water from all layers above. Thus, it is clear that the water storage capacity of a soil is not only related to time but also to the textural composition, sequence of layers of distinct physical properties, initial moisture, etc.

An extreme example is presented in Table 12.1, of a soil with a textural B horizon, of hydraulic conductivity approximately ten times smaller than the surface horizon. As can be verified, three irrigations with different water depths, applied to the same soil in identical initial conditions, lead to three different values of field capacity, all different from the result of the laboratory value, obtained at a pressure head of $-1/3$ atm (-33 kPa), adopted classically for soils of fine texture (argisols).

Table 12.1 Data of three cases of field capacity determination in soil with a structural B horizon

Case	Soil water content θ at field capacity (cm^3 cm^{-3})	Matric potential h at field capacity (atm)
I	0.25	−0.41
II	0.36	−0.12
III	0.32	−0.38
Laboratory	0.30	−0.33

Nonetheless, the concept of field capacity is considered by many to be a practical and useful criterion for the upper limit of water that a soil can store. Under such conditions, field capacity must necessarily be determined in the field and the interested party must be aware of its limitations. Scardua (1972) and Reichardt and Libardi (1974) present field-capacity data determined in the laboratory for the municipality of Piracicaba, SP, Brazil. In fact, no laboratory method is useful for its determination, for the reasons already discussed. The values of the various laboratory methods for field capacity, such as values obtained in the pressure plate at 1/10 or 1/3 atm, cannot represent the field capacity measured in the field. These laboratory criteria are static and the redistribution process in the field is essentially dynamic. How can a 10 g sample, placed on the porous plate under certain pressure, represent a soil profile of heterogeneous layers? Only in some circumstances can it represent the profile, but it is fundamentally wrong to expect such a criterion to be universally valid. In Chap. 14, we will continue to discuss this point, including, further, plant interference. In this case, we will be talking about **available water** (AW), that is, the difference between FC and the permanent wilting point (PWP), measured as a soil water content difference or a soil water storage, in mm, for a given soil profile. Recently, De Jong van Lier (2017) criticizes the FC concept as being the upper limit of the AW. His argument points to the fact that in many situations a fair amount (20–50%) of water is extracted from the soil by the plant, for water contents above FC.

Many authors also use the same connotation of field capacity for potted soils and call it **pot water capacity**. When a soil of a pot is saturated and then left to drain, even when the bottom is drilled with wholes, due to the discontinuity soil/pot bottom, the equilibrium soil water content is much larger than the value of the equivalent value in the field (θ_{FC}) and there are no terms of comparison.

Corsini (1974) shows how soil cultivation can affect the physical–water characteristics, especially water retention and soil hydraulic conductivity. Brunini et al. (1975) present one of the first attempts in Brazil to determine the water retained by the soil under field conditions. Reichardt et al. (1976a, b) discuss the problem of soil spatial variability in relation to its retention and water transmission properties in internal drainage experiments. Reichardt et al. (1980) present and discuss available water data on several soils typical of the Amazon, Brazil.

In general, finer textured soils, with higher proportions of silt and clay, have a greater water storage capacity. As discussed in Chap. 3, the type of clay is of great importance in water retention. Type 2:1 clays, such as **vermiculite** and **montmorillonite**, have excellent water retention properties, and 1:1 types, such as **kaolinite**, have poor water retention properties. Alfisols and Oxisols, which occupy large areas of Brazil, are soils that have low water absorption capacity due to lack of 2:1 clays.

How could their retention properties be improved? This is an old problem in soil physics that has never been solved with complete success. The addition of organic matter, in the form of green manure, manure or compost, is the solution. The problem is always the quantity to be applied, which, in general, is large, making the transport and application operations unfeasible. However, for intensive crops of high economic value, this solution is quite feasible. It also has the advantage of increasing the level of soil nutrients, especially N, P, and S. Reichardt (1987) discusses various ways of conserving water in soil.

Another possibility is the addition of **soil conditioners**, consisting of bitumen or mineral emulsions (Moldenhauer 1975), or as expanded vermiculite, found in large quantities in mines, practically at the surface of the soil. This primary mineral receives a heat treatment (about 700 °C), during which it expands to a volume ten times greater. It has the same structure as the secondary vermiculite of 2:1 found in the soil and, therefore, when added to the soil, increases its water retention capacity. Again, the problem is cost, quantity to be incorporated, etc. Studies, however, show that the effect is as expected, and some of those that address these issues are of Salati et al. (1980), Reichardt (1981), and Libardi et al. (1983). These materials are now widely used in commercial substrate mixtures for seedling production. Currently, agricultural residues have been tested in order to increase the soil ability to store water, such as rice husk ash (RHA). Islabão et al. (2016) carried out an experiment in a Red-Yellow Argisol (Southern Brazil) where different rates of RHA were applied and incorporated in the 0–10 cm soil layer by a rotary hoe. The authors concluded that RHA is an efficient residue to improve soil physical properties, mainly at rates between 40 and 80 Mg ha^{-1}, reducing bulk density and increasing total porosity, macroporosity, and soil aggregation values; however, RHA application does not affect microporosity, field capacity, permanent wilting point, and available water capacity of the soil.

More details on the concept of field capacity can be found in Reichardt (1987) and especially in Reichardt (1988), who presents two examples of soils with data obtained in the field.

12.4 Exercises

12.1. On a soil, an internal drainage experiment was carried out to determine the hydraulic conductivity, according to the methodology presented in Hillel et al. (1972), the following data was obtained:

(a) Hydraulic conductivity of the saturated soil $K_0 = 2.2$ cm day^{-1}, measured during dynamic equilibrium infiltration.

(b) Soil water content (cm^3 cm^{-3}) versus time (days), during the redistribution process.

	Soil water content				
Depth (cm)	$t = 0$	$t = 1$	$t = 3$	$t = 7$	$t = 15$
0	0.500	0.463	0.433	0.413	0.396
30	0.501	0.466	0.432	0.414	0.398
60	0.458	0.405	0.375	0.347	0.307
90	0.475	0.453	0.438	0.423	0.414
120	0.486	0.464	0.452	0.440	0.427

(c) Total water potential (cm H_2O) versus time (days), during redistribution.

	Total water potential				
Depth (cm)	$t = 0$	$t = 1$	$t = 3$	$t = 7$	$t = 15$
15	−18	−38	−69	−100	−135
45	−47	−76	−104	−129	−164
75	−76	−105	−135	−163	−200
105	−108	−141	−172	−206	−229
135	−140	−172	−201	−240	−265

Determine the functions $K(\theta)$, by the method of Hillel et al. (1972), for the depths 30, 60, 90, and 120 cm.

12.2. With the above data, determine $K(\theta)$ by the methods of Libardi et al. (1980) and Sisson et al. (1980).

12.3. With your reasoning, what would be the soil field capacity of the previous problems?

12.5 Answers

12.1. The equation to be used is Eq. (12.7), most conveniently written in Eq. (12.9).

Calculate, therefore, $A_L(t_i)$ for $L = 30$, 60, 90, and 120 and $t_i = 0$, 1, 3, 7, and 15 days.

	$A_L(t_i)$ em mm				
L	$t = 0$	$t = 1$	$t = 3$	$t = 7$	$t = 15$
30	150.1	139.4	129.5	124.1	119.1
60	291.8	266.8	248.0	234.8	220.2
90	435.2	402.8	377.6	359.3	340.9
120	580.8	540.2	511.2	488.9	466.1

The numerator of the second member of Eq. (12.9) is the storage change as a function of time $\partial A/\partial t$. We can then make a linear regression of $A_L(t_i)$ as a function of ln (t_i) for each depth. As $\ln(0) = -\infty$, the first result cannot be used in the regression. If the regression has a high R^2 (which usually happens), we will have an equation of this type for each depth:

$$A_L(t) = a - b \cdot \ln(t)$$

whose derivatives are $\partial A/\partial t = -b/t$. We would, thus, have the values of the flows $q = \partial A/\partial t = -b/t$, for any time between 0 and 15 days.

Another form, which we will use, is to employ finite differences as indicated by Eq. (12.9):

$$q = \frac{[A_L(t_{i+1}) - A_L(t_i)]}{[(t_{i+1}) - (t_i)]}$$

And with the $A_L(t_i)$ data we just calculated, we can construct the table below:

	q (mm day^{-1})			
L	$t = 0.5$	$t = 2$	$t = 5$	$t = 11$
30	10.7	5.0	1.4	0.6
60	25.0	9.4	3.3	1.8
90	32.4	12.6	4.6	2.3
120	40.6	14.5	5.6	2.9

The next step is to divide these values of q by the respective gradients $\partial H/\partial z$ or $(\partial h/\partial z + 1)$ to obtain the values of K.

The problem provides data of H as a function of time, obtained by tensiometry. Note that the depths of the tensiometers are different from the measurements of soil water content. This was made by purpose. For example, to calculate the gradient of H at $L = 60$ cm, we use the tensiometers immediately above and below:

$$(\text{grad } H)_{60} = \frac{H_{45} - H_{75}}{30}$$

As the flux densities q were measured at intermediate times, we should also calculate the gradients at the same times. One way is to calculate the means of H between t_i and t_{i+1} and construct a new table:

	H (cm H_2O)			
L	$t = 0.5$	$t = 2$	$t = 5$	$t = 11$
15	28.0	53.5	84.5	117.5
45	61.5	90.0	116.5	146.5
75	90.5	120.0	149.0	181.5
105	124.5	156.5	189.0	217.5
135	156.0	186.5	220.5	252.5

and calculating the gradients:

	grad H (cm H_2O cm^{-1} soil)			
L	$t = 0.5$	$t = 2$	$t = 5$	$t = 11$
30	1.117	1.217	1.067	0.967
60	0.967	1.000	1.083	1.167
90	1.133	1.217	1.333	1.200
120	1.050	1.000	1.050	1.167

and dividing the flux densities q by the respective gradients, we obtain the values of K:

	K (mm day^{-1})			
L	$t = 0.5$	$t = 2$	$t = 5$	$t = 11$
30	9.58	4.11	1.31	0.62
60	25.85	9.40	3.05	1.54
90	28.60	10.35	3.45	1.92
120	38.67	14.50	5.33	2.48

Then, to establish the functions $K(\theta)$, we need to know which values of θ correspond to the values of K we have just calculated. The data of θ are for $t_i = 0$, 1, 3, 7, and 15 days (Table of water contents given in the problem) and since the values of K are for $t_i = 0.5$, 2, 5, and 11, one way is to calculate the means. For each L, we have four pairs of K and θ that give us the points to establish the functions $K(\theta)$.

$L = 30$		$L = 60$		$L = 90$		$L = 120$	
K	θ	K	θ	K	θ	K	θ
9.58	0.483	25.85	0.431	28.60	0.464	38.67	0.475
4.11	0.449	9.40	0.390	10.35	0.445	14.50	0.458
1.31	0.423	3.05	0.361	3.45	0.430	5.33	0.446
0.62	0.406	1.54	0.327	1.92	0.418	2.48	0.433

The next step is to try linear regressions of $\ln(K)$ versus θ for each L and check the values of R^2. When high, the equations $K(\theta)$ will be of the exponential type. See examples in Chap. 7 that show how to establish the functions $K(\theta)$.

$$L = 30 \quad \ln K = -14.8786 + 35.763\,\theta \quad R^2 = 0.980$$

$$L = 60 \quad \ln K = -8.8030 + 28.000\,\theta \quad R^2 = 0.987$$

$$L = 90 \quad \ln K = -24.5168 + 60.129\,\theta \quad R^2 = 0.995$$

$$L = 120 \quad \ln K = -27.9925 + 66.711\,\theta \quad R^2 = 0.995$$

Since the values of R^2 are very high, the behavior K versus θ can be considered to be exponential, and the final equations are

$$L = 30 \quad K(\theta) = 3.45 \times 10^{-7} \exp(35.763\,\theta)$$

$$L = 60 \quad K(\theta) = 1.50 \times 10^{-4} \exp(28.000\,\theta)$$

$$L = 90 \quad K(\theta) = 2.25 \times 10^{-11} \exp(60.129\,\theta)$$

$$L = 120 \quad K(\theta) = 6.97 \times 10^{-13} \exp(66.711\,\theta)$$

The problem provides the value of $K_0 = 2.2$ cm day^{-1} measured at the soil surface during infiltration. Let us see how this value compares with the values estimated by the previous equations. To do this, simply substitute in them the respective values of θ_0 (saturation), which are the values of θ at $t = 0$:

$$L = 30 \quad \theta_0 = 0.501 \text{ cm}^3 \text{ cm}^{-3}; K_0 = 18.13 \text{ mm day}^{-1}$$

$$L = 60 \quad \theta_0 = 0.458 \text{ cm}^3 \text{ cm}^{-3}; K_0 = 55.65 \text{ mm day}^{-1}$$

$$L = 90 \quad \theta_0 = 0.475 \text{ cm}^3 \text{ cm}^{-3}; K_0 = 57.04 \text{ mm day}^{-1}$$

$$L = 120 \quad \theta_0 = 0.486 \text{ cm}^3 \text{ cm}^{-3}; K_0 = 83.89 \text{ mm day}^{-1}$$

Theoretically, at the end of the infiltration ($t = 0$ for us), water infiltrates under dynamic equilibrium and K_0 should be the same at any depth. The K_0 of the profile would be defined by the lowest conductivity layer. By the equations, we obtained values different from K_0. This was expected since they were calculated in the redistribution, and in this case, the lower conductivity profile affects the whole profile less, especially in our case, where the lower conductivity profile is the upper profile and the deeper layers drain more freely. Another aspect is still the problem of the equations of $K(\theta)$ being exponential, which gives rise to large errors in K for very small errors in θ. For example, for $L = 120$, if θ_0 was 0.485 instead of 0.486, i.e., 0.1% lower, K_0 would be 78.48 instead of 83.89. In general, errors in θ measurement are of the order of 2%!. So if θ_0 would be 0.475 (2% less), K_0 would be 40.27, which is less than half of the value obtained before.

12.2. (a) For Libardi et al. (1980), we make the regressions
$-(\theta - \theta_0)$ versus $\ln(t)$ (Eq. 12.16)
$-\ln(\theta)$ versus $(\theta_0 - \theta)$ through the expression

$$\ln\left(L\frac{\partial\bar{\theta}}{\partial t}\right) = \ln K_0 + \beta(\theta_0 - \theta)$$

In the previous exercise, we have the data. The obtained regressions are

$$L = 30 \quad (\theta - \theta_0)$$
$$= -0.0376 - 0.0250\ \ln(t) \quad R^2 = 0.989$$

$$L = 60 \quad (\theta - \theta_0)$$
$$= -0.0485 - 0.0354\ \ln(t) \quad R^2 = 0.976$$

$$L = 90 \quad (\theta - \theta_0)$$
$$= -0.0218 - 0.0147\ \ln(t) \quad R^2 = 0.996$$

$$L = 120 \quad (\theta - \theta_0)$$
$$= -0.0207 - 0.0136\ \ln(t) \quad R^2 = 0.990$$

$$L = 30 \quad \ln(\theta)$$
$$= 3.2393 - 37.6616\ (\theta_0 - \theta) \quad R^2 = 0.965$$

$$L = 60 \quad \ln(\theta)$$
$$= 3.9207 - 26.1348\ (\theta_0 - \theta) \quad R^2 = 0.987$$

$$L = 90 \quad \ln(\theta)$$
$$= 4.1735 - 58.2022\ (\theta_0 - \theta) \quad R^2 = 0.996$$

$$L = 120 \quad \ln(\theta)$$
$$= 4.4098 - 64.1967\ (\theta_0 - \theta) \quad R^2 = 0.995$$

Since for Eq. (12.16) the linear coefficient is $(1/\beta)[\ln(\beta K_0/aL)]$ and the slope $1/\beta$, we obtain the following values for $a = 1$:

L (cm)	β	K_0 (mm day^{-1})
30	40.000	33.75
60	28.249	83.59
90	68.27	58.29
120	73.529	74.78

and because by equation $\ln\left(L\frac{\partial\bar{\theta}}{\partial t}\right) = \ln K_0 + \beta(\theta_0 - \theta)$ the linear coefficient is $\ln(K_0)$ and the slope β,

L (cm)	β	K_0 (mm day^{-1})
30	37.662	25.52
60	26.135	50.46
90	58.202	64.94
120	64.197	82.25

(b) For Sisson et al. (1980), we make the regressions
$\ln(z^*/t)$ versus $(\theta - \theta_0)$ (Eq. 12.20)
So
for $L = 30$ ($z^* = 30$),

t	z^*/t	$\ln(z^*/t)$	$(\theta - \theta_0)$
1	30	3.4012	−0.035
3	10	2.3026	−0.069
7	4.286	1.4554	−0.087
15	2	0.6931	−0.103

for $L = 60$ ($z^* = 60$),

t	z^*/t	$\ln(z^*/t)$	$(\theta - \theta_0)$
1	60	4.0943	−0.053
3	20	2.9957	−0.083
7	8.571	2.1484	−0.111
15	4	1.3863	−0.151

for $L = 90$ ($z^* = 90$),

t	z^*/t	$\ln(z^*/t)$	$(\theta - \theta_0)$
1	90	4.4998	−0.022
3	30	3.4012	−0.037
7	12.857	2.5539	−0.052
15	6	1.7917	−0.061

and for $L = 120$ ($z^* = 120$),

t	z^*/t	ln (z^*/t)	$(\theta - \theta_0)$
1	120	4.7875	−0.022
3	40	3.6889	−0.034
7	17.143	2.8416	−0.046
15	8	2.0794	−0.059

and the regressions are

$$L = 30 \quad \ln(z^*/t) = 4.8747 + 39.6139\,(\theta - \theta_0) \quad R^2 = 0.989$$

$$L = 60 \quad \ln(z^*/t) = 5.3961 + 27.5365\,(\theta - \theta_0) \quad R^2 = 0.976$$

$$L = 90 \quad \ln(z^*/t) = 5.9706 + 67.6500\,(\theta - \theta_0) \quad R^2 = 0.996$$

$$L = 120 \quad \ln(z^*/t) = 6.2800 + 72.8108\,(\theta - \theta_0) \quad R^2 = 0.990$$

Since the linear coefficient is $\ln(\beta K_0)$ and the slope β,

L (cm)	β	K_0 (mm day^{-1})
30	39.614	33.05
60	27.537	80.09
90	67.650	57.91
120	72.811	73.31

At this point, the reader must have observed the great differences in the results obtained for the different methods. As an example, the table below shows the values obtained for $L = 90$ cm side by side:

Method	β	K_0 (mm day^{-1})
Hillel et al. (1972)	60.13	57.04
Libardi et al. (1980) Theta	68.03	58.20
Libardi et al. (1980) Flux	58.20	64.94
Sisson et al. (1980)	67.65	57.91

From what has been seen, the best criterion for defining the state of field capacity is the analysis of the water flow. In the solution of Exercise 12.1, we present a table of flows, that is, their variation in time. In that table, it is seen that after 5 days the flows are still of the order of 1–6 mm day^{-1}, which is high in comparison to evapotranspiration values, which are usually of this order of magnitude. For 11 days, they are smaller, but perhaps not yet negligible. Since we have data of θ only up to fifteen days, we would take these values as the field capacity θ_{FC} of this soil. It is also important to verify that on the 15th day of drainage the matric potentials h are of the order of −100 cm H_2O (or −0.1 atm) for any depth. In the initial table, the values of H are given and if they are subtracted from the values of the gravitational potential z, there remain h values of the order of −100. It is seen that this soil, after 15 days of drainage, is far from −1/3 atm. Therefore, for most soils, the value of −1/10 atm for θ_{FC} is more recommendable than −1/3 atm, at least for the sandy loams of Brazil.

References

Bacchi OOS, Reichardt K (1988) Escalonamento de propriedades hídricas na avaliação de métodos de determinação da condutividade hidráulica. Rev Bras Cienc Solo 12:217–223

Bacchi OOS, Corrente JE, Reichardt K (1991) Avaliação de dois métodos simples de determinação da condutividade hidráulica do solo. Rev Bras Cienc Solo 15:249–252

Bacchi OOS, Reichardt K, Libardi PL, Moraes SO (1989) Scaling of soil hydraulic properties in the evaluation of hydraulic conductivity determination methods. Soil Technol 2:163–170

Brunini O, Reichardt K, Grohmann F (1975) Determinação da água disponível em Latossolo Roxo em condições de campo. Sociedade Brasileira de Ciência do Solo, Campinas, pp 81–87

Corsini PC (1974) Agregação e fluxo de água do solo. Tese de Doutorado, Faculdade de Agronomia e Veterinária, Universidade do Estado de São Paulo, Jaboticabal

Davidson JM, Nielsen DR, Biggar JW (1963) The measurement and description of water flow through Columbia Silt Loam and Hesperia Sandy Loam. Hilgardia 34:601–617

De Jong van Lier Q (2017) Field capacity, a valid upper limit of crop available water ? Agric Water Manag 193:214–220

Dourado-Neto D, Reichardt K, Silva AL, Bacchi OOS, Timm LC, Oliveira JCM, Nielsen DR (2007) A software to calculate soil hydraulic conductivity in internal drainage experiments. Braz J Soil Sci 31:1219–1222

Gardner WR (1970) Field measurement of soil water diffusivity. Soil Sci Soc Am Proc 34:215–238

Hillel D, Krentos VD, Stylianou Y (1972) Procedure and test of an internal drainage meth-od for measuring soil hydraulic characteristics in situ. Soil Sci 114:395–400

IAEA (1984) Field soil-water properties measured through radiation techniques. International Atomic Energy Agency, Vienna

Islabão GO, Lima CLR, Vahl LC, Timm LC, Teixeira JBS (2016) Hydro-physical properties of a Typic Hapludult under the effect of rice husk ash. Braz J Soil Sci 40: e0150161

LaRue ME, Nielsen DR, Hagan RM (1968) Soil water flux below a ryegrass root zone. Agron J 60:625–629

Libardi PL, Reichardt K (2001) Libardi's method refinement for soil hydraulic conductivity measurement. Austr J Soil Res 3:851–860

Libardi PL, Reichardt K, Nielsen DR, Biggar JW (1980) Simplified field methods for estimating the unsaturated hydraulic conductivity. Soil Sci Soc Am J 44:3–6

Libardi PL, Salati E, Reichardt K (1983) The use of expanded vermiculite as a soil conditioner in the tropics. International Atomic Energy Agency, Vienna

Miller EE, Klute A (1967) Dynamics of soil water. Part I: mechanical forces. In: Hagan RM, Haise HR, Edminster TW (eds) Irrigation of agricultural lands. American Society of Agronomy, Madison, WI, pp 209–244

Moldenhauer WC (1975) Soil conditioners. Soil Science Society of America, Madison, WI

Reichardt K (1993) Unit gradient in internal drainage experiments for the determination of soil hydraulic conductivity. Sci Agric 50:151–153

Reichardt K (1988) Capacidade de campo. Rev Bras Cienc Solo 12:211–216

Reichardt K (1987) A água em sistemas agrícolas. Manole, Barueri

Reichardt K (1981) Uma discussão sobre o conceito de disponibilidade da água às plantas. Congresso Brasileiro de Ciência do Solo, Sociedade Brasileira de Ciência do Solo, Campinas, pp 256–284

Reichardt K (1974) Determinação da condutividade hidráulica em condições de campo para a estimativa da drenagem profunda em balanços hídricos. Centro de Energia Nuclear na Agricultura. Universidade de São Paulo, Piracicaba

Reichardt K, Libardi PL (1974) An analysis of soil-water movement in the field. I. Hydrological field site characterization. Centro de Energia Nuclear na Agricultura, Universidade de São Paulo, Piracicaba

Reichardt K, Timm LC, Bacchi OOS, Oliveira JCM, Dourado-Neto D (2004) A parameter-ized equation to estimate hydraulic conductivity in the field. Austr J Soil Res 42:283–287

Reichardt K, Portezan-Filho O, Libardi PL, Bacchi OOS, Moraes SO, Oliveira JCM, Falleiros MC (1998) Critical analysis of the field determination of soil hydraulic conductivity functions using the flux-gradient approach. Soil Tillage Res 48:81–89

Reichardt K, Ranzani G, Freitas Júnior E, Libardi PL (1980) Aspectos hídricos de alguns solos da Amazônia-Região do Baixo Rio Negro. Acta Amaz 10:43–46

Reichardt K, Libardi PL, Saunders LCU, Cadima A (1979) Dinâmica da água em cultura de milho. Rev Bras Cienc Solo 3:1–5

Reichardt K, Grohmann F, Libardi PL, Queiroz SV (1976a) Spatial variability of physical properties of a tropical soil. I. Geometric properties. Centro de Energia Nuclear na Agricultura, Universidade de São Paulo, Piracicaba

Reichardt K, Grohmann F, Libardi PL, Queiroz SV (1976b) Spatial variability of physical properties of a tropical soil. II. Soil water retention curves and hydraulic conductivity. Centro de Energia Nuclear na Agricultura, Universidade de São Paulo, Piracicaba

Remson I, Fungaroli AA, Hornberger GM (1967) Numerical analysis of soil moisture systems. J Irrig Drain Div Proc 3:153–166

Richards LA (1960) Advances in soil physics. International Society of Soil Science. International Congress of Soil Science, Madison, WI, pp 67–69

Rubin J (1967) Numerical method for analyzing hysteresis affected post-infiltration redistribution of soil moisture. Soil Sci Soc Am Proc 31:13–20

Salati E, Reichardt K, Urquiaga SS (1980) Efeitos da adição de vermiculita na retenção e armazenamento de água por latossolos. Rev Bras Cienc Solo 4:125–131

Scardua R (1972) Porosidade livre de água de dois solos do Município de Piracicaba, SP. Dissertação de Mestrado, Escola Superior de Agricultura Luiz de Queiroz, Universidade de São Paulo, Piracicaba

Silva AL, Bruno IP, Reichardt K, Bacchi OOS, Dourado-Neto D, Favarin JL, Costa FMP, Timm LC (2009) Soil water extraction by roots and Kc for the coffee crop. Agri 13:257–261

Silva AL, Reichardt K, Roveratti R, Bacchi OOS, Timm LC, Oliveira JCM, Dourado-Neto D (2007) On the use of soil hydraulic conductivity functions in the field. Soil Tillage Res 93:162–170

Silva AL, Roveratti R, Reichardt K, Bacchi OOS, Timm LC, Bruno IP, Oliveira JCM, Dourado-Neto D (2006) Variability of water balance components in a coffee crop grown in Brazil. Sci Agric 63:105–114

Sisson JB (1987) Drainage from layered field soils: fixed gradient models. Water Resour Res 23:2071–2075

Sisson JB, Ferguson AH, van Genuchten MT (1980) Simple method for prediction drain-age from field plots. Soil Sci Soc Am J 44:1147–1152

Staple WJ (1969) Comparison of computed and measured moisture redistribution following infiltration. Soil Sci Soc Am Proc 33:328–335

Timm LC, Oliveira JCM, Tominaga TT, Cássaro FAM, Reichardt K, Bacchi OOS (2000) Soil hydraulic conductivity measurement on a sloping field. Agri 4:480–482

Veihmeyer FJ, Hendrickson AH (1949) Methods of measuring field capacity and wilting percentages of soil. Soil Sci 68:75–94

Villagra MM, Michiels P, Hartmann R, Bacchi OOS, Reichardt K (1994) Field determined variation of the unsaturated hydraulic conductivity functions using simplified analysis of internal drainage experiments. Sci Agric 51:113–122

Evaporation and Evapotranspiration: The Vapor Losses to the Atmosphere 13

13.1 Introduction

The term **evaporation** is used when water passes from the liquid to the gaseous state and, in agronomy, includes two distinct processes. One is for inanimate surfaces, such as the water of a moist soil or a reservoir like a river, dam, lake, or sea. This process is governed by purely physical laws. The term evaporation is reserved for this process. For the water evaporation through living surfaces, like those of a plant or of an animal, biological phenomena limit the physical laws. The term **transpiration** is reserved for this process. When both processes occur simultaneously, as it occurs in a crop as a whole, with loss of water by the soil and the plant, the term **evapotranspiration** is used.

Loss of soil water by evaporation through its surface or by transpiration by plants are important parameters in the hydrological cycle, especially in cultivated areas. Remembering that 1 mm of water corresponds to 1 L m^{-2}, an evapotranspiration of 10 mm day^{-1}, common in the Amazon rainforest, corresponds to millions of m^3 of water that pass from this huge area to the atmosphere, a volume much greater than the flow that the Amazon river delivers to the ocean. For every gram of nutrients absorbed from the soil by the plant, hundreds of grams of water must be absorbed and transpired. For this reason, the transpiration is often called productive evaporation in order to contrast it with the soil evaporation, called non-productive evaporation. This evaporation of water by the surface of the soil may, however, and in special cases be of great importance from the quantitative point of view. Peters (1960) verified a case where the loss of soil water by evaporation reached 50% of the evapotranspiration of a crop during a normal vegetative cycle.

The process of water evaporation is its transition from the liquid to the gaseous state, at temperatures below the boiling point of the water. Being a change of state, it is a process that requires energy, in this case the **latent heat of evaporation** L (see Table 2.1, and do not to confuse this L with liter), which is larger the colder the water. Pereira et al. (1997) present the equation of L as a function of T, for T in °C between 0 and 100, in the international system:

$$L = 2497 - 2.37\,T \qquad \left(\mathrm{J\,g^{-1}}\right)$$

As the quantification of the processes is made per unit area, usually m^2 or ha, a definition of "**evaporation equivalent**" is used, which is the value L expressed in J mm^{-1} of evaporated water. In this way, the comparison of evaporation with precipitation, irrigation, and storage is facilitated. The mass of 1 g of water may be represented by a

K. Reichardt, L. C. Timm, *Soil, Plant and Atmosphere*, https://doi.org/10.1007/978-3-030-19322-5_13

cube of 1 cm^3 (1 × 1 × 1 cm) with an upper surface of 1 cm^2. Thus, when 1 g of water from a reservoir is evaporated, a height of 1 cm or 10 mm is lost. Therefore, at 10 °C < T < 30 °C, a mean value of $L = 2450$ J g^{-1} (Table 2.1) is taken, and this yields 245 J mm^{-1}. This energy, the evaporation equivalent, comes from **solar radiation**, which is why it is an important factor in the process. As was seen in Chap. 5 on the atmosphere, the vapor content of the air is also important. If the saturation deficit is large, the evaporation is stimulated and, when zero, which represents a saturated air (RH = 100%), the evaporation process ceases, or rather, enters into dynamic equilibrium, in which the number of molecules of water that passes into the gas phase is equal to the number that returns to the liquid phase. The wind affects evaporation, or accelerates it with the entry of drier air, or slows it down with the entry of a more humid air. The atmospheric movement, therefore, maintains an "evaporating power" (Ea), that is, the ability to dry surfaces, even in the shade without the presence of solar radiation. In this case, the latent heat L is withdrawn, by diffusion (sensible heat), from the air surrounding the surface, or from the surface itself. Because of this, when we step out of a shower, even on a hot day, we feel cold. The following expression is used:

$$\text{Ea} = f(u) \cdot d \tag{13.1}$$

where $f(u)$ is an empirical wind speed u function and d is the saturation deficit (Eq. 5.9). These are the main physical factors that affect evaporation. In the plant, in addition to these, the biological factors enter, especially the control of the stomata.

In this chapter, the evaporation of water from the surface of the soil will be focused first. We will distinguish between two cases: the constant evaporation, in dynamic equilibrium, occurring in the presence of a water table near the surface of the soil, and the variable evaporation, which occurs in a very deep soil without the presence of groundwater near the surface. Afterward, the evapotranspiration will be taken care of, which is well presented by Pereira et al. (1997).

13.2 Evaporation Under Steady State

Many researchers studied the evaporation of soil water under conditions of dynamic equilibrium. Willis (1960) is a classic work that presents an interesting analysis, which we will see below.

Consider a soil column, as shown in Fig. 6.18, which simulates a soil with water table near the surface. In this column, obviously, the matrix potential h is null at the water table position, and at the evaporating surface, because the soil is relatively dry, h will be considered very small, i.e., tending to $-\infty$. Thus, the constant flux of water q, within the soil column, is equal to evaporation at the soil surface, and given by the Darcy–Buckingham equation:

$$q = -K\frac{dH}{dz} = K\frac{dh}{dz} - K \tag{13.2}$$

where $q = E$ and $H = -h + z$.

Note that the differentials are total, because it is a case of dynamic equilibrium in which H is only a function of z and not of t, although there is a flux q. The water content profiles θ and potential H are invariant with time.

At that time, Willis (1960) found that hydraulic conductivity values, obtained by a series of researchers, could be related to the matrix potential h, in the wetter range of the soil, using an expression of the type:

$$K(h) = \frac{a}{(h^n + b)} \tag{13.3}$$

where a, b, and n are constants obtained by fitting experimental data to Eq. (13.3). Even though we have seen in Chap. 6 that the best model for K is the exponential, this relationship up to now is valid, in the wet range, for several soils, so that the results of Willis (1960) are still useful to describe the evaporation process. Clayey soils usually have values of n around 2, while sandy soils may have $n = 4$ or more, which means that for the latter the hydraulic conductivity decreases more drastically with water content than for the clayey soils. Let us imagine first the case of soils for which $n = 2$ fits well for Eq. (13.3). It would be the case of soils of fine texture. In this case,

$$K(h) = \frac{a}{(h^2 + b)} \tag{13.4}$$

Rearranging Eq. (13.2), substituting the K value of Eq. (13.4) and separating the variables, we have

$$dz = \frac{dh}{\left[1 + \frac{q(h^2+b)}{a}\right]}$$

which integrated from 0 to L with respect to z (since this is the field of variation of z in the soil column) and from 0 to ∞ with respect to $-h$ (also the field of variation of h in the column), we have

$$\int_0^L dz = \int_0^\infty \frac{(a/q)dh}{(a/q) + b + h^2} \tag{13.5}$$

and remembering that

$$\int \frac{du}{K^2 + u^2} = \frac{1}{K} \text{arc tan}\,(u/K)$$

and making $u = h$ and $K = \sqrt{(a/q) + b}$:

$$[z]_0^L = \frac{a}{q} \frac{1}{\sqrt{(a/q) + b}} \left\{ \text{arc tan} \left[\frac{h}{\sqrt{(a/q) + b}} \right] \right\}_0^\infty$$

Since the arc for which the tangent is zero is 0°, and for ∞ is $\pi/2$, we have

$$L = \frac{a}{q} \frac{1}{\sqrt{(a/q) + b}} [(\pi/2) - 0] \tag{13.6}$$

In Eq. (13.6), as "a" for most soils is generally greater than b, and q is so small, b can be neglected as the first approximation, that is, $\sqrt{(a/q) + b} \cong \sqrt{a/q}$.

Under these conditions, rearranging (13.6), we will have

$$q = \frac{a\pi^2}{4L^2} \tag{13.7}$$

Equation (13.7) tells us that the evaporation flux density q is proportional to the inverse of L^2. This means that for clay soils (or more correctly, for the soils that Eq. (13.4) is valid, with $n = 2$), the evaporation rate decreases with the square of the depth at which the water table lies. This is an important information in the control of evaporation in the face of the manipulation of the water table in drainage projects.

Let us now consider another soil for which $n = 4$, that is, of sandy texture. For this case, Eq. (13.5) is

$$\int_0^L dz = \int_0^\infty \frac{(a/q)dh}{(a/q) + h^4} \tag{13.8}$$

in which b was already neglected. So,

$$\frac{Lq}{a} = \int_0^\infty \frac{dh}{\left[(a/q)^{1/4}\right]^4 + h^4}$$

This integral is of the type

$$\int \frac{du}{K^4 + u^4}$$

so that

$$\frac{Lq}{a} = \frac{\pi}{2(a/q)^{3/4}\sqrt{2L}}$$

or

$$q = \frac{a\pi^4}{64L^4} \tag{13.9}$$

which means that in the case of a sandy soil ($n = 4$), q is proportional to the inverse of L^4, that is, the flow density decreases much faster than in the clayey soil, in case of a lowering of the water table. These two examples, despite all approximations, give the reader a good idea of the process of constant evaporation from the water table. Obviously, possible values of n other than 2 and 4, were a problem for Willis (1960) in the integration of equations of type (13.5) and (13.8), especially if n is fractional. Today, with the advancement of integration techniques, this is no longer a problem. Other examples can be seen in Gardner (1958) and Gardner and Fireman (1958). Another text that addresses this subject is that of Hillel (1998).

13.3 Evaporation in the Absence of a Water Table

Lemon (1956) divided the process of evaporation of a naked soil into three distinct stages. The first stage is characterized by a constant and independent evaporation rate E of the soil water (whose content should be high), as shown in Fig. 13.1. At this stage, the evaporation depends on the conditions prevailing in the atmosphere near the ground, such as radiant energy, wind speed, temperature, and humidity of the air. The first stage ends when resistance to water flow is established at the soil surface and the rate of evaporation ceases to be constant, decreasing with time. In the second stage, the evaporation rate E is a linear function of the mean soil water content profile θ_m and the formerly dominant conditions lose

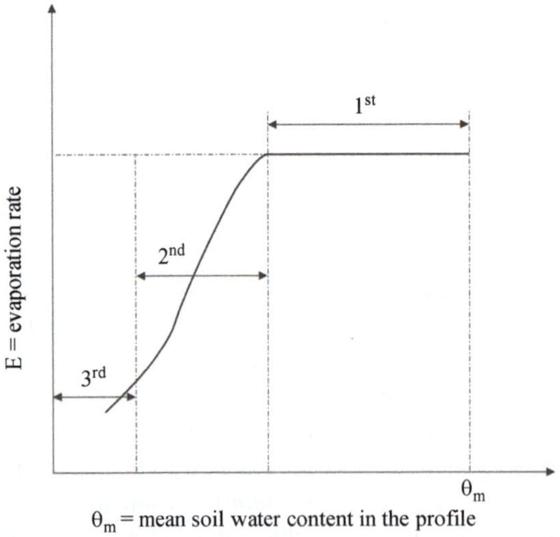

Fig. 13.1 The three phases of soil water evaporation as a function of the mean soil water content of the profile. Start reading the graph from right to left, at the maximum soil water content θ_m. As the soil dries out, it passes from first to second and then to third stages of evaporation

importance, while intrinsic soil conditions begin to govern water transport in the profile and the evaporation rate.

When the function that correlates E with θ_m begins to lose linearity and the soil is already very dry, the third stage of the process begins. This stage is characterized by a very slow movement of the water, due to the low hydraulic conductivity of the soil, being mainly in the vapor phase.

The mathematical analysis of the various processes is based on the solution of the general water-flow equation

$$\frac{\partial \theta}{\partial t} = \frac{\partial}{\partial z}\left[K(\theta)\frac{\partial H}{\partial z}\right] \tag{13.10}$$

subject to the boundary conditions of each particular case. Gardner and Milklich (1962) presented a solution to Eq. (13.10) for the case of constant evaporation in horizontal columns of finite-length soil in which the water is removed at one end by evaporation. This solution fits for soil columns in the first stage of evaporation.

Reichardt (1968) studied the process of evaporation of water from sandier soils, using the gamma radiation attenuation technique described in Chap. 6. In Figs. 13.2 and 13.3 are some of his typical results for columns of a Red-Yellow Podzolic soil (Argisol), from Piracicaba, São Paulo (SP) state, Brazil.

Based on these graphs, it is verified that the first stage of the process of evaporation of the water from the soil is so much more delayed the smaller the speed of evaporation. As a consequence, low evaporation rates deplete more the soil water, that is, the lower the evaporation rate, the lower the average profile water content at the end of the first stage. His column 4 had a layer of aggregates of diameter between 1 and 2 mm, and as can be seen in Fig. 13.2, this column barely noticed the first stage, because a dry crust was immediately formed. In this study, the author also verified that a layer of 0–5 mm of thickness conditioned the evaporation of the water of the soils studied, after the first stage.

In the third stage of evaporation, the water flow is quite slow and a considerable fraction of that flow occurs in the form of vapor. The main

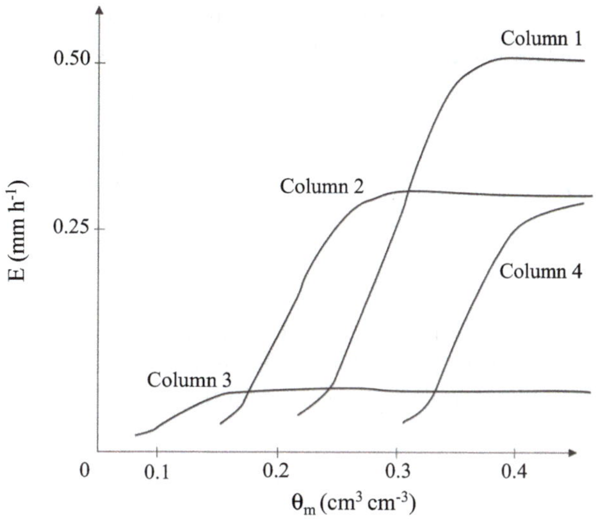

Fig. 13.2 Study of Reichardt (1968) showing that column 1 exposed to highest evaporation rate passes quickly from stage 1 to stage 2. Column 2 exposed to medium evaporation rate extends the first phase, and column 3 exposed to very low evaporation rate extends even further the first stage. Column 4, with aggregates on the surface, dried out so fast that it did not experience the first stage

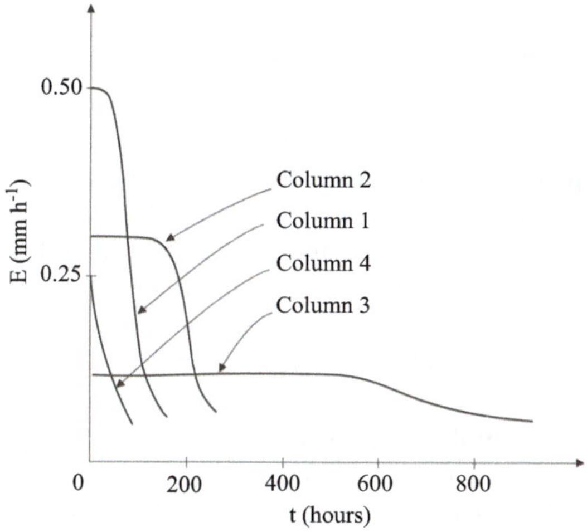

Fig. 13.3 Study of Reichardt (1968) shows the same effect as Fig. 13.2 as a function of time

process responsible for the vapor movement is that of diffusion (Chaps. 8 and 9), due to partial pressure gradients of water vapor and the current soil water content, the partial vapor pressure e_v defined in Chap. 5 being of importance. The vapor movement becomes considerable only for very dry soils. For the vapor flow, we can write, according to **Fick's law of diffusion**,

$$\dot{j}_d = -D_v \frac{de_a}{dx} \tag{13.11}$$

where $\dot{j}_d$ is the vapor flux density and D_v the diffusion coefficient of the vapor in the soil. Since e_a is a function of θ (soil water content which in turn is a function of the matric potential h) and e_0 (the saturation pressure of a pure water-free surface, which in turn is a function of temperature T), we can say that

$$\frac{de_a}{dx} = \frac{\partial e}{\partial e_0}\frac{\partial e_0}{\partial T}\frac{\partial T}{\partial x} + \frac{\partial e}{\partial h}\frac{\partial h}{\partial \theta}\frac{\partial \theta}{\partial x} \tag{13.12}$$

and substituting Eq. (13.12) into (13.11), we have

$$J_d = \underbrace{-D_v \frac{\partial e}{\partial e_0}\frac{\partial e_0}{\partial T}}_{D_{T_v}}\frac{\partial T}{\partial x} \underbrace{-D_v \frac{\partial e}{\partial h}\frac{\partial h}{\partial \theta}}_{D_{\theta_v}}\frac{\partial \theta}{\partial x} \tag{13.13}$$

Since the functions $e = e(T)$ and $\theta = \theta(h)$ are characteristic for a given situation, the partial derivatives of Eq. (13.13) can be grouped with D_v, defining new values of the diffusion coefficients:

$$\dot{j}_d = -D_{T_v}\frac{\partial T}{\partial x} - D_{\theta_v}\frac{\partial \theta}{\partial x} \tag{13.14}$$

D_{T_v} is the diffusion coefficient of vapor due to temperature gradients and D_{θ_v} the diffusion coefficient of vapor due to soil water gradients.

Philip and deVries (1957) assembled Eq. (13.14) with the equation of the water flux density in the liquid phase q in order to obtain the total water flow:

Table 13.1 Diffusivity values as a function of temperature

Diffusivities ($cm^2\ day^{-1}\ °C^{-1}$)	10 °C	20 °C	30 °C
D_{T_v}	0.8×10^{-2}	1.5×10^{-2}	1.7×10^{-2}
D_{T_l}	0.2×10^{-2}	0.4×10^{-2}	0.6×10^{-2}
D_{θ_v}	0.6×10^{-5}	2.0×10^{-5}	4.0×10^{-5}
D_{θ_l}	0.2	0.4	0.6

$$q_t = \dot{J}_d + q \qquad (13.15)$$

in which q can be expressed in the same way as was made to obtain $\dot{J}_d$, which we will see next. The Darcy–Buckingham equation (Eq. 7.3) gives us q:

$$q = -K\frac{\partial h}{\partial x}$$

Considering that the soil water characteristic curve is also a function of the temperature T, we can say that

$$q = \underbrace{-K\frac{\partial h}{\partial \theta}}_{D_{\theta_l}}\frac{\partial \theta}{\partial x} \underbrace{-K\frac{\partial h}{\partial T}}_{D_{T_l}}\frac{\partial T}{\partial x}$$

and so

$$q = -D_{\theta_l}\frac{\partial \theta}{\partial x} - D_{T_l}\frac{\partial T}{\partial x} \qquad (13.16)$$

Substituting Eqs. (13.16) and (13.14) into (13.15), we obtain the total flux of water, in both phases:

$$q = \underbrace{-(D_{T_l} - D_{T_v})}_{D_T}\frac{\partial T}{\partial x} \underbrace{-(D_{\theta_l} - D_{\theta_v})}_{D_\theta}\frac{\partial \theta}{\partial x} \qquad (13.17)$$

The same authors also combined Eq. (13.17) with the continuity equation (Chap. 7) and obtained the differential equation for water movement in the two phases:

$$\frac{\partial \theta}{\partial t} = \nabla\cdot(D_T\nabla T) + \nabla\cdot(D_\theta\nabla\theta) - \frac{\partial K}{\partial z} \qquad (13.18)$$

where ∇ is the divergent operator; $D_T = (D_{T_l} + D_{T_v})$ and $D_\theta = (D_{\theta_l} + D_{\theta_v})$.

In order to exemplify the order of magnitude of the different diffusivities, the data presented in Table 13.1 correspond to a sandy clayey soil, not very humid, $\theta = 0.10\ m^3\ m^{-3}$, taken from Philip and deVries (1957).

These data show the importance of the flow in the liquid phase as a result of the gradient, since D_{θ_l} being 100 times greater than D_{T_v} and D_{T_l} and often greater than D_{θ_v}.

13.4 Potential and Real Evaporation

According to Bernard (1956), we can define as **potential evaporation** the one that occurs on a surface of water, freely exposed to the conditions of solar radiation, air humidity, and wind. If there is great availability of water in the soil, the situation is very similar and its evaporation is also commonly called potential. The potential evaporation E_p of a soil is the maximum loss of water that a soil can present, by evaporation, when subjected to certain meteorological conditions. When there is not enough water available, evaporation ceases to be potential, and it is called **real evaporation** E. In general, we can say that $E \leq E_p$.

For the soil case, E_p occurs during the first stage of evaporation. As already mentioned, this is a function of the prevailing weather conditions on the evaporating surface. If a quantity of energy Q_L per unit area and time is available for the evaporation process at the soil surface, it is easy to verify that the rate of evaporation is given by

$$E = E_p = \frac{Q_L}{L} \qquad (13.19)$$

where L is the **latent heat of evaporation**. When Q_L is given in $J\ m^{-2}\ day^{-1}$ and L in $J\ L^{-1}$, we will have E in $L\ m^{-2}\ day^{-1}$ or $mm\ day^{-1}$.

13.5 Potential and Real Evapotranspiration

As we have seen, **evapotranspiration** depends essentially on the energy available for the water evaporation process. Thus, with available water in the soil, evapotranspiration is directly proportional to the available energy. For example, for a typical day in Piracicaba, SP, Brazil, in which the net radiation Q_L was 337 cal cm^{-2} day^{-1}, 5.8 mm of water could be evaporated. This calculation takes into account that the 337 cal cm^{-2} day^{-1} were totally absorbed by an evaporating surface and that all energy was used in the evaporation process. For a crop, however, the surface exposed to radiation is not flat and has an area larger than its projection on the ground, and the absorption of the radiation is not total. Wind, by turbulence, and the relative humidity of the air, by the potential of water vapor, also interfere in the process, sometimes by accelerating it, or by restricting it.

In order to standardize the evapotranspiration of plant communities, conditions were set in which their measurement should be made. ET_0, the **reference evapotranspiration**, was then defined as "the amount of evapotranspirated water per unit of time and area, by a green crop, completely covering the soil, of uniform height and without water deficiency" (Fig. 13.4). The choice of a lawn of grass (may be *Paspalum notatum* L.) which is common in tropical and subtropical regions is normally taken because it remains practically green over the year and with good development, provided it is irrigated.

For this defined area, the climatic conditions such as net radiation, wind, and relative humidity determine the value of ET_0. In view of this, the reference evapotranspiration is taken as a reference meteorological element for comparative studies of water loss by vegetation in different situations and locations around the globe.

Due to differences in the crop–atmosphere interface between grasslands and other crops, also at different stages of development, the **maximum evapotranspiration** ET_m of a crop, also called **crop evapotranspiration** (ET_c), was defined by a correction of the ET_0 reference evapotranspiration by a **crop coefficient** K_c already presented for a maize crop (Fig. 4.7) and in more detail in Fig. 13.5, so that

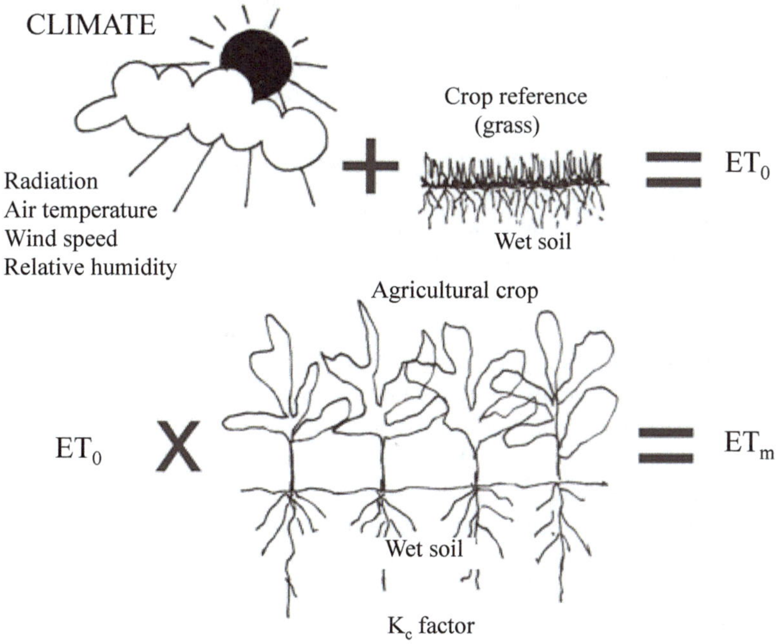

Fig. 13.4 Schematic presentation of the concepts of ET_0 and ET_m

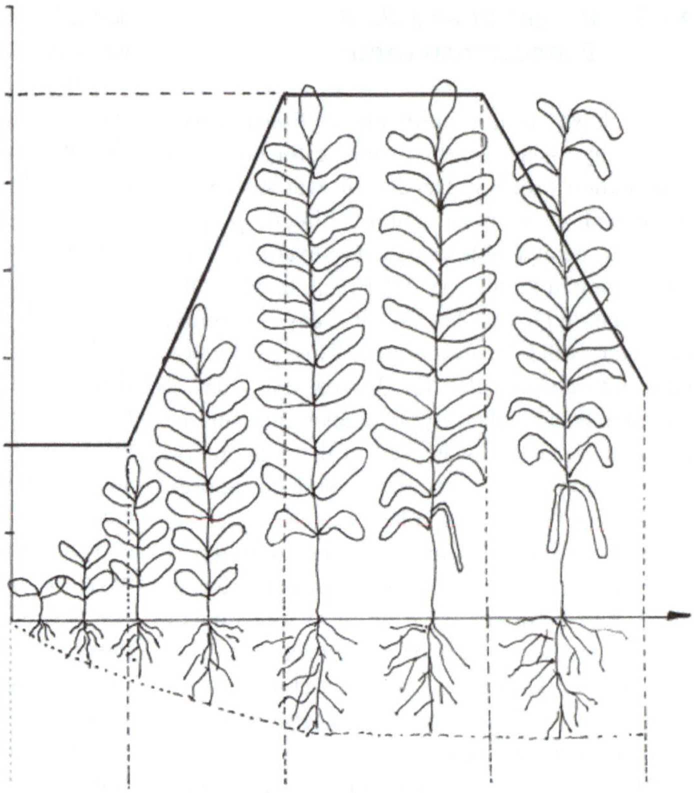

Fig. 13.5 An example of how the crop coefficient K_c is considered along an annual crop cycle

$$ET_c = K_c \cdot ET_0 \qquad (13.20)$$

K_c being evaluated experimentally for different crops in different growth stages, using the ratio ET_c/ET_0.

Figure 13.5 shows how the K_c variation is taken during the cycle of an annual crop (of the type of crop as rainfed rice, beans, corn, or soybeans). At the beginning of the establishment of the crop, K_c is small because a small fraction of the soil is covered by the crop that has a little developed root system. With the crop in full development, the value of K_c is maximum and may even assume values greater than 1. Values greater than 1 indicate that the crop in question loses more water than the grass lawn, both under the same climatic conditions. Table 13.2 also shows K_c values for some crops. The evapotranspiration of the crop ET_c represents, therefore, the maximum loss of water that a certain crop presents, at a given stage of development, when there is no water restriction in the soil. Analyzing the data in Table 13.2, we see that K_c varies more with the stage of development than with the type of crop. This means that the maximum loss of water ET_c, for a given climatic condition, is not very different for a forest, sugarcane, or pasture. It depends, essentially, on the available energy per unit area and time.

Table 13.2 Examples of crop coefficients K_c for some crops

	Stage of development			
Crop	I	II	III	IV
Beans	0.3–0.4	0.7–0.8	1.05–1.2	0.65–0.75
Cotton	0.4–0.5	0.7–0.8	1.05–1.25	0.8–0.9
Peanut	0.4–0.5	0.7–0.8	0.95–1.1	0.75–0.85
Corn	0.3–0.5	0.8–0.85	1.05–1.2	0.8–0.95
Sugarcane	0.4–0.5	0.7–1.0	1.0–1.3	0.75–0.8
Soybean	0.3–0.4	0.7–0.8	1.0–1.15	0.7–0.8
Wheat	0.3–0.4	0.7–0.8	1.05–1.2	0.65–0.75

The **actual evapotranspiration** or current evapotranspiration ET_a is the one that actually occurs in any soil moisture situation. If there is water available in the soil and the water flow in the plant meets the atmospheric demand, ET_a will equal to ET_c. If there is water restriction in the soil and the atmospheric demand is not met, ET_a will be lower than ET_c. In general, we have

$$ET_a \leq ET_c \quad (13.21)$$

As the availability of water affects productivity, the ideal situation for a crop is that ET_a is equal to ET_c. Whenever $ET_a < ET_c$, there is water restriction and productivity may be affected. Therefore, ET_c is used in irrigation projects to calculate the maximum climatic demand of a crop.

The evapotranspiration ET_c of a crop is in general given in mm day^{-1} and, if integrated for a month, crop cycle or year, we will have mm month^{-1}, mm cycle^{-1}, and mm year^{-1}. Some examples for Piracicaba, SP, Brazil, are

- Summer: 3–5 mm day^{-1}, with maximum values up to 8 mm day^{-1}
- Autumn and spring: 2–4 mm day^{-1}
- Winter: 1–3 mm day^{-1}
- Corn crop in spring/summer: 360–600 mm cycle^{-1}

Average annual change:

Jan	Feb	Mar	Apr	May	Jun	Jul	Aug	Sept	Oct	Nov	Dec	
4.1	3.8	3.6	2.6	2.2	2.0	2.2	2.8	3.2	4.0	4.4	4.2	mm day^{-1}
127	107	112	79	68	60	68	87	96	125	132	131	mm month^{-1}
												Yearly total: 1192 mm

13.6 Measurement of the Evapotranspiration

As the difference between crop evapotranspiration ET_c and the real evapotranspiration ET_a is only soil water restriction, direct methods of measurement are the same for both. The most elaborate direct method is the **evapotranspirometer** or **lysimeter**. Figure 13.6 shows schematically a drainage lysimeter. It consists of a tank (of brick, asbestos cement, metal, etc.) filled with a volume of soil up to almost its border, placed within the area for which the evapotranspiration is to be measured. The tank has a drainage system that allows the measurement of the water that percolates through the soil. Its open area A to the atmosphere should not be less than 1 m^2, and can reach 10 m^2. Its depth, h, should be large, from 0.5 m to more, depending on the crop, being the ideal 1–1.2 m for annual crops. When filling the tank with soil, it begins with a layer of gravel and another layer of fine sand. The soil must be placed obeying the layers that occur in the profile, as similar as possible to the natural soil profile. To make a measurement, the soil is wetted until drainage water appears in the collection well. After 1–2 days, the drainage ceases, the soil water is in equilibrium (not the field capacity but the pot capacity). Under these conditions, the measurement period begins, and evapotranspiration is measured by the total water used by the vegetation in a given period, determined by the difference between the amounts of the water added to the lysimeter (I) and the percolated water (D). The operator needs to acquire practice in order to know how to evaluate the amount of water to be added, so that there is not too much percolation. The calculation is made by

$$ET = I - D \quad (13.22)$$

ET will be equal to ET_0, if the crop is the grass lawn, and equal to ET_c for any other crop and shall be ET_a if the measurement period is long and there is water restriction.

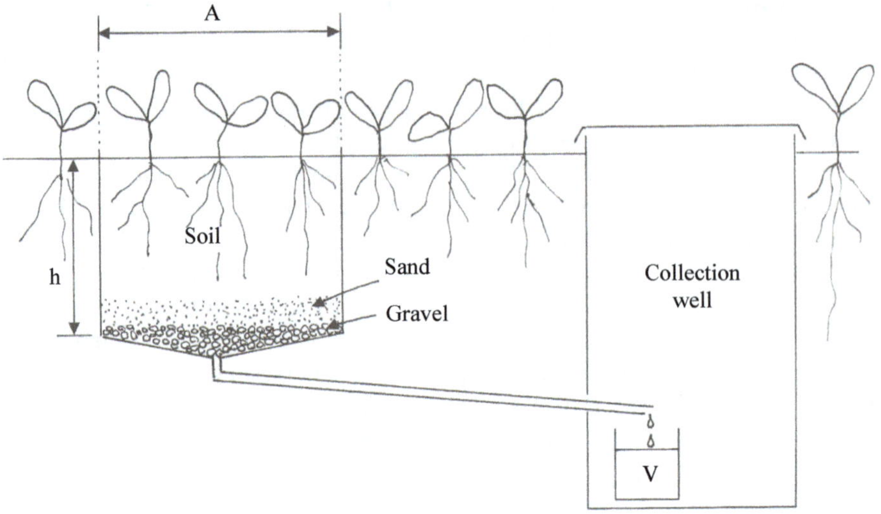

Fig. 13.6 Schematic view of field lysimeter

The soil reaches its maximum storage after wetting and obtaining its equilibrium (drainage ceases). From then on, it loses water by evapotranspiration. After 2–5 days, when water is added to replace the water lost by ET, the amount must be excessive so that the storage returns to the maximum value and the difference percolates. In this way, it is not necessary to know the maximum storage capacity of the soil.

Here is an example of a tank with $A = 4\ \text{m}^2$ planted with beans:

Date (08:00 h)	10/12	10/15	10/18	10/22	10/24
Added water (L)	60	60	60	60	60
Drained water (L)	9.5	14.3	12.1	15.8	18.4

We see that of the 60 L applied in 10/12, 14.3 L were drained until the day 10/15. The remainders 60–14.3 = 45.7 L were evapotranspirated in the period from 12 to 15. As $A = 4\ \text{m}^2$, the total evapotranspiration 45.7/4 = 11.4 L m^{-2} or 11.4 mm. Since the period is 3 days, we have $\text{ET}_\text{c} = 3.8\ \text{mm day}^{-1}$. For the subsequent periods, we have $\text{ET}_\text{c} = 4.0$; 2.8 and 5.2 mm day^{-1}.

Another direct method is the measurement, in the field, of soil water storage to a depth L greater than the root system of the crop, in two consecutive dates. If there is no rainfall in the period and if the downward movement of water is not appreciable, the storage difference is an estimate of evapotranspiration:

$$\text{ET}_\text{a} = \frac{[A_L(t_2) - A_L(t_1)]}{(t_2 - t_1)} \tag{13.23}$$

where A_L (t_2) and A_L (t_1) are the soil water storages at times t_2 and t_1, respectively.

The ET_0 reference evapotranspiration, as the definition itself indicates, depends exclusively on climatic conditions. Based on this, several so-called indirect (or theoretical–empirical) methods provide estimates of ET_0, using only climate data. The Blaney–Criddle and Thornthwaite methods estimate ET_0 from air temperature and day length data. The Penman method uses data from solar radiation, wind, and air humidity.

We will go into more detail for three widely used methods (Thornthwaite, Penman, and Penman–Monteith methods). For more details and other methods, see Pereira et al. (1997) and Allen et al. (1998).

13.6.1 The Thornthwaite Method (Thornthwaite 1948)

The calculation of evapotranspiration by the Thornthwaite method (ET_0) is performed as follows:

Table 13.3 Correction factors for the Thornthwaite method of ET_0 estimation, for the latitude 22°S

Month	f	Month	f
January	1.14	July	0.94
February	1.00	August	0.99
March	1.05	September	1.00
April	0.97	October	1.09
May	0.95	November	1.10
June	0.90	December	1.16

Source: Thornthwaite (1948) and Pereira et al. (2002)

$$ET_0 = f \cdot 16 \cdot \left(10 \cdot \frac{\text{Tn}}{\text{I}}\right)^a \qquad (13.24)$$

where Tn is the mean temperature of month n, in °C; I is the heat index in the region, calculated according to Eq. (13.26); f is a correction factor as a function of latitude and the month of the year, shown in Table 13.3; and a is a regional thermal index, calculated by Eq. (13.27). In this case, the ET_0 result will be in mm month^{-1}, since the reference evapotranspiration is calculated for a 30-day month, considering that each day has a 12-h photoperiod. The factor f is important for its correction and transformation into actual number of days of the month. To obtain ET_0 on a daily scale (mm day^{-1}), it is enough to divide the result in mm month^{-1} by the number of days of the month.

Equation (13.25) is used when $0 \leq \text{Tn} < 26.5$ °C. In the case of Tn ≥ 26.5 °C, ET_0 will be given by

$$ET_0 = -415.85 + 32.24 \times \text{Tn} - 0.43 \times \text{Tn}^2 \qquad (13.25)$$

The value of I depends on the annual oscillation of the air temperature and integrates the thermal effect of each month, being preferably calculated with the normal temperature values for the place in question (Pereira et al. 1997):

$$I = \sum_{i=1}^{12} (0.2 \times \text{Tn})^{1.514} \qquad (13.26)$$

In the same way as I, **a** is calculated with climatological normal period, these coefficients being characteristic of the region and independent of the year of estimation. The exponent **a** is calculated according to Eq. (13.27):

$$a = 6.75 \times 10^{-7} \cdot I^3 - 7.71 \times 10^{-5} \cdot I^2 + 1.7912 \times 10^{-2} \cdot I + 0.49239 \qquad (13.27)$$

As can be seen, the Thornthwaite method uses only air temperature as a temporal variable. It actually encompasses indirectly solar radiation that would be the most important variable in the intensity of evapotranspiration. The method is still widely used, especially when only T is available, which occurs in many regions of the globe.

13.6.2 Penman Method (Penman 1948)

The estimation of ET_0 by the Penman method is given by

$$ET_0 = \frac{W \cdot Q_L}{\lambda} + (1 - W) \cdot E_a \qquad (13.28)$$

where λ is the latent heat of evaporation (MJ kg^{-1}); W is the air temperature dependent weighting factor, obtained by Eq. (13.29); Q_L is the net radiation (MJ m^{-2} day^{-1}) given by Eq. 5.17; and E_a is the air evaporation power (MJ m^{-2} day^{-1}), obtained by Eq. (13.1). So that

$$W = \frac{\Delta}{\Delta + \lambda} \qquad (13.29)$$

where Δ is the slope (derivative) of the saturation vapor pressure versus air temperature curve, in kPa (see Eq. 5.10), which uses the atmospheric pressure value (P_{atm}) for its calculation and is obtained by

$$\lambda = 0.664742 \times 10^{-3} \cdot P_{atm} \qquad (13.30)$$

$$\Delta = \frac{4098 \times e_s}{(T_{mean} + 237.3)^2} \qquad (13.31)$$

e_s being the partial pressure of vapor saturation (kPa) (Eq. 5.10). In this method, $f(U)$ of Eq. (13.1) is given by Eq. (13.32):

$$f(U) = m \cdot (a + bU_2) \qquad (13.32)$$

where m is equal to 6.43 MJ m^{-2} day^{-1} kPa^{-1}; $a = 1$; $b = 0.526$ s m^{-1}; and U_2 is equal to the velocity of the wind measured at 2 m above soil surface (m s^{-1}).

As can be seen above, this method, in addition to air temperature, employs characteristics of air humidity, solar radiation, and wind speed.

13.6.3 Penman–Monteith Method (Allen et al. 1998)

The calculation of evapotranspiration (ET_0) by the Penman–Monteith method can be performed according to Eq. (13.33):

$$ET_0 = \frac{0.408 \times \Delta \times (Q_L - G) + \lambda \frac{900}{T_{mean}+273.16} U_2 \times (e_s - e_a)}{\Delta + \lambda \times (1 + 0.34 \times U_2)} \quad (13.33)$$

where G is the heat flux density by conduction into the soil (MJ m^{-2} day^{-1}) that enters as a new variable (see Chap. 10). This method is the most used today, since its calculation can easily be made with the use of an Excel[R] worksheet.

Another indirect way to measure evapotranspiration is through evaporation tanks. The most common is the **Class A tank**, standardized according to the scheme of Fig. 13.7, also shown in Fig. 13.8, and the result of the measurement is called **tank evaporation** (ECA). It is a galvanized metal tank, filled with water, placed on a wooden platform. The set is preferably placed on a grass lawn, with a large grass border.

With evaporation, the water level in the tank lowers, directly providing the height of the evaporated water. The measurement of the water height can be done with a graduated ruler, but it is difficult to accurately assess the position of the water level in relation to the ruler. Therefore, there are special instruments for this measurement, such as micrometric screws and floating systems. The accuracy of the measurement is important because the water level drops only a few millimeters a day, and ideally, one could measure fractions of millimeters.

The tank should not be completely filled with water to prevent water loss by wind. It is a standard measure to leave 5 cm of edge, being, therefore, the maximum height of water of 25 cm. The minimum level should also not be too low because the volume of water in the tank becomes too small and the water heats up too much, introducing an error in the measurement. The recommended minimum level is 20 cm. The tank, therefore, has a useful height for measures of 25–20 = 5 cm or 50 mm. For evaporations of 5 mm day^{-1}, the water in the tank lasts for 10 days. Normally, the water is replenished once a week, so that the maximum level is maintained.

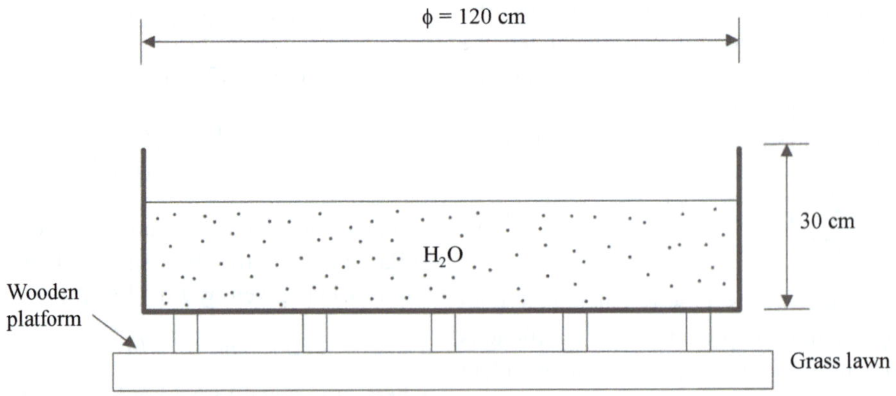

Fig. 13.7 Schematic view of a Class A evaporation tank

The example shown in Table 13.4 shows the data sequence for a Class A tank for a period without rainfall.

If there is rain, the tank acts like a rain gauge and its level rises. One should not, however, rely on this data for rainfall measurement because during rain there may be a lot of water loss, especially due to the wind that usually accompanies the rain. It is common, therefore, to lose readings on the tank on rainy days. This is not a great problem since the evaporation measurement loses its importance on a rainy day.

A free water surface like that of the Class A tank loses more water than a crop. Therefore, the ECA tank evaporation values should be corrected. For this, a **tank coefficient** (K_p) is used:

$$\mathrm{ET}_0 = K_p \times \mathrm{ECA} \tag{13.34}$$

The value of K_p depends on the size of the boundary in which the tank is exposed, the relative humidity of the air, and the wind speed. Table 13.5 gives some K_p values.

The use of Table 13.5 involves measurements of wind and relative humidity. When these are not available, it is common to use an average value of $K_p = 0.8$. If tank data are used directly, which implies $K_p = 1$, there is a safety margin of approximately 20%, since, as already mentioned, the tank always loses more water than a crop.

Regarding the direct measure of evapotranspiration, the studies of Ferraz (1968), Pereira et al. (1974), and Reichardt et al. (1974) are interesting. These authors obtained measurements of real evapotranspiration, for grasses and sugarcane, coffee, and beans. They used the principle of neutron moderation to measure soil water content (Chap. 6).

Fig. 13.8 An evaporation Class A tank with constant water feeding (Source: The authors)

13.7 Exercises

13.1. In the scheme of Fig. 6.18 imagine a clayey soil with $n = 2$ and $a = 2.03 \times 10^3$. The constant a was estimated such that when applied to Eq. (13.7) with L in cm, the result of q is mm day^{-1}. Calculate q when the water table is at $L = 50$, 75, and 100 cm.

13.2. Same exercise as above, for sandy soil, with $n = 4$ and $a = 2.06 \times 10^6$.

13.3. The soil columns used by Reichardt (1968), referring to the graphs in Fig. 13.2, were 30 cm deep. How many mm of water were

Table 13.4 An example of Class A measurement sheet

Day	Hour	Water height (cm)	Evaporation rate (mm day^{-1})
03/05/85	08:00	25.0	
03/06/85	08:00	24.5	5
03/07/85	08:00	24.1	4
03/08/85	08:00	23.9	2
03/09/85	08:00	23.6	3
03/10/85	08:00	23.5	1
03/11/85	08:00	22.9	6
03/12/85	08:00	22.4 (25.0) replenished	5
03/13/85	08:00	24.7	3
03/14/85	08:00	24.1	6

Table 13.5 Tank coefficient as a function of wind speed, boundary distance, and relative humidity

		Relative humidity		
		Low	Moderate	High
Wind speed (km day^{-1})	Boundary (Grass Lawn) m	<40%	40–70%	>70%
<175 low	1	0.55	0.65	0.75
	10	0.65	0.75	0.85
	100	0.70	0.80	0.85
	1000	0.75	0.85	0.85
175–425 moderate	1	0.50	0.60	0.65
	10	0.60	0.70	0.75
	100	0.65	0.75	0.80
	1000	0.70	0.80	0.80
475–700 strong	1	0.45	0.50	0.60
	10	0.55	0.60	0.65
	100	0.60	0.65	0.70
	1000	0.65	0.70	0.75

lost in the first stage of evaporation by columns 1, 2, 3, and 4?

13.4. The previous problem showed that the lower E of the first stage, more water can be removed from the soil. Is not that contradictory?

13.5. How many mm of water evaporate in one day from a Class A tank that received that same day a radiant energy net of 756 cal cm^{-2} day^{-1}, which is fully utilized in the evaporation process? Average tank temperature 30 °C.

13.8 Answers

13.1. 2.00, 0.89, and 0.50 mm day^{-1}.

13.2. 0.50, 0.10, and 0.03 mm day^{-1}.

13.3. From Fig. 13.2, $\Delta\theta$ from column 1 in the first stage is approximately 0.01 cm^3 cm^{-3}. Therefore, $\Delta A = 0.01 \times 30 = 0.3\,cm = 3\,mm$. For columns 2, 3, and 4, we have 6, 9, and 0 mm, respectively.

13.4. It is not contradictory. Although more water can be withdrawn from the soil with a smaller E, the process takes much longer. These times are shown in Fig. 13.3. Column 3, which lost more water, lost that water in about 500 h or 20 days.

13.5. 13.02 mm day^{-1}.

References

Allen RG, Pereira LS, Raes D, Smith M (1998) Crop evapotranspiration – guidelines for computing crop water requirements. FAO, Roma

Bernard EA (1956) Le determinisme de l'evaporation dans la nature. Institut National pour l'Étude Agronomique du Congo Belgue, Leopoldville

Ferraz ESB (1968) Determinação da evapotranspiração real pela moderação de nêutrons. Dissertação de Mestrado, Escola Superior de Agricultura Luiz de Queiroz. Universidade de São Paulo, Piracicaba

Gardner WR (1958) Some steady-state solutions of the unsaturated moisture flow equation with applications to evaporation from a water table. Soil Sci 85:228–232

Gardner WR, Fireman M (1958) Laboratory studies of evaporation from soil columns in the presence of a water table. Soil Sci 85:244–249

Gardner WR, Milklich FJ (1962) Unsaturated conductivity and diffusivity measurement by a constant flux method. Soil Sci 93:271–274

Hillel D (1998) Environmental soil physics. Academic, San Diego, FL

Lemon ER (1956) The potentialities for decreasing soil moisture evaporation loss. Soil Sci Soc Am Proc 20:120–125

Penman HL (1948) Natural evaporation from open water, bare soil and grass. Proc R Soc Ser A 193:120–145

Pereira AR, Angelocci LR, Sentelhas PC (2002) Agrometeorologia: fundamentos e aplicações práticas. Agropecuária, Guaíba

Pereira AR, Ferraz ESB, Reichardt K, Libardi PL (1974) Estimativa da evapotranspiração e da drenagem profunda em cafezais cultivados em solos podzolizados Lins e Marília. Centro de Energia Nuclear na Agricultura. Universidade de São Paulo, Piracicaba

Pereira AR, Villa Nova NA, Sediyama GC (1997) Evapo (transpi)ração. Fundação de Estudos Agrários Luiz de Queiroz, Piracicaba

Peters DB (1960) Relative magnitude of evaporation and transpiration. Agron J 52:536–538

Philip JR, deVries DA (1957) Moisture movement in porous materials under temperature gradients. Trans Am Geoph Union 38:222–232

Reichardt K (1968) Estudo do processo de evaporação da água do solo. Tese de Livre Docência, Escola Superior de Agricultura Luiz de Queiroz. Universidade de São Paulo, Piracicaba

Reichardt K, Libardi PL, Santos JM (1974) An analysis of soil-water movement in the field. II. Water balance in a snap bean crop. Centro de Energia Nuclear na Agricultura, Universidade de São Paulo, Piracicaba

Thornthwaite CW (1948) An approach toward a rational classification of climate. Geogr Rev 38:55–94

Willis WO (1960) Evaporation from layered soil in the presence of a water table. Soil Sci Soc Am Proc 24:239–242

How Do Plants Absorb Soil Water? 14

14.1 Introduction

Plants, in general, absorb hundreds of grams of water for each gram of accumulated dry matter. They have their roots dipped in the soil water reservoir, and their leaves are subject to the action of solar radiation and wind, forcing the plant to transpire incessantly. To grow properly, plants need a "water economy" such that the demand made on them by the atmosphere is balanced by the water supply from the soil. The problem is that the demand for evaporation due to the atmosphere is practically constant, whereas the processes that add water to the soil, such as rain or irrigation, occur only occasionally and, in general, with great irregularity. In order to survive in the intervals between rains, the plant needs to have the reserve contained in the soil. In this chapter, we will briefly see how efficient the soil can be as a reservoir of water for plants, how can they remove the water from the soil, until which soil moisture limit can the plant maintain vegetal growth, and how the rate of transpiration is determined by the interaction between the plant, the soil, and the atmosphere.

14.2 Water Availability for Plants

The concept of availability of water for plants has, for many years, brought controversy among researchers. The main cause of the controversy is probably the lack of a physical definition of the concept. Veihmeyer and Hendrickson (1927, 1949, 1950, 1955) stated that soil water is equally available in a range of water contents extending from an upper limit, the **field capacity** FC (θ_{FC}—defined in Chap. 12), to a lower limit, the **permanent wilting point** PWP (θ_{PWP}). The PWP was defined by Veihmeyer and Hendrickson (1949) as soil water content at which a wilted plant does not restore turgidity, even when placed in a saturated atmosphere for 12 h. It is commonly assumed that this soil water content corresponds to a matrix potential of -15 atm (-1.5 MPa). These authors postulated that the biological functions of plants remain unaffected within this range, abruptly changing beyond the lower limit (curve "a" in Fig. 14.1). Other researchers, especially Richards and Waldleigh (1952), have found evidence that water availability to plants decreases with decreasing soil water content and that the plant may suffer from water deficiency and reduce growth before reaching the permanent wilting point (curve "b" in Fig. 14.1). Other researchers, disagreeing with both views, sought to divide the range of available water into two intervals, one of "immediately available water" and another one of "available water," and looked for a "critical point" between field capacity and permanent wilting point, as an additional criterion for the

K. Reichardt, L. C. Timm, *Soil, Plant and Atmosphere*, https://doi.org/10.1007/978-3-030-19322-5_14

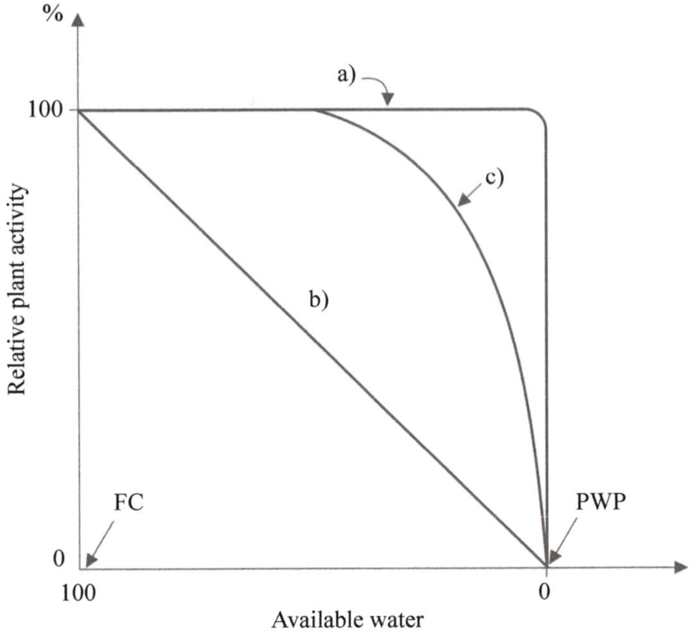

Fig. 14.1 Different points of view of the soil water availability

availability definition (curve "c" in Fig. 14.1). None of these schools was able to base their hypotheses on a well-founded theory. These authors drew generalized conclusions from a small number of experiments conducted under specific conditions.

The problem has become more complex when different plants have been found to respond differently to soil water, which has led researchers to recognize that soil water content alone is not a sufficient criterion for defining soil water availability. The attempt to solve the problem by correlating the energy state of the water in the plant with the state of the water in the soil, in terms of its potential, was an improvement. Therefore, soil "constants" were defined in terms of potential [$-1/3$ atm (-33 kPa) for θ_{FC} and -15 atm (-1.5 MPa) for θ_{PWP}] to be applied universally. However, although the use of these energy concepts represented a considerable advance, there was still a need to consider the Soil-Plant-Atmosphere System as a continuous and extremely dynamic system.

The exact description of water absorption by plants by a well-founded theory is difficult given the inherent complications of the space-time relationships involved in the process. The roots grow in a disorderly way in the most diverse directions and spaces. Conventional methods of measuring θ and Ψ are based on sampling a relatively large volume of soil. As a result of these and many other difficulties, only a semi-quantitative analysis of the phenomenon was possible.

In the second half of the twentieth century, a fundamental change occurred in the interpretation of soil-plant relationships with respect to water. With the development of theoretical knowledge of the state of water in the soil, plant, and atmosphere, and with the development of experimental techniques, a more solid interpretation could be given to the problem. It has become increasingly clear that in a dynamic system like this, static concepts, for example, equivalent water content, permanent wilting point, critical water content, capillary water, gravitational water, and others, are generally meaningless because they are based on the hypothesis that the processes that occur in the field are directed toward such static states.

This development has led us to improve the classical concept of available water in its original sense. Of course, there is no qualitative difference between water retained at different soil potentials. Nor is the amount of water absorbed by plants only a function of their potential in the soil. This quantity depends on the ability of the roots to absorb the water of the soil with which they are in contact, as well as the soil properties in the supply and transfer of this water to the roots, in a proportion that satisfies the demands of transpiration. It is then seen that the phenomenon depends on soil factors (hydraulic conductivity, diffusivity, relationships between water content and potential), plant (root density, depth, root growth rate, root physiology, leaf area), and the atmosphere (saturation deficit, wind, available radiation).

Many researchers, insistent in maintaining the classical concepts involved in the availability of water, argue that the development of soil science has brought the abandonment of useful concepts without replacing them with more accurate ones. It should be clear from the foregoing that it is difficult to find an exact and precise form for describing a phenomenon as complex as that of water dynamics in the Soil-Plant-Atmosphere System. This difficulty is inherent to the process, so complex in its space-time relationships.

This is not to say that the problem is unsolvable. Each case in particular should be studied, taking into account our knowledge of the dynamics of water in the soil, plant, and atmosphere. Each case may have a particular solution, often requiring a series of rational simplifications.

Denmead and Shaw (1962), for example, reached an experimental confirmation of the effect of dynamic conditions on plant water uptake and subsequent transpiration. These authors measured transpiration rates of maize plants grown in pot and field under different irrigation and evaporation conditions. Under evapotranspiration conditions of ET_c equal to 3–4 mm day^{-1}, the actual evapotranspiration rate ET_a fell below the ET_c rate, for mean soil water contents corresponding to soil water potentials of approximately −2 atm. In more extreme weather conditions, ET_c varied between 6 and 7 mm day^{-1}, and ET_a fall was already verified at soil water contents corresponding to potentials of the order of −0.3 atm. On the other hand, for ET_c very low values, less than 1.5 mm day^{-1}, no ET_a drop was observed until potential of −12 atm. A similar result has already been discussed in Chap. 13 for soils without vegetation. In addition to this contribution, the reader can search for studies carried out in Brazil: Brunini et al. (1975), Reichardt (1976), Reichardt (1979, 1981), and Angelocci (2002).

14.3 The Soil-Plant-Atmosphere System Considered as a Whole

Gardner (1960), Cowan (1965), and Philip (1966) were the first to recognize the Soil-Plant-Atmosphere System (SPAS) as a whole, a physical continuum in which the dynamics of different transfer processes occur interdependently. In this system, the water flow occurs in the direction of the decrease of its total energy, being the concept of total potential of water Ψ valid in the soil, plant, and atmosphere. In the past, researchers from different areas (plant physiology, soil physics, and meteorology) expressed the same potential in different ways: **diffusion pressure deficit** (DPD), stress, vapor pressure, etc. This fact, added to the measurements made in different units, did not allow, for a long time, an immediate communication between these researchers and an analysis of the SPAS as a whole. The important principle to be understood is that the various terms used by different disciplines (plant physiology, soil science, agrometeorology) to characterize the state of energy of the water in the different parts of the system, that is, the potential Ψ. Differences in Ψ from one point to another are responsible for the flow of water. This principle applies throughout the soil, the plant, and the atmosphere, continuously. As seen in Chap. 7, the Darcy equation describes water movement, which we rewrite in finite differences:

$$q = \frac{1}{r}\frac{\Delta \Psi}{\Delta x} \tag{14.1}$$

where q is the flux density ($m^3\ m^{-2}\ s^{-1}$); ΔΨ (mH_2O) is the total potential difference between two points (including the potential components that fit for each part of the system) separated by Δx (m); and r is the water resistivity in the medium, equal to the inverse of the hydraulic conductivity K. Δx is difficult to be measured in the different parts of the SPAS. How to measure distances from soil points to points in the root system that develops in disorder? How to measure distances that water travels within the plant? And the atmosphere? One way to solve the problem is to include Δx in the water resistivity r, and Eq. (14.1) simplifies in:

$$q = \frac{\Delta \Psi}{R} \tag{14.2}$$

where R is now the resistance of the system to the flow of water. The reader should compare this equation with Ohm's law of electricity and verify that they are of the same nature. This equation shows us that the flux at any point in the system is inversely proportional to a resistance. The trajectory of the water includes its movement from the soil to the roots, absorption by the roots, transport in the roots to the stems and leaves through the xylem, evaporation of the intercellular spaces of the leaves, diffusion of vapor through the stomata and cuticle, and its movement in the atmosphere near the leaf to the outer atmosphere.

The amount of water transpired daily is large in relation to the variation of the water content of the plant, in such a way that the flow of water by the plant can, for short periods of time, be considered a steady-state equilibrium process. Thus, by Eq. (14.2) with constant q, the potential differences $\Delta\Psi$ in the different parts of the system are proportional to the respective resistances R to the flow. Figure 14.2 shows the Soil-Plant-Atmosphere System, indicating the resistances, just as it is done in electrical circuits. In this figure, R_s represents the resistance of the soil, a variable resistance, depending on θ, Ψ_s, etc., as already seen in Chap. 7, which referred to the flow of water in the soil. R_{co} is the resistance of the radicular cortex, and R_x is the resistance of the xylem. These resistances, although varying for long times, can also be considered constant for shorter times in which plant growth can be neglected (e.g., 1 day). After the xylem, in the

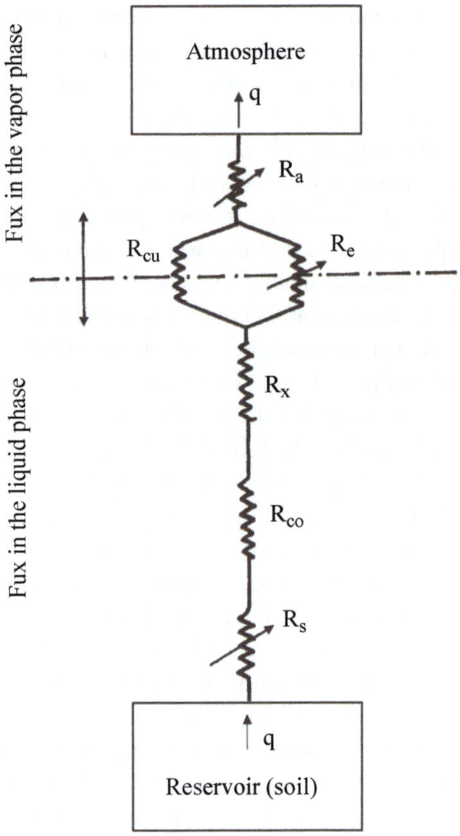

Fig. 14.2 Comparison of the Soil-Plant-Atmosphere System with an electric circuit

leaf, the water flow can take two parallel paths to the atmosphere: either through the cuticle (R_{cu}) or through the stomata (R_e), typically variable resistance due to the changes of the opening of the stomata. After the leaf, the water flow, already in the vapor phase, encounters the resistance of the atmosphere R_a, also variable according to turbulence, solar radiation, etc.

As we consider this process a case of steady-state equilibrium, q must be constant at any point, so that:

$$q = -\frac{\Delta\Psi_s}{R_s} = -\frac{\Delta\Psi_{co}}{R_{co}} = -\frac{\Delta\Psi_x}{R_x} = -\frac{\Delta\Psi_l}{R_l} = -\frac{\Delta\Psi_a}{R_a} \tag{14.3}$$

where $\Delta\Psi_l$ and R_l are, respectively, the potential change in the leaf and the resistance of the leaf. It should be recalled that the negative sign appears in Eq. (14.3) due to the fact that the water flow occurs in the opposite direction of the hydraulic gradient. In the same way, as for parallel resistances in electricity, we have:

$$\frac{1}{R_l} = \frac{1}{R_{cu}} + \frac{1}{R_e}$$

Typical values of $\Delta\Psi$ can be seen in Fig. 14.3 for different soil and air humidity conditions as well as for different parts of the plant.

If in a crop, the plants are turgid and the available net energy is only used in the phase change of the water (liquid → vapor), it is possible to assume that the flow in the plant occurs in dynamic equilibrium. This means that the transpiration rate q is equal to the flow of water in the plant and equal to the absorption by the roots:

$$q = -\frac{\Delta\Psi_{soil}}{R_{soil}} = -\frac{\Delta\Psi_{plant}}{R_{plant}} = -\frac{\Delta\Psi_{atmosphere}}{R_{atmosphere}}$$

The order of magnitude of these potentials changes, under normal conditions, is $\Delta\Psi_{soil} \cong -1$ to -3 atm; $\Delta\Psi_{plant} \cong -1$ to -10 atm; and $\Delta\Psi_{atmosphere} \cong -20$ to -500 atm. It can be concluded that the resistance $R_l + R_a$ between the leaves and the atmosphere can be 50 times greater than the resistance of the plant and the soil. During the hottest hours of the day, when the stomata close, this resistance becomes even greater, resulting in a decrease in the transpiration rate.

14.4 Water Flux from Soil to Root

In Chap. 7 we studied the flow of water in the soil using the **Cartesian coordinate system** of **orthogonal coordinates** (x, y, z), which is revised in Chap. 18. The **equation of continuity** written in this system (Eq. 7.20b) is:

$$\frac{\partial\theta}{\partial t} = \nabla \cdot K\nabla H$$

or

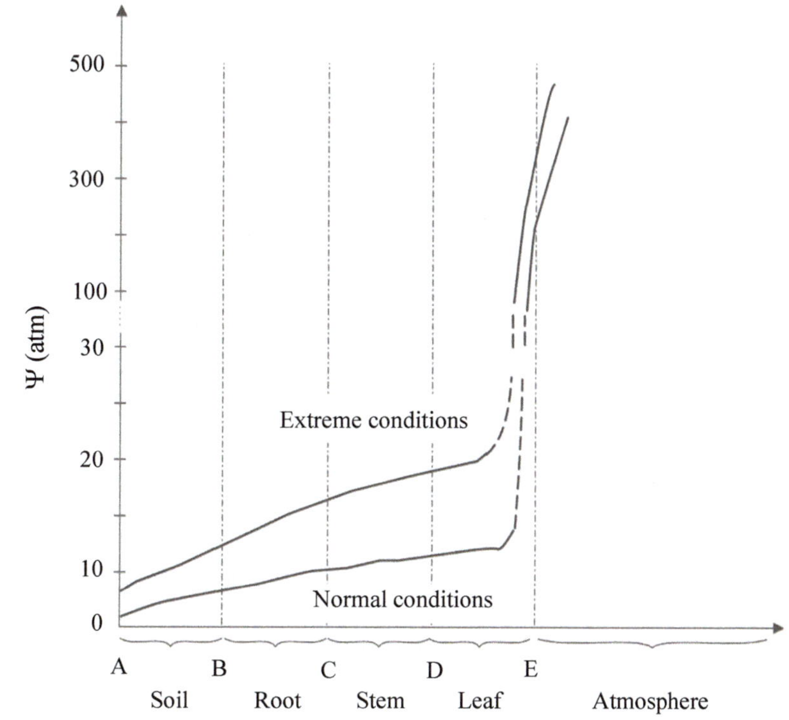

Fig. 14.3 Water potential distributions along the Soil-Plant-Atmosphere System

$$\frac{\partial\theta}{\partial t} = \frac{\partial}{\partial x}\left[K(\theta)_x \frac{\partial H}{\partial x}\right] + \frac{\partial}{\partial y}\left[K(\theta)_y \frac{\partial H}{\partial y}\right] + \frac{\partial}{\partial z}\left[K(\theta)_z \frac{\partial H}{\partial z}\right] \tag{14.4}$$

As a part of a root can be approximated by a cylinder of radius "a" and length L, and the flow of water from the soil to the roots is radial, it can better be described in the **cylindrical coordinate system**, in which the variables r, α, and z are used as indicated in Fig. 14.4.

In this system, r is the straight line that includes the root cylinder radius, α is the angle between the plane containing OM and the coordinate z, and a plane of reference zx and z is the height (in this case, root length). A point M, characterized by the coordinates x, y, and z in the orthogonal Cartesian system, is characterized by the coordinates r, α, and z in the cylindrical system. The **volume element** that in the Cartesian system is a cube, in this system, is shaped as a "piece of cheese" (see Fig. 14.4).

The curved distance $r\Delta\alpha$ comes from a simple rule of three:

$$2\pi r(\text{cm}) \rightarrow 2\pi(\text{radians})$$

$$x \rightarrow \Delta\alpha$$

In this system, we could, as we did in Chap. 7, deduce the equation of continuity. The result would be:

$$\frac{\partial\theta}{\partial t} = -\frac{1}{r}\frac{\partial}{\partial r}(r \cdot q_r) - \frac{1}{r}\frac{\partial q_\alpha}{\partial \alpha} - \frac{\partial q_z}{\partial z} \tag{14.5}$$

The water flux densities in the directions r, α, and z, given by the Darcy-Buckingham equation in this system, are:

$$q_r = -K_r \frac{\partial H}{\partial r}$$

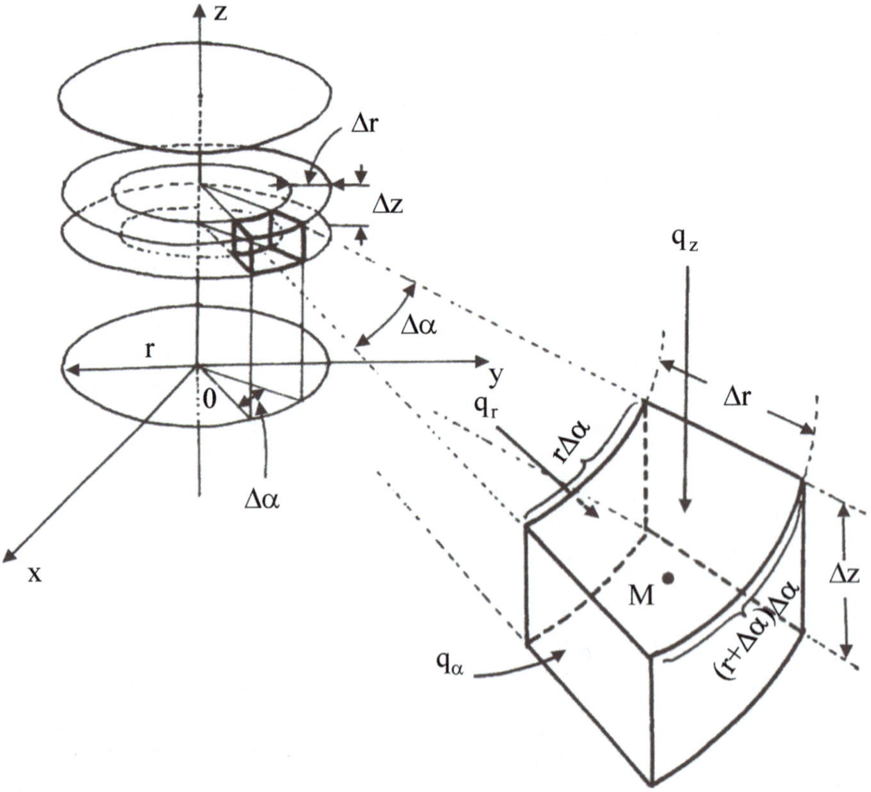

Fig. 14.4 Illustration of the cylindrical coordination system, showing an elementary volume around a point M

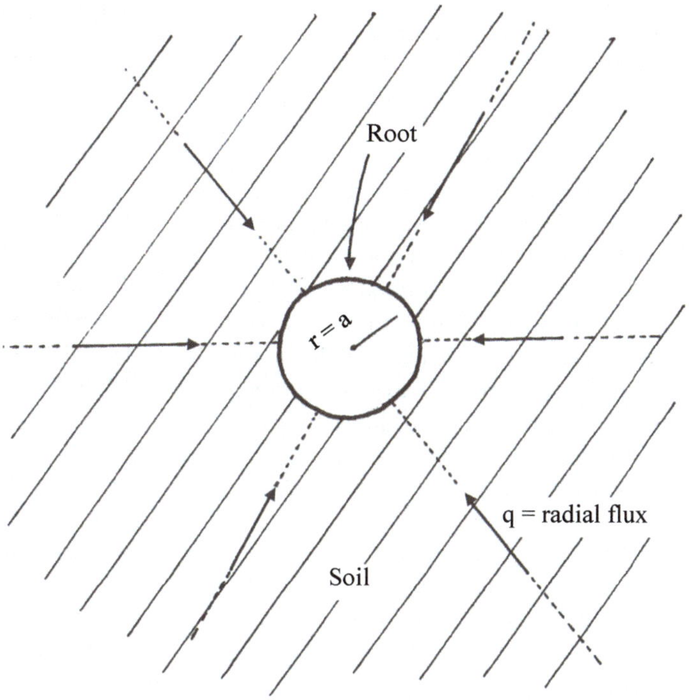

Fig. 14.5 A root cross section showing the radial directions of water flow from the soil to the root

q_r being a radial flux, into or out of the root;

$$q_\alpha = -K_\alpha \frac{\partial H}{\partial (r\alpha)} = -\frac{K_\alpha}{r}\frac{\partial H}{\partial \alpha}$$

q_α being a circular flux, as it were around the root, but not found in our case of water moving to the root;

$$q_z = -K_z \frac{\partial H}{\partial z}$$

q_z being an axial flux, along the root.

Substituting these flows in Eq. (14.5) and considering $K_z = K_\alpha = K_r = K$, we obtain the general equation of the water flow in the cylindrical system:

$$\frac{\partial \theta}{\partial t} = \frac{1}{r}\frac{\partial}{\partial r}\left[r \cdot K \frac{\partial H}{\partial r}\right] + \frac{1}{r^2} \times \frac{\partial}{\partial \alpha}\left[K \frac{\partial H}{\partial \alpha}\right] + \frac{\partial}{\partial z}\left[K \frac{\partial H}{\partial z}\right] \quad (14.6)$$

If z is chosen as the root axis and if we take a cross section of the root and the surrounding soil (see in Sect. 14.4 in Fig. 14.5), we will be only interested in the coordinate r, and the equation is summarized in:

$$\frac{\partial \theta}{\partial t} = \frac{1}{r}\frac{\partial}{\partial r}\left[r \cdot K \frac{\partial H}{\partial r}\right] \quad (14.7)$$

which, by neglecting the gravitational component of water and introducing the concept of **soil water diffusivity**, becomes:

$$\frac{\partial \theta}{\partial t} = \frac{1}{r}\frac{\partial}{\partial r}\left[r \cdot D \frac{\partial \theta}{\partial r}\right] \quad (14.8)$$

The solution of the problems involving Eqs. (14.7) and (14.8) can in general be obtained by the technique of separable variables, but is not found directly, falling generally into Bessel functions. For this reason, we will not provide any detailed examples; we will only mention that Gardner (1964) solved a problem involving Eq. (14.7) for the following boundary conditions:

$$H = H_0 \quad \text{or} \quad \theta = \theta_0, \quad r > 0, \quad t = t_0$$

$$q = 2\pi a K \frac{\partial H}{\partial r}, \quad r = a, \quad t > 0$$

$$\theta = \theta_0, \quad r = \infty, \quad t > 0$$

The first condition tells us that at the beginning, the soil was at the constant water content (θ_0) and potential (H_0), in any position. Second, the flow density at the root surface is given by the Darcy-Buckingham equation, with the factor 2π being introduced to have the flow around the root cross section, considered to be cylindrical. The last condition tells us that soil water content does not vary at a fairly large distance from the root. Its solution is, for K and D constants:

$$H - H_0 = \frac{q}{4\pi K}\left[\ln\left(\frac{4Dt}{r^2}\right) - 0.57722\right]$$

Many other examples of using cylindrical coordinates can be found in the literature. Equation (14.6) is presented here, more to show the reader the existence of this other coordinate system. Besides this system, there is still the spherical coordinate system, much less used in agronomic applications. In this system, the generic point M is characterized by two angles and a radius (see Chap. 19).

For more detailed studies of water flow in the plant, leaf, and atmosphere surrounding the system, we recommend again the text of Angelocci (2002). We also mention the recent work by De Jong Van Lier et al. (2009) that deals with transpiration flows from soil to root using the concept of **matrix flow**.

In terms of modeling in the Soil-Plant-Atmosphere System, excellent programs such as SWAP (Soil, Water, Atmosphere and Plant) were developed by Kroes, van Dam, Groenendijk, Hendriks, and Jacobs (Alterra-report 1649, Wageningen, UR, the Netherlands). This program allows simulations of water flow, solute flow, and heat flux in the soil and can be used by aggregating other models to simulate crop productivity.

14.5 Available Water and Evapotranspiration

Having seen in general terms the complexity of the study of water absorption by plants, we now see practical aspects of this subject, necessary for the management of water in agricultural crops and especially in irrigation planning. In this context, **available water** (AW) (often referred to as the **available water capacity**—AWC—in irrigation) is considered the water between the θ_{FC} (**field capacity**) and θ_{PWP} (**permanent wilting point**), as shown at the lower part of Fig. 14.6. The value to be adopted as θ_{FC} will depend on the objectives of each project, taking into account the type of soil. Classical values of θ_{FC} are taken from the **soil water retention curve** at potentials of $-1/3$ atm (-333 cm of water or -33.3 kPa) for more clayey soils and -0.1 atm (-100 cm of water or -10 kPa) for sandier soils, but any other value measured under field conditions can also be used. For θ_{PWP} there is not much controversy, using the value of the retention curve corresponding to the potential of -15 atm (-150 m of water or -1.5 MPa). So that:

$$\text{AWC} = (\theta_{FC} - \theta_{PWP}) \tag{14.9}$$

For the comparison of the AWC given as above in $m^3\ m^{-3}$ with precipitation, drainage, irrigation, and evapotranspiration data, which are expressed in mm, it is necessary to multiply Eq. (14.9) by the depth of L (in mm) of the considered soil. Since θ is dimensionless ($m^3\ m^{-3}$), the result of the AWC will be mm. For calculation and irrigation design purposes, the water between θ_s (saturation) and θ_{FC} does not enter as useful water for the plants because it is considered to be strongly affected by gravity, percolating to the deepest layers. This is a limitation, because for small precipitation events where infiltration reaches small depths, all water in this range is available to the plants (De Jong Van Lier 2017). In irrigation projects, care is taken not to irrigate to very deep depths, limiting the calculations to the main root zone (to a depth called the **effective root depth** of the root system,

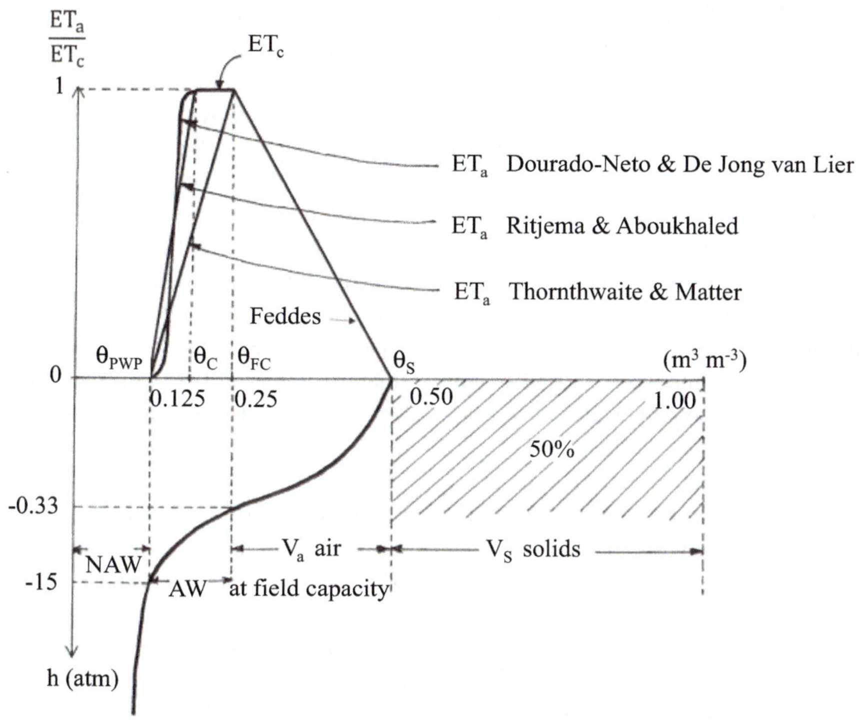

Fig. 14.6 Schematic view of an ideal soil with porosity 50% and water extraction by evapotranspiration through three different evaluation methods

in irrigation texts), so that redistribution of water after irrigation can be extracted by the plant. In irrigations with water of poor quality (e.g., salt water), which should be avoided, irrigations are made in excess to cause salts to be leached out of the root zone. It can be seen, therefore, that considering the interval θ_s–θ_{FC} as lost water for the plants can be problematic, and, therefore, in Fig. 14.6, ET is shown in three of the many available options. Since the amount of water below the PWP, θ_{PWP}, is generally very small, it can safely be considered unavailable to plants (represented by NAW, nonavailable water in Fig. 14.6).

Figure 14.6 shows an average, "ideal" soil with 50% porosity, which results in $\theta_s = 0.500\ m^3\ m^{-3}$ and in the optimum condition has a fraction of 25% with $\theta_{FC} = 0.250\ m^3\ m^{-3}$ and with a $\theta_{PWP} = 0.125\ m^3\ m^{-3}$. For a homogeneous soil profile of 1 m depth, its AWC is 125 mm. A crop developing in this soil will have $ET_a = ET_c$ in θ_{FC}, and its evapotranspiration will decrease thereafter due to the restrictions of soil water flow to the plant, resulting from the decrease of K with θ. According to Thornthwaite and Mather (1955), ET_a linearly decreases to θ_{PWP} as indicated at the top of Fig. 14.6. This linear reduction from ET_a to θ_{PWP} was much criticized in the following years, and several substitutes were proposed. One of the most accepted is that of Rijtema and Aboukhaled (1975) who assume that the equality between ET_a and ET_c continues until a critical value θ_c from which it decreases linearly to θ_{PWP} as indicated in Fig. 14.6 in that its value is $0.1875\ m^3\ m^{-3}$. Feddes et al. (1978) proposed for the water between saturation θ_s and θ_{FC} a line going from ET_c in θ_{FC} to zero in θ_s. Recall that Fig. 14.6 is timeless, the abscissa is θ, and thus the line that Feddes et al. (1978) proposed only indicates that any ET value is possible in this range. The value 0 for θ_s indicates that in the saturation, there is no oxygen in the soil and the cultivated plants do not develop, except for those of the rice type, whose root system receives O_2 through the aerial part of the plant. Dourado Neto and De Jong van Lier (1993), dissatisfied with the discontinuity of the ET_c–ET_a function at points θ_c and zero, proposed a model that follows a branch of a cosine curve, also shown in Fig. 14.6, which gives continuity in the ET curve at points θ_c and zero. The above work discusses ten different models describing the decrease of ET_a from θ_c. More details on these models of ET_a reduction will be discussed in the next chapter, together with the water balance.

As a recent contribution to this subject, we would like to cite a recent work carried out by Monteiro et al. (2018) dealing water absorption by fruit trees.

In order to improve the concept of water availability for plants, Silva et al. (2014) present the concept of **least limiting water range** (LLWR), which is the range of available water within which plant development is limited not only by water potential and AWC but by three factors: (1) the potential of the water in the soil, (2) **soil aeration** level, and (3) soil **resistance to root penetration** (RP). This concept has been approached as an index of **structural quality of soils** for plant growth. To calculate the LLWR, its upper limit (wet limit) takes into account θ_{FC} and the soil water content θ of 10% which is a value that guarantees 10% aeration (threshold value below which the crops are considered to be affected by the low level of O_2 in the soil), both dependent on the soil bulk density d_s. The lower limit of the LLWR takes into account θ_{PWP} and θ_{RP}, the water content limiting root penetration (taken as 2 MPa), also both dependent on d_s. The smallest interval between these four moisture levels is the LLWR which can be presented in the form of a graph of θ versus d_s (Fig. 14.7), hence the name of "least limiting" factor. Figure 14.7 represents schematically the LLWR for two d_s (d_{s1} and d_{s2}), the respective LLWR and AWC for a given soil, and the area indicates the full range of possible LLWRs as a function of soil density. It can be seen that the concept of LLWR is more rigorous regarding the availability of water compared to the classical available water.

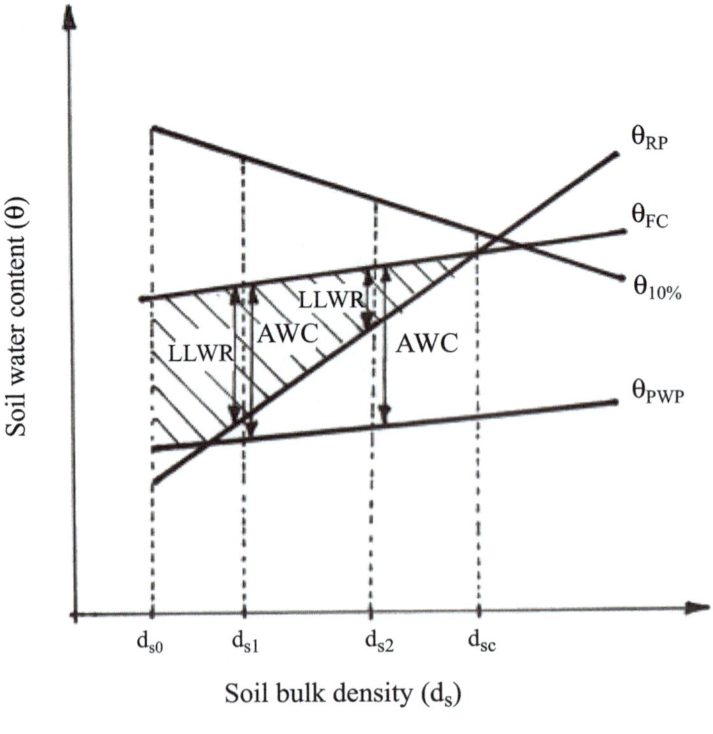

Fig. 14.7 Illustration of the **least limiting water range** (LLWR) correlating soil water content and soil bulk density

The LLWR concept is relatively new and has been applied in the literature for different soil conditions and types of management, such as Silva and Kay (1996, 1997a, b). An example of application of the LLWR concept in a coal mining area (Candiota, RS, Brazil) in order to evaluate different types of vegetation cover during the soil reconstruction process in this area is Lima et al. (2012).

The relation of $\theta_{10\%}$ with d_s can easily be established through Eq. (3.9) of Chap. 3 making $V_{air} = 10\%$ and calculating V_s from values chosen for d_s and d_p. Usually a linear relationship is obtained. The relations θ_{FC} and θ_{PWP} with d_s need to be obtained experimentally. It is common to find slightly increasing relationships. The θ_{RP} relationship with d_s is also experimental but laborious since RP depends on both θ and d_s. An advantage is that the penetrometers (Stolf et al. 2012) are simple and fast to operate. Let's take the opportunity here to describe the **penetrometers**. They consist of a rod with a conical end that penetrates the soil by the fall of a hammer, as shown below (Fig. 14.8):

The expression relating the impact of the hammer with **soil resistance** is obtained from the maximum total potential energy E available in the hammer during the falling process in which it initially moves from a height h generating an energy equal to (Mgh) and then penetrating the soil to depth x with an energy equal to $(M + m)gx$:

$$E = Mgh + (M + m)gx \tag{14.10}$$

Considering losses due to the collision between M and m, we will have the factor f which reduces the final energy of the hammer to:

$$f = \frac{M}{(M + m)} \tag{14.11}$$

so that:

$$E = F \cdot x = fMgh + (M + m)gx \tag{14.12}$$

and expliciting the penetration force F:

$$F = \frac{fMgh}{x} + (M + m)g \tag{14.13}$$

and dividing the force by the contact area A of the cone with the soil, we obtain the soil resistance RP:

$$\text{RP} = \frac{fMgh}{Ax} + \frac{(M + m)g}{A} \tag{14.14}$$

which is an equation of the type $y = a + b/x$ where RP is given in N m^{-2} or pascal (Pa) and x in m, where x is the penetration in m/impact, and we have $1/x$ corresponding to impacts/m. Defining N as the number of impacts to penetrate 10 cm (1 dm), we have:

$$\frac{1}{x} = 10 \cdot N \tag{14.15}$$

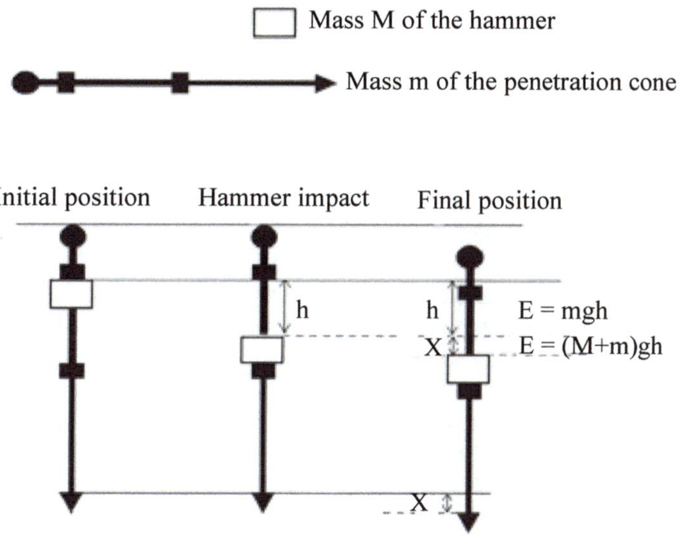

Fig. 14.8 Diagram of an impact penetrometer as developed by Stolf et al. (1983)

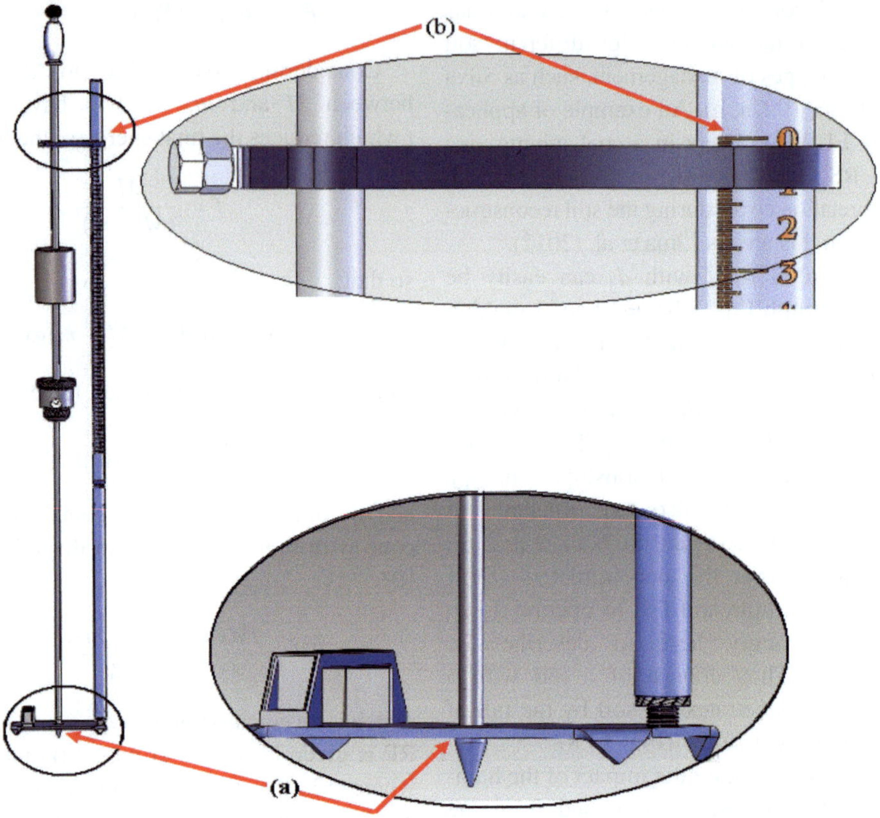

Fig. 14.9 Details of the measurement ruler of an impact penetrometer (Stolf et al. 2012)

resulting in:

$$\mathrm{RP} = c + d \cdot N$$

in which the constants c and d are characteristic of each penetrometer because they involve M, m, h, and A. An illustration of details of a penetrometer is given in Fig. 14.9.

The table below shows, for example, eight replicates of measurements with a penetrometer made in an area of sugarcane.

Soil layer (cm)	Mean depth (cm)	Resistance (MPa)								
		1	2	3	4	5	6	7	8	Mean
0–5	2.50	2.68	4.69	5.38	4.26	2.95	3.19	6.42	6.07	4.46
5–10	7.50	3.29	8.64	8.95	6.71	6.20	4.29	8.55	6.94	6.70
10–15	12.50	4.79	6.60	7.27	5.99	5.82	4.45	6.81	6.19	5.99
15–20	17.50	5.50	2.98	5.40	5.64	4.06	4.58	5.09	5.35	4.82
20–25	22.50	6.52	2.98	4.86	4.59	3.65	4.38	4.07	4.51	4.45
25–30	27.50	5.70	2.84	4.64	4.08	3.89	3.30	3.62	4.66	4.09
30–35	32.50	4.01	3.87	3.37	3.35	3.31	2.09	3.46	4.29	3.47
35–40	37.50	3.09	4.70	2.87	3.00	2.64	1.82	3.76	3.16	3.13
40–45	42.50	2.93	4.14	2.95	3.25	3.00	1.19	3.50	2.78	2.96
45–50	47.50	2.90	2.90	2.93	3.45	3.18	1.28	2.88	2.56	2.76
50–55	52.50	2.85	2.58	3.15	3.21	2.94	2.13	2.62	2.47	2.74
55–60	57.50	2.74	2.53	2.94	2.87	2.70	2.23		2.15	

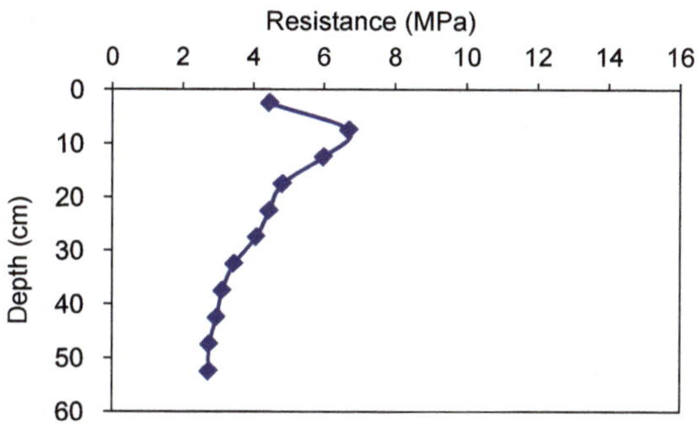

14.6 Exercises

14.1. Certain crop transpires in dynamic equilibrium at a rate of 5.5 mm day^{-1}. The total potentials were measured at several points and obtained:

(a) Soil mean: −0.2 atm
(b) On the root surface: −1.5 atm
(c) In the root xylem: −3 atm
(d) In the leaf xylem: −5 atm
(e) In the leaf chamber: −10 atm
(f) In the atmosphere: −220 atm

What are the resistances of the different parts of the system?

14.2. How do you understand the flows q_r, q_α, and q_z in Eq. (14.5)?

14.3. A soil has water content in the field capacity of 0.325 $m^3\ m^{-3}$ and at the permanent wilting point of 0.205 $m^3\ m^{-3}$. What is its available water capacity for the depths of 20, 40, and 60 cm?

14.7 Answers

14.1. The value of the resistances depend on the units used. We will use the International System MKS, as indicated below from Eq. (14.1):

(a) Flow $q = 5.5$ mm $day^{-1} = 6.37 \times 10^{-8}$ m s^{-1}
(b) $\Delta\Psi$(soil) = −0.2 − (−1.5) = 1.3 atm = 13.4 mH_2O
(c) $\Delta\Psi$(cortex) = −1.5 − (−3) = 1.5 atm = 15.5 mH_2O
(d) $\Delta\Psi$(xylem) = −3 − (−5) = 2.0 atm = 20.7 mH_2O
(e) $\Delta\Psi$ (leaf) = −5 − (−10) = 5.0 atm = 51.7 mH_2O
(f) $\Delta\Psi$ (atmosphere) = −10 − (−220) = 210 atm = 2169 mH_2O

and, according to Eq. (14.3):

$$6.37 \times 10^{-8} = \frac{13.4}{R_s} = \frac{15.5}{R_{co}} = \frac{20.7}{R_x} = \frac{51.7}{R_l} = \frac{2169}{R_a}$$

and as a result, we have $R_s = 2.10 \times 10^8$ s; $R_c = 2.43 \times 10^8$ s; $R_x = 3.25 \times 10^8$ s; $R_l = 8.12 \times 10^8$ s; and $R_a = 3.40 \times 10^{10}$ s.

14.2. If we take a root as a cylinder, at the center of which, the z axis passes:
The flow q_r indicates radial flow, for example, from the soil toward the root, or more rarely from the root to the soil. The flows q_α are along concentric circles around the root. The flows q_z are along the root, mainly upward to the leaf.

14.3. According to Eq. (14.9):
AWC = 0.325 − 0.205 = 0.120 $m^3\ m^{-3}$.
AWC (0 − 200 mm) = 0.120 × 200 = 24 mm
AWC (0 − 400 mm) = 48 mm
AWC (0 − 600 mm) = 72 mm

References

Angelocci LR (2002) Água na planta e trocas gasosas/energéticas com a atmosfera: introdução ao tratamento biofísico. Angelocci LR, Piracicaba

Brunini O, Reichardt K, Grohmann F (1975) Determinação da água disponível em Latosso-lo Roxo em condições de campo. Sociedade Brasileira de Ciência do Solo, Campinas, pp 81–87

Cowan IR (1965) Transport of water in soil-plant-atmosphere system. J Appl Ecol 2:221–229

De Jong van Lier Q (2017) Field capacity, a valid upper limit of crop available water? Agric Water Manag 193:214–220

De Jong van Lier Q, Dourado-Neto D, Metselaar K (2009) Modeling of transpiration reduction in van Genuchten–Mualem type soils. Water Resour Res 45:1–9

Denmead OT, Shaw RH (1962) Availability of soil water to plants as affected by soil moisture content and meteorological conditions. Agron J 54:385–390
Dourado Neto D, De Jong van Lier Q (1993) Estimativa do armazenamento de água no solo para realização de balanço hídrico. Rev Bras Ciênc Solo 17:9–15
Feddes RA, Kowalik PJ, Zaradny H (1978) Simulation of field water use and crop yield. Centre for agricultural publishing and documentation, Wageningen
Gardner WR (1960) Dynamic aspects of water availability to plants. Soil Sci 89:63–73
Gardner WR (1964) Relation of root distribution to water uptake and availability. Agron J 56:35–41
Lima CLR, Miola ECC, Timm LC, Pauletto EA, Silva AP (2012) Soil compressibility and least limiting water range of a constructed soil under cover crops after coal mining in Southern Brazil. Soil Tillage Res 124:190–195
Monteiro AB, Reisser Junior C, Romano LR, Timm LC, Toebe M (2018) Water potential in peach branches as a function of soil water storage and evaporative demand of the atmosphere. Rev Bras Frutic 40:e-403
Philip JR (1966) Plant water relationships: some physical aspects. Annals Rev Plant Phys 17:245–268
Reichardt K (1976) Noções gerais sobre solo. In: Malavolta E (ed.) Manual de química agrícola. Agronômica Ceres, Piracicaba, pp 121–157
Reichardt K (1979) A água: absorção e translocação. In: Ferri MG (ed) Fisiologia vegetal. EPU/EDUSP, São Paulo, pp 3–24
Reichardt K (1981) Uma discussão sobre o conceito de disponibilidade da água às plantas. Sociedade Brasileira de Ciência do Solo, Salvador, pp 256–284
Richards LA, Waldleigh CH (1952) Soil water and plant growth. In: Shaw BT (ed) Soil physical conditions and plant growth. American Society of Agronomy, Madison, WI, pp 73–251
Rijtema PE, Aboukhaled A (1975) Crop water use. In: Aboukhaled A, Arar A, Balba AM, Bishay BG, Kadry LT, Rijtema PE, Taher A (eds) Research on crop water use, salt affected soils and drainage in the Arab Republic of Egypt. FAO Regional Office for the Near East, Cairo, pp 5–61
Silva AP, Kay DB (1996) The sensitivity of shoot growth of corn to the least limiting water range of soils. Plant and Soil 184:323–329
Silva AP, Kay DB (1997a) Estimating the least limiting water range of soils from properties and management. Soil Sci Soc Am J 61:877–883
Silva AP, Kay DB (1997b) Effect of soil water content variation on the least limiting water range. Soil Sci Soc Am J 61:884–888
Silva AP, Bruand A, Tormena CA, Silva EM, Santos GG, Giarola NFB, Guimarães RML, Marchão RL, Klein VA (2014) Indicators of soil physical quality: from simplicity to complexity. In: Teixeira WG, Ceddia MB, Ottoni MV, Donnagema GK (eds) Application of soil physics in environmental analysis: measuring, modelling and data integration. Springer, New York, NY, pp 201–221
Stolf R, Fernandes J, Furlani Neto VL (1983) Penetrômetro de impacto modelo IAA/Planalsucar-Stolf: recomendação para seu uso. STAB 1:18–23
Stolf R, Murakami JH, Maniero MA, Silva LCF, Soares MR (2012) Incorporação de régua para medida de profundidade no projeto do penetrômetro de impacto Stolf. Rev Bras Ciênc Solo 36:1476–1482
Thornthwaite CW, Mather JR (1955) The water balance. Drexel Institute of Technology, Centerton, NJ
Veihmeyer FJ, Hendrickson AH (1927) Soil moisture conditions in relation to plant growth. Plant Physiol 2:71–78
Veihmeyer FJ, Hendrickson AH (1949) Methods of measuring field capacity and wilting percentages of soil. Soil Sci 68:75–94
Veihmeyer FJ, Hendrickson AH (1950) Soil moisture in relation to plant growth. Ann Rev Plant Phys 1:285–304
Veihmeyer FJ, Hendrickson AH (1955) Does transpiration decrease as the soil moisture decreases? Trans Am Geoph U 36:425–448

The Water Balance in Agricultural and Natural Systems 15

15.1 Introduction

The various processes involving water flow described in previous chapters, that is, infiltration, redistribution, evaporation, and absorption by plants, although studied independently, can and do occur simultaneously. To study the water cycle in a crop or in general in any ecosystem, it is necessary to consider the **water balance** (WB) that encompasses these processes. WB is nothing more than the sum of the amounts of water entering and leaving a **soil volume element** in a given time interval, resulting in the net amount of water remaining in it.

The water balance is, in fact, the very law of mass conservation and is closely linked to the energy balance, because the processes involved require energy. The energy balance, in turn, is also the very law of conservation of energy. From the agronomic point of view, the water balance is fundamental, since it defines the water conditions under which a crop has developed or is developing in each of its phenological stages.

15.2 The Balance

Imagine any crop, such as those outlined in Figs. 15.1 and 15.2. The first step for the WB is to choose the soil layer (water reservoir) in which we are interested in determining the water balance. In general terms, this layer should include the entire root absorption zone or at least most of it. Hence, the need to know the distribution of the root system, in its different stages of development. Let the depth of interest be $z = L$. The value of L for crops such as beans and soybeans can vary from 0.40 to 0.50 m, for maize and cotton from 0.50 to 0.80 or 1.00 m, and for sugarcane from 1.0 to 1.5 m and for perennial crops or forests can reach a few meters. The choice of L is quite difficult, but the most widely used criterion is that the 0–L layer should include 95% or more of the active root system.

The volume element, on which the water balance is made, is a prism of unit area (say 1 m^2) and height L (m), which contains the layer of interest mentioned above (Fig. 15.2) and stores the water needed for plant growth. Since the horizontal cross section is 1 m^2, all volumes (litters) of water entering or leaving the element volume are represented by a water height in mm, because 1 $L/m^2 = 1$ mm. In fact, the area of 1 m^2 is "symbolic"; it can be any one, that is, 10 m^2, 1.0 ha, 100 ha, or more. The reader should check on this and recognize that in any case, the result will be a height in mm (see in Chap. 3 the discussion on rainfall and soil water storage). This is because the WB involves water heights, which are independent of the area. In Fig. 15.2 we can

K. Reichardt, L. C. Timm, *Soil, Plant and Atmosphere*, https://doi.org/10.1007/978-3-030-19322-5_15

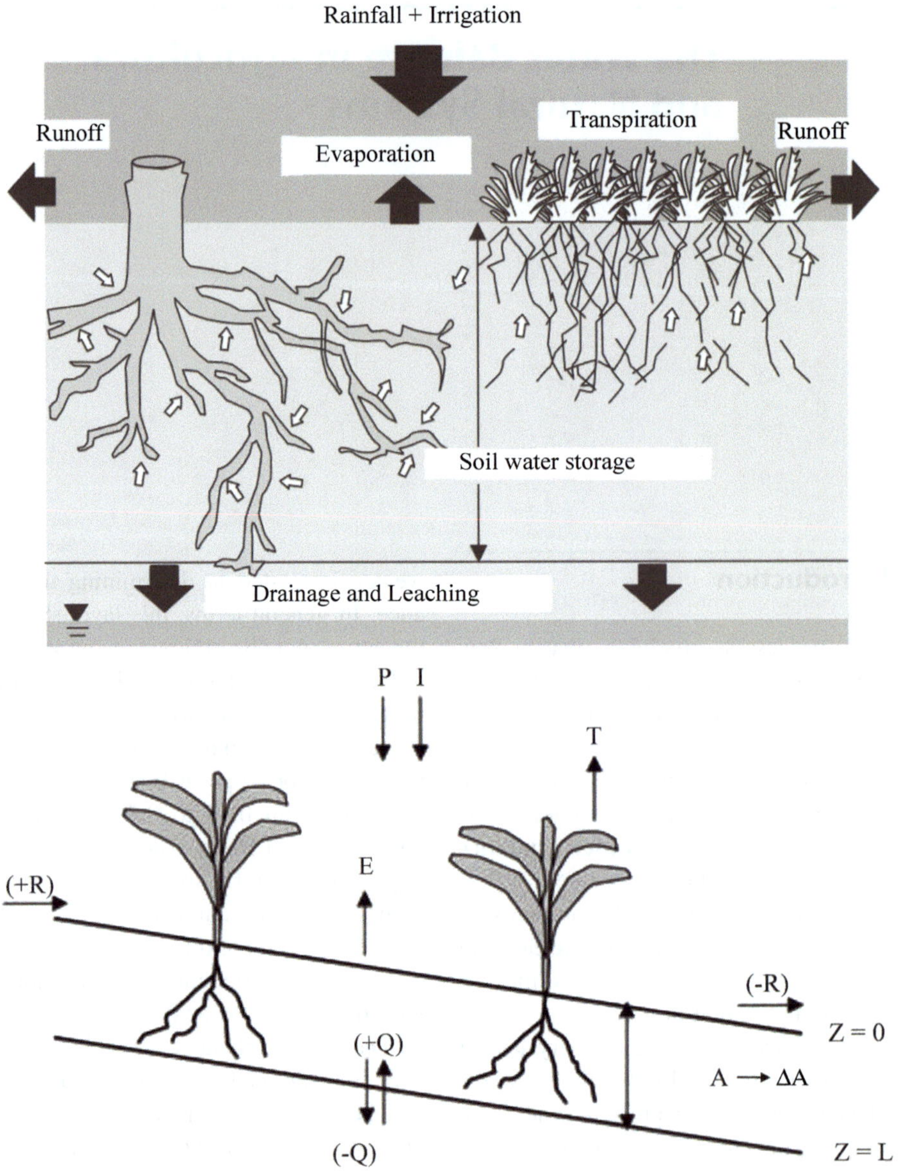

Fig. 15.1 Components of the water balance indicating the soil water storage (*A*, mm) in the elemental volume of thickness 0–*L* m

also see the coordinates *x*, *y*, and *z* and the different layers of soil with their water contents θ_1, θ_2, . . ., θ_n. Despite the three dimensions, the balance is most times calculated in one dimension, in the *z* direction. Only the surface runoff is considered as a horizontal loss of total precipitation (*P*) or irrigation (*I*), is in the *xy* plane, and will be taken into account subtracting it from *P* and *I*.

We will start with the **instantaneous water balance**, in which the inputs and outputs are measured in terms of flow densities (see Darcy equation, Chap. 6), i.e., volumes (L) of water by area (m^2), by time (day), resulting mm day^{-1}. These flow densities are actually intensities; they are vectors and are represented here in a simpler way by lowercase letters: **rainfall** *p*, **irrigation** *i*,

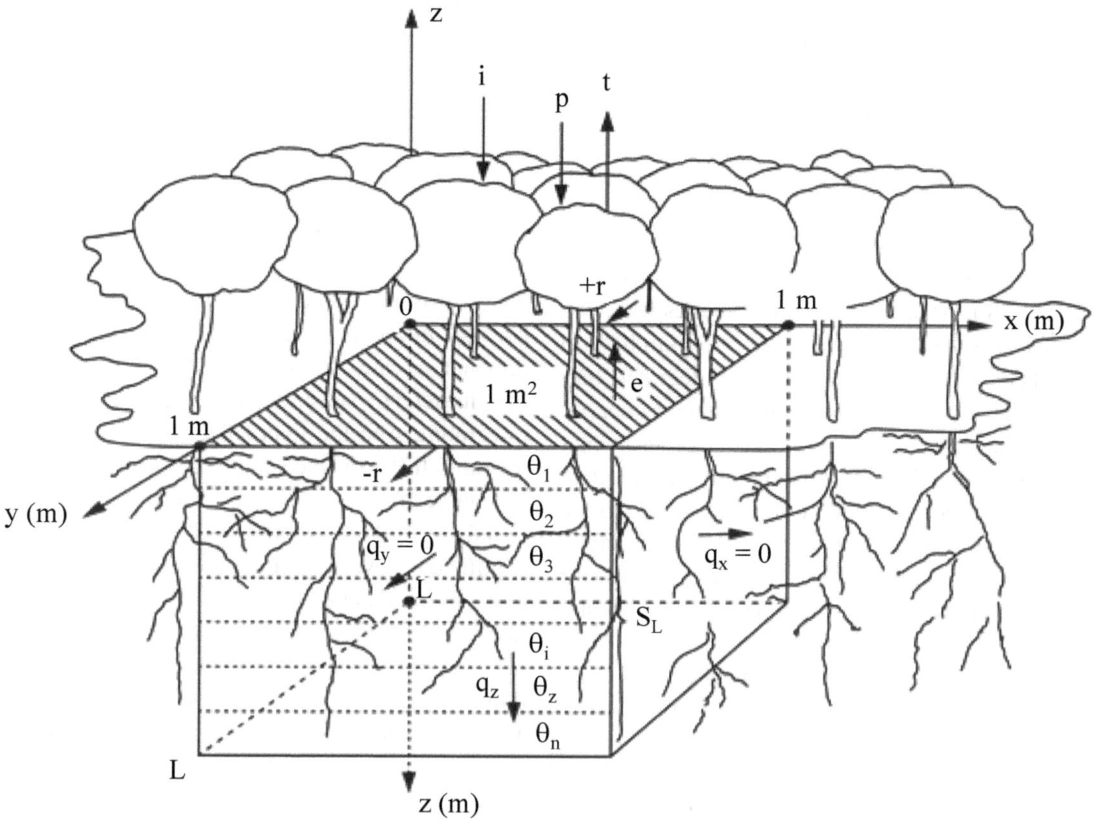

Fig. 15.2 Details of the elemental volume expressing the balance components as rates [i, p, t, $+r$, $-r$, q_z in mm day^{-1}]. The soil water content (θ) distribution is also shown in the elemental volume

surface runoff = runoff = d_s, **evapotranspiration** q_e, and **internal drainage** q_z, all in mm day^{-1}. The positive flows, i.e., which contribute to the increase in the amount of water in the volume element, are, in particular, p and i. As mentioned above, part of this water can, however, be lost due to the surface runoff d_s and is discounted in the calculation of the components of the balance sheet. Surface runoff from upstream areas may constitute water additions. The evapotranspiration flow q_e contributes to the decrease of the amount of available water in the volume element, and, at the depth $z = L$ (the lower boundary of the considered layer), we can have flows of water q_z, in both directions upstream or downstream depending on the water potential gradient at depth L, that is, $(\partial H/\partial z)_{z=L}$. This gradient will determine the direction of the flow q_z and, along with the soil hydraulic conductivity prevailing at L, will determinate its intensity q_z.

The instantaneous water balance can, therefore, be written in the form:

$$p + i \pm d_s - q_e \pm q_z = \int_0^L \left(\frac{\partial \theta}{\partial t}\right) dz \qquad (15.1)$$

which is to say that the addition of the flow densities entering and leaving the volume element is equal to the changes in soil water content θ in the volume element, represented by the integral of the second term of Eq. (15.1). To understand this integral, refer to Eq. (3.19), on water storage in the soil, and to Eq. (3.24), on its variations in time. But we will come back to this integral at the end of this section.

Equation (15.1) is the instantaneous balance with units in mm day^{-1}. The reader might wonder: if it is instantaneous, how can the unit be millimeters evaporated in a day? This is just a matter of unity, in this case the day. If we divide the values by 86,400, which is the number of seconds of 1 day, we will have the result in mm s^{-1}, which would not make much practical sense, because 1 s can also not be considered an instant. It is as in the case of kinematics when we speak at an instantaneous speed of 50 km h^{-1}. In practice, what interests us is the integrated balance sheet over a certain period or time interval. The time interval Δt is our choice. Because water dynamics are relatively slow, periods shorter than a day are not feasible. For short-cycle crops, Δt may be of the order of 3, 7, 10, or 15 days . For long-cycle or perennial crops 10, 15, or 30 days. For ecological purposes, the semester or even the year is often used.

To obtain the integrated balance, it is only necessary to integrate Eq. (15.1) as a function of time, in the chosen interval between t_i and t_j:

$$\int_{t_i}^{t_j} (p + i \pm d_s - q_e \pm q_z)\mathrm{d}t = \int_{t_i}^{t_j}\int_{0}^{L} \left(\frac{\partial \theta}{\partial t}\right) \mathrm{d}z\mathrm{d}t \qquad (15.2)$$

This equation simply tells us that the algebraic sum of the flux densities over a period $\Delta t = t_j - t_i$ is equal to the changes in the quantity of water, within the same interval in the element of the volume of soil of thickness 0–L. Let us study each element from Eqs. (15.1) and (15.2), separately. By the rules of calculus, the first member of Eq. (15.2) can be unfolded into an algebraic sum of integrals. The first one will be:

$$\int_{t_i}^{t_j} p\mathrm{d}t = P \qquad (15.3)$$

where p is the precipitation flux density, called **rainfall intensity** by the meteorologists, and its dimensions are L m^{-2} day^{-1} = mm day^{-1}. **Pluviometers** measure the integral P directly, integrating p in the interval $t_j - t_i$. Graphically, the rainfall P is the area under the curve p versus t, and the result is mm. Figure 15.3 shows this for a rain from 12:00 to 20:00 h. It is seen that p varied significantly over this time interval, with a maximum value of about 10 mm h^{-1} at 16 h. With the integral P, this detailed information is lost, leaving only the possibility to calculate the average rain intensity $i = P/(t_j - t_i) = 40/8 = 5$ mm h^{-1}. The rainfall intensity p can be obtained from rain gauges, which make a graph (pluviograph) similar to that in Fig. 15.3.

The second integral obtained from the unfolding of Eq. (15.2) is:

$$\int_{t_i}^{t_j} i\mathrm{d}t = I \qquad (15.4)$$

The same considerations made for p and P can be made for the **irrigation rate** i and the accumulated irrigation I in the case of sprinkler

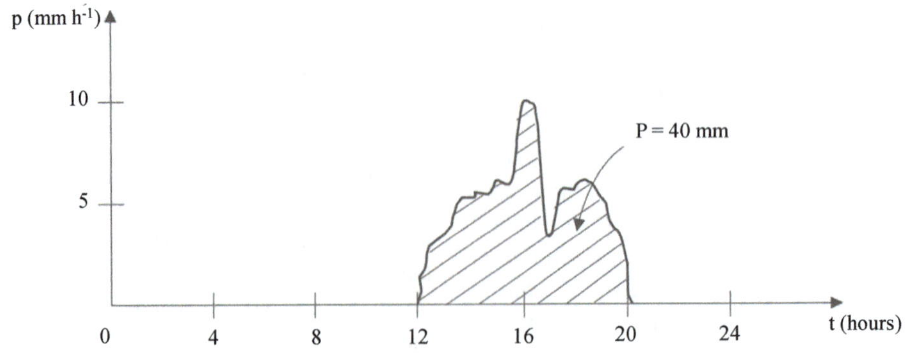

Fig. 15.3 Example of a pluviograph to elucidate the difference between p and P

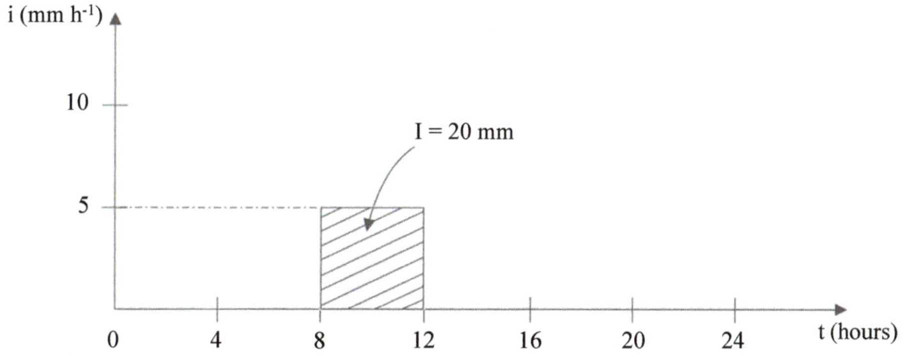

Fig. 15.4 Graph of the irrigation rate i as a function of time, elucidating the difference between i and I

irrigation, which is, in fact, an artificial rainfall. In this case, the irrigation intensity i is constant with time, during the irrigation period. In this way, an irrigation of intensity i (mm h^{-1}), applied during a time t (h), results in $I = i \cdot t$ (mm). Or graphically, as it can be seen in Fig. 15.4.

For drip irrigation and other types of irrigation, the determination of i is more complicated, but the concepts of i and I are the same. Each irrigation method (or system) presents different characteristics in terms of water consumption, water distribution in the soil profile, application efficiency, required labor, and possibilities of rotating irrigated areas in fields, among others. All of these characteristics influence the way irrigation should be managed.

The correct determination of the quantity of water required for irrigation is one of the main parameters for the planning, design, and management of an irrigation system, to be adequately done, as well as for the assessment of the needs for water collection, storage, and conduction and the evaluation of available sources of supply. When the amount of water to be applied by irrigation is overestimated, irrigation systems are oversized, increasing the cost of irrigation per unit area. It may also cause damage to the crop, leaching of nutrients, and elevation of the water table in the area (drainage problem), causing a reduction in productivity. On the other hand, when the quantity is underestimated, there will be an inability of the system to irrigate the entire area. In this way, irrigation management consists of defining the methods that will be used to answer the two main questions of irrigation: (1) When to irrigate? and (2) What is the amount of water to be applied? There are different types of irrigation management, one of which is based on the soil water balance.

The third integral is the **surface runoff**:

$$\int_{t_i}^{t_j} d_s \mathrm{d}t = \pm \mathrm{DS} \tag{15.5}$$

difficult to determine and depends mainly on the rainfall or irrigation intensities, on soil properties, and on the slope of the surface. It is usually measured by collecting the water flowing from a given area, surrounded by a "wall" to restrict and collect the flow. It is positive when it is a contribution coming from areas situated above the WB plot and negative when it flows out of the WB plot. Contributions coming from above and losses downhill should be avoided, and, in general, this is done by means of terracing. The distance between terraces varies mainly with the terrain slope and the soil type, and a common average value for the distance of terraces is between 20 and 30 m.

Therefore, the surface runoff is studied in **standard runoff ramps** measuring 22.4 m in length and 2 m in width. They are surrounded by wooden or metal foil ditches (Fig. 15.5), and the runoff water is collected in tanks that make it possible to measure the volume V (L) of water leaving the parcel. The value of DS (mm) is V/A when the ramp area is given in m^2. If, for example, a runoff volume of 65.8 L was collected from a 44.8 m^2 ramp, DS = 65.8/44.8 = 1.47 mm.

Fig. 15.5 Runoff plots (2×22.5 m^2) of different soil covers on a 12% slope field in Piracicaba, São Paulo (SP), Brazil. At the lower end, runoff water is collected in a type of triangular funnel which directs by gravity the water to the collecting tanks. Neutron probe access tubes and tensiometers are installed inside the plots to monitor soil water content and potential

These ramps are also used to estimate **soil erosion losses**. If, after each event with runoff, the solids that have been flushed into the collection tank are weighed, erosion can be calculated. Assuming that 2.24 kg of solid material was collected over a year of observations on a 44.8 m^2 ramp, the soil erosion value is equivalent to 500 kg ha^{-1} year^{-1}.

Since soil erosion is directly linked to runoff, we will devote in this chapter some of the main aspects of this process. Accelerated soil erosion is a global problem, and, despite the difficulty of accurately calculating soil losses, it is known that the magnitude of these losses has caused serious economic and environmental consequences for agriculture, because the transported material comes from the very surface layers of the soil (Lal 1988, among others). The determination of soil losses by erosion by direct methods, e.g., through ramps as seen above, is time-consuming and expensive, and this is the main cause of the growing interest in erosion prediction models. These models allow to identify areas of greater erosion risk and help in choosing more appropriate management practices (Foster et al. 1985) for the control of erosion. Renard and Mausbach (1990) present a brief history of the evolution of these models. The most widely used model in predicting soil losses is the **universal soil loss equation (USLE)**, used in several regions of the globe and for different purposes (Albaladejo Montoro and Stocking 1989; Toy and Osterkamp 1995). This model was developed in the United States in 1954 at the National Runoff and Soil Loss Data Center (Agricultural Research Service, University of Purdue). Wischmeier and Smith (1978) reviewed the equation that evolved into the model currently employed. USLE calculates the average annual loss of soil per unit area and per unit of time A (Mg ha^{-1} year^{-1}) by the following model:

$$A = R \cdot K \cdot L \cdot S \cdot C \cdot P$$

where: R = rainfall erosivity factor (MJ mm ha^{-1} h^{-1}); K soil erodibility factor (Mg ha h MJ^{-1} mm^{-1} ha^{-1}); L = slope length factor (dimensionless); S = slope degree factor (dimensionless); C = use and management factor (dimensionless); and P = conservationist (dimensionless) practice factor.

The factors R, K, L, and S are dependent on natural conditions, and factors C and P are related to the forms of land occupation and use (anthropic factors).

Rainfall erosivity (R) is a numerical index that represents the potential of rainfall and flooding to cause erosion in an unprotected area (Bertoni and Lombardi Neto 1990). With the other USLE factors constant, soil losses from rainfall are directly proportional to the product of the rainfall kinetic energy at the maximum intensity in 30 min. This product is called the **erosion index** (IE). This method, proposed by Wischmeier and Smith in 1958, is slow and cumbersome and requires information contained in daily rainfalls, which are often scarce or nonexistent (Vieira and Lombardi Neto 1995). Lombardi Neto and Moldenhauer (1992) used long series of data for Campinas (SP), Brazil, and found a high correlation between the monthly mean value of IE and the average monthly rainfall coefficient (p/P), where p = mean monthly precipitation and P = mean annual precipitation, both in mm. Thus, it became possible to calculate the R factor without the use of rain gauge.

The **soil erodibility factor** (K) reflects the differential loss that the soil presents when the other factors that influence erosion remain constant, being influenced, in particular, by the characteristics that affect the infiltration capacity and soil permeability, its ability to resist detachment, and transport of particles by rainfall and runoff (Lombardi Neto and Bertoni 1975). The soil erodibility index is an experimentally determined quantitative value. It consists of the rate of soil loss per unit erosion index measured in a unit plot (Lane et al. 1992). Within this plot, the factors LS, C, and P are unitary (=1), and factor K is given by the slope of the regression line between the erosion index and soil loss.

One established indirect method of erodibility determination is the model proposed by Wischmeier et al. (1971), based on the physical parameters—texture, structure, and permeability classes—and the percentage of organic matter in the soil, combining them graphically in a nomogram. According to these authors, erodibility tends to increase with increasing silt content. The advantages of using these models are the rapidity in the determination of soil erodibility, in comparison with conventional direct methods, costly and requiring several years of replication, besides the possibility of their estimation through parameters obtained by laboratory analyses of easy execution. The ramp slope length (LS) factor participates in the USLE as a combined component, presenting a single value (Bertoni and Lombardi Neto 1990). The LS factor is the expected relationship of soil loss per unit area at any slope relative to the corresponding soil losses of a standard unit plot (Wischmeier and Smith 1978). The USLE is designed to be applied to uniform ramps, not considering the deposition of sediments along the slopes. However, the topography of the terrestrial surface is extremely heterogeneous and discontinuous. The use of an average gradient of ramp length, considering it uniform, can underestimate soil losses on convex slopes and overestimate concave slopes. This is one of the main limitations of USLE's application in watersheds. Some methods that have been used to obtain the LS factor are the Digital Terrain Elevation Model (DTEM), proposed by Hickley et al. (1994), and the model proposed by Bertoni and Lombardi Neto (1990). The **soil use and management factor** (C) is obtained from the soil loss rate during a determined stage of crop development compared to the rate of soil loss in a standard plot during the same period. To determine the factor C, defined stages of crop development and their influence on soil erosion are considered. The crop management practices are variable, implying in differences of the calculations to obtain factor C. These calculations are valid for specific conditions in each region. These considerations represent other limitations of USLE use. The **conservationist practices factor** (P) represents the relationship between the expected intensity of soil losses with a given conservation practice and the losses obtained when the crop is planted in the direction of slope decline (Ranieri 1996). The P factor indicates the effect of conservation practices, such as level planting, terracing, and planting in soil erosion bands (Lane et al. 1992).

The main advantage of USLE is that it is a model composed of a reduced number of components, when compared to more complex models, and because it is well known and well studied. However, because it is an empirical model and is designed based on single parcels of soil erosion evaluation, it presents several limitations. Among them are:

a) The need to work with relatively homogeneous areas with respect to soil type, use, and slope.
b) The equation leaves several parameters and their effects implicit.
c) Calculations for factor *C* are valid only for local conditions.
d) Does not consider areas of deposition or linear erosion.

These limitations hamper the application of USLE at a river basin scale and imply the need to develop a new technique to estimate erosion soil losses (Laflen et al. 1991). Lane et al. (1992) refer to another model of soil loss prediction, the **RUSLE** (**Revised Universal Soil Loss Equation**) model, which was successfully used by Steinmetz et al. (2018) to estimate the magnitude and spatial distribution of soil erosion rates in two data-scarce watersheds in southern Brazil. Knisel (1980) presented a model, the **CREAMS** (**Chemical Runoff and Erosion from Agricultural Management Systems**), containing a sophisticated erosion component based partly on USLE and partly on hydraulic flow and processes of disaggregation, transport, and deposition of sediments. This model allowed a better prediction of soil losses, but it is more complex, which made it difficult to be used in conservation projects.

In 1985, the US Department of Agriculture (USDA), in cooperation with several universities, initiated a national project called the **Water Erosion Prediction Project** (**WEPP**) to develop new technologies for the prediction of soil losses by water erosion. At the same time, a review of the USLE was initiated. Other hydrological models have been recently developed and used to predict variables associated with water, sediment, nutrient transport, etc. Examples are **Limburg Soil Erosion Model** (**LISEM**) (De Roo et al. 1996), **Soil and Water Assessment Tool** (**SWAT**) (Arnold et al. 1998; Gassman et al. 2007), and **Agricultural Non-Point Source Pollution Model** (**AGNPS**) (Young et al. 1986).

Most developing countries, such as Brazil, have data shortages at the basin scale, except when basins are monitored for research purposes or by power generation companies (Beskow et al. 2009). Therefore, in the case of lack of data, it may be impracticable to apply complex hydrologic models that are fed with a large amount of data, such as those previously mentioned (Beskow et al. 2011b). In order to overcome this drawback, it may be advisable to choose hydrological models based on simpler formulations, which use a reduced database (Beskow et al. 2011a). The **Lavras Simulation of Hydrology** (**LASH**) model, developed by Beskow (2009), employs a simplified formulation and was created to simulate total flow in watershed control sections of places where there is a lack of data regarding climate, soil, and land use.

The LASH model (Beskow 2009) was developed at the Federal University of Lavras, Brazil, in a partnership with the National Soil Erosion Research Laboratory (NSERL/USDA)—Purdue University, USA. It is similar to the model developed by Mello et al. (2008). The main difference is that LASH uses a distributed formulation, in addition to having a computational algorithm that is efficient for performing automatic calibration. This new model takes into account the spatial and temporal variability of all input variables used in the hydrological components, in such a way that it divides the basin into cells of uniform size. It is a continuous simulation model classified as deterministic and semi-conceptual, which simulates the following components in daily increments of time: evapotranspiration, foliar interception, capillary rise, soil water availability, direct surface flow, subsurface runoff, and base flow. LASH has been implemented in the Delphi programming language (Windows environment) and provides a graphical user interface. Such an interface allows the user to import maps from different geographic information systems (GIS), thus facilitating the use of the model. In addition, LASH has a built-in autocalibration routine, which is based on the shuffled complex evolution method (SCE-UA, Duan et al. 1992). With this option, the users are able to calibrate as many parameters as necessary. The model is divided into three basic modules: (a) the first module is designed to compute the direct surface flow, subsurface flow, base flow, and capillary rise, which are drained from the soil layer considered in the water balance; (b) the second module generates the flow within each cell to the drainage network taking into account the re-delay effect through the concept of a linear reservoir; and (c) in the third module, LASH employs the Muskingum-Cunge model in order to propagate the flows through the network of channels.

The water balance is calculated at each increment of time for each cell within the river basin. The simulation structure is based on the subdivision of the watersheds into cells of different sizes. This subdivision aims to reduce the problems associated with spatial variability and is based on the digital elevation model of each basin, characterizing the behavior of the drainage network. It is important to mention that this physiographic characterization is of fundamental importance for the physical structuring of the model, to evaluate and obtain its input parameters, whether variable (as initial estimate) or fixed, as canopy characterization, effective depth of the root system, stomatal and aerodynamic resistance, and others.

Although the spatial variability of the input variables of the model should preferably be taken into account, users may also use focused values to represent some variables, depending on the amount of data available.

All necessary maps are derived from maps that represent the **digital elevation model (DEM)**, land use, soil, and drainage network. In addition to the maps, the LASH model also requires two other files in the table format. The first one contains information associated with the climatic data and also the variation in flow observed over time. Another table format file is used to inform the variation of some variables related to soil use in time, such as **leaf area index** ($m^2\ m^{-2}$), height (m), albedo (dimensionless), resistance ($S\ m^{-1}$), rooting depth (mm), and crop coefficient (dimensionless). It should be noted that some of these land use data can be measured in the field or even found in the literature. Further details on the LASH model can be seen in Beskow (2009). Applications of the model in the Brazilian environment can be found in Mello et al. (2008), Beskow et al. (2011a, b, 2016), and Caldeira et al. (2019), in which the model was successfully applied to watersheds of different sizes in the state of Minas Gerais and Rio Grande do Sul, Brazil.

A more sophisticated technique for estimating soil loss rates is that of the radioisotope ^{137}Cs. Since the mid-1970s, a model based on the analysis of redistribution of the **fallout** of the ^{137}Cs has been used in the evaluation of soil losses by erosion. The ^{137}Cs present in the environment is originated from the nuclear tests, transported by the atmosphere, distributed globally, and slowly deposited, a posteriori, on the soil surface by rainfall. This process is called fallout. The spatial variability of fallout is evident on a global scale, with smaller depositions in the southern hemisphere compared to the northern hemisphere, where most of the nuclear tests (United States and former USSR) were located. On a regional scale, a few available data show a correlation between the magnitude of fallout and the annual total rainfall precipitation. It is assumed that on a local scale, the deposition occurred uniformly. Basic studies demonstrate a rapid and strong adsorption of the ^{137}Cs by soil clay minerals, indicating their ready fixation in the upper horizons soon after their deposition and showing a very low rate of vertical migration in undisturbed soils after fallout. This isotope has a half-life of about 30 years, so that any strength of radioactivity measured today will be reduced to half in 30 years. Therefore, this radioisotope can be detected up to date. In the absence of significant lateral or vertical translocation, soil ^{137}Cs content allows to distinguish eroded, non-eroded, and soil deposited sites, based not only on the shape of the ^{137}Cs distribution in the soil profile but also on the total amount of ^{137}Cs at points of interest along the countryside.

The differences in the storage of ^{137}Cs in soil profiles down to the depth reached by agricultural implements, in relation to that observed in reference profiles, not eroded or very little eroded after fallout, allow to evaluate the rates of soil loss under natural conditions of rainfall erosivity at the sampling site. The activities of ^{137}Cs can be converted to erosion rates using the proportional model, according to the methodology described in Walling and Quine (1993). In Brazil, Bacchi et al. (2000, 2003), Correchel et al. (2006), and Pires et al. (2009) used this technique to evaluate erosion in a watershed of Piracicaba, SP.

Coming back to the main subject of this chapter, the fourth integral of the WB is:

$$\int_{t_i}^{t_j} q_e \mathrm{d}t = -\mathrm{ET} \tag{15.6}$$

The **evapotranspiration rate** q_e is the water flow from the soil surface ($z = 0$) added to the plant transpiration. These concepts are discussed in detail in Chaps. 7, 12, and 13. In most cases, mean values of q_e are taken, which, considered as constants, can be taken out from the integration symbol, resulting in:

$$q_e \int_{t_i}^{t_j} \mathrm{d}t = -q_e(t_j - t_i) = -\mathrm{ET}$$

assuming that evapotranspiration does not cease in the interval ($t_j - t_i$), which is reasonable for mean values of q_e.

The fifth integral is:

$$\int_{t_i}^{t_j} q_z \mathrm{d}t = \pm Q_z \tag{15.7}$$

which is the most difficult part to estimate in a water balance. It is the flow of water in the soil at the lower limit of the volume element under consideration. The flow q_z is given by the Darcy-Buckingham equation, applied to the vertical flow (see Chap. 7):

$$q_z = -K(\theta)\frac{\partial H}{\partial z}\bigg|_L = -K(\theta)\frac{\partial h}{\partial z}\bigg|_L - K(\theta) \tag{15.8}$$

Depending on the sign of the gradient $\partial H/\partial z$, at $z = L$, the flow may be upward (**capillary rise**) or downward (**deep drainage**), hence the + and – signs in Eq. (15.7). Thus, for the estimation of Q_z, it is necessary to know the hydraulic characteristics of the soil at $z = L$, that is, its water retention curve $h = h(\theta)$ and its hydraulic conductivity $K = K(\theta)$ or $K(h)$. When L is large ($L > 1.0$ m), the gradient $(\partial h/\partial z)_{z=L}$ may not vary widely in both direction and magnitude. Our experience in Piracicaba, SP, Brazil (Reichardt et al. 1990), for a very deep water table, is that this gradient, at the depth of 1.50 m, oscillates around one and always indicates downward flow. The few times the gradient indicated upward flow, the soil water content was very low, so that $K(\theta)$ became negligible. Of course, in situations with the presence of groundwater, the upward flow can be considerable.

The best way to obtain grad H for the application of the Darcy equation is to measure the $H(z)$ or $h(z)$ profiles with tensiometers or TDR probes (or even the FDR probe), obtaining graphs of the type of Fig. 6.19 of Chap. 6. The gradient $\partial H/\partial z$ indicates the direction of the flow of water. The grad H is the tangent to the curve $H(z)$ at L, which can be obtained graphically or by finite differences. In the upper layers, $0 < z < 0.60$ m, there are roots, and, although the gradient H indicates ascending flows, soil drying occurs mainly by root extraction.

The simplest case would be the gradient measurement H only at $z = L$, considering negligible the presence of roots at this depth. In this case, tensiometers or TDR probes are installed at two depths: $L - \Delta z$ and $L + \Delta z$, where $2\Delta z$ is a much smaller distance that L, but large enough to detect the difference ΔH. In this case, grad $H = \Delta H/2\Delta z$. Let's see an example: be the case of a water balance in which $L = 1.00$ m and $\Delta z = 0.10$ m. For this case, tensiometers were installed at $z = 0.90$ and 1.10 m, as indicated in Fig. 15.6.

In A:

$h(A) = -12.6 \times 22.8 + 30 + 90 = -167.3$ cmH$_2$O

$H(A) = -167.3 - 90 = -257.3$ cmH$_2$O

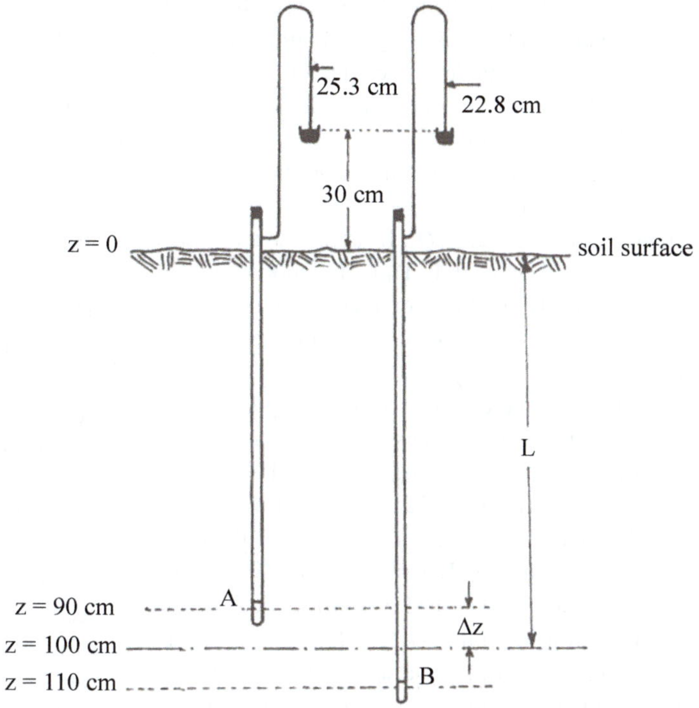

Fig. 15.6 Measurement of the gradient grad $H = \Delta H/2\Delta z$ in a soil with a very deep water table

In *B*:

$h(B) = -12.6 \times 25.3 + 30 + 110 = -178.8$ cmH_2O

$H(B) = -178.8 - 110 = -288.8$ cmH_2O

As $H(A) > H(B)$, the water moves from *A* to *B*, i.e., represents a drainage. The gradient was positive, indicating that the potential field of *H* grows from *B* to *A*. Since the Darcy equation has a negative sign, the flow q_L will be negative, i.e., a loss to the balance (drainage).

For this soil, the equation $K(h) = 171.$ exp (0.036 h) mm day^{-1} was determined at the depth $L = 1.0$ m. Since we have *h* at 0.90 m and 1.10 m and not at 1.00 m, we will use the mean: $h = -173.05$ cmH_2O. Thus, $K = 171.$ exp [0.036 (−173.05)] = 0.36368 mm day^{-1} and the drainage flow q_L will be:

$$q_L = -K(h)\frac{\partial H}{\partial z} = -0.3368 \times 1.575$$
$$= -0.5305 \text{ mm day}^{-1}$$

The ideal is to take daily readings of tensiometers, and, if the Δt of the balance is 7 days, we will have seven values of q_L for the considered interval. Thus, Eq. (15.7) becomes:

$$\int_1^7 q_i \mathrm{d}t \cong \sum_{i=1}^{7} q_i \Delta t' = q_1\Delta t' + q_2\Delta t' + \cdots q_7\Delta t'$$
$$= [q_1 + q_2 + \cdots + q_7]\Delta t'$$

where $\Delta t'$ is the interval of the numerical (trapezoidal) integration, taken equal to 1 day. By dividing and multiplying by $n = 7$, we have:

$$Q_L = \bar{q} \cdot 7\Delta t' = \bar{q} \cdot \Delta t$$

If, in our example, we have $\bar{q} = 0.5815$ mm day^{-1}, it follows that $Q_L = 0.5815 \times 7 = 4.0705$ mm.

The limitations of this calculation of Q_z are enormous, especially for large time intervals $\Delta t = t_j - t_i$. In water balances, often, $\Delta t = 10$, 15, or even 30 days. For these time intervals, one generally does not have complete information on how θ varied with *z*, and it is not known whether gradient and conductivity estimates are acceptable. For short intervals of time, this calculation of Q_z is already much better. In any case, the exact determination of Q_z is very difficult, and this problem is still under investigation. In many water balances, the term Q_z is neglected without plausible justification.

If there is no evapotranspiration or water additions in the period $t_j - t_i$, then the storage change ΔA_L is equal to Q_z. If there is evapotranspiration, $Q_z = A_L - \text{ET}$. Thus, by choosing propitious periods in which some components of the balance are null, we can simplify the determination of Q_z.

More accurate determination attempts at deep drainage Q_z were successfully performed by LaRue et al. (1968), Renger et al. (1970), and Daian and Vachaud (1972). Pereira et al. (1974) and Reichardt et al. (1974) present Q_z data for Brazilian conditions. Reichardt et al. (1990) and, more recently, Silva et al. (2006) discussed the problem of determining Q_z in relation to soil spatial variability. Still, Lima and Reichardt (1977) presented an example of a forest water regime.

Let us now look at the second member of Eq. (15.2). It is a double integral, equal to the time **change of water storage** in the 0–L soil layer (see definition in Chap. 3), which can be more easily understood in the form:

$$\int_0^L \left[\int_{t_i}^{t_j} \left(\frac{\partial \theta}{\partial t} \right) \mathrm{d}t \right] \mathrm{d}z = \Delta A_L \qquad (15.9)$$

The integral between brackets tells us that we must sum all the changes of θ with t in the period $t_j - t_i$. Since these variations of θ with t depend on the depth z, we must integrate them along z, from 0 to L. The final result is the variation of θ in the interval $t_j - t_i$ in the layer 0–L. The value of this integral is equal to the area between the soil water content profiles, measured in t_i and t_j, down to the depth L. This subject has already been approached when defining storage changes by Eq. (3.22).

ΔA_L must have the same dimensions as P, I, ET, and Q_z, i.e., mm of water. If θ is given in $\text{m}^3\,\text{m}^{-3}$ and z in m, any area of the graph θ versus z will have dimensions of meter, which multiplied by 1000 becomes millimeter.

We can now rewrite Eq. (15.1) in the most simplified way, that is, integrated in time Δt, as it commonly is presented by many authors:

$$P + I + \text{DS} + \text{ET} + Q_z + \Delta A_L = 0 \qquad (15.10)$$

in which the sign of each component depends on whether it is a gain or a loss.

The term ΔA_L is the variation of storage, and, logically, it can be + or −, depending on the magnitude of all other components, since the algebraic sum of all must remain null (law of mass conservation).

The following five cases exemplify Eq. (15.10):

1. A soil profile stores 280 mm of water and receives 10 mm of precipitation and 30 mm of irrigation. It loses 40 mm by evapotranspiration. Neglecting the surface runoff and groundwater flow below the root zone, what is the value of your new storage?
2. A soybean crop loses 35 mm by evapotranspiration in a period without rainfall and irrigation. It also loses 8 mm by deep drainage. What is the change in storage?
3. During a rainy season, a plot receives 56 mm of precipitation, of which 14 mm is lost by surface runoff. The value of deep drainage is 5 mm. Neglecting evapotranspiration, what is the change in storage?
4. Calculate the daily evapotranspiration of a bean crop in a period of 10 days, which received 15 mm of precipitation and two irrigations of 10 mm each. At the same time, the deep drainage was of 2 mm, and the change in storage was of −5 mm.
5. What is the amount of water added to a crop by irrigation, knowing that in a dry period its evapotranspiration was 42 mm and the storage variation was −12 mm? Soil water content was in the value of field capacity, and there was no runoff during irrigation.

In practice, the terms that interest us most are ET and ΔA_L. ET has to be known to calculate how much the soil and plant have lost by evapotranspiration in the interval $t_j - t_i$, in order to restore the water lost, and ΔA_L to know the availability of water in the soil for the plants at time t_j. For example, neglecting Q_z, measuring P, and I and estimating ET by theoretical-empirical formulas, we can obtain ΔA_L without measuring soil water changes. The agrometeorologists proceed this way. In this type of water balance, the definition of **available water capacity** (AWC), usually measured in mm, is calculated by:

$$\mathrm{AWC} = (\theta_{\mathrm{FC}} - \theta_{\mathrm{PWP}}) \cdot L \tag{15.11}$$

where L is the considered depth, in mm. It must be clear that in this case, the soil profile is taken as homogeneous from 0 to z. Thus, if for a given soil $\theta_{\mathrm{FC}} = 0.32\ \mathrm{m^3\ m^{-3}}$ and $\theta_{\mathrm{PWP}} = 0.19\ \mathrm{m^3\ m^{-3}}$, then the AWC for a 1.2 m layer is 156 mm. **Available water** (AW) is the quantity of water, in mm, that the soil holds at any given time. In the above example, the soil could have, for example, 95 mm of AW, which corresponds to 61% of the AWC. In many water balances carried out for locations of soils with low AWC, like Brazilian latosols and podsols, and in the absence of available soil data, a layer $L = 1$, m = 1000 mm, and $(\theta_{\mathrm{FC}} - \theta_{\mathrm{PWP}}) = 0.1\ \mathrm{m^3\ m^{-3}}$ are considered, resulting in a AWC of 100 mm.

15.3 Thornthwaite and Mather Water Balance

Thornthwaite and Mather (1955) presented a methodology for the computation of climatological water balances (WB), which until today is still the basis of several balance sheets and which, therefore, will be seen here in detail. It uses meteorological data that are found in most regions, air temperature and precipitation, in addition to soil water characteristics, AWC (Eq. 15.11). Its balance obviously follows Eq. (15.10), neglects I (or includes I in the precipitation), and joins DS and Q_z in a so-called excess component (EXC). It introduces the concept of water deficit (DEF) that appears when the soil restricts the evapotranspiration, being DEF = $\mathrm{ET_c} - \mathrm{ET_a}$. Below (Table 15.1) is an illustrative example, for the city of Franca, SP, Brazil, latitude 21° 10′ S. It is a monthly WB, elaborated with the normal climatological data of the place.

In the second column of Table 15.1, after the presentation of the months, we find the monthly average air temperature, with which $\mathrm{ET_0}$ is calculated by the Thornthwaite method (1948), as shown in Chap. 13, Eq. (13.24), here presented in column 3. In the fourth column, we find the precipitation P, which characterizes a type of

Table 15.1 An example of the Thornthwaite and Mather water balance for Franca, São Paulo state, Brazil

	T	$\mathrm{ET_c}$	P	$P - \mathrm{ET_c}$	ACCUM	A_L	ΔA_L	$\mathrm{ET_a}$	DEF	EXC
Month	°C	mm	mm	mm	NEG	mm	mm	mm	mm	mm
Jan	23.4	117	275	+158	0	125	0	117	0	158
Feb	23.4	102	218	+116	0	125	0	102	0	116
Mar	22.9	104	180	+76	0	125	0	104	0	76
Apr	21.2	79	60	−19	−19	107	−18	78	1	0
May	19.3	60	25	−35	−54	81	−26	51	9	0
Jun	18.2	49	20	−29	−83	64	−17	37	12	0
Jul	18.4	54	15	−39	−122	47	−17	32	22	0
Aug	20.6	74	12	−62	−184	29	−18	30	44	0
Sep	22.4	93	48	−45	−229	20	−9	57	36	0
Oct	23.1	107	113	+6	−196	26	+6	107	0	0
Nov	23.2	108	180	+72	−30	98	+72	108	0	0
Dec	23.3	117	245	+128	0	125	+27	117	0	101
Year	21.6	1064	1391	327	–	–	±105	940	124	451

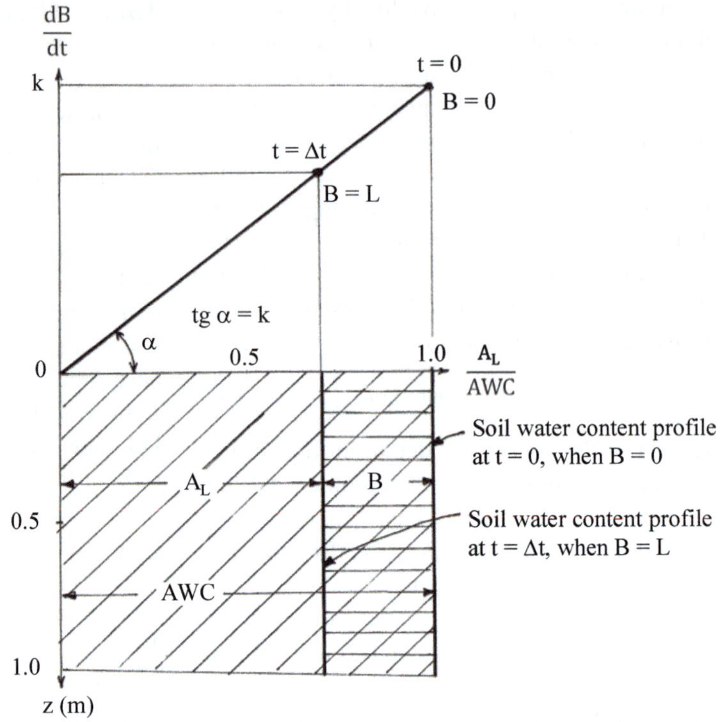

Fig. 15.7 Thornthwaite's model for soil water extraction when $P - ET_c$ is negative

climate with rainy summer. The fifth column shows the difference $P - ET_0$, which in the winter months is negative, indicating a water deficiency that needs to be supplied by the soil or, in many cases, by irrigation. The sixth column is the accumulated negative at the depth of interest L [defined by Camargo 1978] and will be better discussed below. The seventh column is the soil water storage A_L, not the ΔA_L that appears in Eq. (15.10), which is the eighth column. In this example, the value of AWC = 125 mm was assumed, and therefore ΔA_L = AWC in the wettest months. In the sequence $P > ET_c$, EXC appears in the 11th column. The ninth column contains ET_a calculated by the balance as we shall see below and, at the tenth, the DEF water deficit defined above.

Let's start the balance sheet calculation in the month of March, for which we are sure that the soil reservoir is full, that is, $A_L = \text{AWC} = 125$ mm, because in the previous months, it rained a lot. Since $P - ET_0$ is positive, an EXC = 76 mm (11th column) appears. In the following month, April, as $P - ET_0$ becomes negative, the consumption of the stored water begins. Thornthwaite assumes that the actual loss B of soil water occurs at a rate dB/dt which is a linear function of the storage, expressed in a relative manner, i.e., A_L/AWC, ranging from 1 to 0, such that:

$$\frac{dB}{dt} = k\frac{A_L}{\text{AWC}} \tag{15.12}$$

Figure 15.7 shows the relation dB/dt as a function of A_L/AWC. Supposing that at the end of the wet period (March in Piracicaba, SP, Brazil) and beginning of the period (April) in which $P - ET_c$ is negative (−19 mm), the soil is at field capacity (θ_{FC}), the water loss B, in the interval Δt (30 days), is given by:

$$B = \text{AWC} - A_L \tag{15.13}$$

and, substituting Eq. (15.13) into Eq. (15.12), we have:

$$\frac{dB}{dt} = k\frac{(\text{AWC} - B)}{\text{AWC}} \tag{15.14}$$

To obtain the constant k, we use the initial condition, A_L = AWC and B = 0, and Eqs. (15.12) and (15.14) become:

$$\left.\frac{\mathrm{d}B}{\mathrm{d}t}\right|_{t=0} = k \tag{15.15}$$

and, on the other hand, dB/dt for $t = 0$ can only be the negative difference P – ET_c (−19 mm), which is the first contribution to the **accumulated negative** L, of −19 mm. Do not confuse this L with the depth of the soil layer (we only used the same symbol for both). So:

$$\left.\frac{\mathrm{d}B}{\mathrm{d}t}\right|_{t=0} = \frac{L}{\Delta t} \tag{15.12a}$$

Equating Eq. (15.15) with Eq. (15.12a), we have the value of k:

$$k = \frac{L}{\Delta t} \tag{15.16}$$

and introducing k into the general Eq. (15.14), we have:

$$\frac{\mathrm{d}B}{\mathrm{d}t} = \frac{L}{\Delta t}\frac{(\mathrm{AWC} - B)}{\mathrm{AWC}} \tag{15.17}$$

Equation (15.17) can be integrated separating the variables and rearranging:

$$\int_0^L \left[\frac{1}{\mathrm{AWC} - B}\right]\mathrm{d}B = \frac{L}{\Delta t \cdot \mathrm{AWC}}\int_0^{\Delta t}\mathrm{d}t \tag{15.18}$$

or

$$-\ln(\mathrm{AWC} - B)|_0^L = \frac{L}{\Delta t \cdot \mathrm{AWC}}t|_0^{\Delta t} \tag{15.19}$$

and, remembering that AWC – $L = A_L$, results in:

$$A_L = \mathrm{AWC}\exp\left[\frac{-L}{\mathrm{AWC}}\right] \tag{15.20}$$

It can be seen that the result of the changes of A_L by the Thornthwaite method is exponential. This is shown by the data in column 7 of the worksheet (Table 15.1), changing from 125 to 107, 81, 64, 47, 29 mm, etc. showing that the extraction of soil water becomes more and more difficult. The accumulated negative L, used in the deductions from Eqs. (15.12a) to (15.20), becomes more clear by analyzing the column 6 of the worksheet. It is the sum of the negative values of P – ET_c. ET_a is the sum of P plus the water withdrawn from the soil. Thus, for April, $ET_a = 60 + 18 = 78$ mm.

Practical results show that the Thornthwaite linear model for ET_a range from θ_{FC} to θ_{PWP} works very well for high atmospheric demands or high ET_0 values of the order of 8–12 mm day^{-1}. For lower rates, the plant is able to remove water from the soil more easily, and, therefore, Rijtema and Aboukhaled (1975) have suggested a factor p so that ET_a is equal to ET_0 for values of θ below θ_{FC}. This factor p defines a critical value θ_C between θ_{FC} and θ_{PWP}, until which $ET_a = ET_0$. As in their equations a factor (1 – p) appears, for $p = 0.7$, we have for the first 30% of the available water $ET_a = ET_c$. For plants sensitive to water deficit or for conditions of high atmospheric demand, p is of the order of 0.7. For intermediate conditions, $p = 0.5$, and for plants tolerant or low demands, $p = 0.3$. We often speak of a **critical available water capacity** or critical storage [(1 – p). AWC], which represents the part of the AW of greatest difficulty for plants to withdraw soil water.

The method of Thornthwaite and Mather (1955) can be adapted to any method of calculation of ET_0. To exemplify, for the case of Rijtema and Aboukhaled, the storage A_L becomes:

$$A_L = (1 - p)\mathrm{AWC}\exp\left(\frac{\frac{L}{\mathrm{AWC}} - p}{(1-p)}\right) \tag{15.21}$$

and for the case of the cosenoidal method of Dourado-Neto and De Jong van Lier (1993):

$$A_L = (1 - p)\mathrm{AWC} \times \left\{1 - \frac{2}{\pi}\arctan\left[\frac{\pi}{2}\left(\frac{\left(\frac{L}{\mathrm{AWC}} - p\right)}{(1 - p)}\right)\right]\right\} \tag{15.22}$$

The bibliography on water balance is extensive. In Brazil, the first studies of water balances were done by agrometeorologists who, with meteorological data, estimated ET_0 by theoretical-empirical formulas (mainly ET of reference) for the state of São Paulo. Camargo (1961) estimated ET_0 for 24 municipalities in the state of São Paulo, applying the Penman (1948) method. He also presented a comparison of the estimated values with measurements in Class A tanks and evapotranspirometers, suggesting conversion coefficients.

Camargo (1964) also presented a water balance for the state of São Paulo. Ranzani (1971) included in the balance variables inherent to the soil in the storage capacity. Among others, Lima and Reichardt (1977), who were more concerned with the term Q_z, can also be mentioned. Timm et al. (2002) made a quantitative and qualitative analysis of the methodologies used to estimate the components of the water balance in a Rhodic Kandiudalf, cultivated with sugarcane and submitted to different management practices (1) bare soil, (2) the surface of the soil with the presence of straw residues left after the cane harvest, and (3) the soil surface with the presence of the residues of the sugarcane burned before harvest, in the city of Piracicaba, SP, Brazil. The authors conclude that surface runoff, water flows at lower boundaries of soil volume ($z = L = 1.0$ m), and changes in soil water storage were not affected by different management practices and that the runoff and soil water fluxes were strongly affected by the spatial variability of soil physical properties.

A very interesting way to use water balances to monitor soil water content variations in a developing crop is the **sequential water balance** (Rolim et al. 1998). It consists of an Excel sheet, in which current values of air temperature (for the calculation of ET_0 by Thornthwaite) and precipitation (or irrigation) are introduced, and the program calculates the storage of water in the soil, based on a chosen AWC. It provides water deficits and calculates an excess, which corresponds to the sum of the runoff and the deep drainage. Bortolotto et al. (2011) used the sequential water balance for the estimation of deep drainage (and N leaching) in fertirrigated coffee plantations in western Bahia. Silva et al. (2013) controlled the water conditions in vessels cultivated with *Jatropha*, applying the sequential water balance methodology.

There are also detailed programs that make crop water balances, one of them being the SWAP. This program requires, in addition to climatic data, various data on the plant and on the soil water properties. Recently, Pinto et al. (2015) have shown that this program can also be used to simulate the water balance of perennial crops such as coffee.

As far as the different components of the water balance are concerned, there is always the problem of temporal and spatial variability (see Chaps. 17 and 18), sampling, number of replicates, etc. Precipitation, for example, is easily measured by rain gauges, as seen in Chap. 3, but we know from our day to day that it may rain where we are and some hundred meters away, there falls no drop of water. This is typical for tropical very localized showers. Therefore, in establishing water balances, data from distant weather stations should be avoided or not be used, and preferably precipitation should be measured on site. Reichardt et al. (1995) approach this subject, showing the temporal and spatial variability of precipitation over an area of 1000 ha, in Piracicaba, SP, Brazil. The measurement of irrigation also presents difficulties because its distribution is slowly homogeneous. In irrigation by pivots, by furrows, and by drip, the variability from site to site may be very large. Evapotranspiration, if measured by aerodynamic models, does not present serious problems of variability, due to aerodynamic air movements. In the case of Class A tanks or lysimeters, their number and location may be extremely important. The runoff is greatly affected by small variations of slope, vegetation cover, soil preparation, etc. The variability of the deep drainage follows the variability of the water properties along the soil profile. It is therefore necessary that the researcher be attentive to these problems, not having only a general rule for samplings and number of replicates. Each case is a special case and needs to be approached with criterium.

In a coffee experiment conducted in Piracicaba, SP, Brazil, Silva et al. (2006) studied the variability of water balance components, and Timm et al. (2011) analyzed, for the same crop, the time series of 2 years of measurements of ET and ΔA.

Rose and Stern (1967) presented an analysis of the water absorption ratio for different depths in any crop, from the water balance. Equation (15.1) can be rewritten by separating q_e into two parts: q_{es} (evaporation of the soil water) and q_{et} (transpiration of the plants):

$$\underbrace{\int_{t_i}^{t_j} (p + i \pm d_s - q_{es} \pm q_z)\mathrm{d}t}_{P+I\pm DS-E\pm Q_z} - \underbrace{\int_{t_i}^{t_j}\int_0^L \left(\frac{\partial\theta}{\partial t}\right)\mathrm{d}z\mathrm{d}t}_{\Delta A_z} = \underbrace{\int_{t_i}^{t_j}\int_0^z r_z\mathrm{d}z\mathrm{d}t}_{T} \quad (15.23)$$

where the double integral of the second member is the integral of q_{et}, equal to T. In this equation, r_z is the ratio of the decrease in time of the soil water content in the "0–z" layer, due to root activity. When all the terms of the first member of Eq. (15.23) are known, T can be estimated, varying z at predetermined intervals Δz, from $z = 0$ to $z = L$.

Defining a mean absorption ratio—$\overline{r_z}$—in a 0–z layer by:

$$\overline{r_z} = \frac{\int_{t_i}^{t_j} r_z\mathrm{d}t}{(t_1 - t_0)} \quad (15.24)$$

the distribution of root activity can be determined by performing repeated calculations with Eqs. (15.23) and (15.24), varying z in small intervals. Reichardt et al. (1979) applied for the first time the technique of Rose and Stern (1967), under Brazilian conditions, specifically for a maize crop. Silva et al. (2009) evaluated by this technique the root system of a coffee crop.

15.4 Water-Depleted Productivity

The potential productivity of a crop in a given region is determined by plant genetic factors and by the degree of plant adaptation to the environment, with water and nutrients fully available, without limitation by pests and diseases, during all periods of crop development, until maturity (Doorenbos and Kassam 1994; Heifig 2002). The availability of water in the periods of greater demand for the crop is essential to achieve the expected result, since it participates in all the metabolic processes that will establish growth and development, to ending the period of growth of the crop (Doorenbos and Kassam 1994). The adoption of management practices that minimize the impact of external factors on the assimilation of carbon and nitrogen in the reproductive phase is determinant of grain yield. This is the phase that the crop presents high physiological activity, reaching the maximum rate of assimilation of carbon and nitrogen (Fagan 2007; Taiz and Zeiger 2010).

The water deficiency (D, DH, or DEF) is characterized when ET_a is smaller than ET_0, defined by $D = ET_c - ET_a$, that is, the soil restrictions prevent the plant from losing the maximum water by transpiration. It induces physiological and morphological adaptations, such as stomatal closure. As a consequence, the reduction of photosynthesis occurs, affecting growth and productivity (Pereira et al. 2002). According to Marin et al. (2000), the model of Doorenbos and Kassam (1994) is commonly used in estimating the yield of agricultural crops according to a given water condition. The model penalizes the potential productivity (Y_0) for water deficiency, a function of the evapotranspiration deficit and the deficit sensitivity coefficient, resulting in a real or **water-depleted productivity** (Y_r). The coefficient k_y, also known as the response factor of the crop to water availability, relates the relative yield decrease $[1 - (Y_r/Y_0)]$ to the relative evapotranspiration deficit $[1 - (ET_a/ET_c)]$, ET_c being the evapotranspiration of the crop in each phase of development. In the specific periods of growth, flowering and harvesting, the penalty for k_y is relatively large, whereas for the vegetative and ripening periods, it is smaller. Thus, according to Doorenbos and Kassam (1994):

$$Y_r = \left[1 - k_y\left(1 - \frac{ET_a}{ET_c}\right)\right] \cdot Y_0 \qquad (15.25)$$

15.5 A Holistic View of the Agricultural Production System

Having fully approached the behavior of water in the Soil-Plant-Atmosphere System, we will now present an overview of the agricultural production system, necessary for the researcher, modeler, and the farmer himself. For this, we will discuss in detail Fig. 15.8, taken as an example for the case of a 120-day cycle maize crop. It encompasses four quadrants, those of the right (first and fourth) with static aspects of the system, nontemporal, the first referring to the atmosphere and the fourth to the soil and the quadrants of the left, involving the development of the crop along its cycle to maturation (second and third), the second referring to the development of the aerial part of the plant, affected by the soil and the atmosphere, and the third referring to the development of the root system, also affected by the soil and the atmosphere. The third also includes weather information, in order not to overload the second.

The plot of the fourth is the graph $\theta(h)$ for the available water (AW). It is the plot of the soil water retention curve, in the range $h = -0.33$ atm (for $\theta = 0.250\ m^3\ m^{-3}$ as the FC) and $h = -15$ atm

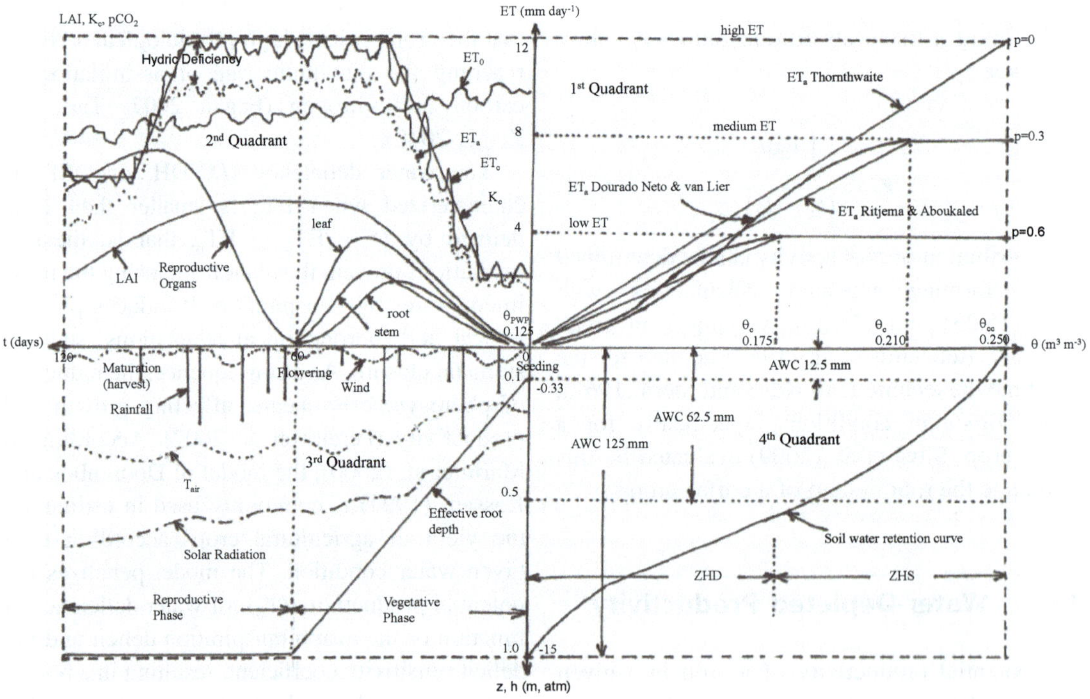

Fig. 15.8 A holistic view of an example of a corn crop. First quadrant shows statically the conditions of evapotranspiration, the second the dynamics of plant growth as affected by climate, the third climate and root development, and the forth soil water characteristics. *LAI*, leaf area index; K_c, crop coefficient; pCO_2, photosynthetic carbon partition of the different plant organs; *ET*, evapotranspiration; θ, soil water content; *AWC*, available water capacity; *ZHD*, zone of hydric deficiency; *ZHS*, zone of hydric sufficiency; *h*, soil water potential; *p*, Rijtema and Aboukhaled's available water factor; z_e, effective root depth; *FC*, field capacity; *PWP*, permanent wilting point. Collaboration of Prof. Durval Dourado Neto

(for $\theta = 0.125$ m^3 m^{-3} as the PWP), already discussed in Fig. 14.6, showing the increase of available soil water as a function of the root development shown in the third quadrant. Here the **effective depth** concept of the root system z_e is introduced, considered growing linearly from the emergence up to the beginning of the appearance of the reproductive organs, afterward remaining constant until the end of the plant cycle. Thus, considering the seeding depth $z_{ei} = 0.1$ m, the AWC_0 of this layer would be 12.5 mm; for a given time t, we could have $AWC_t = 62.5$ mm, and, after 60 days, $AWC_{60–120} = 125$ mm (Fig. 15.8). In modeling it is therefore important to consider the AWC as a variable during crop development, not constant as we have seen in the Thornthwaite and Mather balance sheet.

The first quadrant is the ET(θ) graph, which is tied to the fourth quadrant. Three ET_0 values were chosen arbitrarily: high, above 10 mm day^{-1}; medium, in the order of 5–8 mm day^{-1}; and low, < 5 mm day^{-1}. At high ET_0 values, the Thornthwaite method is the most suitable for the calculation of ET_a, which decreases linearly with θ or AWC from FC (θ_{Fc}) to PWP (θ_{PWP}). For medium or low conditions, the methods of Rijtema and Aboulkhaled and Dourado-Neto and De Jong van Lier are better adapted. In these methods appears the factor p that extends the interval in which $ET_a = ET_0$, defining values of the critical soil water content θ_c.

The second quadrant includes parameters related to the development of the crop. Notice that the time scale is inverted, growing from right to left. The temporal variations of K_c, already discussed in Fig. 13.5 of Chap. 13, accompany the growth of the crop. ET_0 depends on the evolution of the climate and therefore is presented oscillating but slightly increasing since the air temperature (T_{air}) and solar radiation (SR) show this trend. ET_c, which is the product of ET_0 by K_c, accompanies K_c also in an oscillating way. ET_a that is what actually occurs is always less than or equal to ET_c, depending on precipitation P or irrigation I. LAI follows leaf growth, tending to decrease at the end of the cycle when the older leaves senesce. Today, evergreen corn seeds are on the market that keep the leaves green until the end of the cycle. Of importance also is the carbon partition (pCO_2) for the different parts of the plant: root, stem, leaf, and reproductive organs.

The third quadrant shows an example of climate evolution through air temperature (T_{air}), solar radiation (SR), wind, and precipitation ($P + I$).

The elements shown in Fig. 15.8 are essential to establish models, management practices related to water and nutrients for agricultural crops in general. With them, one can predict periods of water deficiency for irrigation and fertilizer application, in order to maximize production. We have already seen that the potential productivity of a crop is determined by a number of factors, among which the genetic (variety), degree of adaptation to the environment (which opens the possibility of varying the number of plants per hectare), availability of water and nutrients, control of pests and diseases, this during all periods of development until maturation. Let us return to the case of corn as an example, for which the potential productivity is quite variable. Let us make a relative and approximate comparison of potential productivity in three major corn-producing countries: the United States, Argentina, and Brazil.

In this comparison we used mean data from soils cultivated to maize, such as the availability of water in terms of AWC/z_e, i.e., millimeter of AW per centimeter soil. Thus, for the three countries chosen and for a soil layer of 1.0 m, we will have AWC of 200 mm, 150 mm, and 100 mm, or 2.0 mm cm^{-1}, 1.5 mm cm^{-1}, and 1.0 mm cm^{-1}, respectively. The CEC in mmol$_c$ dm^3 is very high for the soils of the United States, average for those of Argentina, and very low for Brazil. Mainly due to these two factors, plantation populations per hectare have to be very different, as indicated above. Other factors such as variety, pest control, diseases, weeds, and day length N (Chap. 5) are also different in these countries, resulting in very different potential average yields. Hence, the very high corn yields in the US Corn Belt.

Countries	AWC/z_e (mm cm^{-1})	CEC (mmol$_c$ dm^{-3})	Population × 10^3	Pests, diseases, and weeds	N (h)	Potential productivity (tons ha^{-1})
United States	2.0	300	120	↓	12–15	8–18
Argentina	1.5	120	85	↓	10–14	5–13
Brazil	1.0	50	70	↓	11–13	3–10

15.6 Exercises

15.1. The intensity of a rainfall was 5 mm h^{-1} for 5 min, 15 mm h^{-1} for 10 min, and 2 mm h^{-1} for 3 min. Calculate the total rainfall P by the integral of Eq. (15.3).

15.2. An irrigation system operates for 3 h at a rate of 10 mm h^{-1}. What is the total irrigation I calculated by the integral Eq. (15.4)?

15.3. In a ramp used to measure the runoff, measuring 2 × 22 m, 156 L of water was collected in the collecting tank. The rainfall was 18.6 mm. What is the percentage of runoff in relation to rainfall?

15.4. In Problem (15.3), how does integral Eq. (15.5) fit?

15.5. In one crop, the deep drainage q_z was measured using Eqs. (15.7), (15.8), (15.9), and (15.10), and the following results were obtained:

q_z (mm day^{-1})	Day
0.3	May 3, 1993
1.2	May 7, 1993
0.8	May 10, 1993
0.3	May 14, 1993
0.1	May 18, 1993

Which is the value of Q_z in the period May 1–20, 1993?

15.6. For the same crop of Problem (15.5), the following data were obtained for evapotranspiration:

q_e (mm day^{-1})	Day
5.6	May 10, 1993
7.3	May 11, 1993
6.2	May 12, 1993
4.1	May 13, 1993
5.4	May 14, 1993

Which is the value of ET in the period May 10–14, 1993?

15.7. In a cotton crop, the following water balance components were measured in three distinct periods:
Being $\Delta A_L = A_L(t_f) - A_L(t_i)$, estimate ET for these periods.

Period A	Period B	Period C
$P = 0$	$P = 33$ mm	$P = 30$ mm
$I = 20$ mm	$I = 0$	$I = 0$
DS = 0	DS = −5 mm	DS = 0
$Q_z = -1$ mm	$Q_z = -2$ mm	$Q_z = 0$
$\Delta A_L = -5$ mm	$\Delta A_L = +3$ mm	$\Delta A_L = 0$

15.8. Soil water content measurements were performed in a soybean crop, and the following results were obtained for averages of ten locations (Fig. 15.9):

Depth (cm)	Soil water content (m^3 m^{-3})	
	October 15, 1989	October 22, 1989
0	0.401	0.298
10	0.402	0.305
20	0.410	0.319
30	0.424	0.336
40	0.435	0.375
50	0.449	0.412
60	0.462	0.438
70	0.463	0.455
80	0.461	0.462
90	0.464	0.463
100	0.466	0.466

Apply Rose and Stern's (1967) technique, and estimate the root distribution of the soy plants in this development period. There was no rain, and plants covered totally the soil surface, so that evaporation can be neglected. The soil water contents measured were below field capacity.

15.9. Make a critical analysis of the advantages and disadvantages of the water balance methodology presented in this chapter, when compared to other methodologies for calculating the water balance of a crop.

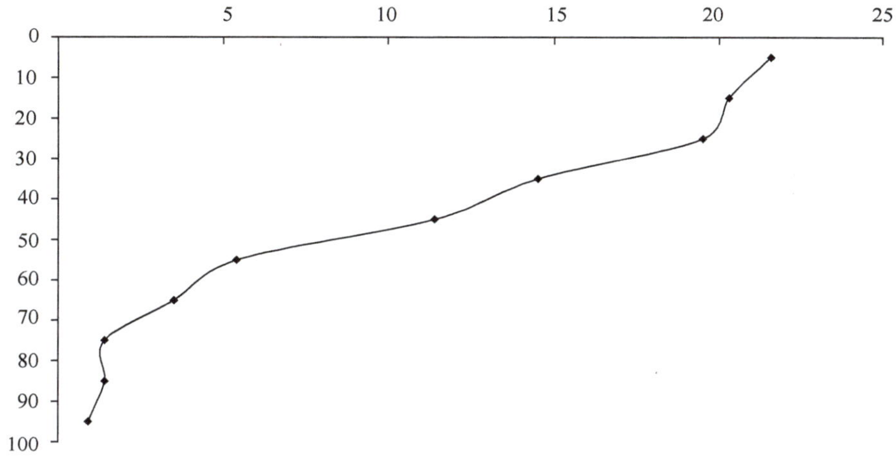

Fig. 15.9 Root distribution of the soybean crop

15.7 Answers

15.1. $P = \int p\mathrm{d}t = p_1\Delta t_1 + p_2\Delta t_2 + p_3\Delta t_3$

$$= 5 \times 0.083 + 10 \times 0.167 + 2 \times 0.05$$
$$= 2.185\,\mathrm{mm}$$

15.2. $I = 10 \times 3 = 30$ mm.

15.3. From the rain that fell over the area of $2 \times 22 = 44$ m^2, 156 L represents runoff. The "height" h of the runoff is calculated as in the case of rain: $h = V/A$; hence $h = 156$ L/44 m^2 = 3.5 mm. The percentage is (3.5/18.6) × 100 = 18.8% of the water that fallen on the ramp runs off.

15.4. The integral Eq. (15.5) is already embedded in the 156 L that left the ramp. They did not flow over soil surface at once, but over time. The collection in the container already does the integration, in the same way as a rain gauge integrates the rain giving the result of P.

15.5. Soil drainage is continuous and the measurements we have are timely over time. Q_z is given by the integral of q_z in time (Eq. 15.7). The best way is to make an approximation:

$$Q_z = \int_{t_i}^{t_f} q_z\mathrm{d}t \cong \bar{q}_z(t_f - t_i) = 0.54 \times (21 - 1)$$
$$= 10.8\,\mathrm{mm}$$

The average of q_z is 0.54 mm day^{-1}. When individual values correspond to very different periods of time, a weighted average is recommended.

15.6. ET = 28.6 mm. The reasoning is the same as above.

$$\mathrm{ET} = \int_{t_i}^{t_f} q_e\mathrm{d}t \cong \bar{q}_e(t_f - t_i) = 5.72 \times (15 - 10)$$
$$= 28.6\,\mathrm{mm}$$

15.7. Using Eq. (15.10) (watch the signs!):
ET(A) = 24 mm
ET(B) = 23 mm
ET(C) = 30 mm

15.8. Equation (15.12) is reduced to:

$$\int_{t_i}^{t_f}\int_0^L \left(\frac{\partial\theta}{\partial t}\right)\mathrm{d}z\mathrm{d}t = \int_{t_i}^{t_j}\int_0^z r_z\mathrm{d}z\mathrm{d}t = T_z$$

because $p = 0$; $i = 0$; $q_{es} = 0$; and $q_z = 0$. The member on the left represents soil water storage variations. In our case we will calculate these variations in layers 0–10; 0–20; 0–30; …; 0–100. Of course, the thicker the layer, the greater the change. The right member can be simplified by assuming that r_z is constant over time, which is reasonable. This means that the rate of root extraction of water is constant in the interval $t_f - t_i$, which should not be too long. Even if r_z varies in time, it could be replaced by an average value in time, and this value is considered constant. Therefore:

$$T_z = \bar{r}_z \cdot z(t_f - t_i)$$

since:

$$\int_0^z dz = z \quad \text{and} \quad \int_{t_i}^{t_f} dt = t_f - t_i$$

and we can calculate:

$$\bar{r}_z = \frac{T_z}{z(t_f - t_i)}$$

remembering that T_z is the term at the left, representing the changes in soil water storage. The table below illustrates this case:

Layer	A_L (mm)		$\Delta A_L = T_z$	r_z
(cm)	October 15, 1989	October 22, 1989	(mm cm^{-1} day^{-1})	
0–10	40.15	30.15	10.00	0.143
0–20	80.87	61.47	19.40	0.139
0–30	122.78	94.35	28.43	0.135
0–40	165.76	130.64	35.12	0.125
0–50	210.08	170.41	40.35	0.115
0–60	255.68	212.83	42.85	0.102
0–70	301.52	257.08	44.44	0.091
0–80	347.29	302.22	45.07	0.080
0–90	393.39	347.67	45.72	0.072
0–100	439.73	393.55	46.18	0.066

Attention, what is the meaning of the value $r_z = 0.115$ mm cm^{-1} day^{-1} for the 0–50 cm layer? It means that the soil layer 0–50 cm loses on average, through root extraction, 0.115 mm in each cm in each day. Logically the layer loses $0.115 \times 50 = 5.75$ mm in each day and in 7 days $5.75 \times 7 = 40.35$ mm.

It would be more interesting to define a $\bar{r}_{z_j - z_i}$ of successive layers, that is, from 0 to 10; 10–20; 20–30; …; 90–100 cm. In this way, we would have:

Layer (cm)	$\Delta A_{z_j - z_i}$ (mm)	$\bar{r}_{z_j - z_i}$ (mm cm^{-1} day^{-1})	$\bar{r}_{z_j - z_i}$ (%)
0–10	10.00	0.143	21.6
10–20	9.40	0.134	20.3
20–30	9.03	0.129	19.5
30–40	6.69	0.096	14.5
40–50	5.23	0.075	11.4
50–60	2.50	0.036	5.4
60–70	1.59	0.023	3.5
70–80	0.63	0.009	1.4
80–90	0.65	0.009	1.4
90–100	0.46	0.006	0.9

References

Albaladejo Montoro J, Stocking MA (1989) Comparative evaluation of two models in predicting storm soil loss from erosion plots in semi-arid Spain. Catena 16:227–236

Arnold JG, Srinivasan R, Muttiah RS, Williams JR (1998) Large area hydrologic modeling and assessment: part I. Model development. J Am Water Resour Assoc 34:73–89

Bacchi OOS, Reichardt K, Sparovek G, Ranieri SBL (2000) Soil erosion evaluation in a small watershed in Brazil through 137-Cs fallout redistribution analysis and conventional models. Acta Geol Hisp 35:251–259

Bacchi OOS, Reichardt K, Sparovek G (2003) Sediment spatial distribution evaluated by three methods and its relation to some soil properties. Soil Tillage Res 69:117–125

Bertoni J, Lombardi Neto F (1990) Conservação do solo. Ícone, São Paulo

Beskow S (2009) LASH model: a hydrological simulation tool in GIS framework. Departamento de Engenharia, Universidade Federal de Lavras, Lavras, Tese de Doutorado

Beskow S, Mello CR, Norton LD, Curi N, Viola MR, Avanzi JC (2009) Soil erosion prediction in the Grande River, Brazil using distributed modeling. Catena 79:49–59

Beskow S, Mello CR, Norton LD, Silva AM (2011a) Performance of a distributed semi-con-ceptual

hydrological model under tropical watershed conditions. Catena 86:160–171
Beskow S, Mello CR, Norton LD (2011b) Development, sensitivity and uncertainty analysis of LASH model. Sci Agric 68:265–274
Beskow S, Timm LC, Tavares VEQ, Caldeira TL, Aquino LS (2016) Potential of the LASH model for water resources management in data-scarce basins: a case study of the Fragata River basin, Southern Brazil. Hydrol Sci J 61:2567–2578
Bortolotto RP, Bruno IP, Dourado-Neto D, Timm LC, Silva AN, Reichardt K (2011) Soil profile internal drainage for a central pivot fertigated coffee crop. Rev Ceres 58:723–728
Caldeira TL, Mello CR, Beskow S, Timm LC, Viola MR (2019) LASH hydrological model: an analysis focused on spatial discretization. Catena 173:183–193
Camargo AP (1961) Contribuição para a determinação da evapotranspiração potencial no Estado de São Paulo. Tese de Doutorado, Escola Superior de Agricultura Luiz de Queiroz. Universidade de São Paulo, Piracicaba
Camargo AP (1964) Balanço hídrico no Estado de São Paulo. Instituto Agronômico de Campinas, Campinas
Camargo AP (1978) Balanço hídrico no Estado de São Paulo. Instituto Agronômico de Campinas, Campinas
Correchel V, Bacchi OOS, Maria IC, Dechen SCF, Reichardt K (2006) Erosion rates evaluated by the 137 Cs technique and direct measurements on long-term runoff plots. Soil Tillage Res 86:199–208
Daian FJ, Vachaud G (1972) Methode d'evaluation du bilan hydrique in situ a partir de la mesure des teneures en eau et des succions. In: Symposium on isotopes and radiation in soil-plant relationships including forestry. International Atomic Energy Agency, Vienna, pp 649–660
De Roo APJ, Wesseling CG, Ritsema CJ (1996) LISEM: a single event physically-based hydrologic and soil erosion model for drainage basins: I. Theory, input and output. Hydrol Process 10:1107–1117
Doorenbos J, Kassam AH (1994) Efeito da água no rendimento das culturas. Tradução de H Ghey HR, Sousa AA, Damasceno FAV, Medeiros JF (tradutores). Universidade Federal da Paraíba, Campina Grande
Dourado-Neto D, De Jong van Lier Q (1993) Estimativa do armazenamento de água no solo para realização de balanço hídrico. Rev Bras Ciênc Solo 17:9–15
Duan Q, Sorooshian S, Gupta V (1992) Effective and efficient global optimization for conceptual rainfall-runoff models. Water Resour Res 28:1015–1031
Fagan EB (2007) A cultura da soja: modelo de crescimento e aplicação da estrobilurina piraclostrobina. Tese de Doutorado, Escola Superior de Agricultura Luiz de Queiroz. Universidade de São Paulo, Piracicaba
Foster GR, Moldenhauer WC, Wischmeier WH (1985) Transferability of US technology for prediction and control of erosion in the tropics. In: Symposium on soil erosion and conservation in the tropics. American Society of Agronomy, Madison, WI, pp 135–149
Gassman PW, Reyes MR, Green CH, Arnold JG (2007) The soil and water assessment tool: historical development, applications, and future research directions. Trans ASABE 50:1211–1250
Heifig LC (2002) Plasticidade da cultura da soja (Glycine max (L.) Merril) em diferentes arranjos espaciais. Dissertação de Mestrado, Escola Superior de Agricultura Luiz de Queiroz, Universidade de São Paulo, Piracicaba, Brazil
Hickley R, Smith A, Jankowski P (1994) Slope length calculations from a DEM within ARC/Info grid. Comp Env Urban Sys 18:365–380
Knisel WG (1980) CREAMS: a field-scale model for chemicals, runoff and erosion from agricultural management systems. United States Department of Agriculture, Washington, DC
Laflen JM, Lane LJ, Foster GR (1991) WEPP: a new generation of erosion prediction technology. J Soil Water Conserv 46:34–38
Lal R (1988) Soil erosion by wind and water: problems and prospects. In: Lal R (ed) Soil erosion research methodology. Soil and Water Conservation Society of America, Ankeny, IA, pp 1–8
Lane LJ, Renard KG, Foster GR, Laflen JM (1992) Development and application of modern soil erosion prediction technology. Aust J Soil Res 30:893–912
LaRue ME, Nielsen DR, Hagan RM (1968) Soil water flux below a ryegrass root zone. Agron J 60:625–629
Lima WP, Reichardt K (1977) Regime de água do solo sob florestas homogêneas de eucalipto e pinheiro. Centro de Energia Nuclear na Agricultura. Universidade de São Paulo, Piracicaba
Lombardi Neto F, Bertoni J (1975) Erodibilidade de solos paulistas. Instituto Agronômico de Campinas, Campinas
Lombardi Neto F, Moldenhauer WC (1992) Erosividade da chuva: sua distribuição e relação com as perdas de solo em Campinas (SP). Bragantia 51:189–196
Marin FR, Sentelhas PC, Ungaro MRG (2000) Perda de rendimento potencial da cultura do girassol por deficiência hídrica, no Estado de São Paulo. Sci Agric 57:1–6
Mello CR, Viola MR, Norton LD, Silva AM, Weimar FA (2008) Development and application of a simple hydrologic model simulation for a Brazilian headwater basin. Catena 75:235–247
Penman HL (1948) Natural evaporation from open water, bare soil and grass. Proc R Soc Ser A 193:120–145
Pereira AR, Ferraz ESB, Reichardt K, Libardi PL (1974) Estimativa da evapotranspiração e da drenagem profunda em cafezais cultivados em solos podzolizados Lins e Marília. Centro de Energia Nuclear na Agricultura. Universidade de São Paulo, Piracicaba
Pereira AR, Angelocci LR, Sentelhas PC (2002) Agrometeorologia: fundamentos e aplicações práticas. Agropecuária, Guaíba
Pinto VM, Reichardt K, van Dam J, Van Lier QDJ, Bruno IP, Durigon A, Dourado-Neto D, Bortolotto RP (2015) Deep drainage modeling for a fertigated coffee

plantation in the Brazilian savanna. Agric Water Manag 140C:130–140
Pires LF, Bacchi OOS, Correchel V, Reichardt K, Filippe J (2009) Riparian forest potential to retain sediment and carbon evaluated by the 137 Cs fallout and carbon isotopic technique. Anais Acad Bras Ci 81:271–279
Ranieri SBL (1996) Avaliação de métodos e escalas de trabalho para determinação de risco de erosão em bacia hidrográfica utilizando Sistema de Informações Geográficas (SIG). Master Dissertation, Escola de Engenharia de São Carlos, Universidade de São Paulo, São Carlos, Brazil
Ranzani G (1971) A marcha anual d'água disponível do solo. Escola Agricultura Luiz de Queiroz/Universidade de São Paulo, Piracicaba
Reichardt K, Libardi PL, Santos JM (1974) An analysis of soil-water movement in the field. II. Water balance in a snap bean crop. Centro de Energia Nuclear na Agricultura, Universidade de São Paulo, Piracicaba
Reichardt K, Libardi PL, Saunders LCU, Cadima A (1979) Dinâmica da água em cultura de milho. Rev Bras Ciênc Solo 3:1–5
Reichardt K, Libardi PL, Moraes SO, Bacchi OOS, Turatti AL, Villagra MM (1990) Soil spatial variability and its implications on the establishment of water balances. International Congress of Soil Science, International Union Soil Science, Kyoto, pp 41–46
Reichardt K, Angelocci LR, Bacchi OOS, Pilotto JE (1995) Daily rainfall variability at a local scale (1,000 ha), in Piracicaba, SP, Brazil, and its implications on soil recharge. Sci Agric 52:43–49
Renard KG, Mausbach MJ (1990) Tools for conservation. In: Larson WE, Foster GR, Allmaras RR, Smith CM (eds) Proceedings of soil erosion and productivity workshop. University of Minnesota, Minneapolis, MN, pp 55–64
Renger M, Giesel W, Strebel O, Lorch S (1970) Erste ergebnisse zur quantitativen erfassung der wasserhaushaltskomponenten in der ungessättigten bodenzone. Z Pflanzenernährung Bod 126:15–35
Rijtema PE, Aboukhaled A (1975) Crop water use. In: Aboukhaled A, Arar A, Balba AM, Bishay BG, Kadry LT, Rijtema PE, Taher A (eds) Research on crop water use, salt affect-ed soils and drainage in the Arab Republic of Egypt. FAO Regional Office for the Near East, Cairo, pp 5–61
Rolim GS, Sentelhas PC, Barbieri V (1998) Planilhas no ambiente Excell para cálculos de balanços hídricos: normal, sequencial, de cultura e de produtividade real e potencial. Rev Bras Agromet 6:133–137
Rose CW, Stern WR (1967) Determination of withdrawal of water from soil by crop roots as function of depth and time. Aust J Soil Res 5:11–19
Silva AL, Roveratti R, Reichardt K, Bacchi OOS, Timm LC, Bruno IP, Oliveira JCM, Dourado-Neto D (2006) Variability of water balance components in a coffee crop grown in Brazil. Sci Agric 63:105–114
Silva AL, Bruno IP, Reichardt K, Bacchi OOS, Dourado-Neto D, Favarin JL, Costa FMP, Timm LC (2009) Soil water extraction by roots and Kc for the coffee crop. Agriambi 13:257–261
Silva AN, Bortolotto RP, Tomaz HVQ, Reis LG, Olinda RA, Heiffig-del-Águila LS, Reichardt K (2013) Pot irrigation control through the climatologic sequential water balance. Rev Agric 88:101–106
Steinmetz AA, Cassalho F, Caldeira TL, Oliveira VA, Beskow S, Timm LC (2018) Assessment of soil loss vulnerability in data-scarce watersheds in southern Brazil. Cienc Agrotecnol 42:575–587
Taiz L, Zeiger E (2010) Plant physiology, 5th edn. Sinauer Associates Inc, Sunderland, MA
Thornthwaite CW (1948) An approach toward a rational classification of climate. Geogr Rev 38:55–94
Thornthwaite CW, Mather JR (1955) The water balance. Drexel Institute of Technology, Centerton, NJ
Timm LC, Oliveira JCM, Tominaga TT, Cássaro FAM, Reichardt K, Bacchi OOS (2002) Water balance of a sugarcane crop: quantitative and qualitative aspects of its measurement. Agriambi 6:57–62
Timm LC, Dourado-Neto D, Bacchi OOS, Hu W, Bortolotto RP, Silva AL, Bruno IP, Reichardt K (2011) Temporal variability of soil water storage evaluated for a coffee field. Aust J Soil Res 49:77–86
Toy TJ, Osterkamp WR (1995) The applicability of RUSLE to geomorphic studies. J Soil Water Conserv 50:498–503
Vieira SR, Lombardi Neto F (1995) Variabilidade espacial do potencial de erosão das chuvas do estado de São Paulo. Bragantia 54:405–412
Walling DE, Quine TA (1993) Use of caesium-137 as a tracer of erosion and sedimentation. In: Handbook for application of the Caesium-137 technique, Exeter. Department of Geography, University of Exeter, Exeter
Wischmeier WH, Smith DD (1978) Predicting rainfall erosion losses - a guide to conservation planning. United States Department of Agriculture, Washington DC
Wischmeier WH, Johnson CB, Cross BW (1971) A soil erodibility nomograph for farmland and construction sites. J Soil Water Conserv 26:189–193
Young R, Onstad C, Bosch D, Anderson W (1986) Agricultural nonpoint source pollution model: a watershed analysis tool, model documentation. Agricultural Research Service, US Department of Agriculture, Morris, MN

How Plants Absorb Nutrients from the Soil

16

16.1 Introduction

In the introduction of Chap. 8 some aspects of the nutrient uptake dynamics of plants were presented, summarized in Eq. (8.1). In that chapter we were concerned with the characterization of *M* (solution) and the processes of nutrient transfer in the soil. In this chapter, we will develop the subject a little further, discussing the flow of soil ions to the roots and we will introduce the mechanisms of nutrient uptake by plant roots.

16.2 Movement of Nutrients from the Soil to the Surface of Roots

The solid fraction of the soil, both mineral and organic, is the reservoir of nutrients for the plant. For a satisfactory development of the plant, it is necessary that the **ion activity** of each nutrient is adequate in the solution of the soil. This activity depends, above all, on the absorption by the roots and their "release" by the solid phase. The release of nutrients from the soil solid fraction is an important chapter in the study of soil fertility. In Chap. 8, we describe the phenomena of adsorption and ion exchange and we will not discuss further this matter. Let's just draw the reader's attention to the fact that nutrients, in their different forms, are bound to the solid phase with different energies. Thus, for example, NO_3^- and Cl^- are practically free of adsorption in most soils that have excess negative charges; K^+, Ca^{2+}, Mg^{2+}, and NH_4^+ are adsorbed electrically by clay minerals and organic matter; Fe^{3+} and Cu^{2+} can form complexes and chelates; the P can also form complexes of high insolubility with the oxides of Al and Fe; etc. The rate of release of these ions into the liquid phase depends on all these forms of adsorption. Once in the liquid phase, each nutrient can be absorbed by the roots, and this absorption depends on a series of problems that will be discussed later in this chapter. To be absorbed by the plant, a nutrient must be present in soil solution, and in contact with the active surface of the root system, in a form that can be absorbed and utilized by the plant. This "available" form of the different nutrients has been the object of attention of the soil chemists, being extensive the literature on the subject. In general, it can be said that the main factors controlling the transition from *M* (solid) to *M* (solution) (see Eq. 8.1) are solubility and oxidation potential. Sposito (1989) is a complete text on these physicochemical aspects of the soil.

Once in solution, two processes are responsible for the transfer of a nutrient from the soil to the plant: **diffusion** and **mass transport**. The diffusion comprises transport due to gradients of

K. Reichardt, L. C. Timm, *Soil, Plant and Atmosphere*, https://doi.org/10.1007/978-3-030-19322-5_16

chemical potential, measured by the activity of the ion in question in the soil solution; and the mass transport refers to all transport of ions carried along by the flow of water in the soil. Olsen and Kemper (1968) present an extensive theoretical discussion about the different factors involved in these ion transport processes.

16.3 Diffusion

The fundamental equation of the diffusion of a solute in the soil in given direction x is the **Equation of Fick**, seen in Chap. 8:

$$j_d = -\underbrace{\left[\theta D_0 \left(\frac{L}{L_e}\right)^2 \alpha \gamma\right]}_{D} \left(\frac{\partial C}{\partial x}\right) \quad (16.1)$$

where j_d is the flux density of a given ion in the soil by diffusion and x is the position coordinate measured directly in the soil. As already mentioned, the volume-based soil water content θ is included in the equation because it measures the available cross-section area for the flow, because the movement occurs only within the solution. In a cross section of soil A, only "$\theta \cdot A$" is available to the solution flow. The other factors mentioned in Eq. (16.1) were discussed in Chap. 8.

Thus, for example, if at a point P of the soil the concentration C_P of NO_3^- is 6.2 mg L^{-1} and at another point M, distant from $\Delta x = 10$ cm of P, the C_M concentration of NO_3^- is 2.9 mg L^{-1}, there will be a flow of NO_3^- from P to M, which can be calculated by the Fick equation. Let the diffusion coefficient D of NO_3^- in the soil be equal to 0.54×10^{-5} cm^2 s^{-1}. The concentration gradient $\partial C/\partial x$ can be approximated by finite differences $(C_M - C_P)/\Delta x$ and thus we have:

$$j_d = \frac{-0.54 \times 10^{-5}\left(2.9 \times 10^{-3} - 6.2 \times 10^{-3}\right)}{10}$$
$$= 1.78 \times 10^{-8}\,\mathrm{mg\,cm^{-2}\,s^{-1}}$$

In this example, P could be a generic point of the soil that is 10 cm from an absorbent root, on which is M. By diffusion, j_d would be the ionic flux density from the soil to the root.

16.4 Mass Flow

The transport of nutrients by **mass flow**, also called convection flow, depends strictly on the flow of water, since it comprises the amount of nutrients carried by the water per unit cross section of the flow per unit time. The water flux density q is conveniently described by the **Darcy-Buckingham equation** discussed in Chap. 7:

$$q = -K(\theta)\frac{\partial H}{\partial x} \quad (16.2)$$

where q is the water flux density (L H_2O m^{-2} day^{-1}), $K(\theta)$ the hydraulic conductivity of the soil (mm day^{-1}), and $\partial H/\partial x$ the hydraulic potential gradient (m m^{-1}).

Given the water flux density q, the mass flow density j_m of a nutrient can be calculated by the expression:

$$j_m = q \cdot C \quad (16.3)$$

C is the nutrient concentration in the water. If, as an example, the water flux is 0.2 L m^{-2} day^{-1}, or 0.2 mm day^{-1}, and the concentration of NO_3^- in the water is 6.2 g L^{-1}, we will have:

$$j_m = 0.2 \times 6.2 = 1.24\,\mathrm{mg\ of\ NO_3^-\,cm^{-2}\,day^{-1}}$$

In this example, if the water flow in the soil toward the roots, provoked by the evaporative demand of the atmosphere, would be q, the mass flow of nitrate into the plant would be j_m.

16.5 Relative Importance of Root Extension in Relation of Nutrient Absorption

The two processes described separately above occur simultaneously in the transport of nutrients from the soil to the plants. The magnitude of each process varies from situation to situation. In addition to these processes, plant nutrition is still

Table 16.1 Barber and Olsen's example of the sharing among root interception, mass flow, and diffusion, for a maize crop

		Quantity supplied by:		
Nutrient	Maize requirement in kg ha^{-1}	Root interception	Mass flow	Diffusion
N	190	2.2	188.0	0
P	39	1.1	2.2	35.8
K	196	4.5	39.2	152.5
Ca	39	67.3	168.2	0
S	22	1.1	21.3	0
Mo	0.01	0.001	0.02	0

affected by the extension of the root system of plants. The question can be posed as follows: either the nutrient moves from the soil to the root (diffusion and mass transport) or the root "directs" its growth to a point where it encounters the nutrient (**root interception**). By such an affirmative, we do not mean that roots have directed growth; they grow randomly and, as they grow, they explore new volumes of soil where they meet nutrients.

The relative importance of root interception, diffusion, and mass flow in maintaining an adequate concentration of a nutrient close to root absorption surfaces is difficult to determine. Even for a given soil-plant condition, the relative importance of each process varies with the time of day and point to point within the soil profile. Barber and Olsen (1968) made the first efforts to determine the contribution of each of these three processes, starting from very simplified hypotheses. Some of their examples are shown in Table 16.1.

16.6 Influence of Soil Physical Condition on the Transport of Nutrients

16.6.1 Soil Water Content

Soil water content varies greatly during the vegetative cycle, decreasing gradually in quantity while evapotranspiration proceeds, increasing steeply with rainfall or irrigation. The description of the water regime in a crop is a problem that requires the knowledge of water content and water potential in space and time, which is done by the water balance (Chap. 15). In the same way, from the nutritional point of view, it is important to know the soil water content and its matrix potential. In general, the greatest use of nutrients by plants occurs when soil water contents are kept as high as possible (close to field capacity), but do not cause aeration and temperature problems.

Soil water content, when appropriate, allows for potential transpiration by plants; the nutrients are absorbed by mass flow to the root surface and, in many cases, entering into the root and then to the upper part by the xylem. The mass flow of nutrients, directly proportional to the flow of water in the soil, described by the Darcy-Buckingham equation (Eq. 7.3, Chap. 7), is extremely affected by soil water content conditions. The hydraulic conductivity $K(\theta)$, which expresses the soil hydro-physical property of transmitting water, is very much affected by soil water contents. It is drastically reduced with relatively small decreases in θ; in general the relation $K(\theta)$ can be described by an exponential function. A K reduction of 100–1000 times is common for a 5% decrease in soil water content. Also, the potential gradient $\partial\Psi/\partial x$ is important. Often a considerable gradient implies in a reasonable flow despite low hydraulic conductivity of the soil. Roots, removing water from the soil, diminish the water potential increasing the gradient and allowing an adequate flow of water and nutrients, even for conditions of small K.

As seen above, the soil nutrient mass flow to the plants is affected by the transpiration, which in turn depends on the atmospheric conditions and the soil water content that affects the hydraulic conductivity of the soil and the potential gradient. With these variations in space and time, the description of the phenomenon becomes complicated. Thus, the results of most research on the subject, such as the early work of Barber and Olsen (1968), presented in Table 16.1, cannot be generalized. An updated text on the dynamics of nutrients in the soil-plant system is that of Havlin et al. (2014). The authors discuss the importance of the presence of **available water** in the soil for the **nutrient mass flow** from the soil to the plants, emphasizing that the **availability of nutrients** is characterized by diffusion and mass flow in the soil, factors directly related to soil water content. Beyond this topic, Havlin et al. (2014) deal in depth with aspects related to soil fertility and plant nutrition. With the same soil and the same plant, under different conditions, opposite results can be obtained. Soil moisture also affects the diffusion of nutrients. The main effect is to reduce the area available to the flow, when θ decreases. In addition, the effective path of diffusion increases with the decrease of θ, noting also significant increases of viscosity and negative adsorption γ. Porter et al. (1960), Partil et al. (1963), and Olsen et al. (1965) were the main authors to initiate studies of the effect of θ on the diffusion coefficient D of different ions in the soil. In general, their data shows reduction of the order of 10 times in the values of D by a reduction of 10% of water content. Under field conditions, the variations of θ in space and time make it very difficult to analyze the phenomena. In the same way as we discuss the case of mass flow, generalizations cannot be made easily, in the case of nutrient diffusion. Each particular situation should be carefully considered, taking into account the influence of all of the factors together.

Soil water content also affects root development and thus "root interception" of nutrients. High levels of water affect the aeration of the soil, damaging root growth and low levels of water; impede the flow of water in the soil, increasing its matrix potential, to prevent water absorption; and also slow down root growth. Indirectly, variations in soil water content also imply variations in soil consistency (mechanical properties) of great importance in radicular penetration and management practices.

16.6.2 Soil Air

The importance of **soil aeration** in the nutrition of a typical upland crop is generally related to the activity of microorganisms and to root respiration. The supply of air to the soil is inversely proportional to the water supply, hence the dilemma: high levels of water are beneficial to nutrition, but may compromise the activity of essential microorganisms and root respiration. It is therefore necessary to determine an optimal point between the supply of water and air to the plants in the root zone. Most of the time, attention is focused on water supply, since aeration problems are often temporary, often unnoticed.

The aeration of the soil occurs mainly by diffusion. The diffusion equation of one gas into another is described by Eq. (9.2) (Chap. 9), varying only the magnitude of the diffusion coefficient. In this case:

$$D = (\alpha - \theta) D_0 \left(\frac{L}{L_e}\right)^2 \qquad (16.4)$$

where α is the total porosity of the soil (Eq. 3.12) and $(\alpha - \theta)$ the water free porosity β (Eq. 3.30).

Under unsaturated soil conditions, water occupies the smaller pores, leaving larger pores for air diffuse. The relationship between the diffusion coefficient D of a gas in the soil and its diffusion coefficient D in the air is not large and, for average conditions of field crops, the ratio is of the order of 0.6 times $(\alpha - \theta)$.

Studies of the effects of soil aeration on plant nutrition are found in a large number of studies. In the field, any restriction on aeration is generally temporary and difficult to diagnose, and its relationship with nutrition is quite complex. Anaerobic conditions can, for example, promote the availability of Fe, Cu, Mo, and Mn. Variations of oxidation potentials and pH, resulting from variations in aeration, increase the complexity of the problem.

16.6.3 Soil Texture

The textural analysis characterizes a soil from the point of view of the distribution of the solid particle size of the constituents. This distribution determines a **porosity** and arrangement of characteristic particles, which, in turn, will determine its water properties, such as hydraulic conductivity and the relationship between the soil water content θ and matrix potential (the **soil water retention curve**). These water properties directly or indirectly affect nutrient uptake processes, that is, diffusion, mass flow, and root interception. In general, however, for a given soil-plant system, the textural features are practically invariant over time. Its influence becomes more indirect, that is, by variations in the water content of the system.

16.6.4 Soil Temperature

The availability and absorption of nutrients are affected by the temperature in all phases of Eq. (8.1). The microbiological activity, solubility of compounds, diffusion coefficients, root absorption, root permeability, metabolic activity, etc., are all affected by **soil temperature** variations. Studies that relate temperature variations with root absorption and nutrient accumulation in plants are numerous and diversified, making a global interpretation difficult. In general, the biological activities in the soil increase with the increase of temperature, until a maximum around 30 °C. It appears that nutrient absorption has different temperature dependence for the various nutrients. For example, Walker (1969) has shown that maize seed growth has a maximum at a soil temperature of 26 °C, decreasing rapidly with the cooling or heating of the soil. The nutrient uptake, expressed as shoot concentration, has different response curves. In general terms, nitrogen uptake increases from 12 to 18 °C, decreases from 18 to 26 °C, and increases again from 26 to 34 °C. The phosphorus concentration decreases when the temperature increased from 12 to 25 °C and then increased when the temperature increased from 25 to 34 °C. Potassium increased more than 100% when soil temperature was raised from 12 to 18 °C, remaining constant at higher temperatures. On the other hand, Knoll et al. (1964) found an increase in the absorption of phosphorus with a temperature increase of 15–25 °C. These results demonstrate the complexity of the relationship between nutrient absorption and temperature. Even under very controlled conditions, it is difficult to determine causes of effects in experiments of this nature, making it virtually impossible to generalize about the effect of temperature on plant nutrition.

16.6.5 Root System

The type of root system and its distribution along the profile are also extremely important factors in nutrient absorption, especially in relation to radicular interception. Root grows mostly in the vegetative growth stages of annual crops because in the reproductive phases the plant devotes most of its energy for flowering and grain maturation. At the end of the vegetative phase, the root size is mostly defined, and in modelling the **effective root depth** is an important parameter. In relation to plant species, grasses, belonging to the family poaceae, for example, have a fasciculate root system, well distributed in the superficial layers of the soil. Dicotyledons have a pivotal root system, with a main root that can reach great depth. In general, its distribution follows an exponential model that decreases with depth. In addition, there are particular cases, such as coffee and orange, which proliferate absorbent roots around the crown skirt. Certain plants of the cerrado (the South American savanah) have a very deep pivotal root that reaches several meters in soil depth to guarantee their survival in prolonged periods of drought. Aspects of the influence of physical factors on root development under tropical conditions are discussed, among others, by Reichardt (1976, 1980).

16.7 Examples of Nutrient Movement in the Soil

Let us first consider the model of a root of radius $r = \text{a}$ in a constant water content soil in which the diffusion coefficient of a nutrient is D. Assume also that the root absorbs the nutrient only by diffusion, at a constant rate, in such a way that its concentration on the root surface is C_a. If the transport of nutrients occurs only by diffusion, what should be the concentration of the soil solution C at any point between A and B, the point B being located at the mean distance between two active roots (see Fig. 16.1).

In Chap. 14, we have already said that for problems of this kind, the cylindrical coordinates (see Fig. 14.4) fit better. In the same way, as it was made for the differential equation of water movement, we can write Eq. (14.8) as follows, using only the coordinate r:

$$\frac{\partial(\theta C)}{\partial t} = \frac{1}{r}\frac{\partial}{\partial r}\left(rD\frac{\partial C}{\partial r}\right) \tag{16.5}$$

Since in the second member we have a derivative of a product of functions of r, that is, r and $D\partial C/\partial r$, by the chain rule we have:

$$\frac{\partial(\theta C)}{\partial t} = \frac{1}{r}\left(rD\frac{\partial^2 C}{\partial r^2} + \frac{\partial r}{\partial r}D\frac{\partial C}{\partial r}\right)$$

$$\frac{\partial(\theta C)}{\partial t} = D\left(\frac{\partial^2 C}{\partial r^2} + \frac{1}{r}\frac{\partial C}{\partial r}\right)$$

or

$$\frac{\partial C}{\partial t} = D'\left(\frac{\partial^2 C}{\partial r^2} + \frac{1}{r}\frac{\partial C}{\partial r}\right) \tag{16.6}$$

where $D' = D/\theta$.

It is opportune at this point to show the reader that Eq. (16.6) (cylindrical coordinates) is identical to Eq. (8.41a) (orthogonal Cartesian coordinates) for r tending to ∞, since in this case for very far distances two close radii become parallel and the coordinate r can be replaced by. If $r \to \infty$, $(1/r) = 0$, and:

$$\frac{1}{r}\frac{\partial C}{\partial r} = 0 \quad \text{and} \quad \frac{\partial C}{\partial t} = D\frac{\partial^2 C}{\partial r^2}$$

Thus, it is verified that when $r \to \infty$, the cylindrical geometry tends toward the Cartesian.

Our problem, however, is a case of typically cylindrical geometry, in dynamic equilibrium, so that $\partial C/\partial t = 0$ and Eq. (16.6) becomes:

$$\frac{\mathrm{d}^2 C}{\mathrm{d}r^2} + \frac{1}{r}\frac{\mathrm{d}C}{\mathrm{d}r} = 0 \tag{16.7}$$

subject to the conditions:

$r = a$, $C = C_a$ and $r = b$, $C = C_b$

Let's assume Eq. (16.7) is of the type:

$$C = k_1 \ln r + k_2 \tag{16.8}$$

and thus it must satisfy Eq. (16.7), which contains a first derivative and a second derivative. Recalling that the derivative of ln r is $1/r$, the derivative of Eq. (16.8) is:

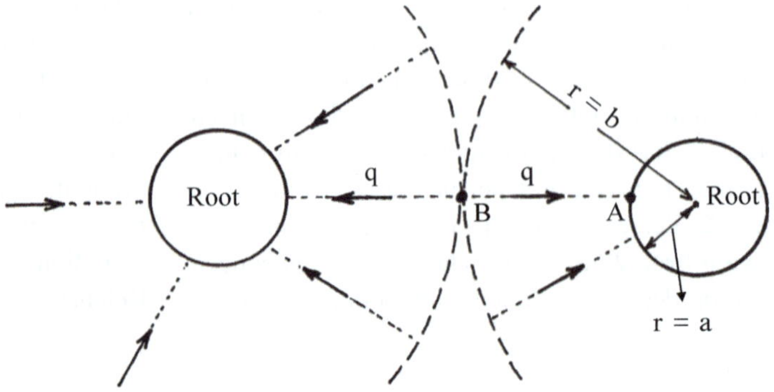

Fig. 16.1 Cut of two parallel roots indicating the radial flux q from point B to point A at the root surface

$$\frac{\partial C}{\partial r} = \frac{k_1}{r} = k_1 \cdot r^{-1}$$

and

$$\frac{d^2C}{dr^2} = -k_1 \cdot r^{-2}$$

substituting these values in Eq. (16.7), we have:

$$-k_1 \cdot r^{-2} + \frac{1}{r} k_1 \cdot r^{-1} = 0$$

which confirms that Eq. (16.8) is the general solution of Eq. (16.7). The particular solution will be obtained using the boundary conditions:

$$C_a = k_1 \ln a + k_2$$

$$C_b = k_1 \ln b + k_2$$

So that:

$$k_1 = \frac{C_a - C_b}{\ln\left(\frac{a}{b}\right)}$$

and

$$k_2 = C_a - \frac{C_a - C_b}{\ln\left(\frac{a}{b}\right)} \times \ln a$$

and the particular solution will be:

$$C = \frac{C_a - C_b}{\ln\left(\frac{a}{b}\right)} \cdot \ln r + C_a - \frac{C_a - C_b}{\ln\left(\frac{a}{b}\right)} \cdot \ln a$$

or

$$C = C_a + \frac{C_a - C_b}{\ln\left(\frac{a}{b}\right)} \cdot \ln\left(\frac{r}{a}\right) \quad (16.9)$$

With data for C_a, C_b, a, and b, it is easy to calculate C for any r between A and B, as the problem asks for besides C, we could still be interested in the input flow of nutrients to the root:

$$j_d = -D\frac{\partial C}{\partial r}$$

Since $\partial C/\partial r = k_1/r$, for $r = a$, we have $\partial C/\partial r = k_1/a$, so that:

$$j_d = -D\frac{k_1}{a}$$

It is often interesting to know the amount that penetrates the root per unit of length and time. The area of 1 cm radius root a is $(2\pi a)$ cm^2. So that:

$$j_d = \frac{Q}{A \cdot t} = \frac{Q}{2\pi a \cdot t} = D\frac{k_1}{a}$$

where:

$$\frac{Q}{t} = 2\pi k_1 D \ \left(\text{g s}^{-1} \text{ per cm of root}\right)$$

Let us now see how the previous example applies to a maize crop, logically with a series of approximations. It is a maize crop that in 3 months produces 8 tons of dry matter per hectare. The dry matter of this maize has, on average, 1.5% of nitrogen N. The number of plants is 20,000 per hectare and each one has 900 m of active roots of average diameter of 0.1 cm. Every N that reaches the surface of the root is absorbed, so in the previous equations, $C_a = 0$, which is a way to say that this soil is deficient in N. Assuming that the only process that transports nutrients until the root is the diffusion, calculate the mean concentration of the soil solution in g N cm^{-3}, at a distance of 5 cm from the roots, so that the absorption occurs constantly, without harm to plant growth. A great restriction of this example is neglecting the N mass flow due to evapotranspiration.

With the data presented, one can easily determine the absorption of N, at least in average terms:

1. Total nitrogen mass in the crop $= 8000 \times 0.015 = 120$ kg N ha^{-1}
2. $\text{kg of N/plant} = \frac{120}{20{,}000} = 6 \times 10^{-3} \frac{\text{kg of N}}{\text{plant}}$
3. $\text{kg of N/cm of root} = \frac{6 \times 10^{-3} \text{ kg of N}}{90{,}000 \text{ cm of root}}$

$$= 6.67 \times 10^{-8} \frac{\text{kg of N}}{\text{cm of root}}$$

4. kg of N/cm of root s

$$= 6.67 \times 10^{-8} \frac{\text{kg of N}}{\text{cm of root}} \times \frac{3 \text{ months}}{90 \text{ days}} \times \frac{1 \text{ day}}{86{,}400 \text{ s}} = 2.57 \times 10^{-14} \frac{\text{kg of N}}{\text{cm of root s}}$$

which means that the flow of N into each centimeter of root must have been, on average, equal to 2.57×10^{-14} kg of N s^{-1}, so that crop would establish itself and produce what it produced.

If we assume the same conditions of the previous problem, the solution will be given by Eq. (16.9) and the flow is given by:

$$j_d = -D\frac{dC}{dr}$$

derivating Eq. (16.9) in relation to r, we obtain the gradient:

$$\frac{dC}{dr} = 0 + \frac{C_a - C_b}{\ln\frac{a}{b}} \cdot \frac{1}{a} \cdot \frac{1}{r}$$

So that the flux in $r = a$ is given by:

$$j_d = -\frac{D(C_a - C_b)}{a^2 \cdot \ln\left(\frac{a}{b}\right)}$$

and for 1 cm of root of area $2\pi a$, we have:

$$\frac{Q}{t} = \frac{2\pi D(C_a - C_b)}{a \cdot \ln\left(\frac{a}{b}\right)}$$

Using the data of the exercise and considering $D = 2 \times 10^{-10}$ m^2 s^{-1}, we have:

$$2.57 \times 10^{-14} = -\frac{2\pi \cdot 2 \times 10^{-10}(0 - C_b)}{0.0001 \times \ln\left(\frac{0.1}{5}\right)}$$

$$C_b = 8.00 \times 10^{-9} \text{ kg N m}^{-3}$$

The reader must have realized that the examples seen are quite simplified and were presented here only for didactic reasons, to gain familiarity with the concepts used. These are cases of dynamic equilibrium. Most of the time, this does not happen and the equations to be solved are much more difficult, depending on each problem.

16.8 Nutrient Root Absorption

In Chap. 4, we said that the entry of nutrients into the plant can be passive (by diffusion through intercellular spaces) and active (by metabolic processes through cell membranes). In Fig. 4.8 we saw a cross section of a root. When this root is in contact with soil solution, the nutrients can diffuse into the intercellular spaces, which include the cell walls. This space is termed "outer space" (or free space) for the processes of diffusion and ion exchange between ions and the electrically charged radii of cell walls. It is separated from "inner space" (inner space or non-free space) by cell membranes, which are barriers of selective permeability. The inner space is accessible only by active transportation of the nutrients, which requires metabolic energy. The first observations that led the researchers to recognize this active transport were:

1. Nutrient uptake depends on the temperature, approximately doubling for every 10 °C increase, just as the whole metabolic process.
2. Oxygen is required.
3. The process is sensitive to inhibitors (e.g., CN^-).
4. Depends on the nature and concentration of the nutrient (e.g., K is absorbed more readily when supplied as KCl than as K_2SO_4).
5. There is interference of one nutrient in the absorption of another.

The absorption process is often adequately described by the **Michaelis-Menten equation**, which is based on the hypothesis of the **ion carriers**. This equation can be written in the form:

$$j = \frac{j_{max} \cdot C}{K_m + C} \tag{16.10}$$

where:

j = absorption rate of a nutrient in kg m^{-2} s^{-1} or mol g^{-1} s^{-1}, that is, the amount of absorbed nutrient in moles per gram of root.
j_{max} = maximum absorption rate.

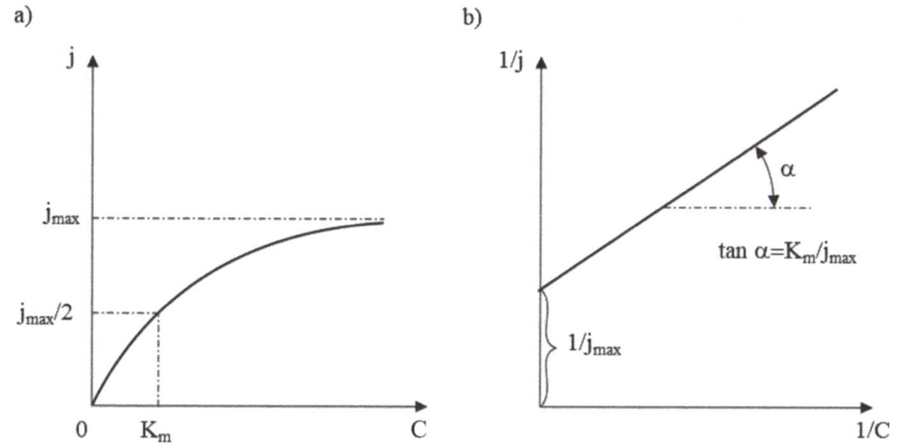

Fig. 16.2 Excised root absorption of an ion in solution, illustrated (**a**) by a graph of the ion flux j as a function of the solution concentration C and (**b**) by a graph of $1/j$ as a function of $1/C$, constructed with the same data of case (**a**)

K_m = **Michaelis-Menten constant** that gives an idea of the affinity between the nutrient and the carrier.

C = concentration of the solution in which the nutrient is absorbed.

Graphically, the phenomenon is presented in Fig. 16.2a. It is seen by this figure that j increases with C, but not linearly, having a j_{max} for high concentrations, that cannot be surpassed by biological restrictions. The value of K_m for a given plant in relation to the absorption of a given nutrient is equal to the value of the concentration for which $j = 1/2\ j_{max}$. To verify this, simply substitute j for $1/2\ j_{max}$ in Eq. (16.10) and verify that, under these conditions, $K_m = C$. The units are also identical to the units of C.

From Eq. (16.10), it can be seen that the absorption of a given nutrient by a given plant depends on the parameters j_{max} and K_m. These can be obtained in experiments in which the absorption j is studied as a function of the solution concentration C, which has been made with plants or roots placed in nutrient solutions of variable concentration. Thus, a curve of the type of Fig. 16.2a is obtained experimentally. Since it is asymptotic for values of j near the j_{max}, the latter becomes difficult to be determined precisely. The problem is circumvented by plotting $1/j$ versus $1/C$, as shown in Fig. 16.2b. In this graph, from Eq. (16.10), it is easy to verify that:

$$\frac{1}{j} = \frac{K_m}{j_{max}} \cdot \frac{1}{C} + \frac{1}{j_{max}}$$

is a straight line equation when $1/j$ is taken as the dependent variable and $1/C$ as the independent variable. Based on such a graph it is very easy to find j_{max} and K_m.

During the absorption of several nutrients, competition phenomena still appear, one element interfering with or inhibiting the absorption of another. On the other hand, for high concentrations, it seems that other absorption mechanisms start to work, since absorption may be larger than j_{max}, with another curve similar to that of Fig. 16.2a, with a second j_{max}, at high concentrations. The subject is complicated and runs away from the goals of this book. Epstein (1972) was one of the pioneers in root absorption and wrote a great text that presents details on the subject.

In the case of potassium, Yamada et al. (1982) present aspects about the availability of K in Brazilian soils, K functions in the plant, absorption mechanisms, and potassium nutrition of the main agricultural crops in Brazil. It is an example of a work that applies the concepts seen in this chapter. Havlin et al. (2014) apply the concepts seen here also for potassium and other elements.

16.9 Nutrient Balance

In the same way as we did for water in Chap. 15, we can make a balance of a given nutrient in a crop, which is the accounting for all additions and subtractions of that nutrient in a soil layer. Balances are made for any nutrient, but given the importance of nitrogen, most of the work found in the literature refers to this nutrient. In this item, we will only introduce the subject and we will talk, in particular, of the nitrogen.

Given a soil layer of thickness L (cm), the following components of the nitrogen balance are important:

1. Addition by mineral fertilizer (MF). In this case, the most common forms of nitrogen fertilizer are urea, ammonium sulfate, ammonium nitrate, sodium nitrate, etc.
2. Addition by organic fertilizer (OF). Sprays, green manures, residues from previous crops, and other organic compounds are the most common.
3. Addition by **biological fixation** of atmospheric nitrogen (BF). The legumes—common beans, soybean, etc.—have the ability, in symbiosis with microorganisms (such as *Rhizobium* spp.), to fix atmospheric nitrogen N_2 and, finally, transform it into protein. Other plants, such as grasses (poaceae), can also fix the atmospheric N_2 by a nonsymbiotic association with microorganisms found in the soil.
4. Addition by rain (AR). Small (but not negligible) amounts of nitrogen (usually in the form of ammonia) can be added to crops by rainfall. It is nitrogen transformed into ammonia during electrical discharges into the atmosphere. Today, depending on the region, air pollution can also contribute with nitrogenous compounds.
5. Addition by irrigation water (AI). Here we include only the N present in the irrigation water. The case of fertigation can be included here or in item 1, MF.
6. Extraction or export by crops (EC). Each crop, at the end of its cycle, contains a reasonable quantity of nitrogen. As the products are taken out of the area (grains, fruits, tubers, green mass, etc.), they represent loss of N to the soil. Hence the need to replenish N through fertilizers in order to maintain soil productivity.
   Even vegetable wastes are often burned in the area and, under these conditions, represent nitrogen loss. Only when incorporated into the soil is it not entirely lost.
7. Losses by volatilization (VL). It is the loss of nitrogen by passing from the solid state to the gaseous phase. It usually occurs with fertilizers: urea and ammonium sulfate can volatilize and a good proportion can be lost. Soil pH, temperature, and fertilizer application are the main factors that affect volatilization.
8. Denitrification losses (DL). Under special conditions, especially lack of oxygen, microorganisms can supply their oxygen needs using nitrates (NO_3^-). The final result of this use is N_2, passing through NO_2^-, NO, and N_2O.
9. Leaching losses (LL). Nitrogen may also be lost through the bottom of the volume element at depth L, being carried by water in the form of NO_3^-, NH_4^-, or humidified organic compounds. It is the mass flow by the deep drainage. One way to estimate the leaching is by sampling the soil solution by means of solution extractors which are made up of porous coups subjected to vacuum (Figs. 16.3 and 16.4).

When the suction probes or soil solution extractors are installed in an unsaturated soil, say with Ψ_m of the order of -100 to -200 cmH_2O, at a depth $z = L$, they do not fill with water because they are at atmospheric pressure, i.e., $\Psi = 0$. When applying vacuum to them (points A, B, and C in Fig. 16.3, the pressure inside the cup will be of the order of -1 atm or -1000 cmH_2O), and because of the potential difference $\Delta\Psi$ between the inside the cup and the soil, the solution of the soil is directed toward the capsule. The solution flow depends on $\Delta\Psi$ and the soil water content, which, in turn, determines the hydraulic conductivity of the

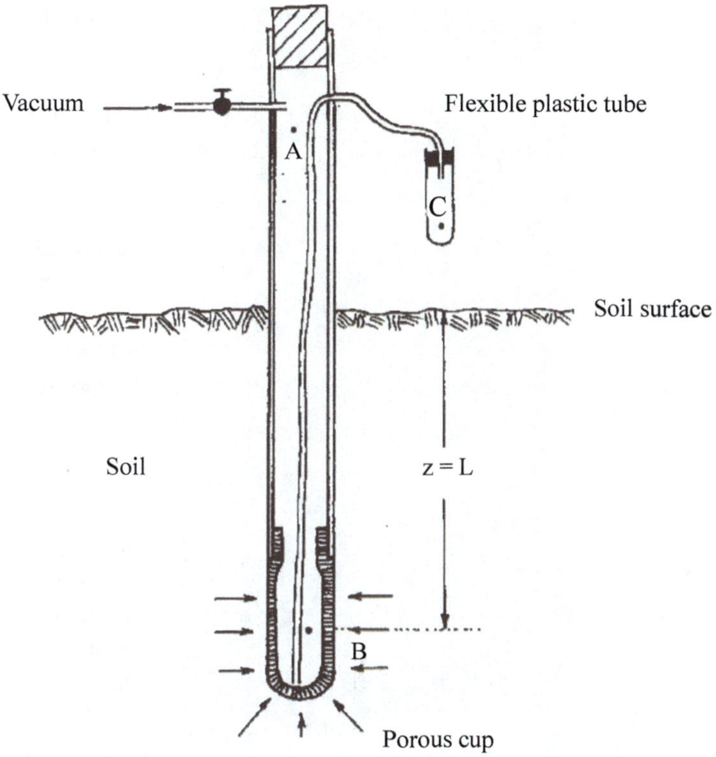

Fig. 16.3 A porous cup soil solution extractor installed at a depth L

soil. The level of the solution in the capsule rises above the point B and leaves the tip of the flexible plastic tube immersed in the extracted soil solution. After 1–2 h of vacuum (depending on the soil water content), the vacuum is disconnected and the atmospheric pressure is automatically exposed to point A. Since C is still under vacuum, the solution rises and will fill the test tube. It is a clear solution, ready for analysis of ionic concentrations C_i. As we saw in Chap. 8, the product of water flow density q, by concentration C_i, gives us the flux density of solute j. If these instruments are used in conjunction with tensiometers, as shown in Chap. 15 and in Fig. 16.4 for the determination of the deep drainage, one can estimate the quantities of leached solute.

From the algebraic sum of the nine previous items results the N storage (or its variation) (ΔA_N) in the considered layer 0–L (see Eq. 3.22, Chap. 3). Even with no additions or losses, the nitrogen in this layer is extremely dynamic. By the process of mineralization, it goes from organic to mineral; by absorption processes by microorganisms it can pass from mineral to the organic phase. NO_3^- can be adsorbed by positive charges of the soil and NH_4^+ by negative charges. Thus, the amount of N available to the plants in the soil solution is quite variable. The processes that determine this availability are complex and interdependent.

The nitrogen balance (Fig. 16.5) can then be written in the form:

$$\begin{aligned} &\mathrm{MF} + \mathrm{OF} + \mathrm{BF} + \mathrm{AR} + \mathrm{AI} - \mathrm{EC} - \mathrm{VL} - \mathrm{DL} \\ &\quad - \mathrm{LL} \pm \Delta A_{\mathrm{N}} = 0 \end{aligned} \tag{16.11}$$

in which it is seen that the number of components is much larger than in the case of the water balance.

The literature on **nitrogen balance** or on the various components of the balance sheet is vast and we will not make a review here. Just as an example, we indicate some work done in Brazil.

Fig. 16.4 A sugarcane crop showing in the front two tensiometers and in the back center a soil solution extractor

Reichardt et al. (1977) discuss the problem of extraction and analysis of nitrates in the soil solution; Libardi and Reichardt (1978) studied the fate of urea applied to a bean culture using ^{15}N as a tracer; Reichardt et al. (1979) presented the dynamics of nitrogen in a corn crop; Cervellini et al. (1980) and Meirelles et al. (1980) studied the fate of ^{15}N applied as ammonium sulfate to a bean crop; and Reichardt et al. (1982) reviewed the nitrogen balance in tropical crops. Reichardt (1990) discussed the problem of soil spatial variability in the estimation of biological nitrogen fixation (FB).

Interesting contributions in this area were given by Fenilli et al. (2007b), who evaluated the losses of N by volatilization of the urea fertilizer in a coffee crop, and Fenilli et al. (2007a), which present a complete balance of a coffee crop conducted in Piracicaba. Bruno et al. (2011) present urea nitrogen uptake data for 1 year in fertigated coffee in western Bahia, Brazil. With the same data from the previous work, Bruno et al. (2015) present the complete balance of the crop of fertigated coffee.

16.10 Use of Isotopes in Agricultural Experimentation

Isotopes, both stable and radioactive, have been successfully used as tracers or markers of their respective elements. Each element of the periodic table has isotopes—atoms of the same species—behaving chemically and biologically in the same way, differing only in some physical properties. They have in the nucleus the same number of protons (which defines the element), but a different number of neutrons, which makes them different in terms of mass. Its majority is natural, being present in nature since the formation of our planet Earth; some are artificial, produced in the laboratory by nucleation reactions. When stable, they differ only in atomic weight and, when radioactive, emit radiation. The important thing is that both differ in a measurable way and can be detected. **Stable isotopes** are detected by **mass spectrometers**, instruments capable of distinguishing atomic weights. **Radioactive isotopes** are detected by radiation detectors, which include a series of instruments, depending on the type of radiation emitted by the radioisotope. Radioisotopes are unstable isotopes that seek their stability by emitting radiation. These can be detected easily and in minimal quantities, i.e., at concentrations as low as 10^{-11} (about 1—100 billion). Each radioisotope has its own characteristics and emits one or more types of radiation (in particular α, β^-, β^+, γ, and neutrons), with one or more energies and at a rate that depends on their **half-life**. As this rate decreases exponentially over time, the half-life $T_{1/2}$ has been defined as the time required for any emission rate to be reduced to its half. Depending on these properties, each radioisotope is suitable (or not) for different uses as a tracer. The type and energy of radiation affect the detection. A short half-life may limit its use in longer duration experiments.

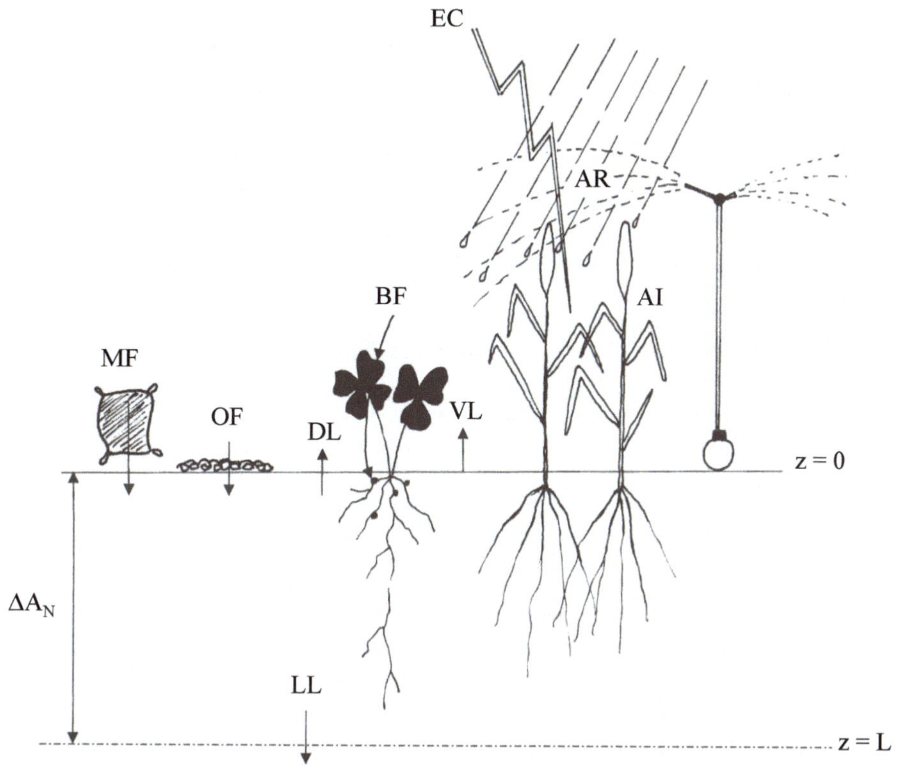

Fig. 16.5 Illustration of a nitrogen balance: *MF*, mineral fertilizer; *OF*, organic fertilizer; *DL*, denitrification loss; *BF*, biological N fixation; volatilization loss; *EC*, electric discharges; *AR*, irrigation water contribution; *AI*, crop export; *LL*, leaching at depth *L*; and ΔA_N, the change in soil N storage during the period of the balance

Radioisotopes are found in nature, as is the case of ^{3}H (tritium), ^{14}C, ^{40}K, ^{226}Ra, ^{235}U, and others; some are produced continuously by nuclear reactions that occur in the atmosphere by interaction of solar or cosmic radiation with dispersed atoms, such that their rate of production is in equilibrium with the rate of radioactive disintegration, thus presenting a constant global content, as is the case with ^{14}C; others have half-lives longer than the age of our planet and are still present in the environment and are made up of radioactive series beginning with uranium, thorium, and actinium, passing, for example, by ^{226}Ra, finishing with some isotope of lead. The vast majority of radioisotopes used as tracer are, however, artificially produced in nuclear reactors. Table 16.2 shows the main radioisotopes used as tracers in the SPAS. To use them as tracers, they are added to their stable isotope in amounts that allow their detection until the end of the experiment. When possible, they should be added in the same chemical form in which the stable compound is present and thus become true tracers. If a phosphorus study involves superphosphate, the ^{32}P radioactive isotope must also be present in the form of superphosphate. This is sometimes more difficult, as in the case of studies with herbicides, in which ^{14}C should be incorporated into the complex molecule of the respective pesticide.

The intensity of radioactive materials is measured in terms of **radioactive activity** (Bequerel, 1 Bq = 1 disintegration/s, Curie, 1 Ci = 3.7×10^{10} disintegrations/s). Since each detected radiation corresponds to a count in the measuring instrument, the activity is also measured in counts per second (cps) or counts per minute (cpm). In addition, because the measured activity depends on the sample size (extensive magnitude), the activities are also

Table 16.2 Most important radioisotopes used in agricultural research

More abundant isotope	Tracer radioisotope	Energy of radiation (MeV)		Half- life
		Beta	Gamma	
Calcium-40	^{45}Ca	0.254		153 days
Carbon-12	^{14}C	0.156		5720 years
Cesium-133	^{137}Cs	0.52; 1.18	0.662	30 years
Cobalt-59	^{60}Co	0.31	1.17; 1.33	5.27 years
Strontium-88	^{89}Sr	1.47		50.4 days
Sulfur-32	^{35}S	0.168		86.7 days
Hydrogen-1	^{3}H	0.0181	0.36; 0.08; 0.72	12.26 years
Iodine-127	^{131}I	0.61; 0.25; 0.85	0.032; 1.35; 0.95	8.05 years
Magnesium-24	^{28}Mg	0.45	0.94; 1.46	21.3 h
Manganese-55	^{52}Mn	0.60		5.7 days
Phosphorus-31	^{32}P	1.71	1.46	14.3 days
Potassium-39	^{40}K	1.32	1.08	1.3×10^9 years
Rubidium-85	^{86}Rb	0.7; 1.77		18.7 days
Zinc-64	^{65}Zn	0.33	1.11	245 days

expressed in terms of the specific activity s_a, which is the activity per unit mass or volume, i.e., cps g^{-1}, cpm g^{-1}. The use of isotopes in experiments covers a wide range of forms, from the simple presence of the tracer at a given point in the system until its follow-up in dynamic studies. The basis of most applications is the principle of **isotopic dilution**, whereby "given a constant amount of radioactivity, the specific activity is inversely proportional to the total element or substance present in the system." This means that for a fixed amount of $^{40}CaCl_2$, the more stable compound $CaCl_2$ is added, the lower the specific activity of the blend (when homogeneous). This principle allows, for example, to estimate the amount of a nutrient in a plant, which came from the fertilizer applied to the soil. If a phosphate fertilization study uses ^{32}P-labeled superphosphate, one can make for the P that is in the plant a distinction between the one that comes from the fertilizer and the one that comes from the soil. In this case, the phosphorus derived from the fertilizer (Pddf) is defined, which can be calculated by the expression:

$$\%\text{Pddf} = \left(\frac{s_a \text{ of}^{32}\text{P in the plant}}{s_a \text{ of}^{32}\text{P in the fertilizer}}\right) \times 100 \quad (16.12)$$

If a soil is fertilized with a fertilizer that contains a specific activity $s_a = 102{,}441$ cps/μ mol P and a plant is grown on it, the plant is exposed to sources of P, the labeled fertilizer, and the native soil P which is not radioactive. It is easy to understand that if the plant only absorbs P from the fertilizer, in the end its specific activity would be equal to that of the fertilizer. On the contrary, if it absorbs only P from the soil, its specific activity would be $s_a = 0$. As it absorbs from both, and according to the availability of each, its value of s_a will be intermediate. Taking the plant and measuring its s_a, for example, 35,758 cps/μmol P, we would have % Pddf = (35,758/102,441) × 100 = 35%, which means that of all P found in the plant, 35% came from the fertilizer and 65% from the soil. If the total extraction of the crop was of 15 kg ha^{-1} of P, 0.35 × 15 = 5.25 kg ha^{-1} of P came from the fertilizer and 9.75 comes from the soil. The efficiency of the fertilization was (5.25/60) × 100 = 8.75%, if the fertilization was 60 kg ha^{-1} of P.

It can be seen that, with the use of markers, fertilization efficiency, fertilizer application optimization, rates, times, crop responses, etc., can all be studied.

Stable isotopes can also be used for the same purposes. Not always a certain element has a

Table 16.3 Most commonly used stable isotopes in agricultural experimentation

Element	Isotope	Isotopic abundance
Carbon	^{12}C	99.985
	^{13}C	0.015
	^{14}C	Traces
Hydrogen	^{1}H	98.89
	^{2}H	1.11
	^{3}H	Traces
Nitrogen	^{14}N	99.635
	^{15}N	0.365
Oxygen	^{16}O	99.759
	^{17}O	0.037
	^{18}O	0.204
Sulfur	^{32}S	95.0
	^{33}S	0.76
	^{34}S	4.22
	^{36}S	0.014

suitable radioisotope and there is made use of the stable isotope, with the advantage of not being radioactive. Table 16.3 shows the most used ones.

The basis of the use of stable isotopes lies in the concept of **isotopic abundance**. Abundance is the proportion in which the isotopes of a given element appear in nature. The nitrogen element, for example, has only two stable and natural isotopes, ^{14}N and ^{15}N. The abundance of ^{15}N in a sample is:

$$\%\ ^{15}\text{N atoms} = \left(\frac{\text{number of}\,^{15}\text{N atoms}}{\text{number of}\,^{15}\text{N}+^{14}\text{N atoms}}\right) \times 100 \tag{16.13}$$

Since N is a very dynamic element in SPAS, abundances of ^{14}N and ^{15}N are practically constant and the same in any biological sample. Their values are 99.635% for ^{14}N (more abundant) and 0.365% for ^{15}N. Samples can be enriched in ^{15}N by special laboratory techniques, and the amounts of ^{15}N in enriched samples can be expressed in terms of % excess atoms over the natural abundance of 0.365%.

Dynamic processes in which the flow of nitrogenous compounds is affected by the atomic weight discriminate the ^{14}N of the ^{15}N. This phenomenon is used in the laboratory to enrich samples in ^{15}N. Using batteries of long ion exchange columns, ammonium sulfate can be enriched in ^{15}N. In this way, we can have a compound like a nitrogen fertilizer, enriched at high levels of ^{15}N, for example, a sample of ammonium sulfate with an abundance of 10.365% of ^{15}N. Its abundance in ^{14}N is 89.635% and it has 10.000% atoms in excess of ^{15}N.

This fertilizer can be used as a tracer in agronomic studies. If a soil is fertilized with this material, the crop established there is exposed to two sources of nitrogen at its disposal: the native N of the soil with abundance of 0.365% and the fertilizer with abundance of 10.365%. As the crop makes use of both sources, in proportion to their availability, we can also calculate the **nitrogen derived from the fertilizer** (Nddf), given by the expression:

$$\%\ \text{Nddf} = \left(\frac{\%\ ^{15}\text{N atoms in excess in the plant}}{\%\ ^{15}\text{N atoms in excess in the fertilizer}}\right) \times 100 \tag{16.14}$$

If, for example, the plant grown on the soil shown above presents an abundance of 7.855%, it will have a value of ^{15}N atoms in excess of 7.855 − 0.365 = 7.490% and Nddf% = (7.490/10.000) × 100 = 65%. Its interpretation is the same as Pddf.

Important studies performed in Brazil that used ^{15}N as a tracer are those of Fenilli et al. (2007a, c), Bortolotto et al. (2011), and Bruno et al. (2011).

Legumes have a third source of nitrogen, the atmospheric N_2, absorbed in symbiosis with bacteria *Rhizobium* spp. Which form nodules at their roots. When it is desired to make a distinction of this **biological nitrogen fixation** (BNF), it is necessary to cultivate a N_2 non-fixing plant next to the legume, in the same soil fertilized with ^{15}N. Although it is not a legume, it should have similar size and development, so that the comparison would be reasonable. In many cases, a grass is used, such as wheat, barley, and oats, but with the discovery of the nonsymbiotic fixation of atmospheric nitrogen by the poaceae (grasses), especially sugarcane, the non-fixing plant should be well thought out. Imagine that in the previous example, the plant was barley and that soybean plants grown in the same situation had an abundance of 5.136%. The abundance of soybeans is lower because it has three N sources and therefore absorbs less labeled fertilizer. In this case, the nitrogen derived from the atmosphere (Ndda) is calculated by the expression:

$$\%\text{Ndda} = \left[1 - \left(\frac{\%\ ^{15}\text{N atoms in excess in the legume}}{\%\ ^{15}\text{N atoms in excess in the non fixing plant}}\right)\right] \times 100 \qquad (16.15)$$

and, in the example, we would have Ndda % = [1 – (4.771/7.490)] × 100 = 36.6%. Thus, from the N contained in soybean, 36.6% came from the atmosphere, and the remaining 63.4% came from both the soil and the fertilizer, in a proportion that cannot be calculated and, of course, different from that of the barley.

The Nddf concept can be applied to any N compartment in the crop, as long as each is sampled separately. In a study with sugarcane with ^{15}N, started in 1997 and in which the tracer could be followed for 5 years, Basanta et al. (2003) separately calculated the Nddf for leaves, straw, stems, soil, and soil solution, following the "pulse" of fertilizer applied to the cane plant, until the third ratoon, succeeding in making a nitrogen balance. The labeled cane plant straw was also used as mulching and could be followed by its mineralization over the years. This study was later compared to the mineralization of crop residues from several tropical regions in the world by Dourado-Neto et al. (2010).

The concept of **A Value** of a soil, which refers to the availability of nutrients, especially in relation to nitrogen and phosphorus, is also linked to the use of isotopes. Very good manuals for these applications of nuclear energy in agriculture are those of Vose (1980), L'Annunziata (1998), Hardarson (1990), and IAEA (2001).

Another way to use stable isotopes as tracers is by means of small (but significant) natural variations in abundance. As we have already said, dynamic processes occurring in the SPAS lead to isotope discrimination. In the first exercise of Chap. 2, we saw the 18 types of water molecules, which differ only in weight. When water is lost from a Class A tank, the phase change process discriminates the 18 types, and the lighter ones evaporate more easily. Thus, after some time, the water remaining in the tank is somewhat enriched in ^{18}O and ^{2}H. In other dynamic processes there is also discrimination of ^{15}N in relation to ^{14}N, ^{13}C in relation to ^{12}C, and so on. As these abundance variations are very small, a more sensitive measure is used, which is the **isotopic ratio** given in δ values. For the $^{13}C/^{12}C$ relationship, for example, the value of δ is calculated by:

$$\delta^{13}\text{C}‰ = \left[\left(\frac{^{13}\text{C}/^{12}\text{C in the sample}}{^{13}\text{C}/^{12}\text{C in the standard}}\right) - 1\right] \times 1000 \qquad (16.16)$$

and standards are reproducible and internationally chosen samples. The most commonly used isotopic ratios are $^{2}H/^{1}H$, $^{13}C/^{12}C$, $^{15}N/^{14}N$, and $^{18}O/^{16}O$ (Table 16.3). The plants discriminate the C isotopes in the photosynthetic process. **C3 plants** (see Chap. 4) have values of $\delta\ ^{13}C‰$ in the range of −11 to −14, while **C4 plants** from −25 to −30 C‰. This fact makes possible the study of interesting aspects in the carbon cycle in different ecosystems, as it is done in the comparison of carbon storage (or carbon stock) in the soil in tropical natural forests (mainly C4 species) and in introduced pastures (mainly with C3 plants).

Even differences in cattle grazing (between grasses and legumes) can be studied by the $\delta\ ^{13}C$ values found in their feces.

The use of the $\delta\ ^{13}C$ ratio is also very useful in studies of the **global carbon balance** in the atmosphere. The increase in CO_2 concentration in atmospheric air is a concern with regard to the **global changes** that occur on planet Earth, including the **greenhouse effect**. In the carbon balance, CO_2 emissions from burning fossil fuels are important and, from the agricultural point of view, the shift from forest areas to pasture and agricultural crops, which also causes emissions by burning the carbon stock contained in the forests of stable form. **Carbon sequestration** is done by photosynthesis, mainly by seaweed, forests, pastures, and agricultural crops. The balance between emission and sequestration defines the levels of CO_2 in the atmosphere. Cerri et al. (1991) is an example of a study of this nature and Rosenzweig and Hillel (1998) deal with the subject in a global way.

The main controversy over the causes of **global warming** is the question if the rise in air temperature recorded in recent decades is caused by natural global processes or **anthropogenic** influence through increases in CO_2 concentration in the atmosphere caused by the burning of petroleum and coal. Kutilek and Nielsen (2010) dealt with this subject with propriety, making an analysis of climate variations from prehistory to now.

Hydrological studies make use of $\delta\ ^{18}O$ in very varied ways. The value of $\delta\ ^{18}O$ is different for seawater (taken as standard) and freshwater to such an extent that different water colors may present consistent differences. As an important example, we cite Matsui et al. (1976) who measured the proportion in which the Solimões and Negro rivers of the Amazon basin contribute to the formation of the Amazon River.

Other applications of nuclear tools were already presented in Chap. 6, like gamma ray attenuation, neutron probes, and soil tomography. An excellent review on **soil tomography** was recently published by Pires (2018).

16.11 Exercises

16.1. If you had to organize an example such as page 319 of this Chapter with a bean crop, what data would you use for dry matter per hectare production, % nitrogen, plants per hectare, root system length, and average root diameter? What would be the flow of nitrogen per centimeter of root per second?

16.2. In a root absorption experiment, the following results were obtained:

j (mol g^{-1} s^{-1} × 10^{15})	C (mol L^{-1} × 10^3)
4.6	0.5
7.8	1.0
11.1	2.0
14.6	4.0
17.5	7.0
19.1	10.0
19.6	15.0
19.8	20.0

Make the graphs of j versus C and $1/j$ versus $1/C$ to estimate the values of K_m and j_{max}.

16.12 Answers

16.1. 5000 kg; 3%; 125,000; 500 m; 0.05 cm; N flux = 4.63×10^{-12} for a 60 day cycle.

16.2. $j_{max} = 2.2 \times 10^{14}$ mol g^{-1} s^{-1} and $K_m = 1.89 \times 10^{-3}$ mol L^{-1}.

References

Barber SA, Olsen RA (1968) Fertilizer use on corn. In: Nelson LB, Mcvickar MH, Munson RD, Seatz LF, Tisdale SL, White WC (eds) Changing patterns in fertilizer use. Soil Science Society of America, Madison, WI, pp 163–188

Basanta MV, Dourado-Neto D, Reichardt K, Bacchi OOS, Oliveira JCM, Trivelin PCO, Timm LC, Tominaga TT, Correchel V, Cassaro FAM, Pires LF, Macedo JR (2003) Quantifying management effects on fertilizer and trash nitrogen recovery in a sugarcane crop grown in Brazil. Geoderma 116:235–248

Bortolotto RP, Bruno IP, Dourado-Neto D, Timm LC, Silva AN, Reichardt K (2011) Soil profile internal drainage for a central pivot fertigated coffee crop. Rev Ceres 58:723–728

Bruno IP, Unkovich MJ, Bortolotto RP, Bacchi OOS, Dourado-Neto D, Reichardt K (2011) Fertilizer nitrogen in fertigated coffee crop: absorption changes in plant compartments over time. Field Crop Res 124:369–377

Bruno IP, Reichardt K, Bortolotto RP, Pinto VM, Bacchi OOS, Dourado-Neto D, Unkovich MJ (2015) Nitrogen balance and fertigation use efficiency in a field coffee crop. J Plant Nutr 38:2055–2076

Cerri CC, Volkoff B, Andreux F (1991) Nature and behavior of organic matter in soils under natural forest, and after deforestation, burning and cultivation, near Manaus. Forest Ecol Manag 38:247–257

Cervellini A, Ruschel AP, Matsui E, Salati E, Zagatto EAG, Ferreyra HH, Krug FJ, Bergamin Filho H, Reichardt K, Meirelles NMF, Libardi PL, Victoria RL, Saito SMT, Nascimento Filho VF (1980) Fate of 15N applied as ammonium sulphate to a bean crop. In: Soil nitrogen as fertilizer or pollutant. International Atomic Energy Agency, Vienna, pp 23–34

Dourado-Neto D, Powlson D, Bakar RA et al (2010) Multiseason recoveries of organic and inorganic nitrogen-15 in tropical cropping systems. Soil Sci Soc Am J 74:139–152

Epstein E (1972) Mineral nutrition of plants: principles and perspectives. John Wiley & Sons, New York, NY

Fenilli TAB, Reichardt K, Dourado-Neto D, Trivelin PCO, Favarin JL, Costa FMP, Bacchi OOS (2007a) Growth, development and fertilizer N-15 recovery by the coffee plant. Sci Agric 64:541–547

Fenilli TAB, Reichardt K, Trivelin PCO, Favarin JL (2007b) Volatization losses of ammonia from fertilizer and its reabsorbtion by coffee plants. Commun Soil Sci Plant Anal 38:1741–1751

Fenilli TAB, Reichardt K, Bacchi OOS, Trivelin PCO, Dourado-Neto D (2007c) The 15N isotope to evaluate fertilizer nitrogen absorbtion efficiency by the coffee plant. An Acad Bras Cienc 79:767–776

Hardarson G (1990) Use of nuclear techniques in studies of soil-plant relationships. International Atomic Energy Agency, Vienna

Havlin JL, Tisdale SL, Nelson WL, Beaton JD (2014) Soil fertility and fertilizers, 8th edn. Prentice Hall, Upper Saddle River, NJ

IAEA (2001) Use of isotope and radiation methods in soil and water management and crop nutrition. International Atomic Energy Agency, Vienna

Knoll HA, Brady NC, Lathwell DJ (1964) Effect of soil temperature and phosphorus fertilization on the growth and phosphorus content of corn. Agron J 56:145–147

Kutilek M, Nielsen DR (2010) Facts about global warming. Catena Verlag, Cremlingen-Destedt

L'Annunziata MF (ed) (1998) Handbook of radioactivity analysis. Academic Press, San Diego, CA

Libardi PL, Reichardt K (1978) Destino da uréia aplicada a um solo tropical. Rev Bras Ciênc Solo 2:34–40

Matsui E, Salati E, Friedman I, Brinkman WLF (1976) Isotopic hydrology in Amazonia. 2. Relative discharges of the Negro and Solimões rivers through ^{18}O concentrations. Water Resour Res 12:781–785

Meirelles NMF, Libardi PL, Reichardt K (1980) Absorção e lixiviação de nitrogênio em cultura de feijão. Rev Bras Ciênc Solo 4:83–88

Olsen SR, Kemper WD (1968) Movement of nutrients to plant roots. Adv Agron 20:91–151

Olsen SR, Kemper WD, van Schaik JC (1965) Self-diffusion coefficients of phosphorus in soil measured by transient and steady-state methods. Soil Sci Soc Am Proc 29:154–158

Partil AS, King KM, Miller MH (1963) Self-diffusion of rubidium as influenced by soil moisture tension. Can J Soil Sci 43:44–51

Pires LF (2018) Soil analysis using nuclear techniques: a literature review of the gamma ray attenuation method. Soil Tillage Res 184:216–234

Porter LK, Kemper WD, Jackson RD, Stewart BA (1960) Chloride diffusion in soils as influenced by moisture content. Soil Sci Soc Am Proc 24:460–463

Reichardt K (1976) Noções gerais sobre solo. In: Malavolta E (ed) Manual de química agrícola. Agronômica Ceres, Piracicaba, pp 121–157

Reichardt K (1980) Physico-chemical conditions and development of roots. In: Symposium on Root/Soil System. Instituto Agronômico do Paraná; Empresa Brasileira de Pesquisa Agropecuária; Conselho Nacional de Desenvolvimento Científico e Tecnológico, Londrina, pp 103–114

Reichardt K (1990) Soil spatial variability and symbiotic nitrogen fixation by legumes. Soil Sci 150:579–587

Reichardt K, Libardi PL, Meirelles NMF, Ferreyra FF, Zagatto EAG, Matsui E (1977) Extração e análise de nitratos em solução do solo. Rev Bras Ciênc Solo 1:130–132

Reichardt K, Libardi PL, Victoria RL, Viegas GP (1979) Dinâmica do nitrogênio em solo cultivado com milho. Rev Bras Ciênc Solo 3:17–20

Reichardt K, Libardi PL, Urquiaga SS (1982) The fate of fertilizer nitrogen in soil-plant systems with emphasis on the tropics. International Atomic Energy Agency, Vienna, pp 277–289

Rosenzweig C, Hillel D (1998) Climate change and the global harvest. Oxford University Press, New York, NY

Sposito G (1989) The chemistry of soils. Oxford University Press, New York, NY

Vose PB (1980) Introduction to nuclear techniques in agronomy and plant biology. Pergamon Press, Oxford

Walker JM (1969) One-degree increments in soil temperatures affect maize seedling behavior. Soil Sci Soc Am Proc 33:729–736

Yamada T, Igue K, Mizilli O, Usherwood NR (1982) Potássio na agricultura brasileira. Instituto da Potassa e Fosfato, Instituto Internacional da Potassa, Instituto Agronômico do Paraná, Londrina

How Soil, Plant, and Atmosphere Properties Vary in Space and Time in the SPAS: An Approach to Geostatistics

17

17.1 Introduction

Observations made in agronomic studies of characteristics in the soil–plant–atmosphere system (SPAS) need to include considerations on the **spatial and temporal variability** of soil and plant attributes, under field conditions, as well as atmospheric parameters. The soil and the distributions of the different parts of the plants, inside and outside the soil, are fundamentally heterogeneous. Variations in the soil are due to variable rates of the processes of soil formation and the various actions of man during cultivation. The root and shoot distribution of the plants depends on the species, genotype, soil properties, management operations, pests, and diseases. Thus, measurements of soil and plant parameters often have irregularities that may or may not be randomly distributed in relation to their spatial distribution in the field. Therefore, it is important to establish criteria to define spacing between samples for the measurements to be made, to define the frequency of observations, and the number of observations needed so that the average value obtained characterizes the considered site. Classically, agronomists have sought to achieve these objectives through the most diverse statistical techniques applied to data obtained, without taking into account their spatial distribution in the field.

Often, homogeneous areas and/or soils are chosen without a well-defined criterion of homogeneity, in which plots are randomly distributed to avoid the effect of existing irregularities. Experiments in **randomized blocks**, factorial, etc. are planned in this way and in the analysis of the data. If the analysis of variance shows a relatively small residual component, conclusions can be drawn on differences between treatments, interactions, etc. If, however, the residual component of the variance is relatively large, which is usually indicated by a high coefficient of variation, the results of the experiment are compromised. The cause may be soil variability, assumed to homogeneous at the beginning of the experiment setup.

If the spatial distribution of measurements is recognized and taken into account in the analysis, in many cases it is even possible to take advantage of the spatial variability. This is another way of planning experiments, relatively new in agronomy, but using recent techniques imported from **geostatistics** and **analysis of time and space series**, the latter being studied in Chap. 18. Textbooks for a first contact with these techniques are those of Journel and Huijbregts (1978), Clark (1979), Isaaks and Srivastava (1989), Cressie (1993), Goovaerts (1997), Nielsen and Wendroth (2003), Webster and Oliver (2007), and Shumway and Stoffer (2017), and

K. Reichardt, L. C. Timm, *Soil, Plant and Atmosphere*, https://doi.org/10.1007/978-3-030-19322-5_17

revisions on spatial variability in soils are given by Trangmar et al. (1985), Reichardt et al. (1986), Wendroth et al. (1997), Goovaerts (1999), Si (2008), among other texts.

The classical or "casual" technique often called **Fisher's statistics** and the technique of regionalized or "spatial" variables simply called geostatistics complement each other. One does not exclude the other and questions answered by one often cannot be answered by the other, but together they work better.

In agronomic experimentation, the methodology of sampling the SPAS, either soil, plant, or atmosphere, is of extreme importance. In "**classical statistics**," random sampling is a must, by lot, distributed randomly within the area generally assumed homogeneous, with no rigid criterion. In such designs, the points on which samplings were made, that is, the coordinates of the sampled sites, are not taken into account in the statistical analysis. In the technique of **regionalized variables**, regional sampling is used, in which the collection positions and consequently the coordinates of the sampled locations participate in the statistical analysis that is very concerned with neighboring samples. In this case, the sampling is done along a transect at equidistant intervals, denoted in the literature as **lag**, which is sometimes called spacing; or in two dimensional designs, as **grid**, also with fixed spacing; or even not in regular positions, but of known coordinates.

In order to make a comparison between classical statistics and the analysis of regionalized variables, we will apply concepts from both to the same set of data, presented in Table 17.1, collected along a **transect**. The soil volumetric water content ($m^3\ m^{-3}$) and clay (%) were measured in the same samples and collected in a fairly homogeneous field along a 150 m spatial transect at a regular interval of 5 m.

17.2 Mean (Average), Variance, Standard Deviation, and Coefficient of Variation

In classical statistics, which is based primarily on the normal distribution, each measurement at a sampling point $[z(x_i)]$ is taken as a random variable $Z(x_i)$, independently of the others $Z(x_j)$, with a total of n measurements. In a new sampling on the same area, what would be the most expected value of Z? It will be around the most likely value, which is the **mean** or **average** $\bar{z}$, also called the **expected value** of Z, given by the expression

$$\bar{z} = \left(\sum_{i=1}^{n} z(x_i) \right) \cdot n^{-1} \qquad (17.1)$$

When the distribution of a data set is asymmetric (this will be seen later in this chapter), besides the mean defined by Eq. (17.1), the **mode** and the **median** are also considered measures of position of the distribution of a set of data. The mode (Mo) represents the most frequent value of Z and the median (M_d) is the value of Z for which the probability is 0.5 or the value that divides the probability distribution curve into two equal areas. In addition to the average value of the variable Z, we are interested in a measure of the dispersion of the data around the mean. The deviations $[z(x_i) - \bar{z}]$ are a measure of this dispersion, but since they are positive and negative, their mean value tends to zero. The variance (s^2) of Z avoids this problem since it is the mean value of the squares of the deviations

$$s^2 = \left(\sum_{i=1}^{n} [z(x_i) - \bar{z}] \right)^2 \cdot (n-1)^{-1} \qquad (17.2)$$

The sum indicated in Eq. (17.2) is divided by $(n - 1)$ because one degree of freedom was lost. As we are interested in the mean value of the deviations, simply extract the square root of the variance to obtain the **standard deviation** s.

A measure of dispersion still widely used when there is interest in comparing variabilities of different data sets is the **coefficient of variation** (CV).

The coefficient of variation is defined as the relationship between the mean and the estimate of the standard deviation of a set of data. If we divide the standard deviation s by the mean $\bar{z}$, we have the proportion of the magnitude of the casual differences of the observations in relation to the mean value. Thus, if the standard deviation of a sample is 20, for a calculated mean value of

Table 17.1 Example of a transect of soil water content and clay content measured at the same points with a lag of 5 m

Distance	Soil water content θ	Clay content a		Position p		Ordered soil water content θ
5	0.390	36.5		1		0.350
10	0.380	35.0		2		0.355
15	0.385	35.0		3		0.360
20	0.375	35.5		4		0.360
25	0.385	34.0		5		0.360
30	0.360	33.0		6		0.370
35	0.350	32.5		7		0.370
40	0.370	34.5	p_1	8	Q_1	0.370
45	0.375	37.0		9		0.370
50	0.375	37.5		10		0.370
55	0.385	37.0		11		0.370
60	0.400	38.0		12		0.375
65	0.390	36.0		13		0.375
70	0.395	38.5		14		0.375
75	0.380	37.5	p_2	15	Q_2	0.375
80	0.385	35.0		16		0.375
85	0.370	35.0		17		0.380
90	0.390	34.0		18		0.380
95	0.370	35.0		19		0.380
100	0.370	34.0		20		0.380
105	0.360	33.0		21		0.385
110	0.370	35.0		22		0.385
115	0.380	35.5	p_3	23	Q_3	0.385
120	0.375	36.0		24		0.385
125	0.370	36.0		25		0.385
130	0.385	35.5		26		0.390
135	0.375	35.0		27		0.390
140	0.360	33.5		28		0.390
145	0.355	32.5		29		0.395
150	0.380	34.5		30		0.400

200, we see that the magnitude of the casual differences is 10%. This is the CV of the sampling that is given by the expression

$$\mathrm{CV}(\%) = \frac{s}{\bar{z}} \cdot 100 \tag{17.3}$$

The advantages of the CV over other dispersion measures (standard deviation, variance, amplitude) are as follows:

- The CV does not have a unit of measurement, since it is expressed in percentage.
- CV is a relative measure, that is, it relates the estimate of the standard deviation (s) of a data set with its respective arithmetic mean.

Because of the ease of calculation of the CV, it has been indiscriminately used to compare the variability of different variables even though they have different orders of magnitude. This, according to Webster (2001), would not be adequate. The same author points out that it can be used to compare variations in two sets of data as long as the variable is the same as, for example, variability of a set of pH data of a Nitosol compared to the pH of an Argisol.

For the data in Table 17.1, we have

Variable	Mean	s^2	s	CV (%)
θ	0.376	0.000141	0.01189	3.2
a	35.2	2.4954	1.5797	4.5

It is important to note that s^2 and s follow the units of the variables θ and a of the data we are using in the example, with the unit of variance being squared. According to the classification proposed by Wilding and Drees (1983), the dispersion of data around the mean in both sets is classified as low (CV $\leq$ 15%), that is, the data present low variability around the mean of the sampled transect in this case.

It is important to point out that for an entire or complete population, we refer to the mean value as the expected or true average μ, and as we actually made only a sample population, which is practically never complete, we refer to the average estimate $\hat{\mu}$ or only $\bar{z}$, as we did in this example, which also applies to variance and standard deviation:

σ^2 = expected variance; s^2 = estimated variance; σ = expected standard deviation; s = estimate of the standard deviation.

As in this chapter, we will restrict ourselves only to sets of never complete samplings and abandon the symbols μ, σ^2, and σ.

17.3 Quartiles and Moments

Quartiles are three measures that divide an ordered data set (from the lowest to the highest values) into four equal parts, as also shown in Table 17.1. They serve to construct box charts for the identification of discrepant values (most often by measurement error) and also to corroborate with the assessment that the data tends to be normal or not. They are as follows:

- First quartile (Q_1): 25% of the values are below and 75% are above the measure that occupies the position (p) within the ordered set.
- Second quartile (Q_2): 50% of the values are below and 50% are above this measure. It corresponds to the median of a set of data.
- Third quartile (Q_3): 75% of the values are below and 25% are above its value.

The determination of the quartiles, first, consists in ordering the data and then determining the position (p) of the quartile in the ordered data set. There are two different cases for determination of p:

First case: the number of data (n) is odd.

- For Q_1

$$p_1 = \frac{n+1}{4} \tag{17.4}$$

- For Q_2

$$p_2 = \frac{2(n+1)}{4} \tag{17.5}$$

- For Q_3

$$p_3 = \frac{3(n+1)}{4} \tag{17.6}$$

Second case: the number of data is even.

- For Q_1

$$p_1 = \frac{n+2}{4} \tag{17.7}$$

- For Q_2

$$p_2 = \frac{2n+2}{4} \tag{17.8}$$

- For Q_3

$$p_3 = \frac{3n+2}{4} \tag{17.9}$$

For the soil water content data of Table 17.1 with $n = 30$, therefore even, we have $p_1 = 8$, $p_2 = 15.5$, and $p_3 = 23$, and, respectively, $Q_1 = 0.385$, $Q_2 = 0.370$, and $Q_3 = 0.375$. For cases where p is not an integer, Montgomery and Runger (1994) point out that the quartile will be the arithmetic mean of the two values that occupy the positions corresponding to the smallest and the closest integers of p. For example, if $p = 5.5$, the quartile will be the mean of the values occupying positions 5 and 6. Thus, for our example with $p_2 = 15.5$, Q_2 is the mean of θ at position 15 and θ at position 16, that is, 0.370.

Each data set follows a distribution when ordered according to their magnitudes and frequencies with which they occur, as shown below in Figs. 17.2, 17.3, and 17.4.

Moments (m_r) are measures calculated in order to analyze these distributions of data sets. In general, a moment of order r, centered on a value b, is given by

$$m_r = \frac{\sum (z_i - b)^r}{n} \quad (17.10)$$

Of special interest are the moments of order $n = 1, 2, 3$, and 4, centered on the mean ($b = \bar{z}$), which are given by

$$m_r = \frac{\sum (z_i - \bar{z})^r}{n} \quad (17.11)$$

and Moment 1 for $r = 1$,

$$m_1 = \frac{\sum (z_i - \bar{z})^1}{n} = 0 \quad (17.11a)$$

which we already mentioned above, that is, the mean of the deviations with a value of zero.

Moment 2 for $r = 2$,

$$m_2 = \frac{\sum (z_i - \bar{z})^2}{n} = \text{variance} \quad (17.11b)$$

already presented in Eq. (17.2), with "$n - 1$" because of the loss of a degree of freedom. We will not enter here in the matter of degrees of freedom since this subject is well discussed in statistical texts, such as Montgomery and Runger (1994).

Moment 3 for $r = 3$,

$$m_3 = \frac{\sum (z_i - \bar{z})^3}{n} \quad (17.11c)$$

which is used to verify the degree of asymmetry of a distribution through the coefficient a_3, which we will see shortly.

Moment 4 for $r = 4$,

$$m_4 = \frac{\sum (z_i - \bar{z})^4}{n} \quad (17.11d)$$

which is used to evaluate the degree of flattening of a distribution through the coefficient a_4 shown below.

17.4 Total Amplitude and Interquartile Range

The total **amplitude** (at) of a data set gives an idea of the dispersion of the values of the variable in question and consists of the difference between the largest and the smallest value of a data set, that is, in its calculation we use only the two most extreme values of a data set. For this reason, it is extremely influenced by discrepant values, or outliers. So, we have

$$\text{at} = \text{Ls} - \text{Li} \quad (17.12)$$

where Ls and Li are the highest and the lowest values of an ordered data set, respectively.

The total amplitude is used when only a rudimentary idea of the variability of the data is sufficient since it is not a very precise measurement.

One measure that is seldomly used, but is not influenced by discrepant values, is the so-called **interquartile range** (q). It is the difference between the third quartile (Q_3) and the first quartile (Q_1). So, we have

$$q = Q_3 - Q_1 \quad (17.13)$$

17.5 Skewness and Kurtosis Coefficients

The **skewness coefficient** (a_3) reports whether most values in a data set lie to the left, or to the right of the mean, or are evenly distributed around the arithmetic mean. It indicates the degree and direction of the data distribution for asymmetry and is obtained by using the second (m_2) and third (m_3) moments centered on the mean. So, we have

$$a_3 = \frac{m_3}{m_2\sqrt{m_2}} \quad (17.14)$$

The classification of the **distribution for asymmetry** is based on the value of a_3:

- If $a_3 < 0$, the distribution is classified as negative asymmetric, that is, most of the values are larger or are located to the right of the arithmetic mean.
- If $a_3 = 0$, the distribution is classified as symmetric, indicating that the values are evenly distributed around the arithmetic mean.
- If $a_3 > 0$, the distribution is classified as positive asymmetric, that is, most values are smaller or are located to the left of the arithmetic mean.

The **kurtosis coefficient**, denoted by a_4, indicates the degree of flattening of a distribution. It is calculated by means of the second (m_2) and fourth (m_4) moments centered on the mean. So, we have

$$a_4 = \frac{m_4}{(m_2)^2} - 3 \quad (17.15)$$

The degree of flattening of a distribution is based on the value of a_4:

- If $a_4 < 0$, the distribution is classified as platicurtic, that is, greater flattening.
- If $a_4 = 0$, the distribution is classified as mesocurtic, indicating average flattening.
- If $a_4 > 0$, the distribution is classified as leptokurtic, that is, less degree of flattening.

For a given distribution, the closer to zero the coefficients a_3 and a_4 are, the more symmetric is the distribution. For the soil water content data set in Table 17.1, the values of the coefficients a_3 and a_4 are −0.159 and −0.382, respectively. Therefore, this distribution is classified as negative asymmetric ($a_3 < 0$) and platicurtic ($a_4 < 0$), and can be practically considered as normal because coefficients a_3 and a_4 are close to zero. Webster and Oliver (2007) commented that in order to overcome the difficulties when the data move away from the normal distribution (symmetry), a transformation of the measured values of the variable to a new scale in which the distribution becomes close to normal is often necessary. Examples of commonly used transformations are the logarithmic and the square root, among others. Webster (2001) discusses the difficulties encountered when the distribution of the data is far from normal. However, there is a controversy on this subject such as that raised by Goovaerts (1997). Goovaerts (1997) indicated that transformation is not ideal if the aim is prediction because in general those who ultimately use the predictions, such as land managers, environmental scientists, and so on, want values on the original scale of measurement which involves a back-transformation.

17.6 Identification of Outliers (Discrepant Values)

An **outlier** is a **discrepant value**, an observation that seems to be suspect to the researcher because its magnitude is very different from the others, within replicates or even treatments. In this way, when a discrepant value in a data set is found, its origin must be investigated. Often, discrepant values, in fact, are part of the data set, but eventually, these values may come from gauging errors or data logging. Careful inspection of the data and the possible causes of the occurrence of discrepant value(s) is always a necessary step before any action is taken in relation to these data. For the identification of outliers in a data set, we used two measures, called lower fence (LF) and upper fence (UF). The lower fence is calculated by subtracting from the first quartile (Q_1) one and a half interquartile range (q), and the upper fence, summing this amount to the third quartile (Q_3). So, we have

$$\mathrm{LF} = Q_1 - 1.5q \quad \text{and} \quad \mathrm{UF} = Q_3 + 1.5q \quad (17.16)$$

Values that are outside the range $[Q_1 - 1.5q; Q_3 + 1.5q]$ are considered outliers.

17.7 Box Plot (Box-and-Whisker Plot)

The **box plot** (also called box-and-whisker plot in the literature) aggregates a series of information about the distribution of a data set, such as position, dispersion, asymmetry, tails, and discrepant data. For its construction, we consider a rectangle in which the quartiles (Q_1 and Q_3) and the median ($M_d = Q_2$) are represented. From the rectangle, up and down, follow lines, called whiskers, that go up to the adjacent values. The lower and higher values of a data set are considered to be adjacent lower and adjacent upper, respectively. They do not exceed the upper fence (UF) and the lower fence (LF). The discrepant values are given an individual representation by means of a letter or symbol. Thus, we obtain a figure that represents many relevant aspects of a data set, as we can see in Fig. 17.1 (Table 17.2).

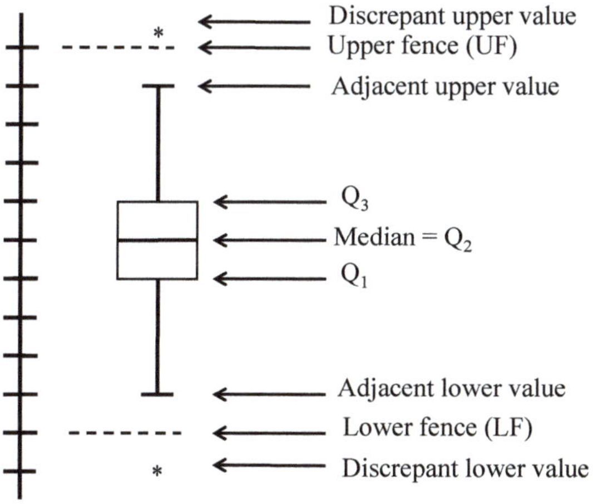

Fig. 17.1 Example of a Box plot

The central position of the values is given by the median ($M_d = Q_2$) and the dispersion by the interquartile range (q). The relative positions of the median and quartiles and the shape of the whiskers give an idea of the symmetry and size of the tails of the distribution.

17.8 Normal Frequency Distribution

Many variables of the soil–plant–atmosphere system follow the **normal frequency distribution**, which is considered one of the most important distributions in statistical theory.

Assuming that a variable Z_i follows the normal distribution, there is a function $h(z_i)$ such that

$$h(z_i) = \frac{1}{s\sqrt{2\pi}} \exp\left[\frac{-(z_i - \bar{z})^2}{2s^2}\right] \tag{17.17}$$

called the **normal probability density function**. The value of h for each z_i is proportional to the probability of occurring within the total of n observations made. The curve $h(z)$ versus z is given in Fig. 17.2.

It is a symmetric bell with a maximum probability value corresponding to $z = \bar{z}$, asymptotic with respect to the z-axis for both sides. The integral of the curve is the probability of occurrence of the values of z, and for the interval of z equal to ($-\infty$ to $+\infty$) that includes all values, the total probability of occurrence is 1% or 100%. Since it is a symmetric curve, the probability of

Table 17.2 Box plot summary for soil water content data of Table 17.1

Summary of the box plot	Soil water content ($m^3\ m^{-3}$)
Lower adjacent value (minimum value)	0.350
Q_1	0.370
Median = Q_2	0.375
Q_3	0.385
Upper adjacent value (maximum value)	0.400
Interquartile range (q)	0.015
Lower fence	0.348
Upper fence	0.408
Number of lower discrepant values	0
Number of upper discrepant values	0

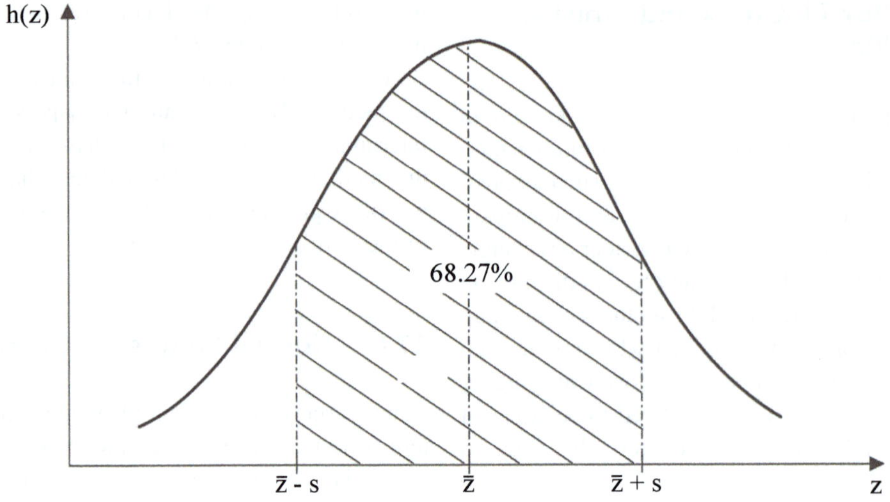

Fig. 17.2 Schematic presentation of Eq. (17.17) of a normal distribution curve

Fig. 17.3 Schematic presentation of a log-normal distribution curve

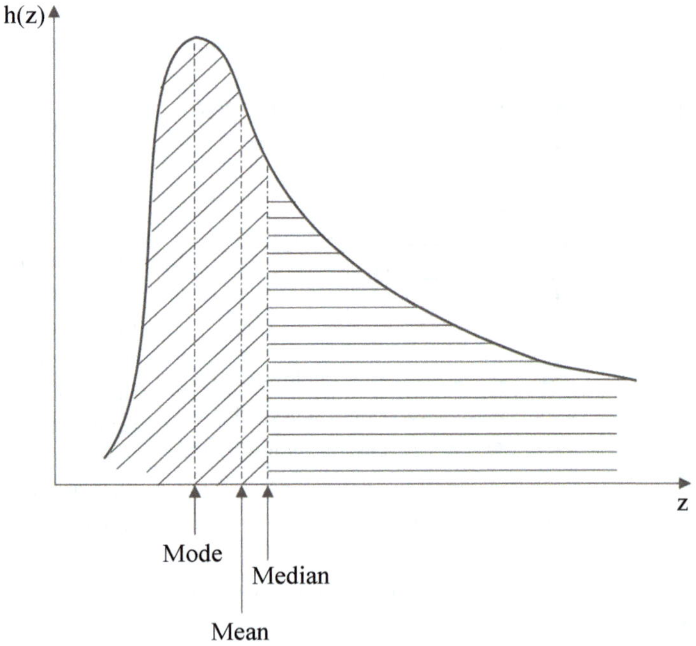

the intervals ($-\infty$ to $\bar{z}$) and ($\bar{z}$ to $+\infty$) is 0.5% or 50%. It can be seen that the two inflection points occur at $z = -s$ and $z = +s$ and that the probability of occurrence of a value of z falling between these inflection points is 0.6827% or 68.27%. Therefore, one of the most common ways of expressing the data dispersion is by the expression $\bar{z} \pm s$. In this case, only 31.73% of the data fall outside this range. Since the normal curve is symmetrical, the **mode** and the **median** are equal to the mean, which is not true for other types of distribution, as shown in Fig. 17.3 for the log-normal distribution.

The normal curve of Fig. 17.2 is theoretical and shows the expected values of z. The more the observed values approach those expected, the

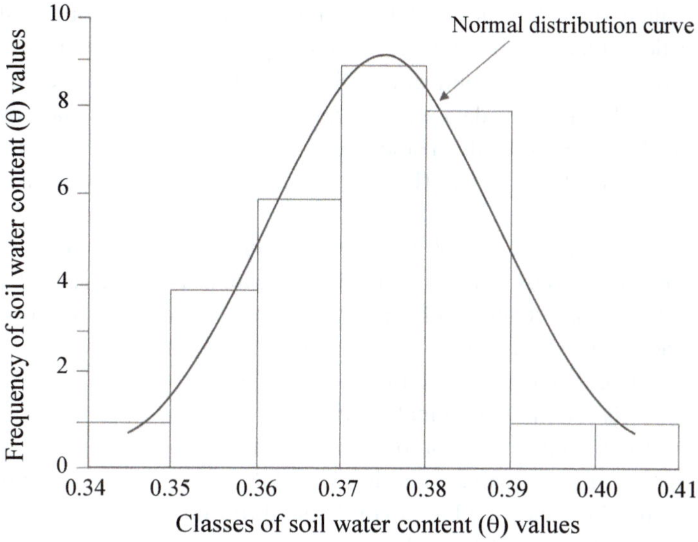

Fig. 17.4 Theoretical curve and histogram of the soil water content data (θ) presented in Table 17.1

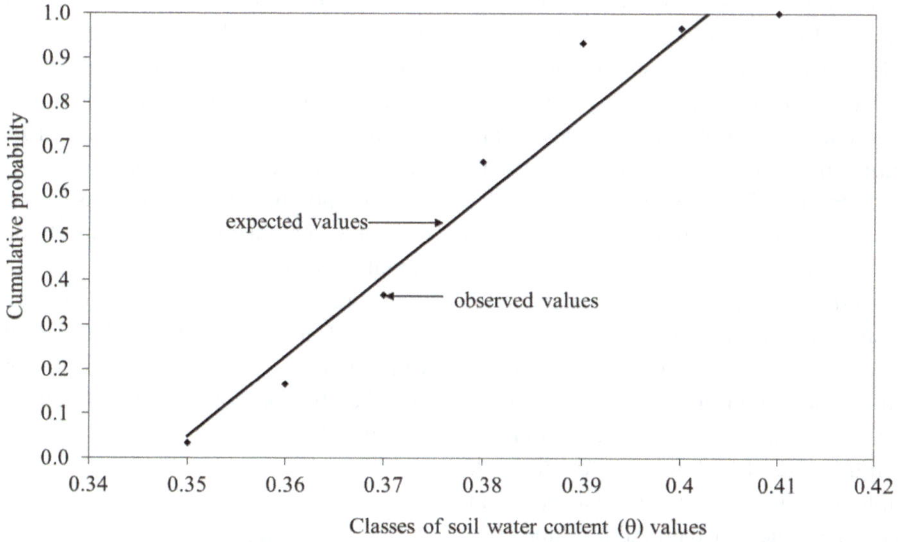

Fig. 17.5 Fractile or normal diagram showing that the graph of the accumulated probability is a straight line

better the adjustment of the population to the normal curve. The observed data are usually grouped into classes and presented as a histogram superimposed on the normal curve. Figure 17.4 shows the theoretical curve and histogram of the data of θ in Table 17.1.

There are several ways to verify if a data set follows the normal distribution. One much used way is through an accumulated probability graph called **fractile diagram** or **normal diagram**, which for an expected (theoretical) population is a straight line. The **cumulative probability** is the integral of Eq. (17.17) from $-\infty$ to z_i, and choosing increasing values of z_i, the result is a straight line. Therefore, the more the integrals of observed data adjust to a straight line, the more the data set tends to normality. Figure 17.5 shows this graph for the θ values of Table 17.1.

The adherence of the data distribution to the normal curve can also be checked using the

Kolmogorov–Smirnov and the Shapiro–Wilk statistics (Jury and Horton 2004; Pachepsky and Rawls 2004). The Kolmogorov–Smirnov statistic is usually used for data sets including more than 50 observations, while in other cases the Shapiro–Wilk statistic is used (Pachepsky and Rawls 2004).

When the data do not fit the normal curve, the population must belong to another distribution. Often, transformations of the data are made that lead to an adjustment to the normal curve. This adjustment is desired because most of the statistical tools refer to the normal distribution.

A common case in soil physics is the transformation of the variable z into log z, and the curve obtained is called the **log-normal distribution** or curve. The natural logarithm (ln) has also been used as a transformation of a variable. The log-normal distribution is an asymmetric distribution, and therefore, the mode and the median appear displaced from the geometric mean defined by Eq. (17.1). As previously stated, the **mode** represents the most probable value of z and the **median** is the value of z for which the probability is 0.5. Figure 17.3 schematizes such a distribution, which, for example, describes soil hydraulic conductivity data very well.

It is important to recognize again that for the normal distribution: mean = mode = median and also that the mean of z is different from the mean of ln z. In practical solutions, it is therefore difficult to decide when to use the mean, mode, or median. In the case of hydraulic conductivity K, for example, it could be assumed that the mean, which represents the most probable value, best represents its distribution in an experimental area. It turns out, however, that some extremely high values of K that make the distribution asymmetric can play a very important role in the places where they were measured, to the point that the places in which these high values occur, they cause a more pronounced drainage in relation to the value represented by mode.

Many other distributions can be used for data adjustment, such as **beta**, **gamma, and kappa distributions**, with applications in rainfall (Marques et al. 2014; Murshed et al. 2018), maximum streamflows (Cassalho et al. 2018), and maximum wind velocity (Gusella 1991).

Of fundamental importance is also the **standard normal distribution**. If the random variable z_i of the normal distribution (Eq. 17.17) is transformed into $z_j = \left(z_i - \bar{z}\right)/s$, the random variable z_j is considered to be a "standardized" version of z_i because it has mean 0 and standard deviation 1 (Montgomery and Runger 1994). Its equation is simplified to

$$h\left(z_j\right) = \frac{1}{\sqrt{2\pi}} \exp\left(\frac{-1}{2z_j^2}\right) \quad (17.17a)$$

It is important to recognize that $(z_i - \bar{z})$ are the deviations from the mean, and that dividing by s we obtain relative deviations. The integral of Eq. (17.17a) is 0 for $z_j = -\infty$; 0.5 for $z_j = 0$ (mean) and 1 for $z_j = +\infty$, which are cumulative probability values. From this, it can easily be seen that for any set of data, this will be the result, and therefore, the standard normal distribution is used to make probability tattles, found in most statistics texts. The transformation of z_i into z_j turns the variable nondimensional, because z_i and s have the same dimensions (see Chap. 19 for dimensional analysis).

Another interesting transformation is that presented in Chap. 18 by Hui et al. (1998), which makes $\bar{z} = 0.5$ also for any population, very helpful for comparisons of data of different magnitudes.

17.9 Covariance

In case of two populations with two different aleatory variables x and y, with some dependent relation, the covariance is used to quantify this dependence. In our example, there is a high suspicion that the soil water content data are related to clay content, if samples were collected at the same positions. The **covariance** (C) is defined by

$$C = \frac{1}{(n-1)}\left[\sum_{i=1}^{n} \left(x_i - \bar{x}\right)\left(y_i - \bar{y}\right)\right] \quad (17.18)$$

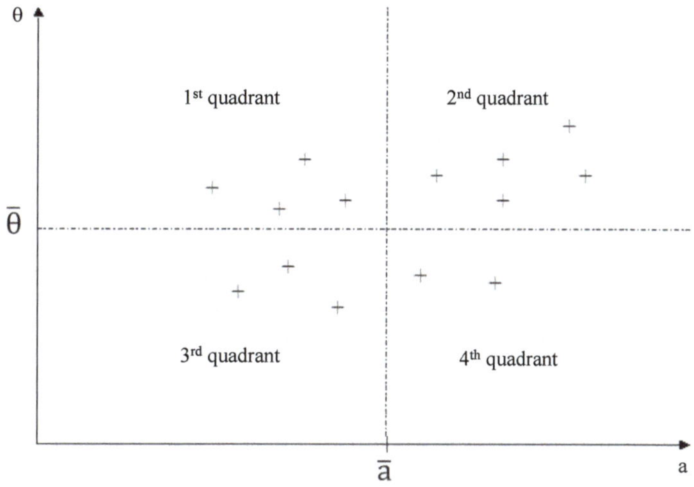

Fig. 17.6 Illustration of the quadrants obtained when separating data greater or smaller in relation to the averages

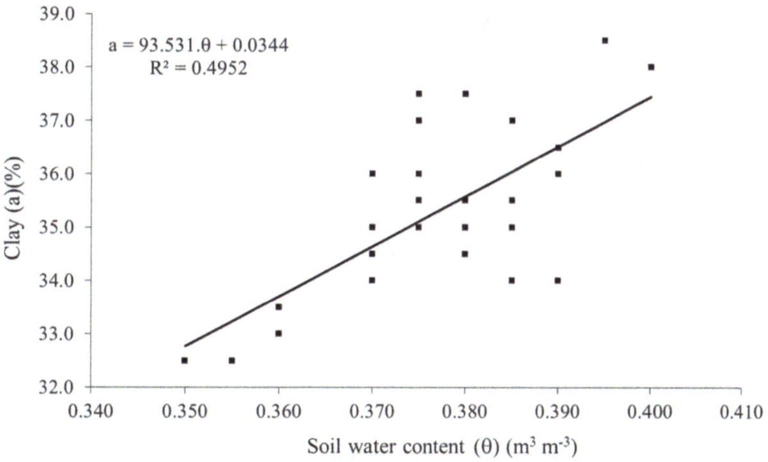

Fig. 17.7 Regression of clay content and soil water content data of Table 17.1

There is covariance between two variables x_i and y_i if there is a relation between these variables. The graph of the couples x_i, y_i or θ_i, a_i shown in Fig. 17.6 (for any x_i, y_i) and in Fig. 17.7 (for θ_i and a_i, prepared for the data in Table 17.1) illustrates the issue.

If we divide the data point area into four quadrants drawing straight lines and $y = \bar{y}$ and $x = \bar{x}$, we have data points in each quadrant. For quadrants 1 and 2, we have points $y_i > \bar{y}$ and the differences $y_i - \bar{y}$ are positive. For quadrants 3 and 4 ($y_i < \bar{y}$), the differences $y_i - \bar{y}$ are negative. For x in quadrants 1 and 3 ($x_i < \bar{x}$), the differences ($x_i - \bar{x}$) are negative, and for quadrants 2 and 4 positive. When summing the products of the differences from Eq. (17.18) for all pairs, we see that these are positive for the first and third quadrants and negative for the second and fourth quadrants. If most points fall in the first and third quadrants, the result of C will be positive and we say that y correlates directly with x. If most fall on the first and fourth, the correlation will be reversed, that is, negative. If the points are equally distributed in the four quadrants, C tends to zero and the variables are not correlated. The higher the C (without considering the signal), the greater the correlation. This evaluation is done by the correlation coefficient r, given by

$$r = \frac{C(x,y)}{\sqrt{s_x \cdot s_y}} \quad (17.19)$$

In practice, we use r or R^2. For our data of θ and a, we have $r = 0.7037$ and $R^2 = 0.4952$ (Fig. 17.7).

We have seen very briefly the application of the concepts most used in **classical statistics**. In addition, this statistic makes use of **experimental designs** (randomized blocks, completely randomized, factorial, etc.) in which means of treatments are compared to each other by tests of significance, such as F, t, Tukey, and Duncan. A full text on the subject is that of Glaz and Yeater (2018), among others. It should be recalled that the positions or locations where the samplings were taken do not participate in the analysis of classical statistics. The statistical picture obtained by classical statistics is generalized over the sampled area, losing details of local variability, that is, we lose the important information of where in the field the lowest and the highest values occur, for example. Therefore, the analysis of regionalized variables, seen below, is an improvement in the statistical treatment of field data.

17.10 Autocorrelogram

We will now introduce the **analysis of regionalized variables**, showing some of the most used concepts applied to the data in Table 17.1. As already said, this analysis is adapted to data sets with samplings on a **transect** or **grid**. The first concept is **autocorrelation**, which leads to the construction of the **autocorrelogram**. Equation (17.18) of the correlation of two variables x and y becomes an autocorrelation if y is exchanged for x itself, but at another position along the transect. Since it is a relationship between a variable and itself in another position, the process is called autocorrelation.

Thus, for variables x_i (at position i) and x_{i+j} [at position $(i + j)$, at a distance from i of jh, where h is the spacing (**lag**) with $j = 0, 1, 2, 3, \ldots$] Eqs. (17.18) and (17.19a) become

$$C(j) = \frac{1}{(n-1-j)}\left[\sum_{i=1}^{n-j}(x_i - \bar{x})(x_{i+jh} - \bar{x})\right] \quad (17.18a)$$

$$r(j) = \frac{C(j)}{s^2} \quad (17.19a)$$

Autocorrelation is, therefore, the correlation between neighbors; between adjacent neighbors for $j = 1$ (x_1 with x_2, x_2 with x_3, x_3 with x_4, ..., x_i with x_{i+1}); between second neighbors for $j = 2$ (x_1 with x_3, x_2 with x_4, ..., x_i with x_{i+2}), and so on. We can see that the coordinates of the sampling positions of x do not enter in the analysis, but their position on the transect and their ordination are of importance. For $j = 1$, we lose one pair of data in the correlation; for $j = 2$, we lose two pairs so that with the increase of j the number of pairs in the sum of Eq. (17.18a) decreases and is equal to the upper index of the summation: $n - j$. Due to this fact, in the use of autocorrelation and also in the use of other tools of the regionalized statistics, we need a large number n of observations. In Eq. (17.19a), we have s^2 because $s_x = s$ and $s_y = s$, so that $s\,.\,s = s^2$. For the calculation of the autocorrelation function, it is assumed that the first two moments (mean and covariance) of the data distribution are invariant under translation, that is, there is an assumption of stationarity of the data. We will not go further in this subject. More details on it can be found in the textbooks of Journel and Huijbregts (1978) and Goovaerts (1997).

If we apply Eq. (17.19a) to values of $j = 0, 1, 2, \ldots, k$ (with k much less than $n - j$), we obtain $r(0)$, $r(1)$, $r(2)$, ..., $r(k)$. The value of $r(0)$ is 1, since it correlates with x_i with itself. If there is correlation between neighbors, we will have real values of $r(1)$, $r(2)$, ... proportional to their correlations, but always smaller than 1. For very distant neighbors, the correlation is expected to decrease tending to zero. The graph of $r(j)$ as a function of j is called **autocorrelogram**. It therefore expresses the variation of the autocorrelation as a function of the distance separating the adjacent observations of the variable. If $r(j)$ drops rapidly to zero, the variable x is not autocorrelated and the values x_i can be considered independent, which is, moreover, required by classical statistics and is rarely observed. If $r(j = 5)$ is still significant (which can be done by means of probability tests), this means that even the fifth neighbor (fifth) is still autocorrelated. The next step is to calculate the confidence intervals of r to check whether it is significant or not and, thus, define the length jh in which there is the spatial dependence between the adjacent observations of the variable under study. Nielsen and Wendroth (2003) mentioned that there are different ways to calculate confidence intervals for determining if an autocorrelation coefficient is significantly different from zero or not. One way, according to those authors, to determine the significant autocorrelation confidence interval (CI) is to use the cumulative probability function p (e.g., ± 1.96 for 95% probability) for a standardized normal distribution function (Davis 1986) and the number of observations (n). In this way,

$$\mathrm{CI} = \pm \frac{p}{\sqrt{n}} \tag{17.20}$$

The autocorrelogram of the θ data of Table 17.1 is shown in Fig. 17.8. In practice, the autocorrelograms can assume varied forms, depending on the spatial variability of the variable. For high values of j, $r(j)$ can even assume negative values and, with the increase of j, return to be positive.

17.11 Semivariogram

The study of the **spatial variability** of soil, plant, or atmosphere attributes, when sampling is done on a transect or in a grid (i.e., in two dimensions) requires the use of the so-called **geostatistics**, which emerged in South Africa. Danie Krige, in 1951, working with gold concentration data on soil/rock (geological) samples, concluded that he could not find meaning in the values of the variances of the data, if the distance between the sampling positions was not taken into account. Based on these observations, Matheron (1963) developed a theory, which he called "**regionalized variable theory**" and which contains the foundations of geostatistics. This theory is based on the fact that the difference of the values of a given variable measured at two points in the field depends on the distance between them. Thus, the difference between the values of the attribute measured at two closest points in space must be smaller than the difference between the values measured at two more

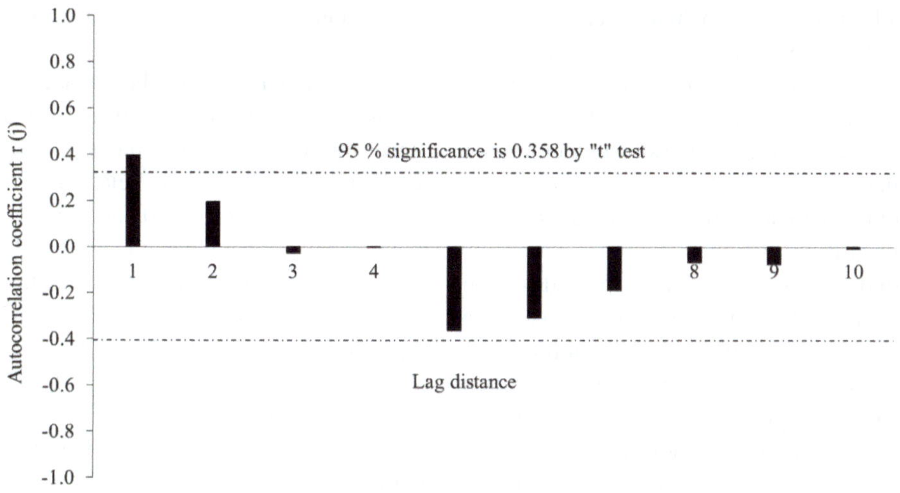

Fig. 17.8 Example of an autocorrelation graph as a function of the distance separating observations in terms of lags

distant points. Therefore, each value carries with it a strong interference of the values of its neighborhood, illustrating the spatial continuity.

A variable is said to be regionalized when it is distributed in space (Journel and Huijbregts 1978). Geostatistics is based on the concept of random function (RF) since this concept allows us to account for structures in the spatial variation of the attribute (Goovaerts 1997). When the geostatistical tool is used to analyze the data, some working hypotheses are assumed, mainly the **intrinsic hypothesis**. An RF $Z(x)$ is said to be intrinsic when the mathematical expectation exists and does not depend on the position (first-order stationarity), and for all vectors h, the increment $[Z(x) - Z(x + h)]$ has a finite variance which does not depend on the position (Journel and Huijbregts 1978). The intrinsic hypothesis can be seen as the limitation of the second-order stationarity to the increments of the RF $Z(x)$. Therefore, second-order stationarity implies the intrinsic hypothesis but the converse is not true (Journel and Huijbregts 1978). The second-order stationarity assumes the existence of the first-order stationarity and the existence of a covariance which depends on the separation distance h. According to Journel and Huijbregts (1978), the stationarity of the covariance implies the stationarity of the variance and the variogram (this will be seen soon in this chapter).

It should be noted that when we are analyzing the data variability along a transect (i.e., in one dimension), the distance h is a scalar being characterized only by its modulus. When we are analyzing data collected in a grid, whether regular or irregular (i.e., in two dimensions), h is also a vector being characterized by its modulus, direction, and sense.

A fundamental step that precedes the geostatistical analysis is the performance of a careful exploratory analysis of the data (classical descriptive statistics, i.e., calculation of position measurements, dispersion, moments, histogram, normal plot, box plot). The normality of the data distribution should be verified, checking if the data need to be transformed. The way to do these calculations has already been presented previously.

By including N pairs of values of $z(x_i)$ separated by a distance ih [$N(h)$] in the equation of the variance (s^2) (Eq. 17.2), we obtain the **semivariance** (γ), given by

$$\gamma(ih) = \frac{1}{2N(h)} \sum_{i=1}^{N(h)} [z(x_i) - z(x_{i+h})]^2 \quad (17.21)$$

where $\gamma(ih)$ is the experimental semivariance value of the data pairs as a function of the distance ih (hereinafter referred to simply as h as is commonly done in the geostatistics literature) and $z(x_i)$ and $z(x_{i+h})$ are the measured values of the variable Z under study at positions x_i and x_{i+h}, respectively. Equation (17.21) is known in the literature as the Matheron's classical semivariance estimator. It should be emphasized that this equation must be used when the distribution of the variable under study follows the normal distribution since it is affected by the presence of discrepant values in the data set. When the variable does not follow the normal distribution, a robust estimator is recommended for the calculation of experimental semivariance, such as that of Cressie and Hawkins (1980). Further details on this topic can be found in Webster and Oliver (2007), who present and discuss the effects of the presence of discrepant values on the Matheron estimator and suggest other robust estimators for the calculation of experimental semivariance if the variable distribution is not normal.

It should be noted that the variance s^2 of a data distribution is calculated with the squares of the deviations of z from the mean $\bar{z}$ and that the semivariance γ is done in relation to data pairs. For $h = 0$, $z(x_i) - z(x_i) = 0$ and $\gamma(h = 0) = 0$. For $h = 1, 2, 3, \ldots$, we have $\gamma(1), \gamma(2), \gamma(3), \ldots, \gamma(k)$ whose values are increasing until data pairs $z(x_i)$ and $z(x_{i+h})$ are so far apart that they have no dependence and $\gamma(k)$ tends (or not) for s^2 which is an estimate of the variability of a data set of independent observations. The graph of γ as a function of the distance h is called **experimental semivariogram**.

The experimental semivariogram is fitted with mathematical models that provide the maximum possible correlation with the points calculated from Eq. (17.21). The fitted model is called the theoretical model of the semivariogram or also called the **theoretical semivariogram**. The adjustment of a theoretical model to the experimental semivariogram is one of the most important aspects of the applications of the theory of regionalized variables (geostatistics) and can be one of the major sources of ambiguity and controversy in these applications. All calculations of geostatistics depend on the adjusted theoretical semivariogram model and its respective parameters (Vieira et al. 1983). Therefore, if the fit quality of the model is not satisfactory based on some statistical measures that evaluate it, all subsequent calculations will be compromised.

The step of fitting the mathematical model to the experimental semivariogram is of great importance, since it can influence the later results. The adjusted model should approximate to the maximum description of the phenomenon in the field, and the verification of the best fit of the theoretical model to the experimental semivariogram can be performed by the cross-validation procedure. In a summarized way, this procedure consists in estimating a value of the variable by means of **kriging** (it will be seen later) for each one of the locations (experimental points) where one has a measured value of the variable. In this way, a scatter plot is constructed between the estimated and measured values of the variable, and statistical measures [coefficient of determination r^2, mean error (ME), mean squared error (MSE), mean squared deviation ratio (MSDR), etc.] can be used to evaluate the quality of the fitted model. Further details on this procedure can be found in Webster and Oliver (2007), among others.

Figure 17.9 presents an example of an experimental semivariogram and a theoretical one (and its parameters) with characteristics very close to the expected. The pattern represents what is intuitively expected of field data, that is, that the differences $[z(x_i) - z(x_{i+h})]$ decrease as h, the distance separating them, decreases.

The parameters of the theoretical semivariogram can be seen directly from the figure above.

- a (**range**): distance within which the observations of the variable are spatially correlated.

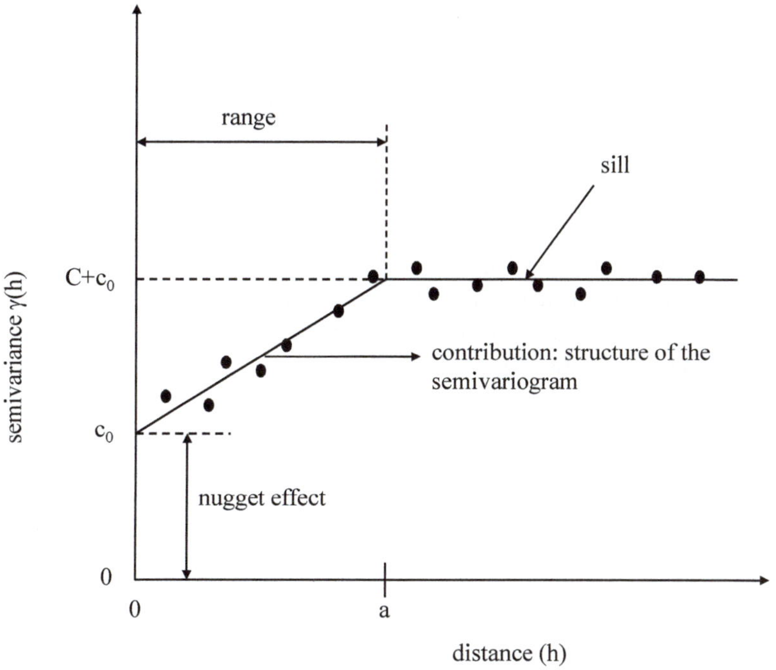

Fig. 17.9 Experimental and theoretical semivariograms with their terms

- $C + c_0$ (**sill**): the value of the semivariance corresponding to its range (a). From this point on, it is considered that there is no more spatial dependence between adjacent observations of the variable, because the variance of the difference between pairs of observations $\{\text{Var}[z(x_i) - z(x_{i+h})]\}$ becomes invariant with the sampling distance.
- c_0 (**Nugget effect**): by definition, $\gamma(h = 0) = 0$. However, in general, as the distance h tends to 0 (zero), $\gamma(h)$ approaches a positive value (c_0). The c_0 value reveals a discontinuity of the semivariogram for distances smaller than the h, the distance between observations. Part of this discontinuity may also be due to measurement errors, but it is impossible to quantify whether the greatest contribution comes from measurement errors or small-scale variability not captured by sampling. It may also be the result of a not very suitable choice of lag when doing field samplings.
- C (contribution): the difference between the sill ($C + c_0$) and the nugget effect (c_0).

It should be emphasized again that the adjustment of a theoretical model to the experimental semivariogram is one of the most important aspects of geostatistics. There are commercial programs [e.g., GS+ software (Robertson 2008)] and free softwares [e.g., R software (R Core Team 2012)] that have routines implemented to make semivariogram adjustments. As a rule, the simpler the adjusted model can be, the better, and excessive importance should not be given to small fluctuations that may be artifacts concerning a small number of data. The condition for fitting models to experimental data is that it represents the tendency of $\gamma(h)$ as a function of h and that the model has a conditionally defined positivity. In general, a model is positively conditional if $\gamma(h) \geq 0$ and $\gamma(-h) = \gamma(h)$, whichever is h (Journel and Huijbregts 1978).

Based on the semivariogram parameters defined above, the main semivariograms models used in geostatistics are as follows (Nielsen and Wendroth 2003).

17.11.1 Models with Defined Sill (Bounded Models)

Pure nugget effect

$$\gamma(h) = \begin{cases} 0 & \text{for } h = 0 \\ C + c_0 & \text{for } h > 0 \end{cases} \tag{17.22}$$

Linear

$$\gamma(h) = \begin{cases} c_0 + \dfrac{C\,h}{a} & \text{for } 0 \leq h \leq a \\ C + c_0 & \text{for } h > 0 \end{cases} \tag{17.23}$$

Spherical

$$\gamma(h) = \begin{cases} c_0 + C\left[\dfrac{3h}{2a} - \dfrac{1}{2}\left(\dfrac{h}{a}\right)^3\right] & \text{for } 0 \leq h \leq a \\ C + c_0 & \text{for } h > a \end{cases} \tag{17.24}$$

Exponential

$$\gamma(h) = c_0 + C[1 - \exp(-h/a)] \text{ for } h \geq 0 \tag{17.25}$$

Gaussian

$$\gamma(h) = c_0 + C\left\{1 - \exp\left[-(h/a)^2\right]\right\} \text{ for } h \geq 0 \tag{17.26}$$

17.11.2 Models Without Defined Sill (Unbounded Models)

Linear

$$\gamma(h) = c_0 + mh \text{ for } h \geq 0 \tag{17.27}$$

Power

$$\gamma(h) = c_0 + mh^{\alpha} \text{ for } h \geq 0;\ 1 < \alpha < 2 \tag{17.28}$$

Nielsen and Wendroth (2003) point out that bounded semivariograms occur when the variance of the data set remains constant throughout the sampled spatial domain, while unbounded semivariograms occur when the variance within the spatial domain is not constant. It is worth to emphasize that an unbounded semivariogram reflects that (1) the intrinsic hypothesis was not reached and we are probably facing a phenomenon with infinite dispersion capacity, and (2) the maximum distance h between the sampling points was not able to display the data variance and there is probably a trend of the data distribution for a given direction. If the trend is identified, this tendency can be removed and it is verified if the transformed variable presents a semivariogram with a defined sill (intrinsic hypothesis); it can be one of the solutions to be adopted to remove the trend of behavior of the variable in question. Another alternative is to work with the original data and to use a robust semivariance estimator (e.g., Cressie and Hawkins estimator). It is worth mentioning that the first alternative is the simplest and most used.

The semivariogram of clay data from Table 17.1 is shown in Fig. 17.10.

Analyzing Fig. 17.10, it is verified that the clay data collected along the 150 m spatial transect have spatial dependence and that it can be described by the spherical model with a range of 30.6 m, that is, pairs of clay observations separated for distances less than 30.6 m are correlated. For distances greater than this range, they are not spatially correlated. Cambardella et al. (1994) proposed a classification for the degree of spatial dependence of a variable based on the nugget effect (c_0) and the sill ($C + c_0$), so that if the relation $(c_0/C + c_0) \leq 25\%$, the degree of dependence is strong; if $25\% < (c_0/C + c_0) \leq 75\%$, moderate; and if $(c_0/C + c_0) > 75\%$, weak. For our clay case, we have 2.6% [100 × (0.09)/(3.446)] indicating that the clay data have a strong dependence degree.

Examples of experimental and theoretical semivariograms were extracted from Parfitt et al. (2009, 2013, 2014). These authors applied geostatistical tools to evaluate the effect of land leveling on 31 chemical, physical, and microbiological attributes determined in the 0–0.20 m depth layer of a lowland soil located in Pelotas (RS), Brazil. A grid of 100 points (10 m × 10 m) was established with georeferenced points in the experimental area where soil samples were collected, at the same points,

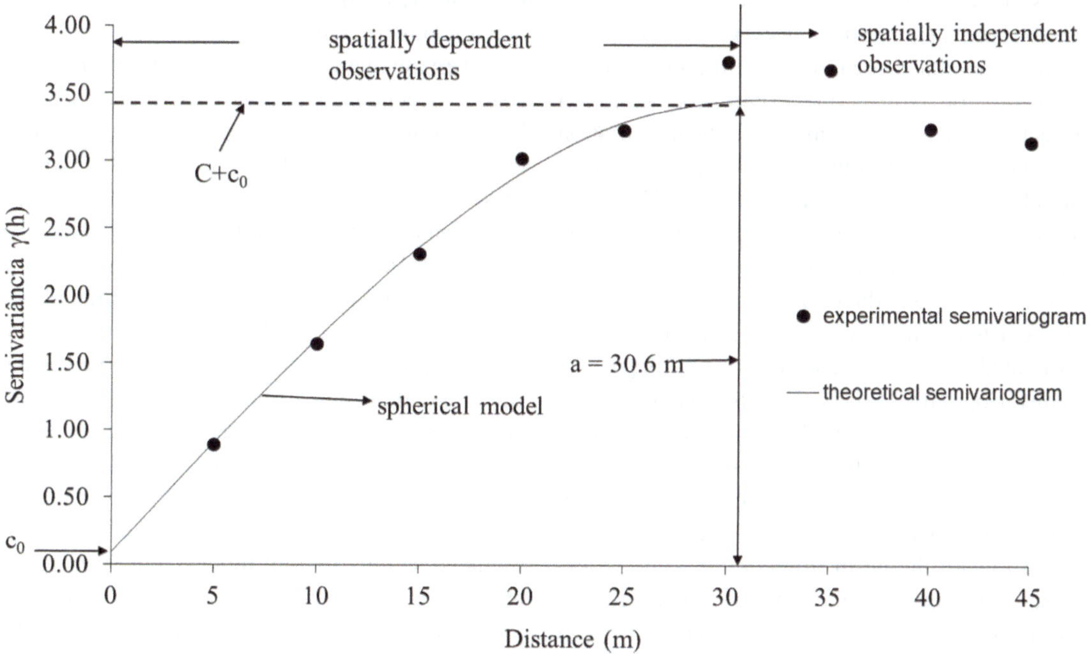

Fig. 17.10 Semivariogram of the clay data of Table 17.1

before and after land leveling. Parfitt et al. (2013, 2014) used geostatistics tools to evaluate the effects of land leveling on the spatial variability of lowland soil physicochemical attributes in the same experimental area as well as established relationships between the magnitude of soil cuts and fills and those of soil attributes by means of linear regressions after leveling (Figs. 17.11 and 17.12).

It is important to mention again that when we are analyzing the variability of data along a transect (i.e., in a linear dimension), the spacing *h* is a scalar being characterized only by its modulus. In this way, the semivariogram is one dimensional and nothing can be said about anisotropy. However, when we are analyzing in a grid (in two dimensions), *h* is a vector being characterized by its modulus, sense, and direction. In this case, the semivariogram also depends on the magnitude and direction of *h*.

As already said, geostatistics tools can quantify the spatial dependence structure of a given variable. If the semivariogram is a function of the Euclidean distance between the observations of the variable only, the semivariogram presents second-order stationarity and the phenomenon is classified as isotropic. Otherwise, it is considered anisotropic (Zimmerman 1993). Anisotropy is usually classified, for models with a defined sill, as geometric, zonal, and combined, based on directional semivariograms that indicate variation in range, sill, and both parameters, respectively (Isaaks and Srivastava 1989). However, Zimmerman (1993) also considers the variation of the nugget effect and slope for models without sill with change in direction. The identification of the existence and type of anisotropy can be evaluated by the construction of directional semivariograms (Isaaks and Srivastava 1989; Eriksson and Siska 2000; Gringarten and Deutsch 2001; Guedes et al. 2008). The directions conventionally used for the construction of semivariograms are 0°, 45°, 90°, and 135° oriented toward the north (direction 0°) and clockwise (Guedes et al. 2013); however, the anisotropic phenomenon can occur in other directions (Isaaks and Srivastava 1989). Figure 17.13 shows experimental semivariograms for two variables (*P* variable—Fig. 17.13a and *K* variable—Fig. 17.13b) illustrating, as examples, directional semivariograms constructed varying the direction by 10 in 10° with angle tolerance of 40° (Isaaks and Srivastava 1989) in order to characterize the anisotropy type (geometric, zonal, or a mixture of both anisotropy types). Figure 17.13A shows that the *P* variable has, from the practical point of view, an isotropic behavior and Fig. 17.13B shows that *K* variable has an anisotropic one. More details about this study can be found in Bitencourt et al. (2015).

Anisotropic phenomena will influence the shape of the estimation window used in the interpolation process, assigning a greater weight to points located closer to the direction of greater spatial continuity of the phenomenon, thus affecting the kriging variance (Guan et al. 2004; Guedes et al. 2013). For this reason, when the influence of anisotropic phenomena on the spatial variability of soil properties is considered, the thematic maps will tend to present greater accuracy. Guedes et al. (2013) evaluated the influence of incorporating geometric anisotropy in the construction of simulated thematic maps of some soil chemical attributes. The authors concluded that there are relevant differences in thematic maps when geometric anisotropy is considered.

When the interest is to evaluate the structure of spatial dependence between two variables collected in an experimental grid or along a spatial transect, we can also calculate the cross-semivariogram between two variables that aims to describe the spatial and/or temporal behaviors between themselves. The cross-semivariogram is only calculated using the existing information for coincident geographical positions, that is, the two variables must necessarily be sampled at the same locations. A **cross-semivariogram** with characteristics that can be identified as ideal would have the appearance of the simple semivariogram (of a single variable, i.e., defined sill, increasing semivariance for small distances, bounded semivariogram model), but with different meanings, simply by involving the product of the differences of two different variables. The theoretical models used for the adjustment process of the cross-semivariogram are the same as

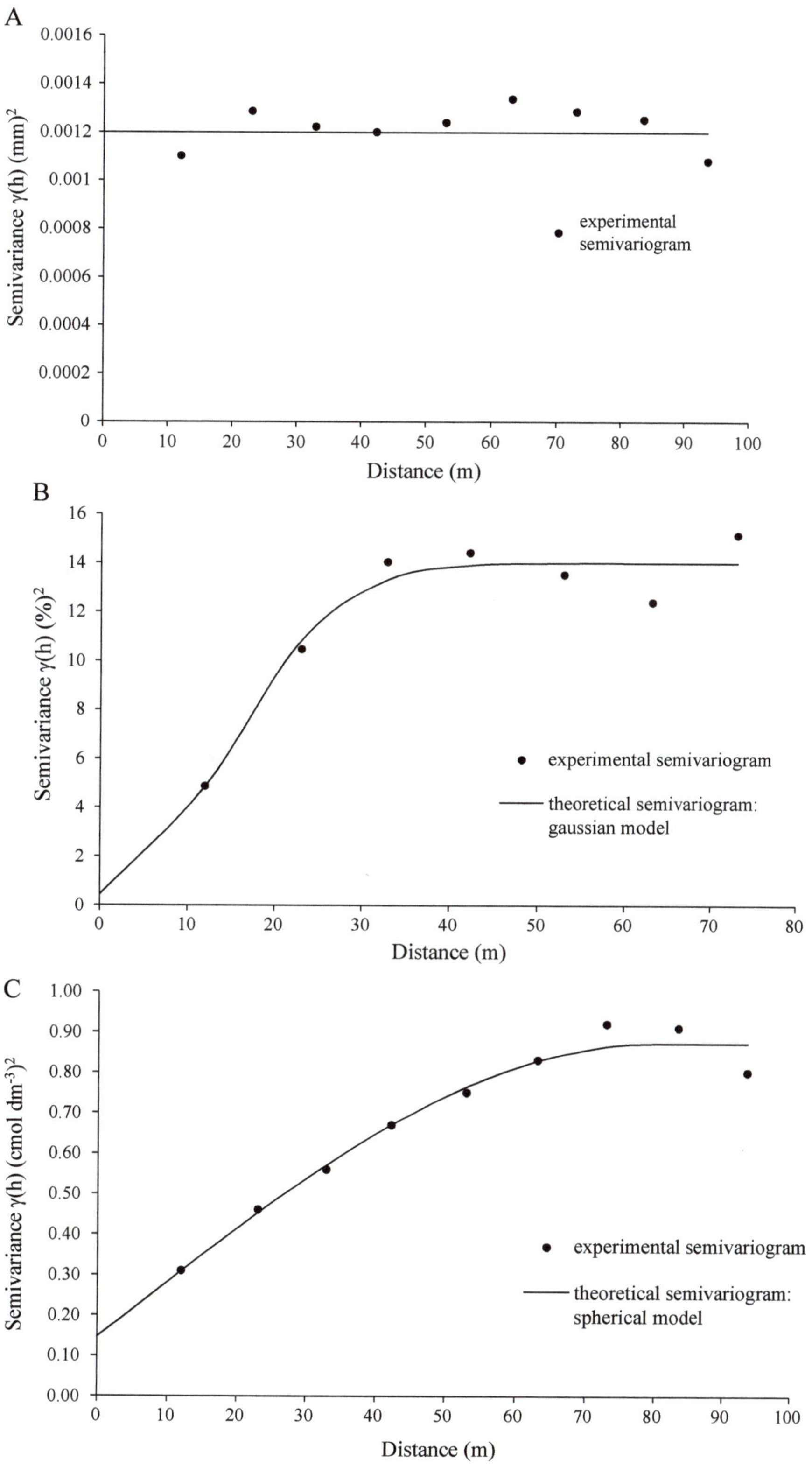

Fig. 17.11 Semivariograms of (**a**) available water capacity variable, (**b**) sand variable, and (**c**) cation exchange capacity variable

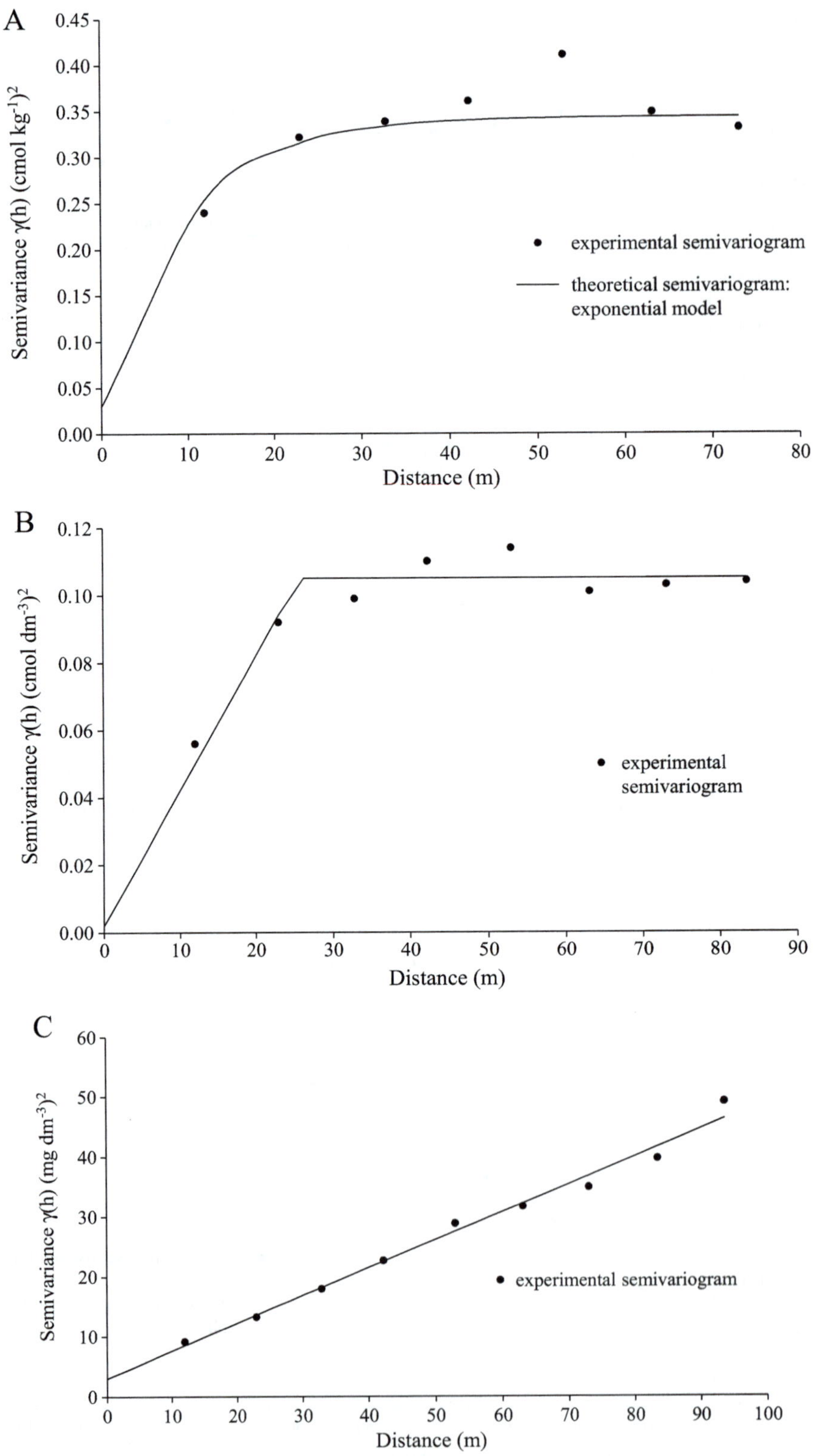

Fig. 17.12 Semivariograms of (**a**) hydrogen + aluminum variable, (**b**) aluminum variable, and (**c**) phosphorus variable

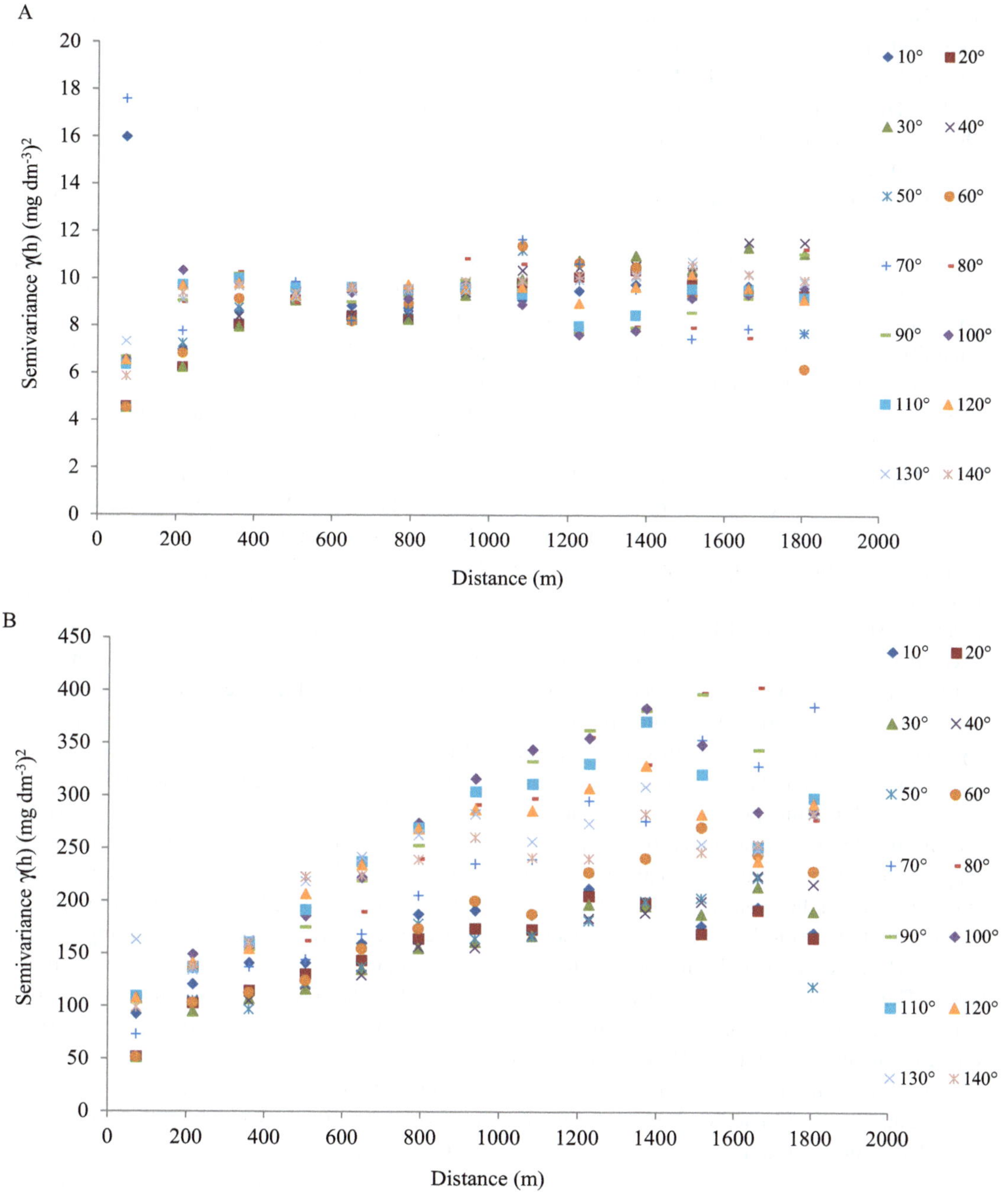

Fig. 17.13 Directional semivariograms illustrating isotropy (**a**) and anisotropy (**b**) phenomena

those previously seen for the individual semivariogram. Carroll and Oliver (2005) emphasized that for a co-regionalization analysis, the semivariograms of the individual variables and the cross-semivariogram between them are modeled individually. However, the same function with the same number of structures must be fitted to each semivariogram; for bounded semivariance functions (theoretical semivariogram models), the distance parameter (lag distance) must be the same, and for unbounded ones, the exponent must be the same.

Journel and Huijbregts (1978), Vieira et al. (1983), Trangmar et al. (1985), Goovaerts (1997), Nielsen and Wendroth (2003), and Webster and Oliver (2007) present a more in-depth and detailed study on isotropic and anisotropic semivariograms and cross-semivariograms. The book of Isaaks and Srivastava (1989) is a very good reference on this subject.

When the objective is to compare the spatial variability structure of the same soil attribute over time, Vieira et al. (1997) proposed the technique of scaling of the semivariogram, where each experimental semivariance value is divided by the most appropriate scale factor (the sample variance of the data set or the sill of the adjusted theoretical model). Several studies have also been published using the scaling technique (technique discussed in Chap. 19) to compare spatial variability structures of different soil attributes collected in the same experimental grid, such as Vieira et al. (2014). With the objective of comparing the spatial variability structure of soil physicochemical attributes (pH in water, organic carbon, phosphorus, potassium, sodium, calcium, magnesium, aluminum, potential acidity, and clay content) in the superficial layer of a Gleysoils mapping unit at reconnaissance scale and taking into consideration the existence of three Gleysoils mapping units at semi-detailed scale, Bitencourt et al. (2015) used the scaled semivariogram technique. The authors concluded that the scaled semivariogram technique revealed that the spatial behavior of the attributes pH in water and exchangeable sodium was similar, regardless of the evaluation scale adopted or the factor used for the scaled semivariogram.

17.12 Ordinary Kriging: A Geostatistical Method of Interpolation

A method of **interpolation** is to make inferences for the points not sampled from the data observed along the grid in the experimental area. There are deterministic interpolation methods, such as polygonal method, triangulation, and inverse functions of distance, among others. However, these methods do not estimate the error associated with each interpolated value, which can be obtained through the kriging geostatistical interpolator (Webster and Oliver 2007).

The semivariogram is the tool of geostatistics that allows to verify and to model the spatial dependence of a variable as previously seen. The existence of a semivariogram model of spatial dependence allows estimating attribute values at unsampled locations (Goovaerts 1997). An immediate application of the semivariogram is the use of its available spatial variance structure for estimating an unmeasured value of a given variable from weighted values measured in a local neighborhood and, later, mapping of the variable (Journel and Huijbregts 1978; Nielsen and Wendroth 2003). The interpolator that uses the semivariogram in its modeling is called **kriging**. According to Goovaerts (1997), kriging is a generic name adopted by geostatisticians for a family of generalized least-squares regression algorithms in recognition of the pioneering work of Danie Krige in 1951.

Kriging is considered the best unbiased linear interpolation method with minimum variance, considering the semivariogram parameters (Nielsen and Wendroth 2003). No other interpolation method is based on the minimum variance between measured values. The kriging weights are associated to the sample data within the neighborhood of the position to be estimated in such a way as to minimize the estimation or kriging variance, and the estimates are unbiased (Webster and Oliver 2007). These weights vary as a function of the distance separating the unmeasured value at position $x_0[z^*(x_0)]$ to be estimated and the measured values at positions $x_i[z(x_i)]$ in its neighborhood. The kriging weights are calculated considering the spatial structure available in the theoretical semivariogram model, that is, that an attribute varies in space described by the semivariogram model. The unmeasured value at position x_0 is then calculated by the solution of a kriging matrix system (Journel and Huijbregts 1978; Isaaks and Srivastava 1989).

For the application of kriging, it is assumed that the realizations of $z(x_i)$ [$z(x_1)$, $z(x_2)$, $z(x_3)$, …, $z(x_n)$] of the random variable Z at the positions x_i ($i = 1, 2, \ldots, n$) are known, and that the semivariogram of the variable already has been correctly determined. Thus, the objective is to determine z^* in the position $x_0[z^*(x_0)]$ of interest, for which there is no measure.

The estimated value $z^*(x_0)$ is given by

$$z^*(x_0) = \sum_{i=1}^{N} \lambda_i z(x_i) \quad (17.29)$$

where N is the number of measured data of the variable Z within the neighborhood involved in the estimate of $z^*(x_0)$ and λ_i are the weights associated with each measured value $z(x_i)$.

If there is spatial structure adequately described by the semivariogram of the variable, the kriging weights are variable according to the distance between the unmeasured value at position $x_0[z^*(x_0)]$ to be estimated and the measured values at positions $x_i[z(x_i)]$ in its neighborhood involved in the estimate.

The best estimate of $z^*(x_0)$ is obtained when

(a) The estimated value is unbiased

$$E[z^*(x_0) - z(x_0)] = 0$$

(b) The kriging variance of the estimated value is minimized

$$\mathrm{Var}[z^*(x_0) - z(x_0)] = \text{minimized kriging variance}$$

For z^* to be an unbiased estimate of z, the sum of the kriging weights of the sample data involved in the estimate has to be 1 (Nielsen and Wendroth 2003; Webster and Oliver 2007).

$$\sum_{i=1}^{N} \lambda_i = 1 \quad (17.30)$$

To obtain the kriging weights that minimize the kriging variance, subject to the constraint Eq. (17.30), we use the method of Lagrange multiplier. The minimized kriging variance is obtained when

$$\sum_{i=1}^{N} \lambda_i \gamma\left[z(x_i), z(x_j)\right] + \psi = \gamma[z(x_i), z(x_0)], \quad i = 1 \text{ to } N, \quad j = 0,1,2,\ldots k \quad (17.31)$$

in which ψ is the Lagrange multiplier. This is the ordinary kriging system (Eqs. 17.30 and 17.31) for punctual kriging (Webster and Oliver 2007).

The set of $N + 1$ equations (Eqs. 17.30 and 17.31) are solved to find the $N + 1$ unknown weights and the Lagrange multiplier (Nielsen and Wendroth 2003). This is the ordinary kriging system for points. From which, the estimation variance (prediction variance or specifically kriging variance) can be calculated as

$$s_E^2[z^*(x_0)] = \psi + \sum_{i=1}^{N} \lambda_i \gamma[z(x_i), z(x_0)] \quad (17.32)$$

In matrix notation, we call $[A]$ the **semivariance matrix** of the sample data involved in the estimate of $z^*(x_0)$; $[\lambda]$ the column matrix containing the kriging weights and the Lagrange multiplier; and $[b]$ the column matrix of the semivariance values between the sample data and the unmeasured value at position $x_0[z^*(x_0)]$ to be estimated. It can be written as follows:

$$[A] \cdot [\lambda] = [b] \quad (17.33)$$

and so

$$[\lambda] = [A]^{-1} \cdot [b] \quad (17.34)$$

$[A]^{-1}$ being the inverse (inverted) matrix of the semivariances $[A]$.

The kriging variance s_E^2, in matrix notation, is written as

$$s_E^2[z^*(x_0)] = [\lambda][b]^t \quad (17.35)$$

the matrix $[b]^t$ is the transposed matrix of $[b]$.

The matrixes $[A]$, $[b]$, and $[\lambda]$, can be written as

$$[A] = \begin{bmatrix} y[z(x_1), z(x_1)] & y[z(x_1), z(x_2)] & \dots & \dots & y[z(x_1), z(x_N)] & 1 \\ y[z(x_2), z(x_1)] & y[z(x_2), z(x_2)] & \dots & \dots & y[z(x_2), z(x_N)] & 1 \\ y[z(x_3), z(x_1)] & y[z(x_3), z(x_2)] & \dots & \dots & y[z(x_3), z(x_N)] & 1 \\ \vdots & \vdots & & & \vdots & \\ \vdots & \vdots & & & \vdots & \\ \vdots & \vdots & & & \vdots & \\ y[z(x_N), z(x_1)] & y[z(x_N), z(x_2)] & \dots & \dots & y[z(x_N), z(x_N)] & 1 \\ 1 & 1 & \dots & \dots & 1 & 0 \end{bmatrix};$$

$$[b] = \begin{bmatrix} \gamma[z(x_1), z(x_0)] \\ \gamma[z(x_2), z(x_0)] \\ \vdots \\ \vdots \\ \gamma[z(x_N), z(x_0)] \\ 1 \end{bmatrix}; \quad [\lambda] = \begin{bmatrix} \lambda_1 \\ \lambda_2 \\ \vdots \\ \vdots \\ \lambda_N \\ \psi \end{bmatrix} \tag{17.36}$$

It is worth remembering that the semivariance value for $\gamma[z(x_1),z(x_1)]$, ..., $\gamma[z(x_N),z(x_N)]$ corresponds to the value of the semivariance between the pairs of values of the variable Z separated by a lag h equal to zero and because of this the main diagonal is equal to zero or equal to the value of the nugget effect.

Some issues should be highlighted about the matrix systems for performing kriging:

(a) The matrix $[A]$ is symmetric.
(b) The values that appear in the matrices $[A]$ and $[b]$ are consequences of the Lagrange multiplier.
(c) The kriging system must be solved for each estimate of z^* and for each variation of the number of sample data involved in its estimate.

For a further stepwise understanding of the calculation sequence of the ordinary kriging interpolation method, an example based on the adjusted spherical semivariogram (Fig. 17.10) is presented for the clay data (Table 17.1).

The clay content of the soil was measured at points of an experimental transect of 150 m, spaced 5 by 5 m (Table 17.1). It is desired to estimate the value of the clay content (not measured) of a point at the distance of 12.5 m. The measured clay contents were 36.5%, 35.0%, 35.0%, and 35.5% at points x_1, x_2, x_3, and x_4, respectively (Fig. 17.14).

Distances d between data pairs:

$$d[z(x_1) - z(x_2)] = d[z(x_3) - z(x_4)] = 5 \text{ m or 1 lag};$$

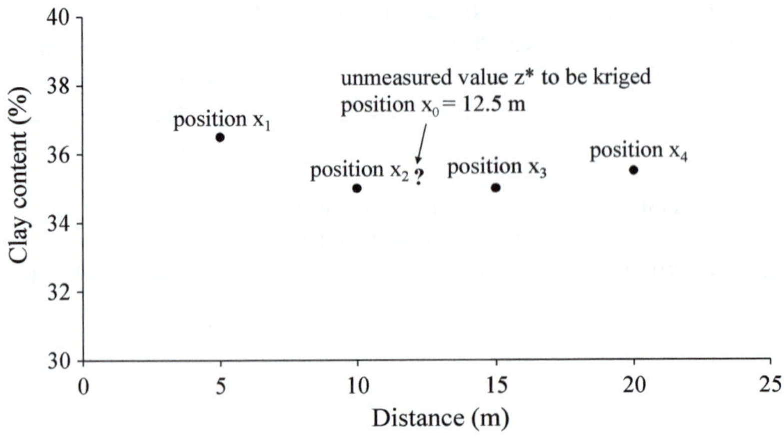

Fig. 17.14 Illustration of the kriging procedure

$d[z(x_1) - z(x_3)] = d[z(x_3) - z(x_4)] = 10$ m or 2 lags;
$d[z(x_1) - z(x_4)] = 15$ m or 3 lags;
$d[z(x_1) - z^*(x_0)] = d[z(x_4) - z^*(x_0)] = 7.5$ m;
$d[z(x_2) - z^*(x_0)] = d[z(x_3) - z^*(x_0)] = 2.5$ m.

Through the spherical semivariogram model (Fig. 17.10), we have

$$\gamma(h) = 0.09 + 3.356\left[\frac{3h}{61.2} - \frac{1}{2}\left(\frac{h}{30.6}\right)^3\right] \quad \text{for } 0 \le h \le a$$

The calculated semivariances for these distances are

$\gamma(h{=}2.5\text{ m}) = 0.50\ \%^2$.
$\gamma(h{=}5.0\text{ m}) = 0.91\ \%^2$;
$\gamma(h{=}7.5\text{ m}) = 1.30\ \%^2$.
$\gamma(h{=}10\text{ m}) = 1.68\ \%^2$;
$\gamma(h{=}15\text{ m}) = 2.36\ \%^2$.

Thus, one can construct the system of equations (in analogy to the system 17.36) for estimation of the unmeasured value at position x_0 by ordinary kriging:

$$[A] = \begin{bmatrix} \gamma[z(x_1),z(x_1)] & \gamma[z(x_1),z(x_2)] & \gamma[z(x_1),z(x_3)] & \gamma[z(x_1),z(x_4)] & 1 \\ \gamma[z(x_2),z(x_1)] & \gamma[z(x_2),z(x_2)] & \gamma[z(x_2),z(x_3)] & \gamma[z(x_2),z(x_4)] & 1 \\ \gamma[z(x_3),z(x_1)] & \gamma[z(x_3),z(x_2)] & \gamma[z(x_3),z(x_3)] & \gamma[z(x_3),z(x_4)] & 1 \\ \gamma[z(x_4),z(x_1)] & \gamma[z(x_4),z(x_2)] & \gamma[z(x_4),z(x_3)] & \gamma[z(x_4),z(x_4)] & 1 \\ 1 & 1 & 1 & 1 & 0 \end{bmatrix}$$

$$[b] = \begin{bmatrix} \gamma[z(x_1),z(x_0)] \\ \gamma[z(x_2),z(x_0)] \\ \gamma[z(x_3),z(x_0)] \\ \gamma[z(x_4),z(x_0)] \\ 1 \end{bmatrix}; \qquad [\lambda] = \begin{bmatrix} \lambda_1 \\ \lambda_2 \\ \lambda_3 \\ \lambda_4 \\ \psi \end{bmatrix}$$

$$\underset{[A]}{\begin{bmatrix} 0.09 & 0.91 & 1.68 & 2.36 & 1 \\ 0.91 & 0.09 & 0.91 & 1.68 & 1 \\ 1.68 & 0.91 & 0.09 & 0.91 & 1 \\ 2.36 & 1.68 & 0.91 & 0.09 & 1 \\ 1 & 1 & 1 & 1 & 0 \end{bmatrix}} \underset{[\lambda]}{\begin{bmatrix} \lambda_1 \\ \lambda_2 \\ \lambda_3 \\ \lambda_4 \\ \psi \end{bmatrix}} = \underset{[b]}{\begin{bmatrix} 1.30 \\ 0.50 \\ 0.50 \\ 1.30 \\ 1 \end{bmatrix}}$$

which is solved according to

$$[\lambda] = [A]^{-1} \cdot [b]$$

where the matrix $[A]^{-1}$ is the inverse matrix of $[A]$. A good text on the operation of matrixes is presented by Jeffrey (2010).

$$[A]^{-1} = \begin{bmatrix} -0.6156 & 0.5981 & -0.0162 & 0.0336 & 0.4617 \\ 0.5981 & -1.1949 & 0.6129 & -0.0162 & 0.0383 \\ -0.0162 & 0.6129 & -1.1949 & 0.5981 & 0.0383 \\ 0.0336 & -0.0162 & 0.5981 & -0.6156 & 0.4617 \\ 0.4617 & 0.0383 & 0.0383 & 0.4617 & -1.2301 \end{bmatrix}$$

Resulting

$$[\lambda] = \begin{bmatrix} -0.003195 \\ 0.503195 \\ 0.503195 \\ -0.003195 \\ 0.00781 \end{bmatrix}$$

As can be seen in the result of the kriging weights matrix $[\lambda]$, negative values of λ_1 and λ_4 were found. Journel and Rao (1996) have proposed to correct the negative kriging weights by translation and rescaling. The largest negative weight value is added to all kriging weights, which are then rescaled to sum up to one. This correction is only possible if all weights of ordinary kriging are positive and have a sum equal to 1 (Eq. 17.30). Therefore, this means that if a negative weight is found, it must be replaced. The algorithm for substitution of negative kriging weights, proposed by Journel and Rao (1996), consists in adding a constant (W), equal to the module of the largest negative kriging weight, to all weights. Then, the weights are normalized (τ_i) again to sum equal to 1, according to the expression

$$\tau_i = \frac{\lambda_i + W}{\sum_{i=1}^{N} (\lambda_i + W)} \quad \text{for } i = 1,2,3, \ldots N$$

where $W = |\lambda_i|$ is the largest negative module kriging weight. In this way, Journel and Rao (1996) mentioned that the algorithm eliminates only the sample with the highest negative weight in the module, and if there are more samples with negative weights, its weights will be replaced by the addition of the constant W. The estimate of ordinary kriging after the correction of the negative kriging weights, at the unsampled location x_0, becomes

$$z^*(x_0) = \sum_{i=1}^{N} \tau_i z(x_i) \tag{17.29a}$$

For our example,

λ_i	W	$\lambda_i + W$	τ_i
−0.003195	0.003195	0	0
0.503195		0.50639	0.5
0.503195		0.50639	0.5
−0.003195		−2.8E−16	0
0.007814			

Applying Eq. (17.29a), we have

$$\begin{aligned} z^*(x_0 = 12.5\,\text{m}) &= (0 \times 36.5) + (0.5 \times 35.0) \\ &\quad + (0.5 \times 35.0) + (0 \times 35.5) \\ &= 35.0\% \end{aligned}$$

This means that $z(x_1)$ and $z(x_4)$ have the same weight 0 and $z(x_2)$ and $z(x_3)$ have kriging weights 0.5 in the estimation of $z^*(x_0)$. The kriging variance (Eq. 17.32) associated to the estimation is

$$\begin{aligned} s_E^2[z^*(x_0)] &= 0.00781 + (0 \times 1.30) \\ &\quad + (0.5 \times 0.5) + (0.5 \times 0.5) \\ &\quad + (0 \times 1.30) \\ &= 0.5078\ (\%)^2 \end{aligned}$$

or in matricial form, Eq. (17.35):

$$\underset{}{s_E^2\ z^*(x_0)} = \underset{[\lambda]}{\begin{bmatrix} 0 \\ 0.5 \\ 0.5 \\ 0 \\ 0.00781 \end{bmatrix}} \underset{\times}{\times} \underset{[b]^t}{[1.30 \quad 0.5 \quad 0.5 \quad 1.30 \quad 1]} = \underset{= s_E^2[z^*(x_0)]}{0.5078(\%)^2}$$

Thus, the clay content estimated at the distance of 12.5 m is 35% with a standard deviation (remembering that it is equal to the square root of the variance) of 0.713%. It may be noted that the two neighboring points contributed equal weight to the estimate of the new point, which is correct since the two are located at the same distance from that point. By choice of example,

$z^*(x_0)$ equals the mean of $z(x_2)$ and $z(x_3)$, but in general, this does not happen. If the question would have been to calculate $z^*(x_0)$ at 11.0 m, a larger contribution of $z(x_2)$ would probably appear.

There are other types of kriging besides ordinary kriging, such as universal, lognormal, probability, factorial, indicator, and disjunctive. Further details on these types can be found in the texts of Journel and Huijbregts (1978), Nielsen and Wendroth (2003), and Webster and Oliver (2007), among others.

The determination of variables in some studies may be costly and have a difficult methodology, and may compromise the study of their temporal or spatial variabilities. However, if the variable costly and/or difficult to be determined and has a spatial correlation with another variable of simple determination and/or low cost, we can make its estimate using information from the cross-semivariogram, by means of a technique called **co-kriging**. Further details of this technique can be found in Cressie (1993), Nielsen and Wendroth (2003), and Webster and Oliver (2007), among other geostatistical texts already mentioned above.

17.13 Pedotransfer Functions

Although not formally recognized and denominated until 1989, the concept of **pedotransfer function** (PTF) has been applied to estimate soil attributes that are time-consuming, have a complex methodology, and use expensive equipment for their determination. The term "pedotransfer function" was formally defined by Bouma (1989) as "the conversion of the data we have in data we need," that is, transform data that are part of the routine analysis in (e.g., soil texture, soil organic matter, soil color, soil bulk density, soil porosity) in properties that spend more time and cost for their determination (e.g., soil water content, hydraulic conductivity of the soil, water retention curve in the soil).

McBratney et al. (2002) cite that many PTFs are being developed for predicting certain soil attributes for a geographic area and that literature reviews on their development and use, particularly for predicting soil hydraulic attributes, can be seen in Rawls et al. (1991), Wösten (1997), Pachepsky et al. (1999a), Wösten et al. (2001), Botula et al. (2014), and more recently van Looy et al. (2017). Pachepsky and Rawls (2004) have published a comprehensive text on the theory surrounding this subject as well as examples of different PTFs developed in different parts of the world. These authors also commented that the development of new PTFs is an arduous task and that the use of already developed functions would be more sensible. However, a given PTF should not be extrapolated beyond the geomorphic region or type of soil for which it was developed.

According to Nemes et al. (2003), most of the pedotransfer functions available in the literature use soil texture, soil density, and soil organic matter as explanatory variables and that other variables are seldom used (Rawls et al. 1991; Wösten et al. 2001). The same authors also comment that many PTFs have been developed in the last decades; among them are as follows: Tietje and Tapkenhinrichs (1993) and Kern (1995) evaluated different PTFs to estimate water retention; Tietje and Hennings (1996) tested PTFs to estimate soil hydraulic conductivity; Imam et al. (1999) compared three PTFs to calculate water retention capacity in inorganic soils; Cornelis et al. (2001) compared nine PTFs to estimate the water retention curve of the soil; Wagner et al. (2001) evaluated the performance of eight PTFs to estimate the hydraulic conductivity of unsaturated soil. Revisions on pedotransfer functions can be seen in Leij et al. (2002), Tomasella and Hodnett (2004), Pachepsky et al. (2006), Vereecken et al. (2010), and Botula et al. (2014).

In Brazil, there are still few references of research results in pedotransfer functions. Tomasella and Hodnett (1998) developed PTF equations to predict parameters of the equation developed by Brooks and Corey (1964) to estimate the soil water retention curve using soil texture, soil density, porosity, and water content, using multiple linear regression. The authors observed high significance in the correlation between the values of the measured and estimated parameters of the equation. Tomasella et al. (2000) used soil texture, organic carbon, equivalent moisture, and soil density, available in soil survey reports, and established relationships between the van Genuchten (1980) equation water retention parameters. Other examples of studies that have developed PTFs for Brazilian soils are Arruda et al. (1987), van den Berg et al. (1997), Oliveira et al. (2002), Hodnett and Tomasella (2002), Giarola et al. (2002), and Silva et al. (2008). Barros and De Jong van Lier (2014) present a review of the state-of-the-art PTFs in Brazil and offer suggestions for the development of future work on this topic. Botula et al. (2014) published a review on the use of PTFs to estimate soil water retention in the humid tropics, including Brazil, mentioning that among the 35 publications found in the literature on these PTFs, 91% are based on an empirical approach and only 9% in a semi-physical approach. Nebel et al. (2010) evaluated eight point-specific gravimetric PTFs developed by some authors using soils of different characteristics to estimate water retention in the soil, depth of 0–0.20 m, in 100 points of an experimental grid established in a lowland soil in Pelotas (RS), Brazil. Comparisons were made between the data measured and estimated by the PTFs, using statistical and geostatistical tools, such as mean error, square root mean error, semivariograms, cross-validation, and regression coefficient. The eight PTFs tested to evaluate gravitational soil water (Ug) under the potentials of 33 kPa and 1500 kPa showed a tendency to overestimate Ug33kPa and to underestimate Ug1500kPa. PTFs were classified according to their performance and also in relation to their potential in describing the spatial variability structure of the measured data set.

Wösten et al. (2001) and Pachepsky and Rawls (2004) distinguish three different types of PTFs to estimate hydraulic properties:

1. Type 1—prediction of hydraulic properties based on soil structure model: the models presented by Bloemen (1980) and Arya and Paris (1981) are examples of this type of PTF where the water retention curve was estimated from particle size distribution, soil density, and particle density. Another example is that of Silva et al. (2017) who applied the Splintex model, which is a physico-empirical model, based on Arya and Paris (1981).
2. Type 2—prediction of one specific point (e.g., soil water content at −10 kPa) in the soil water retention curve (referred to as punctual PTF): in this type of PTF, classical regression equations are used which estimate specific points of interest of the soil water retention curve (Gupta and Larson 1979; Rawls et al. 1982; Ahuja et al. 1985, among others). Consequently, these functions have the following form:

$$\begin{aligned}\theta_h &= a \times \text{sand content} + b \times \text{clay content} \\ &\quad + c \times \text{organic carbon content} + \cdots \\ &\quad + x \times X \ \text{variable}\end{aligned}$$

   where θ_h is the value of the soil volumetric water content at the matrix potential h; a, b, c, and x are the coefficients of the classical multiple regression. The variable X is any other basic property easily obtained for the soil.
3. Type 3—prediction of parameters used to describe soil hydraulic properties (termed parametric PTF): PTFs are functional relationships that transform available soil properties (e.g., texture, structure, soil organic matter) into soil properties not available (e.g., soil water retention curve). In contrast to type 2, type 3 PTFs usually estimate model parameters describing the complete relationship between soil water content–matric potential–hydraulic conductivity. Type 3 PTF is simpler and more straightforward than type 2 due to the fact that the results are directly applicable in model simulation [e.g., PTFs to estimate the parameters of the Brooks and Corey (1964) and Van Genuchten (1980) models to estimate the soil water retention curve].

Minasny et al. (2003) point out that although most PTFs have been developed to predict hydraulic attributes of soils, it also works to estimate other physical and chemical properties. They also point out that while only a few PTFs were developed for chemical attributes, citing as an example the research developed by Janik et al. (1995), it allows the prediction of various chemical attributes from measurements of spectroscopy.

In addition to the methodologies described above, the pedotransfer functions also make use of **neural networks**, which constitute a method of solving artificial intelligence problems, constructing a system that has circuits that simulate the human brain, even without behavior (Pachepsky et al. 1996; Pachepsky and Rawls 2004). In the agronomic area, neural networks have been used to develop pedotransfer functions (Schaap and Bouten 1996; Pachepsky et al. 1996, 1999b; Tamari et al. 1996; Schaap et al. 1998), which have the purpose of estimating soil hydraulic properties that demand time and cost.

An example of PTF that uses neural networks is the Rosetta computer program described in detail in Schaap et al. (2001) and with examples of applications in Radcliffe and Simunek (2010) and Silva et al. (2017). In this program, five pedotransfer functions based on neural network models combined with the bootstrap method are implemented, thus allowing estimation of the uncertainties in the estimated values of the hydraulic properties. The Rosetta program allows estimating the parameters of the van Genuchten (1980) retention curve model and the saturated soil hydraulic conductivity values using five different input levels: the simplest (Model 1) uses the average hydraulic parameters adjusted within a textural class of soil based on the USDA textural triangle and the other four models progressively use more input data, starting with the sand, silt, and clay fractions (Model 2) (Model 3) and water content retained at 33 kPa (Model 4) and at 1500 kPa (Model 5).

According to Warner and Misra (1996), neural networks have been used in a wide variety of applications where statistical methods are traditionally employed, such as classification problems related to the identification of marine sonars (Gorman and Sejnowski 1988), prediction of cardiac problems in patients (Baxt 1991; Fujita et al. 1992), speech recognition (Lippmann 1989), and applications in time series for predicting supermarket stocks (Hutchinson 1994). The same authors also comment that for statisticians, these problems mentioned above would normally be solved by classical statistical models such as discriminant analysis (Flury and Riedwyl 1990), logistic regression (Studenmund 1992), Bayes classifier and other types of classifiers (Duda and Hart 1973), multiple regression (Neter et al. 1990), and time series models (ARIMA models, etc.) (Studenmund 1992).

Timm et al. (2006) is an example of the use of neural networks in the agronomic area. They evaluated the relationship between more expensive, labor-intensive, and time-consuming variables (e.g., total soil nitrogen) and other variables that are cheaper and faster (e.g., soil organic carbon, pH) using feed-forward (Fig. 17.15) and recurrent (Fig. 17.16) neural

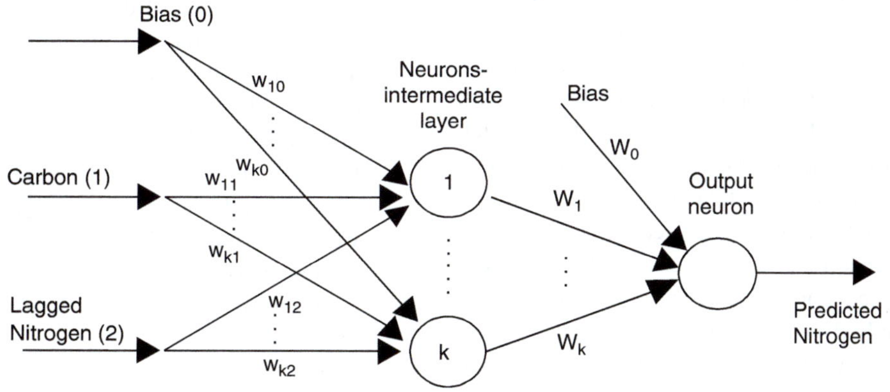

Fig. 17.15 Feed-forward neural network

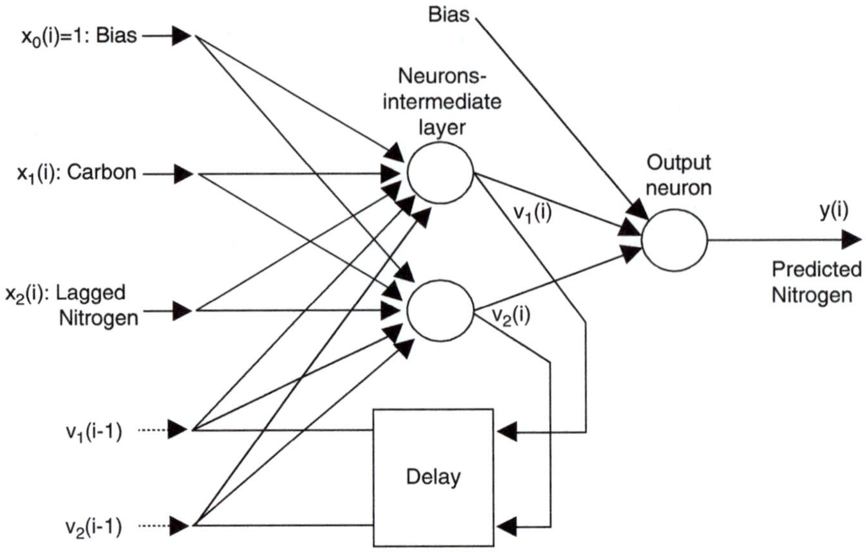

Fig. 17.16 Recurrent neural network

network models and state-space models (Shumway 1988; West and Harrison 1997), which will be addressed in Chap. 18. The predictive capacity of these two important classes of models was compared with standard regression models used as reference. Samples of a soil classified as Latossol in an experimental area cultivated with oats (Embrapa-CNPMA—Jaguariúna, SP, Brazil) were collected in the 0–0.20 m depth layer along a spatial transect of 194 m, points equidistant from each other by 2 m, totaling 97 observations of each variable. The results showed that the models of recurrent neural network and standard state space had a better predictive performance of the last ten and first ten observations of the total soil nitrogen variable along the transect when compared to the standard regression models, independently of the statistical criterion used. For the prediction of the last ten observations, the standard state-space model had a better performance when compared to all other models used, while for the prediction of the first ten, the recurrent neural network had a better performance compared to all models. Among the standard regression models, the autoregressive vector had a better predictive performance in estimating the total soil nitrogen variable for both cases.

Examples of texts with conceptual fundaments, operational principles, and modeling methods with artificial neural networks are those by Haykin (1999) and Hastie et al. (2016) which are very broad texts on this subject.

17.14 Exercises

17.1. In an experimental area, a sampling grid of 10 m × 10 m was established, totalizing 100 sample points [more details about the experimental area and the established grid are found in the work of Parfitt et al. (2009) (2013) (2014)]. At each sampling point, undisturbed soil samples were collected in the 0–0.20 m deep layer, where the sand contents and microporosity values (micro) were determined (here considered as the values of soil water content retained at the potential of $\geq$6 kPa), which are presented in the table below.

Point	*X*	*y*	Sand (%)	Micro (%)	Point	*x*	*y*	Sand (%)	Micro (%)
1	0	90	46	28	51	50	90	44	27
2	0	80	45	29	52	50	80	48	28
3	0	70	43	34	53	50	70	45	31
4	0	60	44	34	54	50	60	39	32
5	0	50	45	34	55	50	50	40	35
6	0	40	43	35	56	50	40	38	36
7	0	30	43	36	57	50	30	40	33
8	0	20	42	36	58	50	20	42	33
9	0	10	42	35	59	50	10	41	33
10	0	0	41	34	60	50	0	41	32
11	10	90	48	31	61	60	90	44	28
12	10	80	47	33	62	60	80	48	28
13	10	70	44	29	63	60	70	48	28
14	10	60	44	28	64	60	60	47	18
15	10	50	43	30	65	60	50	39	31
16	10	40	44	36	66	60	40	40	32
17	10	30	43	37	67	60	30	42	32
18	10	20	41	35	68	60	20	45	30
19	10	10	41	35	69	60	10	44	29
20	10	0	42	33	70	60	0	41	31
21	20	90	46	29	71	70	90	48	28
22	20	80	48	27	72	70	80	49	29
23	20	70	47	31	73	70	70	51	27
24	20	60	46	30	74	70	60	48	29
25	20	50	44	31	75	70	50	40	32
26	20	40	44	33	76	70	40	43	33
27	20	30	44	36	77	70	30	47	28
28	20	20	43	36	78	70	20	49	30
29	20	10	42	32	79	70	10	52	29
30	20	0	42	33	80	70	0	47	31
31	30	90	50	28	81	80	90	48	28
32	30	80	48	28	82	80	80	52	26
33	30	70	49	28	83	80	70	51	28
34	30	60	48	27	84	80	60	53	28
35	30	50	48	28	85	80	50	46	31
36	30	40	46	33	86	80	40	47	30
37	30	30	45	28	87	80	30	49	28
38	30	20	46	33	88	80	20	54	28
39	30	10	44	31	89	80	10	49	30
40	30	0	43	31	90	80	0	50	27
41	40	90	48	28	91	90	90	49	31
42	40	80	47	29	92	90	80	51	27
43	40	70	46	28	93	90	70	49	38
44	40	60	46	27	94	90	60	53	30
45	40	50	46	33	95	90	50	52	32
46	40	40	42	31	96	90	40	51	29
47	40	30	43	33	97	90	30	52	31
48	40	20	43	30	98	90	20	54	28
49	40	10	45	30	99	90	10	52	31
50	40	0	47	30	100	90	0	49	34

It is requested:

(a) Perform the classical statistical analysis of each series determining mean, standard deviation, variance, coefficient of variation (CV), skewness, and kurtosis coefficients.

(b) Calculate the experimental and theoretical semivariograms with respective parameters of adjustments (c_0: nugget effect, $c_0 + C$: sill; a: range) for each variable, the degree of spatial dependence (DSD according to Cambardella et al. 1994) and check the quality of semivariogram adjustment using the cross-validation procedure.

(c) Is it possible to map each variable using the kriging technique? If so, do the spatial distribution map for each variable?

17.15 Answers

17.1. (a)

Attribute	Mean	Standard Deviation	Variance	CV (%)	Skewness Coefficient	Kurtosis Coefficient
Sand	46	3.741	13.992	8.1	0.19	−0.67
Micro	31	3.095	9.577	10.0	−0.26	1.55

(b)

								Cross-Validation	
Attribute	Model	c_0	$c_0 + C$	a (m)	R^2	RSS	DSD (%)	R^{2*}	CR
Sand	Gaussian	0.31	14.17	33.3	0.98	1.27	2.2	0.78	0.91
Micro	Gaussian	4.81	9.62	41.7	0.97	0.33	50.0	0.36	0.92

c_0: nugget effect; $c_0 + C$: sill; a: range; R^2: determination coefficient; RSS: residual sum of squares; DSD = degree of spatial dependency $[c_0/(c_0 + C)] \times 100$; R^2 and CR: slope of the linear regression (measured versus estimated values) made for the cross-validation.

(c) Yes, it is possible.

- Map of the spatial distribution of sand content.

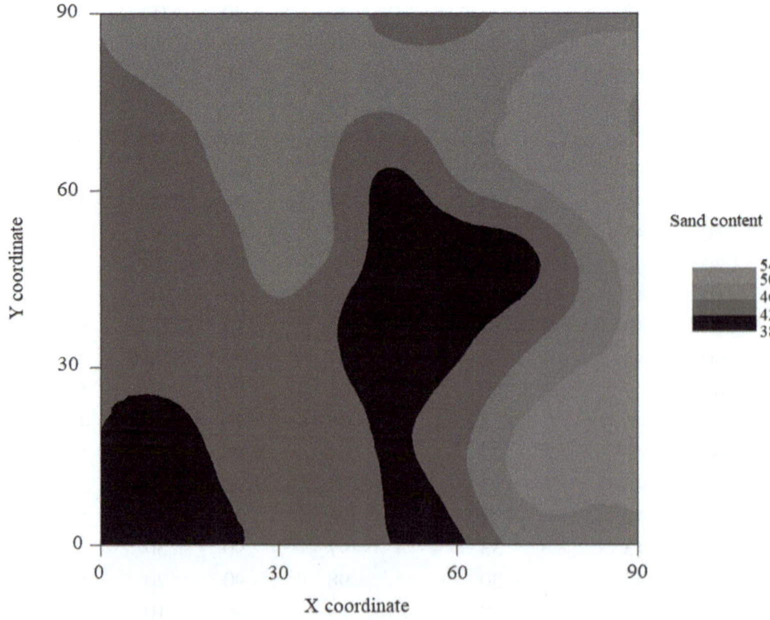

- Map of the spatial distribution of soil microporosity.

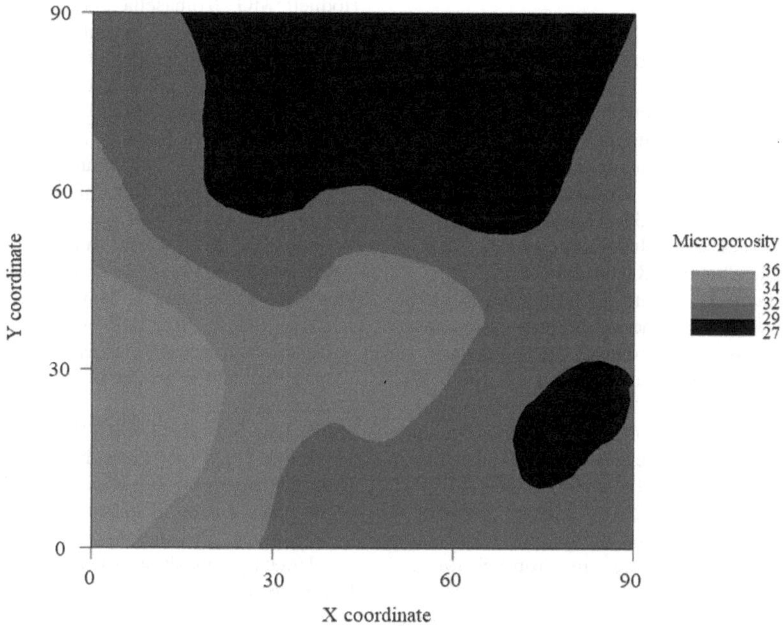

References

Ahuja LR, Naney JW, Williams RD (1985) Estimating soil water characteristics from simpler properties or limited data. Soil Sci Soc Am J 49:1100–1105

Arruda FB, Zullo Junior J, Oliveira JB (1987) Parâmetros de solo para o cálculo da água disponível com base na textura do solo. Rev Bras Ciênc Solo 11:11–15

Arya LM, Paris JF (1981) A physico-empirical model to predict the soil moisture characteristic from particle-size distribution and bulk density data. Soil Sci Soc Am J 45:1023–1030

Barros AHC, De Jong van Lier Q (2014) Pedotransfer functions for Brazilian soils. In: Teixeira WG, Ceddia MB, Ottoni MV, Donnagema GK (eds) Application of soil physics in environmental analysis: measuring, modeling and data integration. Springer, New York, NY, pp 131–162

Baxt WG (1991) Use of an artificial neural network for the diagnosis of myocardial infarction. Ann Intern Med 115:843–848

Bitencourt DGB, Timm LC, Guimarães EC, Pinto LFS, Pauletto EA, Penning LH (2015) Spatial variability structure of the surface layer attributes of Gleysols from the Coastal Plain of Rio Grande do Sul. Biosci J 31:1711–1721

Bloemen GW (1980) Calculation of hydraulic conductivities from texture and organic matter content. Z Pflanz Bod 143:581–605

Botula YD, Van Ranst E, Cornelis WM (2014) Pedotransfer functions to predict water retention for soils of the humid tropics: a review. Braz J Soil Sci 38:679–698

Bouma J (1989) Using soil survey data for quantitative land evaluation. Adv Soil Sci 9:177–213

Brooks RH, Corey AT (1964) Hydraulic properties of porous media. Colorado State University, Fort Collins, CO, pp 1–27

Cambardella CA, Moorman TB, Novak JM, Parkin TB, Karlen DL, Turco RF, Konopka AE (1994) Field-scale variability of soil properties in Central Iowa soils. Soil Sci Soc Am J 58:1501–1511

Carroll ZL, Oliver MA (2005) Exploring the spatial relations between soil physical properties and apparent electrical conductivity. Geoderma 128:354–374

Cassalho F, Beskow S, Mello CR, Moura MM, Kerstner L, Ávila LF (2018) At-site flood frequency analysis coupled with multiparameter probability distributions. Water Resour Manag 32:285–300

Clark I (1979) Practical geostatistics. Applied Science Publications, London

Cornelis WM, Ronsyn J, Van Meirvenne M, Hartmann R (2001) Evaluation of pedotransfer functions for

predicting the soil moisture retention curve. Soil Sci Soc Am J 65:638–648
Cressie NAC (1993) Statistics for spatial data. John Wiley & Sons Inc., New York, NY
Cressie NAC, Hawkins DM (1980) Robust estimation of the variogram: I. Math Geol 12:115–125
Davis JC (1986) Statistics and data analysis in geology, 2nd edn. John Wiley & Sons Inc., New York, NY
Duda RO, Hart PE (1973) Pattern classification and scene analysis. John Wiley & Sons Inc., New York, NY
Eriksson M, Siska P (2000) Understanding anisotropy computations. Math Geol 326:683–700
Flury B, Riedwyl H (1990) Multivariate statistics: a practical approach. Chapman & Hall, London
Fujita H, Katafuchi T, Uehara T, Nishimura T (1992) Application of artificial neural network to computer aided diagnosis of coronary artery disease in myocardial spect bull's-eye images. J Nucl Med 33:272–276
Giarola NFB, Silva AP, Imhoff S (2002) Relações entre propriedades físicas e características de solos da região sul do Brasil. Rev Bras Ciênc Solo 26:885–893
Glaz B, Yeater KM (eds) (2018) Applied statistics in agricultural, biological, and environmental sciences. Am Soc Agron, Soil Sci Soc Am, Crop Sci Soc Am, Madison, WI
Goovaerts P (1997) Geostatistics for natural resources evaluation. Oxford University Press Inc., New York, NY
Goovaerts P (1999) Geostatistics in soil science: state-of-the-art and perspectives. Geoderma 89:1–45
Gorman RP, Sejnowski TJ (1988) Analysis of hidden units in a layered network to classify sonar targets. Neural Netw 1:75–89
Gringarten E, Deutsch CV (2001) Teacher's aide, variogram interpretation and modeling. Math Geol 33:507–534
Guan Y, Sherman M, Calvin JA (2004) A nonparametric test for spatial isotropy using subsampling. J Am Stat Assoc 99:810–821
Guedes LPC, Uribe-Opazo MA, Johann JA, Souza EG (2008) Anisotropia no estudo da variabilidade espacial de algumas variáveis químicas do solo. Rev Bras Ciênc Solo 32:2217–2226
Guedes LPC, Uribe-Opazo MA, Ribeiro Junior PJ (2013) Influence of incorporating geometric anisotropy on the construction of thematic maps of simulated data and chemical attributes of soil. Chil J Agric Res 73:414–423
Gupta SC, Larson WE (1979) Estimating soil water characteristic from particle size distribution, organic matter percent, and bulk density. Water Resour Res 15:1633–1635
Gusella V (1991) Estimation of extreme winds from short-term records. J Struct Eng 117:375–390
Hastie T, Tibshirani R, Friedman J (2016) The elements of statistical learning: data, mining, inference, and prediction, 2nd edn. Springer, New York, NY
Haykin S (1999) Neural networks – a comprehensive foundation, 2nd edn. Prentice Hall, Englewood Cliffs, NJ
Hodnett MG, Tomasella J (2002) Marked differences between van Genuchten soil water-retention parameters for temperate and tropical soils: a new water-retention pedotransfer function developed for tropical soils. Geoderma 108:155–180
Hui S, Wendroth O, Parlange MB, Nielsen DR (1998) Soil variability – infiltration relationships of agroecosystems. J Balkan Ecol 1:21–40
Hutchinson JM (1994) A radial basis function approach to financial time series analysis. PhD Dissertation, Massachusetts Institute of Technology, Massachusetts
Imam B, Sorooshian S, Mayr T, Schaap MG, Wösten JHM, Scholes RJ (1999) Comparison of pedotransfer functions to compute water holding capacity using the van Genuchten model in inorganic soils. IGBP-DIS Report, Toulouse
Isaaks EH, Srivastava RM (1989) Applied geostatistics. Oxford University Press, New York, NY
Janik LJ, Skjemstad JO, Raven MD (1995) Characterization and analysis of soils using midinfrared partial least squares. I. Correlations with XRF-determined major element composition. Aust J Soil Res 33:621–636
Jeffrey A (2010) Matrix operations for engineers and scientists: an essential guide in linear algebra. Springer, New York, NY
Journel AG, Huijbregts CHJ (1978) Mining geoestatistics. Academic Press Inc., New York, NY
Journel AG, Rao SE (1996) Deriving conditional distributions from ordinary kriging. Stanford Center for Reservoir Forecasting, Stanford, CA
Jury WA, Horton R (2004) Soil physics, 6th edn. John Wiley & Sons, Hoboken, NJ
Kern JS (1995) Evaluation of soil water retention models based on basic soil physical properties. Soil Sci Soc Am J 59:1134–1141
Leij FJ, Schaap MG, Arya MP (2002) Indirect methods. In: Dane JH, Topp GC (eds) Methods of soil analysis: Part 4, Physical methods, 3rd edn. Soil Science Society of Agronomy, Madison, WI, pp 1009–1045
Lippmann RP (1989) Review of neural networks for speech recognition. Neural Comput 1:1–38
Marques RFPV, Mello CR, Silva AM, Franco CS, Oliveira AS (2014) Performance of the probability distribution models applied to heavy rainfall daily events. Cienc Agrotec 38:335–342
Matheron G (1963) Principles of geostatistics. Econ Geol 58:1246–1266
McBratney AB, Minasny B, Cattle SR, Vervoort RW (2002) From pedotransfer functions to soil inference systems. Geoderma 109:41–73
Minasny B, McBratney AB, Mendonça-Santos ML, Santos HG (2003) Revisão sobre funções de pedotransferência (PTFs) e novos métodos de predição de classes e atributos do solo. Embrapa Solos, Rio de Janeiro

Montgomery DC, Runger GC (1994) Applied statistics and probability for engineers. John Wiley & Sons Inc., Crawfordsville, IN

Murshed MS, Seob YA, Parkc J-S, Lee Y (2018) Use of beta-P distribution for modeling hydrologic events. Commun Statist Appl Meth 25:15–27

Nebel ALC, Timm LC, Cornelis WM, Gabriels D, Reichardt K, Aquino LS, Pauletto EA, Reinert DJ (2010) Pedotransfer functions related to spatial variability of water retention attributes for lowland soils. Braz J Soil Sci 34:669–680

Nemes A, Schaap MG, Wösten JHM (2003) Functional evaluation of pedotransfer functions derived from different scales of data collection. Soil Sci Soc Am J 67:1093–1102

Neter J, Wasserman W, Kutner MH (1990) Applied linear statistical models, 3rd edn. Richard D. Irwin, Homewood, IL

Nielsen DR, Wendroth O (2003) Spatial and temporal statistics – sampling field soils and their vegetation. Catena Verlag, Cremlingen-Desdedt

Oliveira LB, Ribeiro MR, Jacomine PKT, Rodrigues JVV, Marques FA (2002) Funções de Pedotransferência para predição da umidade retida a potenciais específicos em solos do Estado de Pernambuco. Rev Bras Ciênc Solo 26:315–323

Pachepsky YA, Rawls WJ (eds) (2004) Development of pedotransfer functions in soil hydrology. Elsevier, Amsterdam

Pachepsky Y, Timlin D, Varallyay G (1996) Artificial neural networks to estimate soil water retention from easily measurable data. Soil Sci Soc Am J 60:727–733

Pachepsky YA, Timlin DJ, Ahuja LR (1999a) The current status of pedotransfer functions: their accuracy, reliability and utility in field and regional scale modeling. In: Corwin DL, Loage K, Ells-Worth TR (eds) Assessment of non-point source pollution in vadose zone. American Geophysical Union, Washington, DC, pp 223–234

Pachepsky YA, Timlin DJ, Ahuja LR (1999b) Estimated saturated soil hydraulic conductivity using water retention data and neural networks. Soil Sci 164:552–560

Pachepsky YA, Rawls WJ, Lin HS (2006) Hydropedology and pedotransfer functions. Geoderma 131:308–316

Parfitt JMB, Timm LC, Pauletto EA, Sousa RO, Castilhos DD, Ávila CL, Reckziegel NL (2009) Spatial variability of the chemical, physical and biological properties in lowland cultivated with irrigated rice. Braz J Soil Sci 33:819–830

Parfitt JMB, Timm LC, Reichardt K, Pinto LFS, Pauletto EA, Castilhos DD (2013) Chemical and biological attributes of a lowland soil affected by land leveling. Pesq Agropec Bras 48:1489–1497

Parfitt JMB, Timm LC, Reichardt K, Pauletto EA (2014) Impacts of land levelling on lowland soil physical properties. Braz J Soil Sci 38:315–326

Radcliffe DE, Simunek J (2010) Soil physics with Hydrus: modeling and applications. CRC Press Taylor & Francis Group, Boca Raton, FL

Rawls WJ, Brakensiek DL, Saxton KE (1982) Estimation of soil water properties. Trans ASAE 25:1316–1320

Rawls WJ, Gish TJ, Brakensiek DL (1991) Estimating soil water retention from soil physical properties and characteristics. Adv Soil Sci 16:213–234

R Core Team (2012) R: a language and environment for statistical computing. R Foundation for Statistical Computing, Vienna. https://www.R-project.org

Reichardt K, Vieira SR, Libardi PL (1986) Variabilidade espacial de solos e experimentação de campo. Rev Bras Ciênc Solo 10:1–6

Robertson GP (2008) GS: geostatistics for the environmental sciences. Gamma Design Software, Plainwell, MI

Schaap MG, Bouten W (1996) Modeling water retention curves of sandy soils using neural networks. Water Resour Res 32:3033–3040

Schaap MG, Leij FJ, Van Genuchten MT (1998) Neural network analysis for hierarchical prediction of soil hydraulic properties. Soil Sci Soc Am J 62:847–855

Schaap MG, Leij FJ, Van Genuchten MT (2001) ROSETTA: a computer program for estimating soil hydraulic parameters with hierarchical pedotransfer functions. J Hydrol 251:163–176

Shumway RH (1988) Applied statistical time series analyses. Prentice Halll, Englewood Cliffs, NJ

Shumway RH, Stoffer DS (2017) Time series analysis and its applications with R examples, 4th edn. Springer, New York, NY

Si BC (2008) Spatial scaling analyses of soil physical properties: a review of spectral and wavelet methods. Vadose Zone J 7:547–562

Silva AP, Tormena CA, Fidalski J, Imhoff S (2008) Funções de pedotransferência para as curvas de retenção de água e de resistência do solo à penetração. Rev Bras Ciênc Solo 32:1–10

Silva AC, Armindo RA, Brito AS, Schaap MG (2017) An assessment of pedotransfer function performance for the estimation of spatial variability of key soil hydraulic properties. Vadose Zone J 16:1–10

Studenmund AH (1992) Using econometrics: a practical guide. Harper Collins, New York, NY

Tamari S, Wösten JHM, Ruiz-Suárez JC (1996) Testing an artificial neural network for predicting soil hydraulic conductivity. Soil Sci Soc Am J 60:771–774

Tietje O, Hennings V (1996) Accuracy of the saturated hydraulic conductivity prediction by pedotransfer functions compared to the variability within FAO textural classes. Geoderma 69:71–84

Tietje O, Tapkenhinrichs M (1993) Evaluation of pedotransfer functions. Soil Sci Soc Am J 57:1088–1095

Timm LC, Gomes DT, Barbosa EP, Reichardt K, Souza MD, Dynia JF (2006) Neural network and state-space models for studying relationships among soil properties. Sci Agric 63:386–395

Tomasella J, Hodnett MG (1998) Estimating soil water characteristics from limited data in Brazilian Amazonia. Soil Sci 163:190–202

Tomasella J, Hodnett MG (2004) Pedotransfer functions for tropical soils. In: Pachepsky YA, Rawls WJ (eds) Development of pedotransfer functions in soil hydrology. Elsevier, Amsterdam, pp 415–430

Tomasella J, Hodnett MG, Rossato L (2000) Pedotransfer functions for the estimation of soil water retention in Brazilian soils. Soil Sci Soc Am J 64:327–338

Trangmar BB, Yost RS, Uehara G (1985) Application of geostatistics to spatial studies of soil properties. Adv Agron 38:45–93

Van Den Berg M, Klamt E, Van Reeuwijk LP, Sombroek WG (1997) Pedotransfer functions for the estimation of moisture retention characteristics of Ferralsols and related soils. Geoderma 78:161–180

Van Genuchten MT (1980) A closed-form equation for predicting the conductivity of un-saturated soils. Soil Sci Soc Am J 44:892–898

Van Looy K, Bouma J, Herbst M et al (2017) Pedotransfer functions in Earth system science: challenges and perspectives. Rev Geophys 55:1199–1256

Vereecken H, Weynants M, Javaux M, Pachepsky Y, Schaap MG, Van Genuchten MT (2010) Using pedotransfer functions to estimate the van Genuchten–Mualem soil hydraulic properties: a review. Vadose Zone J 9:795–820

Vieira SR, Hatfield JL, Nielsen DR, Biggar JW (1983) Geostatistical theory and application to variability of some agronomical properties. Hilgardia 51:1–75

Vieira SR, Tillotson PM, Biggar JW, Nielsen DR (1997) Scaling of semivariograms and the kriging estimation of field-measured properties. Braz J Soil Sci 21:525–533

Vieira SR, Grego CR, Topp GC, Reynolds WD (2014) Spatial relationships between soil water content and hydraulic conductivity in a highly structured clay soils. In: Teixeira WG, Ceddia MB, Ottoni MV, Donnagema GK (eds) Application of Soil Physics in Environmental Analysis: measuring, modeling and data integration. Springer, New York, NY, pp 75–90

Wagner B, Tarnawski VR, Hennings V, Müller U, Wessolek G, Plagge R (2001) Evaluation of pedotransfer functions for unsaturated soil hydraulic conductivity using an independent data set. Geoderma 102:275–297

Warner B, Misra M (1996) Understanding neural networks as statistical tools. Am Statist 50:284–293

Webster R (2001) Statistics to support soil research and their presentation. Eur J Soil Sci 52:331–340

Webster R, Oliver MA (2007) Geostatistics for environmental scientists, 2nd edn. John Wiley & Sons, Chichester

Wendroth O, Reynolds WD, Vieira SR, Reichardt K, Wirth S (1997) Statistical approaches to the analysis of soil quality data. In: Gregorich EG, Carter MR (eds) Soil quality for crop production and ecosystem health. Elsevier Science, Amsterdam, pp 247–276

West M, Harrison J (1997) Bayesian forecasting and dynamic models, 2nd edn. Springer, London

Wilding LP, Drees LR (1983) Spatial variability and pedology. In: Wilding LP, Smeck NE, Hall GF (eds) Pedogenesis and soil taxonomy: concepts and interactions. Elsevier, New York, NY, pp 83–116

Wösten JHM (1997) Pedotransfer functions to evaluate soil quality. In: Gregorich EG, Carter MR (eds) Soil quality for crop production and ecosystem health. Elsevier Science, Amsterdam, pp 221–245

Wösten JHM, Pachepsky YA, Rawls WJ (2001) Pedotransfer functions: bridging the gap between available basic soil data and missing soil hydraulic characteristics. J Hydrol 251:123–150

Zimmerman D (1993) Another look at anisotropy in geostatistics. Math Geol 25:453–470

Spatial and Temporal Variability of SPAS Attributes: Analysis of Spatial and Temporal Series

18

18.1 Introduction

We have seen in the previous chapter that SPAS attributes vary in space and time. This variability, however, in many cases has built in a structure that might tell us interesting features of a system. If the temporal and/or spatial distribution of the attributes is taken into account in agronomic studies, tools of **Geostatistics** can be used to better understand the relationship between SPAS attributes and how to map their spatial and temporal distributions. In this chapter, we will see another way of planning experiments, still new in agronomy, but using techniques imported from **time series analysis**. Texts for a first contact with this technique are Shumway (1988), Nielsen and Wendroth (2003), and Shumway and Stoffer, (2017) and reviews on their application in agronomic studies are given by Wendroth et al. (1997, 2014) and Timm et al. (2014). It should be emphasized again that Fisher's classical technique is complemented by the technique of regionalized variables (Geostatistics) and the time series analysis. One does not exclude the other and questions answered by one often cannot be answered by the other.

It has been said in the previous chapter that in agronomic experimentation, the methodology of sampling, whether soil, plant, or atmosphere, is fundamental. It was emphasized that in "classical or Fisher's statistics," **random sampling** implies in randomly distributed samples within the system, and the coordinates of the sampled locations are not taken into account in the statistical analysis. In the technique of regionalized variables, the coordinates of the sampled locations are of importance in the statistical analysis, that is, very concerned with neighboring samples. In this case, the sampling is done along a **transect** at equidistant intervals, denoted as **lag**, also called spacing; or in a **grid**, also with fixed spacings; or even in positions of any kind, but of known coordinates. In the time series analysis, we use the sampling in equidistant times that could be called "temporal transects" in analogy to the spatial transects already discussed in Chap. 17. Thus, several statistical tools (mean, variance, standard deviation, coefficient of variation, quartiles, box plot, skewness and kurtosis coefficients, probability distribution, covariance function, and autocorrelation function) already presented in the previous chapter are useful in time series analysis. In this way, the classical statistical and autocorrelation tools applied to the data sets in Table 17.1 are also valid in this chapter and will obviously not be presented again. Based on the soil water content data sets θ (m^3 m^{-3}) and clay (%) already presented in Table 17.1, the concept of cross-

K. Reichardt, L. C. Timm, *Soil, Plant and Atmosphere*, https://doi.org/10.1007/978-3-030-19322-5_18

correlation function (or also called cross-correlation) will be introduced.

18.2 Cross-Correlogram

With the notion of autocorrelation (Chap. 17), the simple correlation (x_i with y_i) given by Eq. 17.18 can be extended to neighbors. Thus, the **cross-correlogram** appears when making correlation of x_i with y_{i+j}:

$$r_c(h) = C(x_i, y_{i+j})\left(\sqrt{s_x^2 \cdot s_y^2}\right)^{-1} \tag{18.1}$$

The cross-correlogram makes the correlation between x and y more consistent, since it also relies on neighbors. Figure 18.1 shows the cross-correlogram between θ and a data from Table 17.1.

It should be noted that the correlations between x_i and x_{i+j} are identical to those of x_{i+j} and x_i, so the autocorrelogram of Fig. 17.8 has been presented only for positive j. In the case of cross-correlogram, the correlation between x_i and y_{i+j} is different from y_{i+j} and x_i and, therefore, it is presented as in Fig. 18.1.

18.3 Temporal and Spatial Series: Definition and Examples

There is a great class of phenomena (physical, chemical, and biological) whose observational process and consequent numerical quantification produce a sequence of data distributed over time or space. The sequence of data collected along time is called **time series**. Examples of time series are:

1. monthly values of air temperature at a given location,
2. daily values of precipitation at a given location,
3. data on annual production of sugarcane in a given area, and
4. annual soil organic matter content at a given site.

Similarly, a sequence of data arranged in a spatial order is called **spatial series**. Some examples are:

1. soil temperature values collected along a transect,

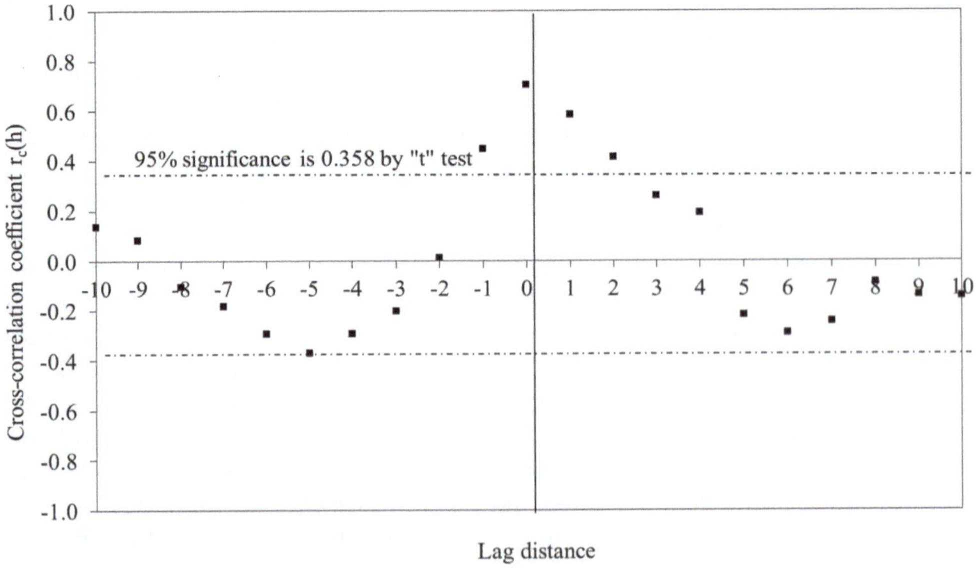

Fig. 18.1 Cross-correlogram between soil water content (θ) and clay content (a) from Table 17.1

2. soil water content values collected along a maize crop line,
3. sugarcane production data measured along a strip, and
4. soil pH values measured along a transect.

Time series can be discrete or continuous, and the simplest form of conceptualization is given by a function of the type $Z(t_i)$, $i = 1, 2, \ldots, n$, composed of a set of n discrete observations observed in $t_i - t_{i-1} = \alpha$ equidistant times that have a serial dependence between them. Even if a series is continuously obtained during a time interval of amplitude T, which is done by continuous recording instruments, it will be necessary to transform it into a discrete series, by means of sampling in equidistant time intervals α. The interval of time between successive observations is sometimes determined by the researcher, but in many situations, it is determined by the availability of the data, and the smaller the sampling interval and the greater the number of observations, consequently, the better the data analysis. According to Tukey (1980), the basic goals in mind when analyzing a time series are:

1. modeling of the phenomena under consideration,
2. obtaining statistical conclusions, and
3. evaluation of the ability of the model in terms of forecasting.

In all research involving statistical methodology, one of the first steps to be taken in analyzing a series is sample planning and data preparation. Depending on the objectives of the analysis, several problems with the observations may occur, and measures must be taken to avoid them or at least to minimize their effect. Among these measures, we can mention stationarity, transformations, lost observations, outliers, and short registers.

The models used to describe time series are stochastic processes, that is, processes controlled by probabilistic laws. The choice of these models depends on several factors, such as the behavior of the phenomenon or the "a priori" knowledge that we have of its nature and the purpose of the analysis. From the practical point of view, it also depends on the existence of good estimation methods and the availability of adequate softwares.

A time series can be analyzed in two ways: (1) **time domain analysis** and (2) **frequency domain analysis**. In both cases, the goal is to construct models for the series of certain purposes. In the first case, the objective of the analysis is to identify the models for the stationary components (random variables) and nonstationary (mean function), in which case the proposed models are parametric models (with finite number of parameters). Among the parametric models, we have, for example, the AR (autoregressive), MA (moving average) models, ARMA (mobile average autoregressive), ARIMA (integrated autoregressive moving average), and state-space models. In the second case, the proposed models are nonparametric models and consist of decomposing the given series into frequency components, in which the existence of the spectrum is the fundamental characteristic. Among the nonparametric models, we can mention the spectral analysis, in which phenomena that involve periodicity of the data are studied, having, therefore, numerous applications in all the areas of science. Recently, the **wavelet analysis** has been used in the soil science area to analyze a spatial (or temporal) series in both domains.

When we are interested in analyzing a series in the time domain, one of the most frequent assumptions is that this series is stationary, that is, it develops in time randomly where the statistical properties (mean and variance) do not vary, reflecting some sort of stable equilibrium. However, most of the series that we find in practice have some form of nonstationarity (mean and variance vary), requiring or not, in this way, a transformation of the original data since most of the analysis procedures suppose that they are stationary. Nevertheless, there are analysis procedures that are applied when the series is nonstationary, such as the wavelet analysis (Si 2008) already mentioned previously.

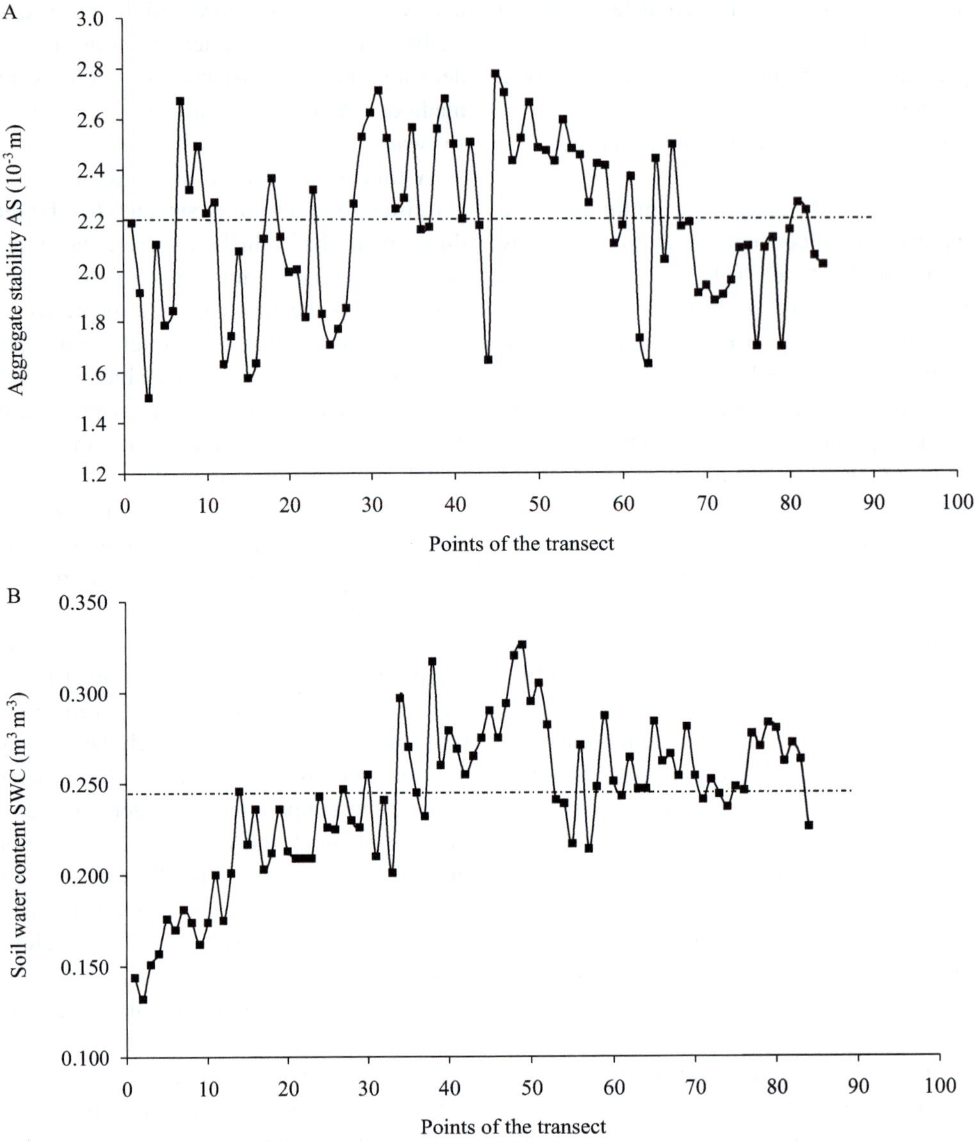

Fig. 18.2 Examples of time series: (**A**) stationary; (**B**) nonstationary

Examples of a stationary and nonstationary series, at a practical point of view, are shown in Fig. 18.2.

The definition of time series presented previously, although simple, evidences to some extent the "analysis of time series" as well-defined area in Statistics, since we are clearly discarding the independent and identically distributed data (classic statistics) commonly used in the various statistical models.

As mentioned, until recently, scientists linked to agricultural sciences studied the variability of soil properties through classical statistics (variance analysis, mean, coefficient of variation, regression analysis, etc.), which presupposes that the observations of a given property are independent of each other, regardless of their location in the area. In this case, the experiments are carried out to minimize the impact of spatial or temporal variability, thus ignoring the fact that observations can be spatially (or temporarily) dependent. However, it has been emphasized that adjacent observations of given soil property are not completely independent and that spatial variability should be considered in the statistical analysis of the data. Nielsen and Alemi (1989) observed that the measurements inside and between treatments may not be independent of one another, which makes the experimental arrangement in the field inadequate.

Spatial variability of soil properties can occur at different levels and may be related to several factors: variation of the parent material (rocks), climate, relief, organisms, and time, that is, of the genetic processes of soil formation and/or effects of soil management techniques resulting from agricultural uses (McGraw 1994). Statistical tools such as auto-and cross-correlation functions, wavelet analysis, multivariate empirical mode decomposition (MEMD), autoregressive models, state-space models, etc. have been used to study spatial variability of soil attributes and may potentially lead to a management that provides a better understanding of soil-plant-atmosphere interaction processes (Wendroth et al. 1992, 2014; Dourado-Neto et al. 1999; Timm et al. 2000, 2001, 2003a, b, c, 2004, 2006a; Grinsted et al. 2004; Si and Zeleke 2005; Ogunwole et al. 2014a; Aquino et al. 2015; Awe et al. 2015; She et al. 2015, 2016, 2017; Hu and Si 2016a; Yang et al. 2016). A brief introduction on **spectral** analysis, **wavelet** analysis, and the **multivariate empirical mode decomposition method** (MEMD) will be seen in the following section. Further research on these topics can be found in Huang et al. (1998), Hubbard (1998), Brillinger (2001), Nielsen and Wendroth (2003), Chatfield (2004), Rehman and Mandic (2010), Addison (2017), Shumway and Stoffer (2017), and Flandrin (2018). Applications in the field of soil science can be found in Shumway et al. (1989), Si (2008), She et al. (2017), Biswas (2018), among others. An important issue is related to the number of samples required for a particular attribute to be representative of a particular area. According to Warrick and Nielsen (1980), when a soil attribute follows the normal distribution and the samples are independent, it is possible to calculate the number of samples needed in future samplings so that prediction with a desired probability level is obtained:

$$N = \frac{t^2 \cdot s^2}{d^2} \qquad (18.2)$$

in which N is the number of samples needed in future samplings, and the value of t is obtained from the t student distribution, with infinite degrees of freedom (DF) and probability given by $(1 - \beta/2)$, where β is the desired confidence level. The standard deviation of the n known data from previous samples is represented by s and d is the acceptable or desired variation around the mean.

Example: It is known that in one area, the average soil water content θ at the point of permanent wilt (PWP) is 9.5% by volume, calculated with a standard deviation $s = 3.1\%$. In a new sampling, how many measurements (N) must be made so that 95% of them are within a range d corresponding to 15% of the mean?

Solution:

At a 5% probability level, the value of t for DF $= \infty$ is 1.96; 15% of the mean $= 0.15 \times 9.5 = 1.4$.

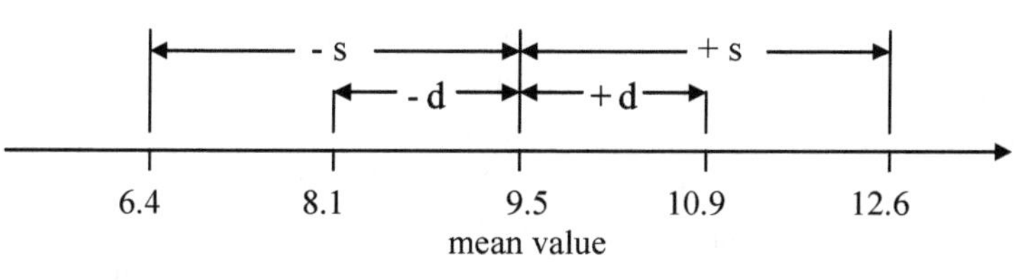

$$N = \frac{(1.96)^2 \times (3.1)^2}{(1.4)^2} = 19 \text{ measurements}$$

According to this approach to the problem, the appropriate sampling procedure for the characterization of the behavior of a soil attribute Z in a given area is to perform random sampling, assuming the independence of the data. The sampling intensity will depend on the variability of the attribute in the area, as well as on the level of accuracy desired around the average and the amount of financial resources available in the project, since the cost of each sampling depends on the attribute to be measured.

The concern with the spatial variability and, more recently, with the temporal variability of the attributes of the soil, is expressed by the volume of research related to the agronomic area. Until recently, more detailed studies of this variability have revealed the limitations of Fisher's classical statistical methods. In general, the assumptions of normality and independence of the data are not tested, and furthermore, independence must be assumed a priori before being sampled. All the variability presented by the values of the variable is ascribed to the residue, that is, to uncontrolled factors. The use of statistical tools that take into account the structure of spatial and/or temporal dependence between observations has contributed to the adoption of a better management of agricultural practices as well as the impacts caused by these practices on the environment.

18.4 Spectral Analysis

In many time and space series, the values of the observed property oscillate around an average value evidencing a periodic process or a cyclical behavior that from time to time (or spaces to spaces) return to past (or previous) values. The annual minimum temperature of Piracicaba, for example, reaches values below zero, provoking frost every 7–10 years. If soil density is measured along a transect perpendicular to the crop lines, each nth line will show a higher bulk density caused by vehicles used for cultivation and harvesting. For these series, the autocorrelogram also presents a cyclic behavior, remembering a cosine that oscillates around the h axis, with both positive and sometimes negative values, often within the nonsignificant level. The **power spectrum (spectral function)** $S(f)$ is obtained from the integrated autocorrelation function $r(h)$ (Chap. 17, Eqs. 17.18a and 17.19a) (Nielsen and Wendroth 2003):

$$S(f) = 2\int_0^{\infty} r(h)\cos(2\pi \cdot f \cdot h)\,dh \qquad (18.3)$$

The graph of S as a function of the frequency (f) gives us the spectrum of variability of the variable under study (remember that the frequency is the inverse of the period of one wavelength). Si (2008) cites that the spectral analysis transforms values of the variable in the space domain to the frequency domain. As a result, the total variance of the variable is divided into scales of spatial frequencies, which allows the identification of the dominant spatial scales in the variability spectrum of the variable. The same author also points out that due to the fact that many patterns of spatial behavior are a combination of variations in different scales, the spectral analysis divides the total variance into different spatial scales defined by the frequency f (i.e., high frequencies = small scale of variation and low frequencies = large scale of variation). A schematic example of the variability spectrum of a property x (soil bulk density d_s in this case) is that of Fig. 18.3, where two frequency peaks of distinctly different d_s are present: one for $f_1 = 0.083\ m^{-1}$ and another for $f_2 = 1.25\ m^{-1}$. This means that the variable under study at every $1/f_1 = 12$ m and every $1/f_2 = 0.8$ m has a maximum value that is a reflection of operations performed every 0.8 m (e.g., line of planting) and every 12 m (vehicle passage).

Figure 18.4 shows, as examples, that the spectrum $S(f)$ (Eq. 18.3) for saturated soil hydraulic conductivity (expressed as ln K_0) (Fig. 18.4a) and

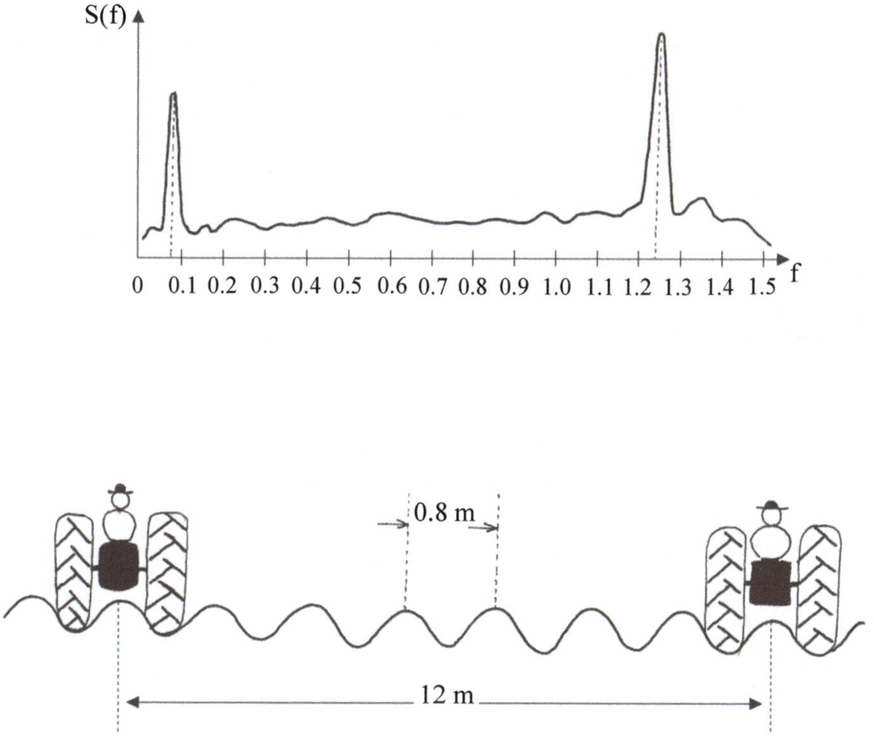

Fig. 18.3 Power spectrum of the frequency of soil bulk density for a field with planting lines every 0.8 m and tractor passage every 12 m

soil macroporosity (Fig. 18.4b) has several peaks at approximately the same frequencies. Saturated soil hydraulic conductivity and macroporosity were measured at 100 points equidistantly spaced along a 15 km spatial transect established in the Fragata River Watershed, Southern Brazil (Beskow et al. 2016).

It was seen that when we are interested in evaluating the spatial correlation structure between two variables, the **cross-correlation function** (Eq. 18.1) can be used. Likewise, we may be interested in evaluating whether the variability spectra of two attributes of the soil (or plant or atmosphere) are spatially (or temporally) correlated with each other. Nielsen and Wendroth (2003) cite that the spectral correlation analysis consists of two components: the **co-spectrum** and the **quadrature spectrum**. In a similar way to the use of the autocorrelation coefficient in Eq. 18.3, the cross-correlation coefficient $r_c(h)$ is used to divide the total covariance between two sets of observations x and y collected over time or space. The co-spectrum function [Co(f)] is given by Wendroth et al. (2014):

$$\mathrm{Co}(f) = 2\int_{0}^{\infty} r_c(h)\cos(2\pi \cdot f \cdot h)\mathrm{d}h \tag{18.4}$$

where Co is the co-spectrum value as a function of frequency f. The quadrature spectrum [$Q(f)$] is a measure of the contribution of the different frequencies in the total covariance between x and y when all the cyclic variations of a set of observations are lagged in a quarter of a period. According to Nielsen and Wendroth (2003), $Q(f)$ is important because it identifies the lag between two sets of data that are correlated at the same frequency. It is given by Wendroth (2013):

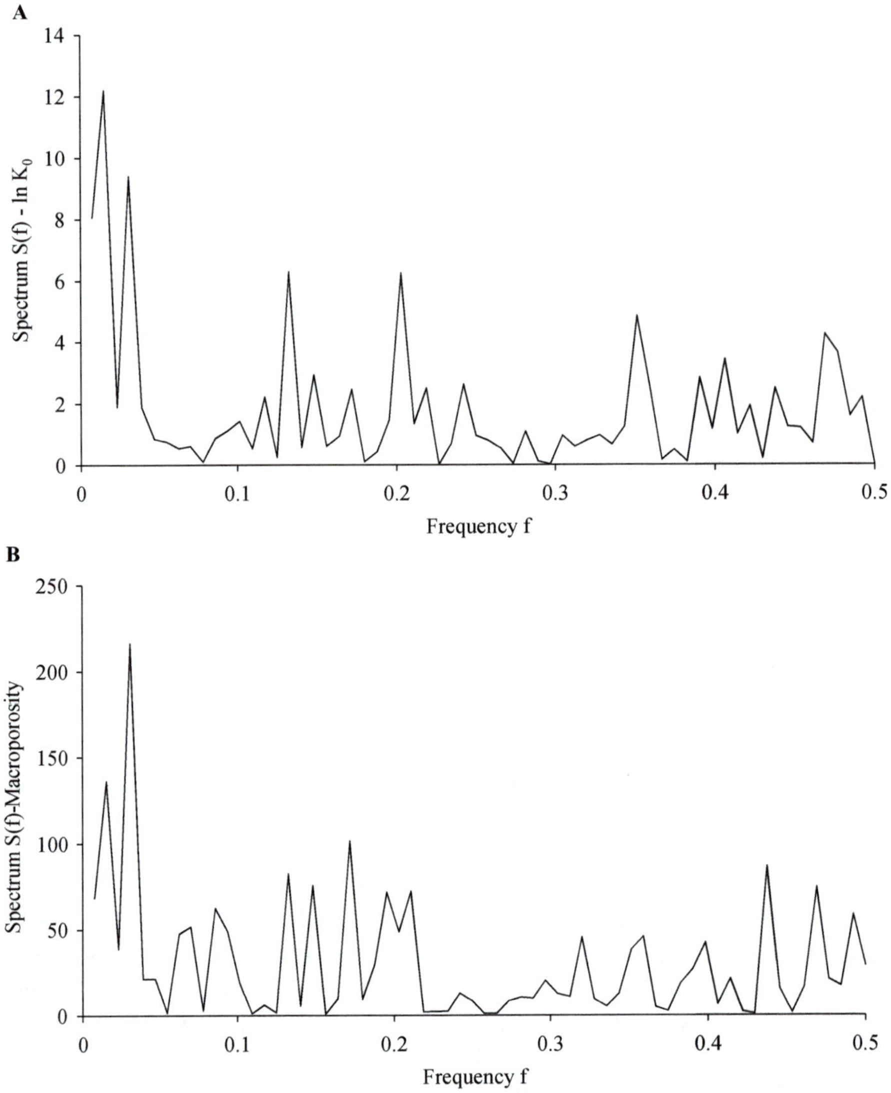

Fig. 18.4 Spectra for saturated soil hydraulic conductivity (expressed as ln K_0) (**A**) and soil macroporosity (**B**)

$$Q(f) = 2\int_0^\infty r_c'(h)\sin(2\pi\cdot f\cdot h)dh \quad (18.5)$$

where $r_c' = 0.5[r_c(h > 0) - r_c(h < 0)]$. Because the sine is an odd function $[r_c(-h) = -r_c(h)]$, this subtraction process reinforces any cyclic variations described by the sine function and eliminates cyclic variations described by the cosine function (Nielsen and Wendroth 2003). From Eqs. 18.3–18.5, the **coherency function** [Coh(f)], which is a quantitative measure of the correlation between two sets of data x and y for various frequencies f, can be calculated as follows:

$$\text{Coh}(f) = \frac{\text{Co}^2(f) + Q^2(f)}{S_x(f)\cdot S_y(f)} \quad (18.6)$$

where $S_x(f)$ and $S_y(f)$ are the spectral functions of the variables x and y, respectively, calculated from Eq. 18.3. Coh(f) values vary from 0 to 1 and can be interpreted as a measure of the degree of correlation between x and y similar to

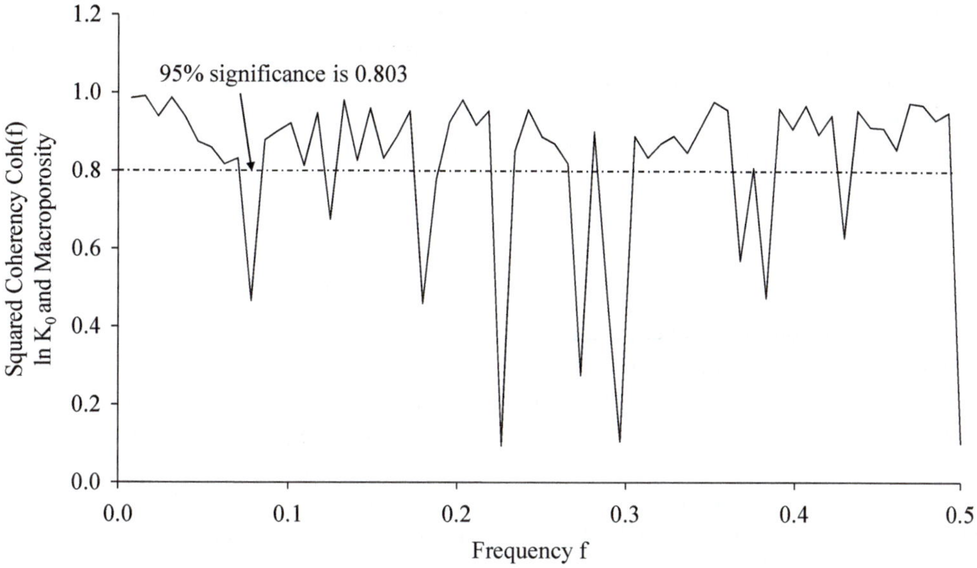

Fig. 18.5 Coherence of saturated soil hydraulic conductivity (expressed as ln K_0) and soil macroporosity

the coefficient of determination r^2 in the case of a linear regression between x and y. If x and y are strongly correlated for a given (f), the magnitude of Coh(f) approaches 1, and if they are not correlated it approaches 0. The significance of the Coh(f) value for a given probability level p (Coh$_p$) is approximately

$$\text{Coh}_p = \sqrt{1 - p^{1/(\text{df}-1)}} \qquad (18.7)$$

where df in the above equation is the number of degrees of freedom. The coherence graph (Fig. 18.5) is displaying several significant peaks at approximately the same frequencies between ln K_0 (Fig. 18.4a) and soil macroporosity (Fig. 18.4b) variables.

Spectral and coherence analyses require second-order stationarity of soil processes under study and deal with global information or mean states (Si 2008). Often, two completely different spatial series with different local information may result in very similar mean states. Therefore, spatial information is completely lost in spectral and coherence analyses. According to Si (2008), more often than not, soil spatial variation is nonstationary, consisting of a variety of frequency regimes that may be localized in space (relative to the entire spatial domain). Therefore, methods, such as wavelet and multivariate empirical mode decomposition analyses (they will be seen soon), that could examine these natural trends, localized features, or transient features of the soil processes are needed.

As the interest here is not to exhaust the subject, theoretical aspects in more details on spectral analysis and its applications in the agronomic area can be found in Brillinger (2001), Nielsen and Wendroth (2003), Chatfield (2004) and James (2011), Shumway and Stoffer (2017), among other texts. An interesting application is that of Bazza et al. (1988), which deals with the use of two-dimensional spectral analysis in space. A review on the application of spectral analysis in the study of spatial variability of soil hydrophysical properties is presented in Si (2008).

18.5 Wavelet Analysis

The basic tool in **wavelet analysis** is the wavelet transform, which can be classified as continuous (CWT) or as discrete (DWT). The basis for the use of CWT in scale analysis is that the wavelet function (wavelet mother) can be stretched or contracted in space (location) and at different scales (Biswas 2018). Among the mother wavelets, the Morlet wavelet is commonly used in CWT since it is a complex symmetric function that retains the real and imaginary component of wavelet coefficients (Biswas and Si 2011). Morlet wavelet can be represented as (Grinsted et al. 2004):

$$\psi(\eta) = \pi^{-1/4} e^{i\omega\eta - 0.5\eta^2} \tag{18.8}$$

where ω is dimensionless frequency and η is dimensionless space ($\eta = s/x$; s is the scale and x is the distance) and $i = (-1)^{1/2}$. The Morlet wavelet ($\omega = 6$) is good for feature extraction because it provides a good balance between space and frequency location (Farge 1992; Wendroth et al. 2011; Biswas 2018) and has been used in soil science applications.

The CWT can be defined as the convolution of a spatial series Y_i of length N ($i = 1, 2, 3, \ldots, N$) along a transect with equal increments of distance δx (Torrence and Compo 1998; Biswas 2018):

$$W_i^Y(s) = \sqrt{\frac{\delta x}{s}} \sum_{j=1}^{N} Y_j \psi\left[(j - i)\frac{\delta x}{s}\right] \tag{18.9}$$

where $W_i^Y(s)$ are the wavelet coefficients being expressed as complex numbers $a + ib$, a, and b are, respectively, the real and imaginary components of $W_i^Y(s)$. Equation 18.9 can be implemented through a series of Fast Fourier Transform. The function $\psi[]$ is the mother wavelet function (Eq. 18.8). The parameter s is the dilation-contraction factor, and it is associated with scales (Biswas 2018). The energy (strength of the variance) associated with each scale (s) and location (x) can be calculated from the magnitude of wavelet coefficients (Eq. 18.9). Therefore, the wavelet power spectrum can be calculated as a function of scales and locations and can be used to examine the scale-location-specific variations in soil properties. To evaluate if the wavelet coefficients associated at each scale and location were significant, a red noise test of statistical significance (5% of significance level) can be used. More details on theoretical aspects about the red noise test of statistical significance can be found in Si and Zeleke (2005) and Si (2008). Briefly, if the wavelet power of a spatial series at a certain scale falls in the 95% confidence interval, we say that the wavelet power at this scale and location is not significantly different from the background red noise.

Before calculating the wavelet coherency between two spatial series (X and Y), with wavelet transforms $W_i^X(s)$ and $W_i^Y(s)$, the cross-wavelet power spectrum has to be calculated and can be defined as $\left|W_i^{XY}(s)\right| = \left|W_i^Y(s).\overline{W_i^X(s)}\right|$, where $\overline{W_i^X}$ is the complex conjugate of W_i^x. The complex argument of W_i^{XY} can be interpreted as the local relative phase between X_i and Y_i in the spatial frequency domain (Si and Zeleke 2005). Then, the squared wavelet coherence between X and Y spatial series is calculated as (Grinsted et al. 2004):

$$R_i^2(s) = \frac{\left|S\left(s^{-1}\, W_i^{XY}(s)\right)\right|^2}{S\left(s^{-1}\left|\, W_i^X(s)\right|^2\right) \cdot S\left(s^{-1}\left|\, W_i^Y(s)\right|^2\right)} \tag{18.10}$$

in which S is a smoothing operator and can be written as:

$$S(W) = S_{\text{scale}}\left(S_{\text{space}}\left(W(s, \tau)\right)\right) \tag{18.11}$$

where S_{scale} denotes smoothing along the wavelet scale axis and S_{space} smoothing in spatial domain. More details on the smoothing function can be found in Si and Zeleke (2005), among others. To examine if the squared wavelet coherency coefficients associated at each scale and location were significant, a red noise test of statistical significance (at 5% of significance level) can also be used.

The wavelet power spectrum (Eq. 18.9) for saturated soil hydraulic conductivity and soil macroporosity is shown in Fig. 18.6, illustrating applications of both wavelet analyses. Data are the same used in Fig. 18.4.

At small scales (300–450 m), K_0 showed significantly high variances at locations that are 8.5–10 km from the origin of the transect (Fig. 18.6A). At the scales of 450–750 m and 670–950 m, high and significant variances of K_0 are exhibited at locations around 9.2–10.7 km and

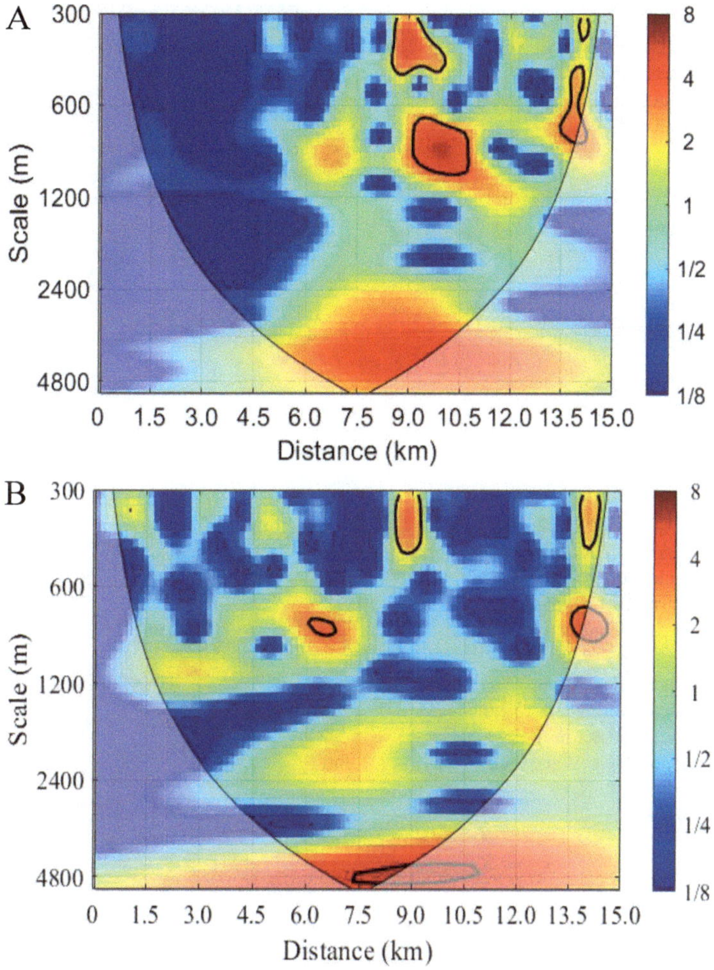

Fig. 18.6 Wavelet spectrum for saturated soil hydraulic conductivity (K_0) (**A**) and soil macroporosity (**B**)

13.5–14.0 km, respectively. It can also be seen that K_0 shows high variances at the scales of 2400–4800 m at locations that are 4.7–10.7 km from the origin of the transect. However, they are not significantly different from that of red noise at this scale. Figure 18.6B shows that high and significant variances of soil macroporosity are exhibited at small scales (300–450 m) at locations that are 8.5–9.2 and 14.0–14.3 km from the origin of the transect. High wavelet spectra of macroporosity are also observed at the scale of 4300–4800 m in the middle of the transect.

Figure 18.7 shows the wavelet coherence spectrum for saturated hydraulic conductivity and soil macroporosity, indicating a strong and significant coherence between K_0 and macroporosity along the entire transect. At all scales, high coherency exists between K_0 and macroporosity along the entire transect. The positive (right-directed arrow) relationship between K_0 and macroporosity prevails along the transect, that is, higher K_0 corresponded to higher macroporosity values in the transect.

More recently, Hu and Si (2016a) developed a multiple wavelet coherence (MWC) method for examining scale-specific and localized multivariate relationships. The authors mentioned that similar to bivariate wavelet coherence (Eq. 18.10), MWC is based on a series of auto- and cross-wavelet power spectra, at different scales and

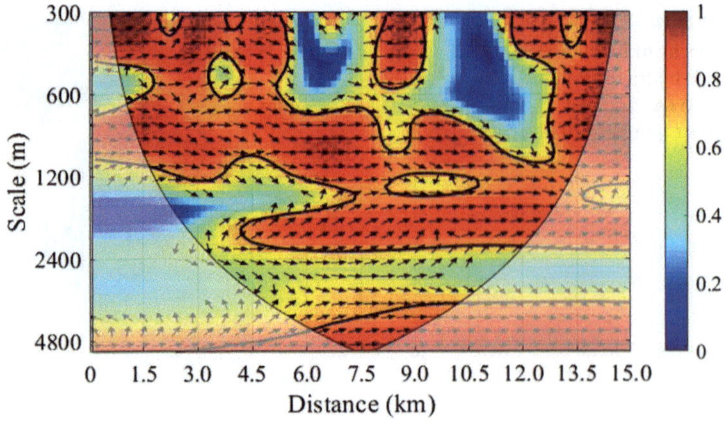

Fig. 18.7 Wavelet coherence spectrum for saturated hydraulic conductivity and soil macroporosity

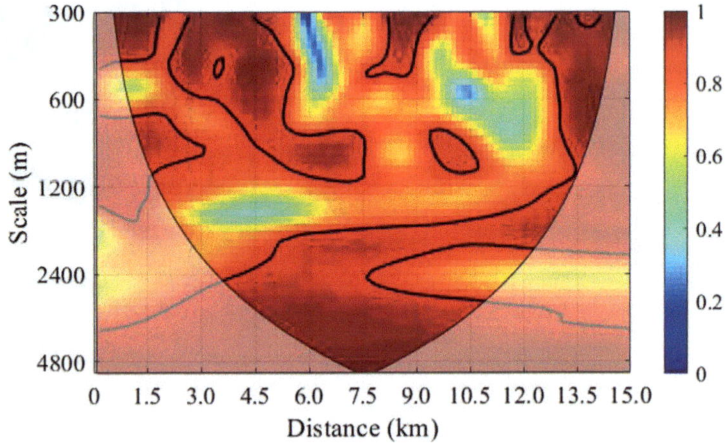

Fig. 18.8 Multiple wavelet coherence spectrum for saturated hydraulic conductivity, soil macroporosity, and elevation data series collected along the 15 km spatial transect in the Fragata River Watershed

spatial (temporal) locations, for the response variable and all predictor variables. The MATLAB code for calculating multiple wavelet coherence can be found in Hu and Si (2016b). Figure 18.8 shows, as an example, the multiple wavelet coherence spectrum for saturated hydraulic conductivity, elevation, and soil macroporosity. Elevation values were also obtained for the 15 km spatial transect established in the Fragata River Watershed.

Further details and comprehensive reviews on theoretical aspects and application of wavelet analyses are given by Farge (1992), Graps (1995), Hubbard (1998), Torrence and Compo (1998), Si (2008), Hu and Si (2016a), Addison (2017), Biswas (2018), among others.

18.6 Multivariate Empirical Mode Decomposition (MEMD) Analysis

In nature, the effects of different processes as represented by different frequency components are not cumulative and do not follow the principle of superposition, thus indicating that the system is nonstationary and nonlinear. A relatively new multiscale analysis method, that is, the empirical mode decomposition (EMD), which was developed by Huang et al. (1998), and its extension algorithm, the multivariate empirical mode decomposition (MEMD) (Rehman and Mandic 2010), are now available. This method has been applied to characterize the scale-dependent spatial relationships between environmental factors and soil properties of nonstationary and nonlinear systems. Example of application of the MEMD analysis is of the She et al. (2017) study. The authors used the MEMD method to investigate the multiscale correlations between two soil hydraulic properties and two topographic attributes and three soil factors along a transectin the Pelotas River Watershed, situated in Southern Brazil. The multivariate empirical mode decomposition separated the overall variation in soil hydraulic properties (soil water content at field capacity and saturated soil hydraulic conductivity—ln K_0) and associated factors (elevation, slope, sand content, soil bulk density, and soil organic content) into six intrinsic mode functions according to the scale of occurrence. The authors concluded that the overall ln K_0 prediction using MEMD outperformed the prediction using the original data alone, which illustrated the importance of taking scale dependence into account when investigating the spatial distribution of K_0 in the Pelotas River Watershed.

The MEMD was pioneered by Huang et al. (1998) who developed it to empirically assess data as a form of wavelet analysis. The MEMD algorithm created by Rehman and Mandic (2009) was used in the She et al. (2017) study. Briefly, this algorithm enabled projections of multiple input data to be made with various directions in an n-dimensional space, thereby making it possible to produce multiple n-dimensional envelopes, aligning "common scales" present within multivariate data. Assuming that the n spatial data sets denoted by $D(v) = \{d_1(v), d_1(v), \ldots, d_n(v)\}$, as a function of space (v), represent a multivariate signal with n components, and $K^{f_x} = \{k_1^x, k_2^x, \ldots, k_n^x\}$ is the direction vector along the direction given by angles $f^x = \{f_1^x, f_2^x, \ldots, f_{n-1}^x\}$ ($x = 1, 2, \ldots, m$, m is the total number of directions), then the "common scales" IMFs of the n-depth data sets can be obtained by MEMD algorithm (Rehman and Mandic 2010; She et al. 2014a). The MEMD algorithm is a fully data-driven method designed for multiscale decomposition and time-frequency analysis of real-world signals, which has already been detailed in the literature (Huang et al. 1998; Rehman and Mandic 2010; Hu and Si 2013; She et al. 2013; and more recently by Flandrin 2018).

18.7 The State-Space Approach

The **state-space model** of a stationary or nonstationary stochastic process is based on the property of Markovian systems that establishes the independence in the future of the process in relation to its past, once given the present state. In these systems, the state of the process under consideration condenses all information of the past needed to forecast the future.

The formulation of a state-space model is a way of representing a linear system (or not) from two dynamic equations. For the case of a linear system, we will have:

1. the way in which the vector of the observations $Y_j(x_i)$ of the process is generated as a function of the state vector $Z_j(x_i)$, called the equation of the observations (Eq. 18.12), and
2. the dynamic evolution of the unobserved state vector $Z_j(x_i)$, called the state or system equation (Eq. 18.13).

In an analytical way, we have the following:

$$Y_j(x_i) = M_{jj}(x_i)Z_j(x_i) + \nu_{Y_j}(x_i) \quad (18.12)$$

$$Z_j(x_i) = \phi_{jj}Z_j(x_{i-1}) + u_{Z_j}(x_i) \quad (18.13)$$

The **matrix of observations M_{jj}** in Eq. 18.12 is originated from the following set of observation equations:

$$
\begin{aligned}
Y_1(x_i) &= m_{11}Z_1(x_i) + m_{12}Z_2(x_i) + \cdots \cdots + m_{1j}Z_j(x_i) + v_{Y_1} \\
Y_2(x_i) &= m_{21}Z_1(x_i) + m_{22}Z_2(x_i) + \cdots \cdots + m_{2j}Z_j(x_i) + v_{Y_2} \\
Y_3(x_i) &= m_{31}Z_1(x_i) + m_{32}Z_2(x_i) + \cdots \cdots + m_{3j}Z_j(x_i) + v_{Y_3} \\
&\vdots \\
Y_j(x_i) &= m_{j1}Z_1(x_i) + m_{j2}Z_2(x_i) + \cdots \cdots + m_{jj}Z_j(x_i) + v_{Y_j}
\end{aligned}
$$

which can be written in a matrix form:

$$
\begin{bmatrix} Y_1(x_i) \\ Y_2(x_i) \\ \vdots \\ \vdots \\ Y_j(x_i) \end{bmatrix} = \underbrace{\begin{bmatrix} m_{11} & m_{12} & m_{13} & \cdots\cdots & m_{1j} \\ m_{21} & m_{22} & m_{23} & \cdots\cdots & m_{2j} \\ \vdots & \vdots & \vdots & \cdots\cdots & \vdots \\ \vdots & \vdots & \vdots & \cdots\cdots & \vdots \\ m_{j1} & m_{j2} & m_{j3} & \cdots\cdots & m_{jj} \end{bmatrix}}_{M_{jj}} \times \begin{bmatrix} Z_1(x_i) \\ Z_2(x_i) \\ \vdots \\ \vdots \\ Z_j(x_i) \end{bmatrix} + \begin{bmatrix} v_{Y_1}(x_i) \\ v_{Y_2}(x_i) \\ \vdots \\ \vdots \\ v_{Y_j}(x_i) \end{bmatrix}
$$

A transition matrix (or coefficient matrix) ϕ_{jj} in Eq. 18.13 is originated from the following set of state equations:

$$
\begin{aligned}
Z_1(x_i) &= \phi_{11}Z_1(x_{i-1}) + \phi_{12}Z_2(x_{i-1}) + \cdots \cdots + \phi_{1j}Z_j(x_{i-1}) + u_{Z_1}(x_i) \\
Z_2(x_i) &= \phi_{21}Z_1(x_{i-1}) + \phi_{22}Z_2(x_{i-1}) + \cdots \cdots + \phi_{2j}Z_j(x_{i-1}) + u_{Z_2}(x_i) \\
Z_3(x_i) &= \phi_{31}Z_1(x_{i-1}) + \phi_{32}Z_2(x_{i-1}) + \cdots \cdots + \phi_{3j}Z_j(x_{i-1}) + u_{Z_3}(x_i) \\
&\vdots \\
Z_j(x_i) &= \phi_{j1}Z_1(x_{i-1}) + \phi_{j2}Z_2(x_{i-1}) + \cdots \cdots + \phi_{jj}Z_j(x_{i-1}) + u_{Z_j}(x_i)
\end{aligned}
$$

Or in matrix form:

$$
\begin{bmatrix} Z_1(x_i) \\ Z_2(x_i) \\ \vdots \\ \vdots \\ Z_j(x_i) \end{bmatrix} = \underbrace{\begin{bmatrix} \phi_{11} & \phi_{12} & \phi_{13} & \cdots\cdots & \phi_{1j} \\ \phi_{21} & \phi_{22} & \phi_{23} & \cdots\cdots & \phi_{2j} \\ \vdots & \vdots & \vdots & \cdots\cdots & \vdots \\ \vdots & \vdots & \vdots & \cdots\cdots & \vdots \\ \phi_{j1} & \phi_{j2} & \phi_{j3} & \cdots\cdots & \phi_{jj} \end{bmatrix}}_{\phi_{jj}} \times \begin{bmatrix} Z_1(x_{i-1}) \\ Z_2(x_{i-1}) \\ \vdots \\ \vdots \\ Z_j(x_{i-1}) \end{bmatrix} + \begin{bmatrix} u_{Z_1}(x_i) \\ u_{Z_2}(x_i) \\ \vdots \\ \vdots \\ u_{Z_j}(x_i) \end{bmatrix}
$$

The observation vector $Y_j(x_i)$ is related to the state vector $Z_j(x_i)$ through the observations matrix $M_{jj}(x_i)$ and by an observation error (or noise) $v_{Yj}(x_i)$ (Eq. 18.12). This means that observed values Y_j (which can be any variables such as soil water content θ, pH, organic matter, soil bulk density, etc. measured along a transect of n points spaced by a lag h, i.e., $i-(i-1)=h$) are not taken as true but are considered as the indirect measures of Y_j, reflecting the true state of the variable Z_j (non observed value of the variable), added to an observation error $v_{Yj}(x_i)$. On the other hand, the state vector $Z_j(x_i)$ at position i is related to the same vector at position $i - 1$ by means of the matrix of the state coefficients $\phi_{jj}(x_i)$ (transition matrix) and an error (or noise) associated with the state $u_{Zj}(x_i)$ with the structure of a first-order autoregressive model. It is assumed that $v_{Yj}(x_i)$ and $u_{Zj}(x_i)$ are normally distributed, independent, and uncorrelated to each other for all lags.

Equations 18.12 and 18.13 contain distinct perturbations (or noises), one being associated with the observations (Eq. 18.12) and the other with the state (Eq. 18.13). According to Gelb (1974), the development of methods for processing noise-contaminated observations can be attributed to the works of Gauss and Legendre (around 1800), which independently developed the least squares method for linear models. Later, a recursive solution for the least squares method in linear models was obtained by Plackett (1950). Kalman (1960), using a state-space formulation, developed an optimal recursive filter for estimation in stochastic dynamic linear systems, which is currently known in the literature as **Kalman Filter** (FK). According to Gelb (1974), an optimal estimator is a computational algorithm that processes the observations to deduce a minimum estimate (according to some optimization criterion) of the state error of a system using (1) knowledge of the dynamics of observations and system, (2) assumed statistical inferences to the noises associated with observations and those associated with the state, and (3) knowledge of the initial condition of the information. In summary, given the dynamic system of equations that describes the behavior of the state vector and observations, the statistical models that characterize the observational and state errors, and the initial condition of the information, the Kalman Filter performs the sequential update of the state vector in time (or space) $i - 1$ for time (or space) i. In fact, it can be argued that FK is, in essence, a recursive solution (a solution that allows sequential processing of observations) to the original Gauss minimum squares method. However, it is necessary to use another algorithm (e.g., the maximum likelihood algorithm, widely discussed by Shumway and Stoffer 2000, 2011) so that together with the FK, the problem can be solved of observations contaminated by noise, that is, presence of uncertainty parameters (Gelb 1974).

According to the purpose of the study involving the state-space methodology, one can have different types of estimates: (a) when the time (or space) in which an estimate is desired coincides with the last observed data (t or $x = n$), the problem is said of filtering; (b) when the time (or space) of interest lies within the whole set of observed data, that is, the entire set of observed data is used to estimate the point of interest (t or $x < n$), the problem is of smoothing; and (c) when the time (or space) of interest lies beyond the last observed data (t or $x > n$), the problem is said to be prediction.

State-space formulation can be used, such as kriging and co-kriging (Alemi et al. 1988; Deutsch and Journel 1992), for spatial data interpolation, but the philosophy behind these tools differs. For example, for the application of kriging and co-kriging techniques, the condition of stationarity of the data is required, differing from the state-space approach in which this condition is not a limiting factor.

Until now, the system of dynamic linear equations (Eqs. 18.12 and 18.13), which describes the state-space formulation, was presented in a general way. However, the purpose of this chapter is to present the use of state-space formulation under two different approaches: the first one presented in Shumway (1988) and Shumway and Stoffer (2000, 2011, 2017) that has been used by several researchers in the agronomic area (Eq. 18.13), and the second one presented in West and Harrison (1989, 1997), which has still been little explored in the agronomic area, where greater emphasis is given to the equation of observations (Eq. 18.12). It is intended to achieve this objective by illustrating examples of application in the agronomic area where such approaches have been used.

18.8 Shumway's Approach to State-Space

This approach, presented in Shumway (1988) and Shumway and Stoffer (2000, 2011) and more recently in Shumway and Stoffer (2017), places greater emphasis on the state equation of the system in which the transition coefficient matrix ϕ (Eq. 18.13) is a matrix of dimension jxj, which indicates the spatial extent of the linear association between the variables of interest. These coefficients are optimized by a recursive procedure using Kalman's filtering algorithm (Shumway and Stoffer 1982) in which the maximum likelihood estimation method is used in conjunction with the expectation-maximization algorithm of Dempster et al. (1977). In this case, Eqs. 18.12 and 18.13 are solved by assuming initial values for the mean and variance of each variable and for the noise covariance matrices: of the observations R, of noise covariance associated to the state vector Q, of the transition coefficients ϕ, and of observation M. Shumway (1988) considers the matrix M as unitary (identity). In this way, Eq. 18.12 becomes:

$$Y_j(x_i) = Z_j(x_i) + \ \nu_{Y_j}(x_i) \qquad (18.12a)$$

which means that Y differs from Z only by an observation error.

We will illustrate some applications of this approach in an experiment conducted in an area cultivated with sugarcane located in Piracicaba, SP, Brazil. The field experiment with sugarcane cultivation, with 1.4 m line spacing, was installed according to a randomized complete block design with four treatments and four replicates per treatment, with each replicate being subdivided into 1 m (lag = 1 m) strips, composing, in total, of a spatial transect of 84 plots, including the borders placed at the ends and between each treatment. Figure 18.9 shows the experimental scheme used.

In total, 15 lines of sugarcane were planted with 100 m of length each. Treatment T_1 received N^{15}-labeled fertilizer at planting, and in the first cut (November 1998), it received unmarked straw from treatment T_2 that was fertilized with the same but unlabeled fertilizer. Treatment T_2 received on the first straw scored T_1. The T_3 treatment was used for the production of unmarked straw, the soil surface being maintained without straw, that is, bare. The T_4 treatment received the same labeled fertilizer that was applied on T_1, but the sugarcane was burned before the harvest, leaving the residues of the burning on the surface of the soil. These treatments were based on the tendency to change the management practices of sugarcane, replacing the traditional burning of sugarcane with raw cane harvesting, where the vegetation cover is left on the soil surface.

18.8.1 Analysis of the Behavior of Soil Water Content and Temperature

This example aims to illustrate the use of state-space methodology to better understand the behavior of soil water content and soil temperature in the sugarcane experiment just described. The soil moisture θ in the 0–0.15 m layer along the 84 points of the spatial transect was measured using a neutron-gamma surface probe (Fig. 18.10). Simultaneously, soil temperature was measured at depths of 0.03, 0.06, and 0.09 m in the same transect with a digital thermometer. The mean soil temperature value of the three depths was used in this study. The analysis was performed using the ASTSA software (Shumway 1988).

From Eq. 18.13, it can be seen that the transition coefficient matrix ϕ_{jj} relates the state vector Z_j at position i to its value at position $i - 1$. When the original data are normalized [$z_j(x_i)$] before the application of the state-space methodology, the magnitude of the coefficients ϕ becomes directly proportional to the contribution of each variable in the estimate of $Z_j(x_i)$. For this, the following transformation was used:

$$z_j(x_i) = \frac{Z_j(x_i) - (\bar{Z}_j - 2 \cdot s)}{4 \cdot s} \qquad (18.14)$$

The observed soil water content and temperature data used in this example are shown in

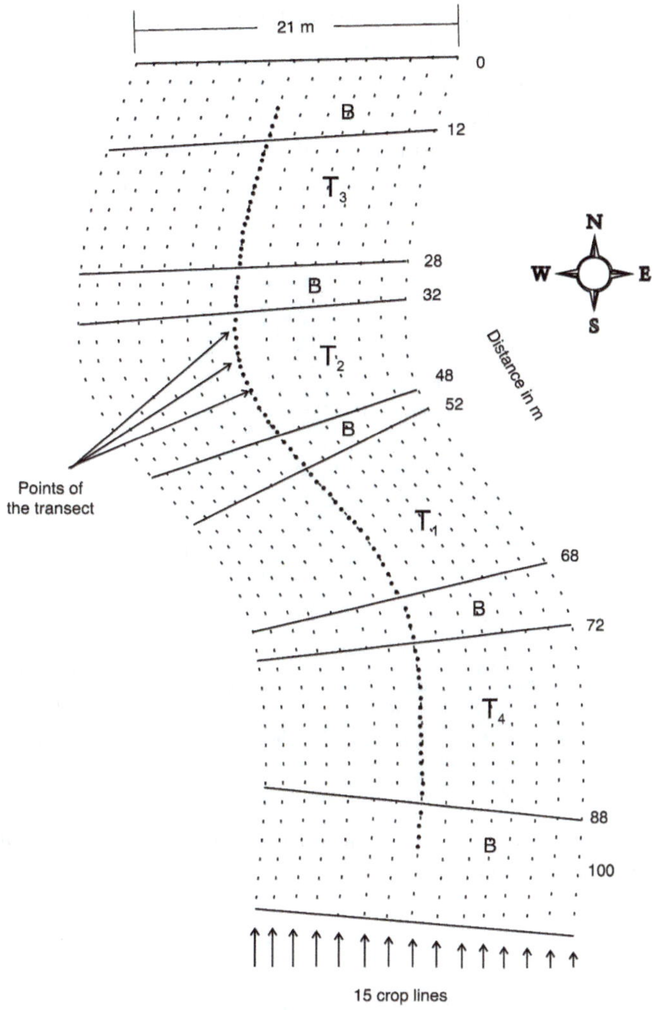

Fig. 18.9 Schematic experimental design showing the 15 cane lines, each 100 m long, indicating the central line used to measure physical and chemical soil properties. B = border; T = treatments; R = replicate

Fig. 18.11; both variables present a marked variation along the transect.

Analyzing together Figs. 18.9 and 18.11, the effect of the presence of the straw mulch cover on the soil surface (T_1 and T_2) with higher values of soil water content and lower soil temperature can be clearly seen. This sharp variation among the values is due to the fact that the sugarcane harvest was carried out in October 1998 and the measurements made on November 20 of the same year, that is, the sugarcane ratoon was still small in size, not covering soil surface, exposing the surface to climatic adversities in a more pronounced way. The linear regression shown in Fig. 18.12 ($R^2 = 0.45$) shows that soil water content is inversely related to soil temperature.

Figure 18.13A, B presents the autocorrelograms for soil water content (Fig. 18.13A) and soil temperature (Fig. 18.13B). Analyzing these figures, it can be verified that both soil moisture and temperature have a spatial dependence of up to 8 lags, at the level of 5% of significance. This indicates that there is spatial dependence, in this case up to 8 m between the adjacent observations of both variables. The cross-correlogram between soil water content and soil temperature is shown in Fig. 18.14, showing the strong spatial dependence between these variables up to a distance of 6 m, in both directions in this case.

The results of the application of the state-space approach, justified by Figs. 18.13A, B, and 18.14, in data $z_j(x_i)$ (transformed by Eq. 18.14) of moisture θ and soil temperature T are shown in Figs. 18.15 and 18.16, respectively. For this example, Eq. 18.13 becomes:

$$\theta_i = 0.8810 \times \theta_{i-1} + 0.1148 \times T_{i-1} + u_{\theta_i} \tag{18.13a}$$

and

$$T_i = 0.0615 \times \theta_{i-1} + 0.9272 \times T_{i-1} + u_{T_i} \tag{18.13b}$$

or in matrix form:

$$\begin{pmatrix} \theta_i \\ T_i \end{pmatrix} = \begin{pmatrix} 0.8810 & 0.1148 \\ 0.0615 & 0.9272 \end{pmatrix} \times \begin{pmatrix} \theta_{i-1} \\ T_{i-1} \end{pmatrix} + \begin{pmatrix} u_{\theta_i} \\ u_{T_i} \end{pmatrix} \tag{18.13c}$$

Fig. 18.10 View of the gamma-neutron surface probe used in the experiment

In Figs. 18.15 and 18.16, the observed values of the variables are represented by the square symbol. The estimated values of the variable through the state-space model (Eqs. 18.13a and 18.13b) are represented by the middle line. The estimated values plus and minus the standard deviation of the estimate at each position i are represented by the upper and lower lines corresponding to the area that the model

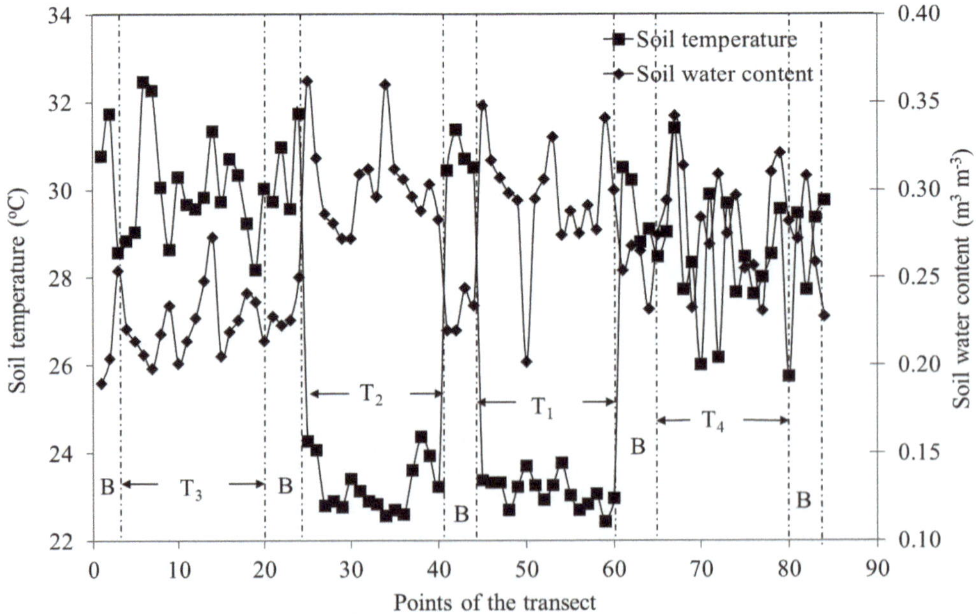

Fig. 18.11 Soil water content and temperature distributions along a sugarcane transect with several treatments T_i

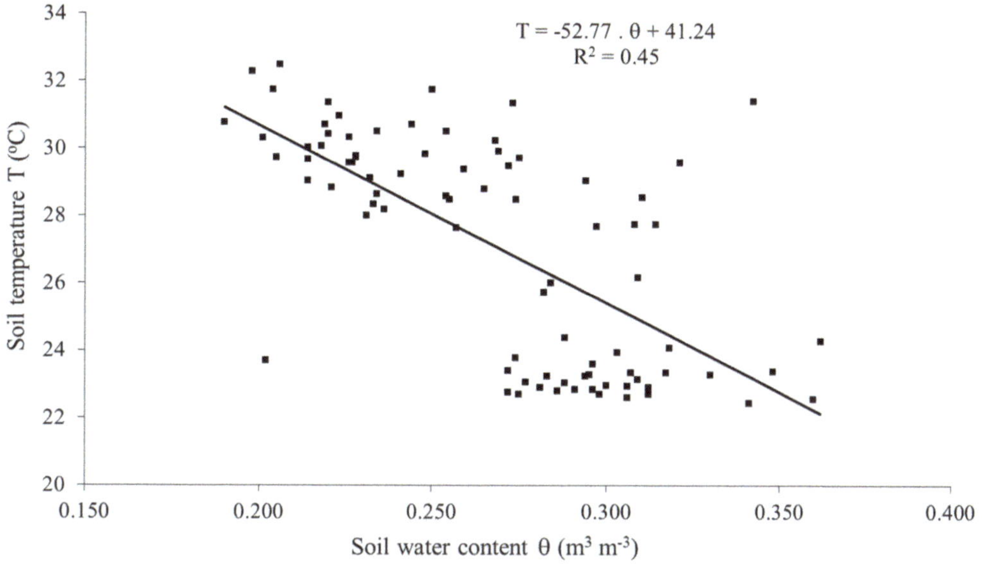

Fig. 18.12 Correlation between soil temperature and soil water content

presented good performance. It should be noted that the standard deviations are estimated for each position i. Examining Eq. 18.13a, it can be seen that the soil water content at position $i - 1$ contributes 88% to the estimate at position i, while the soil temperature at point $i - 1$ contributes only 11.5%.

Analyzing Eq. 18.13b, soil water content at position $i - 1$ contributes 6.15% to the estimate of soil temperature at position i, which means that the contribution of the first neighbor is greater in the case of temperature (92.7%) when compared to soil water content (88.1%).

Figure 18.17A, B shows the linear regression between observed and estimated soil water content values (Fig. 18.17A) and observed and estimated soil temperature values (Fig. 18.17B), with $R^2 = 0.91$ for soil water content and 0.96 for temperature, that is, the model had better performance in soil temperature estimates.

It is important to note that in this example, analyzed with a first-order model, it could also be analyzed with higher-order models. Hui et al. (1998) used a second-order model relating the infiltration rate of water at position i with the rate of infiltration and electrical conductivity of the extract at position $i - 1$ and with the infiltration rate at the second $i - 2$ neighbor. In this example, a two-variable system was used (soil water content and soil temperature), but we could use a system relating more variables, such as clay content, soil bulk density, mineralogy, organic matter, and soil surface microtopography.

This example (extracted from Dourado-Neto et al. 1999) is the first application of the state-space methodology in Brazil using spatial series, aiming to introduce this approach in Latin American soil science literature as a tool for better understanding of soil-plant-atmosphere system variables.

18.8.2 Relation Between Physical and Chemical Soil Properties

This second example (extracted from Timm et al. 2004) illustrates the application of the state-space approach in four spatial series (soil water content [θ], organic matter [OM], clay content [CC], and aggregate stability [AS]) collected along the same transection of the 84 points in the same experimental area of sugarcane. Soil water content was measured as previously described during the dry season (September 6, 1999). For the determination of OM, CC, and AS, disturbed soil samples were collected in the 0–0.15 m depth layer along

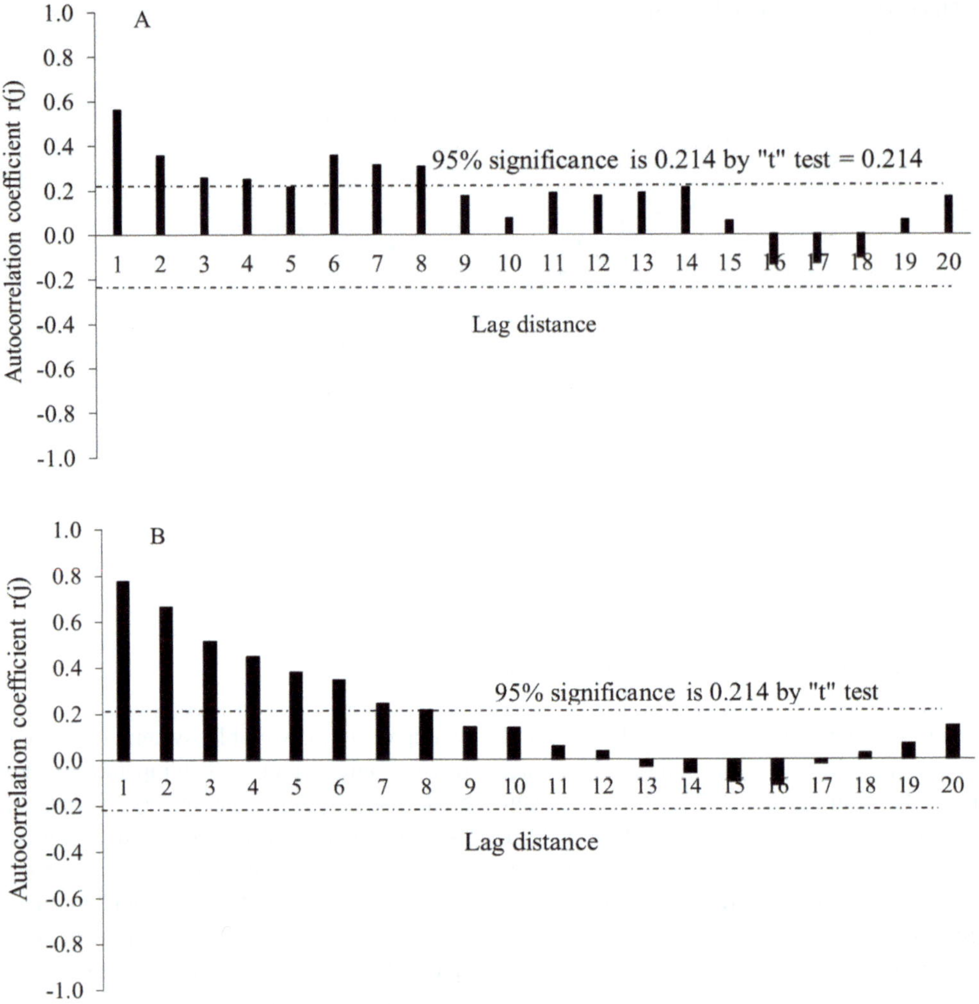

Fig. 18.13 Autocorrelograms: (**A**) soil water content; (**B**) soil temperature

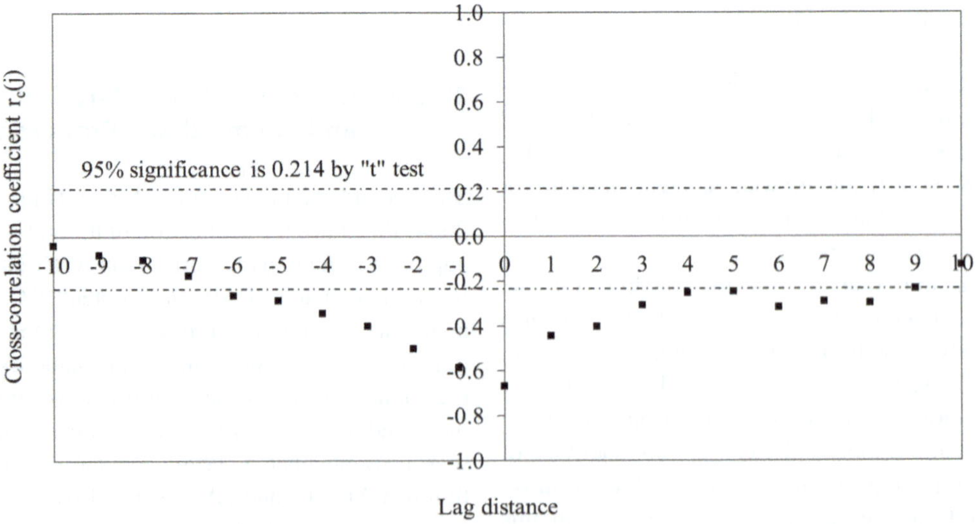

Fig. 18.14 Cross-correlogram between soil water content and soil temperature

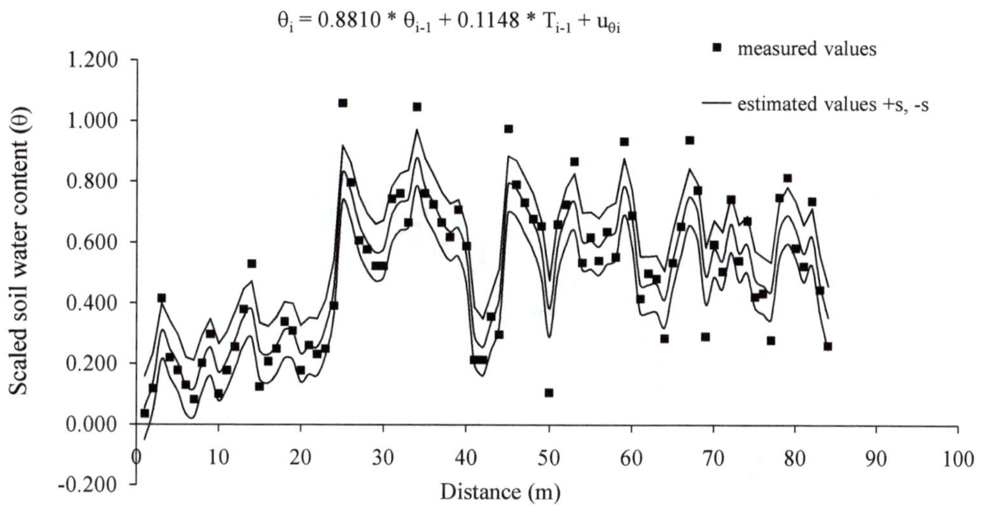

Fig. 18.15 Soil water content as described by state-space using the transformation of Eq. 18.14

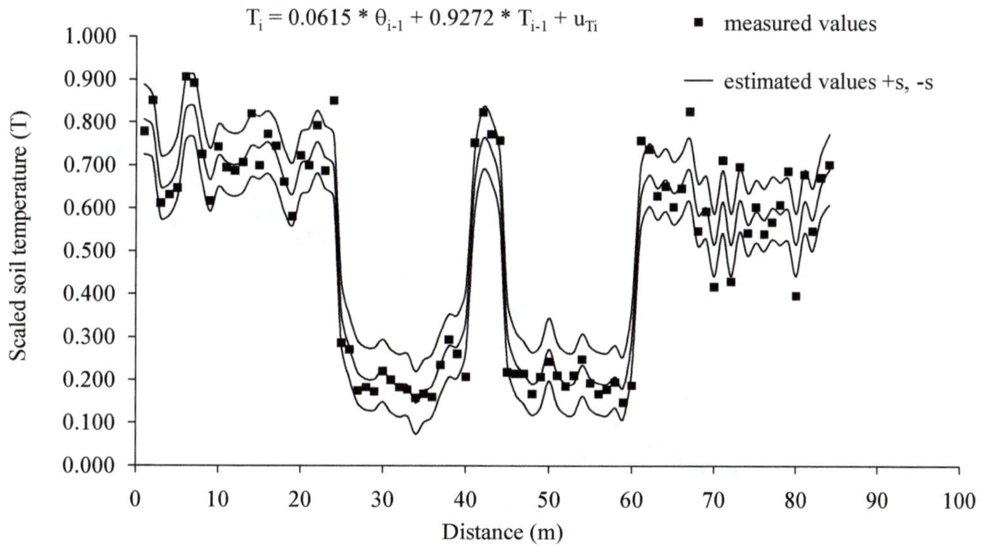

Fig. 18.16 Soil temperature as described by state-space using the transformation of Eq. 18.14

the same experimental transect. After a long period without rain, it was expected that there was a relationship between soil water content and other variables. In the same way as in the previous example, the data were transformed by means of Eq. 18.14 before using the ASTSA software.

The spatial distribution of soil water content along the transect is shown in Fig. 18.18A with coefficient of variation (CV) of 13.4%, indicating that the variability of the data in relation to the mean is relatively small, although the point-to-point fluctuation seems to be large when compared to the total variation of the data along the transect. Although the effect of the treatments is not visually perceived, it is possible to observe a clear tendency of an increase of the soil water content along the transect. The autocorrelation function, calculated through Eq. 17.19a (Chap. 17), for soil water content is shown in

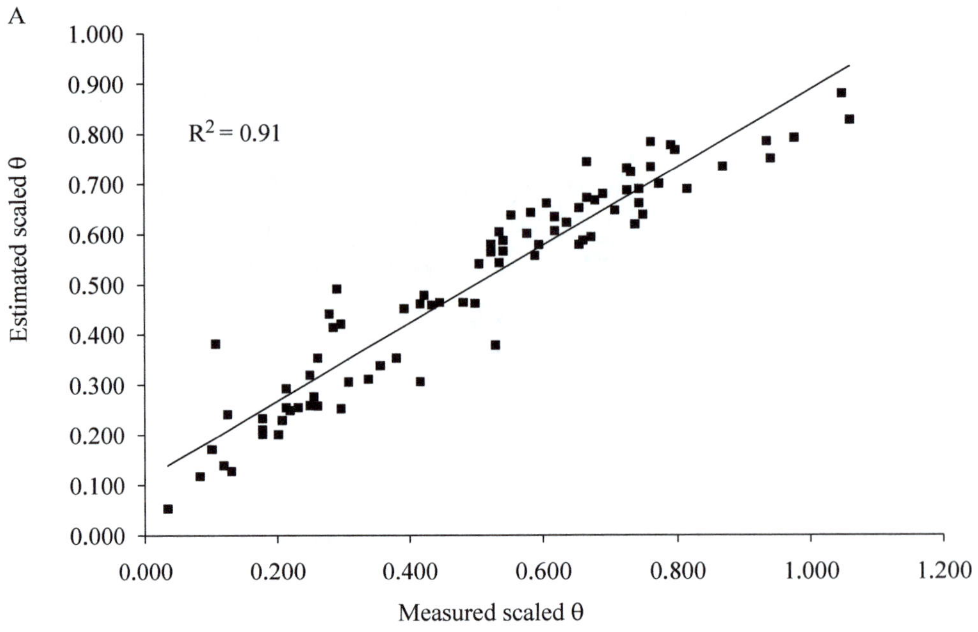

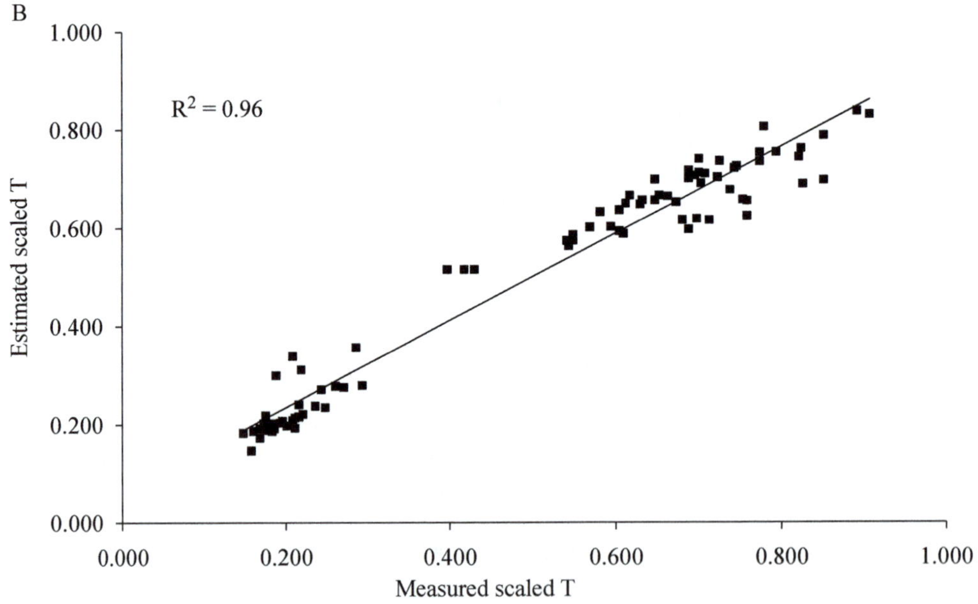

Fig. 18.17 Linear regression between estimated and measured data: (**A**) soil water content; (**B**) soil temperature

Fig. 18.18B. Analyzing this figure, it is verified that there is a strong spatial correlation structure up to 14 lags (in this case 14 m), at the 5% probability level using the *t* test (Eq. 17.20—Chap. 17), which is expected due to the spatial trend in the data (Fig. 18.18A).

Figure 18.19A, B shows the spatial distribution of organic matter values along the 84-point transect and the autocorrelogram of these data, respectively. The CV value is 7.8% and the spatial dependence reaches 10 lags (10 m). In this case, it is observed that OM values show a

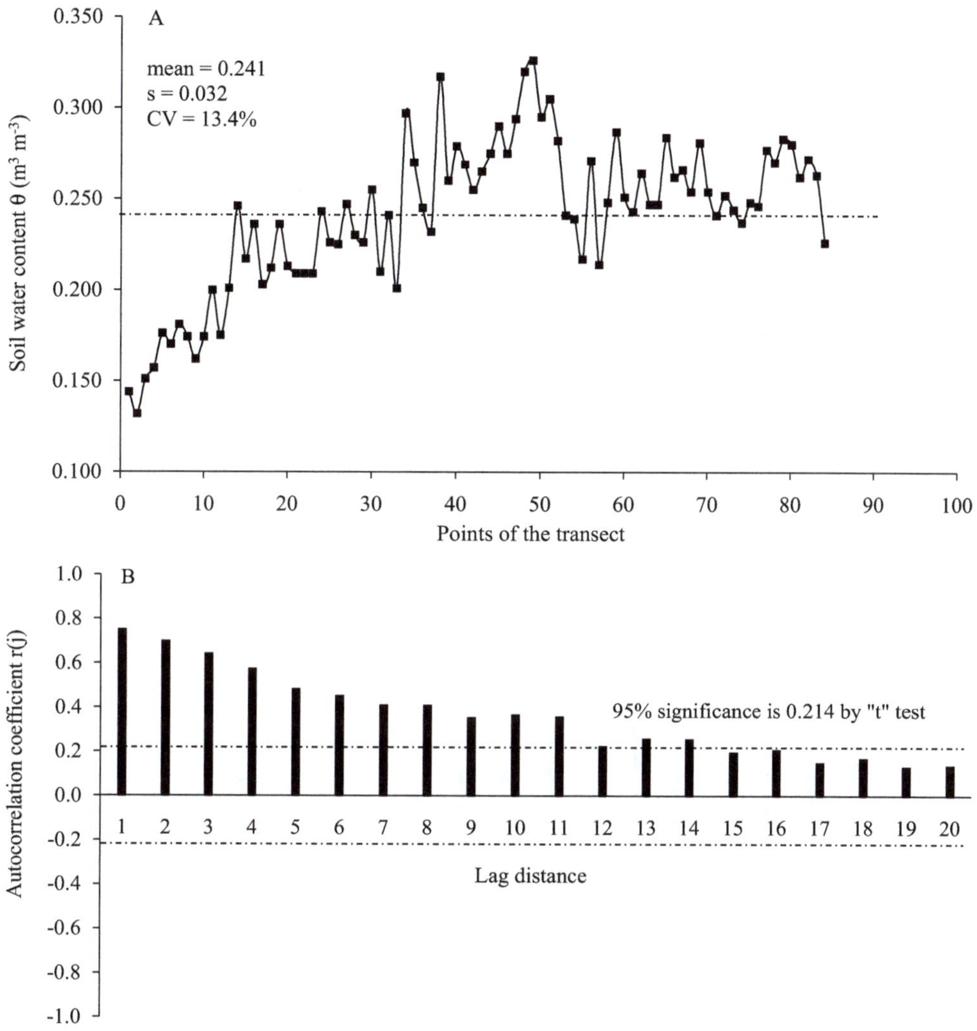

Fig. 18.18 (**A**) Spatial distribution of soil water content data along the transect on Sept 06, 1999; (**B**) autocorrelogram of the data presented in (**A**)

decreasing tendency along the transect (Fig. 18.19A).

The spatial variation of the clay content (CC) and its autocorrelation function are shown in Fig. 18.20A, B, respectively, also reflecting the presence of a spatial trend along the transect. The point-to-point variations are also large when compared to the total variation of CC throughout the transection (CV = 8.7%). This behavior, according to Wendroth et al. (1998), is best identified when statistical tools are used that consider local behavioral trends, which is not the case for classical statistics.

Analyzing Figs. 18.18A, 18.19A, and 18.20A, it can be verified that the spatial distribution of observations of soil water content, organic matter, and clay content show a trend along the transect. This tendency causes strong spatial dependence of each variable as evidenced by the autocorrelation function in Figs. 18.18B, 18.19B, and 18.20B. Figure 18.21A, B presents the spatial variability and autocorrelation function for the stability aggregate observations along the transect. Unlike the other three data sets, aggregate stability does not have a trend and manifests spatial dependence up to 3 m (Fig. 18.21B).

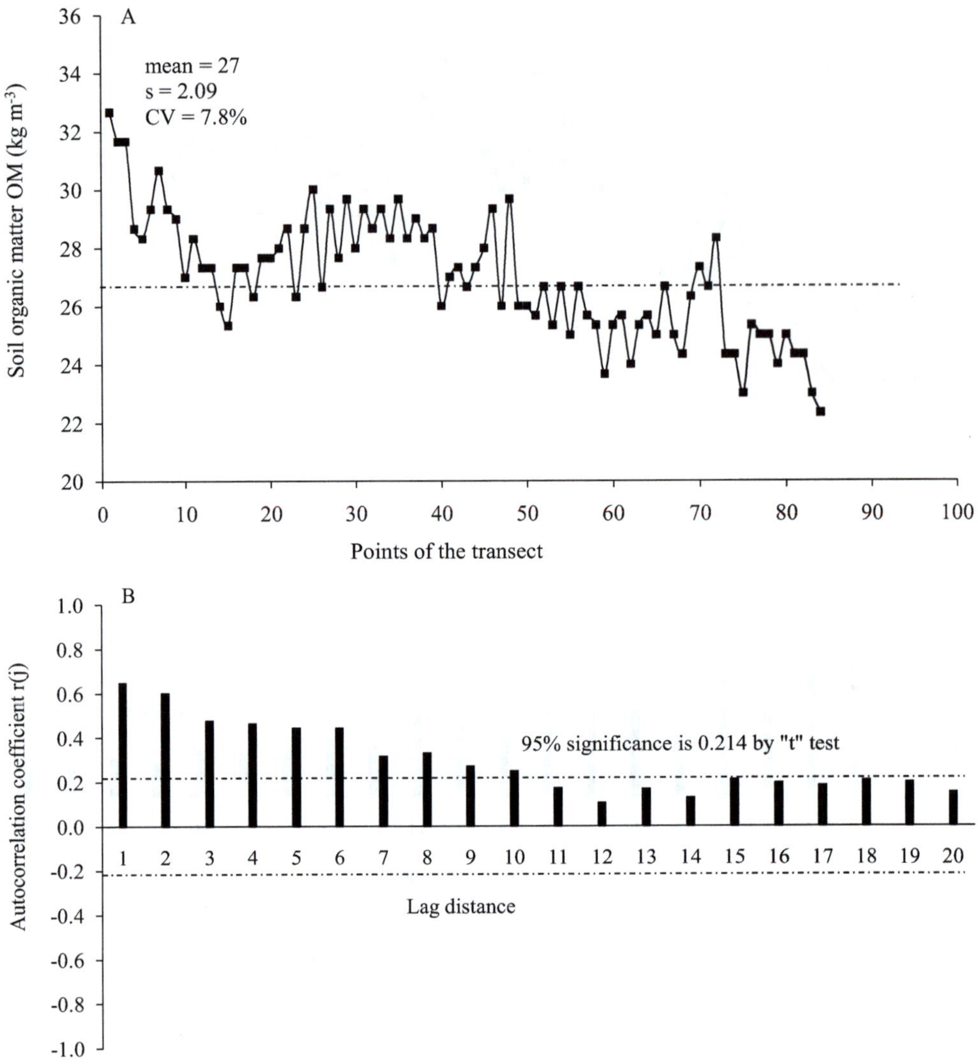

Fig. 18.19 (**A**) Spatial distribution of soil organic matter data along the transect; (**B**) autocorrelogram of the data presented in (**A**)

Traditionally, most agronomic investigations have used random sampling techniques assuming that the samples are independent of each other. Thus, classical statistical analyses are used, such as analysis of variance (ANOVA) and regression analyses, to describe the changes observed within and between different treatments. In cases where observations within and between treatments are not independent, Nielsen and Alemi (1989) argue that experimental design in the field does not provide for the application of classical statistical analysis.

In this second example, the relationship among soil water content (Fig. 18.18A) and organic matter (Fig. 18.19B) and clay content (Fig. 18.20A) and aggregate stability (Fig. 18.21A) data measured along the transect (Fig. 18.9) are first evaluated by the classical regression equations. Not taking into account the positions of the observations, no more than 55% of the soil water content variance is explained by linear and multiple regression analysis using any combination of the observed series. In Table 18.1, it can be noted that the best regression result was obtained

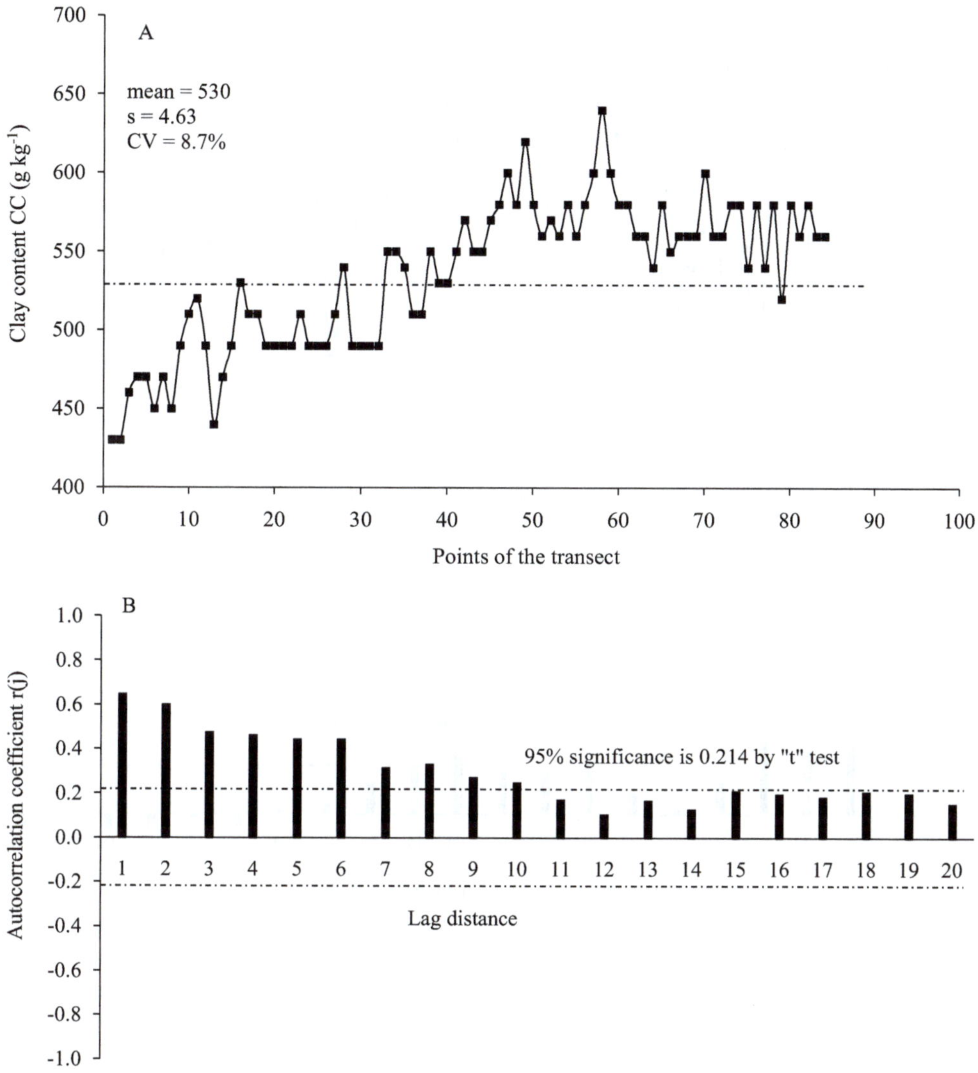

Fig. 18.20 (**A**) Spatial distribution of soil clay content data along the transect; (**B**) autocorrelogram of the data presented in (**A**)

using the organic matter, clay content, and aggregate stability series, and the worst is obtained using only aggregate stability. Using the series of clay content and stability of aggregates, a coefficient of determination (R^2) was obtained close to that in which the three series are used as dependent variables.

These classical regression analyses are based on the assumptions that each series manifests a constant mean along the transect and ignores the local spatial cross-correlation in the transect. The AS observations manifesting only local spatial autocorrelation and not having a trend to match those of the other sets of observations account for its poor regression with soil water content.

To verify that the use of time series analysis leads to additional information on the spatial variability of soil properties and, together with the classical statistics analysis (mean, standard deviation, and coefficient of variation), provides better soil management for a rational use of natural resources, we proceeded with the study calculating the spatial correlation structure between the series by means of cross-correlograms (Eq. 18.1).

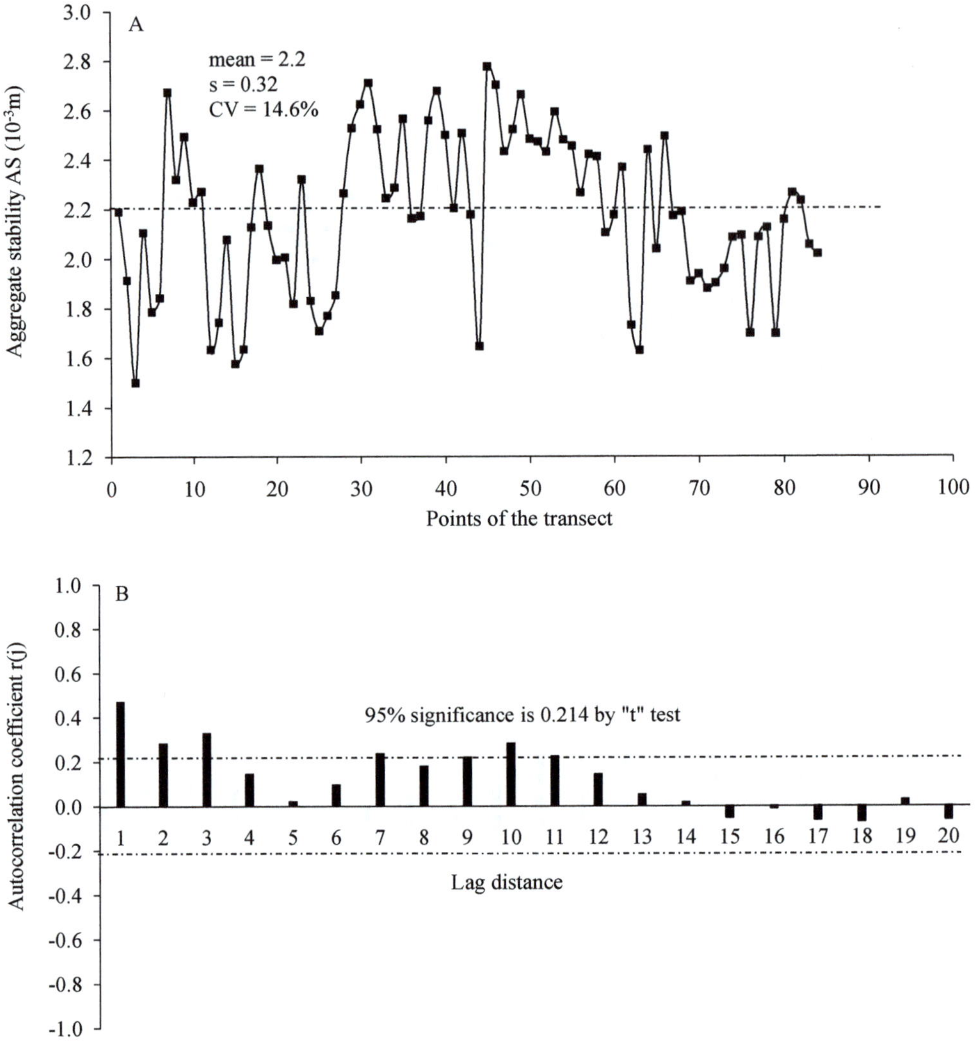

Fig. 18.21 (**A**) Spatial distribution of aggregate stability data along the transect; (**B**) autocorrelogram of the data presented in (**A**)

Table 18.1 Classical linear and multiple regression analyses of the four sets of observations

Equation	Coefficient of determination R^2
Multiple regression	
$\theta = -0.073 - 0.00128 \times OM + 0.000591 \times CC + 0.0150 \times AS$	0.54
$\theta = -0.096 - 0.000322 \times OM + 0.000645 \times CC$	0.53
$\theta = 0.397 - 0.00942 \times OM + 0.0448 \times AS$	0.32
$\theta = -0.124 + 0.000632 \times CC + 0.0122 \times AS$	0.54
Linear regression	
$\theta = 0.474 - 0.0087 \times OM$	0.21
$\theta = -0.109 + 0.000655 \times CC$	0.53
$\theta = 0.159 + 0.375 \times AS$	0.08

The cross-correlograms shown in Fig. 18.22A, B show stronger spatial dependence between soil water content and organic matter and clay content, respectively, than that between soil water content and stability of aggregates (Fig. 18.22C). With such magnitudes for the cross-correlation functions, the potential to describe the spatial relationship among SWC and the other variables along the transect is recognized using the state-space analysis. Initially, in the presence of the trends noted above, we evaluated how well the application of the time series analysis can describe the soil water content series using various combinations between the OM, CC, and AS series. Each original series was transformed by Eq. 18.14.

Figure 18.23A presents the state-space analysis applied to θ, OM, CC, and AS with the square symbol representing the measured values of soil water content. The middle line represents the estimated values of θ using the state equation (Eq. 18.13). The upper and lower lines represent the confidence limits considering plus or minus two standard deviations, respectively. It can be verified from the state equation that the soil water content at position $i - 1$ contributes approximately 91% to the estimate at point i, while OM, CC, and AS at position $i - 1$ contribute with 7%, 6%, and 2.4%, respectively. Figure 18.23B indicates that this state-space model using all four series describes soil water content ($R^2 = 0.80$) better than the equivalent using the multiple regression equation ($R^2 = 0.54$).

Figure 18.24A, B presents the state-space analysis applied to soil water content without the aggregate stability series. Comparing with the graphs in Fig. 18.23, the contribution of soil water content at position $i - 1$ is lower, confidence limits are higher, and R^2 is higher (0.84).

Table 18.2 shows that state-space equations describe soil water content better than any equivalent classical regression equation (Table 18.1).

Examining the results of Table 18.2, it is found that the best performance of all state-space equations was using clay content and aggregate stability. For this equation, the contribution of soil water content at position $i - 1$ was the lowest and R^2 was the highest, that is, local and regional variations of CC and AS along the transect were the most important variations related to the spatial distribution of θ.

The spatial trends (noted in Figs. 18.18A, 18.19A, and 18.20A) were removed using second-order polynomial regression in each series. Subtracting these trends from each respective series of observations, the residues were obtained along the transect (Figs. 18.25A, 18.26A, and 18.27A). The autocorrelation functions of these residues (Figs. 18.25B, 18.26B, and 18.27B) reveal that the spatial dependence along the transect persists at a minimum of 2 lags after the trend has been withdrawn, that is, the observations are still related to the adjacent observations.

Additional analyses did not reveal spatial cross-correlation between soil water content and organic matter and clay content series after the trend removal.

The results presented in Tables 18.1 and 18.2 reveal that the spatial variations of OM, CC, and AS are significantly related to the variations of θ along the transect. Because these space-state equations are empirical, we know that the spatial variations of the series are related to each other, but we have to identify why they are related, that is, how physical and chemical laws can be incorporated into equations of state-space allowing more rigorous selection of the input variables, providing a more realistic explanation of the local and regional variations of the processes that occur at the field level.

18.8.3 Soil-Plant System Evaluation

This third example (extracted from Timm et al. 2003a) illustrates the application of the state-space methodology to evaluate the same soil-plant system described above, using six variables measured along the 84 points of the spatial transect (Fig. 18.9). The number of cane stalks per meter NCS of cane line (Fig. 18.28A) is related to soil chemical properties—phosphorus content (P) (Fig. 18.29A), calcium (Ca) (Fig. 18.30A), and magnesium (Fig. 18.31A)—and with soil

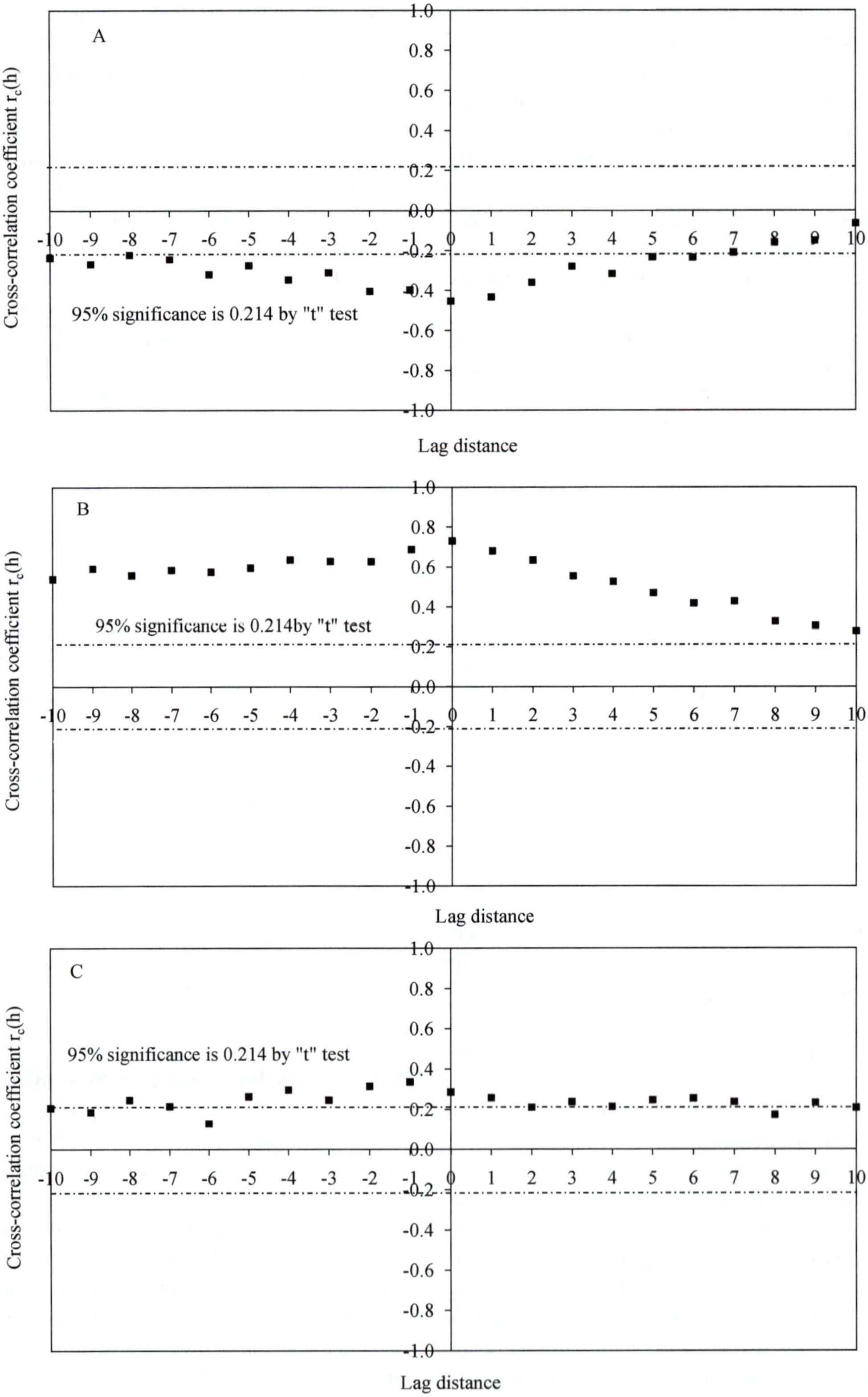

Fig. 18.22 Cross-correlograms between (**A**) soil water content and organic matter, (**B**) soil water content and clay content, and (**C**) soil water content and aggregate stability

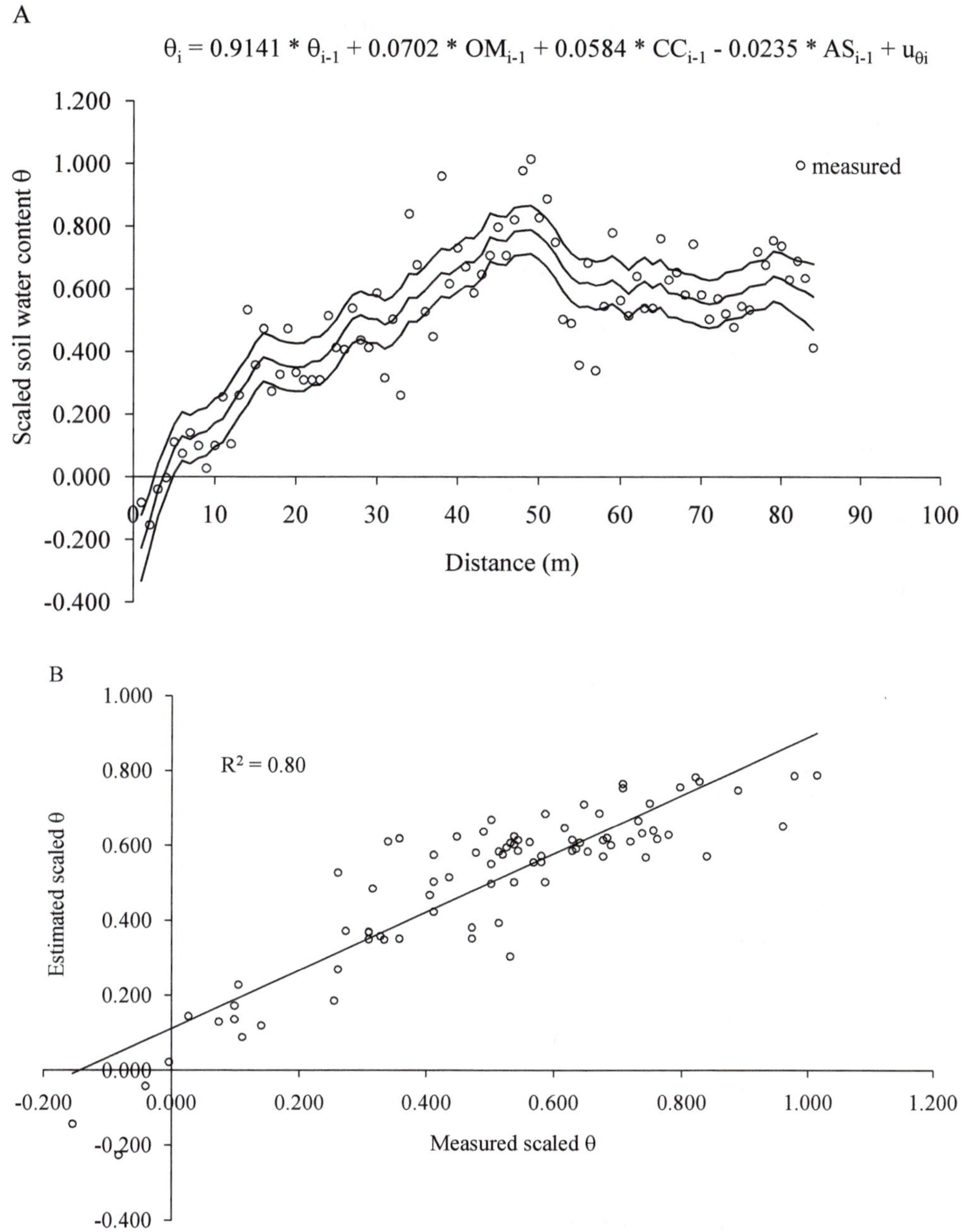

Fig. 18.23 (**A**) State-space analysis of scaled soil water content as a function of data of organic matter, clay content, and aggregate stability at position $i - 1$; (**B**) correlation between estimated and measured values of soil water content

physical properties (clay content [CC, Fig. 18.20A] and stability of aggregates [AS, Fig. 18.21A]). In order to evaluate the spatial correlation structure of adjacent observations of each variable, that is, if each variable was monitored at a sufficient distance to identify its spatial structure, the autocorrelation function (ACF) was calculated (Fig. 18.28B), P (Fig. 18.29B), Ca (Fig. 18.30B), and Mg (Fig. 18.31B). Using the *t* test (Eq. 17.20, Chap. 17) at the 5% probability level, the NCS autocorrelation structure is shown to be 10 lags (Fig. 18.28B) associated with its

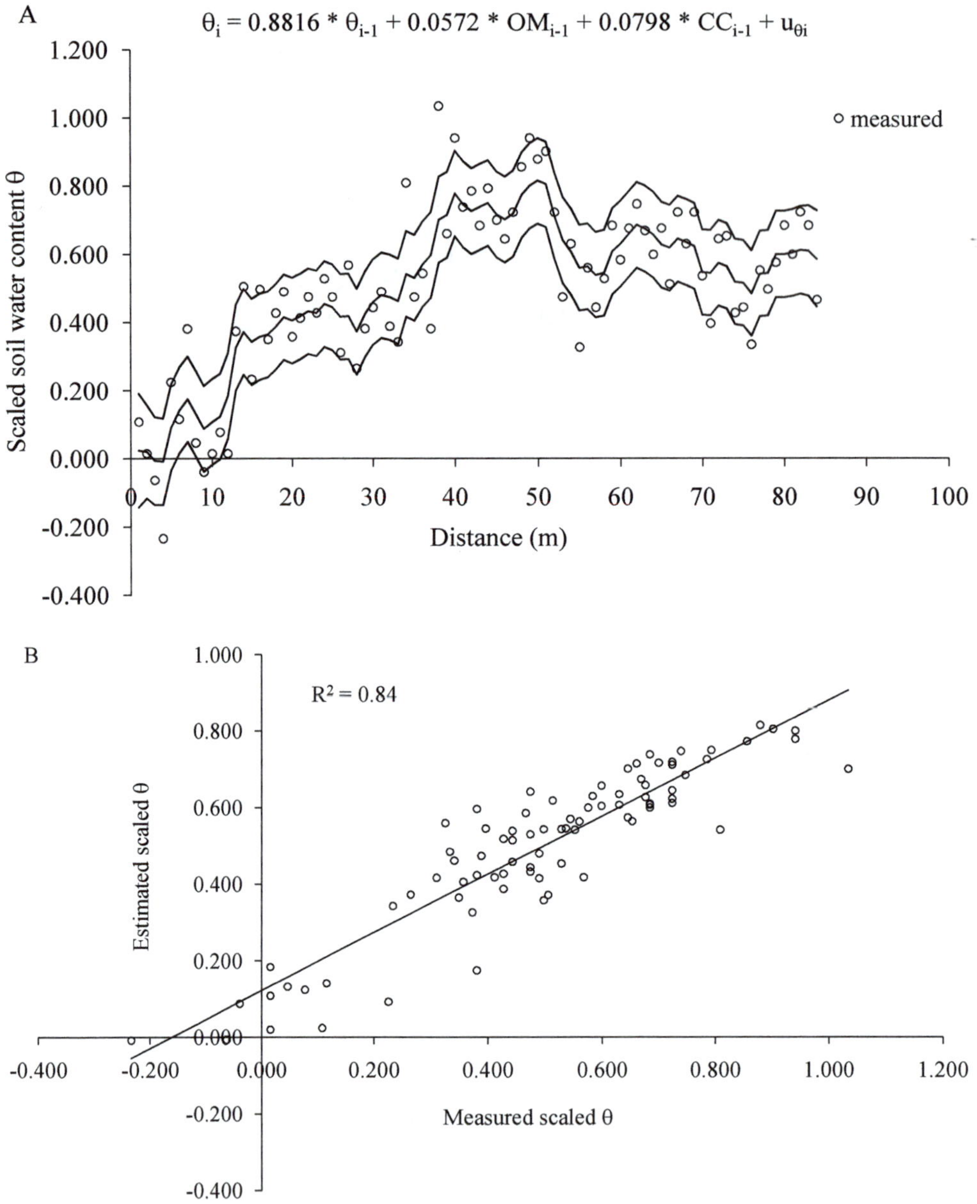

Fig. 18.24 (**A**) State-space analysis of scaled soil water content as a function of data of organic matter and clay content at position $i - 1$; (**B**) correlation between estimated and measured values of soil water content

spatial tendency shown in Fig. 18.28A. Figures 18.29B, 18.30B, and 18.31B show that the spatial dependence of the adjacent observations of P, Ca, and Mg is significant up to 6 lags (6 m in this study). The autocorrelograms for CC and AS observations can be found in Figs. 18.20B and 18.21B, respectively. The cross-correlation function CCF (Eq. 18.1) used to analyze the spatial correlation structure between the number of NCS and (1) P (Fig. 18.32A), (2) Ca (Fig. 18.32B), (3) Mg (Fig. 18.32C), (4) CC (Fig. 18.32D), and (5) AS (Fig. 18.32E) shows a poor spatial dependence between NCS and P and NCS and Ca. The cross-correlogram of Fig. 18.32D shows that the spatial dependence between the number of cane stalks

Table 18.2 State-space equations of soil water content (Fig. 18.18A) using the data presented in Figs. 18.19A, 18.20A, and 18.21A and values of R^2 from linear regression between estimated and measured values of θ. All observations have been scaled using Eq. 18.14

Equation	Coefficient of determination R^2
$\theta_i = 0.914 \times \theta_{i-1} + 0.070 \times OM_{i-1} + 0.058 \times CC_{i-1} - 0.024 \times AS_{i-1} + u_{\theta_i}$	0.80
$\theta_i = 0.882\theta_{i-1} + 0.057 \times OM_{i-1} + 0.080 \times CC_{i-1} + u_{\theta_i}$	0.84
$\theta_i = 0.971\theta_{i-1} + 0.061 \times OM_{i-1} - 0.014 \times AS_{i-1} + u_{\theta_i}$	0.80
$\theta_i = 0.768 \times \theta_{i-1} + 0.146 \times CC_{i-1} - 0.096 \times AS_{i-1} + u_{\theta_i}$	0.91
$\theta_i = 0.961 \times \theta_{i-1} + 0.053 \times OM_{i-1} + u_{\theta_i}$	0.85
$\theta_i = 0.900 \times \theta_{i-1} + 0.100 \times CC_{i-1} + u_{\theta_i}$	0.89
$\theta_i = 0.924 \times \theta_{i-1} + 0.083 \times AS_{i-1} + u_{\theta_i}$	0.88

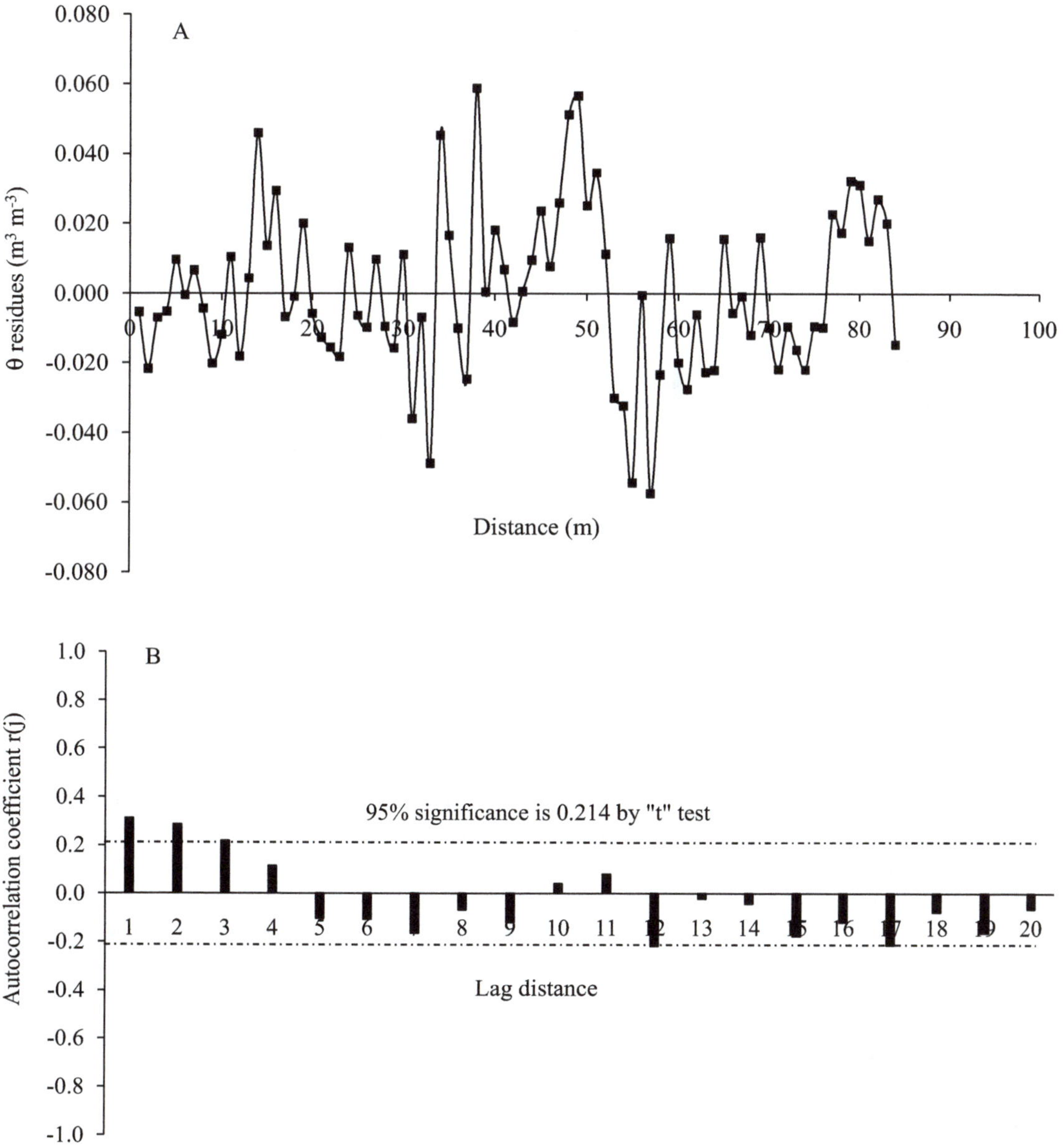

Fig. 18.25 (**A**) Soil water content residues distribution, meter by meter, along the 84 point transect. (**B**) Calculated autocorrelation function (ACF) for soil water content residues data of (**A**)

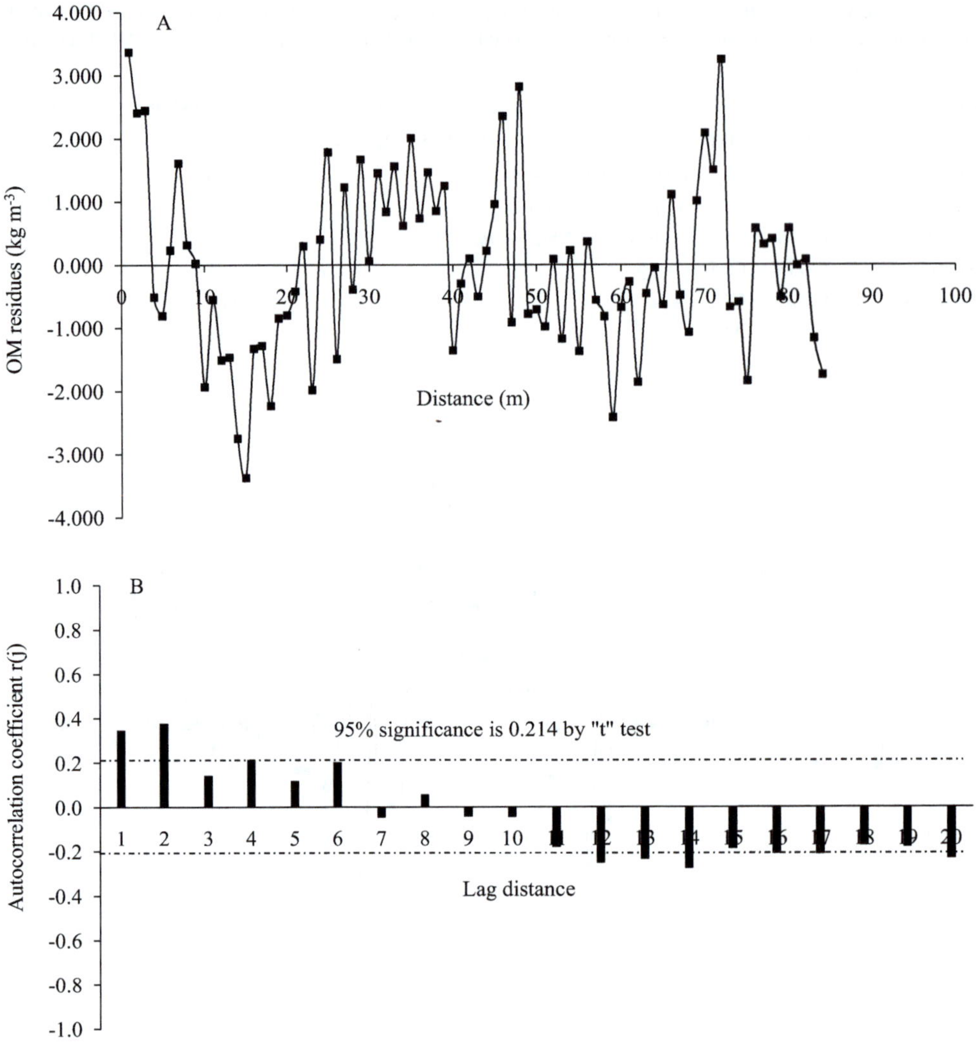

Fig. 18.26 (**A**) Soil organic matter residues distribution, meter by meter, along the 84 point transect. (**B**) Calculated autocorrelation function (ACF) for soil organic matter residues data of (**A**)

and the clay content is stronger than that between the number of stalks and the content of magnesium (Fig. 18.32C). Similar results were found for the cross-correlation between the number of cane stalks and the stability of aggregates (Fig. 18.32E).

Based on the magnitudes of the autocorrelation and cross-correlation functions, the state equation (Eq. 18.13) was used in 31 different combinations of the variables under study to evaluate the behavior of the model with respect to the estimates of the observed values of NCS/m. The results are presented in Table 18.3.

The number of cane stalks per meter (NCS/m) estimated with all the variables used in the state-space analysis is shown in Fig. 18.33. The overall trend of NCS is captured by the analysis, but not the local variation of the data. Analyzing the figure, it can be seen that many observed NCS values fall outside the area corresponding to the lower and upper limits of the estimate and the R^2 value is only 0.50 (Table 18.3). NCS values and the other variables at position $i - 1$ contribute with 86% and between ±10% and 20%, respectively, in the estimate of NCS at position i. As a result, the Kalman Filter excludes many measured

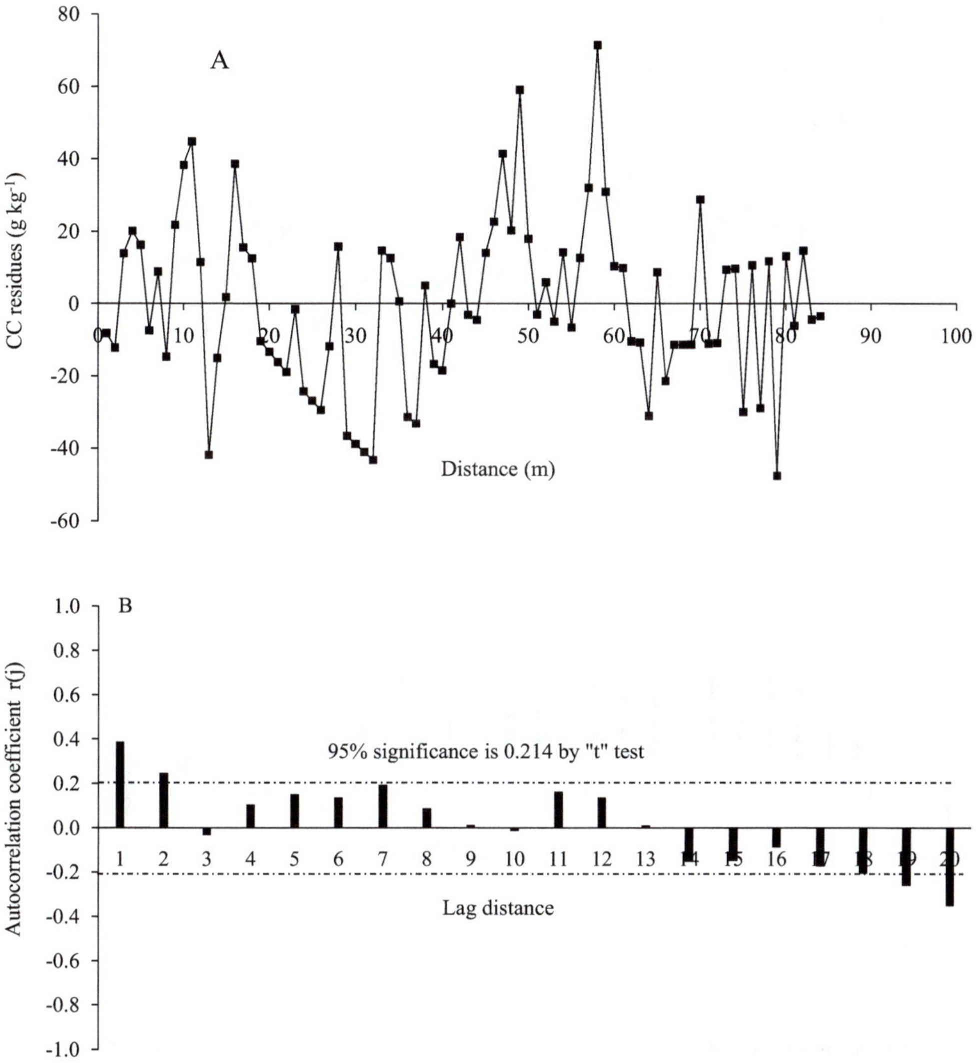

Fig. 18.27 (**A**) Clay content residues distribution, meter by meter, along the 84 point transect. (**B**) Calculated autocorrelation function (ACF) for clay content residues data of (**A**)

NCS values from the narrow area corresponding to the lower and upper limits of the estimates derived from the use of the six variables in the state equation.

In Fig. 18.32A, B, it was found that P and Ca manifested weak spatial structure of cross-correlation with NCS; thus, P and Ca were omitted from the state-space analysis in Fig. 18.34. Although the area corresponding to the confidence limits is slightly larger than in Fig. 18.33, the results of the state-space analysis remain practically the same, that is, the general behavior trend of NCS is captured, but its local variation not.

The largest spatial cross-correlation distance between NCS and the other five variables was obtained with the clay content (CC) (Fig. 18.32D). State-space analysis using only the clay content with NCS is shown in Fig. 18.35. The value of R^2 increased slightly (=0.53), with only 5% of the NCS estimate coming from the neighbor clay content value.

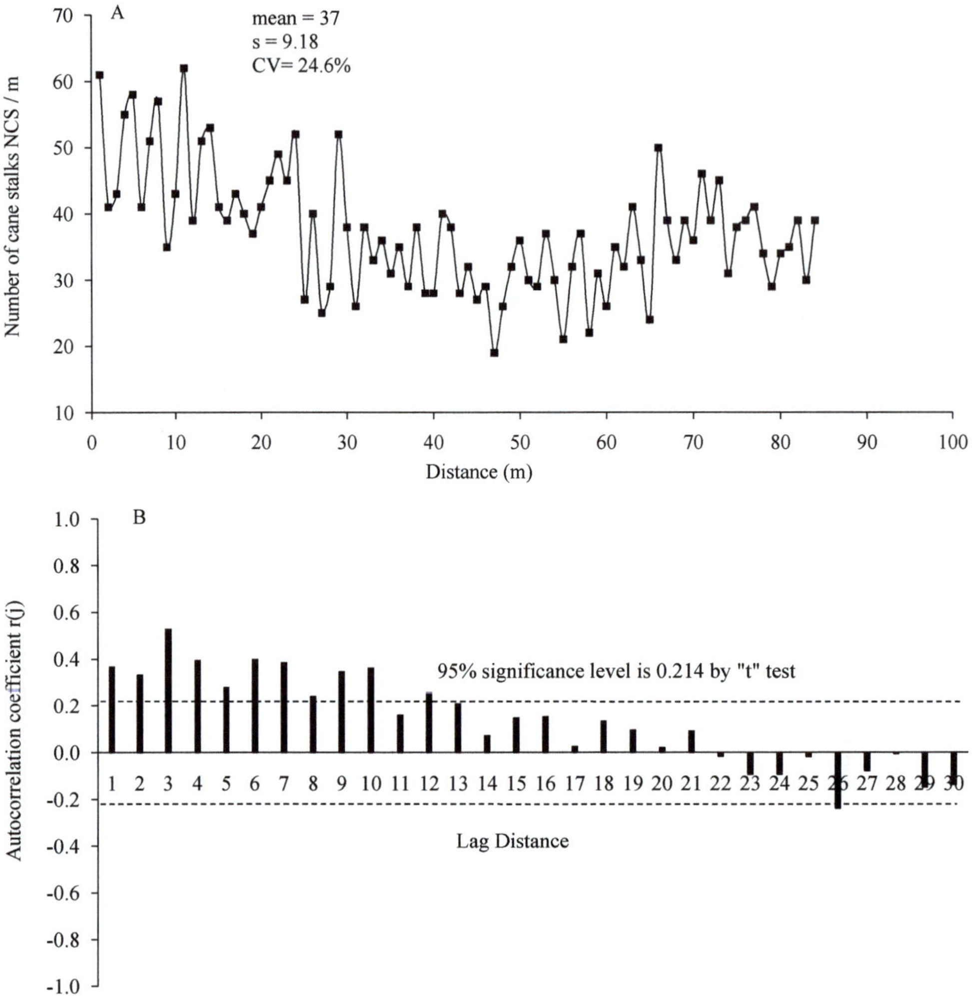

Fig. 18.28 (**A**) Number of cane stalks per meter (NCS/m) distribution, meter by meter, along the 84 point transect. (**B**) Calculated autocorrelation function (ACF) for NCS data of (**A**)

In this example, it was found that local variations in the number of cane stalks were not adequately described by the state-space model using any combination of measured variables. Thus, the cause of variations of NCS remains unknown, which could be solved using other variables or the selection of a different model.

As stated earlier, the state-space approach can be used in spatial or temporal interpolation of data. As an example, the state-space model described in Eqs. 18.12a and 18.13 was used for spatial interpolation of NCS data in two different scenarios: (1) when two of four NCS observations were not considered in the estimation, that is, only using 50% of NCS observations and all complete sets of P, Ca, Mg, CC, and AS (84 points each) (Fig. 18.36A), and (2) when three of four NCS observations were not considered in the estimation, that is, only using 25% of the NCS observations and all the complete sets of P, Ca, Mg, CC, and AS (84 points each) (Fig. 18.36B). According to Wendroth et al. (2001), this implies that the weight of different variables that contribute to the NCS estimation changed and depended on the available data for the Kalman Filter update. When 50% of the NCS observations are available, the update step becomes possible only at the positions where the NCS is available, that is, in

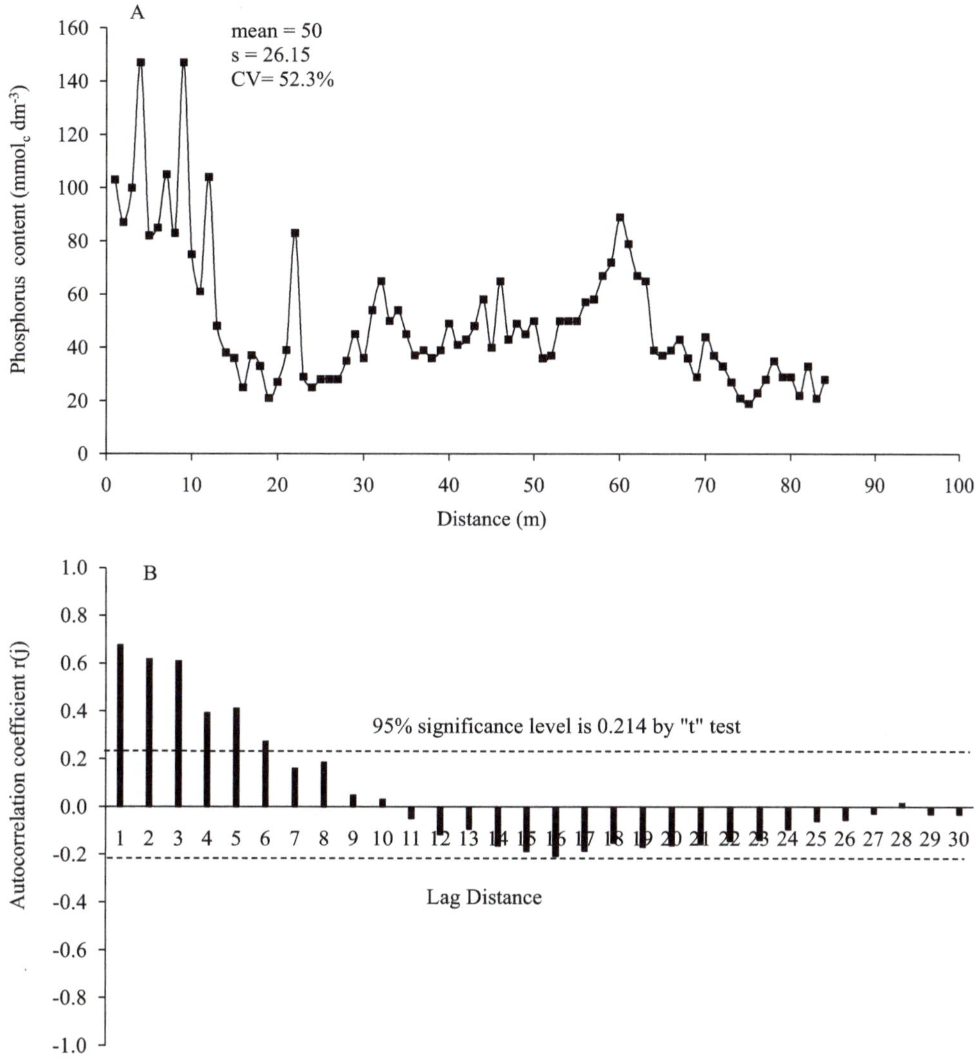

Fig. 18.29 (**A**) Phosphorus (P) content distribution, meter by meter, along the 84 point transect. (**B**) Calculated autocorrelation function (ACF) for P data of (**A**)

two positions. Already when 25% of the NCS observations are available, the update step becomes possible only in four positions. This is the reason why the width of the confidence interval is larger the smaller the number of data available for the NCS estimate (Fig. 18.36C).

Although the state-space approach described in Shumway (1988) comes from time series analysis, it has been successfully applied to evaluate the spatial variability of soil properties collected along spatial transects (e.g., Timm et al. 2003a; Yang et al. 2013; Ogunwole et al. 2014a, b; Yang and Wendroth 2014). However, some applications of this approach can also be found for data collected in experimental grids, such as Stevenson et al. (2001), Liu et al. (2012), and She et al. (2014b). In this sense, Aquino et al. (2015) used the state-space approach, in data collected in an experimental grid, to evaluate the effects of land leveling on the spatial relationship among soil properties in eight different scenarios (four vertical and four horizontal). The authors quantified the relationships between soil water content at field capacity (θ_{FC}) and the permanent

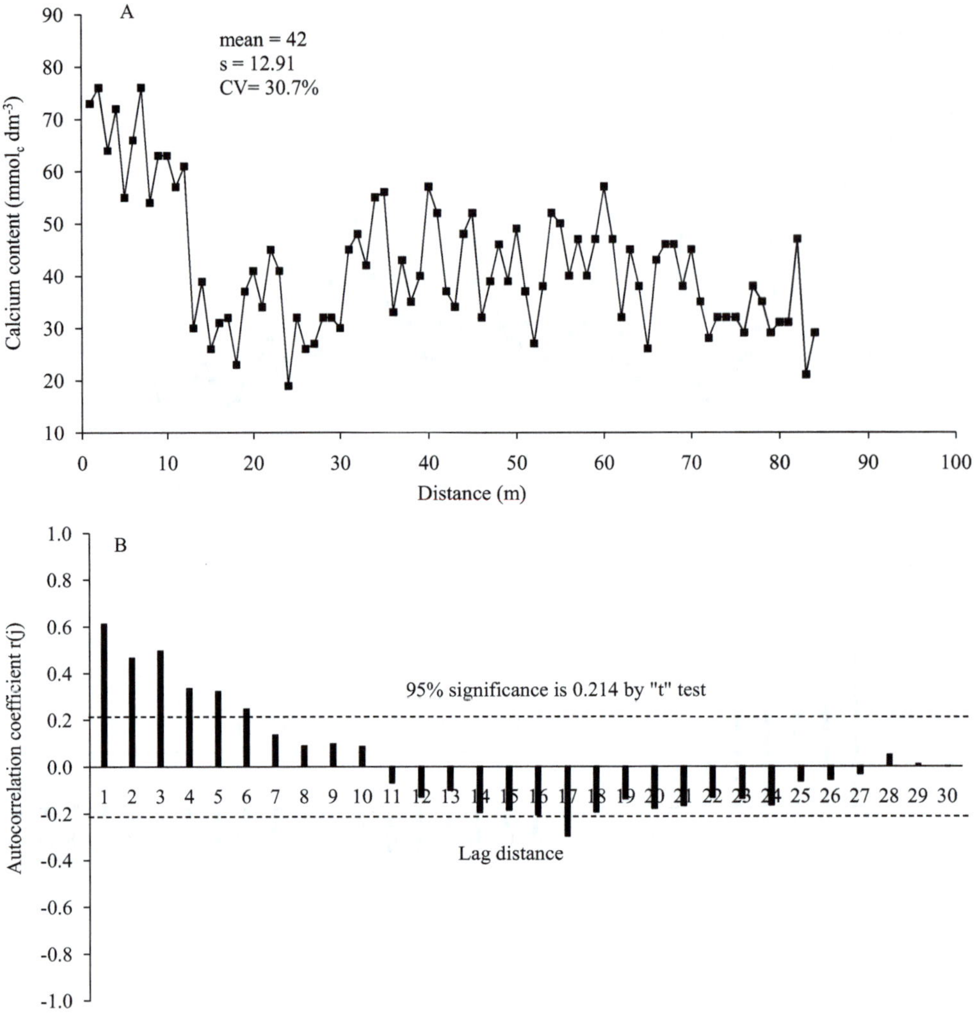

Fig. 18.30 (**A**) Calcium (Ca) content distribution, meter by meter, along the 84 point transect. (**B**) Calculated autocorrelation function (ACF) for Ca data of (**A**)

wilting point (θ_{PWP}) adopted as response variables, before and after land leveling, with sand, silt, and clay contents, microporosity of the soil, soil bulk density, cation exchange capacity, organic carbon content, and depth of the top of the B horizon in relation to the soil surface, also determined before and after leveling and adopted as co-variables in the state-space models. Aquino et al. (2015) concluded that the performance of the state-space models in estimating θ_{FC} and θ_{PWP} is affected by the scenario after leveling, being the best performance of the models for the vertical scenarios. They also concluded that B-horizon depth and cation exchange capacity covariates contributed to the estimation of θ_{FC} and θ_{PWP} for all scenarios evaluated after leveling.

An application of the state-space approach, described by Shumway (1988) and Shumway and Stoffer (2000, 2011, 2017), in series collected over time (time series) was presented by Timm et al. (2011). The authors applied the state-space model for the state-time analysis between water balance components (precipitation [*P*], evapotranspiration [ET], and soil water storage [*S*]) in an established coffee crop in an experimental area

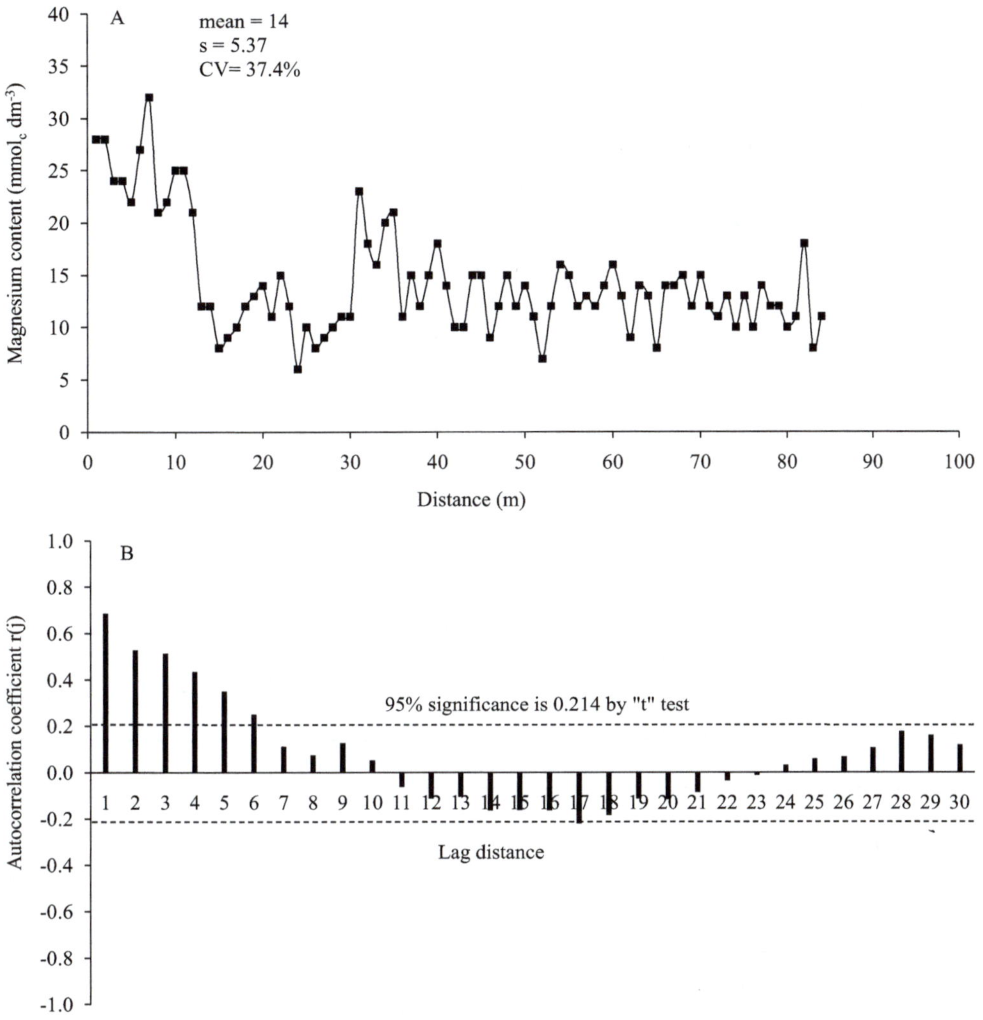

Fig. 18.31 (**A**) Magnesium (Mg) content distribution, meter by meter, along the 84 point transect. (**B**) Calculated autocorrelation function (ACF) for Mg data of (**A**)

in Piracicaba, SP, Brazil. The application of state-time models showed that estimates of S at a time t depended more on precipitation measurements (52%) than on ET (28%) and S (20%) measurements at a time $t - 1$. Analyses also showed that ET estimates at time t were more dependent on ET measurements (59%) than P (30%) and S (9%) at a time $t - 1$. It is worth mentioning that this is the first application of state-time analysis in the Latin American soil science literature as a tool for a better understanding of the temporal relationships of the soil-plant-atmosphere system variables. Awe et al. (2015) also applied the state-space approach to data collected over time in order to evaluate soil water storage in an area planted with sugarcane and the presence of plant residues on the soil surface. The authors monitor soil water storage and matrix potential in the soil layers 0–10 cm, 10–20 cm, and 40–60 cm using time domain reflectometry sensors and tensiometers, respectively. Daily precipitation was also monitored by means of rain gauges, and the actual evapotranspiration of the crop was calculated from data collected at a meteorological station. Awe et al. (2015) compared the performance of the state-space models in estimating soil water storage with the equivalent classical linear regression

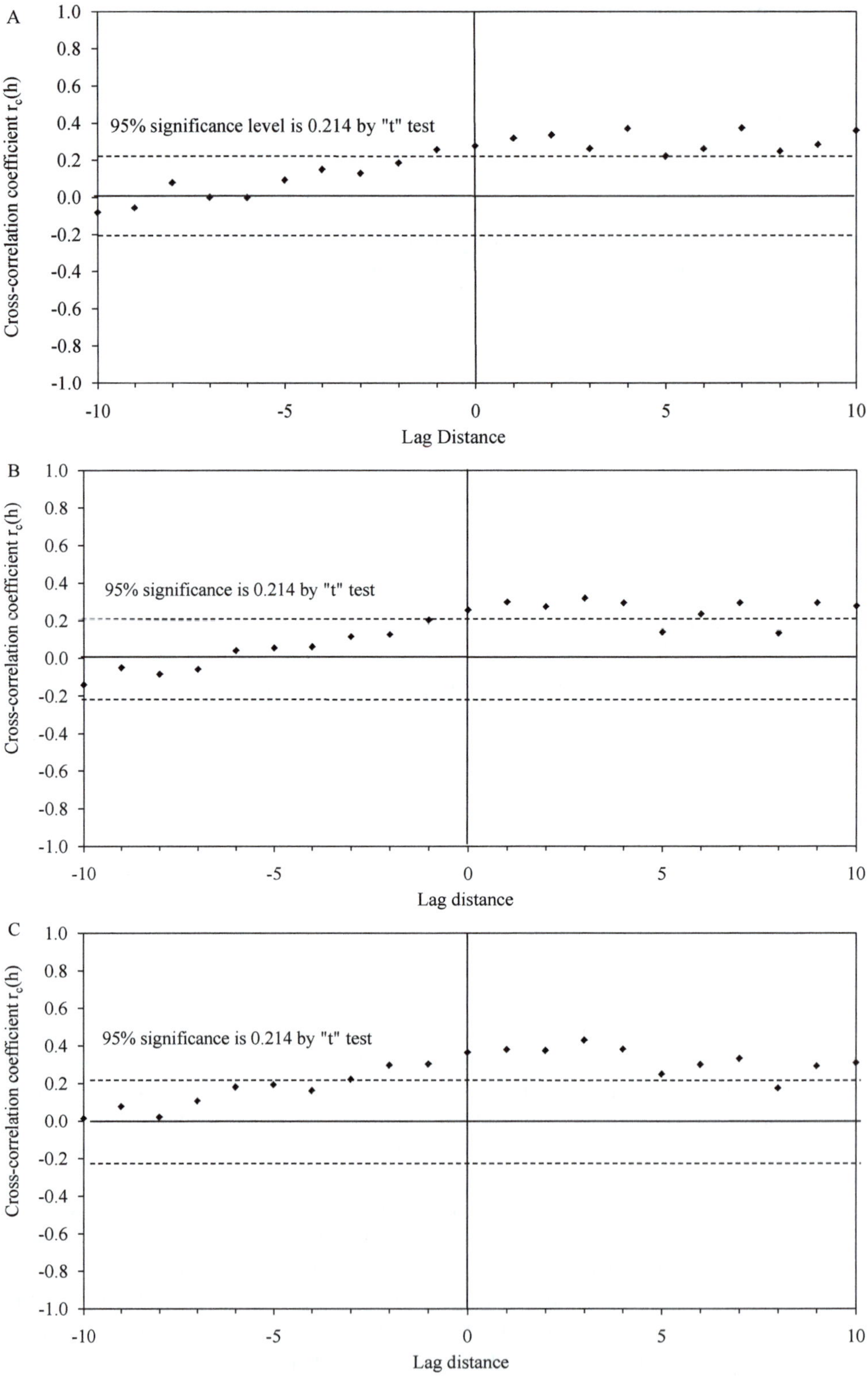

Fig. 18.32 Cross-correlation function for (**A**) NCS/m and phosphorus, (**B**) NCS/m and calcium, (**C**) NCS/m and magnesium, (**D**) NCS/m and clay content, and (**E**) NCS/m and aggregate stability

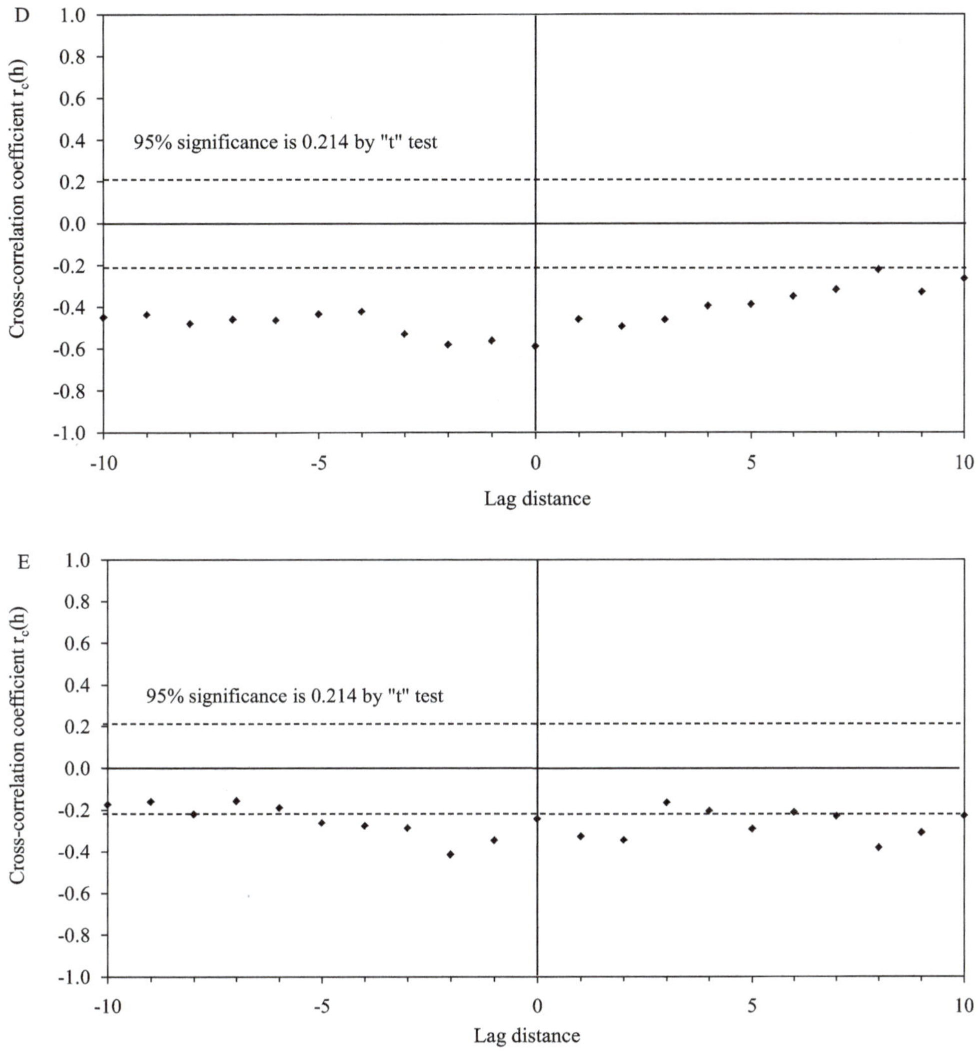

Fig. 18.32 (continued)

models and concluded that the former had better performance in estimating soil water storage over the time in all combinations evaluated using the logarithm of the matrix potential, actual crop evapotranspiration, and precipitation as covariates in the equations.

18.9 State-Space Approach as Described by West and Harrison

The Bayesian formulation presented by West and Harrison (1989, 1997) and originally published by Harrison and Stevens (1976) has not frequently been used in agronomy. In this case, a general parametric formulation is used by which the observations are linearly related to parameters (Eq. 18.12) that have a dynamic evolution according to a random walk (Eq. 18.13), with the possibility of the incorporation of uncertainties associated to the model itself and to the parameters of the model. The probabilities of the model and its parameters are continuously updated in time/space using the Bayes theorem. The acceptance and use of this approach in agronomy have not been as quick as expected, particularly by those without a deep knowledge in

Table 18.3 State-space equations of number of cane stalks per meter (NCS/m) using soil phosphorus (P), soil calcium (Ca), soil magnesium (Mg), soil clay content (CC), and soil aggregate stability (AS) data and values of R^2 from linear regression between estimated and measured values of NCS

Equation	R^2
$NCS_i = 0.857NCS_{i-1} - 0.106P_{i-1} + 0.026Ca_{i-1} + 0.267Mg_{i-1} + 0.163CC_{i-1} - 0.221AS_{i-1} + u_{NCSi}$	0.50
$NCS_i = 0.876NCS_{i-1} + 0.098P_{i-1} - 0.144Ca_{i-1} + 0.133Mg_{i-1} + 0.0286CC_{i-1} + u_{NCSi}$	0.53
$NCS_i = 0.971NCS_{i-1} - 0.005P_{i-1} + 0.148Ca_{i-1} - 0.108Mg_{i-1} - 0.016AS_{i-1} + u_{NCSi}$	0.50
$NCS_i = 0.898NCS_{i-1} + 0.352P_{i-1} - 0.279Ca_{i-1} + 0.040CC_{i-1} - 0.028AS_{i-1} + u_{NCSi}$	0.52
$NCS_i = 0.817NCS_{i-1} - 0.094P_{i-1} + 0.314Mg_{i-1} + 0.165CC_{i-1} - 0.218AS_{i-1} + u_{NCSi}$	0.58
$NCS_i = 0.785NCS_{i-1} - 0.197Ca_{i-1} + 0.420Mg_{i-1} + 0.159CC_{i-1} - 0.182AS_{i-1} + u_{NCSi}$	0.52
$NCS_i = 0.951NCS_{i-1} + 0.068P_{i-1} + 0.048Ca_{i-1} - 0.079Mg_{i-1} + u_{NCSi}$	0.48
$NCS_i = 0.920NCS_{i-1} + 0.050P_{i-1} + 0.002Ca_{i-1} + 0.018CC_{i-1} + u_{NCSi}$	0.49
$NCS_i = 0.902NCS_{i-1} + 0.250P_{i-1} - 0.192Ca_{i-1} + 0.027AS_{i-1} + u_{NCSi}$	0.48
$NCS_i = 0.926NCS_{i-1} + 0.069P_{i-1} - 0.025Mg_{i-1} + 0.019CC_{i-1} + u_{NCSi}$	0.51
$NCS_i = 0.968NCS_{i-1} + 0.123P_{i-1} - 0.084Mg_{i-1} - 0.018AS_{i-1} + u_{NCSi}$	0.46
$NCS_i = 0.925NCS_{i-1} + 0.095P_{i-1} + 0.082CC_{i-1} - 0.115AS_{i-1} + u_{NCSi}$	0.50
$NCS_i = 0.879NCS_{i-1} - 0.012Ca_{i-1} + 0.097Mg_{i-1} + 0.027CC_{i-1} + u_{NCSi}$	0.50
$NCS_i = 0.937NCS_{i-1} + 0.160Ca_{i-1} - 0.094Mg_{i-1} - 0.018AS_{i-1} + u_{NCSi}$	0.49
$NCS_i = 0.959NCS_{i-1} + 0.056Ca_{i-1} + 0.062CC_{i-1} - 0.088AS_{i-1} + u_{NCSi}$	0.51
$NCS_i = 0.918NCS_{i-1} + 0.103Mg_{i-1} + 0.099CC_{i-1} - 0.134AS_{i-1} + u_{NCSi}$	0.49
$NCS_i = 0.942NCS_{i-1} - 0.039P_{i-1} + 0.087Ca_{i-1} + u_{NCSi}$	0.47
$NCS_i = 0.946NCS_{i-1} + 0.127P_{i-1} - 0.085Mg_{i-1} + u_{NCSi}$	0.49
$NCS_i = 0.912NCS_{i-1} + 0.058P_{i-1} + 0.019CC_{i-1} + u_{NCSi}$	0.53
$NCS_i = 0.923NCS_{i-1} + 0.062P_{i-1} + 0.004AS_{i-1} + u_{NCSi}$	0.48
$NCS_i = 0.938NCS_{i-1} + 0.148Ca_{i-1} - 0.097Mg_{i-1} + u_{NCSi}$	0.49
$NCS_i = 0.923NCS_{i-1} + 0.050Ca_{i-1} + 0.018CC_{i-1} + u_{NCSi}$	0.52
$NCS_i = 0.929NCS_{i-1} + 0.065Ca_{i-1} - 0.003AS_{i-1} + u_{NCSi}$	0.45
$NCS_i = 0.904NCS_{i-1} + 0.063Mg_{i-1} + 0.023CC_{i-1} + u_{NCSi}$	0.53
$NCS_i = 0.914NCS_{i-1} + 0.060Mg_{i-1} + 0.015AS_{i-1} + u_{NCSi}$	0.49
$NCS_i = 0.972NCS_{i-1} + 0.035CC_{i-1} - 0.018AS_{i-1} + u_{NCSi}$	0.48
$NCS_i = 0.915NCS_{i-1} + 0.073P_{i-1} + u_{NCSi}$	0.51
$NCS_i = 0.924NCS_{i-1} + 0.065Ca_{i-1} + u_{NCSi}$	0.51
$NCS_i = 0.920NCS_{i-1} + 0.068Mg_{i-1} + u_{NCSi}$	0.51
$NCS_i = 0.958NCS_{i-1} + 0.030CC_{i-1} + u_{NCSi}$	0.53
$NCS_i = 0.962NCS_{i-1} + 0.026AS_{i-1} + u_{NCSi}$	0.53

All observations have been scaled using Eq. 18.14.

statistics, due to the difficulties in establishing values (or their law of variation) for the noise parameters "$v(x_i)$ (Eq. 18.12)" and "$u(x_i)$ (Eq. 18.13)". To make this approach more accessible, Ameen and Harrison (1984) used discount factors to calculate the covariance matrix of the noise parameters $u(x_i)$. Discount factors relate to the relevance of the observations during the evolution of time/space—with the most recent information usually being more relevant in the modeling process. The smaller the discount factor, the less importance is given to previous information. Hence, the use of these factors assures that the stochastic influence on the evolution of the parameters (Eq. 18.13) is not directly made explicit through the noise $u(x_i)$. The stochastic influence is derived by the combination of a relation that establishes only the deterministic evolution of $Z_j(x_i)$ and the random process guaranteed by the discount matrix.

In this state-space approach, the state equation describes the evolution of the regression coefficients β through a random walk vector

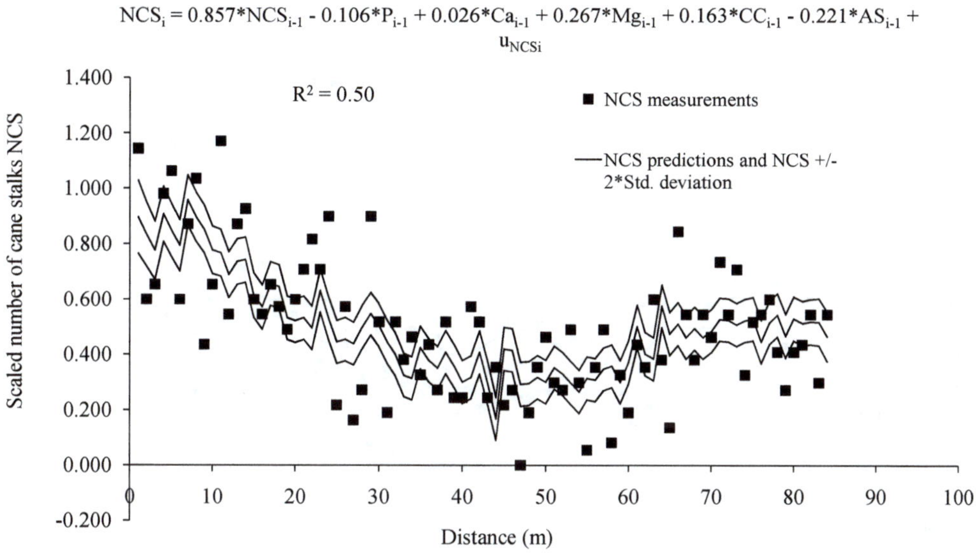

Fig. 18.33 State-space analysis of scaled number of cane stalks data at position i as a function of scaled number of cane stalks, scaled phosphorus content, scaled calcium content, scaled magnesium content, scaled clay content, and scaled aggregate stability, all at position $i - 1$

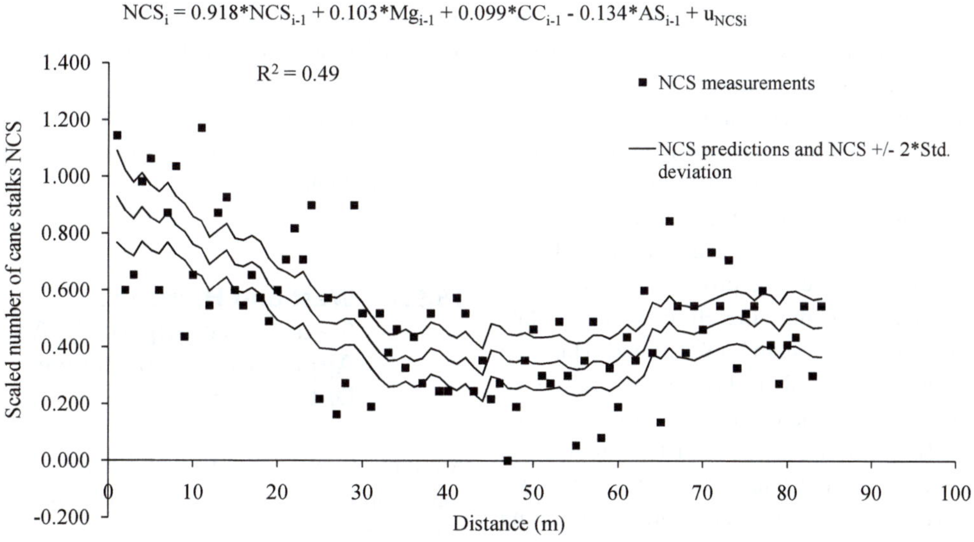

Fig. 18.34 State-space analysis of scaled number of cane stalks data at position i as a function of scaled number of cane stalks, scaled magnesium content, scaled clay content, and scaled aggregate stability, all at position $i - 1$

$$\beta_j(x_i) = \beta_j(x_{i-1}) + u(x_i) \tag{18.15}$$

where $u(x_i)$ are normally distributed with zero mean and constant variance W [$u(x_i) \sim N(0, W)$] and are noncorrelated among themselves (white noise) (Timm et al. 2003b). The regression coefficients vector $\beta_j(x_i)$ is related to the observable response variable $Y(x_i)$ through the observation equation

$$Y(x_i) = F_j(x_i)\beta_j(x_i) + v(x_i) \tag{18.16}$$

where $F_j(x_i)$ is a known matrix containing the regressors, which reduces to a vector for

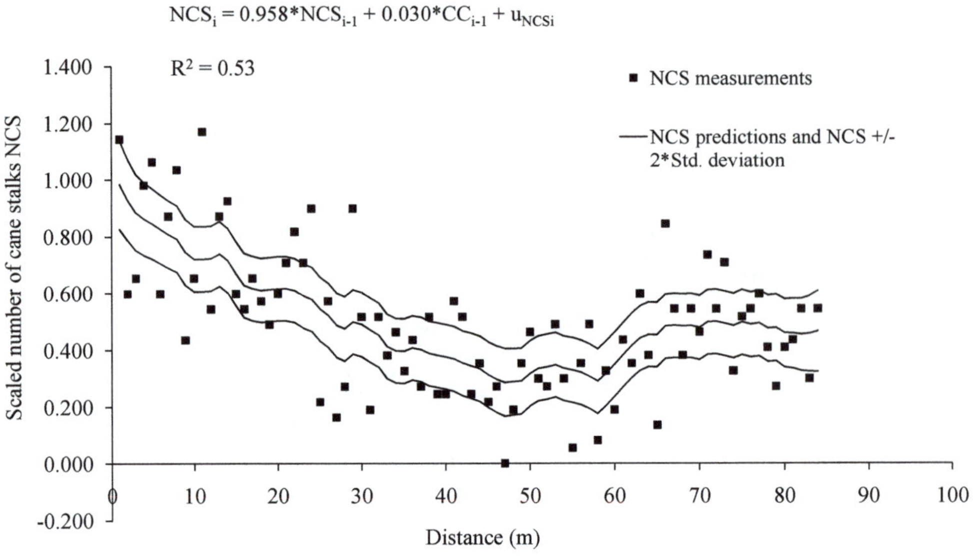

Fig. 18.35 State-space analysis of scaled number of cane stalks data at position i as a function of scaled number of cane stalks and scaled clay content, all at position $i - 1$

unidimensional responses, and $v(x_i)$ are noncorrelated errors with zero mean, constant variance, and normal distribution.

The dynamic regression model (Eqs. 18.15 and 18.16) is a local and not a global model, because it contains variable $\beta(x_i)$ coefficients having the subscript j. These coefficients vary along space/time according to a Markovian evolution (first-order autoregressive process, not having to be stationary), being therefore called "state variables of the system" (West and Harrison 1989, 1997; Pole et al. 1994). Hence, we have a basic regression equation (observation equation, Eq. 18.16) and a second equation (evolution equation, Eq. 18.15) that characterizes the form of the variation of these state parameters along space. Parameters are estimated in an optimal way through algorithms of the Kalman Filter type (extensions of the basic Kalman Filter). The equations of estimation are sequential, comprising the observational equations of actualization (via Bayes theorem and observation equation) and the spatial/time actualization equations (consequence of the evolution equation).

In this approach, more emphasis is given to Eq. 18.15. In the dynamic regression, the β coefficients are considered as a state vector (Eq. 18.16) following a random walk process. The transition matrix ϕ is unity and the observation matrix M that relates the state vector to the observation vectors, being formed by the regression coefficients.

18.9.1 West and Harrison's State-Space Approach: An Application Using Space Data Series

An application of the West and Harrison's state-space approach will be presented here using space data series. Space data series are from an oat crop (*Avena strigosa*) field experiment carried out in Jaguariúna, SP, Brazil. As mentioned before, the objective of this study was to propose another methodology using a dynamic regression model with coefficients changing along the space as an alternative to the standard state-space model.

Soil samples used in this study were collected in Jaguariúna (22° 41′ S and 47° W), SP, Brazil, on a Typic Haplustox in May 1999. Samples were taken along a line (transect) located in the middle

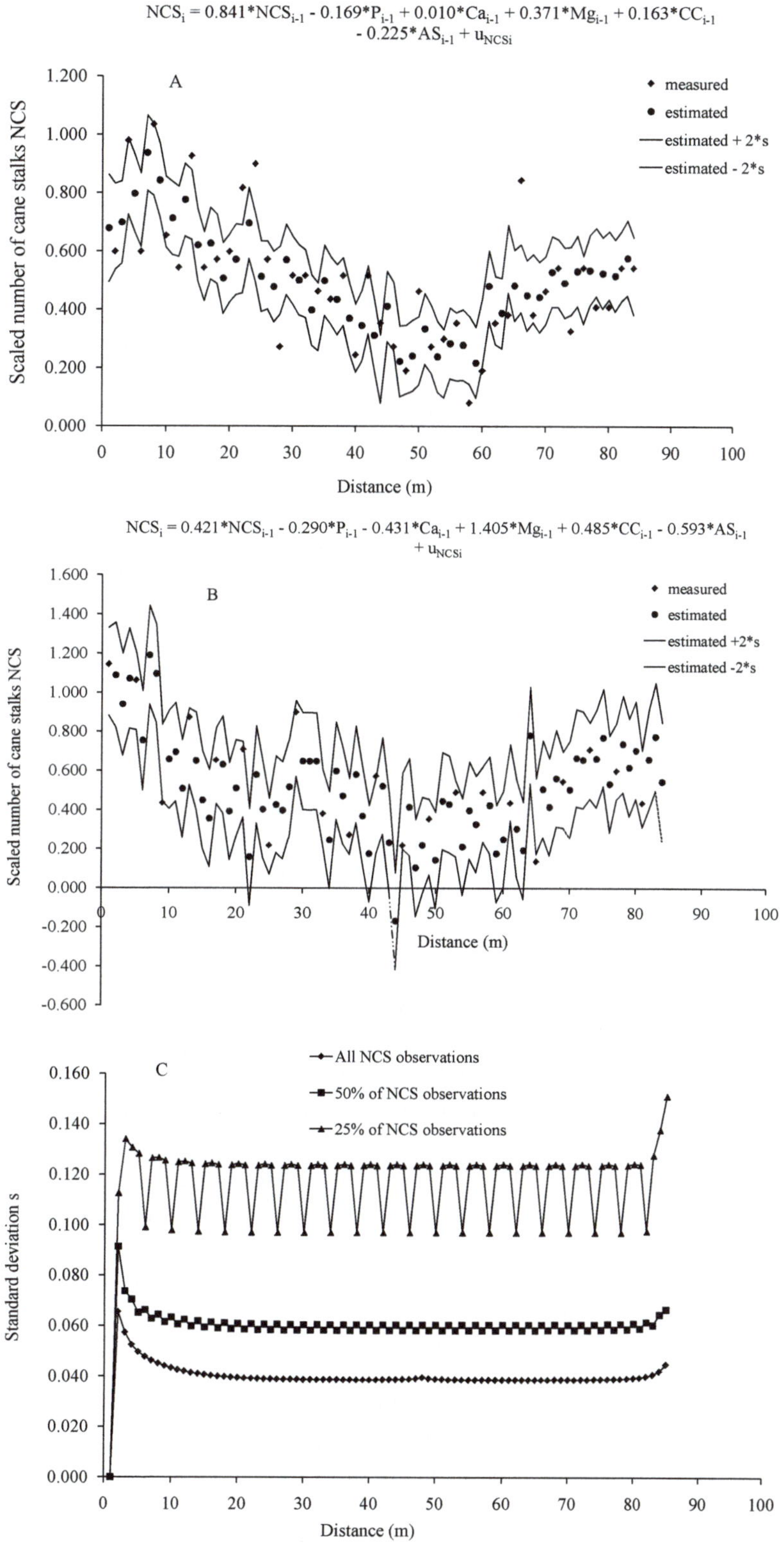

Fig. 18.36 (**A**) Scaled NCS estimated with a state-space model with one out of two observations considered in the estimation; (**B**) scaled NCS estimated with a state-space model with only one out of four observations considered in the estimation; (**C**) standard deviation behavior along 84 point transect when all observations are included in the estimation of a state-space model with 50% of all observations (data from **A**) and with 25% of all observations (data from **B**)

of two adjacent contour lines. The transect samples, totaling 97, were collected in the plow layer (0–0.20 m) at points spaced 2 m apart. The transect soil had been limed, received phosphate (broadcasted and incorporated), and was planted to an oat crop, 3 months before soil sampling. Samples were air-dried, granted to pass a 2 mm sieve, and analyzed for soil organic carbon (SOC) by the Walkley-Black method (Walkey and Black 1934), and for total soil nitrogen (TSN) by the Kjeldahl method (Bremner 1960).

The proposed method to build and implement a model for soil quality data analysis (along a spatial transect) was based on the following main assumptions:

1. Along the transect with inherent soil variation, local characteristics are better represented by a local model (and not a global one) with space-varying coefficients expressing the heterogeneity of the site.
2. Different variables measured by different scales or units, in order to be analyzed or modeled in conjunction and in an efficient way, could be conveniently transformed into a dimension-free scale using Eq. 18.14. However, a method that does not involve data transformation is preferable since it allows an easier interpretation.
3. Some soil quality variables (as for instance, total soil nitrogen) are time-consuming and expensive to be measured but can be well correlated to other variables easier to be measured (as, e.g., the soil organic carbon).

The proposed method for soil quality data analysis through space-varying regression models was based on the following procedure:

First stage: Data transformation
The data should be transformed using Eq. 18.14.
Second stage: Model Building and Fitting
Step 2.1—Define which variable was the basic regressor (in this case, the "easy to measure" variable was soil organic carbon in its filtered version) and which one was the response or dependent variable (in this case, the "expensive to measure" variable was total soil nitrogen in its original version). The lagged version of the dependent variable can eventually be used as a second regressor.
Step 2.2—Once the model variables have been previously defined and prepared (transformation, etc.), then fit the space-varying coefficient regression model as a special dynamic model with regression components, using any implementation of the dynamic linear model (DLM), as for instance, the BATS system (**B**ayesian **A**nalysis of **T**ime **S**eries; West and Harrison 1997).

The alternative modeling approach to the soil data was the space-varying coefficient regression (state-space) model in which the state vector was formed by the dynamic regression coefficients β_i, as follows,

$$\text{Observation equation} : Y_i = F_i\beta_i + v_i, \quad v_i \sim N(0; V)$$

$$\text{State equation} : \beta_i = \beta_{i-1} + u_i, \quad u_i \sim N(0; W),$$

where $Y_i = \text{TSN}_i$ is the response variable, $F_i = (1, \text{TSN}_{i-1}, \text{SOC}_i)$ are the considered regressors, v_i and u_i are defined as before (with the exception that v_i is now a scalar), and the regression coefficient state vector follows a random-walk type of evolution. This nonstandard state-space model, known as dynamic regression model (which is a special case of the so-called dynamic linear model), was implemented as a DLM using the BATS system (Pole et al. 1994; West and Harrison 1997).

The available data consisted of two spatial series with 97 observations each: the total soil nitrogen series and the soil organic carbon series (Fig. 18.37). The series do not present any exceptional or extreme points such as outliers or likewise (Fig. 18.37A, B). For both series along the transect points, there is a detectable soil heterogeneity previously assumed (assumption **1**), since the process level and the process variability are not totally stable through the space. Therefore, the

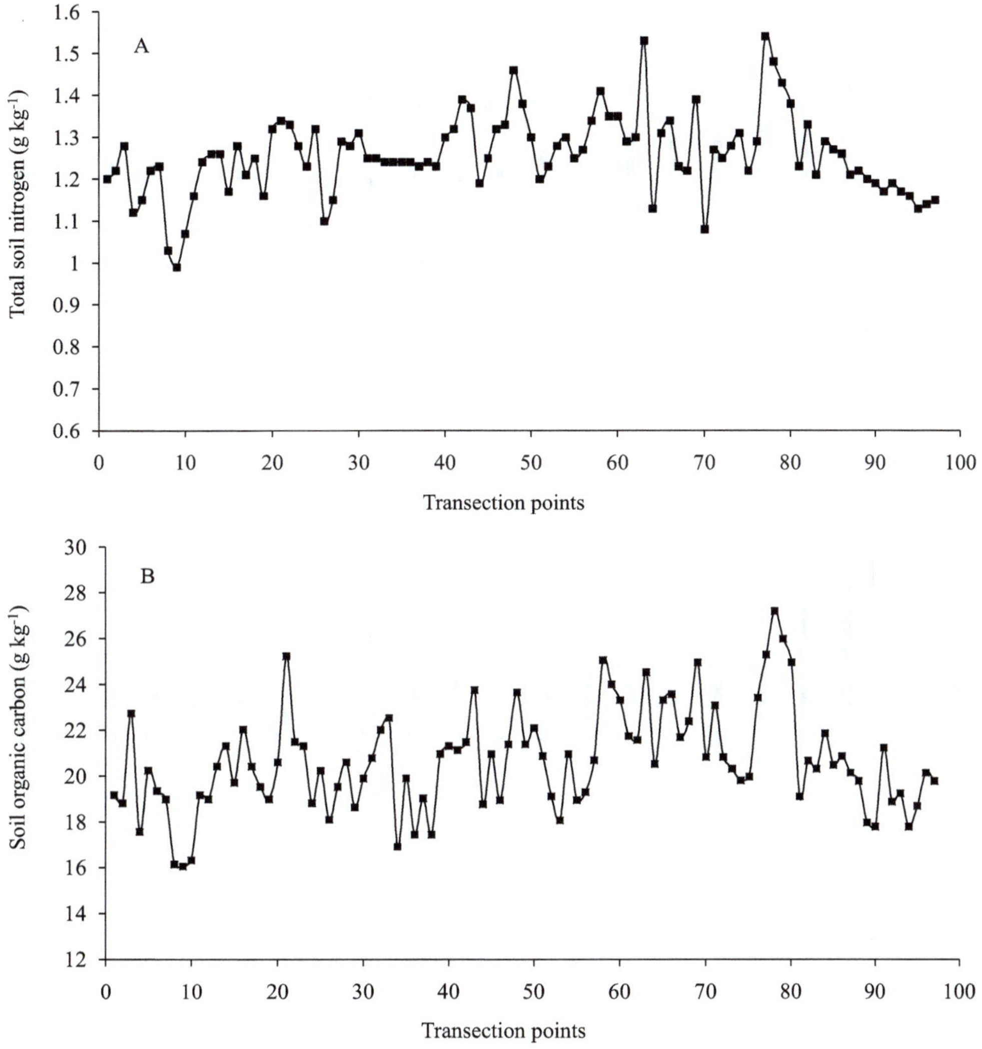

Fig. 18.37 Spatial data series: (**A**) total soil nitrogen (TSN) and (**B**) soil organic carbon (SOC) along the 97 transect points

assumption **1** that the process presents local characteristics changing in space seems to be reasonable.

The data information was initially explored through an analysis of its correlation structure. The correlation structure between the two series [cross-correlation function (CCF), Eq. 18.1] is shown at Fig. 18.38C, in which the high value (near 0.80) for CCF at lag 0 suggests that the carbon series can be a good predictor for the nitrogen series. This is in accordance with our assumption **3** about using a variable easy to measure such as the soil organic carbon to predict a more expensive one such as the soil total nitrogen. Also, the significant values for the autocorrelation function (ACF, Eq. 17.19a) at the first three lags for both series (Fig. 18.38A, B) show the presence of spatial correlation and an autoregressive (AR) structure for observations distant until 6 m in this study. This data can be analyzed through a state-space model, since it is based on a vector AR structure.

The dynamic (space-varying) regression model for the nitrogen series is here implemented

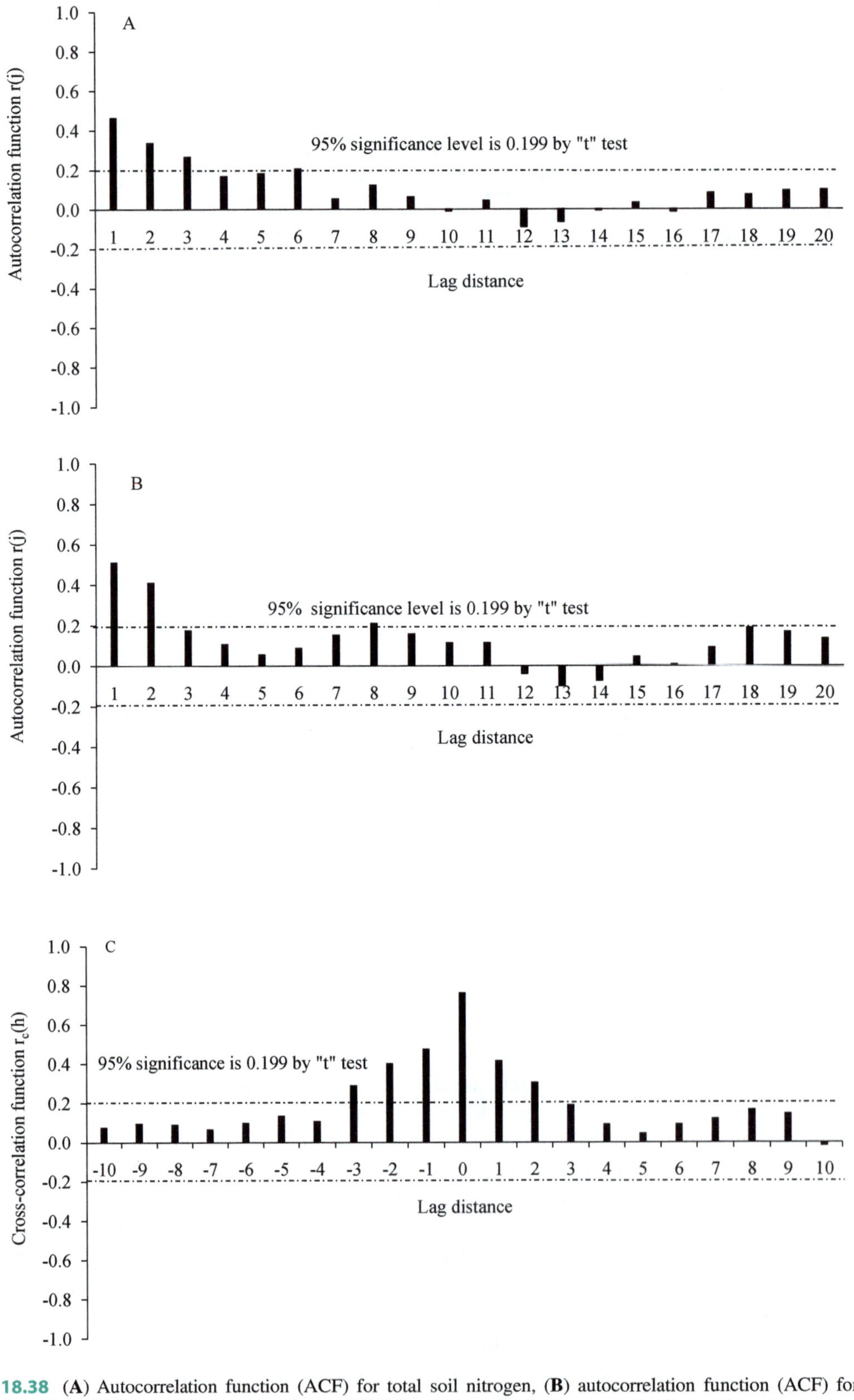

Fig. 18.38 (**A**) Autocorrelation function (ACF) for total soil nitrogen, (**B**) autocorrelation function (ACF) for soil organic carbon (SOC), and (**C**) cross-correlation (CCF) between total soil nitrogen and soil organic carbon

in two different versions: in the first one (called Model I), the regressors are the lagged nitrogen and lagged carbon; and in the second one (called Model II), the regressors are carbon and lagged nitrogen. Since soil carbon is more correlated to soil nitrogen than lagged carbon (as shown in Fig. 18.38C), we expect that Model II will fit better to the data than Model I.

The space-varying regression model has been fitted to the nitrogen-carbon data, and some parameter estimation (point and interval) and goodness of fit results are obtained. For the Model I, R^2 measure is 0.943, and for the proposed Model II is 0.997, that is, the proposed method (Model II) has better-fitting performance when compared to Model I. There is an advantage of qualitative nature for the proposed method since the data present local characteristics (level and variability) that change along the transect points. It is not surprising that the soil nitrogen-carbon model parameter estimates change in space (Fig. 18.39). In this figure, the interval parameter estimates are also presented for each point, giving more complete information on the soil carbon-nitrogen relationship.

The model II for the nitrogen series with its two standard deviation confidence intervals is presented in Fig. 18.40, together with the original data, showing clearly the good-fitting performance, with all the data points inside the confidence intervals.

The fact that the alternative state-space approach provides very good-fitting performance (R^2 coefficient greater than 0.99) is not surprising since it has a local characteristic, having the state variable adapted to the data at each point via Kalman Filter. On the other hand, it is very different as compared to the standard state-space approach presented by Shumway (1988) and Shumway and Stoffer (2000, 2011, 2017) regarding the extraction of qualitative information from the data, where the two main differences are related to parameter definition and interpretation. As explained here, the key parameters for the standard state-space are the autoregressive coefficients (with the mentioned shortcomings), and the key quantities for the alternative state-space approach are the space-varying soil carbon-nitrogen relationship expressing the soil spatial heterogeneity along the transect.

This above example is more completely reported in Timm et al. (2003b). An earlier example of the application of the West and Harrison's state-space approach can be found in Timm et al. (2000).

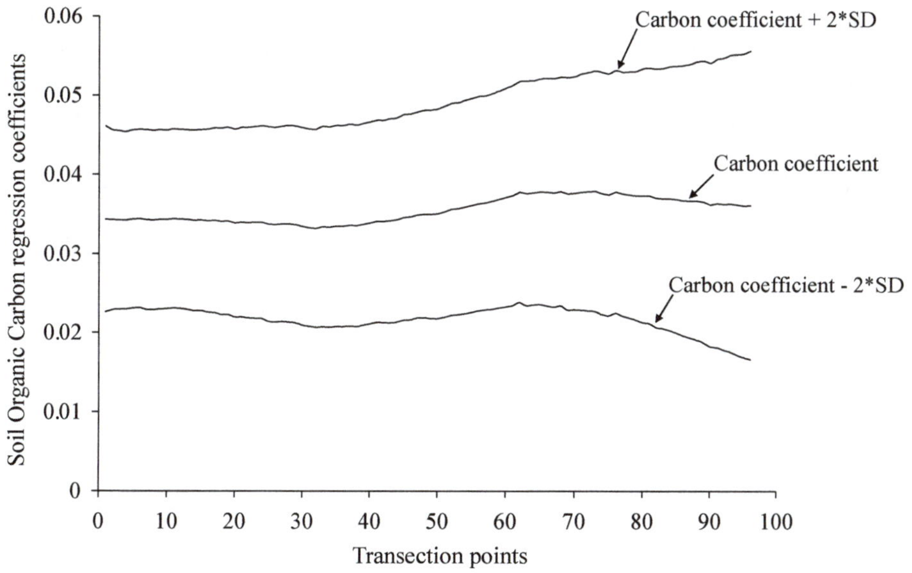

Fig. 18.39 Space-varying parameter estimates (soil organic carbon coefficient)

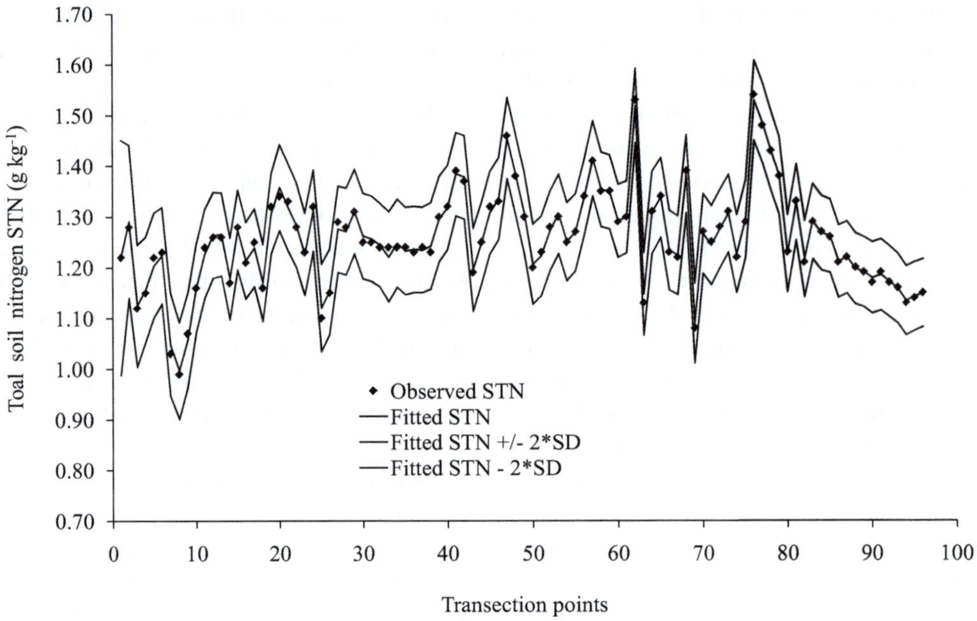

Fig. 18.40 Observed total soil nitrogen series and its model fitting with 2 × standard deviation (SD) confidence interval from Model II

18.9.2 Application of Both State-Space Approaches to Forecast Space Data Series

As mentioned by Tukey (1980), one of the basic objectives for analyzing a time series is to evaluate the ability of the model in terms of forecast. Based on this and on the fact that there were no published manuscripts in the literature to evaluate the potential of both state-space approaches to forecast spatial series in the agronomic area, Timm et al. (2006b) evaluated the potential of both state-space approaches to forecast total soil nitrogen (TSN) from soil organic carbon (SOC) collected along a spatial transect, to compare their performances between themselves, and to compare both state-space approaches with standard uni- and multivariate regression models used as reference. Artificial neural networks [both feed-forward (Fig. 17.13) and recurrent neural (Fig. 17.14) networks] were also used in this comparison. The prediction performance of all models was evaluated in terms of the distance between the observed and predicted values of total soil nitrogen (TSN), and the statistical measures considered were the Mean Square Error (MSE) and the Mean Absolute Percentage Error (MAPE). More details about the theoretical aspects of each class of model can be found in Timm et al. (2006b).

The TSN and SOC data sets used in Timm et al. (2006b) come from Timm et al. (2003b) and consisted of the same two spatial series with 97 observations of each variable (Fig. 18.37). The spatial structure of each variable was evaluated using the autocorrelation function (Fig. 18.38A, B), and the spatial dependency between the TSN and SOC data sets was calculated by the cross-correlation function (Fig. 18.38C). Figure 18.41 shows a linear relation (R^2 coefficient of 0.58) between the TSN and SOC collected at the same spatial point of the transect. Additionally, the analysis of Fig. 18.41 looks promising for the use of neural network models, suggested by a possible nonlinear relation between the TSN and SOC. From this figure, it can be seen that the slope in the extremes of the dispersion diagram is higher than the slope in the middle, and

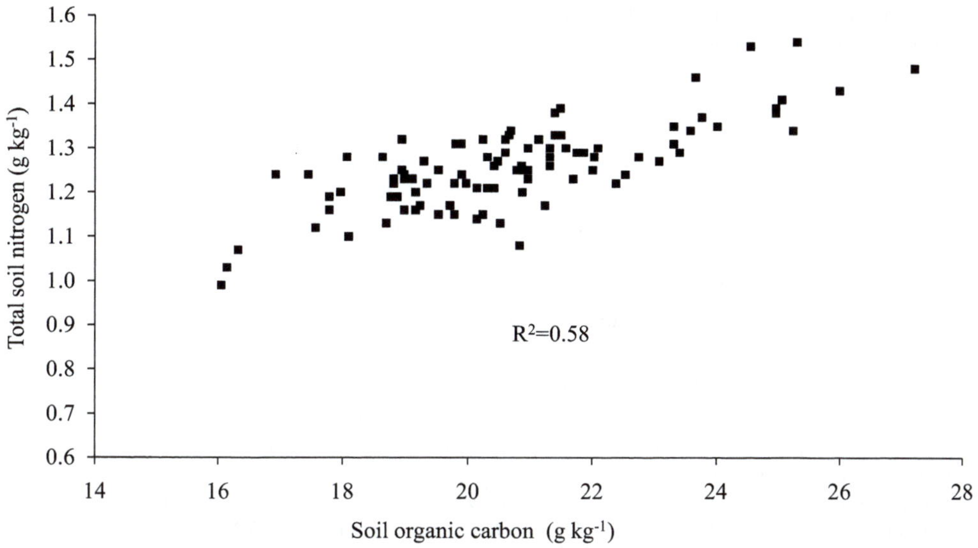

Fig. 18.41 Dispersion diagram between the total soil nitrogen (TSN) and soil organic carbon (SOC) data set showing a strong linear dependence between the variables

Table 18.4 Predictive performance (ten last transect points) of standard regression, of state-space, and of neural network models for total soil nitrogen (TSN)

Prediction models			Statistical measures	
			MSE	MAPE
Without latent variable	Scalar regression	Standard linear	0.00388	0.04301
		AR (1) error	0.00389	0.04279
	Vector autoregression	Standard VAR	0.00713	0.06390
		Corrected VAR	0.00350	0.03905
	Nonparametric regression	GAM/splines	0.00435	0.04359
		GAM/lowess	0.00361	0.04084
With latent variable	Artificial neural networks	Feed-forward	0.00313	0.03727
		Recurrent	0.00279	0.03599
	State-space models	Standard	0.00096	0.02302
		Dynamic	0.00288	0.03960

AR (1) error = regression model with first-order autoregressive error; Standard VAR = Standard Vector Autoregressive model; Corrected VAR = Corrected Vector Autoregressive model; GAM = Generalized Additive Model; Standard = Standard state-space model (Shumway 1988; Shumway and Stoffer 2000, 2011, 2017); Dynamic = Dynamic state-space model (West and Harrison 1989, 1997); MSE = Mean Square Error; and MAPE = Mean Absolute Percentage Error

therefore, the relation can be expressed as a linear segment, to represent the global nonlinear relation. Therefore, the analysis of Figs. 18.37, 18.38, and 18.41 suggests that the total soil nitrogen at each point could be reasonably predicted by the soil organic carbon at the same spatial point and by the total soil nitrogen at the nearest neighbor.

Timm et al. (2006b) evaluated the prediction performance of all adjusted models in two versions: (a) for the first, the last ten transect points of TSN were omitted in order to make their prediction (Table 18.4), and (b) for the second, the first ten points of TSN were omitted with the same objective (Table 18.5). As already

Table 18.5 Predictive performance (ten first transect points) of standard regression, of state-space, and of neural network models for total soil nitrogen (TSN)

Prediction models			Statistical measures	
			MSE	MAPE
Without latent variable	Scalar regression	Standard linear	0.00483	0.04665
		AR (1) error	0.00475	0.04601
	Vector autoregression	Standard VAR	0.00713	0.06390
		Corrected VAR	0.00423	0.04358
	Nonparametric regression	GAM/splines	0.00793	0.05583
With latent variable	Artificial neural networks	Feed-forward	0.00344	0.03898
		Recurrent	0.00213	0.02827
	State-space models	Standard	0.00314	0.04192
		Dynamic	0.00407	0.04589

AR (1) error = regression model with first-order autoregressive error; Standard VAR = Standard Vector Autoregressive model; Corrected VAR = Corrected Vector Autoregressive model; GAM = Generalized Additive Model; Standard = Standard state-space model (Shumway 1988; Shumway and Stoffer 2000, 2011); Dynamic = Dynamic state-space model (West and Harrison 1989, 1997); MSE = Mean Square Error; and MAPE = Mean Absolute Percentage Error

mentioned, the statistical measures considered for comparisons between models were the MSE and the MAPE.

Table 18.4 shows that among the models without latent variables, that is, among the true regression models, the original VAR model gives the worst results (independent of the statistical measure considered, MSE = 0.00713 and MAPE = 0.0639). In this case, unlike other models, it uses the lagged SOC as a regressor variable and not the SOC value at the same point, which has a stronger linear relation with TSN as shown in the Figs. 18.38C and 18.41. The corrected VAR shows the best results among the regression models, for which the minimum values of MSE (=0.00350) and MAPE (=0.0395) were found. This model has the SOC as a more appropriate predictor measured at the same point in space. This is consistent because the model, although being a global one, is presented as a bidimensional system. It is composed of two equations that treat the dynamics of the relation between TSN and SOC in the soil in a more adequate way, that is, there is no a hierarchy between variables. They are both treated in the same way, considered as random variables. The standard linear regression model (scalar model) is, also, a global model. It, however, is presented as a unidimensional system with a hierarchical treatment between TSN and SOC variables (only the variable TSN is considered a random variable). Therefore, both statistical performance measures (MSE = 0.00388 and MAPE = 0.04031) gave higher values as compared to the corrected VAR model. The GAM model (in its best performance version, i.e., the lowess) presented an MSE of 0.00361, slightly higher than the corrected VAR (MSE of 0.00350), however similar values as compared in terms of the MAPE measure (=0.04084). Such performance of this model is due to the fact that it incorporates nonlinear (Fig. 18.41) and local characteristics, that is, it does not ignore the spatial correlation structure between the TSN and SOC variables along the transect, taking into consideration the information the variables carry in function of their neighborhood.

Among the models with latent variables, the standard state-space model (linear and local characteristics), independent of the statistical measure considered, presented the best prediction performance for the ten last transect points (MSE = 0.00096 and MAPE = 0.02302). It was followed by the nonlinear recurrent neural network model (MSE = 0.00279 and MAPE = 0.03599). Such performance can be due to the fact that the state-space model expresses the local linear behavior of the ten last

TSN values of the transect (Fig. 18.37A). The recurrent neural network takes into account the local characteristic of the spatial dependence structure between TSN and SOC data (Fig. 18.38C) by data feedback expressing in this way the point-to-point spatial variability of TSN. On the other hand, the feed-forward neural network is a nonlinear and global model, which does not express this point-to-point fluctuation (Fig. 18.37A). In the same way, the static linear regression models, intensively used in agronomy, are global models whose regression coefficients are mean values that do not change along space, therefore not expressing the point-to-point fluctuations of the variable under study. This can lead to misunderstandings that might induce inadequate procedures of soil management (Nielsen and Alemi 1989). Beyond this, the response of the variable is not unique along the experimental transect, frequently yielding low coefficients of determination when compared to the dynamic models, as shown by Timm et al. (2003a, 2004).

In Table 18.5, we considered only one version of the GAM models (GAM/splines) because the other version (GAM/lowess, implemented by the SAS software) has a restriction with respect to the regression value for prediction (it must be inside the interval of used data), which is not satisfied for these particular data sets. Analyzing Table 18.5, the best predictive performance of the corrected VAR model (MSE = 0.00423 and MAPE = 0.04358) can also be seen for the ten first TSN values of the spatial transect in relation to the other regression models without latent variables. The best performance, however, to predict TSN among the latent variable models (and among all models) was given by the recurrent neural network (MSE = 0.00213 and MAPE = 0.02827).

Tables 18.4 and 18.5 also indicate that the use of dynamic linear models (state-space models), which take into account the local spatial dependence structure, as well as the feed-forward (nonlinear and global characteristics) and recurrent (nonlinear and local characteristics) neural networks gives the best TSN predictions of the ten last and ten first TSN values of the spatial transect, that is, the statistical measured values of MSE and MAPE were lower when compared to the regression models (without latent variables) considered as standard methods for comparison studies. Both state-space and neural network models have, in their essence, the philosophy of the use of state variables that are not observed directly during the different processes that occur simultaneously in the complex soil-plant-atmosphere system. They, however, belong to the used algorithms for practical implementation of these models.

Dynamic regression models, represented in the form of state-space, are relatively recent tools and have not been used frequently to quantify the relations of the soil-plant-atmosphere system. Although originally introduced in the 1960s, only a practical implementation approach was introduced in the 1980s (West and Harrison 1989, 1997). As local adjustment models, it becomes possible to more accurately estimate the regression coefficients at each sampling point, which alleviates the problem of spatial variability found in precision agriculture. The local adjustment models provide an opportunity to describe the spatial or temporal association between different measured variables in the field and may, therefore, be better suited to the study of the complex relations of the soil-plant-atmosphere system. On the other hand, the commonly used static regression models are global fit models, that is, their regression coefficients are mean values that do not vary over time or space. In addition, the response of a variable is not unique across an area, and the application of static multiple regression models frequently gives low values of the coefficient of determination. Because these types of analyses provide global estimates of regression coefficients, they do not represent the point-to-point variations of the variables and may induce inadequate management procedures. However, the dynamic regression models, represented in the form of state-space, are more adequate for the study of soil-plant relationships, especially since they take into account the local spatial and temporal character of the agronomic processes. Finally, we can say that in this chapter, we presented two types of representation in state-space models. One gives emphasis on the state equation (Shumway 1988 and Shumway and Stoffer 2000, 2011, 2017) and the other the emphasis on the equation of observations (West and Harrison 1989, 1997). Both are presented as a dynamic system composed of two equations, differing only in the way in which they are implemented in practice. In the first, the system is treated as a multidimensional spatial process in which there is no hierarchy between the variables, that is, all variables are treated in the same way. In the second, the process is treated as one-dimensional, giving a hierarchical treatment of variables, that is, variables are treated differently from the point of view of practical implementation. Plant and soil attributes were sampled at different positions and times and correlated with each other. In nature, the processes can be physical, chemical, and/or biological, such as infiltration, absorption of nutrients by the plant, etc. These types of models are also valuable when deterministic equations based on a physical process are used in combination with observations that are correlated in time or space as long as the equation is transformed into a state-space formulation and its parameters are theoretically well known (see, e.g., Wendroth et al. 1993 and Katul et al. 1993). Hence, the state-space model can provide the uncertainties associated with observations due, for example, to soil heterogeneity, instrument calibration, etc.

The representation of state-space models in the agronomic area has been recent and indicates that they are potentially valuable tools to study the relationship of several variables involving the soil-plant-atmosphere system. One case is its application in the development of so-called precision agriculture. This modern concept of agriculture is characterized by the differentiated application of agricultural practices (such as fertilization, liming, plowing depth, application of pesticides) in each plot of a crop field, meeting its specific requirements. This type of agriculture requires sophisticated instrumentation, controlled by computers and guided by GPS. The tractor and agricultural implements, when traveling during a cultivation or soil preparation operation, follow spatial instructions of speed and intensity, controlled by the PC that is based on maps of the various physical, chemical, biological, and agronomic properties of the area.

18.10 Exercises

18.1. In an area cultivated with sugarcane on a Nitrosol, four spatial series were collected in a space transect of 84 points on a given day: (a) soil water content (θ) ($m^3\ m^{-3}$), (b) soil organic matter (OM) (kg m^{-3}), (c) clay content (CC) (g kg^{-1}), and (d) aggregate stability (AS) (mm). The measured data are shown in the table below, which is part of what was seen in this chapter. We ask the following:

(a) Make a qualitative analysis of each series regarding the presence of trend and outliers.
(b) Make the classic statistical analysis of each series, determining mean, standard deviation, variance, and coefficient of variation.
(c) Check the normality of the data of each series by means of histograms and normal graph plot. Is there a need for data transformation?
(d) Calculate the autocorrelograms and semivariograms of each series, verifying if there are similarities between these tools and spatial dependence between adjacent observations of each series.
(e) Calculate the cross-correlograms between all four series. Is there a cross-correlation between variables?
(f) On the basis of what has been theoretically set out in this chapter, is it possible to apply the state-space methodology to study the relationship between these series? In affirmative, between which series?

Points	θ	OM	CC	AS	Points	θ	OM	CC	AS
1	0.144	33	430	2.19	43	0.265	27	550	2.18
2	0.132	32	430	1.91	44	0.275	27	550	1.65
3	0.151	32	460	1.50	45	0.290	28	570	2.78
4	0.157	29	470	2.11	46	0.275	29	580	2.70
5	0.176	28	470	1.79	47	0.294	26	600	2.43
6	0.170	29	450	1.84	48	0.320	30	580	2.52
7	0.181	31	470	2.67	49	0.326	26	620	2.66
8	0.174	29	450	2.32	50	0.295	26	580	2.48
9	0.162	29	490	2.49	51	0.305	26	560	2.47
10	0.174	27	510	2.23	52	0.282	27	570	2.43
11	0.200	28	520	2.27	53	0.241	25	560	2.59
12	0.175	27	490	1.63	54	0.239	27	580	2.48
13	0.201	27	440	1.74	55	0.217	25	560	2.46
14	0.246	26	470	2.08	56	0.271	27	580	2.27
15	0.217	25	490	1.58	57	0.214	26	600	2.42
16	0.236	27	530	1.64	58	0.248	25	640	2.41
17	0.203	27	510	2.13	59	0.287	24	600	2.10
18	0.212	26	510	2.37	60	0.251	25	580	2.18
19	0.236	28	490	2.14	61	0.243	26	580	2.37
20	0.213	28	490	2.00	62	0.264	24	560	1.73
21	0.209	28	490	2.01	63	0.247	25	560	1.63
22	0.209	29	490	1.82	64	0.247	26	540	2.44
23	0.209	26	510	2.32	65	0.284	25	580	2.04
24	0.243	29	490	1.83	66	0.262	27	550	2.50
25	0.226	30	490	1.71	67	0.266	25	560	2.17
26	0.225	27	490	1.77	68	0.254	24	560	2.19
27	0.247	29	510	1.85	69	0.281	26	560	1.91
28	0.230	28	540	2.26	70	0.254	27	600	1.94
29	0.226	30	490	2.53	71	0.241	27	560	1.88
30	0.255	28	490	2.62	72	0.252	28	560	1.90
31	0.210	29	490	2.71	73	0.244	24	580	1.96
32	0.241	29	490	2.52	74	0.237	24	580	2.09
33	0.201	29	550	2.24	75	0.248	23	540	2.09
34	0.297	28	550	2.29	76	0.246	25	580	1.70
35	0.270	30	540	2.57	77	0.277	25	540	2.09
36	0.245	28	510	2.16	78	0.270	25	580	2.13
37	0.232	29	510	2.17	79	0.283	24	520	1.70
38	0.317	28	550	2.56	80	0.280	25	580	2.16
39	0.260	29	530	2.68	81	0.262	24	560	2.27
40	0.279	26	530	2.50	82	0.272	24	580	2.23
41	0.269	27	550	2.20	83	0.263	23	560	2.06
42	0.255	27	570	2.51	84	0.226	22	560	2.02

18.11 Answers

18.1. The answers of this exercise can be found in Timm et al. (2004).

Timm LC, Reichardt K, Oliveira JCM et al. (2004) State-space approach to evaluate the relation between soil physical and chemical properties. Braz J Soil Sci 28:49–58.

References

Addison PS (2017) The illustrated wavelet transform handbook: introduction, theory and applications in science, engineering, medicine and finance, 2nd edn. CRC Press, Boca Raton, FL

Alemi MH, Shahriari MR, Nielsen DR (1988) Kriging and cokriging of soil water properties. Soil Technol 1:117–132

Ameen JRM, Harrison PJ (1984) Discount weighted estimation. J Forec 3:285–296

Aquino LS, Timm LC, Reichardt K, Barbosa EP, Parfitt JMB, Nebel ALC, Penning LH (2015) State-space approach to evaluate effects of land levelling on the spatial relationships of soil properties of a lowland area. Soil Tillage Res 145:135–147

Awe GO, Reichert JM, Timm LC, Wendroth O (2015) Temporal processes of soil water status in a sugarcane field under residue management. Plant and Soil 387:395–411

Bazza M, Shumway RH, Nielsen DR (1988) Two-dimensional spectral analyses of soil surface temperature. Hilgardia 56:1–28

Beskow S, Timm LC, Tavares VEQ, Caldeira TL, Aquino LS (2016) Potential of the LASH model for water resources management in data-scarce basins: a case study of the Fragata river basin, southern Brazil. Hydrol Sci J 61:2567–2578

Biswas A (2018) Scale–location specific soil spatial variability: a comparison of continuous wavelet transform and Hilbert–Huang transform. Catena 160:24–31

Biswas A, Si BC (2011) Application of continuous wavelet transform in examining soil spatial variation: a review. Math Geosci 43:379–396

Bremner JM (1960) Determination of nitrogen in soil by the Kjeldahl method. J Agric Sci 55:11–33

Brillinger DR (2001) Time series: data analysis and theory. Society for Industrial and Applied Mathematics, Philadelphia, PA

Chatfield C (2004) The analysis of time series: an introduction, 6th edn. Chapman & Hall/CRC, Boca Raton, FL

Dempster AP, Laird NM, Rubin DB (1977) Maximum likelihood from incomplete data via the EM algorithm. J R Statist Soc Ser B 39:1–38

Deutsch CV, Journel AG (1992) GSLIB. Geostatistical software library and user's guide. Oxford University Press, New York, NY

Dourado-Neto D, Timm LC, Oliveira JCM, Reichardt K, Bacchi OOS, Tominaga TT, Cassaro FAM (1999) State-space approach for the analysis of soil water content and temperature in a sugarcane crop. Sci Agric 56:1215–1221

Farge M (1992) Wavelet transforms and their applications to turbulence. Annu Rev Fluid Mech 24:395–457

Flandrin P (2018) Explorations in time-frequency analysis. Cambridge University Press, Padstow Cornwall

Gelb A (1974) Applied optimal estimation. Massachusetts Institute of Technology Press, Cambridge, MA

Graps A (1995) An introduction to wavelets. IEEE Comp Sci Eng 2:50–61

Grinsted A, Moore JC, Jevrejeva S (2004) Application of cross wavelet transform and wavelet coherence to geophysical time series. Nonlinear Processes Geophys 11:561–566

Harrison PJ, Stevens CF (1976) Bayesian forecasting (with discussion). J R Statist Soc Ser B 38:205–267

Hu W, Si BC (2013) Soil water prediction based on its scale-specific control using multivariate empirical mode decomposition. Geoderma 193-194:180–188

Hu W, Si BC (2016a) Multiple wavelet coherence for untangling scale-specific and localized multivariate relationships in geosciences. Hydrol Earth Syst Sci 20:3183–3191

Hu W, Si BC (2016b) Multiple wavelet coherence for untangling scale-specific and localized multivariate relationships in geosciences. Suppl Hydrol Earth Syst Sci 20:3183–3191

Huang NE, Shen Z, Long SR, Wu MLC, Shih HH, Zheng QN, Yen NC, Tung CC, Liu HH (1998) The empirical mode decomposition and the Hilbert spectrum for nonlinear and non-stationary time series analysis. Proc R Soc Lond Ser AA 454:903–995

Hubbard BB (1998) The world according to wavelets: the story of a mathematical technique in the making, 2nd edn. A K Peters Ltd, Natick, MA

Hui S, Wendroth O, Parlange MB, Nielsen DR (1998) Soil variability – infiltration relationships of agroecosystems. J Balkan Ecol 1:21–40

James JF (2011) A student's guide to Fourier transforms with applications in physics and engineering, 3rd edn. Cambridge University Press, Cambridge

Kalman RE (1960) A new approach to linear filtering and prediction theory. Trans ASME J Basic Eng 8:35–45

Katul GG, Wendroth O, Parlange MB, Puente CE, Folegatti MV, Nielsen DR (1993) Estimation of in situ hydraulic conductivity function from nonlinear filtering theory. Water Resour Res 29:1063–1070

Liu ZP, Shao MA, Wang YQ (2012) Estimating soil organic carbon across a large-scale region: a state-space modeling approach. Soil Sci 177:607–618

McGraw T (1994) Soil test level variability in Southern Minnesota. Better Crops Pot Phosp Inst 78:24–25

Nielsen DR, Alemi MH (1989) Statistical opportunities for analyzing spatial and temporal heterogeneity of field soils. Plant and Soil 115:285–296

Nielsen DR, Wendroth O (2003) Spatial and temporal statistics – sampling field soils and their vegetation. Catena Verlag, Cremlingen-Desdedt

Ogunwole JO, Timm LC, Ugwu-Obidike EO, Gabriels DM (2014a) State-space estimation of soil organic carbon stock. Int Agroph 28:185–194

Ogunwole JO, Obidike EO, Timm LC, Odunze AC, Gabriels DM (2014b) Assessment of spatial distribution of selected soil properties using geospatial statistical tools. Commun Soil Sci Plant Anal 45:2182–2200

Plackett RL (1950) Some theorems in least squares. Biometrika 37:149–157

Pole A, West M, Harrison J (1994) Applied Bayesian forecasting and time series analysis. Chapman & Hall, London

Rehman N, Mandic DP (2009) Empirical mode decomposition. Matlab code and data. http://www.commsp.ee.ic.ac.uk/~mandic/research/emd.htm.

Rehman N, Mandic DP (2010) Multivariate empirical mode decomposition. Proc R Soc A 466:1291–1302

She DL, Liu DD, Peng SZ, Shao MA (2013) Multiscale influences of soil properties on soil water content distribution in a watershed on the Chinese Loess Plateau. Soil Sci 178:530–539

She DL, Tang SQ, Shao MA, Yu SE, Xia YQ (2014a) Characterizing scale specific depth persistence of soil water content along two landscape transects. J Hydrol 519:1149–1161

She DL, Xuemei G, Jingru S, Timm LC, Hu W (2014b) Soil organic carbon estimation with topographic properties in artificial grassland using a state-space modeling approach. Can J Soil Sci 94:503–514

She DL, Zheng JX, Shao MA, Timm LC, Xia YQ (2015) Multivariate empirical mode decomposition derived multi-scale spatial relationships between saturated hydraulic conductivity and basic soil properties. Clean Soil Air Water 43:910–918

She DL, Fei YH, Chen Q, Timm LC (2016) Spatial scaling of soil salinity indices along a temporal coastal reclamation area transect in China using wavelet analysis. Arch Agron Soil Sci 62:1625–1639

She DL, Qiana C, Timm LC, Beskow S, Hu W, Caldeira TL, Oliveira LM (2017) Multi-scale correlations between soil hydraulic properties and associated factors along a Brazilian watershed transect. Geoderma 286:15–24

Shumway RH (1988) Applied statistical time series analyses. Prentice Hall, Englewood Cliffs, NJ

Shumway RH, Stoffer DS (1982) An approach to time series smoothing and forecasting using the EM algorithm. J Time Ser Anal 3:253–264

Shumway RH, Stoffer DS (2000) Time series analysis and its applications. Springer, New York, NY

Shumway RH, Stoffer DS (2011) Time series analysis and its applications with R examples, 3rd edn. Springer, New York, NY

Shumway RH, Stoffer DS (2017) Time series analysis and its applications with R examples, 4th edn. Springer, New York, NY

Shumway RH, Biggar JW, Morkoc F, Bazza M, Nielsen DR (1989) Time-and frequency-domain analyses of field observations. Soil Sci 147:286–298

Si BC (2008) Spatial scaling analyses of soil physical properties: a review of spectral and wavelet methods. Vadose Zone J 7:547–562

Si BC, Zeleke TB (2005) Wavelet coherency analysis to relate saturated hydraulic properties to soil physical properties. Water Resour Res 41:W11424

Stevenson FC, Knight JD, Wendroth O, Van Kessel C, Nielsen DR (2001) A comparison of two methods to predict the landscape-scale variation of crop yield. Soil Tillage Res 58:163–181

Timm LC, Fante Júnior L, Barbosa EP, Reichardt K, Bacchi OOS (2000) Interação solo-planta avaliada por modelagem estatística de espaço de estados. Sci Agric 57:751–760

Timm LC, Reichardt K, Oliveira JCM, Cassaro FAM, Tominaga TT, Bacchi OOS, Dourado-Neto D, Nielsen DR (2001) State-space approach to evaluate the relation between soil physical and chemical properties. Czech Society of Soil Science and Soil Science Society of America, Prague

Timm LC, Reichardt K, Oliveira JCM, Cassaro FAM, Tominaga TT, Bacchi OOS, Dourado-Neto D (2003a) Sugarcane production evaluated by the state–space approach. J Hydrol 272:226–237

Timm LC, Barbosa EP, Souza MD, Dynia JF, Reichardt K (2003b) State-space analysis of soil data: an approach based on space-varying regression models. Sci Agric 60:371–376

Timm LC, Reichardt K, Oliveira JCM, Cassaro FAM, Tominaga TT, Bacchi OOS, Dourado-Neto D (2003c) State-space approach for evaluating the soil–plant–atmosphere system. In: Achyuthan H (ed) Soil and soil physics in continental environment. Allied Publishers Private Limited, Chennai, pp 23–81

Timm LC, Reichardt K, Oliveira JCM, Cassaro FAM, Tominaga TT, Bacchi OOS, Dourado-Neto D, Nielsen DR (2004) State-space approach to evaluate the relation between soil physical and chemical properties. Braz J Soil Sci 28:49–58

Timm LC, Pires LF, Roveratti R, Arthur RCJ, Reichardt K, Oliveira JCM, Bacchi OOS (2006a) Field spatial and temporal patterns of soil water content and bulk density changes. Sci Agric 63:55–64

Timm LC, Gomes DT, Barbosa EP, Reichardt K, Souza MD, Dynia JF (2006b) Neural network and state-space models for studying relationships among soil properties. Sci Agric 63:386–395

Timm LC, Dourado-Neto D, Bacchi OOS, Hu W, Bortolotto RP, Silva AL, Bruno IP, Reichardt K

(2011) Temporal variability of soil water storage evaluated for a coffee field. Aust J Soil Res 49:77–86

Timm LC, Reichardt K, Lima CLR, Aquino LA, Penning LH, Dourado-Neto D (2014) State-space approach to understand Soil-Plant-Atmosphere relationships. In: Teixeira WG, Ceddia MB, Ottoni MV, Donnagema GK (eds) Application of soil physics in environmental analysis: measuring, modelling and data integration. Springer, New York, NY, pp 91–129

Torrence C, Compo GP (1998) A practical guide to wavelet analysis. Bull Am Meteorol Soc 79:61–78

Tukey JW (1980) Can we predict where 'Time Series' should go next? In: Brillinger DR, Tiao GC (eds) Directions in time series. Institute of Mathematical Statistics, Hayward, CA, pp 1–31

Walkey A, Black IA (1934) An examination of the Degtjareff method for determining soil organic matter and a proposed modifications of the chromic acid titration method. Soil Sci 37:29–38

Warrick AW, Nielsen DR (1980) Spatial variability of soil physical properties in the field. In: Hillel D (ed) Applications of soil physics. Academic Press, New York, NY, pp 319–344

Wendroth O (2013) Soil variability. In: Lazarovitch N, Warrick AW (eds) Exercises in soil physics. Catena Verlag, Reiskirchen, pp 292–332

Wendroth O, Al Omran AM, Kirda K, Reichardt K, Nielsen DR (1992) State-space approach to spatial variability of crop yield. Soil Sci Soc Am J 56:801–807

Wendroth O, Katul GG, Parlange MB, Puente CE, Nielsen DR (1993) A nonlinear filtering approach for determining hydraulic conductivity functions. Soil Sci 156:293–301

Wendroth O, Reynolds WD, Vieira SR, Reichardt K, Wirth S (1997) Statistical approaches to the analysis of soil quality data. In: Gregorich EG, Carter MR (eds) Soil quality for crop production and ecosystem health. Elsevier Science, Amsterdam, pp 247–276

Wendroth O, Jürschik P, Giebel A, Nielsen DR (1998) Spatial statistical analysis of on-site-crop yield and soil observations for site-specific management. In: International Conference on Precision Agriculture. American Society of Agronomy/Crop Science Society of America/Soil Science Society of America, Saint Paul, MN, pp 159–170

Wendroth O, Jürschik P, Kersebaum KC, Reuter H, Van Kessel C, Nielsen DR (2001) Identifying, understanding, and describing spatial processes in agricultural landscapes – four case studies. Soil Tillage Res 58:113–127

Wendroth O, Koszinski S, Vasquez V (2011) Soil spatial variability. In: Huang PM, Li Y, Sumner ME (eds) Handbook of soil science, 2nd edn. CRC Press, Boca Raton, FL, pp 10.1–10.25

Wendroth O, Yang Y, Timm LC (2014) State-space analysis in soil physics. In: Teixeira WG, Ceddia MB, Ottoni MV, Donnagema GK (eds) Application of soil physics in environmental analysis: measuring, modelling and data integration. Springer, New York, NY, pp 53–74

West M, Harrison J (1989) Bayesian forecasting and dynamic models, 1st edn. Springer, London

West M, Harrison J (1997) Bayesian forecasting and dynamic models, 2nd edn. Springer, London

Yang Y, Wendroth O (2014) State-space approach to field-scale bromide leaching. Geoderma 217-218:161–172

Yang Y, Wendroth O, Walton RJ (2013) Field-scale bromide leaching as affected by land use and rain characteristics. Soil Sci Soc Am J 77:1157–1167

Yang Y, Wendroth O, Walton RJ (2016) Temporal dynamics and stability of spatial soil matric potential in two land use systems. Vadose Zone J 15:1–15

Dimensional Analysis, Scaling, and Fractals

19

19.1 Introduction

Dimensional analysis refers to the study of **dimensions** that characterize physical quantities, such as mass, strength, or energy. It is a very important issue related to all chapters of this book seen up to now. That's why we left it as the last formal chapter. Classical Mechanics is based on three **fundamental quantities**, with MLT dimensions (mass M, length L, and time T). From the combination of these, derived quantities such as volume, velocity, and force arise, of dimensions L^3, LT^{-1}, and MLT^{-2}, respectively. In the other areas of Physics, four other fundamental dimensions are defined, among them the temperature θ (do not confuse this symbol with that of the soil water content used all along this book) and the electric current I.

To introduce and illustrate the subject of **dimensional analysis**, let's look at a classic example of the romantic literature, which deals with the sizes of objects. Dean Swift in the Adventures of Gulliver describes Lemuel Gulliver's imaginary voyages to the kingdoms of Lilliput and Brobdingnag. In these two places, life was perfectly identical to that of normal men (Gulliver), but its geometrical dimensions were different. In Lilliput, the men, the houses, the cattle, and the trees were 12 times smaller than in Gulliver's place; and in Brobdingnag, all was 12 times larger. The Lilliput man was a Gulliver geometric model on a 1:12 scale, and the Brobdingnag man was a 12:1 scale model (Fig. 19.1).

One can arrive at interesting observations about these two kingdoms, making a dimensional analysis. Long before Gulliver's adventures were written, Galileo had already stated that enlarged or reduced models of men could not be as we are. The human body consists of columns, rods, bones, and muscles. The weight of the body that the skeleton must sustain is proportional to its own volume, that is, proportional to L^3, whereas the resistance of a bone to compression or of a muscle to the traction is proportional to an area (cross section) L^2.

Let us compare Gulliver with the giant of Brobdingnag, which has each of its linear dimensions 12 times larger (Fig. 19.2). The resistance of his legs would be $12^2 = 144$ times greater than that of Gulliver; his weight $12^3 = 1728$ times greater. The giant's resistance/weight ratio would be 12 times smaller than that of Gulliver. To sustain his own weight, he would have to make an effort equivalent to what we would have to do to carry 11 men on our back.

Galileo dealt with these problems clearly and used arguments that refute the possibility of normal-looking giants. If we wanted to keep the

K. Reichardt, L. C. Timm, *Soil, Plant and Atmosphere*, https://doi.org/10.1007/978-3-030-19322-5_19

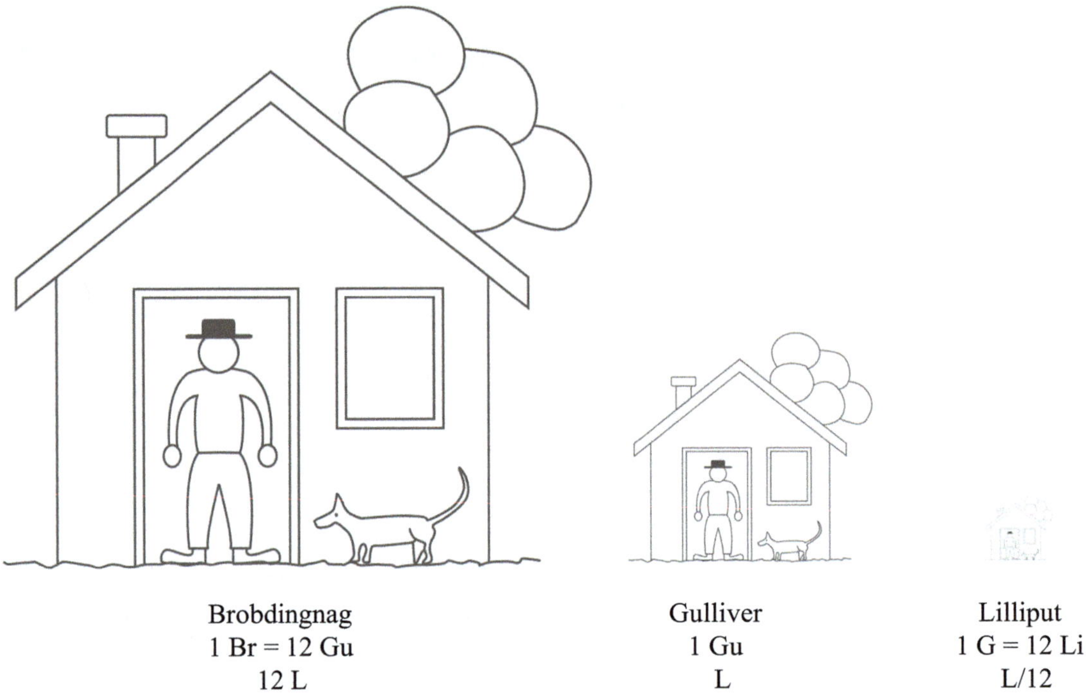

Fig. 19.1 Out of scale illustration of the imaginary voyages of Lemuel Gulliver to the kingdoms of Lilliput (12 times smaller than Gulliver) and Brobdingnag (12 times larger than Gulliver)

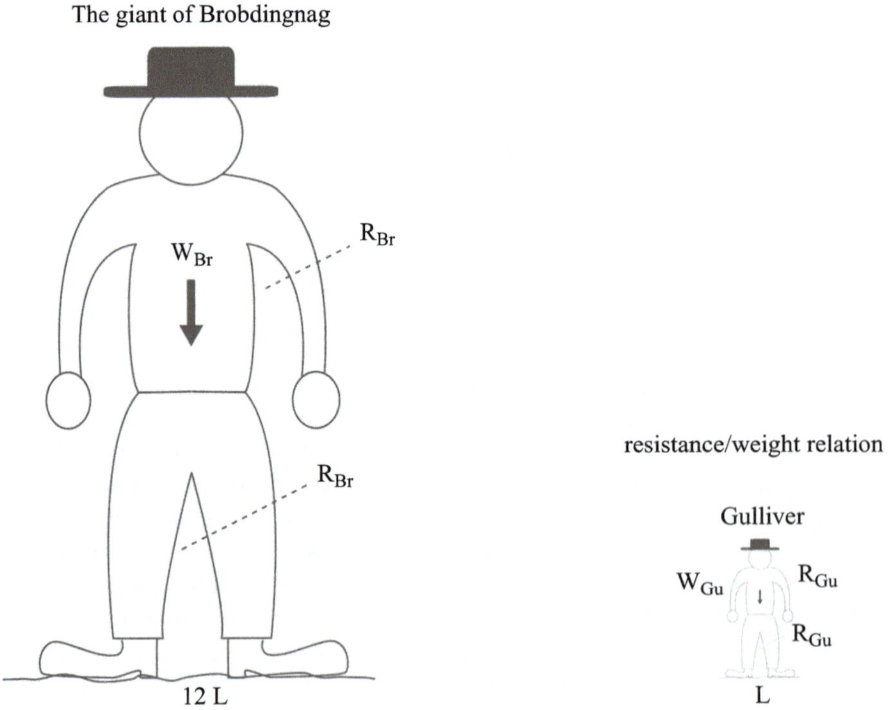

Fig. 19.2 The giant of Brobdingnag (Br), which has each of its linear dimensions 12 times larger than Gulliver (Gu). W stands for weight and R for resistance

same proportion of limbs in a giant as in a normal man, we would have to use a harder and stronger material to make the bones, or we would have to admit an increase in resistance compared to a man of normal stature. On the other hand, if the size of a body is decreased, its resistivity would not decrease in the same proportion; the smaller the body, the greater its relative strength. Thus, a puppy could probably carry on its back two or three puppies of its own size; an elephant could probably not carry even another elephant of its own size.

Let us now look at a problem of the Lilliputians (Fig. 19.3), 12 times smaller than Gulliver. The heat that a living body loses to the environment occurs mainly through the skin. This heat flow is proportional to the surface area covered by the skin, that is, to L^2, provided that the body temperature, the characteristics of the skin, etc. are maintained constant. This dissipated energy, like the energy expended in body movements, comes from the ingested food. Therefore, the minimum amount of food to be consumed by a body should be proportional to L^2. If a man like Gulliver could eat during a day, say, a chicken, a loaf of bread, and a fruit, a Lilliputian would need a volume of food $(1/12)^2$ times lower. But a chicken, a bread, and a fruit, reduced to the scale of its world, would have a volume $(1/12)^3$ times smaller. So to meet his energy needs, he would need a dozen chickens, a dozen loaves, and a dozen fruits a day and feel as well fed as Gulliver was feeding on one of each of them. The Lilliputians should be an irriquiet and hungry people. These qualities are found in many small mammals, such as rats. It is interesting to note that there are no warm-blooded animals much smaller than the rats, perhaps because according to the laws of scale here discussed, such animals would be obliged to ingest such a large quantity of food that it would become impossible to obtain it, or their digestion should be much faster.

From all that we have seen, it is important to note that although Brobdingnag and Lilliput are geometric models of our world, they could not be physical and biological models. To make the models viable, it would be necessary for the variables to be adjusted accordingly. In the case of Brobdingnag, for example, the giant might well

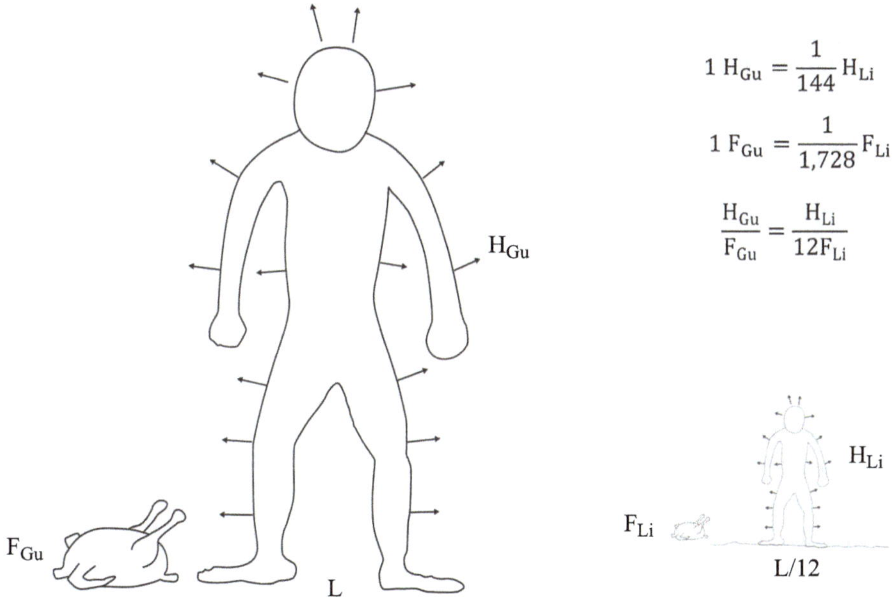

Fig. 19.3 The dwarf of Lilliput (Li), which has each of its linear dimensions 12 times smaller than Gulliver (Gu). F stands for food supply and H for heat exchange at the skin

support his weight even if he had the structure of humans and if he were living on a planet where the gravitational acceleration was $1/12 \times g$.

19.2 Physical Quantities and Dimensional Analysis

The parameters that characterize the physical phenomena are related by laws, in general, quantitative, in which they appear as measures of the **physical quantities**. The measurement of a quantity results from comparing it with another of the same type, called unity. Thus, a quantity (G) is given by two factors, where one is the ratio between the values of the quantities considered or measured (M) and the other is the unit (U). Thus, when we write $V = 50\ \mathrm{m}^3$, the ratio between the quantities is 50, the quantity considered is the volume V, and the unit is m^3. A quantity G can therefore be generalized by the expression:

$$G = M(G) \cdot \mathrm{U}(G)$$

(G) being the **measure** of G and U (G) is the **unit** of G. In addition, the **dimensional symbol** of G is a combination of fundamental quantities, symbolized by capital letters. Some examples, using the **fundamental quantities MLT** (mass, length, time) are shown below.

Quantity or measure	Dimensional symbol
Area	L^2
Speed or velocity	LT^{-1}
Force	MLT^{-2}
Pressure	$\mathrm{ML}^{-1}\mathrm{T}^{-2}$
Flow	$\mathrm{L}^3\mathrm{T}^{-1}$

Unit systems contain fundamental and derived units that are consistently established. The **International System of Units** (ISU) is a coherent system used officially around the world. The seven fundamental units of this system and their respective standards are:

(a) **Mass** (M), kilogram (kg), is the mass of the international prototype of the kilogram, constructed of irradiated platinum and stored in the International Bureau of Weights and Measures in Sèvres, France;

(b) **Length** (L), meter (m), is the length equal to 1,650,763.73 wavelengths of the radiation corresponding to the transition between levels $2p_{10}$ and $5d_5$ of the atom ^{86}Kr, in vacuum;

(c) **Time** (T), second (s), is the duration of 9,192,631,770 periods (travel time of one wavelength) of the radiation corresponding to the transition between the two hyperfine levels of the ground state of the ^{133}Cs;

(d) **Electric current** (I), ampere (A), is the intensity of a constant electric current maintained in two parallel, rectilinear conductors, of infinite length, and of negligible circular section and situated at a distance of 1 m between these conductors in the vacuum, a force equal to 2×10^{-7} Newton per meter of length;

(e) **Thermodynamic temperature** (θ), Kelvin (K), is the fraction 1/273.16 of the thermodynamic temperature of the triple point of water;

(f) **Luminous intensity** (Iv), candela (cd), is the luminous intensity, in the perpendicular direction, of a surface of 1/600,000 m^2 of a black body at a pressure of 101,325 N m^{-2};

(g) **Amount of matter** (N), mol (mol), is the amount of matter of a system containing as many elementary units as there are atoms in 0.012 kg of ^{12}C.

The physical quantities are quantities that relate to each other in such a way that the same types of relationships occur with the units of these quantities, since these are particular values of those quantities. Thus, for example, if we consider Newton's Second Law, we can write:

$$F = m \cdot a$$

where F is the force acting on a mass m of a particle, with the consequent acceleration a. For the units of these quantities, we can write that:

$$\mathrm{U}(F) = \mathrm{U}(m) \times \mathrm{U}(a)$$

where U(F), U(m), and U(a) are the units of force, mass, and acceleration, respectively.

The previous equation, which relates dimensional symbols, is dimensional, and the exponents of m and a, respectively, 1 and 1, define the size of the force in relation to mass and acceleration.

In a general way, if G is a quantity that has exponents, a with respect to X, b with respect to Y, c with respect to Z, etc., we can write:

$$G = kX^a \cdot Y^b \cdot Z^c \ldots.$$

where k is a dimensionless constant.

A true physical equation must be homogeneous in relation to the exponents of each member of the equation so that they represent the relations that really exist between the considered quantities. This criterion represents a necessary condition for any physical equation to be true and is called the **principle of dimensional homogeneity**: "**a physical equation cannot be true unless it is dimensionally homogeneous.**"

If, for example, we were not sure of the formula $F = m \cdot a$, we could do the proof. At least we must admit that F is a function of m and a, like this:

$$G = kX^a \cdot Y^b \quad \text{or} \quad F = km^a \cdot a^b$$

Because F has dimensions MLT^{-2} and the second member has also to have dimensions MLT^{-2}, according to the homogeneity principle seen above:

$$\mathrm{MLT}^{-2} = k\mathrm{M}^a \cdot \left(\mathrm{LT}^{-2}\right)^b$$

remembering that the dimensions of a are LT^{-2}. So $\mathrm{MLT}^{-2} = k\,\mathrm{M}^a\,\mathrm{L}^b\,\mathrm{T}^{-2b}$, and we can see that the only possibility is $k = 1$, $a = 1$, and $b = 1$, resulting to $\mathrm{MLT}^{-2} = \mathrm{MLT}^{-2}$ and, consequently, $F = m \cdot a$.

Let us take another example: what would be the equation of the space S traveled by a free falling body from rest, assuming that S is a function of the body weight p (a force!), of the acceleration of gravity, and of the time t. In this case,

$$S = k \cdot p^a \cdot g^b \cdot t^c$$

$$\mathrm{L} = k \cdot \left(\mathrm{M} \cdot \mathrm{L} \cdot \mathrm{T}^{-2}\right)^a \cdot \left(\mathrm{L} \cdot \mathrm{T}^{-2}\right)^b \cdot (\mathrm{T})^c$$

$$\mathrm{L} = k \cdot \mathrm{M}^a \cdot \mathrm{L}^{(a+b)} \cdot \mathrm{T}^{(-2a-2b+c)}$$

and so $a = 0$; $(a + b) = 1$; $b = 1$; $(-2a - 2b + c) = 0$; $c = 2$ so that the second member has also dimension L. Finally,

$$S = k \cdot p^0 \cdot g^1 \cdot t^2 = \frac{1}{2} g \cdot t^2$$

which is the well-known formula of body fall in the vacuum, with $k = 1/2$. Note that we assumed erroneously that s is a function of p, and because this is false, p disappeared: $p^0 = 1$.

The **products of variables** P are any products of variables involving a phenomenon, each variable raised to an integer exponent. We have just seen that the fall of bodies in the vacuum involves S, g, and t. With these variables, we can make several dimensional products, such as:

$P_1 = S^2 t^{-2} g$, with dimensions $\mathrm{L}^2\mathrm{T}^{-2}\mathrm{LT}^{-2} = \mathrm{L}^3\mathrm{T}^{-5}$

$P_2 = S^0 t^2 g$, with dimensions $1\mathrm{T}^2\mathrm{LT}^{-2} = \mathrm{L}$

$P_3 = S^{-3} t^4 g$, with dimensions $\mathrm{L}^{-3}\mathrm{T}^4\mathrm{LT}^{-2} = \mathrm{L}^{-2}\mathrm{T}^2$

$P_4 = S^{-2} t^4 g^2$, with dimensions $\mathrm{L}^{-2}\mathrm{T}^4\left(\mathrm{LT}^{-2}\right)^2 = \mathrm{L}^0\mathrm{T}^0 = 1$

Whenever a chosen product is dimensionless, as P_4, it is called a **dimensionless product**, symbolized by π, in this case, $P_4 = \pi_4$. By the **theorem of the** π, given n dimensional quantities G_1, G_2, …, G_n, obtained by products of k fundamental quantities, if a phenomenon can be expressed by a function $F(G_1, G_2, \ldots, G_n) = 0$, it can also be described by a function $\phi(\pi_1, \pi_2, \ldots, \pi_{n-k}) = 0$, that is, by a smaller number of variables n–k. In Fluid Mechanics, for example, when studying the flow of a liquid around a fixed obstacle, we have five variables:

$G_1 = \rho$, specific mass of the liquid;

$G_2 = v$, velocity of the liquid;

$G_3 = D$, diameter of the obstacle;

$G_4 = \mu$, viscosity of the liquid;

$G_5 = F$, force on the obstacle,

which describe the phenomenon by an equation of type $F(G_1, G_2, G_3, G_4, G_5) = 0$, which involves the five variables. Since the fundamental units k are three, the same phenomenon can be described by a function $\phi(\pi_1, \pi_2) = 0$, with two variables only. This means that of five G variables, we reduce to $5 - 3 = 2$ variables π, which simplifies the description of the phenomenon. In this case, the two most adequate dimensionless products are:

$$\pi_1 = \frac{\rho \cdot v \cdot D}{\mu} \rightarrow \mathrm{M^0L^0T^0} = 1$$

$\rightarrow$ **Reynolds number**

$$\pi_2 = \frac{F}{\rho/2v^2D^2} \rightarrow \mathrm{M^0L^0T^0} = 1$$

$\rightarrow$ **Drag coefficient**

Besides these nondimensional numbers, **Bond's number** is important in Soil Physics, in the balance between gravitational and matrix forces (Ryan and Dhir 1993).

19.3 Physical Similarity

The problem addressed in the introduction about the kingdoms of Lilliput and Brobdingnag is of **physical similarity**. Whenever you work with object models on different scales, it is necessary that there is physical similarity between the model (prototype, usually smaller) and the actual object under study. Depending on the case, we speak of **kinematic similarity**, which involves relations of velocity and acceleration between the model and the object or of **dynamic similarity**, involving relations between the forces acting on the model and the object. In the similarity analysis, dimensionless products π are used, such as the Euler, Reynolds, Froude, and Mach "numbers."

Thus, for object and prototype, we have:

Object

$$F(G_1, G_2, \ldots, G_n) = 0 \rightarrow \phi(\pi_1, \pi_2, \ldots, \pi_{n-k}) = 0$$

Prototype

$$F(G'_1, G'_2, \ldots, G'_n) = 0 \rightarrow \phi(\pi'_1, \pi'_2, \ldots, \pi'_{n-k}) = 0$$

the G_i may be different from G'_i. There will be only physical similarity between the object and the prototype, if $\pi_1 = \pi'_1$; $\pi_2 = \pi'_2$; ...; $\pi_{n-k} = \pi'_{n-k}$. In our example of a liquid flowing around obstacle, there will be similarity between the object and a possible prototype when:

Reynolds number of the object = Reynolds number of the prototype

Drag coefficient of the object = Drag coefficient of the prototype

This similarity analysis is widely used in hydrodynamics, machines, etc., and there is not much application in our soil-plant-atmosphere system. Exception is the work of Shukla et al. (2002), which uses the dimensionless products π in a work of miscible displacement of solutes in soils.

19.4 Nondimensional Quantities

Nondimensional quantities are of spread use in science and, therefore, also in Soil Physics. They are quantities obtained from nondimensional products π, which have a numeric value k of dimension 1:

$$\mathrm{M^0L^0T^0K^0} = 1$$

In addition to the cases already seen, dimensionless quantities can appear by means of the ratio between two quantities G_1 and G_2 of the same dimension: $G_1/G_2 = \pi$. This is the case for the classic number $\pi = 3.1416\ldots$, the result of dividing the length πD of any circumference (dimension L) by the respective diameter D (dimension L).

In the SPAS, several quantities are dimensionless, represented in percentage (%) or parts per million (ppm, in disuse today). A percentage is also a ratio $G_1/G_2 = \pi$. The soil water contents u, θ and porosities α, β defined in Chap. 3,

Eqs. 3.12, 3.14, 3.15, and 3.30, respectively, are examples of quantities π. In Chap. 3, it was said that it is important to maintain the units (kg kg^{-1}, m^3 m^{-3}) so that the difference between them can be observed.

Important is to recognize that the transformation of a quantity into dimensionless units has several advantages and many times meeting a defined purpose. The simplest case is to divide the quantity by one of its own values, which can be done in two different ways. For example, experiments with soil columns are very common and each researcher uses a different column length L m (note: this L is not the fundamental unit length L). How to compare or generalize their results? If the coordinate x or z (distance along the column) is divided by the maximum length L, a new dimensionless variable $X = x/L$ appears, with the advantage that for any length L, at $x = 0$, $X = 0$; at $x = L$, $X = 1$, thus varying in the range 0–1, which is a great advantage.

This same procedure can be used for quantities that are already dimensionless, such as soil moisture θ. If we divide $(\theta - \theta_s)$ by its range of variation $(\theta_0 - \theta_s)$, where θ_s and θ_0 are the residual water contents, when dry and on saturation, respectively, we will have a new variable $\Theta = (\theta - \theta_s)/(\theta_0 - \theta_s)$, whose value are $\Theta = 0$ for $\theta = \theta_s$ (dry soil) and $\Theta = 1$ for $\theta = \theta_0$ (saturated soil). Thus, for any soil, Θ ranges from 0 to 1 and comparisons between them can be made more adequately.

Dividing a variable G by its maximum value G_{max} (or its range of variation) is a very employed technique. For example, in Fig. 4.2 (sigmoidal model for dry matter accumulation) of Chap. 4, both the ordinate y as the abscissa x could be dimensionalized by $y = DM/DM_{max}$ and $x = DD/DD_{max}$, and Fig. 4.1 (Chap. 4) would be generalized, opening the possibility of comparing growth curves of different crops.

19.5 Main Variables Used to Quantify the SPAS

In this item, we will list the main physical-chemical quantities used in the description of the SPAS, soil-plant-atmosphere system, indicating their dimensional formulas and units in the **International System of Units**. As already mentioned, the use of ISU is mandatory, but even so, we will present other units in routine use by the agronomic scientific community. For length, for example, the unit is the meter m, but in many cases, when the values are too small or too large, multiples and submultiples are used, which are allowed: km, mm, μm, nm, etc. Strictly prohibited is the use of units outside the metric system, such as inch (inch), mile, and angstrom (Å). In the submultiples of the meter, the use of the centimeter (cm) is problematic because it belongs to another unit system, the CGS, and is a submultiple of order 10^{-2}. Even so, for convenience, it is widely used, even in this text.

In the case of time, the official unit is the second (s) and only the submultiples belong to the decimal system, such as μs, ns, etc. The upper multiples we rarely use are ks and Ms. We use the multiples derived from our "calendar": year, month, day, hour, and minute. In our case, as agricultural crops follow the calendar, these units will be heavily employed, especially the day. Another factor that leads to their use is the relatively slow movement of water, whose velocity (or rate) is best described in mm day^{-1} than in m s^{-1}. For example, a typical rate of evapotranspiration is:

$$5\,\mathrm{mm\,day^{-1}} = \frac{5 \times 10^{-3}\,\mathrm{m}}{86{,}400\,\mathrm{s}} = 5.79 \times 10^{-8}\,\mathrm{m\,s^{-1}}$$

In the case of mass, the unit is kg, which is already a multiple of the gram (g). In any case, multiples and submultiples can be used as Mg, mg, μg, etc. Again, the use of the gram is problematic because it belongs to the CGS. Even so, its use is often convenient and, therefore, is widely used in this text. Table 19.1 shows the most commonly used multiples and submultiples of the SI International System.

Table 19.2 presents the main values used in the SPAS, with their dimensional formulas. It facilitates the transformation of units. For example, we are going to transform Newtons (force MLT^{-2}) into dynes (d):

Table 19.1 Factors, prefixes, and symbols for multiplex and submultiples of entities

Factor	Prefix	Symbol
10^{18}	exa-	E
10^{15}	peta-	P
10^{12}	tera-	T
10^{9}	giga-	G
10^{6}	mega-	M
10^{3}	quilo-	k
10^{-3}	mili-	m
10^{-6}	micro-	μ
10^{-9}	nano-	n
10^{-12}	pico-	p
10^{-15}	fento-	f
10^{-18}	atto-	a

$$1\,\mathrm{N} = \frac{1\,\mathrm{kg} \times 1\,\mathrm{m}}{1\,\mathrm{s}^2} = \frac{10^3\,\mathrm{g} \times 10^2\,\mathrm{cm}}{1\,\mathrm{s}^2}$$
$$= 10^5\,\mathrm{g\,cm\,s^{-2}} = 10^5\,\mathrm{d}$$

The fundamental molar entity to a number is equivalent to Avogadro's number: 6.02×10^{23}. This entity is usually employed to quantify chemicals. Thus, 1 mol of any substance corresponds to the mass of 6.02×10^{23} units of that substance. We say that 1 mol of $CaCl_2$ is 75.5 g of this salt, and this mass contains 6.02×10^{23} $CaCl_2$ molecules, or Ca^{2+} ions and twice the Cl^-. A 1 M solution (one molar) has 75.5 g of $CaCl_2$ per liter of solution and is

Table 19.2 The most used units of entities of the soil-plant-atmosphere system

Entity	Name	Dimension	Unity ISU	Other units and multiples
Mass	Kilogram	M	kg	Mg, mg, μg
Length	Meter	L	m	cm, mm, μm, km
Time	Second	T	s	min, h, day
Temperature	Kelvin	θ	K	°C, °F, °R
Absolute viscosity		$ML^{-1}T^{-1}$	$N\,m^{-2}\,s^{-1}$	
Acceleration	a	LT^{-2}	$m\,s^{-2}$	
Angular displacement	Plane angle		rad	degree °
	Solid angle	$L^2\,L^{-2}$	steradian (sr)	
Angular velocity	ω	T^{-1}	$rad\,s^{-1}$	$degree°\,h^{-1}$
Area	Square meter	L^2	m^2	ha (1 ha = 10,000 m^2)
Concentration of a chemical element in the soil		NL^{-3} NM^{-1}	$cmol_c\,dm^{-3}$ $cmol_c\,kg^{-1}$	meq/100 g
Depth of water: precipitation, irrigation depth, soil water, soil water storage	P I S_L	L	m	cm, mm
Electric charge	Coulomb		C = A s	
Electrical conductivity of water	K		$s\,m^{-1}$	$mmho\,cm^{-1}$
Entropy	S	$ML^2T^{-2}\theta^{-1}$	$J\,K^{-1}$	$cal\,°C^{-1}$
Flux	Q	L^3T^{-1}	$m^3\,s^{-1}$	$L\,h^{-1}$
Flux density of nutrients, ions, gases	j	$ML^{-2}T^{-1}$	$kg\,m^{-2}\,s^{-1}$	$mg\,cm^{-2}\,day^{-1}$ $kg\,ha^{-1}\,year^{-1}$
Flux density of water, of precipitation, of irrigation, of evapotranspiration, hydraulic conductivity	q p i q_{et} $K(\theta)$	$L^3L^{-2}T^{-1}$	$m^3\,m^{-2}\,s^{-1}$ ($m\,s^{-1}$)	$mm\,day^{-1}$ $mm\,h^{-1}$
Force	Newton	MLT^{-2}	N	kgf, dyne
Frequency	Hertz	T^{-1}	Hz	cps, cpm

(continued)

Table 19.2 (continued)

Entity	Name	Dimension	Unity ISU	Other units and multiples
Heat flux density or radiant energy	q	MT^{-3}	$J\ s^{-1}\ m^{-2}$ or $W\ m^{-2}$	$cal\ cm^{-2}\ min^{-1}$
Heat capacity	C	$ML^2T^{-2}\theta^{-1}$	$J\ K^{-1}$	$cal\ °C^{-1}$
Intrinsic permeability	k	L^2	m^2	cm^2
Kinematic viscosity		L^2T^{-1}	$m^2\ s^{-1}$	
Latent heat	L	L^2T^{-2}	$J\ kg^{-1}$	$cal\ g^{-1}$
Plant or part of plant dry matter	DM	ML^{-2} MM^{-1}	$kg\ m^{-2}$ $kg\ kg^{-1}$	$kg\ ha^{-1}$, $Mg\ ha^{-1}$, $g\ g^{-1}$, %
Power	Watt	ML^2T^{-3}	$W = J\ s^{-1}$	$cal\ cm^{-2}$
Pressure	Pascal	$ML^{-1}T^{-2}$	$Pa = N\ m^{-2}$	$b =$ dyne cm^{-2}, atm
Quantity of atoms	Mol	N	mol	mmol, μmol
Soil bulk density Soil particle density	d_s d_p	ML^{-3}	$kg\ m^{-3}$	$Mg\ m^{-3}$, $g\ cm^{-3}$
Soil total porosity Water-free porosity	α β	L^3L^{-3}	$m^3\ m^{-3}$	$cm^3\ cm^{-3}$, %
Specific heat	c	$L^2T^{-2}\theta^{-1}$	$J\ kg^{-1}\ K^{-1}$	$cal\ g^{-1}\ °C^{-1}$
Soil water content % weight	u	MM^{-1}	$kg\ kg^{-1}$	$g\ g^{-1}$, %
Soil water content % volume	θ	L^3L^{-3}	$m^3\ m^{-3}$	$cm^3\ cm^{-3}$, %
Soil water diffusivity	$D(\theta)$	$L^{-2}T^{-1}$	$m^{-2}\ s^{-1}$	$cm^{-2}\ s^{-1}$
Surface tension	σ	MT^{-2}	$J\ m^{-2} = N\ m^{-1}$	
Temperature gradient	Grad T	θL^{-1}	$k\ m^{-1}$	$°C\ cm^{-1}$
Thermal conductivity	K	$MLT^{-3}\theta^{-1}$	$J\ m^{-1}\ K^{-1}$ $(W\ m^{-1}\ K^{-1})$	$cal\ cm^{-1}\ °C^{-1}$
Thermal diffusivity	D_T	L^2T^{-1}	$m^2\ s^{-1}$	$cm^2\ s^{-1}$
Tortuosity		LL^{-1}	$m\ m^{-1}$	$cm\ cm^{-1}$
Velocity or speed	v	LT^{-1}	$m\ s^{-1}$	
Volume	Cubic meter	L^3	m^3	L (liter), mL, μL
Water potential gradient	Grad Ψ	$L\ L^{-1}$	$m\ m^{-1}$	cmH_2O/cm
Water potential Ψ	Energy/V Energy/M Energy/weight	MLT^{-2} LT^{-2} L	$J\ m^{-3} = Pa$ $J\ kg^{-1}$ $J\ N^{-1} = m$	atm $erg\ g^{-1}$ mH_2O, $cm\ H_2O$, mmHg
Work Energy Heat	Joule	ML^2T^{-2}	$J = N\ m$	erg = dyne cm, cal (1 cal = 4.18 J)

equivalent to a 1 N solution (one normal or one equivalent per liter) in calcium and 2 N in chlorine. In the evaluation of ionic concentrations (see Table 19.2), the unit meq 100 g^{-1} of soil is used, which has now been changed to $cmol_c\ dm^{-3}$ or $cmol_c\ kg^{-1}$. It therefore equates to a number of moles of charge of the element in question, per unit volume or mass of soil. It should be noted that the soil mass is not equal its volume, and the difference is in the soil bulk density (d_s), which implies a factor that varies from soil to soil, of the order of 1.5. In the evaluation of this magnitude, one method recommends the use of a volume (a reference container of about 10 cm^3) of dry soil sieved through a 2 mm sieve and another method, the use of a soil mass, say 50 g.

Mole is also used to quantify beams of radiation, so, for example, a beam of 6.02×10^{23} photons of wavelength 555 nm (yellow color) has an energy of 21.56×10^4 J and equals to 1 Einstein of that radiation.

19.6 Coordinate Systems

The most common coordinate system is the **Orthogonal Cartesian System** of Euclidean Geometry, in which the three linear dimensions x, y, and z are arranged perpendicular to each other, as was done for the continuity equation (Eq. 7.20, Chap. 7). This system involves three coordinates of dimension L, resulting in a length $L(x, y, \text{or } z)$, an area $L^2(xy, xz, \text{or } zy)$, and a volume $L^3(xyz)$. The exponents of L indicate the "dimension," that is, dimension 1 = linear; dimension 2 = plane; and dimension 3 = volume, and fractional dimensions such as 1.6 or 2.4 are not allowed. As we shall see in the following section of fractal geometry where dimensions are fractional, it is difficult to visualize them in terms of what we are faced daily: line, plane, volume. Even the fourth dimension L^4 becomes quite "virtual" to our perception. In Modern Physics, Einstein uses four dimensions: x, y, z, and t.

In the three-dimensional system, the position of a point is fully defined by the three linear coordinates x, y, and z, that is, there is only one point A in the space with coordinates x_A, y_A, and z_A. In addition to this system, we have several others, some useful in describing the SPAS. In the **cylindrical coordinate system**, a point B in space is defined by two linear coordinates (a height z and a radius r) and an angular coordinate α. In Chap. 14, Fig. 14.4, this system is schematized. In the **spherical coordinate system**, a point C in space is defined by a linear coordinate (radius r) and two angles β and γ. When the object under study is spherical, this coordinate system is advantageous.

19.7 Scales and Scaling

We have already talked about scales at the beginning of this chapter when presenting the problem of physical similarity between the object being studied and the model. Maps are also drawn to scale. For example, on a scale of 1:10,000, 1 cm^2 of paper can represent 10,000 m^2 in the field, that is, 1 ha. Quantities that have differences in scale cannot simply be compared. As we have seen, there is the problem of physical similarity, but what if we want to compare without changing the scale of each? One of the techniques proposed is that of **scaling**, much used in Soil Physics. The technique was introduced in Soil Science by Miller and Miller (1956), by the concept of **similar media** applied to the "capillary" flow of fluids in porous media. According to them, two systems, M_1 and M_2, are similar when the quantities that describe the physical processes that occur in them differ by a linear factor λ, called **microscopic characteristic length**, which relates their physical characteristics. The best way to visualize the concept is to consider M_2 as an enlarged photograph of M_1 by a factor λ (Fig. 19.4). For these media, the diameter D of a particle of one medium would be related to the other by the relation $D_2 = \lambda D_1$, the surface S of this particle by $S_2 = \lambda^2 S_1$, and its volume V by $V_2 = \lambda^3 V_1$ (Fig. 19.5).

Under these conditions, if we know the water flow in M_1, could it be estimated in M_2, based on λ? Klute and Wilkinson (1958) and Wilkinson and Klute (1959), using media consisting of homogeneous glass beads, obtained results on hydraulic conductivity and retention curve of these media, which supported the theory of the similar media very well. Afterward, contributions on this subject did not appear in the literature to develop this concept. More than 10 years later, Reichardt et al. (1972) took up the theme, achieving success, even with natural porous media, that is, soils of different textures. They were based on the fact that soils can be considered similar media, each with its factor λ which, initially, they did not know how to determine. They have taken the horizontal infiltration process, discussed in Chap. 11, whose BVP is repeated here:

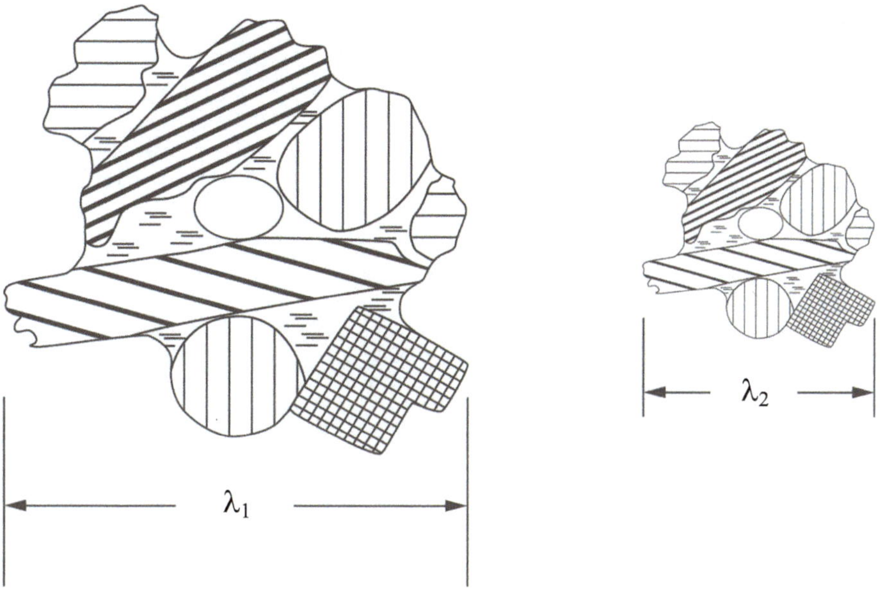

Fig. 19.4 Illustration of physical similarity of two soil particle-water-air media of characteristic lengths λ_1 and of λ_2

$r_1 = 3$ cm

$A_1 = \pi r_1^2 = 28.27\ \text{cm}^2$

$V_1 = \frac{3\pi r_1^3}{4} = 63.62\ \text{cm}^3$

$r_2 = 4.5$ cm

$A_2 = \pi r_2^2 = 63.62\ \text{cm}^2$

$V_1 = \frac{3\pi r_2^3}{4} = 214.71\ \text{cm}^3$

$$\frac{A_2}{A_1} = 2.25 \longrightarrow \sqrt{2.25} = 1.5 \text{ or } A_2 = (1.5)^2 A_1$$

$$\frac{V_2}{V_1} = 3.37 \longrightarrow \sqrt[3]{3.37} = 1.5 \text{ or } V_2 = (1.5)^3 V_1$$

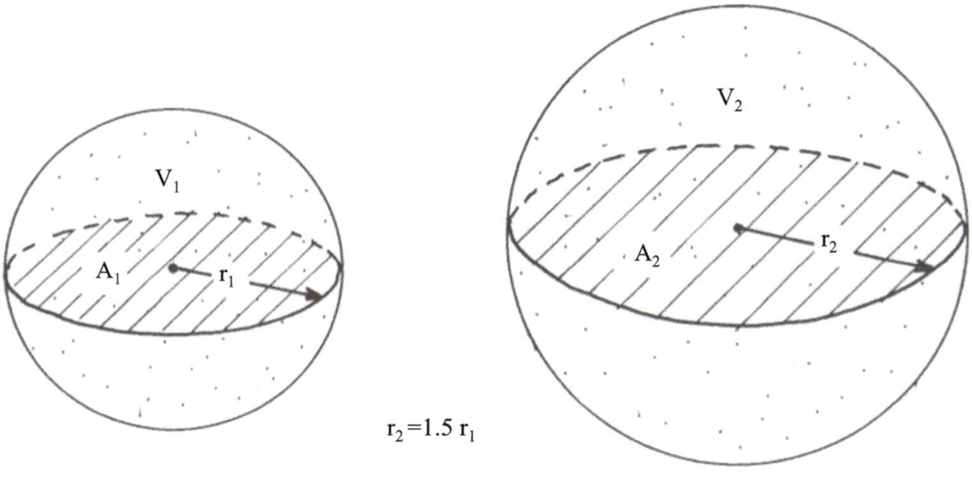

Fig. 19.5 Perfectly similar spheres (by nature)

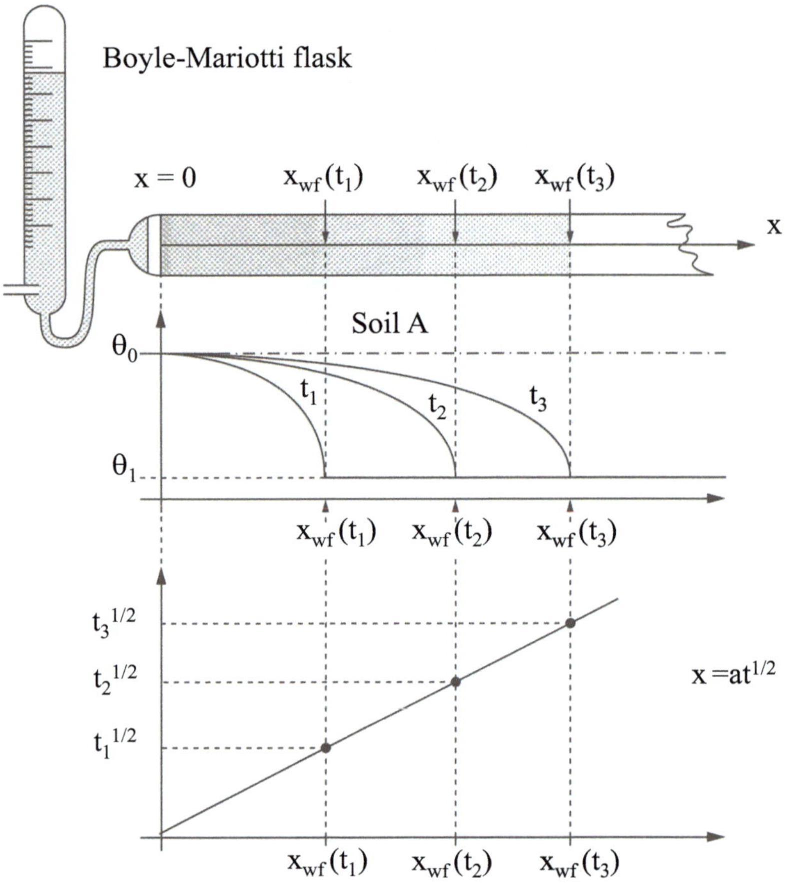

Fig. 19.6 Schematic presentation of an experiment of horizontal infiltration into a homogeneous soil, showing the advance of the wetting front x_{wf} at three times t_1, t_2, and t_3

$$\theta = \theta_i, \quad x > 0, \quad t = 0 \tag{19.1}$$

$$\theta = \theta_0, \quad x = 0, \quad t > 0 \tag{19.2}$$

$$\frac{\partial \theta}{\partial t} = \frac{\partial}{\partial x}\left[D(\theta)\frac{\partial \theta}{\partial x}\right] \tag{19.3}$$

where $D(\theta) = K(\theta) \cdot \mathrm{d}h/\mathrm{d}\theta$. Figure 19.6 illustrates this BVP for the horizontal water flow into homogeneous soils.

As for any soil, the solution of this BVP is of the same type: $x = \phi(\theta) \cdot t^{1/2}$, in which $\phi(\theta)$ depends on the characteristics of each porous medium, would it not be possible to obtain a generalized solution for all (considered similar) media, provided the characteristic λ of each is known? The procedure they used was to dimensionalize all variables, also using the similar media theory applied to i soils, each with its $\lambda_1, \lambda_2, \ldots, \lambda_i$. The soil water content θ and the x coordinate were turned dimensionless as has been seen previously in this chapter:

$$\Theta = \frac{(\theta - \theta_i)}{(\theta_0 - \theta_i)} \tag{19.4}$$

$$X = \frac{x}{x_{\max}} \tag{19.5}$$

With respect to the matrix potential h, this potential was considered only as the result of capillary forces, that is, $h = 2\sigma/\rho g r$ (Eq. 6.18, Chap. 6) or $hr = 2\sigma/\rho g$ = constant. If every soil i were made only of capillaries of radius r_i and if the characteristic lengths λ_i were proportional to the radius r_i, we would have:

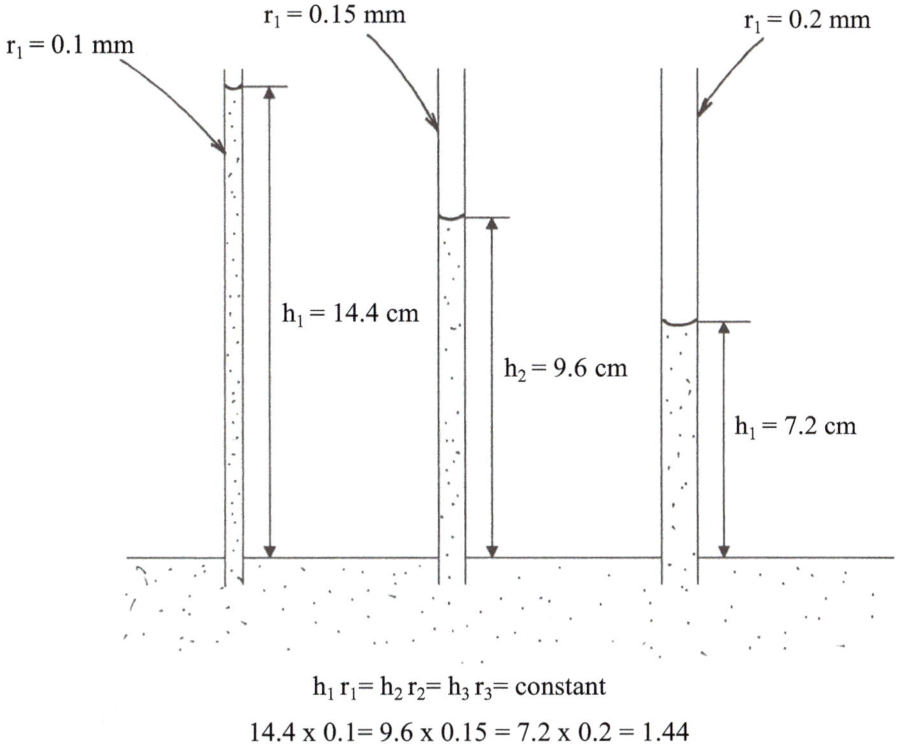

Fig. 19.7 Three glass capillaries considered as similar media

$$h_1 r_1 = h_2 r_2 = \cdots = h_i r_i = \text{constant}$$

If we choose among the soils one considered as standard (could even be a hypothetical soil) for which $\lambda^* = r^* = 1$ (one μm, m, or any other unit), the above constant becomes equal to $h^* r^* = h^*$, which would be the potential matrix h^* of the standard soil (Fig. 19.7). By the dimensional analysis shown earlier, we can also make h^* dimensionless by choosing related constants:

$$h^* = \frac{\lambda_1 \rho g h_1}{\sigma} = \frac{\lambda_2 \rho g h_2}{\sigma} = \cdots = \frac{\lambda_i \rho g h_i}{\sigma} \quad (19.6)$$

With regard to the hydraulic conductivity K, as it is proportional to the area (λ^2) available for the flow (k = intrinsic permeability, L^2), we have the same reasoning $K = k\rho g/\eta$ (Eq. 7.9, Chap. 7) or $K/k = \rho g/\eta$ = constant:

$$\frac{K_1}{k_1} = \frac{K_2}{k_2} = \cdots = \frac{K_i}{k_i} = \text{constant}$$

$$K^* = \frac{\eta K_1}{\lambda_1^2 \rho g} = \frac{\eta K_2}{\lambda_2^2 \rho g} = \cdots = \frac{\eta K_i}{\lambda_i^2 \rho g} \quad (19.7)$$

where K^* is the hydraulic conductivity of the standard soil, assuming $\lambda^* = r^* = k^* = 1$ (Fig.19.8).

Through the definition of $D = K \cdot dh/d\theta$, we can also see that the soil water diffusivity of the standard soil D^* is given by:

$$D^* = \frac{\eta D_1}{\lambda_1 \sigma} = \frac{\eta D_2}{\lambda_2 \sigma} = \cdots = \frac{\eta D_i}{\lambda_i \sigma} \quad (19.8)$$

Of all the variables of Eq. 19.3, we now need only the dimension of the time. If this is done in order to make Eq. 19.3 dimensionless, we would have a time t^* for the standard soil, given by:

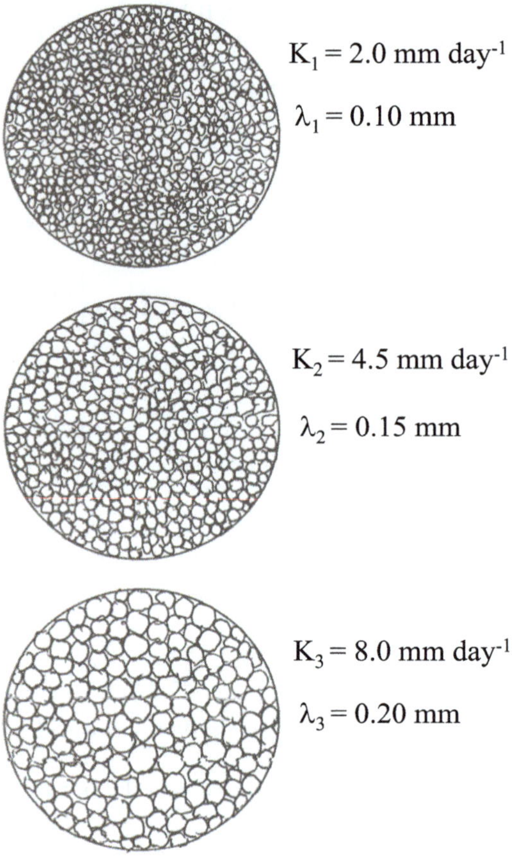

$$\frac{K_1}{\lambda_1^2} = \frac{K_2}{\lambda_2^2} = \frac{K_3}{\lambda_3^2} = \text{constant}$$

$$\frac{2}{(0.10)^2} = \frac{4.5}{(0.15)^2} = \frac{8}{(0.20)^2} = 200$$

Fig. 19.8 Three soil cross-sections considered as similar media

$$t^* = \frac{\lambda_1 \sigma t_1}{\eta (x_{1\max})^2} = \frac{\lambda_2 \sigma t_2}{\eta (x_{2\max})^2} = \cdots = \frac{\lambda_i \sigma t_i}{\eta (x_{i\max})^2} \quad (19.9)$$

Under these conditions, the reader can verify that by replacing θ by Θ, t by t^*, x by X, and D by D^* in Eq. 19.3, we obtain the dimensionless differential equation of the standard soil. This equation differs from the ones of each individual soil only by the respective scaling factors λ_i, hidden in Eq. 19.10 but which are built in the definitions of t^* and D^*:

$$\frac{\partial \Theta}{\partial t^*} = \frac{\partial}{\partial X}\left[D^*(\Theta)\frac{\partial \Theta}{\partial X}\right] \quad (19.10)$$

subject to the conditions:

$$\Theta = 0, \quad X \geq 0, \quad t^* = 0 \quad (19.11)$$

$$\Theta = 1, \quad X = 0, \quad t^* > 0 \quad (19.12)$$

whose solution, by analogy to Eq. 11.17, is:

$$X = \phi^*(\Theta) \cdot (t^*)^{1/2} \quad (19.13)$$

It is opportune to analyze Eq. 19.9 of the dimensionless times in the light of physical resemblance and the kingdoms of Lilliput and Brobdingnag, which shows that in order to compare different soils (but considered similar media), their times must be different and dependent on λ, which is a length. We might even suggest that this fact helps explain how in Modern Physics time enters as a fourth coordinate along with x, y, and z. By analogy to what was done with h and K, we can write:

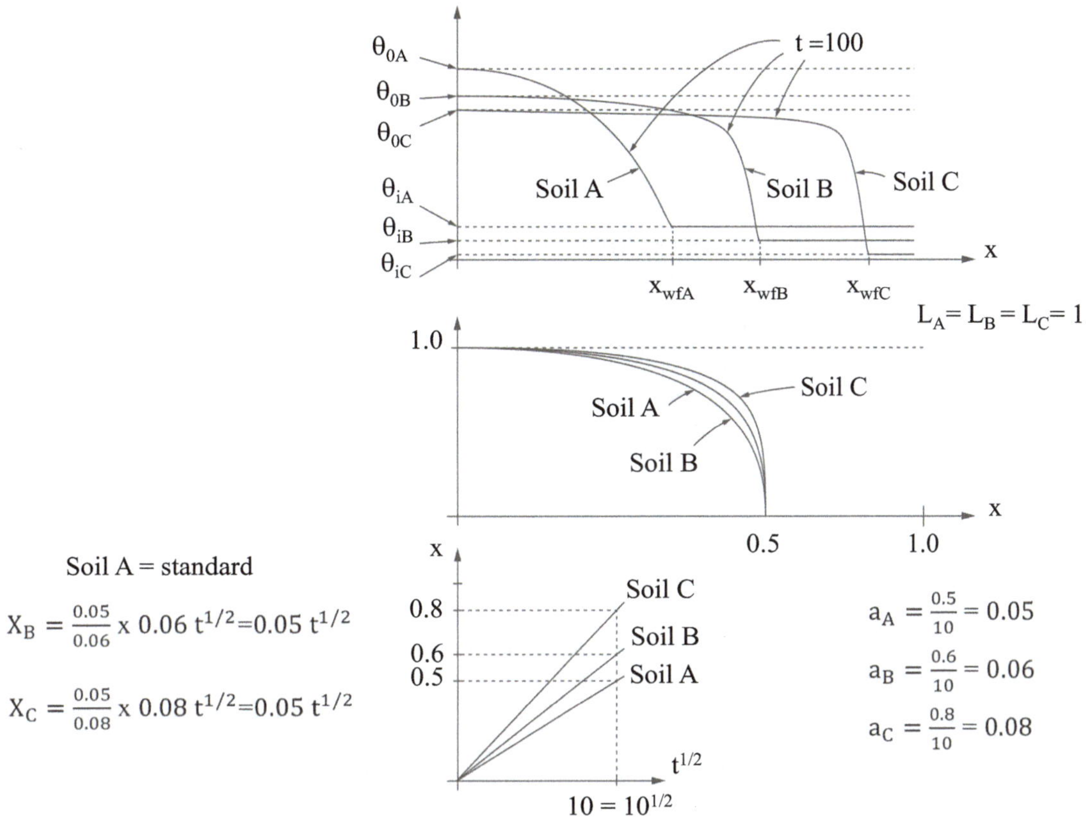

Fig. 19.9 Three soils considered as similar media. (Above) Their soil water content profiles at time 100. (At the center) Scaled graphs. (Below) The way the scaling factors are obtained from the x versus $t^{1/2}$ graphs

$$t_1\lambda_1 = t_2\lambda_2 = \cdots = t_i\lambda_i = \frac{t^*\eta(x_{\max})^2}{\sigma} = \text{constant}$$

Once the theory was established, Reichardt et al. (1972) looked for ways to measure λ for the different soils. The "Columbus egg" arose when they realized that if the straight lines x_f versus $t^{1/2}$ (see Fig. 11.7, Chap. 11) are characteristics for each soil i, they should be reduced to a single line X_f versus $t^{*1/2}$ according to Eq. 19.13, for the standard soil. The factors that make the overlapping of the lines could be served as λ_i of each soil. We know that straight lines that pass through the origin $y = a_i.\ x$ can be overlapped over each other by the relation a_i/a_j of the respective slopes (angular coefficients), as exemplified in Fig. 19.9. Since the line in question involves square root, the relation to be used was:

$$\frac{\lambda_i}{\lambda^*} = \left(\frac{a_i}{a^*}\right)^2 \tag{19.14}$$

and with this relation, Reichardt et al. (1972) determined the values of λ_i for each soil, taking arbitrarily the soil of faster infiltration as the standard soil, for which they postulated $\lambda^* = 1$. Thus, the slower the infiltration of soil i, the smaller its λ_i. This procedure of determining a relative scaling factor rather than a characteristic microscopic length, as suggested by Miller and Miller (1956), facilitated the experimental part and opened the doors to the use of the scaling technique in several other areas of Soil Physics. In conclusion, Reichardt et al. (1972) were able to perfectly scale $D(\theta)$ and with constraints $h(\theta)$ and $K(\theta)$, because the soils were not really similar media. The fact that they succeeded in scaling $D(\theta)$ led Reichardt and Libardi (1973) to establish a general equation for the determination of $D(\theta)$ of a

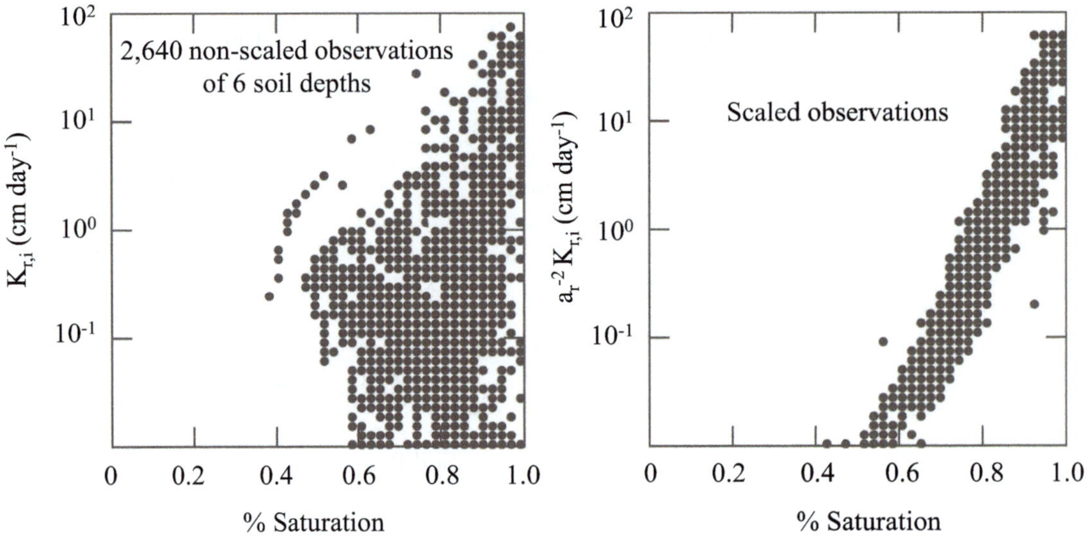

Fig. 19.10 An example of how the scatter of raw data can be improved by scaling

soil, measuring only the linear coefficient a_i of its horizontal infiltration curve x_f versus $t^{1/2}$ obtained in an easy horizontal infiltration experiment:

$$D(\Theta) = 1.462 \times 10^{-5} a_i^2 \exp(8.087 \times \Theta) \quad (19.15)$$

Furthermore, Reichardt et al. (1975) presented a method of determining $K(\theta)$ by means of a_i; Bacchi and Reichardt (1988) used the scaling technique to evaluate the efficacy of $K(\theta)$ determination methods, and Shukla et al. (2002) scaled miscible displacement experiments. In addition, the scaling technique was much used in studies of spatial variability of soils, assuming a characteristic λ for each point of a transect (Fig. 19.10). A good review on scaling was done by Tillotson and Nielsen (1984), Kutilek and Nielsen (1994), and Nielsen et al. (1998). A new form of scaling for dissimilar soils was presented by Sadeghi et al. (2011).

19.8 Fractal Geometry and Dimension

Fractal geometry, unlike the Euclidean geometry, admits fractional dimensions. The term ***fractal*** was defined by Mandelbrot (1982), derived from the Latin adjective *fractus*, whose verb *frangere* means to break, to create irregular fragments. Etymologically, the term fractal is the opposite of the term algebra (from Arabic *jabara*), which means to join, to connect the parts. According to Mandelbrot, fractals are nontopological objects, that is, objects for which their dimension is a noninteger real number, which exceeds the value of the topological dimension. For objects called topological, or Euclidean geometric forms, the dimension is an integer (0 for a point, 1 for any segment, 2 for any surface, 3 for volumes). The dimension Mandelbrot called the fractal dimension is a measure of the degree of irregularity of the object considered at all scales of observation. The fractal dimension is related to the rapidity with which the estimated measure of the object increases as the scale of measurement decreases. The autosimilarity or scaling properties of objects is one of the central concepts of the fractal geometry and allows a better understanding of the concept of fractional dimension. An object normally considered unidimensional (Fig. 19.11) as a straight segment can be divided into N identical parts, such that each part is a new straight segment represented on a scale $r = 1/N$ of the original

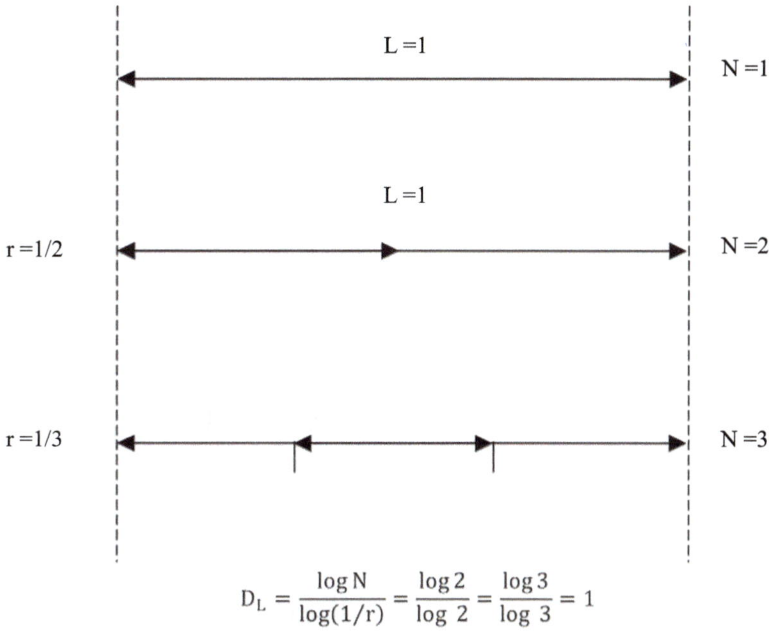

Fig. 19.11 Generalization of the relation $Nr^D = 1$ for the case $D = 1$, resulting $Nr^1 = 1$

segment so that $Nr^1 = 1$. Similarly, a two-dimensional object (Fig. 19.12), as a square area on a plane, can be divided into N identical square areas on a scale of the original area so that $Nr^2 = 1$.

Such a scaling can be extended to three-dimensional objects (Fig. 19.13) and the relationship between the number of similar fragments (N) and their scale in relation to the original object (r) can be generalized by $Nr^D = 1$, in which D defines the dimension of similarity or fractal dimension.

Therefore, the Euclidean geometric forms, with dimensions 0, 1, 2, and 3, with which we are more familiar, can be seen as particular cases of the numerous forms and dimensions that occur in nature. Figure 19.14, adapted from Barnsley et al. (1988), known as the Von Koch's curve, is constructed in an iterative or recursive fashion, starting from a straight segment (a) divided into three equal parts and the central segment replaced by two equal segments, forming part of an equilateral triangle (b). In the next stage, each of these four segments is again divided into three parts and each is replaced by four new length segments equal to 1/3 of the original and arranged according to the same pattern shown in (b) and so on. From stage b, at each stage change the total length L of the figure increases by a factor 4/3, the number N of elements similar to that of stage a increases by a factor of 4, and its dimensions are in scale $r = 1/3$ from the previous stage. At each stage, the figure can be divided into N similar elements, such that $N \cdot r^D = 1$, where D is the fractal dimension of the object. This curve has an approximate fractal dimension $D = 1.26$, which is greater than 1 and less than 2, which means that it fills more space than a single line ($D = 1$) and less than one Euclidean area of a plane ($D = 2$).

Highly complex and irregular shapes and structures, common in nature, can be reproduced in rich detail by similar procedures, indicating that behind an apparent disorder of these forms, dynamic structures, and processes occurring in nature, there is some regularity that can be better understood. Physicists, astrologers, biologists, and scientists in many other areas have been developing in the last decades a new approach to dealing with the complexity of nature, called "Theory of Chaos," and mathematically defining the causality generated by simple deterministic dynamical systems. Such an approach allows the description of a certain order in dynamic

Fig. 19.12 Two-dimensional objects $Nr^2 = 1$

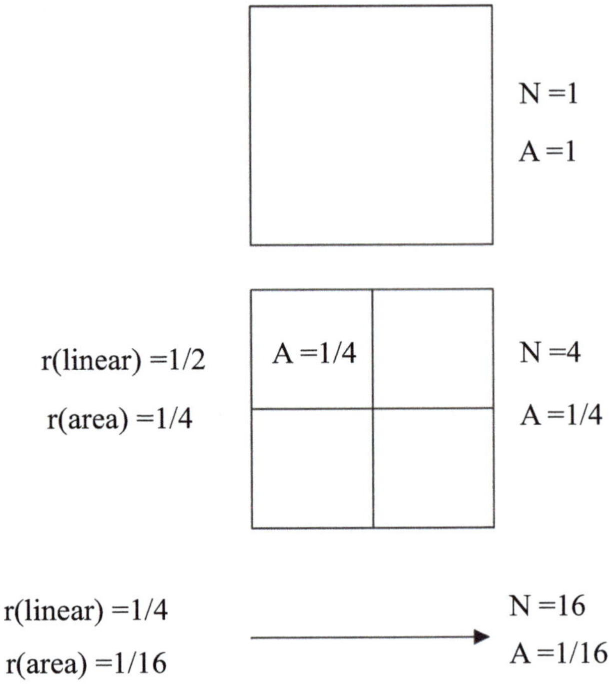

$$D_A = \frac{\log N}{\log(1/r)} = \frac{\log 4}{\log 2} = \frac{\log 16}{\log 4} = 2$$

$$N.r^2 = 1$$

$$r = \frac{1}{\sqrt{N}} \qquad D_L = D_A - 1$$

processes that were formerly defined as completely random.

With the indispensable help of computers, fractal geometry has been taking shape in the most diverse areas of knowledge, including the arts, as a new working tool for a better understanding of nature. Agronomic research, which deals basically with processes and objects of nature, follows this trend and has applied this new approach in several situations, such as the study of the dynamic processes that occur in the soil (water, gas, and solutes), soil structure, architecture and development of plants, drainage processes in hydrographic basins, etc.

Let us now clarify in more detail Figs. 19.11, 19.12, 19.13, and 19.14. When we measure a length L, which can be a line segment, a curve, or the coastal contour of a map, we use as a unit a linear ruler of "size" $\in$, smaller than L. If $\in$ fits N times in L, we have:

$$L(r) = N(r)r, \quad \text{where } r = \frac{\in}{L}$$

We write $L(r)$ because a tortuous length L, measured with the linear ruler, depends on the size of the ruler, since "arcs" are measured rectilinearly. The smaller the ruler, the better the measure. In Fig. 19.11, L is a straight. In the first case, $L = 1$, $N = 1$, and $r = 1$, that is, the ruler is L. If the ruler is half of L, we have $N = 2$ and $r = 1/2$. If it is one third, $N = 3$ and $r = 1/3$.

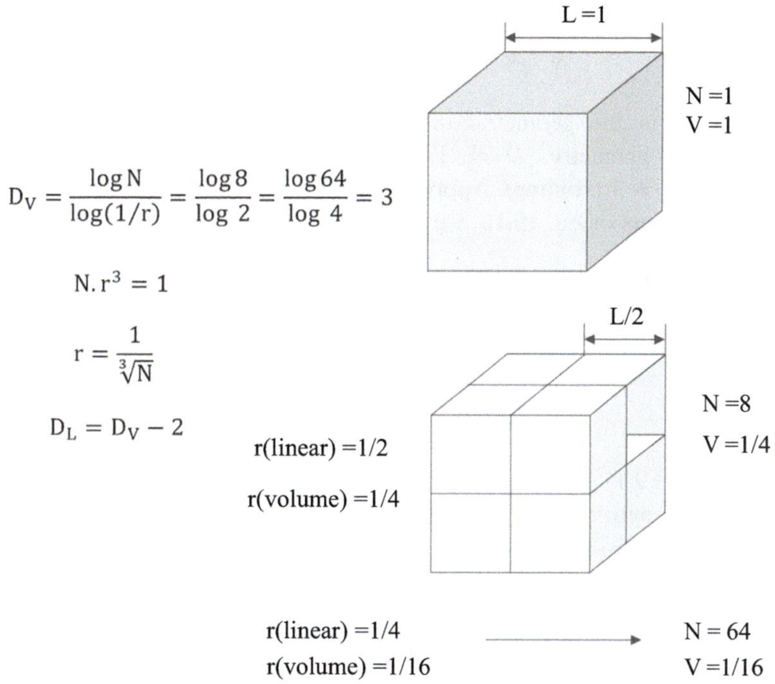

Fig. 19.13 Three-dimensional objects $Nr^3 = 1$

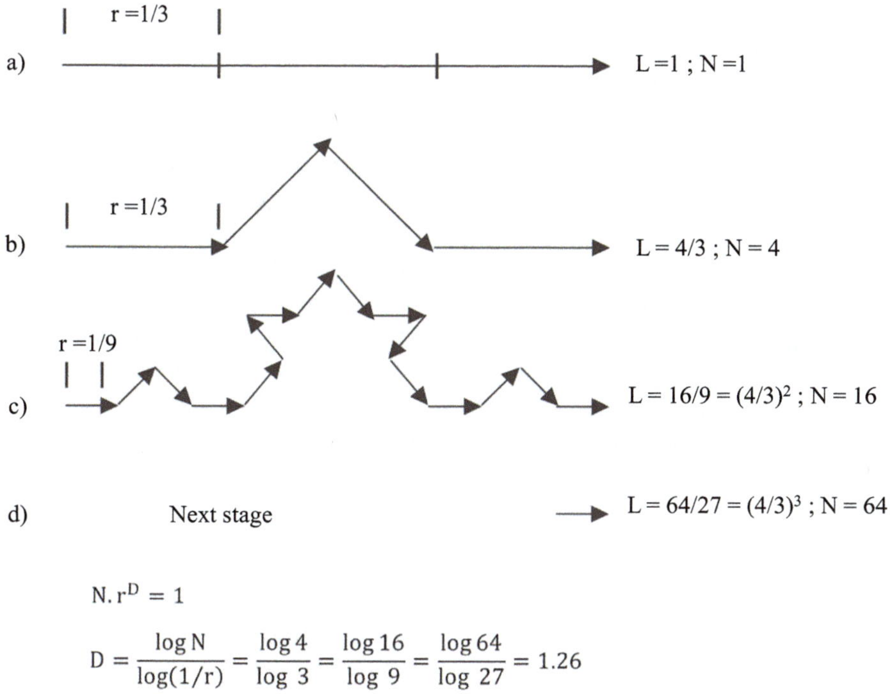

Fig. 19.14 (**a–d**) Von Koch's curve with $D = 1.26$

It can be shown that:

$$N \cdot r^D = 1 \qquad (19.16)$$

where D is the geometric dimension. In the Euclidean geometry, $D = 1$ (line); $D = 2$ (plane); $D = 3$ (volume). Applying logarithm to both members of Eq. 19.16, we have:

$$N = r^{-D}, \quad \text{or} \quad \log N = -D \cdot \log r,$$

or still log $N = D \cdot \log(1/r)$ and so

$$D = \frac{\log N}{\log(1/r)} \qquad (19.17)$$

In Fig. 19.11, we use the symbol D_L for linear dimension and it can be seen that by Eq. 19.17, the measure is linear: $D_L = 1$, agreeing with Euclidean geometry.

In Fig. 19.12, we measure two-dimensional objects, that is, areas and the Euclidean dimension is $D_A = 2$, where $D_L = D_A - 1$. For three-dimensional objects (volumes), the Euclidean dimension is $D_v = 3$, where $D_L = D_v - 2$ (Fig. 19.13).

Equation 19.16 also admits fractional dimensions, the fractal dimensions, which appear when we measure tortuous L-contours, areas A, and irregular V-volumes. In Fig. 19.14, the tortuosity is shown progressively in: (a) a basic length L_0 is given; in (b) is added 1/3 of L_0 and to fit in the same space the next stage (d) is shown. If the ruler is of L_0 length, it does not measure L_1, which is 4/3 L_0; in (c), for each stretch of b, the same assembly is made and a larger length $L_2 = 16/9$ L_0, which would not be observed with a ruler of length L_0.

Equation 19.17 gives the dimension $D = 1.26$ …, greater than 1 and less than 2 of the Euclidean geometry. It is not a straight line or an area; it is a "crooked line."

In the case of Fig. 19.14, if we add two parts, we will have:

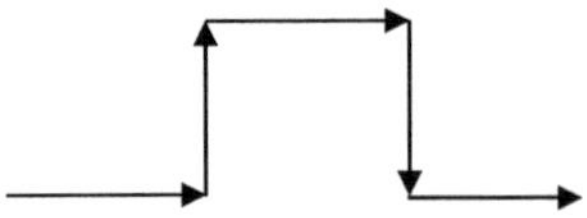

$$D = \frac{\log 6}{\log 3} = 1.63$$

and if add three parts:

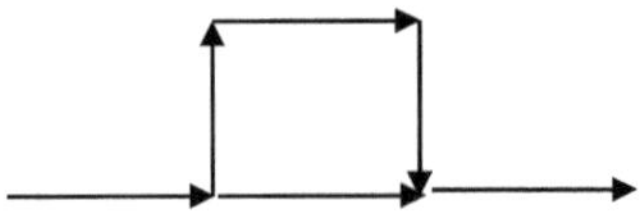

$$D = \frac{\log 7}{\log 3} = 1.77$$

or still, adding six parts:

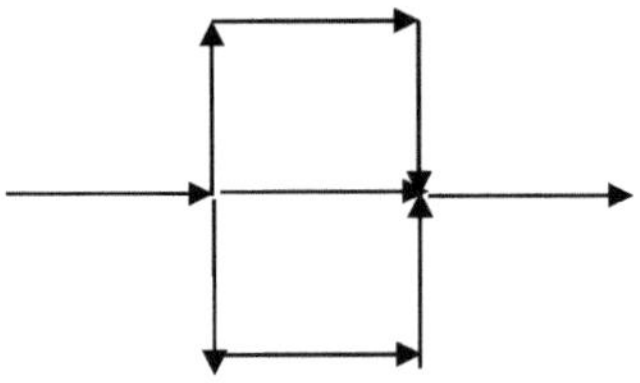

$$D = \frac{\log 9}{\log 3} = 2$$

that is, $D = D_A = 2$, which means that the tortuosity is so large that the "curve" tends toward an area.

In Soil Physics, as the path traveled by water, ions, and gases and as the distribution of particles are all tortuous, fractals appear to be a good option for modeling. In this research line, Tyler and Wheatcraft (1989) measured the volumetric fractal dimension of the soil particle distribution (Chap. 3) by the angular coefficient of log N versus log R graph, where N is the number of particles of radius smaller than R. Later, Tyler and Wheatcraft (1992) recognized the difficulty of measuring the number of particles N and used mass of particles, in dimensionless form $M(R < R_i)/M_t$ and also in dimensionless form R_i/R_t.

Bacchi and Reichardt (1993) used these concepts in the modeling of water retention curves, estimating the length of L_i pores, corresponding to a given textural class, by the empirical expression of Arya and Paris (1981): $L_i = 2R_iN_i^{\alpha}$, where $2R_i$ is the particle diameter of class i and N_i the number of particles of the same class. There was no success and this topic is open for research. Bacchi et al. (1996) compared the use of particle distributions and pore distributions in obtaining D_v and studied their effects on soil hydraulic conductivity data. For those interested in fractal geometry, the basic text is Mandelbrot (1982); and in addition to the works already cited, the following are of interest: Puckett et al. (1985), Turcotte (1986), Tyler and Wheatcraft (1990), Guerrini and Swartzendruber (1994, 1997), and Perfect and Kay (1995).

19.9 Exercises

19.1. In the example in Fig. 19.5, show that the surface of the spheres is also related by similarity.

19.2. The surface tension is given as force per unit length or energy per unit area. Show that the two forms have the same dimension.

19.3. Knowing that the hydraulic conductivity is a function of the intrinsic permeability k (cm^2 or m^2), the fluid density ρ (g cm^{-3} or kg m^{-3}), the acceleration of gravity g (cm s^{-2} or m s^{-2}), and the fluid viscosity η (g cm^{-1} s^{-1} or kg m^{-1} s^{-1}), determine the function $K = K(k, \rho, g, \eta)$.

19.4. In Eq. 19.9, show that t^* is dimensionless.

19.5. What is the relationship between cal cm^{-2} min^{-1} and W m^{-2}?

19.6. How to dimensionalize $K(\theta)$?

19.7. In the equation $N \cdot r^D = 1$, show that $D = \log N/\log(1/r)$.

19.10 Answers

19.1. $4\pi r_2^2 = 4\pi(1.5\, r_1)^2$

19.2. $MT^{-2} = MT^{-2}$

19.3. $K = k\rho g/\eta$

19.4. $t^* = \pi$

19.5. 1 cal cm^{-2} min^{-1} = W m^{-2}

19.6. $\pi = K(\theta)/K_0$, when $K(\theta) = K_0$, $\pi = 1$; when $K(\theta) = 0$, $\pi = 0$

19.7. ___________

References

Arya LM, Paris JF (1981) A physico-empirical model to predict the soil moisture characteristic from particle-size distribution and bulk density data. Soil Sci Soc Am J 45:1023–1030

Bacchi OOS, Reichardt K (1988) Escalonamento de propriedades hídricas na avaliação de métodos de determinação da condutividade hidráulica. Rev Bras Ciênc Solo 12:217–223

Bacchi OOS, Reichardt K (1993) Geometria fractal em física de solo. Sci Agric 50:321–325

Bacchi OOS, Reichardt K, Villa Nova NA (1996) Fractal scaling of particle and pore size distributions and its relation to soil hydraulic conductivity. Sci Agric 53:356–361

Barnsley MF, Devaney RL, Mandelbrot BB, Peitgen HO, Saupe D, Voss RF (1988) The science of fractal images. Springer, New York, NY

Guerrini IA, Swartzendruber D (1994) Fractal characteristics of the horizontal movement of water in soil. Fractal 2:465–468

Guerrini IA, Swartzendruber D (1997) Fractal concepts in relation to soil-water diffusivity. Soil Sci 162:778–784

Klute A, Wilkinson GE (1958) Some tests of the similar media concept of capillary flow: I. Reduced capillary conductivity and moisture characteristic data. Soil Sci Soc Am Proc 22:278–281

Kutilek M, Nielsen DR (1994) Soil hydrology. Catena Verlag, Cremlingen-Destedt

Mandelbrot BB (1982) The fractal geometry of nature. W.H. Freeman and Company, New York, NY

Miller EE, Miller RD (1956) Physical theory of capillary flow phenomena. J Appl Phys 27:324–332

Nielsen DR, Hopmans J, Reichardt K (1998) An emerging technology for scaling field soil water behavior. In: Sposito G (ed) Scale dependence and scale invariance in hydrology. Cambridge University Press, New York, NY, pp 136–166

Perfect E, Kay BD (1995) Applications of fractals in soil and tillage research: a review. Soil Tillage Res 36:1–20

Puckett WE, Dane JH, Hajek BF (1985) Physical and mineralogical data to determine soil hydraulic properties. Soil Sci Soc Am J 49:831–836

Reichardt K, Libardi PL (1973) A new equation for the estimation of soil water diffusivity. In: Symposium on isotopes and radiation techniques in studies of soil physics, irrigation and drainage in relation to crop production. FAO/IAEA, Vienna, pp 45–51

Reichardt K, Nielsen DR, Biggar JW (1972) Scaling of horizontal infiltration into homogeneous soils. Soil Sci Soc Am Proc 36:241–245

Reichardt K, Libardi PL, Nielsen DR (1975) Unsaturated hydraulic conductivity determination by a scaling technique. Soil Sci 120:165–168

Ryan RG, Dhir VK (1993) The effect of soil-particle size on hydrocarbon entrapment near a dynamic water table. J Soil Cont 2:59–92

Sadeghi M, Grahraman B, Davary K, Hasheminia SM, Reichardt K (2011) Scaling to generalize a single solution of Richards' equation for soil water redistribution. Sci Agric 68:582–591

Shukla MK, Kastanek FJ, Nielsen DR (2002) Inspectional analysis of convective-dispersion equation and application on measured breakthrough curves. Soil Sci Soc Am J 66:1087–1094

Tillotson PM, Nielsen DR (1984) Scale factors in soil science. Soil Sci Soc Am J 48:953–959

Turcotte DL (1986) Fractals and fragmentation. J Geophys Res 91:1921–1926

Tyler WS, Wheatcraft SW (1989) Application of fractal mathematics to soil water retention estimation. Soil Sci Soc Am J 3:987–996

Tyler WS, Wheatcraft SW (1990) Fractal processes in soil water retention. Water Resour Res 26:1047–1054

Tyler WS, Wheatcraft SW (1992) Fractal scaling of soil particle-size distributions: analysis and limitations. Soil Sci Soc Am J 56:362–369

Wilkinson GE, Klute A (1959) Some tests of the similar media concept of capillary flow: II. Flow systems data. Soil Sci Soc Am Proc 23:434–437

Index

K. Reichardt, L. C. Timm, *Soil, Plant and Atmosphere*, https://doi.org/10.1007/978-3-030-19322-5

MIX
Papier aus verantwortungsvollen Quellen
Paper from responsible sources
FSC® C105338

If you have any concerns about our products,
you can contact us on
ProductSafety@springernature.com

In case Publisher is established outside the EU,
the EU authorized representative is:
Springer Nature Customer Service Center GmbH
Europaplatz 3, 69115 Heidelberg, Germany

Printed by Libri Plureos GmbH
in Hamburg, Germany